Chapter 4 Systems of Linear Equations and Inequalities

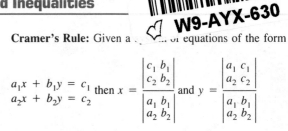

Consistent system   Inconsistent system   Dependent system

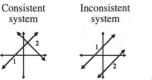

A system of linear equations may be solved: (a) graphically: (b) by the substitution method, (c) by the addition or elimination method, or (d) by determinants.

$$\begin{vmatrix} a_1 & b_1 \\ a_2 & b_2 \end{vmatrix} = a_1 b_2 - a_2 b_1$$

Cramer's Rule: Given a _____ of equations of the form

$$\begin{array}{l} a_1 x + b_1 y = c_1 \\ a_2 x + b_2 y = c_2 \end{array} \text{ then } x = \dfrac{\begin{vmatrix} c_1 & b_1 \\ c_2 & b_2 \end{vmatrix}}{\begin{vmatrix} a_1 & b_1 \\ a_2 & b_2 \end{vmatrix}} \text{ and } y = \dfrac{\begin{vmatrix} a_1 & c_1 \\ a_2 & c_2 \end{vmatrix}}{\begin{vmatrix} a_1 & b_1 \\ a_2 & b_2 \end{vmatrix}}$$

Chapter 5 Polynomials

$a^m \cdot a^n = a^{m+n}$

$a^m / a^n = a^{m-n}, a \neq 0$

$a^{-m} = \dfrac{1}{a^m}, a \neq 0$

$a^0 = 1, a \neq 0$

$(a^m)^n = a^{mn}$

$(ab)^m = a^m b^m$

$\left(\dfrac{a}{b}\right)^m = \dfrac{a^m}{b^m}, b \neq 0$

$\left(\dfrac{a}{b}\right)^{-m} = \left(\dfrac{b}{a}\right)^m, a \neq 0, b \neq 0$

FOIL method to multiply two binomials

F   O   I   L

$(a + b)(c + d) = a \cdot c + a \cdot d + b \cdot c + b \cdot d$

Square of a binomial:

$(a + b)^2 = a^2 + 2ab + b^2,$

$(a - b)^2 = a^2 - 2ab + b^2$

Product of the sum and difference of the same two terms (also called the difference of two squares):

$(a + b)(a - b) = a^2 - b^2$

Polynomial function:

$f(x) = a_n x^n + a_{n-1} x^{n-1} + a_{n-2} x^{n-2} + \cdots + a_1 x + a_0$

Quadratic function: $f(x) = ax^2 + bx + c, a \neq 0$

Chapter 6 Factoring

Difference of two squares: $a^2 - b^2 = (a + b)(a - b)$

Note: the sum of two squares $a^2 + b^2$ cannot be factored over the set of real numbers

Perfect square trinomials:

$a^2 + 2ab + b^2 = (a + b)^2, a^2 - 2ab + b^2 = (a - b)^2$

Sum of two cubes:

$a^3 + b^3 = (a + b)(a^2 - ab + b^2)$

Difference of two cubes:

$a^3 - b^3 = (a - b)(a^2 + ab + b^2)$

To Factor a Polynomial

1. Determine if the polynomial has a greatest common factor other than 1. If so factor out the GCF from every term in the polynomial.
2. If the polynomial has two terms (or is a binomial), determine if it is a difference of two squares or a sum or difference of two cubes. If so factor using the appropriate formula.
3. If the polynomial has 3 terms (or is a trinomial), determine if it is a perfect square trinomial. If so factor accordingly. If it is not, then factor the trinomial using the method discussed in Section 6.2.
4. If the polynomial has more than 3 terms, then try factoring by grouping.
5. As a final step examine your factored polynomial to see if any factors listed have a common factor and can be factored further. If you find a common factor, factor it out at this point.

Standard form of a quadratic equation:

$ax^2 + bx + c = 0, a \neq 0$

Zero-factor property: If $a \cdot b = 0$, then either $a = 0$ or $b = 0$, or both a and $b = 0$.

Intermediate Algebra for College Students

THIRD EDITION

Intermediate Algebra for College Students

Allen R. Angel

Monroe Community College

PRENTICE HALL
Englewood Cliffs, New Jersey 07632

Library of Congress Cataloging-in-Publication Data

Angel, Allen R.
 Intermediate algebra for college students / Allen R. Angel — 3rd
 ed., Annotated instructor's ed.
   p.   cm.
  Includes index.
  ISBN 0-13-478736-6
  1. Algebra.   I. Title.
QA154.2.A53 1992
512.9—dc20   91-22951
                                                    CIP

Editor-in-chief: Tim Bozik

Executive editor: Priscilla McGeehon

Editorial/production supervision: Rachel J. Witty, Letter Perfect, Inc.,
    Edward Thomas, and Thomas Aloisi

Interior design: Lee Goldstein

Design director: Florence Dara Silverman

Cover design: Lee Goldstein

Cover photo: © 1992 by Adam Peiperl

Marketing manager: Gary June

Prepress buyer: Paula Massenaro

Manufacturing buyer: Lori Bulwin

Supplement editor: Susan Black

Copy editor: Bill Thomas

Editorial assistant: Marisol L. Torres

Printed in the United States of America
10 9 8 7 6 5 4

ISBN 0-13-478736-6

Prentice-Hall International (UK) Limited, *London*
Prentice-Hall of Australia Pty. Limited, *Sydney*
Prentice-Hall Canada Inc., *Toronto*
Prentice-Hall Hispanoamericana, S.A., *Mexico*
Prentice-Hall of India Private Limited, *New Delhi*
Prentice-Hall of Japan, Inc., *Tokyo*
Simon & Schuster Asia Pte. Ltd., *Singapore*
Editora Prentice-Hall do Brasil, Ltda., *Rio de Janeiro*

To my mother,
Sylvia Angel-Baumgarten

and

to the memory of my father,
Isaac Angel

Contents

6 Factoring   258

7 Rational Expressions and Equations   298

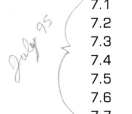

8 Roots, Radicals, and Complex Numbers   352

9 Quadratic Functions and the Algebra of Functions   412

10 Conic Sections   466

11 Exponontial and Logarithmic Functions   504

12 Sequences, Series, and the Binomial Theorem   541

Appendices

Answers   583

Preface

This is the second book in a two-volume algebra series. This book was written for college students who have successfully completed elementary algebra and wish to take a second course in algebra.

My primary goals in writing this book were to write a book that students could read, understand, and enjoy while acquiring the necessary skills to be successful in future mathematics courses. To achieve these goals, I have used short sentences, clear explanations, and many detailed, worked-out examples. I have tried to make the book relevant to college students by using practical applications of algebra throughout the text. For consistency in the series, I have used the same pedagogical features in this book as in *Elementary Algebra for College Students*. Some of these features are outlined below.

Features of the Text

Four-color Format: Color is used pedagogically in the following ways:

- Important definitions and procedures are color screened.
- Color screening or color type is used to make other important items stand out.
- Errors that students commonly make are given in colored boxes as warnings for students.
- Artwork is enhanced and clarified with use of multiple colors.

Readability: One of the most important features of the text is its readability. The book is very readable, even for those with weak reading skills. Short clear sentences are used, and words that are more easily recognized and understood are used whenever possible. With so many of our students from different countries now taking algebra, this feature has become increasingly important.

Accuracy: Accuracy in a mathematics text is essential. To insure accuracy in this book, mathematicians from around the country have read the galleys carefully for typographical errors and have checked all the answers.

Spiral Approach to Learning: Many of our students do not thoroughly grasp new concepts the first time they are presented. In this text we use the spiral approach to learning. That is, we introduce a concept, then later in the text briefly reintroduce it and build upon it. Often an important concept is used in many sections of the text. Students are often reminded where the material was seen before, or where it will be used again. This also serves to emphasize the importance of the concept. Important concepts are also reinforced throughout the text in the Cumulative Review Exercises and Cumulative Review Test.

Keyed Section Objectives: Each section opens with a list of skills that the student should learn in that section. The objectives are then keyed to the appropriate portions of the sections with symbols such as ▶1.

Practical Applications: Practical applications of algebra are stressed throughout the text. Students need to learn how to translate application problems into algebraic symbols. The problem-solving approach used throughout this text gives students ample practice in setting up and solving application problems. The use of practical, real-life applications motivate the students.

Detailed Worked-Out Examples: A wealth of examples have been worked out in a step-by-step, detailed manner. Important steps are highlighted in color, and no steps are omitted until after the student has seen a sufficient number of similar examples.

Study Skills Section: Many students taking this course have poor study skills in mathematics. Section 1.1, the first section of this text, discusses the study skills needed to be successful in mathematics. This section should be very beneficial for your students, and should help them to achieve success in mathematics.

Common Student Errors: Errors that students often make are illustrated. The reasons why certain procedures are wrong are explained, and the correct procedure for working the problem is illustrated. These common student error boxes will help prevent your students from making those errors we see so often.

Helpful Hints: The helpful hint boxes offer useful suggestions for problem solving and other varied topics. They are set off in a special manner so that students will be sure to read them.

Calculator Corners: The Calculator Corners, placed at appropriate intervals in the text, are written to reinforce the algebraic topics presented in the section and to give students pertinent information on using the calculator to solve algebraic problems. No new algebraic information is given in the Calculator Corners.

Exercise Sets: Each exercise set is graded in difficulty. The early problems help develop the student's confidence; then the students are eased gradually into the more difficult problems. A sufficient number and variety of examples are given in the section for the student to successfully complete even the more difficult exercises. The number of exercises in each section is more than ample for student assignments and practice.

Writing Exercises: Many exercise sets now include exercises that require students to write out the answers in words. These exercises improve students understanding and comprehension of the material and help develop better reasoning and critical thinking skills. Writing exercises are indicated by the symbol ✎.

Cumulative Review Exercises: The cumulative review exercises that appear at the end of each exercise set contain questions from previous sections in the chapter and from previous chapters. These exercises will reinforce topics that were previously taught and help students retain the old material, while they are learning the new material. For the students' benefit the cumulative review exercises are keyed to the section where the material is covered.

Just for Fun Problems: At the end of many exercise sets are Just for Fun problems. These problems offer more challenging problems for the bright students in your class who want something extra. These problems present additional applications of algebra, material to be presented later in the text, or material to be covered in a later mathematics course. These exercises lend themselves nicely to group work in the classroom.

Chapter Summary: At the end of each chapter is a chapter summary which includes a glossary and important chapter facts.

Review Exercises: At the end of each chapter are review exercises that cover all types of exercises presented in the chapter. The review exercises are keyed to the sections where the material was first introduced.

Practice Tests: The comprehensive end-of-chapter practice test will enable the students to see how well they are prepared for the actual class test. The Instructor's Resource Manual includes 5 forms of each chapter test that are similar to the students' practice test (multiple-choice tests are also included in the Instructor's Resource Manual.)

Cumulative Review Tests: These tests, which appear at the end of each even-numbered chapter, test the students' knowledge of material from the beginning of the book to the end of that chapter. Students can use these tests for review, as well as for preparation for the final exam. These exams, like the cumulative review exercises, will serve to reinforce topics taught earlier.

Answers: Answers are provided to the following exercises: odd-numbered problems in the exercise sets, all cumulative review exercises, all Just for Fun exercises, all review exercises, all practice tests, and all cumulative review tests.

Prerequisite

The prerequisite for this course is a working knowledge of elementary algebra. Although some elementary algebra topics are briefly reviewed in the text, students should have a basic understanding of elementary algebra before taking this course.

Mode of Instruction

The format of this book lends itself to many different modes of instruction. For your students to be able to understand the material presented, the text must be readable. Short, clear sentences are used to make this text readable for students with weak reading skills. Wherever possible, common, easy-to-understand words are used.

The spiral approach, cumulative review exercises, and the cumulative review tests will continually reinforce important concepts and topics. The approach and the features of the text outlined earlier will result in greater understanding and retention of the material by your students.

The features of the text and the large variety of supplements available make this text suitable for many types of instructional modes including:

- lecture
- modified lecture
- learning laboratory
- self-paced instruction
- cooperative or group study

Changes in the Third Edition

When I wrote the third edition, I considered the many letters and reviews I got from students and faculty. I would like to thank all of you who made suggestions for improving the third edition. I would also like to thank the many instructors and students who wrote to inform me of how much they enjoyed and appreciated the text.

Some of the changes made in the third edition of the text include:

- More detailed examples have been added throughout the book.
- More difficult exercises have been added to the graded exercises sets.

- More exercises and examples with decimal numbers and fractions have been added.
- Cumulative Review Exercises have been added to each exercise set.
- The addition of new sections on:
    Study Skills Needed for Success in Mathematics
    Using Factoring to Solve for a Variable in a Formula or Equation
    Rational Exponents
    The Algebra of Functions
    The Natural Exponential Function and Natural Logarithms
- More and new practical applications of algebra are spread throughout the book.
- Addition of new Helpful Hints, Common Student Errors, and Calculator Corners.
- Material on linear programming has been added.
- More exercises that require written responses by students have been added.
- More problem-solving exercises have been added.
- Cumulative Review Tests have been added.
- There is a greater emphasis on the spiral approach to learning.
- Certain sections have been rewritten or reorganized for greater clarity.
- There is a general fine-tuning of the book for greater clarity.

Supplements to the Third Edition

For Instructors

Annotated Instructor's Edition: Includes answers to every exercise on the same page.

Instructor's Resource Manual: Contains solutions to even-numbered exercises and eight tests per chapter (three are multiple choice)

PH Test Manager: Allows users to generate tests by chapter or section number, choosing from thousands of test questions and hundreds of algorithms which generate different numbers for the same item. Editing and graphing capability are included.

Test Item File: Contains thousands of test items for use with PH Test Manager.

Syllabus and Teaching Outlines (with Instructor's Disk): Contains suggested homework assignments keyed to objectives and teaching outlines integrating supplements into the course. All available on ASCII disk for individual customization in your course.

For Students
Tutorial Software
Math Master Tutor Software: Carefully keyed to the book, with page references; includes four modes of instruction: *Explorations* (including detailed, worked-out examples with explanation); *Summary*; *Exercises* (open-ended, algorithmically generated with step-by-step solutions); and *Quiz* (with a printout option). Available free with a qualified adoption for IBM and Macintosh.

Interactive Algebra Tutor: An alternative, generic software with multiple choice questions, available for Apple, IBM or Macintosh. Contact College Editorial or Marketing.

Videotapes: Closely tied to the book, these instructional tapes feature a lecture format with worked-out examples and exercises from each section of the book. A video on study skills is also included. One master set available with each adoption of 100 or more copies.

Student's Study Guide. Includes additional worked-out examples, additional drill problems, and Practice Tests, and their answers. Important concepts are emphasized.

Student's Solution Manual. Includes detailed step-by-step solutions to odd-numbered problems in the exercise sets.

Acknowledgments

Writing a textbook is a long and time-consuming project. Many people deserve thanks for encouraging and assisting me with this project. Most importantly I would like to thank my wife, Kathy, and sons, Robert and Steven. Without their constant encouragement and understanding, this project would not have become a reality.

I would like to thank my colleagues at Monroe Community College for helping with this project, especially Peter Collinge and Annette Leopard. I would like to thank Richard Semmler of Northern Virginia Community College for his many valuable suggestions. Judith Conturo Karas did an excellent job of typing the manuscript.

I would like to thank my students, and students and faculty from around the country, for using the second edition and offering valuable suggestions for the third edition.

I would like to thank my editors at Prentice Hall, Priscilla McGeehon and Christine Peckaitis, and production editor, Rachel J. Witty, Letter Perfect, Inc.

I would like to thank the following reviewers and proofreaders for their thoughtful comments and suggestions:

WAYNE BARBER, *Chemekata Community College;* JACK BARONE, *Baruch College;* RONALD BOHUSLOV, *Merritt College;* BETH BOREL, *University of S. W. Louisiana;* FRANCINE BORTZEL, *Seton Hall University;* HELEN BURRIER, *Kirkwood Community College;* FRANK CERRATO, *City College of San Francisco;* LAURA CLARKE, *Milwaukee Area Tech.;* BEN CORNELIUS, *Oregon Institute of Technology;* ARTHUR DULL, *Diablo Valley Community College;* DALE EWEN, *Parkland College;* PETER FREEDHAND, *New York University;* ROBERT GESELL, *Cleary College;* MARK GIDNEY, *Lees McRae College;* JAY GRAENING, *University of Arkansas—Main;* MARGARET GREENE, *Florida Community College at Jacksonville;* KEN HODGE, *Rose State College*; LARRY HOEHN, *Austin Peay State University;* JUDY KASABIAN, *El Camino College;* HERBERT KASUBE, *Bradley University;* MELVIN KIRKPATRICK, *Roane State Community College;* ADELE LEGERE, *Oakton Community College*; GLENN LIPELY, *Malone College;* CHARLES LUTTRELL, *Frederick Community College;* MERWIN LYNG, *Mayville State College;* P. WILLIAM MAGLIARO, *Bucks County Community College;* JACK McCOWN, *Central Oregon Community College;* JOHN MICHAELS, *SUNY at Brockport;* LOIS MILLER, *Golden West College;* JULIE MONTE, *Daytona Beach Junior College;* CATHY PACE, *Louisiana Tech University;* C. V. PEELE, *Marshall University;* JAMES PERKINS, *Piedmont Community College;* MATTHEW PICKARD, *University of Puget Sound;* JON PLACHY, *Metropolitan State College;* RAYMOND PLUTA, *Castleton State College;* DOLORES SCHAFFNER, *University of South Dakota;* RICHARD SEMMLER, *Northern Virginia Community College;* KEN SEYDEL, *Skyline College;* EDITH SILVER, *Mercer County Community College;* FAY THAMES, *Lamar University;* TOMMY THOMPSON, *Brookhaven College;* LEE-ING TONG, *Southeastern Massachusetts University;* JOHN WENGER, *Loop College;* BRENDA WOOD, *Florida Community College at Jacksonville;* KARL ZILM, *Lewis and Clark Community College.*

To the Student

Algebra is a course that cannot be learned by observation. To learn algebra you must become an active participant. You must read the text, pay attention in class, and, most importantly, you must work the exercises. The more exercises you work, the better.

This text was written with you in mind. Short, clear sentences are used, and many examples are given to illustrate specific points. The text stresses useful applications of algebra. Hopefully, as you progress through the course, you will come to realize that algebra is not just another math course that you are required to take, but a course that offers a wealth of useful information and applications.

This text makes use of 4 different colors. The different colors are used to highlight important information. Important procedures, definitions, and formulas are placed within colored boxes.

The boxes marked **Common Student Errors** should be studied carefully. These boxes point out errors that students commonly make, and provide the correct procedures for doing these problems. The boxes marked **Helpful Hints** should also be studied carefully, for they also stress important information.

Ask your professor early in the course to explain the policy on when the calculator may be used. If your professor allows you to use a calculator, then pay particular attention to the **Calculator Corners.**

Other questions you should ask your professor early in the course include: What supplements are available for use? Where can help be obtained when the professor is not available? Supplements that may be available include: student's study guide, student's solutions manual, tutorial software, and video tapes, including a tape on the study skills needed for success in mathematics.

You may wish to form a study group with other students in your class. Many students find that working in small groups provides an excellent way to learn the material. By discussing and explaining the concepts and exercises to one another you reinforce your own understanding. Once guidelines and procedures are determined by your group, make sure to follow them.

One of the first things you should do is to read Section 1.1, Study Skills Needed for Success in Mathematics. Read this section slowly and carefully, and pay particular attention to the advice and information given. Occasionally, refer back to this section. This could be the most important section of the book. Carefully read the material on doing your homework and on attending class.

At the end of all exercise sets (after the first two) are **Cumulative Review Exercises.** You should work these problems on a regular basis, even if they are not assigned. These problems are from earlier sections and chapters of the book, and they will refresh your memory and reinforce those topics. If you have a problem when working these exercises, read the appropriate section of the text or study your notes that correspond to that material. The section of the text where the Cumulative Review Exercises were introduced is indicated in brackets, [ ], to the left of the exercise. After reviewing the material, if you still have a problem, make an appointment to see your professor. Working the Cumulative Review Exercises throughout the semester will also help prepare you to take your final exam.

At the end of many exercise sets are **Just for Fun** problems. These exercises are not for everyone. They are for those of you who are doing well in the course and are looking for more of a challenge. These exercises often present additional applications of algebra, material that will be presented in a later section, or material that will be presented in a later course.

At the end of each chapter are a **chapter summary,** a set of **review exercises,** and a **chapter practice test.** Before each examination you should review these sections carefully and take the practice test. If you do well on the practice test, you should do well on the class test. The questions in the review exercises are marked to indicate the section in which that material was first introduced. If you have a problem with a review exercise question turn to and reread the section indicated. You may also wish to take the **Cumulative Review Test** that appears at the end of every even-numbered chapter.

In the back of the text there is an **answer section** which contains the answers to the odd-numbered exercises, all cumulative review exercises, Just for Fun exercises, review exercises, practice tests, and cumulative review tests. The answers should be used only to check your work.

I have tried to make this text as clear and error free as possible. No text is perfect, however. If you find an error in the text, or an example or section that you believe can be improved, I would greatly appreciate hearing from you. If you enjoy the text, I would also appreciate hearing from you.

ALLEN R. ANGEL

CHAPTER

1

Basic Concepts

See Section 1.5, Exercise 83.

1.1

Study Skills for Success in Mathematics

▸**1**   Recognize the goals of the text.

▸**2**   Prepare for class effectively.

▸**3**   Preparing for and taking examinations.

▸**4**   Determine how to find help.

You need to acquire certain study skills that will help you to successfully complete this course. These study techniques will also help you succeed in any other mathematics course you take.

Goals of the Text

▸**1**   1. Teaching you traditional algebra topics.
2. Preparing you to take a more advanced mathematics course.
3. Building your confidence to allow you to enjoy mathematics.
4. Improving your reasoning and critical thinking skills.
5. Increasing your understanding of how important mathematics is in solving real-life problems.
6. Getting you to think mathematically so that you can translate a real-life problem into a mathematical equation and then solve the problem.

It is important for you to realize that this course is the foundation for more advanced mathematics courses. If you have a thorough understanding of algebra, you will find it easier to be successful in later mathematics courses.

Have a Positive Attitude

You may be thinking to yourself, "I hate math" or "I wish I did not have to take this class." You may have heard the term "math anxiety" and feel that you fall in this category. The first thing you need to do to be successful in this course is to change your attitude to a more positive one. You must be willing to give this course and yourself a fair chance.

Based on past experiences in mathematics, you may feel this is difficult. However, mathematics is something you need to work at. Many of you taking this course are more mature now than when you took previous mathematics courses. This maturity factor and the desire to learn are extremely important and can make a tremendous difference in your ability to succeed in mathematics. I believe you can be successful in this course, but you also need to believe it.

Prepare for Class Effectively

▸**2**   To be prepared for class, you need to do your homework. If you have difficulty with the homework or some of the concepts, write down questions to ask your professor. Prior to class, you should spend a few minutes previewing any new material in the textbook. At this point, you do not have to understand everything you read. Just get a feeling for the definitions and concepts that will be discussed. This quick preview will help you to understand what your instructor is explaining during class.

After the material is explained in class, read the corresponding sections of the text slowly and carefully, word by word.

Reading the Text

A mathematics text is not a novel. Mathematics textbooks should be read slowly and carefully. If you do not understand what you are reading, reread the material. When you come across a new concept or definition, you may wish to underline it so that it stands out. This way, when looking for it later, it will be easier to find. When you come across a worked out example, read and follow the example very carefully. Do not just skim it. Try working out the example yourself on another sheet of paper. Make notes of anything you do not understand to ask your instructor.

Doing Homework

Two very important commitments that you must make to be successful in this course are attending class and doing your homework regularly. Your assignments must be worked conscientiously and completely. Mathematics cannot be learned by observation. You need to practice at what you have heard in class. It is through doing homework that you truly learn the material.

Don't forget to check the answers to your homework assignments. This book contains the answers to the odd exercises in the back of the book. In addition, the answers to all the cumulative review exercises, Just for Fun exercises, end-of-chapter review exercises, practice tests, and cumulative review tests are in the back of the book.

Ask questions about homework problems you do not understand in class. You should not feel comfortable until you understand all the concepts needed to successfully work every assigned problem.

When you do your homework, make sure that you write it neatly and carefully. Pay particular attention to copying down signs and exponents correctly. Do your homework in a step-by-step manner. This way you can refer back to it later and still understand what was written.

Attending and Participating in Class

You should attend every class possible. Most instructors will agree that there is an inverse relationship between absences and grades. That is, the more absences you have, the lower your grade will be. Every time you miss a class, you miss important information. If you need to miss a class, contact your instructor ahead of time and get the reading assignment and homework.

While in class, pay attention to what your instructor is saying. If you do not understand something, ask your instructor to repeat or reexplain the material. If you have read the upcoming material before class and have questions that have not been answered, ask your instructor. If you do not ask questions, your instructor will not know that you have a problem in understanding the material.

In class, take careful notes. Write numbers and letters clearly so that you can read them later. It is not necessary to write down every word your instructor says. Copy down the major points and the examples that do not appear in the text. You should not be taking notes so frantically that you lose track of what your instructor is saying. It is a mistake to believe that you can copy down material in class without understanding it and then figure it out when you get home.

Studying

Study in the proper atmosphere. Study in an area where you are not constantly disturbed so that your attention can be devoted to what you are reading. The area where you study should be well ventilated and well lit. You should have sufficient desk space to spread out all your materials. Your chair should be comfortable. There should be no loud music to distract you from studying.

When studying, you should not only understand how to work a problem, you should also know why you follow the specific steps you do to work the problem. If you do not have an understanding of why you follow the specific process, you will not be able to transfer the process to solve similar problems.

Time Management

It is recommended that students spend at least 2 hours per hour of class time studying and doing homework. Some students require more time than others. Finding the necessary time to study is not always easy. The following are some suggestions that you may find helpful.

1. Plan ahead. Determine when you will have time to study and do your homework. Do not schedule other activities for this time period. Try to space these times evenly over the week.
2. Be organized so that you will not have to waste time looking for your books, pen, calculator, or notes.
3. If you are allowed to use calculators, use a calculator to perform tedious calculations.
4. When you stop studying, clearly mark where you stopped in the text.
5. Try not to take on added responsibilities. You must set your priorities. If your education is a top priority, as it should be, then you may have to cut the time spent on other activities.
6. If time is a problem, do not overburden yourself with too many courses. Consider taking fewer credits. If you do not have sufficient time to study, your understanding and your grade in all of your courses could be affected.

Preparing for and Taking Examinations

► 3

Studying for an Exam

If you study a little bit each day, you should not need to cram the night before an exam. If you wait until the last minute, you will not have time to seek the help you may need. To review for an exam:

1. Read your class notes.
2. Review your homework assignments.
3. Study the formulas, definitions, and procedures given in the text.
4. Read the Common Student Error boxes and Helpful Hint boxes carefully.
5. Read the summary at the end of each chapter.
6. Work the review exercises at the end of each chapter. If you have difficulties, restudy those sections. If you still have trouble, seek help.
7. Work the practice chapter test.

Taking an Exam

Make sure you get a good night's sleep the day before the test. If you studied properly, you should not have to stay up late the night before preparing for a test. Arrive at the exam site early so that you have a few minutes to relax before the exam. If you need to rush to get to the exam, you will start out nervous and anxious. After you are given the exam, do the following:

1. Carefully write down any formulas or ideas that you need to remember.
2. Look over the entire exam quickly to get an idea of its length. You will need

to pace yourself to make sure you complete the entire exam. Be prepared to spend more time on problems worth more points.

3. Read the test directions carefully.

4. Read each question carefully. Answer each question completely and make sure you have answered the specific question asked.

5. Work the questions you understand best first; then come back and work those problems you are not sure of. Do not spend too much time on any one problem.

6. Attempt each problem. You may be able to get at least partial credit.

7. Work carefully and write clearly so that your instructor can read your work. Also, it is easy to make mistakes when your writing is unclear.

8. Check your work and your answers if you have time.

9. Do not be concerned if others finish the test before you. Do not be disturbed if you are the last to finish. Use all your extra time to check your work.

How to Find Help ▶ **4**

Using Supplements

This text comes with a large variety of supplements. Find out from your instructor early in the semester which supplements are available and which supplements might be beneficial for you to use. Supplements should not replace reading the text, but should be used to enhance your understanding of the material.

Seeking Help

One thing I stress with my own student is to *get help as soon as you need it!* Do not wait! In mathematics, one day's material is often based on the previous day's material. So, if you don't understand the material today, you may not be able to understand the material tomorrow.

Where should you seek help? There are often a number of places to obtain help on campus. You should try to make a friend in the class with whom you can study. Often you can help one another. You may wish to form a study group with other students in your class. Discussing the concepts and homework with your peers will reinforce your own understanding of the material.

You should not hesitate to visit your instructor when you are having problems with the material. Be sure you read the assigned material and attempt the homework before meeting with your instructor. Come prepared with specific questions to ask.

Often other sources of help are available. Many colleges have a Mathematics Laboratory or a Mathematics Learning Center where tutors are available to help students. Ask your instructor early in the semester if any tutors are available, and find out where the tutors are located. Then use these tutors as needed.

Exercise Set 1.1

Do you know all of the following information? If not, ask your instructor as soon as possible.

1. What are your instructor's office hours?

2. Where is your instructor's office located?

3. How can you best reach your instructor?

4. Where can you obtain help if your instructor is not available?

5. What supplements are available to assist you in learning?

6. When can you use a calculator? Can it be used in class, on homework, on tests?

7. Do you know the name and phone number of a friend in class?

8. For each hour of class time, how many hours outside of class are recommended for homework and studying?

9. List what you should do to be properly prepared for class.

10. Explain how a mathematics textbook should be read.

11. Write a summary of the steps you should follow when taking an exam.

12. Having a positive attitude is very important for success in this course. Are you beginning this course with a positive attitude? It is important that you do!

13. You need to make a commitment to spend the time necessary to learn the material, to do the homework, and to attend class regularly. Explain why you believe this commitment is necessary to be successful in this course.

14. What are your reasons for taking this course?

15. What are your goals for this course?

16. Have you given any thought to studying with a friend, or a group of friends? Can you see any advantages in doing so? Can you see any disadvantages in doing so?

1.2

Sets and the Real Number System

▸**1** Know the meaning of a set.

▸**2** Identify subsets.

▸**3** Perform set operations: union and intersection.

▸**4** Identify various important sets of numbers.

▸**5** Know the real numbers.

▸**1** Sets are used in many areas of mathematics, so an understanding of sets and set notation is important. A **set** is a collection of objects. The objects in a set are called **elements** of the sets. Sets are indicated by means of braces, { }, and are often named with capital letters. When the elements of a set are listed within the braces, as illustrated below, the set is said to be in **roster form.**

$$A = \{a, b, c\}$$
$$B = \{\text{yellow, green, blue, red}\}$$
$$C = \{1, 2, 3, 4, 5\}$$

Set A has three elements, set B has four elements, and set C has five elements. The symbol $\in$ is used to indicate that an item is an element of a set. Since 2 is an element of set C, we may write $2 \in C$; this is read "2 is an element of set C." Note that 6 is not an element of set C. We may therefore write $6 \notin C$, which is read "6 is not an element of set C."

A set may be finite or infinite. Sets A, B, and C each have a finite number of elements and are therefore *finite sets*. In some sets it is impossible to list all the elements. These are *infinite sets*. The following set is an example of an infinite set.

$$N = \{1, 2, 3, 4, 5, \ldots\}$$

The three dots after the last comma indicate that the set continues in the same manner indefinitely. Set N is called the set of **natural numbers,** or **counting numbers.**

If we write

$$D = \{1, 2, 3, 4, 5, \ldots, 280\}$$

it means that the set continues in the same manner until the number 280. Set D is the set of the first 280 natural numbers.

A special set that contains no elements is called the *null set*, or *empty set*, written { } or ∅. For example, the set of students in your class over the age of 150 is the null or empty set.

A second method of writing a set is with **set builder notation.** An example of a set given in set builder notation is

$$E = \{x \,|\, x \text{ is a natural number greater than 6}\}$$

This is read "Set E is the set of all x such that x is a natural number greater than 6." In roster form this set would be

$$E = \{7, 8, 9, 10, 11, \ldots\}$$

The general form of set builder notation is

$$\{x \,|\, x \text{ (followed by properties to be met)}\}$$

is the set   such x (then state properties)
of all x that

Consider

$$F = \{x \,|\, x \text{ is a natural number between 2 and 7}\}$$

This is read "Set F is the set of all x such that x is a natural number between 2 and 7." In roster form the set is

$$F = \{3, 4, 5, 6\}$$

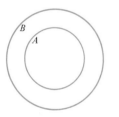

FIGURE 1.1

▶ **2** Set A is a **subset** of set B, written $A \subseteq B$, if every element of set A is also an element of set B. Figure 1.1 illustrates two sets A and B, where set A is a subset of set B. *Note:* Every element that is in set A must also be in set B.

EXAMPLE 1   If the first set is a subset of the second set, insert $\subseteq$ in the shaded area between the two sets. If the first set is not a subset of the second set, insert the symbol $\nsubseteq$ (read "is not a subset of") between the two sets.
(a) $\{1, 2, 3\}$ ▧ $\{1, 2, 3, 4\}$
(b) $\{a, b, c, d\}$ ▧ $\{b, c, d, e, f, g\}$

Solution: (a) $\{1, 2, 3\} \subseteq \{1, 2, 3, 4\}$
(b) $\{a, b, c, d\} \nsubseteq \{b, c, d, e, f, g\}$ Notice that a is in the first set but not the second set. ▪

EXAMPLE 2   Determine if the set of natural numbers, $N = \{1, 2, 3, 4, 5, \ldots\}$ is a subset of the set of whole numbers, $W = \{0, 1, 2, 3, 4, 5, \ldots\}$.

Solution: Since every element in the set of natural numbers is also an element in the set of whole numbers, we may write

$$N \subseteq W$$

The natural numbers are a subset of the set of whole numbers. ▪

▶ **3** Just as operations such as addition and multiplication are performed on numbers, operations are performed on sets. Two operations we will discuss are *union* and *in-*

tersection. The **union** of set A and set B, written $A \cup B$, is the set of elements that belong to either set A *or* set B. The union is formed by combining, or joining together, the elements in set A with those in set B.

Examples of Union of Sets

$A = \{1, 2, 3, 4, 5\}, \quad B = \{3, 4, 5, 6, 7\}, \quad A \cup B = \{1, 2, 3, 4, 5, 6, 7\}$
$A = \{a, b, c, d, e\}, \quad B = \{x, y, z\}, \quad A \cup B = \{a, b, c, d, e, x, y, z\}$

In set builder notation we can express $A \cup B$ as

$$A \cup B = \{x \,|\, x \in A \quad or \quad x \in B\}$$

The **intersection** of set A and set B, written $A \cap B$, is the set of all elements that are common to both set A *and* set B.

Examples of Intersection of Sets

$A = \{1, 2, 3, 4, 5\}, \quad B = \{3, 4, 5, 6, 7\}, \quad A \cap B = \{3, 4, 5\}$
$A = \{a, b, c, d, e\}, \quad B = \{x, y, z\}, \quad A \cap B = \{\ \}$

Note that in the last example sets A and B have no elements in common. Therefore, their intersection is the empty set. In set builder notation we can express $A \cap B$ as

$$A \cap B = \{x \,|\, x \in A \quad and \quad x \in B\}$$

COMMON STUDENT ERROR

Students commonly make the following errors when working with sets.

Correct	*Wrong*
1. $\{\ \}$ or $\varnothing$ is the empty set.	$\{\varnothing\}$ is not the empty set.
2. $3 \in \{1, 2, 3\}$	~~$3 \subseteq \{1, 2, 3\}$~~
	Since there are no braces around the 3, it is not a set. 3 is an element of the set, not a subset.
3. $\{3\} \subseteq \{1, 2, 3\}$	~~$\{3\} \in \{1, 2, 3\}$~~
	Since there are braces around the 3, $\{3\}$ is a set. Thus $\{3\}$ is a subset, not an element, of the set $\{1, 2, 3\}$.

▶**4**   A number line (see Fig. 1.2) can be used to illustrate sets of numbers.

FIGURE 1.2

Some important sets of numbers follow.

Important Sets of Numbers	
Real numbers	$\{x \mid x \text{ is a point on the number line}\}$
Natural or counting numbers	$\{1, 2, 3, 4, 5, \ldots\}$
Whole numbers	$\{0, 1, 2, 3, 4, 5, \ldots\}$
Integers	$\{\ldots, -3, -2, -1, 0, 1, 2, 3, \ldots\}$
Rational numbers	$\left\{\dfrac{p}{q} \,\middle|\, p \text{ and } q \text{ are integers, } q \neq 0\right\}$
Irrational numbers	$\{x \mid x \text{ is a real number that is not rational}\}$

Let us briefly look at the rational, irrational, and real numbers. A rational number is any number that can be represented as a quotient of two integers, with the denominator not zero.

Examples of Rational Numbers

$$\frac{3}{5}, \quad \frac{-2}{3}, \quad 0, \quad 1.63, \quad 7, \quad -12, \quad \sqrt{4}$$

Notice that 0, or any other integer, is also a rational number since it can be written as a fraction with a denominator of 1.

$$0 = \frac{0}{1}, \qquad 7 = \frac{7}{1}, \qquad -12 = \frac{-12}{1}$$

The number 1.63 can be written $\frac{163}{100}$ and is thus a quotient of two integers. Since $\sqrt{4} = 2$ and 2 is an integer, $\sqrt{4}$ is a rational number. *Every rational number when changed to a decimal number will be either a repeating or a terminating decimal number.*

Examples of Repeating Decimals   *Examples of Terminating Decimals*

$\dfrac{2}{3} = 0.6666\ldots$ $2 = 2.0$

$\dfrac{7}{3} = 2.3333\ldots$ $\dfrac{1}{2} = 0.5$

$\dfrac{1}{7} = 0.142857142857\ldots$ $\dfrac{7}{4} = 1.75$
(the block 142857 repeats)

Although $\sqrt{4}$ is a rational number, the square roots of most integers are not. Most square roots will be neither terminating nor repeating decimals when expressed as a decimal number, and are irrational numbers. Some irrational numbers are $\sqrt{2}$, $\sqrt{3}$, $\sqrt{5}$, and $\sqrt{6}$. Another irrational number is pi, π. When we give a decimal value for an irrational number, we are giving only an *approximation* of the value of the irrational number. The symbol $\approx$ means "is approximately equal to."

$$\pi \approx 3.14, \qquad \sqrt{2} \approx 1.41$$

▶**5** The **real numbers** are formed by taking the union of the rational numbers with the irrational numbers. Therefore, any real number must be either a rational number or an irrational number. The symbol $\mathbb{R}$ is often used to represent the set of real numbers. Figure 1.3 illustrates various real numbers on the number line.

FIGURE 1.3

Figure 1.4 illustrates the relationship between the various sets of numbers. Earlier we stated that the natural numbers were a subset of the set of whole numbers. In Figure 1.4, you see that the set of natural numbers is a subset of the set of whole numbers, the set of integers, and the set of rational numbers. Therefore, every natural number must also be a whole number, integer, and rational number. Using the same reasoning, we can see that the set of whole numbers is a subset of the set of integers and the set of rational numbers; and the set of integers is a subset of the set of rational numbers.

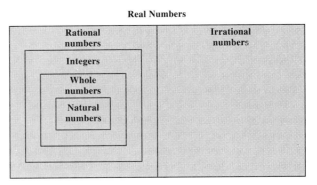

FIGURE 1.4

EXAMPLE 3   Consider the following set of elements.

$$\left\{-4, 0, \frac{3}{5}, 1.8, \sqrt{3}, -\sqrt{5}, \frac{19}{6}, 18, 4.62, -23, \pi\right\}$$

List the elements of the set that are:

(a) Natural numbers   (b) Whole numbers   (c) Integers

(d) Rational numbers   (e) Irrational numbers   (f) Real numbers

Solution: (a) 18
(b) 0, 18
(c) $-4, 0, 18, -23$
(d) $-4, 0, \frac{3}{5}, 1.8, \frac{19}{6}, 18, 4.62, -23$
(e) $\sqrt{3}, -\sqrt{5}, \pi$
(f) $-4, 0, \frac{3}{5}, 1.8, \sqrt{3}, -\sqrt{5}, \frac{19}{6}, 18, 4.62, -23, \pi$ ■

Not all numbers are real numbers. Some numbers that we discuss later in the text that are not real numbers include complex numbers and imaginary numbers.

Exercise Set 1.2

List each set in roster form.

1. $A = \{x \mid x$ is a natural number between 3 and 8$\}$
2. $B = \{x \mid x$ is an even integer between 5 and 10$\}$
3. $C = \{x \mid x$ is an even integer greater than or equal to 6 and less than or equal to 10$\}$
4. $D = \{x \mid x$ is a natural number greater than 5$\}$
5. $E = \{x \mid x$ is a whole number less than 7$\}$
6. $F = \{x \mid x$ is a whole number less than or equal to 7$\}$
7. $G = \{x \mid x$ is a natural number less than 0$\}$
8. $H = \{x \mid x$ is a whole number multiple of 5$\}$
9. $I = \{x \mid x$ is an integer greater than $-5\}$
10. $J = \{x \mid x$ is an integer between -6 and $-1\}$
11. $K = \{x \mid x$ is a whole number between 3 and 4$\}$

Insert either $\in$ or $\notin$ in the shaded area, to make a true statement.

12. $6 \ \ \{1, 3, 6, 9,\}$
13. $4 \ \ \{1, 2, 3\}$
14. $0 \ \ \{-1, 1, 3, 5\}$
15. $72 \ \ \{1, 2, 3, 4, \ldots, 80\}$
16. $-12 \ \ \{1, 2, 3, \ldots\}$
17. $\{3\} \ \ \{1, 2, 3, 4\}$
18. $\{5\} \ \ \{1, 2, 3, 4, 5, 6\}$
19. $\{0\} \ \ \{-3, -2, -1, 0, 1, 2, 3\}$
20. $\{-5\} \ \ \{-4, -3, -2\}$

Insert either $\subseteq$ or $\nsubseteq$ in the shaded area to make a true statement.

21. $\{3\} \ \ \{1, 2, 3, 4\}$
22. $\{0\} \ \ \{0, 1\}$
23. $\{5\} \ \ \{-1, -2, -3\}$
24. $\{4\} \ \ \{3, 4, 5, 6, \ldots\}$
25. $\{72\} \ \ \{6, 7, 8, \ldots, 70\}$
26. $\{-1\} \ \ \{-1, -2, -3\}$
27. $\{-1\} \ \ \{-4, -3, -2\}$
28. $3 \ \ \{1, 2, 3\}$
29. $0 \ \ \{-1, 0, 1\}$
30. $5 \ \ \{1, 2, 3, 4\}$

Let $N =$ the set of natural numbers, $W =$ the set of whole numbers, $I =$ the set of integers, $Q =$ the set of rational numbers, $H =$ the set of irrational numbers, and $\mathbb{R} =$ the set of real numbers. Insert either $\subseteq$ or $\nsubseteq$ in the shaded area to make a true statement.

31. $N \ \ W$
32. $W \ \ Q$
33. $I \ \ Q$
34. $W \ \ N$
35. $Q \ \ H$
36. $Q \ \ \mathbb{R}$
37. $H \ \ \mathbb{R}$
38. $Q \ \ I$
39. $Q \ \ N$
40. $\mathbb{R} \ \ Q$
41. $H \ \ Q$
42. $I \ \ N$
43. $N \ \ I$
44. $N \ \ \mathbb{R}$

Answer true or false.

45. 0 is a real number.
46. 0 is a rational number.
47. 0 is a natural number.
48. 0 is a whole number.
49. Some rational numbers are integers.
50. Some irrational numbers are rational numbers.
51. Some natural numbers are negative numbers.
52. Some whole numbers are natural numbers.
53. Every natural number is a whole number.
54. Every whole number is a natural number.
55. Every integer is a rational number.
56. Every rational number is an integer.
57. The union of the set of rational numbers with the set of irrational numbers forms the set of real numbers.
58. The intersection of the set of rational numbers with the set of irrational numbers is the empty set.
59. The set of natural numbers is a finite set.
60. The set of whole numbers is an infinite set.
61. The set of integers between 1 and 2 is the null set.
62. The set of rational numbers between 1 and 2 is an infinite set.

Consider the set of elements $\{-6, 4, \frac{1}{2}, \frac{5}{9}, 0, \sqrt{7}, \sqrt{5}, -1.23, \frac{99}{100}\}$. List the elements of the set that are:

63. Natural numbers
64. Whole numbers
65. Integers
66. Rational numbers
67. Irrational numbers
68. Real numbers

Consider the set of elements $\{2, 4, -5.33, \frac{9}{2}, \sqrt{7}, \sqrt{2}, -100, -7, 4.7\}$. *List the elements of the set that are:*

69. Whole numbers

70. Natural numbers

71. Rational numbers

72. Integers

73. Real numbers

74. Irrational numbers

Find $A \cup B$ *and* $A \cap B$ *for each set A and B.*

75. $A = \{1, 2, 3\}, B = \{2, 3, 4\}$

76. $A = \{2, 4, 6, 8\}, B = \{1, 3, 5, 7\}$

77. $A = \{-2, -4, -5\}, B = \{-1, -2, -4, -6\}$

78. $A = \{\ \}, B = \{1, 2, 3\}$

79. $A = \{\ \}, B = \{0, 1, 2, 3\}$

80. $A = \{-1, 0, 1\}, B = \{0, 2, 4, 6\}$

81. $A = \{2, 4, 6\}, B = \{2, 4, 6, 8, \ldots\}$

82. $A = \{1, 3, 5\}, B = \{1, 3, 5, 7, \ldots\}$

83. $A = \{0, 2, 4, 6, 8\}, B = \{1, 3, 5, 7\}$

84. $A = \{1, 2, 3, 4, \ldots\}, B = \{0, 1, 2, 3, 4, \ldots\}$

85. $A = \{1, 2, 3, 4, \ldots\}, B = \{2, 4, 6, 8, \ldots\}$

Describe each of the following sets.

86. $A = \{1, 2, 3, 4, \ldots\}$

87. $B = \{5, 7, 9, 11, \ldots\}$

88. $C = \{8, 10, 12, \ldots, 30\}$

89. $A = \{a, b, c, d, \ldots, z\}$

90. $B = \{\ldots, -5, -3, -1, 1, 3, 5, \ldots\}$

91. $C = \{\text{Alabama, Alaska, Arizona}, \ldots, \text{Wyoming}\}$

For exercises 92 and 93, (a) write out how you would read each set; (b) write the set in roster form.

92. $A = \{x \mid x \text{ is a natural number less than } 8\}$

93. $B = \{x \mid x \text{ is one of the last five capital letters in the English alphabet}\}$

94. Describe the counting numbers, whole numbers, integers, rational numbers, irrational numbers and real numbers. Explain the relationships among the sets of numbers.

1.3

Properties of the Real Numbers

▶ **1** Know the properties of the real numbers.

▶ **2** Know the multiplication property of 0 and the double negative property.

The properties discussed in this section are important for an understanding of algebra. You should understand them thoroughly before moving on to the next section. When introducing the properties, letters called **variables** will be used to represent numbers. In this section the letters a, b, and c will be used. In algebra the letters x, y, and z are most often used to represent variables, as will be seen later in the text.

In algebra we do not use an $\times$ to indicate multiplication because it may be confused with the variable x. To indicate multiplication a raised dot may be used. When a number and a variable or two variables are placed next to one another, multiplication is also indicated. For example, both $2 \cdot x$ and $2x$ mean two times x, and both $x \cdot y$ and xy mean x times y.

The term **algebraic expression,** or simply **expression** will be used often in the text. An expression is any combination of numbers, variables, exponents, mathematical symbols, and mathematical operations.

TABLE 1.1 Properties of Real Numbers

For Real Numbers a, b, and c:	Addition	Multiplication
Commutative property	$a + b = b + a$	$ab = ba$
Associative property	$(a + b) + c = a + (b + c)$	$(ab)c = a(bc)$
Identity property	$a + 0 = 0 + a = a$ $\left(\begin{array}{c}\text{0 is called the } \textbf{additive} \\ \textbf{identity element}\end{array}\right)$	$a \cdot 1 = 1 \cdot a = a$ $\left(\begin{array}{c}\text{1 is called the } \textbf{multiplicative} \\ \textbf{identity element}\end{array}\right)$
Inverse property	$a + (-a) = (-a) + a = 0$ $\left(\begin{array}{c}-a \text{ is called the } \textbf{additive} \\ \textbf{inverse} \text{ or } \textbf{opposite} \text{ of } a\end{array}\right)$	$a \cdot \dfrac{1}{a} = \dfrac{1}{a} \cdot a = 1$ $\left(\begin{array}{c}1/a \text{ is called the } \textbf{multiplicative} \\ \textbf{inverse} \text{ or } \textbf{reciprocal} \text{ of } a\end{array}\right)$
Distributive property (of multiplication over addition)	$a(b + c) = ab + ac$	

▶ **1** Table 1.1 lists the basic properties for the operations of addition and multiplication on the real numbers.

The distributive property applies when there are more than two numbers within the parentheses.

$$a(b + c + d + \cdots + n) = ab + ac + ad + \cdots + an$$

This expanded form of the distributive property is called the **extended distributive property.**

Note that the commutative property involves a change in *order*, and the associative property involves a change in *grouping*.

EXAMPLE 1 Name each property illustrated.

(a) $6 \cdot x = x \cdot 6$ (b) $(x + 2) + 3 = x + (2 + 3)$
(c) $x + 3 = 3 + x$ (d) $3(x + 2) = 3x + 3(2) = 3x + 6$

Solution: (a) Commutative property of multiplication.
(b) Associative property of addition.
(c) Commutative property of addition.
(d) Distributive property. ∎

EXAMPLE 2 Name each property illustrated.

(a) $4 \cdot 1 = 4$ (b) $x + 0 = x$
(c) $4 + (-4) = 0$ (d) $1(x + y) = x + y$

Solution: (a) Identity property of multiplication.
(b) Identity property of addition.
(c) Inverse property of addition.
(d) Identity property of multiplication. ∎

EXAMPLE 3 Write the additive inverse (or opposite) and multiplicative inverse (or reciprocal) of the following.

(a) -3 (b) $\frac{2}{3}$

Solution: (a) The additive inverse is 3. The multiplicative inverse is $\dfrac{1}{-3} = -\dfrac{1}{3}$.

(b) The additive inverse is $\dfrac{-2}{3}$. The multiplicative inverse is $\dfrac{1}{\frac{2}{3}} = \dfrac{3}{2}$.

▶ **2**   Two other important properties follow.

For any real number a:

Multiplication Property of 0

$$a \cdot 0 = 0 \cdot a = 0$$

Double Negative Property

$$-(-a) = a$$

EXAMPLE 4   Name the following properties.

(a) $3 \cdot 0 = 0$ (b) $-(-4) = 4$ (c) $-(-x) = x$

Solution: (a) Multiplication property of 0.
(b) Double negative property.
(c) Double negative property.

Exercise Set 1.3

Name the property.

1. $4 + 2 = 2 + 4$

2. $x + y = y + x$

3. $3(4 + 5) = 3 \cdot 4 + 3 \cdot 5$

4. $3(x + 2) = 3x + 6$

5. $(3 + 6) + 2 = 3 + (6 + 2)$

6. $(3 + x) + 5 = 3 + (x + 5)$

7. $x + 2 = 2 + x$

8. $(x + 3) + 6 = x + (3 + 6)$

9. $3 \cdot x = x \cdot 3$

10. $x = 1 \cdot x$

11. $x + 0 = x$

12. $2(x + 2) = 2x + 4$

13. $2 + (-4) = (-4) + 2$

14. $x(y + z) = xy + xz$

15. $3y + 4 = 4 + 3y$

16. $(2x \cdot 3y) \cdot 4y = 2x \cdot (3y \cdot 4y)$

17. $3x + 2y = 2y + 3x$

18. $3(x + y) = 3x + 3y$

19. $4(x + y + 2) = 4x + 4y + 8$

20. $3(2b) = (3 \cdot 2)b$

21. $-(-1) = 1$

22. $5 + 0 = 5$

23. $5 \cdot 1 = 5$

24. $4 \cdot \dfrac{1}{4} = 1$

25. $3 + (-3) = 0$

26. $6 \cdot 0 = 0$

27. $x + (-x) = 0$

28. $x \cdot \dfrac{1}{x} = 1$

29. $-4 + 4 = 0$

30. $(x + y) = 1(x + y)$

31. $(x + 2) = 1(x + 2)$

32. $\left(-\dfrac{1}{2}\right)(-2) = 1$

33. $3 \cdot \dfrac{1}{3} = 1$

34. $1(x + y) = 1x + 1y$

35. $-\left(-\dfrac{1}{2}\right) = \dfrac{1}{2}$

36. $-5 + 5 = 0$

37. $3 \cdot 0 = 0$

38. $x \cdot 0 = 0$

39. $-(-3) = 3$

40. $-(-x) = x$

Fill in the statement on the right side of the equal sign using the property indicated.

41. $x + 3 =$ 　　　　　Commutative property
　　　　　　　　　　　of addition

42. $2 \cdot y =$ 　　　　　Commutative property
　　　　　　　　　　　of multiplication

43. $(x + 2) + 3 =$ 　　Associative property
　　　　　　　　　　　of addition

44. $3(2x + 5) =$ 　　　Distributive property

45. $x + 0 =$ 　　　　　Identity property of
　　　　　　　　　　　addition

46. $3(x + y + 4) =$ 　　Distributive property

47. $1 \cdot x =$ 　　　　　Identity property of
　　　　　　　　　　　multiplication

48. $(x \cdot 3) \cdot 4 =$ 　　Associative property of
　　　　　　　　　　　multiplication

49. $-(-x) =$ 　　　　　Double negative
　　　　　　　　　　　property

50. $5x + (2y + 3x) =$ 　Associative property
　　　　　　　　　　　of addition

51. $a \cdot 0 =$ 　　　　　Multiplication property
　　　　　　　　　　　of zero

52. $5(x + y) =$ 　　　　Distributive property

53. $1(x + y) =$ 　　　　Distributive property

54. $0 + x =$ 　　　　　Identity property
　　　　　　　　　　　of addition

55. $3 \cdot 1 =$ 　　　　　Identity property of
　　　　　　　　　　　multiplication

56. $a \cdot \dfrac{1}{a} =$ 　　　　Inverse property
　　　　　　　　　　　of multiplication

57. $a + (-a) =$ 　　　　Inverse property of
　　　　　　　　　　　addition

58. $-(-2) =$ 　　　　　Double negative
　　　　　　　　　　　property

List both the additive inverse and the multiplicative inverse for each of the following.

59. 4

60. 6

61. -3

62. -5

63. $\dfrac{2}{3}$

64. $\dfrac{5}{2}$

65. -6

66. -12

67. $-\dfrac{3}{7}$

68. $-\dfrac{2}{9}$

69. (a) List the commutative property of addition.
　　(b) In your own words write the meaning of the property.

70. (a) List the associative property of multiplication.
　　(b) In your own words write the meaning of the property.

71. (a) List the distributive property of multiplication over addition.
　　(b) In your own words write the meaning of the property.

72. Using an example, explain why addition is not distributive over multiplication. That is, explain why $a + (b \cdot c) \neq (a + b) \cdot (b + c)$.

Cumulative Review Exercises

[1.2] **73.** Answer true or false. Every irrational number is a real number.

74. Insert either $\subseteq$ or $\nsubseteq$ in the shaded area to make a true statement. $\{3\}$ ▓ $\{3, 4, 5\}$

75. Consider the set $\{3, 4, -2, \frac{5}{6}, \sqrt{3}, 0\}$. List the elements that are (a) integers, (b) rational numbers, (c) irrational numbers, (d) real numbers

76. $A = \{4, 7, 9, 12\}$; $B = \{1, 4, 7, 15\}$. Find (a) $A \cup B$; (b) $A \cap B$.

1.4

Inequalities and Absolute Value

▸ **1** Identify and use inequality symbols.

▸ **2** Evaluate expressions containing absolute value.

▸ **1** The symbols used to indicate an inequality are $>$, $\geq$, $<$, $\leq$, and $\neq$.

Inequality Symbols

$>$ is read "is greater than."
$\geq$ is read "is greater than or equal to."
$<$ is read "is less than."
$\leq$ is read "is less than or equal to."
$\neq$ is read "is not equal to."

The number a is greater than the number b, $a > b$, when a is to the right of b on the number line (Fig. 1.5). We can also state that the number b is less than a, $b < a$, when b is to the left of a on the number line. The inequality $a \neq b$ means either $a < b$ or $a > b$.

FIGURE 1.5

EXAMPLE 1   Insert either $>$ or $<$ in the shaded area between the numbers to make the statement true.

(a) 7 ▨ 3   (b) -4 ▨ -2 (c) 0 ▨ -2

Solution: Draw a number line illustrating the location of all the given points (Fig. 1.6).

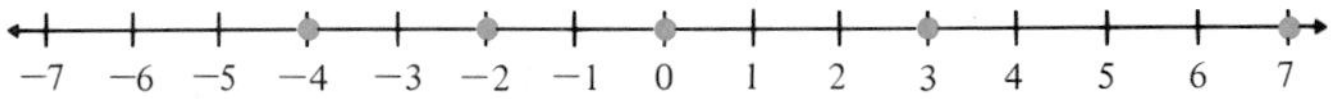

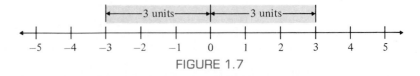

FIGURE 1.6

(a) $7 > 3$ Note 7 is to the right of 3 on the number line.
(b) $-4 < -2$ Note -4 is to the left of -2 on the number line.
(c) $0 > -2$ Note 0 is to the right of -2 on the number line. ■

Remember that the symbol used in an inequality, if it is true, always points to the smaller of the two numbers.

Absolute Value

▸ **2** The **absolute value** of a number is its distance from the number zero on the number line. The symbol $||$ is used to indicate absolute value.

FIGURE 1.7

Consider the numbers 3 and -3 (see Fig. 1.7). Both numbers are 3 units from zero on the number line. Thus

$$|3| = 3$$
$$|-3| = 3$$

EXAMPLE 2   Evaluate each of the following.

(a) $|4|$ (b) $|-8|$ (c) $|0|$ (d) $\left|\dfrac{-5}{3}\right|$

Solution: (a) $|4| = 4$ since 4 is 4 units from zero on the number line
(b) $|-8| = 8$ since -8 is 8 units from zero on the number line
(c) $|0| = 0$
(d) $\left|-\dfrac{5}{3}\right| = \dfrac{5}{3}$

The following is the definition of absolute value.

Absolute Value

If a represents any real number, then

$$|a| = \begin{cases} a & \text{if } a \geq 0 \\ -a & \text{if } a < 0 \end{cases}$$

The definition of absolute value indicates that the absolute value of any nonnegative number is the number itself, and the absolute value of any negative number is the additive inverse (or opposite) of the number. The absolute value of a number can be found by using the definition as illustrated in the following examples.

$$|3| = 3 \qquad \text{since 3 is greater than or equal to 0}$$
$$|8| = 8 \qquad \text{since 8 is greater than or equal to 0}$$
$$|0| = 0 \qquad \text{since 0 is greater than or equal to 0}$$
$$|-4| = -(-4) = 4 \qquad \text{since } -4 \text{ is less than 0}$$
$$|-12| = -(-12) = 12 \qquad \text{since } -12 \text{ is less than 0}$$

The absolute value of any nonzero number will always be a positive number, and the absolute value of zero is zero.

EXAMPLE 3   Insert $>$, $<$, or $=$ in the shaded area between the two values to make the statement true.

(a) $|-3| \blacksquare |3|$ (b) $-3 \blacksquare |-4|$ (c) $|6| \blacksquare |8|$
(d) $-|4| \blacksquare |-3|$ (e) $-|-6| \blacksquare |-5|$

Solution: (a) $|-3| = |3|$ since both $|-3|$ and $|3|$ equal 3.
(b) $-3 < |-4|$ since $|-4| = 4$ and $-3 < 4$.
(c) $|6| < |8|$ since $|6| = 6$ and $|8| = 8$.
(d) $-|4| < |-3|$ since $-|4| = -4$ and $|-3| = 3$.
(e) $-|-6| < |-5|$ since $-|-6| = -6$ and $|-5| = 5$.

Exercise Set 1.4

Insert either < or > in the shaded area to make the statement true.

1. 4 ▒ 2

2. 6 ▒ 8

3. 5 ▒ 3

4. −2 ▒ 4

5. 5 ▒ −3

6. 0 ▒ −1

7. 0 ▒ 4

8. 9 ▒ −2

9. −1 ▒ 1

10. $-\dfrac{1}{2}$ ▒ −1

11. −3 ▒ −3.5

12. 1 ▒ −3

13. −4 ▒ −2

14. −6 ▒ −1

15. −2 ▒ −5

16. −1 ▒ −4

17. 4 ▒ −4

18. 3 ▒ −3.5

19. 1.1 ▒ 1.9

20. −1.1 ▒ −1.9

21. −52 ▒ −55

22. −78 ▒ −65

23. $-\dfrac{7}{8}$ ▒ $-\dfrac{8}{9}$

24. $-\dfrac{4}{7}$ ▒ $-\dfrac{5}{9}$

Evaluate the absolute value expression.

25. $|6|$

26. $|3|$

27. $|-4|$

28. $|-6|$

29. $|-2|$

30. $|0.5|$

31. $\left|-\dfrac{1}{2}\right|$

32. $|-3|$

33. $|0|$

34. $|-22|$

35. $|45|$

36. $|9.6|$

37. $|-13.84|$

38. $|-0.7|$

39. $-|7|$

40. $-|-7|$

41. $-|-3|$

42. $-|-8|$

43. $-\left|\dfrac{5}{9}\right|$

44. $-\left|-\dfrac{5}{7}\right|$

Insert <, >, or = in the shaded area to make the statement true.

45. $|6|$ ▒ $|-6|$

46. $|9|$ ▒ $|3|$

47. $|-9|$ ▒ $|3|$

48. -4 ▒ $|-4|$

49. $|-10|$ ▒ -5

50. -4 ▒ $-|4|$

51. $|-3|$ ▒ $|-2|$

52. $|-4|$ ▒ 6

53. $|-20|$ ▒ $-|24|$

54. $|-16|$ ▒ $-|30|$

55. $-|4|$ ▒ $-|8|$

56. $-|-31|$ ▒ $|-5|$

57. 6 ▒ $|-12|$

58. $|25|$ ▒ $|-23|$

59. $|19|$ ▒ $|-25|$

60. $-|-3|$ ▒ $|-9|$

List the values from smallest to largest.

61. $6, 2, -1, |3|, |-5|$

62. $0, -4, -8, -|12|, |-10|$

63. $4, -2, 8, |-6|, -|3|$

64. $|9|, |4|, |-12|, |3|, |-5|$

65. $-3, |0|, |-5|, |7|, |-12|$

66. $5, -|7|, -9, |15|, |-1|$

67. $12, 24, |36|, |-9|, |-45|$

68. $-8, -12, -|9|, -|20|, -|-18|$

69. $6, -4, -|-6|, |-8|, -2$

70. $3, 3.1, 3.4, |3.9|, |-3.6|$

71. $-2.1, -2, -2.4, |-2.8|, -|2.9|$

72. $-7, -7.1, -7.8, -|7.3|, |-7.4|$

73. $\frac{1}{3}, |-\frac{1}{2}|, -2, |\frac{3}{5}|, |-\frac{3}{4}|$

74. $|-\frac{5}{2}|, \frac{3}{5}, |-3|, |-\frac{5}{3}|, |-\frac{2}{3}|$

Find the unknown number(s) in each of the following.

75. The absolute value of the number 6

76. The absolute value of the number −8

77. All numbers whose absolute value is 4

78. All numbers whose absolute value is 10

79. All numbers such that, if 6 is added to the number, the absolute value of the result is 11

80. All numbers such that, if 2 is added to the number, the absolute value of the result is 5

81. All numbers a such that $|a| = |-a|$

82. All numbers a such that $|a| = a$

83. All numbers a such that $|a| = -a$

84. All numbers a such that $|a| = -3$

85. All numbers a such that $|a| = 5$

86. All numbers x such that $|x - 3| = |3 - x|$

Write out each of the following statements in words.

87. $|7| > 4$

88. $|-2| < |-5|$

89. $-4 = -|4|$

90. $|-5| > -|5|$

91. (a) Give the definition of absolute value.
(b) In your own words, explain what the definition means.

92. If the absolute value of one number is added to the absolute value of another number, can the sum ever be a negative number? Explain your answer.

93. If the absolute value of one number is subtracted from the absolute value of another number, will the difference always be a positive number? Explain your answer.

Cumulative Review Exercises

[1.2] **94.** Explain the difference between a rational number and an irrational number.

[1.3] *Name each of the following properties:*
95. $3x + 4 = 4 + 3x$

96. $5(x + y - 3) = 5x + 5y - 15$
97. $(4x + y) + 3 = 4x + (y + 3)$

1.5

Addition, Subtraction, Multiplication, and Division of Real Numbers

▶ **1** Add real numbers.

▶ **2** Subtract real numbers.

▶ **3** Multiply real numbers.

▶ **4** Divide real numbers.

▶ **5** Write fractions with positive denominators.

Addition of Real Numbers

▶**1** To be successful in algebra, you must have an understanding of how to add and subtract real numbers. We will use absolute value in explaining how real numbers may be added or subtracted. We first discuss how to add two numbers with the same sign, either both positive or both negative, and then how to add two numbers with different signs, one positive and the other negative.

> **To Add Two Numbers with the Same Sign** (Both Positive or Both Negative)
>
> Add their absolute values and place the common sign before the sum.

The sum of two positive numbers will be a positive number, and the sum of two negative numbers will be a negative number.

EXAMPLE 1   Add $-2 + (-5)$.

Solution: Since both numbers being added are negative, the sum will be negative. To find the sum, add the absolute values of these numbers and place a negative sign before the answer.

$$|-2| = 2$$
$$|-5| = \underline{5}$$
$$7$$

Since both numbers are negative, the sum must be negative. Thus

$$-2 + (-5) = -7$$ ∎

EXAMPLE 2   Add $-16 + (-25)$.

Solution:

$$|-16| = 16$$
$$|-25| = \underline{25}$$
$$41$$

Since both numbers are negative, the sum must be negative.

$$-16 + (-25) = -41$$ ∎

> **To Add Two Numbers with Different Signs** (One Positive and the Other Negative)
>
> Take the difference between the absolute values. The answer is positive if the positive number has the larger absolute value. The answer is negative if the negative number has the larger absolute value.

The sum of a positive number and a negative number may be either positive or negative. The sign of the answer will be the same as the sign of the number with the larger absolute value.

EXAMPLE 3   Add $5 + (-2)$.

Solution: Since the numbers being added are of opposite signs, we take the difference between the smaller absolute value and the larger. First evaluate each absolute value.

$$|5| = 5$$
$$|-2| = 2$$

Now find the difference

$$\begin{array}{r} 5 \\ \underline{-2} \\ 3 \end{array}$$

The number 5 has a larger absolute value than the number -2, so the answer is positive.

$$5 + (-2) = 3$$ ∎

EXAMPLE 4   Add $-5 + 2$.

Solution:

$$\begin{array}{lr} |-5| = 5 & 5 \\ |2| = 2 & \underline{-2} \\ & 3 \end{array}$$

The number -5 has a larger absolute value than the number 2, so the answer is negative.

$$-5 + 2 = -3$$ ∎

EXAMPLE 5   Evaluate $6 + (-8)$.

Solution:

$$|6| = 6 \qquad 8$$
$$|-8| = 8 \qquad \frac{-6}{2}$$

The number -8 has a larger absolute value than the number 6, so the answer is negative.

$$6 + (-8) = -2$$ ∎

EXAMPLE 6   A submarine descends 400 feet below sea level. A short while later it descends another 250 feet. What is the submarine's depth with respect to sea level?

Solution: Consider motion in the downward direction to be negative and motion in the upward direction to be positive.

$$\text{Distance} = -400 + (-250) = -650 \text{ feet}$$

The submarine is 650 feet below sea level. ∎

Subtraction of Real Numbers

▶ **2**   Consider the subtraction problem $5 - 2$. To evaluate this problem, we must subtract a positive 2 from a positive 5.

$$\underset{\uparrow\ \text{subtract}}{5 - 2} \quad \text{means} \quad 5 - \overset{\overset{\textbf{positive 2}}{\downarrow}}{\underset{\uparrow\ \text{subtract}}{(+2)}}$$

Every subtraction problem can be expressed as an addition problem using the following rule.

Subtraction of Real Numbers

$$a - b = a + (-b)$$

This rule says that **to subtract b from a, add the opposite (or additive inverse) of b to a.**

In the problem $5 - 2$, which means $5 - (+2)$, the opposite of $+2$ is -2. Thus

$$5 - 2 = 5 + (-2)$$

subtract   positive   add   negative
$\qquad\qquad$ 2 $\qquad\qquad$ 2

We can now find the sum of $5 + (-2)$ to be 3 using the method for adding real numbers presented earlier in this section. Therefore, $5 - 2 = 3$.

EXAMPLE 7   Evaluate $6 - 10$.

Solution: $6 - 10 = 6 + (-10) = -4$ ∎

EXAMPLE 8   Evaluate $-8 - 4$.

Solution: $-8 - 4 = -8 + (-4) = -12$ ∎

EXAMPLE 9   Evaluate $8 - (-10)$.

Solution: This problem is somewhat different since we are subtracting a negative number. The procedure to evaluate remains the same.

$$8 - (-10) = 8 + 10 = 18$$

subtract negative   add   positive
            10   10

Thus $8 - (-10) = 18$. ∎

By studying Example 9 and similar problems, we can see that for any real numbers a and b

$$a - (-b) = a + b$$

We can use this principle to evaluate such problems as $8 - (-10)$ and other problems where we *subtract a negative quantity*.

$$8 - (-10) = 8 + 10 = 18$$

EXAMPLE 10   Evaluate $-4 - (-12)$.

Solution: $-4 - (-12) = -4 + 12 = 8$ ∎

EXAMPLE 11   Subtract 35 from -42.

Solution: $-42 - 35 = -77$ ∎

In Example 12 we will subtract fractions. To add or subtract fractions, the fractions must have a common denominator. When adding or subtracting fractions that do not have a common denominator, rewrite each fraction with the least common denominator, LCD. The **least common denominator** of a set of denominators is the smallest number divisible by all the denominators. For example, if the denominators of two fractions being added or subtracted are 6 and 10, then the LCD is 30, since 30 is the smallest number divisible by both 6 and 10. If you have forgotten how to find the LCD, review an arithmetic book or see your instructor. If you have forgotten the procedures to add, subtract, multiply, or divide fractions, you should review Appendix A now.

EXAMPLE 12   Subtract $-\dfrac{3}{5}$ from $-\dfrac{5}{9}$.

Solution: $-\dfrac{5}{9} - \left(-\dfrac{3}{5}\right) = -\dfrac{5}{9} + \dfrac{3}{5} = -\dfrac{25}{45} + \dfrac{27}{45} = \dfrac{2}{45}$ ∎

EXAMPLE 13   The highest point in the United States, Mt. McKinley in Alaska, is 20,320 feet above sea level. The lowest point in the United States, the Verdigris River in Kansas, is 680 feet below sea level. Find the vertical height difference between the two locations.

Solution: $20{,}320 - (-680) = 20{,}320 + 680 = 21{,}000$ feet ∎

Addition and subtraction are often combined in the same problems, as the following examples illustrate. Unless parentheses are present, if the expression involves only addition and subtraction we evaluate the expression from left to right. When parentheses are used, we evaluate the expression within the parentheses first; then we evaluate from left to right.

EXAMPLE 14 Evaluate $-2 - (4 - 8) - 3$.

Solution:
$$
\begin{aligned}
-2 - (4 - 8) - 3 &= -2 - (-4) - 3 \\
&= -2 + 4 - 3 \\
&= 2 - 3 \\
&= -1
\end{aligned}
$$
∎

EXAMPLE 15 Evaluate $6 - (-2) + (7 - 13) - 9$.

Solution:
$$
\begin{aligned}
6 - (-2) + (7 - 13) - 9 &= 6 + 2 + (-6) - 9 \\
&= 8 + (-6) - 9 \\
&= 2 - 9 \\
&= -7
\end{aligned}
$$
∎

EXAMPLE 16 Evaluate $2 - |-3| + 4 - (6 - |-3|)$.

Solution: Begin by replacing the numbers in absolute value signs with their numerical equivalents; then evaluate.
$$
\begin{aligned}
2 - |-3| + 4 - (6 - |-3|) &= 2 - 3 + 4 - (6 - 3) \\
&= 2 - 3 + 4 - (3) \\
&= -1 + 4 - 3 \\
&= 3 - 3 \\
&= 0
\end{aligned}
$$
∎

Multiplication of Real Numbers

▶ **3** The following rules are used in determining the sign of the product when two numbers are multiplied.

Multiplication of Real Numbers

1. The product of two numbers with **like** signs is a **positive** number.
2. The product of two numbers with **unlike** signs is a **negative** number.

EXAMPLE 17 Evaluate (a) $(4)(-3)$; (b) $(-16)\left(-\dfrac{1}{2}\right)$.

Solution: (a) $(4)(-3) = -12$ The numbers have unlike signs.

(b) $(-16)\left(-\dfrac{1}{2}\right) = 8$ The numbers have like signs, both negative. ∎

EXAMPLE 18   Evaluate $\left(\dfrac{-3}{5}\right)\left|\dfrac{-6}{7}\right|$.

Solution: $\left|\dfrac{-6}{7}\right| = \dfrac{6}{7}$; therefore,

$$\left(\dfrac{-3}{5}\right)\left|\dfrac{-6}{7}\right| = \left(\dfrac{-3}{5}\right)\left(\dfrac{6}{7}\right) = \dfrac{-3 \cdot 6}{5 \cdot 7} = \dfrac{-18}{35}$$ ∎

EXAMPLE 19   Evaluate $4(-2)(-3)(1)$.

Solution: $4(-2)(-3)(1) = (-8)(-3)(1) = 24(1) = 24$ ∎

When multiplying more than two numbers, the product will be *negative* when there is an *odd* number of negative numbers. The product will be *positive* when there is an *even* number of negative numbers.

Division of Real Numbers

▶ **4**   The rules for the division of real numbers are very similar to those for multiplication of real numbers.

> **Division of Real Numbers**
>
> 1. The quotient of two numbers with **like** signs is a **positive** number.
> 2. The quotient of two numbers with **unlike** signs is a **negative** number.

EXAMPLE 20   Evaluate (a) $-24 \div 6$;   (b) $-6.4 \div (-0.4)$.

Solution: (a) $\dfrac{-24}{6} = -4$ The numbers have unlike signs.

(b) $\dfrac{-6.4}{-0.4} = 16$ The numbers have like signs.   ∎

EXAMPLE 21   Evaluate $\dfrac{-3}{8} \div \left|\dfrac{-2}{5}\right|$.

Solution: Since $\left|\dfrac{-2}{5}\right|$ is equal to $\dfrac{2}{5}$, we write

$$\dfrac{-3}{8} \div \left|\dfrac{-2}{5}\right| = \dfrac{-3}{8} \div \dfrac{2}{5}$$

Now invert the divisor and proceed as in multiplication.

$$\dfrac{-3}{8} \div \dfrac{2}{5} = \dfrac{-3}{8} \cdot \dfrac{5}{2} = \dfrac{-3 \cdot 5}{8 \cdot 2} = -\dfrac{15}{16}$$ ∎

▶ **5**   When the denominator of a fraction is a negative number, we usually rewrite the fraction with a positive denominator. To do this, we make use of the following fact.

If a and b represent any real numbers, $b \neq 0$, then

$$\frac{a}{-b} = \frac{-a}{b} = -\frac{a}{b}$$

Thus, when we have a quotient of $\dfrac{1}{-2}$, we rewrite it as either $\dfrac{-1}{2}$ or $-\dfrac{1}{2}$.

Exercise Set 1.5

Evaluate.

1. $4 + (-3)$ **2.** $9 + (-8)$ **3.** $12 + (-2)$ **4.** $14 + (-7)$

5. $-3 + 8$ **6.** $-4 + 12$ **7.** $-9 + 17$ **8.** $-36 + 19$

9. $-16 - (-5)$ **10.** $-32 - (-14)$ **11.** $35 - (-4)$ **12.** $-6.28 - 3.14$

13. $-9.5 - (-3.72)$ **14.** $\dfrac{5}{6} - \dfrac{4}{5}$ **15.** $-\dfrac{3}{8} - \left(-\dfrac{5}{7}\right)$ **16.** $-3 - \dfrac{5}{12}$

Evaluate.

17. $4 + 6 - 3$ **18.** $7 - 4 - 8$ **19.** $6.23 - 4.5 - (-9.67)$

20. $5 + (-0.43) - 6.97$ **21.** $-6 - 4 - \frac{1}{2}$ **22.** $(4 - \frac{2}{3}) + (6 - 8)$

23. $-3 + (4 - 9) + 3$ **24.** $-2 + (4 - 6) - (3 - 8)$ **25.** $-(-4 + 2) + (-6 + 3) + 2$

26. $4 - (8 - 9) + (-6 + 8)$ **27.** $|4| - |3| + |1|$ **28.** $6 - |3| + 4$

29. $3 - |-8| - 5$ **30.** $|9 - 4| - 6$ **31.** $|6 - 9| - 5$

32. $|12 - 5| - |5 - 12|$ **33.** $-|-3| - |7| + (6 + |-2|)$ **34.** $|-4| - |-4| - |-4 - 4|$

Evaluate.

35. $-4 \cdot 12$ **36.** $(-8)(-9)$ **37.** $-4\left(-\dfrac{5}{16}\right)$ **38.** $-4\left(-\dfrac{3}{4}\right)\left(-\dfrac{1}{2}\right)$

39. $(-1)(-1)(-1)(2)(-3)$ **40.** $(4)(-2)(-3)(4)(5)$ **41.** $-6 \div 2$ **42.** $6 \div (-2)$

43. $-3 \div (-3)$ **44.** $-16 \div 8$ **45.** $36 \div \left(-\dfrac{1}{4}\right)$ **46.** $-80 \div (-10)$

47. $-\dfrac{5}{9} \div \dfrac{-5}{9}$ **48.** $-2|4|$ **49.** $-3|8|$ **50.** $\left|-\dfrac{1}{2}\right| \cdot \left|\dfrac{-3}{4}\right|$

51. $\left|\dfrac{3}{5}\right| \cdot \left|\dfrac{-10}{6}\right|$ **52.** $\left|\dfrac{-4}{7}\right| \div \dfrac{1}{14}$ **53.** $\left|\dfrac{3}{8}\right| \div (-2)$ **54.** $\left|\dfrac{-2}{3}\right| \div \left|\dfrac{-1}{2}\right|$

55. $\dfrac{-5}{9} \div |-5|$ **56.** $\dfrac{-3}{8} \div \left|\dfrac{-5}{4}\right|$

Evaluate.

57. $5 - 7$ **58.** $-16 - 8$ **59.** $-64 \div 4$

60. $-20 \div (-2)$ **61.** $-\dfrac{3}{5} - \dfrac{5}{9}$ **62.** $4\left(-\dfrac{8}{5}\right)\left(\dfrac{5}{2}\right)$

63. $3 - (-4) + 6 - 3$ **64.** $8 + (-4) + (-3 + 1)$ **65.** $(4)(-1)(6)(-2)(-2)$

66. $(1)(-3)(-4)(2)$ **67.** $-6 - 6 - (6 + 6) - 3$ **68.** $9 - (4 - 3) - (-2 - 1)$

69. $-|4| \cdot \left|\dfrac{-1}{2}\right|$ **70.** $-\left|\dfrac{-12}{5}\right| \cdot \left|\dfrac{3}{4}\right|$ **71.** $|-1| \div \dfrac{5}{12}$

72. $\left|\dfrac{-9}{4}\right| \div \left|\dfrac{-4}{9}\right|$ **73.** $(-|3| + |5|) - (6 - |-9|)$ **74.** $5 - |-2| + 3 - |-5|$

75. $4 - |8| + (4 - 6) - |12|$ **76.** $-(-9 - 4) - |-3| + 2$ **77.** $(|-4| - 3) - (3 \cdot |-5|)$

78. $(|-16| \div |-4|) \div (-4)$ **79.** $(25 - |36|)(-6 - 5)$ **80.** $\left[(-2)\left|-\dfrac{1}{2}\right|\right] \div \left|-\dfrac{1}{4}\right|$

81. A submarine dives 350 feet. A short time later the submarine comes up 180 feet. Find the submarine's distance with respect to its starting point (consider distance in a downward direction as negative).

82. In New York City, the temperature during a 24-hour period dropped from 46°F to −12°F. Find the change in temperature.

83. Sue made a profit of $3225 in the stock market over a given period. During the same period, Nelson lost $1088. Find the difference in their performances in the stock market.

84. On their first play, the Brown College football team lost 32 yards. On their second play, they gained 25 yards. What is the gain or loss for the two plays?

85. When Jon Bon Jovi signed a contract to record his latest album, he received an advance payment of $350,000. When the album is released and sales begin, the recording company will automatically deduct this $350,000 advance from his royalties.

   (a) If two weeks after the album's release the royalties for the album total $267,000 before the advance is deducted, find the amount of money Jon Bon Jovi will receive from or owe to the record company.

   (b) If two months after the release of the album, his royalties total $1,400,000, find the amount of money he will receive from or owe to the recording company.

See Exercise 84.

86. Mr. Adams had $4264 income tax withheld in 1992 by his employer. Mr. Adams also received income from a second job from which no income tax was withheld. If Mr. Adams' total income tax for the 1992 year was $6053, determine the amount of Mr. Adams' income tax refund, or the balance he owed to the Internal Revenue Service.

Answer each true or false. If false, give an example to illustrate that the statement is false. Such an example is often called a **counterexample.**

87. The product of an odd number of negative numbers is always a negative number.

88. The sum of two positive numbers is always a positive number.

89. The difference of two positive numbers is always a positive number.

90. The product of two positive numbers is always a positive number.

91. The quotient of two positive numbers is always a positive number.

92. The product of two negative numbers is always a negative number.

93. The product of two negative numbers is always a positive number.

94. The sum of two negative numbers is always a positive number.

95. The sum of a positive number and a negative number is always a negative number.

96. The product of a positive number and a negative number is always a negative number.

97. The difference of two negative numbers is always a negative number.

98. The difference of a positive number and a negative number is always a negative number.

99. The quotient of a positive number and a negative number is always a negative number.

100. The product of an even number of negative numbers is always a negative number.

101. Explain in your own words how to add two numbers with different signs.

102. Explain how the rules for multiplication and division of real numbers are similar.

103. Write your own realistic word problem that involves subtracting a positive number from a negative number. Indicate the answer to your word problem.

104. Write your own realistic word problem that involves subtracting a negative number from a negative number. Indicate the answer to your problem.

Cumulative Review Exercises

Name each of the following properties.

[1.3] **105.** $x \cdot 1 = x$

106. $x + 0 = x$

[1.4] **107.** Insert either $>$, $<$, or $=$ in the shaded area to make the statement true: $|-4|$ ▨ -3.

108. List from smallest to largest: $|-4|, |6|, |-5|, -5, -1, -|-7|$.

JUST FOR FUN

1. Evaluate $1 - 2 + 3 - 4 + \cdots + 99 - 100$.
(*Hint:* Group in pairs of two numbers.)

2. Evaluate $1 + 2 - 3 + 4 + 5 - 6 + 7 + 8 - 9 + 10 + 11 - 12 + \cdots + 22 + 23 - 24$.
(*Hint:* Examine in groups of three numbers.)

3. Evaluate $\dfrac{(1) \cdot |-2| \cdot (-3) \cdot |4| \cdot (-5)}{|-1| \cdot (-2) \cdot |-3| \cdot (4) \cdot |-5|}$.

4. Evaluate $\dfrac{(1)(-2)(3)(-4)(5) \ldots (97)(-98)}{(-1)(2)(-3)(4)(-5) \ldots (-97)(98)}$.

1.6

Exponents and Roots

▸**1** Use exponents.

▸**2** Use the zero exponent property.

▸**3** Use roots.

To understand certain topics in algebra, you must know exponents and roots. We discuss exponents in more depth in Sections 5.1 and 5.2 and roots (or radicals) in Chapter 8. We introduce them here so that we can discuss two important topics, the use of parentheses and order of operations.

▸**1** If a, b, and c are integers, and $a \cdot b = c$, then a and b are said to be **factors** of c. If a and b are factors of c, then both a and b will divide c without remainder. For example,

$$\underset{\substack{\uparrow \\ \text{factor}}}{2} \cdot \underset{\substack{\uparrow \\ \text{factor}}}{3} = \underset{\substack{\uparrow \\ \text{product}}}{6}$$

Both 2 and 3 are factors of 6 since each divides 6 without a remainder.

$$\text{product} \rightarrow \frac{6}{2} = \underset{\substack{\uparrow \\ \text{factor}}}{3} \qquad \text{product} \rightarrow \frac{6}{3} = \underset{\substack{\uparrow \\ \text{factor}}}{2}$$

In the expression 3^2, the 3 is called the **base** and the 2 is called the **exponent.** The expression 3^2 is read "three squared" or "three to the second power" and means

$$3^2 = \underbrace{3 \cdot 3}_{\text{2 factors of 3}}$$

The expression 5^3 is read "five cubed" or "five to the third power" and means

$$5^3 = \underbrace{5 \cdot 5 \cdot 5}_{\text{3 factors of 5}}$$

In general, the number b to the nth power, written b^n, means

$$b^n = \underbrace{b \cdot b \cdot b \cdot b \cdot \cdots \cdot b}_{\text{n factors of b}}$$

where n is the exponent and b is the base.

EXAMPLE 1 Evaluate each of the following:

(a) 5^3 (b) $(-2)^5$ (c) 1^{10} (d) $\left(\dfrac{-3}{4}\right)^3$ (e) $(0.6)^3$

Solution: (a) $5^3 = 5 \cdot 5 \cdot 5 = 125$
(b) $(-2)^5 = (-2)(-2)(-2)(-2)(-2) = -32$
(c) $1^{10} = 1$; 1 raised to any power will equal 1. Why?
(d) $\left(\dfrac{-3}{4}\right)^3 = \left(\dfrac{-3}{4}\right)\left(\dfrac{-3}{4}\right)\left(\dfrac{-3}{4}\right) = \dfrac{-27}{64}$
(e) $(0.6)^3 = (0.6)(0.6)(0.6) = 0.216$ ∎

COMMON STUDENT ERROR

Students should realize that $a^b \neq a \cdot b$.

Correct	*Wrong*
$2^4 = 2 \cdot 2 \cdot 2 \cdot 2 = 16$	~~$2^4 = 2 \cdot 4 = 8$~~

It is not necessary to write exponents of 1. Whenever we encounter a numerical value or a variable without an exponent, we assume that it has an exponent of 1. Thus, 3 means 3^1, x means x^1, x^3y means x^3y^1, and $-xy$ means $-x^1y^1$.

▶ **2** A rule that we will be using in Chapter 5 is that any nonzero number, or letter, raised to the 0 power has a value of 1.

Zero Exponent Property

$$a^0 = 1, \qquad \text{for } a \neq 0$$

Examples of the Zero Exponent Property

$$3^0 = 1, \qquad x^0 = 1$$

When a term contains an exponent, that exponent acts only on the one variable or number that directly precedes it, unless parentheses are used. For example, in the expression $2x^0$ the exponent acts only on the variable x: $2x^0 = 2(x^0) = 2(1) = 2$. In the expression $(2x)^0$, since the 2 and the x are in parentheses, the exponent acts on both the 2 and the x: $(2x)^0 = 1$.

EXAMPLE 2 Evaluate.

(a) 5^0 (b) $(2x)^0$ (c) $\frac{5}{9}x^0$ (d) $-4x^0$ (e) -2^0

Solution: (a) $5^0 = 1$

(b) $(2x)^0 = 1$.   Notice that the entire expression is raised to the 0 power because of the parentheses.

(c) $\frac{5}{9}x^0 = \frac{5}{9}(1) = \frac{5}{9}$.   Notice that only the variable x is raised to the 0 power.

(d) $-4x^0 = -4(1) = -4$

(e) $-2^0 = -(2^0) = -1$.   Notice that the exponent refers only to the 2 and not the negative sign preceding it. ∎

Students often evaluate the expression $-x^2$ incorrectly. Study the following example and Common Student Error carefully.

EXAMPLE 3 Evaluate $-x^2$ for the following values of x.

(a) 3   (b) -3

Solution: (a) $-x^2 = -(3^2) = -9$ (b) $-x^2 = -(-3)^2 = -(9) = -9$ ∎

COMMON STUDENT ERROR

A negative sign directly preceding an expression that is raised to a power has the effect of negating that expression.

$$-3^2 \quad \text{means} \quad -(3^2) \quad \text{and not} \quad (-3)^2$$
$$-x^2 \quad \text{means} \quad -(x^2) \quad \text{and not} \quad (-x)^2.$$

Example: Evaluate.

(a) -5^2 (b) $(-5)^2$

Solution:

(a) $-5^2 = -(5^2) = -25$ (b) $(-5)^2 = (-5)(-5) = 25$

Note that $-5^2 \neq (-5)^2$ since $-25 \neq 25$. Note also that $-x^2$ will always be a negative number for any nonzero value of x and that $(-x)^2$ will always be a positive number for any nonzero value of x.

EXAMPLE 4 Evaluate $-3^2 + (-1)^3 - 4^3 + (-2)^0$.

Solution: We evaluate each exponential expression. Then we add or subtract, working from left to right, to obtain the answer.

$$-3^2 + (-1)^3 - 4^3 + (-2)^0 = -(3^2) + (-1)^3 - (4^3) + (-2)^0$$
$$= -9 + (-1) - 64 + 1$$
$$= -9 - 1 - 64 + 1$$
$$= -73$$

∎

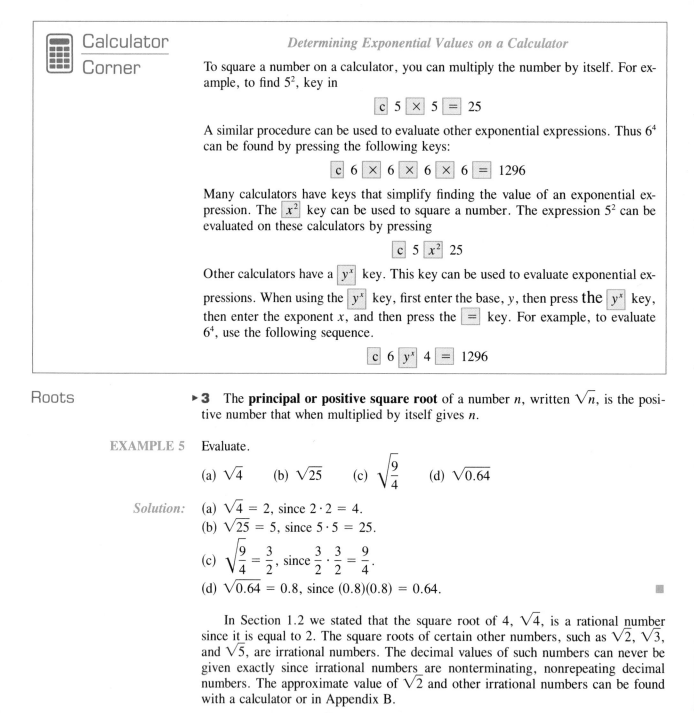

Calculator Corner

Determining Exponential Values on a Calculator

To square a number on a calculator, you can multiply the number by itself. For example, to find 5^2, key in

| c | 5 | × | 5 | = | 25

A similar procedure can be used to evaluate other exponential expressions. Thus 6^4 can be found by pressing the following keys:

| c | 6 | × | 6 | × | 6 | × | 6 | = | 1296

Many calculators have keys that simplify finding the value of an exponential expression. The $\boxed{x^2}$ key can be used to square a number. The expression 5^2 can be evaluated on these calculators by pressing

| c | 5 | x^2 | 25

Other calculators have a $\boxed{y^x}$ key. This key can be used to evaluate exponential expressions. When using the $\boxed{y^x}$ key, first enter the base, y, then press the $\boxed{y^x}$ key, then enter the exponent x, and then press the $\boxed{=}$ key. For example, to evaluate 6^4, use the following sequence.

| c | 6 | y^x | 4 | = | 1296

Roots

▶ **3**   The **principal or positive square root** of a number n, written $\sqrt{n}$, is the positive number that when multiplied by itself gives n.

EXAMPLE 5   Evaluate.

    (a) $\sqrt{4}$ (b) $\sqrt{25}$ (c) $\sqrt{\dfrac{9}{4}}$ (d) $\sqrt{0.64}$

Solution: (a) $\sqrt{4} = 2$, since $2 \cdot 2 = 4$.
    (b) $\sqrt{25} = 5$, since $5 \cdot 5 = 25$.
    (c) $\sqrt{\dfrac{9}{4}} = \dfrac{3}{2}$, since $\dfrac{3}{2} \cdot \dfrac{3}{2} = \dfrac{9}{4}$.
    (d) $\sqrt{0.64} = 0.8$, since $(0.8)(0.8) = 0.64$.   ■

    In Section 1.2 we stated that the square root of 4, $\sqrt{4}$, is a rational number since it is equal to 2. The square roots of certain other numbers, such as $\sqrt{2}$, $\sqrt{3}$, and $\sqrt{5}$, are irrational numbers. The decimal values of such numbers can never be given exactly since irrational numbers are nonterminating, nonrepeating decimal numbers. The approximate value of $\sqrt{2}$ and other irrational numbers can be found with a calculator or in Appendix B.

$$\sqrt{2} \approx 1.414213562 \qquad \text{from a calculator}$$
$$\sqrt{2} \approx 1.41 \qquad \text{from Appendix B}$$

    The concept used to explain square root can be expanded to explain cube roots and higher roots. The cube root of a number n is written $\sqrt[3]{n}$.

$$\sqrt[3]{n} = b \qquad \text{if} \qquad \underbrace{b \cdot b \cdot b}_{\text{3 factors of } b} = n$$

For example, $\sqrt[3]{8} = 2$, because $2 \cdot 2 \cdot 2 = 8$. The expression $\sqrt[m]{n}$ is read the *m*th root of *n*.

$$\sqrt[m]{n} = b \quad \text{if} \quad \underbrace{b \cdot b \cdot b \cdot \ \cdots \ \cdot b}_{m \text{ factors of } b} = n$$

EXAMPLE 6   Evaluate.

    (a) $\sqrt[3]{27}$ (b) $\sqrt[3]{64}$ (c) $\sqrt[4]{16}$

Solution: (a) $\sqrt[3]{27} = 3$, since $3 \cdot 3 \cdot 3 = 27$.
           (b) $\sqrt[3]{64} = 4$, since $4 \cdot 4 \cdot 4 = 64$.
           (c) $\sqrt[4]{16} = 2$, since $2 \cdot 2 \cdot 2 \cdot 2 = 16$.   ■

EXAMPLE 7   Evaluate.

    (a) $\sqrt[4]{81}$ (b) $\sqrt[3]{\dfrac{1}{27}}$ (c) $\sqrt[3]{-8}$ (d) $\sqrt[3]{-1}$

Solution: (a) $\sqrt[4]{81} = 3$, since $3 \cdot 3 \cdot 3 \cdot 3 = 81$.

           (b) $\sqrt[3]{\dfrac{1}{27}} = \dfrac{1}{3}$, since $\left(\dfrac{1}{3}\right)\left(\dfrac{1}{3}\right)\left(\dfrac{1}{3}\right) = \dfrac{1}{27}$.

           (c) $\sqrt[3]{-8} = -2$, since $(-2)(-2)(-2) = -8$.
           (d) $\sqrt[3]{-1} = -1$, since $(-1)(-1)(-1) = -1$.   ■

     Note that in Example 7 (c) and (d) the cube root of a negative number is negative. Why is this so?

Calculator Corner

Finding Roots on a Calculator

The square roots of numbers can be found on calculators with a square-root key, $\boxed{\sqrt{x}}$. To evaluate $\sqrt{25}$ on calculators that have this key, press

$$\boxed{c} \ \ 25 \ \ \boxed{\sqrt{x}} \ \ 5$$

Higher roots can be found on calculators that contain either one of two keys, $\boxed{y^x}$ or $\boxed{\sqrt[x]{y}}$. To evaluate $\sqrt[4]{625}$ on a calculator having a $\boxed{\sqrt[x]{y}}$ key, do the following:

$$\boxed{c} \ \ 625 \ \ \boxed{\sqrt[x]{y}} \ \ 4 \ \ \boxed{=} \ \ 5$$

Note that the number within the radical sign (the radicand) 625 is entered, then the $\boxed{\sqrt[x]{y}}$ key is pressed, and then the root (or index) 4 is entered. When the $\boxed{=}$ key is pressed, the answer 5 is displayed.
    To evaluate $\sqrt[4]{625}$ on a calculator with a $\boxed{y^x}$ key, follow this procedure:

$$\boxed{c} \ \ 625 \ \ \boxed{INV} \ \ \boxed{y^x} \ \ 4 \ \ \boxed{=} \ \ 5$$

The radicand 625 is entered, then the inverse key $\boxed{INV}$ is pressed, then the $\boxed{y^x}$ key is pressed, and then the root 4 is entered. After the $\boxed{=}$ key is pressed, the answer 5 is displayed.

Exercise Set 1.6

Evaluate.

1. 3^2

2. 3^3

3. 5^2

4. -2^4

5. $(-2)^4$

6. $(-2)^3$

7. $(-3)^4$

8. -1^4

9. -2^5

10. $\left(\frac{1}{3}\right)^4$

11. $\left(\frac{2}{3}\right)^4$

12. $\left(\frac{-3}{5}\right)^5$

13. $(0.3)^2$

14. $(0.8)^2$

15. $(0.2)^3$

16. $(0.4)^3$

Evaluate.

17. 6^0

18. x^0

19. $4x^0$

20. $(2x)^0$

21. $-3y^0$

22. $\left(\frac{1}{2}x\right)^0$

23. -7^0

24. $8x^0$

Evaluate.

25. $\sqrt{16}$

26. $\sqrt{25}$

27. $\sqrt{64}$

28. $\sqrt{169}$

29. $\sqrt{\frac{25}{36}}$

30. $\sqrt{\frac{1}{4}}$

31. $\sqrt{\frac{225}{81}}$

32. $\sqrt{\frac{9}{100}}$

33. $\sqrt{.04}$

34. $\sqrt{0.16}$

35. $\sqrt{0.25}$

36. $\sqrt{0.01}$

Evaluate.

37. $\sqrt[3]{64}$

38. $\sqrt[3]{1}$

39. $\sqrt[3]{-8}$

40. $\sqrt[3]{-27}$

41. $\sqrt[3]{-64}$

42. $\sqrt[4]{16}$

43. $\sqrt[4]{1}$

44. $\sqrt[3]{8}$

45. $\sqrt[3]{125}$

46. $\sqrt[3]{-125}$

47. $\sqrt[3]{-216}$

48. $\sqrt[3]{\frac{1}{8}}$

49. $\sqrt[3]{\frac{1}{64}}$

50. $\sqrt[3]{\frac{1}{27}}$

51. $\sqrt[3]{0.001}$

52. $\sqrt[3]{0.064}$

Evaluate.

53. 2^5

54. -4^0

55. $\sqrt[3]{27}$

56. $\sqrt[3]{-216}$

57. $\sqrt{\frac{4}{9}}$

58. $\sqrt{\frac{81}{4}}$

59. $-3x^0$

60. $\sqrt[4]{\frac{1}{16}}$

61. $\sqrt[3]{-125}$

62. $\left(-\frac{5}{9}\right)^3$

63. $\left(-\frac{1}{4}\right)^4$

64. -3^4

Evaluate (a) x^2 and (b) $-x^2$ for the given value of x.

65. 3

66. 4

67. 1

68. -2

69. -1

70. -5

71. $\frac{1}{3}$

72. $-\frac{2}{5}$

Evaluate (a) x^3 and (b) $-x^3$ for the given value of x.

73. 3

74. -5

75. -3

76. -1

77. -2

78. -4

79. $\frac{2}{3}$

80. $-\frac{3}{4}$

Evaluate.

81. $4^2 + 3^2 - 2^2$

82. $(4-1)^2 + 2^3 - (-1)^2$

83. $2^2 + 3^2 + (-4)^2$

84. $(-1)^3 + 1^3 + 1^{10} + (-1)^{12}$

85. $(3 - 2)^3 + (2 - 3)^3$

86. $(-3)^3 - 2^2 - (-2)^2 + (4 - 4)^2$

87. $-2^2 - (2)^3 + 1^0 + (-2)^3$

88. $(-2)^2 + (-3)^2 + (-3)^3 - 4^2$

89. $|-3| + 4^2 - |-2| - 3^0$

90. $-|5| - 4^2 - 1^2 - 3^2$

91. $|5| - |3| - |-8| - 6$

92. $4 - |-9| - (-3)^3 - 4^2$

93. $(0.2)^2 - (1.6)^2 - (3.2)^2$

94. $(3.7)^2 - (0.8)^2 + (2.4)^3$

95. $\left(-\dfrac{1}{2}\right)^3 - \left(\dfrac{1}{3}\right)^2 - \left(-\dfrac{2}{3}\right)^2$

96. $\left(\dfrac{3}{4}\right)^2 - \dfrac{1}{4} - \left(-\dfrac{3}{8}\right) + \left(\dfrac{1}{2}\right)^0$

97. Explain why $\sqrt{-4}$ cannot be a real number.

98. Explain why an odd root of a positive number will be positive.

99. Explain why an odd root of a negative number will be negative.

Cumulative Review Exercises

[1.4] *If the letter* a *represents a real number, for what values of* a *will each of the following be true?*

100. $|a| = |-a|$ **101.** $|a| = a$ **102.** $|a| = 4$

[1.5] **103.** Evaluate $4 - (-6) + 3 - 7$.

1.7

Order of Operations

▶ **1**   Know the order of operations.

▶ **2**   Use parentheses or brackets correctly.

▶ **3**   Evaluate expressions for specific values of the variable.

▶ **1**   What is $2 + 3 \cdot 4$ equal to? Is it 20? Is it 14? To be able to answer questions of this type, we must know the order of operations when evaluating a mathematical expression. You will often have to evaluate expressions containing multiple operations. To do so, follow the order of operations indicated below.

To Evaluate Mathematical Expressions

Use the following order:

1. First, evaluate the information within grouping symbols, including parentheses, ( ), brackets, [ ], or braces, { }. If the expression contains nested parentheses (one pair of parentheses within another pair), evaluate the information in the innermost parentheses first.

2. Next, evaluate all terms containing exponents and roots.

3. Next, evaluate all multiplications or divisions in the order in which they occur, working from left to right.

4. Finally, evaluate all additions or subtractions in the order in which they occur, working from left to right.

It should be noted that a fraction bar acts as a grouping symbol. Thus, when evaluating expressions containing a fraction bar, we work separately above and below the fraction bar.

We can now answer the question posed earlier. Since multiplications are performed before additions,

$$2 + 3 \cdot 4 \quad \text{means} \quad 2 + (3 \cdot 4) = 2 + 12 = 14$$

▸ **2** Parentheses or brackets may be used (1) to change the order of operations to be followed in evaluating an algebraic expression or (2) to help clarify the understanding of an expression.

In the example above, $2 + 3 \cdot 4$, if we wished to have the addition performed before the multiplication, we could indicate this by placing parentheses about the $2 + 3$.

$$(2 + 3) \cdot 4 = 5 \cdot 4 = 20$$

Consider the expression $1 \cdot 3 + 2 \cdot 4$. According to the order of operations, multiplications are to be performed before additions. We can rewrite this expression as $(1 \cdot 3) + (2 \cdot 4)$. Notice that we did not change the order of operations. The parentheses only help clarify the order to be followed.

Brackets are sometimes used in place of parentheses to help avoid confusion. For example, the expression $7((5 \cdot 3) + 6)$ may be easier to follow when written $7[(5 \cdot 3) + 6]$.

EXAMPLE 1   Evaluate $8 + 3 \cdot 5^2 - 7$.

Solution: Colored shading will be used to indicate the order in which the operations are to be evaluated. Since there are no parentheses, we first evaluate 5^2.

$$8 + 3 \cdot 5^2 - 7 = 8 + 3 \cdot 25 - 7$$

Next, perform multiplications or divisions from left to right.

$$= 8 + 75 - 7$$

Next, perform additions or subtractions from left to right.

$$= 83 - 7$$
$$= 76 \qquad \blacksquare$$

EXAMPLE 2   Evaluate $36 + 3[(12 - 4) \div \frac{1}{2}]$.

Solution: First, evaluate the information in the innermost parentheses.

$$36 + 3[(12 - 4) \div \tfrac{1}{2}] = 36 + 3[8 \div \tfrac{1}{2}]$$

Next, evaluate the information within the brackets. To do this we invert the $\frac{1}{2}$ and multiply.

$$= 36 + 3[8 \cdot 2]$$
$$= 36 + 3(16)$$

Now perform the remaining multiplication, and then add.

$$= 36 + 48$$
$$= 84 \qquad \blacksquare$$

EXAMPLE 3   Evaluate $10 + [6 - [4(5 - 2)]]^2$.

Solution: First, evaluate the information within the innermost parentheses.

$$10 + [6 - [4(5 - 2)]]^2 = 10 + [6 - [4(3)]]^2$$
$$= 10 + [6 - (12)]^2$$
$$= 10 + (-6)^2$$
$$= 10 + 36$$
$$= 46$$

EXAMPLE 4 Evaluate $16 \div 8 \cdot 4 - 6^2 \div 2^2$.

Solution: Begin by squaring the 6 and the 2. Then perform the multiplications or divisions, moving from left to right.

$$16 \div 8 \cdot 4 - 6^2 \div 2^2 = 16 \div 8 \cdot 4 - 36 \div 4$$
$$= 2 \cdot 4 - 36 \div 4$$
$$= 8 - 36 \div 4$$
$$= 8 - 9$$
$$= -1$$

EXAMPLE 5 Evaluate $\dfrac{6 \div 2 + 5|7 - 3|}{2 + (3 - 5) \div 2}$.

Solution: Work separately above the fraction bar and below the fraction bar.

$$\frac{6 \div 2 + 5|7 - 3|}{2 + (3 - 5) \div 2} = \frac{6 \div 2 + 5|4|}{2 + (-2) \div 2}$$

$$= \frac{3 + 20}{2 + (-1)}$$

$$= \frac{23}{1} = 23$$

▶ **3** Now we will evaluate expressions for specific values of the variable.

EXAMPLE 6 Evaluate $4x^2 - 2$ when (a) $x = 3$ and (b) $x = -\frac{3}{4}$.

Solution: In part (a), we substitute 3 for each x in the expression. In part (b), we substitute $-\frac{3}{4}$ for each x in the expression.

(a) $4x^2 - 2 = 4(3)^2 - 2$ (b) $4x^2 - 2 = 4\left(-\dfrac{3}{4}\right)^2 - 2$

$\qquad\qquad = 4(9) - 2$ $\qquad\qquad\qquad = \overset{1}{4}\left(\dfrac{9}{\overset{\cancel{16}}{4}}\right) - 2$

$\qquad\qquad = 36 - 2$

$\qquad\qquad = 34$ $\qquad\qquad\qquad\quad = \dfrac{9}{4} - 2$

$\qquad\qquad\qquad\qquad\qquad = \dfrac{9}{4} - \dfrac{8}{4}$

$\qquad\qquad\qquad\qquad\qquad = \dfrac{1}{4}$

EXAMPLE 7   Evaluate $6 - (3x + 1) + 2x^2$ when $x = 4$.

Solution: Substitute 4 for each x in the expression; then evaluate.

$$\begin{aligned}
6 - (3x + 1) + 2x^2 &= 6 - [3(4) + 1] + 2(4)^2 \\
&= 6 - [12 + 1] + 2(16) \\
&= 6 - (13) + 32 \\
&= -7 + 32 \\
&= 25
\end{aligned}$$

EXAMPLE 8   Evaluate $-x^3 - xy - y^2$ when $x = -2$ and $y = 5$.

Solution: Substitute -2 for each x and 5 for each y in the expression; then evaluate.

$$\begin{aligned}
-x^3 - xy - y^2 &= -(-2)^3 - (-2)(5) - (5)^2 \\
&= -(-8) - (-10) - 25 \\
&= 8 + 10 - 25 \\
&= -7
\end{aligned}$$

Calculator Corner

Order of Operations on a Calculator

We now know that $2 + 3 \cdot 4$ means $2 + (3 \cdot 4)$ and has a value of 14. What will a calculator display if you key in the following?

| c | 2 | + | 3 | × | 4 | = |

The answer depends on your calculator. Scientific calculators evaluate problems using the order of operations discussed in this section. Nonscientific calculators will perform the operations in the order they are entered.

Scientific calculator | c | 2 | + | 3 | × | 4 | = | 14

Nonscientific calculator | c | 2 | + | 3 | × | 4 | = | 20

Remember that in algebra, unless otherwise instructed by parentheses, we always perform multiplications and divisions before additions and subtractions.

Do you have a scientific or nonscientific calculator? The cost of a scientific calculator is not much more than that of a nonscientific calculator. You may wish to give some thought to purchasing a scientific calculator, especially if you plan on taking other mathematics and science courses.

Exercise Set 1.7

Evaluate.

1. $6 + 4 \cdot 5$

2. $2 - 2^2 + 4$

3. $2 + 3 \cdot 4^2$

4. $(6^2 - 2) \div (\sqrt{36} - 4)$

5. $6 \div 2 + 5 \cdot \dfrac{3}{4}$

6. $4 \cdot 3 - 4 \cdot 5$

7. $24 \cdot 2 \div \dfrac{1}{3} \div 6$

8. $6 \div 2 + 36 \cdot 3$

9. $(\sqrt{4} - 3) \cdot (5 - 1)^3$

10. $20 - 6 \div 3 - 4$

11. $\dfrac{3}{4} \div \dfrac{5}{6} + \dfrac{1}{2} \cdot \dfrac{9}{4}$

12. $6 + \sqrt{9}(3 + 4)$

13. $2[1 - (4 \cdot 5)] + 6^3$

14. $[12 \div (4 \div 2)] - 5$

15. $(3^2 - 1) \div (3 + 1)$

16. $-4(5 - 2)^3 + \dfrac{5}{2}$

17. $3[(4 + 6)^2 - \sqrt[3]{8}]$

18. $2[3(8 - 2) \div 6]^3$

19. $[[3(14 \div 7)]^2 - 2]^2$

20. $[[(12 - 15) - 3] - 2]^2$

21. $3[6 - [(25 \div 5) - 2]]^3$

22. $4[3(2 - 6) \div (10 \div 5)^2]^2$

23. $\dfrac{\frac{1}{6} - 4 \div 2}{8 - 3 + 6}$

24. $\dfrac{15 \div 3 + 2 \cdot 2}{\sqrt{25} \div 5 + 8 \div 2}$

25. $\dfrac{\frac{1}{2} \cdot \frac{1}{3} \div 4 - 2}{3^2 - 4 \cdot 2 + 3}$

26. $\dfrac{4 \div 2 \cdot 3^2 - 1}{5 - (3 + 4)^2}$

27. $\dfrac{4 - (2 + 3)^2 - 6}{4(3 - 2) - 3^2}$

28. $\dfrac{6 \div 2 + 5^2 + 3}{4 - (-3 + 5) - 4^2}$

29. $\dfrac{2(-3) + 4 \cdot 5 - 3^2}{5 + \sqrt{4}(2^2 - 1)}$

30. $\dfrac{-(-3) + (-2) - (-4) - 3}{6 - (-2) + 3(5) - 4}$

31. $\dfrac{8 - 4 \div 2 \cdot 3 - 4}{5^2 - 3^2 \cdot 2 - 6}$

32. $\dfrac{8 - [4 - (3 - 1)^2]}{5 - (-3)^2 + 4 \div 2}$

33. $-2\left|-3 - \dfrac{2}{3}\right| + 4$

34. $-3|-4 + 2| - 6 \cdot 3$

35. $3|4 - 6| - 2|-4 - 2| + 3^2$

36. $-2|-3| - 6 \div |2| + 3^2$

37. $12 - 15 \div |5| + 2(|4| - 2)^2$

38. $\dfrac{4 - |-12| \div |3|}{2(4 - |5|) + 9}$

39. $\dfrac{6 - 2|9 - 4| + 8}{4 - |-4| + 4^2 \div 2^2}$

40. $\dfrac{6 - |-3| - 4|6 - 2|}{5 - 6 \cdot 2 \div |-4|}$

41. $\dfrac{24 - 5 - 4^2}{|-8| - 4 \div 2(3)} + \dfrac{4 - (-3)^2 - |4|}{3^2 - 4 \cdot 3 + |-7|}$

42. $\dfrac{\sqrt[3]{-27} - 3^2 - 4}{|-6| - 2|-5|} \cdot \dfrac{4 - [6 \div 2 - 2]^2}{5 - [3^2 - (5 - 3)]^2}$

Evaluate.

43. $-3x^2 - 4$ when $x = 1$

44. $2x^2 + x$ when $x = 3$

45. $5x^2 - 2x + 5$ when $x = 3$

46. $-3x^2 + 6x + 5$ when $x = 5$

47. $3(x - 2)^2$ when $x = \dfrac{1}{4}$

48. $4(x + 1)^2 - 6x$ when $x = -\dfrac{5}{6}$

49. $4(x - 3)(x + 4)$ when $x = 1$

50. $3x(x - 1) + 5$ when $x = -4$

51. $-6x + 3y$ when $x = 2, y = 4$

52. $6x + 3y - 5$ when $x = 1, y = -3$

53. $4(x + y)^2 + 4x - 3y$ when $x = 2, y = -3$

54. $(4x^2 - 3y) - 5$ when $x = 4, y = -2$

55. $3(a + b)^2 + 4(a + b) - 6$ when $a = 4, b = -1$

56. $4xy - 6x^2 + 3$ when $x = 5, y = 2$

57. $x^3y^2 - 6xy + 3x$ when $x = 2, y = 3$

58. $\dfrac{6x^2}{3} + \dfrac{2x}{2}$ when $x = 2$

59. $\dfrac{1}{2}(x^2 + y^2 - 2xy)$ when $x = 2, y = -3$

60. $\dfrac{5x}{3} - \dfrac{6y}{4} + 3$ when $x = 6, y = 3$

61. $x^2y^4 - y^3 + 3(x + y)$ when $x = 2, y = -1$

62. $(x - 3)^2 + (y - 5)^2$ when $x = -2, y = 3$

63. $\dfrac{x^2}{25} + \dfrac{y^2}{9}$ when $x = 0, y = 2$

64. $\dfrac{(x - 3)^2}{9} + \dfrac{(y + 5)^2}{16}$ when $x = 4, y = 3$

65. Evaluate each of the following expressions for $a = 3$, $b = 5$, and $c = -12$.

(a) $b^2 - 4ac$

(b) $\sqrt{b^2 - 4ac}$

(c) $-b + \sqrt{b^2 - 4ac}$

(d) $\dfrac{-b + \sqrt{b^2 - 4ac}}{2a}$

(e) $\dfrac{-b - \sqrt{b^2 - 4ac}}{2a}$

66. Evaluate (a) $\dfrac{-b + \sqrt{b^2 - 4ac}}{2a}$ and

(b) $\dfrac{-b - \sqrt{b^2 - 4ac}}{2a}$ for $a = 6$, $b = -11$, and $c = 3$.

Write an algebraic expression for each of the following. Then evaluate the expression for the given value of the variable.

67. Multiply the variable x by 3. To this product add 6. Now square this sum. Find the value of this expression when $x = 3$.

68. Subtract 3 from x. Square this difference. Subtract 5 from this value. Now square this value. Find the value of this expression when $x = -1$.

69. Six is added to the product of 3 and x. This expression is then multiplied by 6. Nine is then subtracted from this product. Find the value of the expression when $x = 3$.

70. The sum of x and y is multiplied by 2. Then 5 is subtracted from this product. This expression is then squared. Find the value of the expression when $x = 2$ and $y = -3$.

71. Three is added to x. This sum is divided by twice y. This quotient is then squared. Finally, 3 is subtracted from this expression. Find the value of the expression when $x = 5$ and $y = 2$.

72. Explain in your own words the order of operations you follow when you evaluate a mathematical expression.

73. (a) In your own words, explain in a step-by-step manner how you would evaluate $\dfrac{5 - 18 \div 3^2}{4 - 3 \cdot 2}$.

(b) Evaluate the expression.

74. (a) In your own words, explain in a step-by-step manner how you would evaluate $[5 - [4 - (3 - 8)]]^2$.

(b) Evaluate the expression.

75. (a) In your own words, explain step-by-step how you would evaluate $16 \div 2^2 + 6 \cdot 4 - 24 \div 6$.

(b) Evaluate the expression.

Cumulative Review Exercises

[1.2] **76.** $A = \{a, b, c, d, f\}$, $B = \{b, c, f, g, h\}$. Find (a) $A \cap B$; (b) $A \cup B$.

[1.3] **77.** Name the following property: $(2 + 3) + 5 = 2 + (3 + 5)$.

[1.4] **78.** List from smallest to largest: $-|6|, -4, |-5|, -|-2|, 0$.

[1.5] **79.** Evaluate $8 - (-4) + (7 - 5) - 10$.

[1.6] **80.** Evaluate $|6 - 3| - 4^2 + \sqrt{25} - (3 - 1)^2$.

JUST FOR FUN

1. Evaluate $[(3 \div 6)^2 + 4]^2 + 3 \cdot 4 \div 12 \div 3$.

2. Evaluate $[-2(3x^2 + 4)^2]^2 \div (3x^2 - 2)$ when $x = -2$.

3. Evaluate $\dfrac{2x + 4 - y\left(2 + \dfrac{3}{x}\right)}{\dfrac{y - 2}{6} + \dfrac{3x^2}{4}}$ when $x = 2$, $y = 3$.

SUMMARY

GLOSSARY

Absolute value *(16):* The distance of a number from 0 on the number line. The absolute value of any real number will be greater than or equal to 0.

Additive identity element *(13):* 0.

Additive inverse *(13):* For any number a, its additive inverse is $-a$.

Algebraic expression (or expression) *(12):* Any combination of numbers, variables, exponents, mathematical symbols, and operations.
Empty or null set *(7):* A set containing no elements, symbolized ∅ or { }.
Factors *(27):* If $a \cdot b = c$ then a and b are factors of c.
Finite set *(6):* A set that contains a finite number of elements.
Inequality symbols *(16):* $<, \leq, >, \geq, \neq$.
Infinite set *(6):* A set that has an infinite number of elements.
Intersection of sets *(8):* The intersection of set A and set B, $A \cap B$, is the set of elements that belongs to both set A and set B.

Multiplicative identity element *(13):* 1.
Multiplicative inverse *(13):* For any number a, $a \neq 0$, its multiplication inverse is $1/a$.
Principal (or positive) square root *(30):* The principal square root of a number n, written $\sqrt{n}$, is the positive number that when multiplied by itself gives n.
Set *(6):* A collection of objects or elements.
Subset *(7):* Set A is a subset of set B, $A \subseteq B$, if every element of set A is also an element of set B.
Union of sets *(8):* The union of sets A and B, $A \cup B$, is the set of elements that belongs to either set A or set B.
Variable *(12):* A letter used to represent a number.

IMPORTANT FACTS

Sets of Numbers

Real numbers	$\{x \mid x$ is a point on the number line$\}$
Natural or counting numbers	$\{1, 2, 3, 4, 5, \ldots\}$
Whole numbers	$\{0, 1, 2, 3, 4, \ldots\}$
Integers	$\{\ldots, -3, -2, -1, 0, 1, 2, 3, \ldots\}$
Rational numbers	$\left\{\dfrac{p}{q} \mid p$ and q are integers, $q \neq 0\right\}$
Irrational numbers	$\{x \mid x$ is a real number that is not rational$\}$

Properties of the Real Number System

Commutative properties	$a + b = b + a, \quad ab = ba$
Associative properties	$(a + b) + c = a + (b + c), \quad (ab)c = a(bc)$
Identity properties	$a + 0 = 0 + a = a, \quad a \cdot 1 = 1 \cdot a = a$
Inverse properties	$a + (-a) = (-a) + a = 0, \quad a \cdot \dfrac{1}{a} = \dfrac{1}{a} \cdot a = 1$
Distributive property	$a(b + c) = ab + ac$
Multiplication property of 0	$a \cdot 0 = 0 \cdot a = 0$
Double-negative property	$-(-a) = a$

Absolute value: $|a| = \begin{cases} a, & a \geq 0 \\ -a, & a < 0 \end{cases}$

Zero exponent property: $a^0 = 1, \quad a \neq 0$

Review Exercises

[1.2] *List each set in roster form.*

1. $A = \{x \mid x$ is a natural number between 2 and 7$\}$

2. $B = \{x \mid x$ is a whole number multiple of 3$\}$

Place either ∈ or ∉ in the shaded area to make a true statement.

3. 0 ▦ $\{0, 1, 2, 3\}$ **4.** $\{3\}$ ▦ $\{0, 1, 2, 3\}$ **5.** $\{5\}$ ▦ $\{4, 5, 6\}$ **6.** 8 ▦ $\{1, 2, 3, 4, \ldots\}$

Place either ⊆ or ⊄ in the shaded area to make a true statement.

7. {3} ▨ {1, 2, 3}   **8.** 0 ▨ {0, 1, 2, 3, . . .}   **9.** 5 ▨ {3, 4, 5, 6}   **10.** {8} ▨ {1, 2, 3, 4, . . .}

Let N = set of natural numbers, W = set of whole numbers, I = set of integers, Q = set of rational numbers, H = set of irra-tional numbers, ℝ = set of real numbers. Insert either ⊆ or ⊄ in the shaded area to make a true statement.

11. N ▨ W **12.** Q ▨ $\mathbb{R}$ **13.** I ▨ Q **14.** N ▨ I

15. H ▨ $\mathbb{R}$ **16.** Q ▨ H

Consider the set of numbers $\{-3, 4, 6, \frac{1}{2}, \sqrt{5}, \sqrt{3}, 0, \frac{15}{27}, -\frac{1}{5}, 1.47\}$. List the elements of the set that are:

17. Natural numbers   **18.** Whole numbers   **19.** Integers

20. Rational numbers   **21.** Irrational numbers   **22.** Real numbers

Answer true or false.

23. $\sqrt{3}$ is an irrational number.

24. A real number cannot be divided by 0.

25. $\dfrac{0}{1}$ is a not a real number.

26. Every rational number and every irrational number is a real number.

27. $0, \frac{3}{5}, -2$, and 4 are all rational numbers.

Find $A \cup B$ and $A \cap B$ for each set A and B.

28. $A = \{1, 2, 3, 4, 5\}, B = \{2, 3, 4, 5\}$ **29.** $A = \{3, 5, 7, 9\}, B = \{2, 4, 6, 8\}$

30. $A = \{1, 2, 3, 4, . . .\}, B = \{2, 4, 6, . . .\}$ **31.** $A = \{4, 6, 9, 10, 11\}, B = \{3, 5, 9, 10, 12\}$

[1.3] *Name the given property.*

32. $3 + 4 = 4 + 3$ **33.** $x + 4 = 4 + x$

34. $3(x + 2) = 3x + 6$ **35.** $xy = yx$

36. $(x + 3) + 2 = x + (3 + 2)$ **37.** $a + 0 = a$

38. $3(x)y = 3(xy)$ **39.** $a \cdot 1 = a$

40. $-(-5) = 5$ **41.** $3(0) = 0$

42. $4 + 0 = 4$ **43.** $5 \cdot 1 = 5$

44. $x + (-x) = 0$ **45.** $x \cdot \dfrac{1}{x} = 1$

46. $(-2)\left(-\dfrac{1}{2}\right) = 1$ **47.** $(x + y) = 1(x + y)$

Fill in the statement on the right side of the equal sign using the property given.

48. $x + 3 =$ commutative property   **49.** $3(x + 5) =$ distributive property

50. $(x + 6) + (-4) =$ associative property   **51.** $3 \cdot x =$ commutative property

52. $(9 \cdot x) \cdot y =$ associative property   **53.** $4(x - y + 5) =$ distributive property

54. $a + 0 =$ identity property   **55.** $1 \cdot a =$ identity property

56. $a + (-a) =$ inverse property   **57.** $a \cdot \dfrac{1}{a} =$ inverse property

58. $-(-a) =$ double-negative property

[1.4] *Insert either $<, >,$ or $=$ in the shaded area between the two numbers to make a true statement.*

59. 3 ▨ 2 **60.** 1 ▨ 4 **61.** -2 ▨ 3 **62.** -4 ▨ -6

63. -8 ▨ 0 **64.** -4 ▨ -3.9 **65.** 1.06 ▨ 1.6 **66.** -1.06 ▨ -1.6

67. $|3|$ ▨ 3 **68.** $|-3|$ ▨ 3 **69.** $|4|$ ▨ $|6|$ **70.** $|-4|$ ▨ $|-6|$

71. 13 ▨ $|-5|$ **72.** $|-12|$ ▨ 4 **73.** $\left|-\frac{2}{3}\right|$ ▨ $\frac{3}{5}$ **74.** $-|-2|$ ▨ -5

Write the numbers from smallest to largest.

75. $4, -2, -5, |7|$

76. $0, \frac{3}{5}, 2.3, |-3|$

77. $|-7|, |-5|, 3, -2$

78. $-4, -2, -2.1, -|3|$

79. $-4, 6, -|-3|, 5$

80. $|1.6|, |-2.3|, -3, 0$

[1.5–1.7] *Evaluate.*

81. $4 - 2 + 3 - \dfrac{3}{5}$

82. $(3)(-4) - 6(8) - 3$

83. $-4|6| - 3(-4)$

84. $(4 - 6) - (-3 + 5) + 12$

85. $3|-2| - (4 - 3) + 2(-3)$

86. $|-16| \div (-4) + 2$

87. $(6 - 9) \div (9 - 6)$

88. $(4^2 - 6) - \sqrt{4} + 8$

89. $|6 - 3| \div 3 + 4 \cdot 8 - 12$

90. $\sqrt{36} \div 2 + |4 - 2| + 4^2$

91. $3^2 - 6 \cdot 9 + 4 \div 2^2 - 3$

92. $4 - (2 - 9)^0 + 3^2 \div 1 + 3$

93. $4^2 - (2 - 3^2)^2 + 4^3$

94. $-3^2 + 14 \div 2 \cdot 3 - 6$

95. $[[(9 \div 3)^2 - 1]^2 \div 8]^3$

96. $\dfrac{8 - 4 \div 2 + 3 \cdot 2}{\sqrt{36} \div 2^2 - 9}$

97. $\dfrac{-(4 - 6)^2 - 3(-2) + |-6|}{18 - 9 \div 3 \cdot 5}$

98. $\dfrac{8 - [5 - (-3 + 2)] \div 2}{|5 - 3| - |5 - 8| \div 3}$

99. $\dfrac{9|3 - 5| - 5|4| \div 10}{-3(5) - 2 \cdot 4 \div 2}$

Evaluate each expression for the values given.

100. $2x^2 + 3x + 1$ when $x = 2$

101. $(x - 2)^2 + 3x$ when $x = -2$

102. $4x^2 - 3y^2 + 5$ when $x = 1$ and $y = -\dfrac{1}{3}$

103. $-3x^2y + 6xy^2 - 2xy$ when $x = 1$ and $y = 3$

104. $(x - 2)^2 + (y - 4)^2 + 3$ when $x = 2$ and $y = 2$

105. $4(x - 3) + 5(y - 3) - 4$ when $x = -1$ and $y = -3$

106. $-x^2 + 3xy^2 - 6y^3$ when $x = 3$ and $y = 4$

107. $-x^2y - 6xy^2 + 4y^3$ when $x = -2$ and $y = 3$

Practice Test

1. List $A = \{x \mid x \text{ is a natural number greater than } 5\}$ in roster form.

Insert either $\subseteq$ or $\not\subseteq$ in the shaded area to make a true statement.

2. $3 \ \blacksquare \ \{1, 2, 3, 4\}$

3. $\{5\} \ \blacksquare \ \{1, 2, 3, 4, 5\}$

Answer true or false.

4. Every rational number is a real number.

5. Every whole number is a natural number.

6. The union of the set of rational numbers and the set of irrational numbers is the set of real numbers.

Consider the set of numbers $\{-\frac{3}{5}, 2, -4, 0, \frac{19}{12}, 2.57, \sqrt{8}, \sqrt{2}, -1.92.\}$ List the elements of the set that are:

7. Rational numbers

8. Real numbers

Find $A \cup B$ and $A \cap B$ for sets A and B.

9. $A = \{8, 10, 11, 14\}, \quad B = \{5, 7, 8, 9, 10\}$

10. $A = \{1, 3, 5, 7, \ldots\}, B = \{3, 5, 7, 9, 11\}$

Insert either $>$, $<$ or $=$ in the shaded area to make a true statement.

11. $-4 \ \blacksquare \ |-9|$

12. $|-3| \ \blacksquare \ -|5|$

13. List from smallest to largest: $|3|, -|4|, -2, 6$.

Name the property.

14. $3(x + 4) = 3x + 12$

15. $(x + y) + 3 = x + (y + 3)$

16. $3x + 4y = 4y + 3x$

17. $-4\left(-\dfrac{1}{4}\right) = 1$

18. $a + 0 = a$

Evaluate each of the following.

19. $[4 - [6 - 3(4 - 5)]]^2 \div (-5)$

20. $5^2 + 16 \div 4 - 3 \cdot 2$

21. $\dfrac{-3|4 - 8| \div 2 + 4}{\sqrt{36} + 18 \div 3^2}$

22. $\dfrac{-6^2 + 3(4 - |6|) \div 6}{4 - (-3) + 12 \div 4 \cdot 5}$

23. $\dfrac{[4 - (2 - 5)]^2 + 6 \div 2 \cdot 5}{|4 - 6| + |-6| \div 2}$

Evaluate each expression for the given values of x and y.

24. $-x^2 + 2xy + y^2$　when $x = 2$　and $y = 3$

25. $(x - 5)^2 + 2xy^2 - 6$　when $x = 2$　and $y = -3$

2

Linear Equations and Inequalities

See Section 2.4, Exercise 14.

2.1

Solving Linear Equations

▶**1** Identify the reflexive, symmetric, and transitive properties.

▶**2** Combine like terms.

▶**3** Solve equations.

▶**4** Solve equations containing fractions.

▶**5** Solve proportions.

▶**6** Identify conditional equations, inconsistent equations, and identities.

▶**1** In elementary algebra you learned how to solve linear equations. We briefly review these procedures in this section. Before we do so, we need to introduce three useful properties of equalities: the reflexive property, the symmetric property, and the transitive property.

Properties of Equalities

For all real numbers a, b and c:

1. $a = a$.   *reflexive property*
2. If $a = b$, then $b = a$.   *symmetric property*
3. If $a = b$ and $b = c$, then $a = c$.   *transitive property*

Examples of the Reflexive Property

$$3 = 3$$
$$x + 5 = x + 5$$
$$x^2 + 2x - 3 = x^2 + 2x - 3$$

Examples of the Symmetric Property

If $x = 3$, then $3 = x$
If $y = x + 4$, then $x + 4 = y$
If $y = x^2 + 2x - 3$, then $x^2 + 2x - 3 = y$

Examples of the Transitive Property

If $x = a$ and $a = 4y$, then $x = 4y$
If $a + b = c$ and $c = 4r$, then $a + b = 4r$
If $4k + 3r = 2m$ and $2m = 5w + 3$, then $4k + 3r = 5w + 3$

You will use these three properties often in this text.

Combining Terms

▶**2** When an algebraic expression consists of several parts, the parts that are added or subtracted are called the **terms** of the expression. The expression

$$6x^2 - 3(x + y) - 4 + \frac{x + 2}{5}$$

has four terms: $6x^2$, $-3(x + y)$, -4, and $\dfrac{x + 2}{5}$.

The $+$ and $-$ signs that break up the expression into terms are a part of a term. However, when listing the terms of an expression, it is not necessary to list the $+$ sign at the beginning of a term.

Expression	*Terms*
$\dfrac{1}{2}x^2 - 3x - 7$	$\dfrac{1}{2}x^2, \quad -3x, \quad -7$
$-5x^3 + 3x^2y - 2$	$-5x^3, \quad 3x^2y, \quad -2$
$4(x + 3) + 2x + 5(x - 2) + 1$	$4(x + 3), \quad 2x, \quad 5(x - 2), \quad 1$

The numerical part of a term is called its **numerical coefficient** or simply its **coefficient.** In the term $6x^2$, the 6 is the numerical coefficient. When the numerical coefficient is 1 or -1, we generally do not write the numeral 1. For example, x means $1x$, $-x^2y$ means $-1x^2y$, and $(x + y)$ means $1(x + y)$.

Term	*Numerical Coefficient*
$\dfrac{2x}{3}$	$\dfrac{2}{3}$
$-4(x + 2)$	-4
$\dfrac{x - 2}{3}$	$\dfrac{1}{3}$
$-(x + y)$	-1

Note that $\dfrac{x - 2}{3}$ means $\dfrac{1}{3}(x - 2)$ and $-(x + y)$ means $-1(x + y)$.

When a term consists of only a number, that number is often referred to as a **constant.** For example, in the expression $x^2 - 4$, the -4 is a constant.

The **degree of a term** is the sum of the exponents on the variable factors in the term. For example, $3x^2$ is a second-degree term, and $-4x$ is a first-degree term ($-4x$ means $-4x^1$). The number 3 can be written as $3x^0$, so the number 3 (and every other constant) is a zero-degree term. The term $6x^2y^3$ is a fifth-degree term since $2 + 3 = 5$. The term $4xy^5$ is a sixth-degree term since the sum of the exponents is $1 + 5$ or 6.

Like terms are terms that have the same variables with the same exponents. For example, $3x$ and $5x$ are like terms, $2x^2$ and $-3x^2$ are like terms, and $3x^2y$ and $-2x^2y$ are like terms. Terms that are not like terms are said to be **unlike terms.** Some unlike terms are $2x$ and $4x^2$, and $3xy$ and $4x$.

To **simplify an expression** means to combine all like terms in the expression. To combine like terms, we can use the distributive property.

Examples of Combining Like Terms

$$5x - 2x = (5 - 2)x = 3x$$

$$3x^2 - 5x^2 = (3 - 5)x^2 = -2x^2$$

$$-7x^2y + 3x^2y = (-7 + 3)x^2y = -4x^2y$$

$$4(x - y) - (x - y) = 4(x - y) - 1(x - y) = (4 - 1)(x - y) = 3(x - y)$$

When simplifying expressions, we are permitted to rearrange the terms because of the commutative and associative properties discussed earlier.

EXAMPLE 1   Simplify each of the following. If an expression cannot be simplified, so state.

(a) $-2x + 5 + 3x - 7$ (b) $7x^2 - 2x^2 + 3x + 4$

(c) $2x - 3y + 5x - 6y + 3$

Solution: (a) $-2x + 5 + 3x - 7 = \underbrace{-2x + 3x}_{x} \underbrace{+ 5 - 7}_{-2}$ Place like terms together.

This expression simplifies to $x - 2$.

(b) $7x^2 - 2x^2 + 3x + 4 = 5x^2 + 3x + 4$

(c) $2x - 3y + 5x - 6y + 3 = 2x + 5x - 3y - 6y + 3$ Place like terms together.

$= 7x - 9y + 3$

■

EXAMPLE 2   Simplify $7 - (2x - 5) - 3(2x + 4)$.

Solution: $7 - (2x - 5) - 3(2x + 4) = 7 - 1(2x - 5) - 3(2x + 4)$

$= 7 - 2x + 5 - 6x - 12$ Distributive property was used.

$= -2x - 6x + 7 + 5 - 12$ Rearrange terms.

$= -8x$ Combine like terms.

Thus $7 - (2x - 5) - 3(2x + 4) = -8x$.

■

Solving Equations

▶ **3**   An **equation** is a mathematical statement of equality. *An equation must contain an equal sign* and a mathematical expression on each side of the equal sign.

Examples of Equations

$$x + 4 = -7$$
$$2x^2 - 4 = -3x + 5$$

The **solution** of an equation is the number or numbers that make the equation a true statement. The solution to the equation $x + 2 = 5$ is 3. The **solution set** of an equation is the set of real numbers that make the equation true. The solution set for the equation $x + 2 = 5$ is {3}.

Two or more equations with the same solution set are called **equivalent equations.** The equations $2x + 3 = 9$, $x + 2 = 5$, and $x = 3$ are all equivalent equations since the solution set for each is {3}. Equations are generally solved by starting with a given equation and producing a series of simpler equivalent equations.

In this chapter we will discuss how to solve **linear equations in one variable.** A linear equation is an equation that can be written in the form $ax + b = c$, $a \neq 0$. Notice that the degree of the highest-powered term in a linear equation is 1; for this reason, linear equations are also called **first-degree equations.**

To solve equations, we use the addition and multiplication properties to isolate the variable on one side of the equal sign.

Addition Property

If $a = b$, then $a + c = b + c$ for any a, b, and c.

The addition property states that the same number can be added to both sides of an equation without changing the solution to the original equation. Since subtraction is defined in terms of addition, *the addition property also allows us to subtract the same number from both sides of an equation.*

Multiplication Property

If $a = b$, then $a \cdot c = b \cdot c$ for any a, b, and c.

The multiplication property states that both sides of an equation can be multiplied by the same number without changing the solution. Since division is defined in terms of multiplication, *the multiplication property also allows us to divide both sides of an equation by the same nonzero number.*

To solve an equation, we will often have to use a combination of properties to isolate the variable. Our goal is to get the variable all by itself on one side of the equation. A general procedure to solve linear equations follows.

To Solve Linear Equations

1. If the equation contains fractions, eliminate all fractions by multiplying both sides of the equation by the least common denominator.
2. Use the distributive property to remove any parentheses.
3. Combine like terms on each side of the equal sign.
4. Use the addition property to rewrite the equation with all terms containing the variable on one side of the equal sign and all terms not containing the variable on the other side of the equal sign. It may be necessary to use the addition property a number of times to accomplish this. Repeated use of the addition property will eventually result in an equation of the form $ax = b$.
5. Use the multiplication property to isolate the variable. This will give an answer of the form $x =$ some number.
6. Check the solution in the original equation.

EXAMPLE 3   Solve the equation $2x + 4 = 9$.

Solution:

$$2x + 4 = 9$$
$$2x + 4 - 4 = 9 - 4 \qquad \text{Subtract 4 from both sides of the equation.}$$
$$2x = 5$$
$$\frac{\overset{1}{\cancel{2}}x}{\underset{1}{\cancel{2}}} = \frac{5}{2} \qquad \text{Divide both sides of the equation by 2.}$$
$$x = \frac{5}{2}$$

Check:
$$2x + 4 = 9$$

$$2\left(\frac{5}{2}\right) + 4 = 9$$

$$5 + 4 = 9$$

$$9 = 9 \qquad \text{true}$$

Since the answer checks, the solution is $\frac{5}{2}$.   ◼

Whenever an equation contains like terms on the same side of the equal sign, combine the like terms before using the addition or multiplication properties.

EXAMPLE 4   Solve the equation $-4.2 = 2(x - 3.5) + 3x - 1.75$.

Solution:

$-4.2 = 2x - 7.0 + 3x - 1.75$	Distributive property was used.
$-4.2 = 5x - 8.75$	Combine like terms.
$-4.2 + 8.75 = 5x - 8.75 + 8.75$	Add 8.75 to both sides of the equation.
$4.55 = 5x$	
$\dfrac{4.55}{5} = \dfrac{5x}{5}$	Divide both sides of the equation by 5.
$0.91 = x$	

The solution is 0.91.   ◼

To save space, we will not always show the check of our answers. You should, however, check all your answers. When the problem contains decimal numbers, using a calculator to work and check the problem may save you some time.

EXAMPLE 5   Solve the equation $2x + 8 - 3x = -2(3x - 5) - 12$.

Solution: First, use the distributive property and then combine like terms.

$2x + 8 - 3x = -2(3x - 5) - 12$	
$2x + 8 - 3x = -6x + 10 - 12$	
$-x + 8 = -6x - 2$	
$6x - x + 8 = -6x + 6x - 2$	Add $6x$ to both sides of the equation.
$5x + 8 = -2$	
$5x + 8 - 8 = -2 - 8$	Subtract 8 from both sides of the equation.
$5x = -10$	
$\dfrac{5x}{5} = \dfrac{-10}{5}$	Divide both sides of the equation by 5.
$x = -2$	◼

In some of the following examples, we will omit some of the intermediate steps in determining the solution to an equation. Now we will illustrate how the solution may be shortened.

Solution   *Shortened Form of Solution*

(a) $x + 4 = 6$ (a) $x + 4 = 6$

       $x + 4 \boxed{- 4} = 6 \boxed{- 4}$ ⟵ Do this step mentally. $x = 2$

              $x = 2$

(b) $3x = 6$ (b) $3x = 6$

       $\dfrac{3x}{3} = \dfrac{6}{3}$ ⟵ Do this step mentally. $x = 2$

              $x = 2$

▶ **4**   Step 1 in the procedure to solve linear equations indicates that, when an equation contains fractions, we begin by multiplying *both* sides of the equation by the least common denominator. The **least common denominator of a set of numbers,** the LCD (also called the **least common multiple,** LCM), is the smallest number that each of the numbers divide (without remainder). For example, if the denominators of two fractions are 4 and 8, then the least common denominator is 8 since 8 is the smallest number that both 4 and 8 divide. If the denominators of two fractions are 5 and 6, then 30 is the least common denominator since 30 is the smallest number that both 5 and 6 divide.

When you multiply both sides of the equation by the LCD, *each term* in the equation will be multiplied by the least common denominator. *After this step is performed, the resulting equation should not contain any fractions.*

EXAMPLE 6   Solve the equation $5 - \dfrac{2x}{3} = -9$.

Solution: The least common denominator is 3. Multiply both sides of the equation by 3 and then use the distributive property. This process will eliminate all fractions from the equation.

$$5 - \frac{2x}{3} = -9$$

$$3\left(5 - \frac{2x}{3}\right) = 3(-9)$$

$$3(5) - \overset{1}{\cancel{3}}\left(\frac{2x}{\underset{1}{\cancel{3}}}\right) = -27$$

$$15 - 2x = -27$$

$$-2x = -42$$

$$x = 21$$ ∎

EXAMPLE 7   Solve the equation $\dfrac{x}{4} + 3 = 2x - \dfrac{5}{3}$.

Solution: Multiply both sides of the equation by the least common denominator 12. Then use the distributive property.

$$12\left(\frac{x}{4} + 3\right) = 12\left(2x - \frac{5}{3}\right)$$

$$\overset{3}{\cancel{12}}\left(\frac{x}{4}\right) + 12(3) = 12(2x) - \overset{4}{\cancel{12}}\left(\frac{5}{3}\right)$$

$$3x + 36 = 24x - 20$$
$$36 = 21x - 20$$
$$56 = 21x$$
$$x = \frac{56}{21} = \frac{8}{3}$$

EXAMPLE 8 Solve the equation $-\frac{2}{5}(x + 3) + 4 = \frac{1}{3}(x - 4)$.

Solution: Multiply both sides of the equation by the least common denominator 15 to remove fractions. Then use the distributive property on the left side of the equation.

$$-\frac{2}{5}(x + 3) + 4 = \frac{1}{3}(x - 4)$$

$$15\left[-\frac{2}{5}(x + 3) + 4\right] = 15\left[\frac{1}{3}(x - 4)\right]$$

Multiply both sides of the equation by 15.

$$\overset{3}{\cancel{15}}\left(-\frac{2}{\cancel{5}}\right)(x + 3) + 15\,(4) = \overset{5}{\cancel{15}}\left(\frac{1}{\cancel{3}}\right)(x - 4)$$

Use the distributive property on the left side of the equation.

$$-6(x + 3) + 60 = 5(x - 4)$$
$$-6x - 18 + 60 = 5x - 20$$
$$-6x + 42 = 5x - 20$$
$$42 = 11x - 20$$
$$62 = 11x$$
$$\frac{62}{11} = x$$

In the third step of the solution to Example 8, we used the distributive property on the *left* side of the equation, but not on the right side. Can you explain why? How many terms appear on the left side of the original equation, and how many terms appear on the right side of the original equation? The left side of the equation contains two terms, while the right side contains only one term. Therefore, we need to use the distributive property on the left side of the equation, but not the right side.

▶ **5** Equations of the form $\frac{a}{b} = \frac{c}{d}$ are called **proportions**. Proportions can often be solved by *cross multiplication,* using the procedure that follows.

Proportions

If $\frac{a}{b} = \frac{c}{d}$, then $ad = bc$, $b \neq 0$, $d \neq 0$.

We will use cross multiplication to solve Example 9.

EXAMPLE 9 Solve the equation $\dfrac{\frac{1}{2}x - 4}{3} = \dfrac{x + 8}{5}$.

Solution:

$$\dfrac{\frac{1}{2}x - 4}{3} = \dfrac{x + 8}{5}$$

$$5\left(\dfrac{1}{2}x - 4\right) = 3(x + 8) \qquad \text{Perform cross multiplication.}$$

$$\dfrac{5}{2}x - 20 = 3x + 24$$

$$2\left(\dfrac{5}{2}x - 20\right) = 2(3x + 24) \qquad \text{Multiply both sides of the equation by 2.}$$

$$5x - 40 = 6x + 48$$

$$-40 = x + 48$$

$$-88 = x$$

A check will show that the answer is correct.

Example 9 could also have been solved by multiplying both sides of the equation by the least common denominator, 15. Try this now to see that you obtain the same solution.

Equations that contain fractions, like those given in Examples 6 through 9, will be discussed in more detail in Section 7.5.

Identities and Inconsistent Equations

▶**6** All the equations discussed thus far have been **conditional equations.** They are true only under specific conditions. For example, in Example 9 the equation was true only when $x = -88$.

Consider the following equation: $2x + 1 = 5x + 1 - 3x$. Solving the equation, we obtain

$$2x + 1 = 5x + 1 - 3x$$

$$2x + 1 = 2x + 1$$

$$2x + 1 - 1 = 2x + 1 - 1 \qquad \text{Subtract 1 from both sides of the equation.}$$

$$2x = 2x$$

$$2x - 2x = 2x - 2x \qquad \text{Subtract } 2x \text{ from both sides of the equation.}$$

$$0 = 0 \qquad \text{A true statement.}$$

This equation, $2x + 1 = 5x + 1 - 3x$, is an example of an identity. An **identity** is an equation that is true for all real numbers. If at any point while solving an equation you realize that both sides of the equations are identical, as in

$$2x + 1 = 2x + 1$$

the equation is an identity. The solution to $2x + 1 = 5x + 1 - 3x$ is all real numbers. **The solution to any identity is all real numbers.** If you continue to solve an equation that is an identity, you will end up with $0 = 0$, a true statement.

Now consider the equation $2(3x + 1) = 9x + 3 - 3x$.

$$2(3x + 1) = 9x + 3 - 3x$$
$$6x + 2 = 6x + 3$$

$6x + 2 - 2 = 6x + 3 - 2$	Subtract 2 from both sides of the equation.
$6x = 6x + 1$	
$6x - 6x = 6x - 6x + 1$	Subtract 6x from both sides of the equation.
$0 = 1$	A false statement.

Since $0 = 1$ is never a true statement, this equation has no solution. An equation that has no solution is said to be an **inconsistent equation.** When solving an equation that turns out to be inconsistent, do not leave the answer blank. Write "no solution" as the answer. **An inconsistent equation has no solution.**

Every linear equation is either a conditional equation with exactly one solution, an identity with an infinite number of solutions, or inconsistent with no solution. Table 2.1 summarizes this information.

TABLE 2.1.

Type of Linear Equation	Solution
Conditional equation	Has exactly one real solution
Identity	Is true for every real number; has an infinite number of solutions
Inconsistent equation	Has no solution

Exercise Set 2.1

Name the property indicated.

1. $2 = 2$

2. If $x = 5$, then $5 = x$

3. If $x + 2 = 3$, then $3 = x + 2$

4. If $x = 3$, and $3 = y$, then $x = y$

5. If $x + 1 = a$ and $a = 2y$, then $x + 1 = 2y$

6. $-3 = -3$

7. $x + 2 = x + 2$

8. If $x = 4$, then $x + 3 = 4 + 3$

9. If $x = 2$, then $x - 2 = 2 - 2$

10. If $x = 3$, then $3 = x$

11. If $2x = 4$, then $3(2x) = 3(4)$

12. If $5x = 4$, then $\frac{1}{5}(5x) = \frac{1}{5}(4)$

13. If $2x = 7$, then $\frac{1}{2}(2x) = \frac{1}{2}(7)$

14. If $x = 5$, then $x - 2 = 5 - 2$

15. If $x = 3$, then $x - 3 = 3 - 3$

16. If $x + 2 = 4$, then $x + 2 - 2 = 4 - 2$

17. If $5x = 3$, then $\frac{5x}{2} = \frac{3}{2}$

18. If $2x = 6$, then $\frac{2x}{2} = \frac{6}{2}$

19. If $x = 2$, then $2 = x$

20. If $x - 3 = x + y$, and $x + y = z$, then $x - 3 = z$

Give the degree of the term.

21. $4x$

22. $-6x^2$

23. $3xy$

24. $18x^2y^3$

25. $\frac{1}{2}x^4y$

26. 7

27. -3

28. $-5x$

29. $3x^4y^6z^3$

30. x^4y^6

31. $3x^5y^6z$

32. $-2x^4y^7z^8$

Simplify each expression. If an expression cannot be simplified, so state.

33. $8x + 7 + 7x - 12$

34. $3x^2 + 4x + 5$

35. $5x^2 - 3x + 2x - 5$

36. $6x^2 - 9x + 3 - 4x - 7$

37. $-4x^2 - 3x - 5x + 7$

38. $7y + 3x - 7 + 4x - 2y$

39. $6y^2 + 6xy + 3$

40. $4x^2 - x^2 - 3x - 3x^2 + 4$

41. $xy + 3xy + y^2 - 2$

42. $3x^2y + 4xy^2 - 2x^2$

43. $4(x + 3) - 7(2x - 5)$

44. $6(x + 5) + 2(x + \frac{2}{3})$

45. $3(x + \frac{1}{2}) - \frac{1}{3}x + 5$

46. $0.4(x - 3) + 6.5(x - 3) + 4x - 2.3$

47. $4 - [6(3x + 2) - x] + 4$

48. $3(x + y) - 4(x + y) - 3$

49. $4x - [3x - (5x - 4y)] + y$

50. $-2[3x - (2y - 1) - 5x] + y$

Solve each equation. If an equation has no solution, so state.

51. $2x + 3 = 5$

52. $4 = 6 - 5x$

53. $4x + 3 = -12$

54. $-\frac{2}{3}x = -12$

55. $-\frac{x}{4} = 8$

56. $6 = 4 - 5x$

57. $\frac{10}{3} = x + 6$

58. $-2 = \frac{-4x}{5} + 9$

59. $3.2(x - 1.6) = 5.88 + 0.2x$

60. $\frac{x - 4}{3} = \frac{x + 4}{2}$

61. $\frac{3x - 9}{3} = \frac{2x - 6}{6}$

62. $\frac{-x + 4}{3} = \frac{\frac{2}{3}x - 1}{4}$

63. $\frac{\frac{1}{4}x - 2}{5} = \frac{3(x - 2)}{4}$

64. $\frac{4x - 1.2}{0.8} = \frac{6x - 3}{1.4}$

65. $\frac{1.5x}{5} = \frac{x - 4.2}{8}$

66. $3(2x - 4) + 3(x + 1) = 9$

67. $-(x - 4) + 3x = -12$

68. $6 - (x + 2) = -3$

69. $\frac{1}{2}(x - 4) = 8$

70. $\frac{2}{3}(2x - 6) + 4 = 8$

71. $6 - 2(x - 3) + 5 = 15$

72. $\frac{4x + 3}{4} = x + 6$

73. $\frac{2x - 5}{3} = -5$

74. $\frac{3x - 7}{5} + 2 = 9$

75. $\frac{4 - 3x}{2} + x = 4$

76. $\frac{1}{2}(3x - 5) = 6$

77. $-\frac{3}{5}(15 - 2x) = -3$

78. $2.5(1.6x - 3) = 4.6x$

79. $-4.2(3.2x - 4) = 2.56x$

80. $6(x - 1) = -3(2 - x) + 3x$

81. $4x - 5(x + 3) = 2x - 3$

82. $\frac{4}{3} + \frac{x}{4} = x$

83. $\frac{7x}{5} = 2x + 3$

84. $\frac{x}{3} + 2 = x + 4$

85. $2(x - 3) + 2x = 4x - 5$

86. $\frac{11 - 3x}{4} = 2x - 4$

87. $\frac{x - 25}{3} = 2x - \frac{2}{5}$

88. $\frac{4}{3} + 4x = 2 - x$

89. $4(2 - 3x) = -[6x - (8 - 6x)]$

90. $\frac{2}{3}x = \frac{9}{4}x$

91. $\frac{1}{2}(2x + 1) = \frac{1}{4}(x - 4)$

92. $4x - 3 = -5x - (x - 1)$

93. $4[x - (5x - 2)] = 2(x - 3)$

94. $-[4 - (3x - 5) + 2x] = -3(x - 2)$

95. $\frac{2}{3}(x - 4) = \frac{2}{3}(4 - x)$

96. $\frac{2(x + 2)}{5} = \frac{x}{3} - 1$

97. $\frac{3x}{4} - 2 = 6 + \frac{x}{3}$

98. $\frac{x - 8}{5} + \frac{x}{3} = \frac{-8}{5}$

99. $\frac{x + 1}{4} = \frac{x - 4}{2} - \frac{2x - 3}{4}$

100. (a) In your own words, explain in a step-by-step manner how you would solve the equation $5x + 2x - 5 = 3(x - 7)$.

(b) Solve the equation.

101. (a) In your own words, explain in a step-by-step manner how you would solve the equation $2x - \dfrac{2}{5} = \dfrac{2}{3}(x + 5)$.

(b) Solve the equation.

102. (a) In your own words, explain in a step-by-step manner how you would solve the equation

$$\frac{3}{4}(2x - 3) + \frac{1}{2} = \frac{2}{3}(x - 4).$$

(b) Solve the equation.

103. What is an identity?

104. What is a conditional equation?

105. What is an inconsistent equation?

106. What are equivalent equations?

107. Consider the equation $x = 4$. Give three equivalent equations.

108. Consider the equation $2x = 5$. Give three equivalent equations.

Cumulative Review Exercises

[1.4] **109.** Write the definition of absolute value.

[1.6] *Evaluate each of the following expressions:*

110. -3^2

111. $\left(-\dfrac{3}{4}\right)^3$

112. $\sqrt[3]{-64}$

JUST FOR FUN

Solve each of the following equations for x.

1. $-\dfrac{3}{5}(x + 2) - \dfrac{4}{3}(2x - 3) + 4 = \dfrac{1}{2}(x + 4) - 6x + 3(5 - x)$

2. $2[-3[4(x + 3) + 2] + 4] = 3[2(x + 3) + 5] + x + 6$

3. $\dfrac{x}{3} + \dfrac{x - 2}{4} + \dfrac{2x - 3}{5} = \dfrac{x - 3}{6} + 4(x - 7) - x + 2$

2.2

Formulas

▶**1** Use subscripts and Greek letters in formulas.

▶**2** Evaluate formulas.

▶**3** Solve for a variable in an equation or formula.

▶**1** To give you more practice in solving equations, we will now discuss literal equations and formulas. **Literal equations** are equations that have more than one letter. **Formulas** are literal equations that are used to represent a scientific or real-life principal in mathematical terms.

Examples of Literal Equations

$$5y = 2x + 3$$
$$x + 2y + 3z = 5$$

Examples of Formulas

$A = P(1 + rt)$ from business

$V = \frac{1}{2}at^2$ from physics

Often a formula will contain subscripts. **Subscripts** are numbers (or other variables) placed below and to the right of variables. They are used to help clarify a formula. For example, if a formula contains two velocities, the original velocity and the final velocity, these velocities might be symbolized as V_0 and V_f, respectively. Subscripts are read using the word "sub"; for example, V_f is read "V sub f" and x_2 is read "x sub 2."

Many mathematical and scientific formulas use *Greek letters*. Examples of the use of Greek letters are given in Table 2.2. *It is not necessary for you to know by name any of the Greek letters except pi, π, delta, Δ, and sigma, Σ, in this course.* Pi is used in finding the circumference and area of a circle. Delta is used when studying slope in Chapter 3, and sigma is used when discussing series in Chapter 12.

TABLE 2.2 Use of Greek Letters in Selected Formulas

Formula	Greek Letter	Greek Letter Represents:
$A = \pi r^2$	π (pi)	A constant
$S = r\theta$	θ (theta)	Angle measurement
$\bar{x} = \dfrac{\Sigma x}{n}$	Σ (sigma)	Summation
$z = \dfrac{x - \mu}{\sigma}$	μ (mu)	Mean of set of data
	σ (sigma)	Standard deviation of a set of data
$m = \dfrac{\Delta y}{\Delta x}$	Δ (delta)	A change in value

Other Greek letters commonly used are γ (gamma), α (alpha), ϵ (epsilon), ρ (rho), λ (lambda), ω (omega), δ (delta), ϕ (phi), β (beta), and χ (chi). The Greek alphabet, like our own, has both upper- and lowercase letters. For example, σ represents the lowercase sigma and Σ represents the uppercase sigma.

▶**2** Many students take algebra in preparation for another mathematics course or a science course. Evaluation of formulas plays an important role in this and other courses. To **evaluate a formula** means to find the value of one of the variables, when you are given the values of the other variables in the formula.

To evaluate a formula follow the order of operations presented in Section 1.7. The order of operations is summarized and included here for you to review.

Order of operations

1. First, evaluate the information within parentheses, ( ), or brackets, [ ].
2. Next, evaluate all terms containing exponents and roots.
3. Next, evaluate all multiplications or divisions, working from left to right.
4. Finally, evaluate all addition or subtractions, working from left to right.

If your instructor permits the use of a calculator, then use it to evaluate formulas. In the following examples, *do not be concerned if you are not familiar with the formulas or symbols.*

EXAMPLE 1   Consider the compound interest formula $A = p\left(1 + \dfrac{r}{n}\right)^{nt}$. The compound interest formula is used by banks to compute the amount (or balance), A, in savings accounts that give compound interest. In the formula, p represents the principal, r represents the interest rate, n represents the number of compounding periods (the number of times the interest is paid annually), and t represents time in years.

Mr. Johnson invests $1000 in a savings account which yields 8% interest, compounded semiannually for a period of one year. Use the compound interest formula to find the amount in the account at the end of one year.

Solution: The principal, p, is $1000. The rate, r, is 8% or 0.08 in decimal form. Since the interest is compounded semiannually, n is 2. The time, t, is 1 year. Substitute the appropriate values in the compound interest formula. Note that percents are always converted to decimal form before being substituted into any formula.

$$A = p\left(1 + \frac{r}{n}\right)^{nt}$$

$$= 1000\left(1 + \frac{0.08}{2}\right)^{2(1)}$$

First, we evaluate the information within parentheses.

$$= 1000(1 + 0.04)^{2(1)}$$

$$= 1000(1.04)^{2(1)}$$

Next we raise the value within parentheses to the second power.

$$= 1000(1.04)^2$$

$$= 1000(1.0816)$$

$$= 1081.60$$

Mr. Johnson has a balance of $1081.60 in his account after one year. This $1081.60 consists of the $1000 principal plus $81.60 in interest.   ∎

EXAMPLE 2   A formula used in the study of statistics is

$$Z = \frac{\bar{x} - \mu}{\dfrac{\sigma}{\sqrt{n}}}$$

Find Z when $\bar{x} = 100.4$, $\mu = 102.8$, $\sigma = 4.2$, and $n = 36$.

Solution: $$Z = \frac{100.4 - 102.8}{\dfrac{4.2}{\sqrt{36}}} = \frac{-2.4}{\dfrac{4.2}{6}} = \frac{-2.4}{0.7} \approx -3.43$$ ∎

▶**3**   Often in science or mathematics courses we are given a formula or equation solved for one variable and asked to solve it for a different variable. To do this, treat each variable in the equation, except the one you are solving for, as if it were a constant. Then solve for the desired variable using the properties discussed previously. To solve for a given variable, it is necessary to get that variable all by itself on one side of the equal sign.

In Chapter 3 we will graph equations. To graph an equation, it is sometimes necessary to solve the equation for the variable y. Examples 3 and 4 illustrate a procedure for doing this.

EXAMPLE 3 Solve the equation $2x - 3y = 6$ for y.

Solution: We must isolate the term containing y.

$$2x - 3y = 6$$
$$2x \boxed{-\ 2x} - 3y = \boxed{-2x} + 6 \qquad \text{Subtract } 2x \text{ from both sides of the equation.}$$
$$-3y = -2x + 6$$
$$\frac{-3y}{\boxed{-3}} = \frac{-2x + 6}{\boxed{-3}} \qquad \text{Divide both sides of equation by } -3.$$
$$y = \frac{-2x + 6}{-3}$$

Since we do not want to leave the answer with a negative number in the denominator, we multiply both the numerator and the denominator by -1 to get

$$y = \frac{2x - 6}{3} \quad \text{or} \quad y = \frac{2}{3}x - 2 \qquad \blacksquare$$

EXAMPLE 4 Solve the equation $\dfrac{2}{3}(2x - y) = \dfrac{5}{7}(x - 3y) + 4$ for y.

Solution: We must isolate the variable y. To do this, collect all terms containing the variable y on one side of the equation and all terms not containing the variable y on the other side of the equation. Since this equation contains fractions, we begin by multiplying both sides of the equation by the least common denominator 21 to eliminate the fractions.

$$\boxed{21}\left[\tfrac{2}{3}(2x - y)\right] = \boxed{21}\left[\tfrac{5}{7}(x - 3y) + 4\right] \qquad \begin{array}{l}\text{Multiply both sides of equation by} \\ \text{the LCD, 21.}\end{array}$$
$$\boxed{21}\left[\tfrac{2}{3}(2x - y)\right] = \boxed{21}\left[\tfrac{5}{7}(x - 3y)\right] + \boxed{21} \cdot 4 \qquad \text{Distributive property was used.}$$
$$14(2x - y) = 15(x - 3y) + 84$$
$$28x - 14y = 15x - 45y + 84$$
$$28x - 14y \boxed{+\ 45y} = 15x - 45y \boxed{+\ 45y} + 84 \qquad \begin{array}{l}\text{Add } 45y \text{ to both sides of the} \\ \text{equation.}\end{array}$$
$$28x + 31y = 15x + 84$$
$$28x \boxed{-\ 28x} + 31y = 15x \boxed{-\ 28x} + 84 \qquad \begin{array}{l}\text{Subtract } 28x \text{ from both sides of} \\ \text{the equation.}\end{array}$$
$$31y = -13x + 84$$
$$\frac{31y}{\boxed{31}} = \frac{-13x + 84}{\boxed{31}} \qquad \begin{array}{l}\text{Divide both sides of the} \\ \text{equation by 31.}\end{array}$$
$$y = \frac{-13x + 84}{31} \quad \text{or} \quad y = -\frac{13}{31}x + \frac{84}{31} \qquad \blacksquare$$

EXAMPLE 5 The formula for the area, A, of a triangle $A = \frac{1}{2}bh$, where b is the length of the base and h is the height. Solve this formula for the height, h.

Solution: We are asked to express the height, h, of the triangle in terms of the triangle's area, A, and base, b. Since we are solving for h, we must isolate the h on one side of the equation. We use the appropriate properties to remove the $\frac{1}{2}$ and the b from the right side of the equation.

$$A = \frac{1}{2}bh$$

$$2A = 2\left(\frac{1}{2}\right)bh \qquad \text{Multiply both sides of the equation by 2.}$$

$$2A = bh$$

$$\frac{2A}{b} = \frac{bh}{b} \qquad \text{Divide both sides of the equation by } b.$$

$$\frac{2A}{b} = h \quad \text{or} \quad h = \frac{2A}{b} \qquad \blacksquare$$

EXAMPLE 6 A formula used in the study of statistics is $z = (x - \mu)/\sigma$. Solve this equation for x.

Solution: Note that this formula contains the Greek letters μ (mu) and σ (sigma). Treat them the same as you would any other letter.

$$z = \frac{x - \mu}{\sigma}$$

$$\sigma \cdot z = \frac{x - \mu}{\sigma} \cdot \sigma \qquad \text{Multiply both sides of the equation by } \sigma.$$

$$\sigma z = x - \mu$$

$$\sigma z + \mu = x - \mu + \mu \qquad \text{Add } \mu \text{ to both sides of the equation.}$$

$$\sigma z + \mu = x$$

The answer may be written in a number of other forms. Other acceptable answers include $x = \mu + \sigma z$ and $x = \mu + z\sigma$. $\qquad \blacksquare$

EXAMPLE 7 A formula that may be important to you now, or in the future, is the *tax-free yield formula*, $T_f = T_a(1 - F)$. This formula can be used to convert a taxable yield, T_a, into its equivalent tax-free yield, T_f, where F is the federal income tax bracket of the individual.

(a) Mary is in a 28% income tax bracket. Find the equivalent tax-free yield of a 12% taxable investment.
(b) Solve this equation for T_a, that is, write an equation for taxable yield in terms of tax-free yield.

Solution: (a) $T_f = 0.12(1 - 0.28)$

$ = 0.12(0.72)$

$ = 0.0864 \quad \text{or} \quad 8.64\%$

Thus, for Mary, or anyone else in a 28% income tax bracket, a 12% taxable yield is equivalent to an 8.64% tax-free yield.

(b) $\qquad T_f = T_a(1 - f)$

$$\frac{T_f}{1 - f} = \frac{T_a(1 - f)}{1 - f} \qquad \text{Divide both sides of the equation by } 1 - f.$$

$$\frac{T_f}{1 - f} = T_a \quad \text{or} \quad T_a = \frac{T_f}{1 - f} \qquad \blacksquare$$

EXAMPLE 8   A formula used in banking is $A = P(1 + rt)$. A represents the amount that must be repaid to the bank when P dollars are borrowed at simple interest rate, r, for time, t. Solve this equation for time, t.

Solution:

$$A = P(1 + rt)$$
$$A = P + Prt \qquad \text{Distributive property was used.}$$
$$A - P = P - P + Prt \qquad \text{Subtract } P \text{ from both sides of the equation to isolate the term containing the variable } t.$$
$$A - P = Prt$$
$$\frac{A - P}{Pr} = \frac{Prt}{Pr} \qquad \text{Divide both sides of equation by } Pr \text{ to isolate } t.$$
$$\frac{A - P}{Pr} = t \qquad\qquad \blacksquare$$

Solving for a variable in a formula will be discussed again in sections 6.6, 7.5 and 8.6.

Exercise Set 2.2

Evaluate the formula for the values given. Round answers to the nearest hundredth.

1. $P = 2l + 2w$; $l = 15, w = 6$ (mathematics)

2. $\bar{x} = \dfrac{x_1 + x_2 + x_3}{3}$: $x_1 = 40, x_2 = 120, x_3 = 80,$ (statistics)

3. $A = \dfrac{1}{2}h(b_1 + b_2)$; $h = 10, b_1 = 20, b_2 = 30$ (mathematics)

4. $P = \dfrac{nRT}{V}$: $n = 10, R = 50, T = 12, V = 500$ (chemistry)

5. $C = a + by_D$: $a = 500, b = 10, y_D = 60$ (economics)

6. $N = 16\left(\dfrac{t_g}{w}\right)^2$: $t_g = 50, w = 10$ (chemistry)

7. $E = a_1 p_1 + a_2 p_2 + a_3 p_3$: $a_1 = 10, p_1 = 0.2, a_2 = 100, p_2 = 0.3, a_3 = 1000, p_3 = 0.5$ (probability)

8. $a_{av} = \dfrac{v_2 - v_1}{t_2 - t_1}$: $v_2 = 80, v_1 = 40, t_2 = 3, t_1 = 0$ (physics)

9. $K = \dfrac{F - 32}{1.8} + 273.1$: $F = 100$ (chemistry)

10. $F = G\dfrac{m_1 m_2}{r^2}$: $G = 0.5, m_1 = 100, m_2 = 200, r = 4$ (physics)

11. $m = \dfrac{y_2 - y_1}{x_2 - x_1}$: $y_2 = 4, y_1 = -3, x_2 = -2, x_1 = -6$ (mathematics)

12. $p = 1 - \dfrac{1}{k^2}$: $k = 1.6$ (statistics)

13. $z = \dfrac{\bar{x} - \mu}{\dfrac{\sigma}{\sqrt{n}}}$: $\bar{x} = 80, \mu = 70, \sigma = 15, n = 25$ (statistics)

14. $a_n = a_1 + (n - 1)d$: $a_1 = -3, n = 12, d = 4$ (mathematics)

15. $d = \sqrt{(x_2 - x_1)^2 + (y_2 - y_1)^2}$: $x_2 = 5, x_1 = -3, y_2 = -6, y_1 = 3$ (mathematics)

16. $n = \left(\dfrac{z\sigma}{E}\right)^2$: $z = 1.96, \sigma = 2, E = 0.2$ (statistics)

17. $R_T = \dfrac{R_1 R_2}{R_1 + R_2}$: $R_1 = 100, R_2 = 200$ (electronics)

18. $V = \sqrt{V_x^2 + V_y^2}$: $V_x = 3, V_y = 4$ (physics)

19. $x = \dfrac{-b + \sqrt{b^2 - 4ac}}{2a}$: $a = 2, b = -5, c = -12$ (mathematics)

20. $x = \dfrac{-b - \sqrt{b^2 - 4ac}}{2a}$: $a = 2, b = -5, c = -12$ (mathematics)

21. $R = O + (V - D)r$: $O = 500, V = 200, D = 12, r = 4$ (economics)

22. $V = \dfrac{1}{3}\pi r^2 h$: $\pi = 3.14, r = 5, h = 10$ (mathematics)

23. $S = \pi r^2 + \pi rs$: $\pi = 3.14, r = 3, s = 4$ (mathematics)
24. $H = (0.14 + 0.47\sqrt{v})(36.5 - T)$: $v = 25, T = 5$ (meteorology)

25. $Y = \dfrac{\dfrac{F}{A}}{\dfrac{\Delta \ell}{\ell_0}}$: $F = 20, A = 2, \Delta \ell = 5, \ell_0 = 10$ (physics)

26. $C = \bar{x} + z\dfrac{\sigma}{\sqrt{n}}$: $\bar{x} = 80, z = 2.05, \sigma = 4, n = 36$ (statistics)

27. $P_c = \dfrac{P_1 - P_2}{\dfrac{P_1 + P_2}{2}}$: $P_1 = 800, P_2 = 600$ (economics)

28. $Z = \dfrac{p' - p}{\sqrt{\dfrac{pq}{n}}}$: $p' = 0.6, p = 0.5, q = 0.5, n = 10$ (statistics)

29. $\bar{x}_w = \dfrac{w_1 x_1 + w_2 x_2 + w_3 x_3}{w_1 + w_2 + w_3}$: $x_1 = 60, x_2 = 80, x_3 = 96, w_1 = 4, w_2 = 6, w_3 = 10$ (statistics)

30. $S_n = \dfrac{a_1(1 - r^n)}{1 - r}$: $a_1 = 6, r = \dfrac{1}{2}, n = 3$ (mathematics)

31. $A = P\left(1 + \dfrac{r}{n}\right)^{nt}$: $P = 100, r = 0.06, n = 1, t = 3$ (banking)

32. $r_{DD} = \dfrac{\dfrac{r_{xx} + r_{yy}}{2} - r_{xy}}{1 - r_{xy}}$: $r_{xx} = 0.7, r_{yy} = 0.9, r_{xy} = 0.3$ (psychology)

Solve each equation for y.

33. $2x + y = 3$
34. $3x + y = 5$
35. $2x + 3y = 6$
36. $3x + 2y = 8$
37. $x - y = 8$
38. $2x - y = -5$
39. $2x - 4y = 6$
40. $5x - 3y = -4$
41. $2y = 8x - 3$
42. $\frac{1}{2}x + 2y = 6$
43. $\frac{3}{5}x + \frac{1}{3}y = 1$
44. $3(x - 2) + 3y = 6x$
45. $2(x + 3y) = 4(x - y) + 5$
46. $3x - 5 = 2(3y + 6)$
47. $\frac{1}{5}(x - 2y) = \frac{3}{4}(y + 2) + 3$

Solve for the variable indicated.

48. $d = rt$, for t

49. $d = rt$, for r

50. $C = \pi d$, for d

51. $A = \dfrac{1}{2}bh$, for b

52. $C = 2\pi r$, for r

53. $i = prt$, for t

54. $P = 2l + 2w$, for w

55. $P = 2l + 2w$, for l

56. $V = \dfrac{1}{3}Bh$, for B

57. $V = \dfrac{1}{3}Bh$, for h

58. $V = lwh$, for h

59. $V = \pi r^2 h$, for h

60. $V = \dfrac{1}{3}lwh$, for l

61. $z = \dfrac{x - \mu}{\sigma}$, for σ

62. $z = \dfrac{x - \mu}{\sigma}$, for μ

63. $P = I^2 R$, for R

64. $I = P + Prt$, for r

65. $A = \dfrac{1}{2}h(b_1 + b_2)$, for h

66. $A = \dfrac{1}{2}h(b_1 + b_2)$, for b_1

67. $y = mx + b$, for m

68. $IR + Ir = E$, for R

69. $y - y_1 = m(x - x_1)$, for m

70. $R_T = \dfrac{R_1 + R_2}{2}$, for R_1

71. $S = \dfrac{n}{2}(f + l)$, for n

72. $S = \dfrac{n}{2}(f + l)$, for l

73. $C = \dfrac{5}{9}(F - 32)$, for F

74. $F = \dfrac{9}{5}C + 32$, for C

75. $y = \dfrac{kx}{z}$, for z

76. $\dfrac{P_1}{T_1} = \dfrac{P_2}{T_2}$, for T_2

77. $F = \dfrac{km_1 m_2}{d^2}$, for m_1

78. $A = \dfrac{r^2 \theta}{2}$, for θ

79. (a) In your own words, explain the procedure you would use to solve the physics formula $\mu = \dfrac{0.5cV^2}{Ad}$ for A.

(b) Solve the formula for A.

80. Ships at sea give their speed in terms of knots. One knot is one nautical mile per hour, where a nautical mile is about 6076 feet. When measuring a car's speed in miles per hour, we consider a mile to contain 5280 ft.

(a) Determine a formula for converting a speed in knots (k) to a speed in miles per hour (m).

(b) Explain how you determined this formula.

Cumulative Review Exercises

[1.6] **81.** Evaluate $-(5 - 8)^2 + |5 - 8| - 4^2$.

[1.7] **82.** Evaluate $\dfrac{4 - 6 \div 3 + 5^2 - 6 \cdot 4}{5 - |6 \div (-2)|}$

83. Evaluate $6x^2 - 3xy + y^2$ when $x = 2$, $y = 3$.

[2.1] **84.** Solve the equation $\dfrac{1}{3}x + 4 = \dfrac{2}{5}(x - 3)$.

2.3

Applications of Algebra

▶ **1** Translate a statement into an algebraic expression.

▶ **2** Solve application problems.

▶ **1** The next few sections will give you an idea of some of the many uses of algebra in real-life situations. Whenever possible, we include other relevant applications throughout the text.

Perhaps the most difficult part of solving a word problem is transforming the problem into an equation. Here are some examples of phrases represented as algebraic expressions.

Phrase	*Algebraic Expression*
a number increased by 4	$x + 4$
twice a number	$2x$
5 less than a number	$x - 5$
a number subtracted from 9	$9 - x$
6 subtracted from a number	$x - 6$
one-eighth of a number	$\frac{1}{8}x$ or $\frac{x}{8}$
2 more than 3 times a number	$3x + 2$
4 less than 6 times a number	$6x - 4$
3 times the sum of a number and 5	$3(x + 5)$

The variable x was used in the algebraic expressions, but any variable could have been used to represent the unknown quantity.

EXAMPLE 1   Express each phrase as an algebraic expression.

(a) The distance, d, increased by 15 miles.
(b) 3 less than 4 times the area.
(c) 4 times a number decreased by 5.

Solution: (a) $d + 15$ (b) $4A - 3$ (c) $4n - 5$ ∎

EXAMPLE 2   Write each of the following as an algebraic expression.

(a) The cost of purchasing x shirts at \$4 each.
(b) The distance traveled in t hours at 55 miles per hour.
(c) The number of calories in x potato chips if each potato chip has 10 calories.
(d) The number of cents in n nickels.
(e) Eight percent commission on x dollars.

Solution: (a) We can reason like this: one shirt would cost 1(4) dollars; two shirts, 2(4) dollars; three shirts, 3(4) dollars; four shirts, 4(4) dollars, and so on. Continuing this reasoning process, we can see that x shirts would cost $x(4)$ or $4x$ dollars. We can use the same reasoning process to complete each of the following.
(b) $55t$
(c) $10x$
(d) $5n$
(e) $0.08x$

Note in part (e) that 8% is written as 0.08 in decimal form.   ∎

HELPFUL HINT

A percent is always a percent of some quantity. Therefore, when a percent is listed, it is *always* multiplied by a number or a variable. In the following examples we use the variable x, but any letter could be used to represent the variable.

Phrase	*How Written*
6% of a number	$0.06n$
the cost of an item increased by a 7% tax	$c + 0.07c$
25% off the cost of an item	$c - 0.25c$

Sometimes in a problem two numbers are related to each other in a certain way. We often represent one of the numbers as a variable and the other as an expression containing that variable. We generally let the less complicated description be represented by the variable and write the second (more complex expression) in terms of the variable. Some examples follow.

Phrase	*One Number*	*Second Number*
Peter's age now and Peter's age in 5 years	x	$x + 5$
one number is 3 times the other	x	$3x$
one number is 7 less than the other	x	$x - 7$
two consecutive integers	x	$x + 1$
two consecutive odd (or even) integers	x	$x + 2$
the sum of two numbers is 10	x	$10 - x$
a 6-foot piece of lumber cut in two pieces	x	$6 - x$
a number and the number increased by 7%	x	$x + 0.07x$
a number and the number decreased by 10%	x	$x - 0.10x$
$10,000 divided between 2 people	x	$10,000 - x$

COMMON STUDENT ERROR

Consider the phrase "$10,000 divided between two people." You might think about dividing the $10,000 by 2 to obtain the answer. But this statement *does not* indicate that the $10,000 is being divided into two *equal* parts.

If one person were to receive x dollars, then the other person would receive $10,000 - x$ dollars. For example, if one person received $2000, then the other person would receive $10,000 - 2000 = 8000$. Note that $2000 + $8000 = $10,000.

EXAMPLE 3 For each of the following relationships, select a variable to represent one quantity and express the second quantity in terms of the first.
(a) Jan is 12 years older than her sister.
(b) The speed of the second train is 1.2 times the speed of the first.
(c) $90 is split between David and his brother.
(d) It takes Tom 3 hours longer than Roberta to complete the task.
(e) Hilda has $4 more than twice the amount of money Hector has.
(f) The length of a rectangle is 2 units less than 3 times its width.

Solution: (a) Sister, x; Jan, $x + 12$
(b) First train, x; second train, $1.2x$
(c) Amount David has, x; amount brother has, $90 - x$
(d) Roberta, x; Tom, $x + 3$
(e) Hector, x; Hilda, $2x + 4$
(f) Width, x; length, $3x - 2$

▶ **2** The word **is** in a word problem often means **is equal to** and is represented by an equal sign, $=$.

Verbal Statement	*Algebraic Equation*
3 more than a number *is* 9	$x + 3 = 9$
4 less than 3 times a number *is* 5	$3x - 4 = 5$
a number decreased by 4 *is* 3 more than twice the number	$x - 4 = 2x + 3$
the product of two consecutive integers *is* 20	$x(x + 1) = 20$
one number is 2 more than 5 times the other number; the sum of the two numbers *is* 62	$x + (5x + 2) = 62$
a number increased by 15% *is* 90	$x + 0.15x = 90$
a number decreased by 12% *is* 38	$x - 0.12x = 38$
the sum of a number and the number increased by 4% *is* 204	$x + (x + 0.04x) = 204$
the cost of renting a VCR for x days at $15 per day *is* $120	$15x = 120$
the distance traveled in n days when 600 miles are traveled per day *is* 1500 miles	$600n = 1500$

Although there are many types of word problems, the general procedure used to solve all word problems is basically the same.

To Solve a Word Problem

1. Read the problem carefully.
2. If possible, draw a sketch to illustrate the problem.
3. Identity the quantity or quantities you are being asked to find.
4. Choose a variable to represent one quantity, *and write down exactly what it represents*. Represent any other quantities to be found in terms of the variable chosen.
5. Write the word problem as an equation.
6. Solve the equation for the unknown quantity.
7. Answer the question asked.
8. Check the solution in the *original word problem*.

Many students solve the equation for the variable, but forget to answer the original question or questions asked.

EXAMPLE 4   Four subtracted from 3 times a number is 29. Find the number.

Solution: *Step 3:*   We are asked to find the unknown number.

Step 4:   Let x = unknown number.

Step 5:   Write the equation.

$$3x - 4 = 29$$

Step 6:   Solve the equation.

$$3x - 4 = 29$$
$$3x = 33$$
$$x = 11$$

Step 7: Answer the question.

The number is 11.

Step 8: Check the solution in the original word problem.

Four subtracted from 3 times a number is 29.

$$3(11) - 4 = 29$$
$$33 - 4 = 29$$
$$29 = 29 \qquad \text{true}$$ ∎

EXAMPLE 5 The sum of two numbers is 47. Find the two numbers if one number is 2 more than 4 times the other number.

Solution:

Let x = smaller number

then $4x + 2$ = larger number

smaller number + larger number = sum of two numbers

$$x + 4x + 2 = 47$$
$$5x + 2 = 47$$
$$5x = 45$$
$$x = 9$$

The smaller number is 9. The larger number is $4x + 2 = 4(9) + 2 = 38$.

Check: The sum of the two numbers is 47.

$$9 + 38 = 47$$
$$47 = 47 \qquad \text{true}$$ ∎

In Example 5, if you gave the answer 9, you would not be correct. The question asked you to find *both* numbers. The two numbers are 9 and 38.

EXAMPLE 6 The population of Springfield is 96,000. If the population is increasing by 5000 per year, how many years will it take to reach 125,500?

Solution:

Let n = number of years

then $5000n$ = population increase in n years

$$\left(\begin{array}{c} \text{present} \\ \text{population} \end{array} \right) + \left(\begin{array}{c} \text{increase in population} \\ \text{over } n \text{ years} \end{array} \right) = \left(\begin{array}{c} \text{future} \\ \text{population} \end{array} \right)$$

$$96{,}000 \qquad + \qquad 5000n \qquad = \qquad 125{,}000$$

$$96{,}000 + 5000n = 125{,}500$$
$$5000n = 29{,}500$$
$$n = 5.9 \text{ years}$$

The population of Springfield will increase to 125,500 in 5.9 years. ∎

EXAMPLE 7   For two consecutive numbers, one less than twice the smaller subtracted from 4 times the larger is 19. Find the two numbers.

Solution:

$$\text{Let } x = \text{smaller of two consecutive numbers}$$
$$\text{then } x + 1 = \text{larger of two consecutive numbers}$$

$$\text{one less than twice the smaller} = 2x - 1$$
$$\text{four times the larger} = 4(x + 1)$$

One less than twice the smaller *subtracted from* four times the larger *is* 19.

$$4(x + 1) - (2x - 1) = 19$$
$$4x + 4 - 2x + 1 = 19$$
$$2x + 5 = 19$$
$$2x = 14$$
$$x = 7$$
$$\text{smaller number} = x = 7$$
$$\text{larger number} = x + 1 = 8$$

Check:

$$\left(\begin{array}{c}\text{four times the}\\\text{larger number}\end{array}\right) - \left(\begin{array}{c}\text{one less than twice}\\\text{the smaller number}\end{array}\right) = 19$$
$$4(8) - [2(7) - 1] = 19$$
$$32 - (14 - 1) = 19$$
$$32 - 13 = 19$$
$$19 = 19 \qquad \text{true} \qquad ■$$

COMMON STUDENT ERROR

When an expression containing more than one term is to be subtracted from another, the entire expression must be subtracted, not just the first term. In Example 7, had we written

$$4(x + 1) - 2x - 1 = 19$$

we would have obtained the wrong answer. The parentheses are needed around the entire expression $2x - 1$ to show that the entire expression is to be subtracted.

Example: Subtract $3x + 4$ from $2x$.

Correct	*Wrong*
$2x - (3x + 4)$	~~$2x - 3x + 4$~~

Example: Subtract $2x - 5$ from $x - 3$.

Correct	*Wrong*
$x - 3 - (2x - 5)$	~~$x - 3 - 2x - 5$~~

EXAMPLE 8   A chemical company's bonus plan states that employees will receive a bonus of 6% of their yearly salary at the end of the calendar year. If Mr. Riordan received a total of $23,320 for the year, including the bonus, what was his yearly salary?

Solution:

$$\text{Let } x = \text{yearly salary}$$
$$\text{then } 0.06x = \text{bonus}$$
$$\text{yearly salary} + \text{bonus} = \text{total received}$$
$$x + 0.06x = 23{,}320$$
$$1.06x = 23{,}320$$
$$x = \frac{23{,}320}{1.06} = \$22{,}000$$

Mr. Riordan's yearly salary is $22,000. ■

EXAMPLE 9   Budget Rent-a-Car offers cars at $30 a day plus 24 cents a mile. Hertz rents the same car for $44 a day plus 16 cents a mile. How many miles would you have to drive in 1 day so that the cost of renting from Hertz would equal the cost of renting from Budget?

Solution:

$$\text{Let } x = \text{number of miles}$$
$$\text{then } 0.24x = \text{cost of traveling } x \text{ miles in Budget's car}$$
$$\text{and } 0.16x = \text{cost of traveling } x \text{ miles in Hertz's car}$$
$$\text{Budget's cost} = \text{Hertz's cost}$$
$$\text{daily fee} + \text{mileage cost} = \text{daily fee} + \text{mileage cost}$$
$$30 + 0.24x = 44 + 0.16x$$
$$0.24x = 14 + 0.16x$$
$$0.08x = 14$$
$$x = \frac{14}{0.08} = 175 \text{ miles}$$

If you plan to drive less than 175 miles, the Budget car will be less expensive. ■

EXAMPLE 10   Mr. Lopez, a farmer, wishes to put up a fence in a rectangular shape to contain his cows. He wishes the length of the rectangle to be 150 feet more than its width. If Mr. Lopez has 2700 feet of fencing, what should be the dimensions of the rectangle?

Solution: The formula for the perimeter of a rectangle is $P = 2l + 2w$. We are told that Mr. Lopez has 2700 feet to construct the fence; therefore, the perimeter, P, must be 2700. This gives $2700 = 2l + 2w$. Note that this equation contains two variables, l and w. To solve the equation, the equation must be expressed in terms of a single variable. We must therefore try to eliminate one of the variables by expressing the length in terms of the width or the width in terms of the length.

$$\text{Let } x = \text{width of rectangle,}$$
$$\text{then } x + 150 = \text{length of rectangle}$$

Note that the length of the rectangle is expressed in terms of its width (see Fig. 2.1).

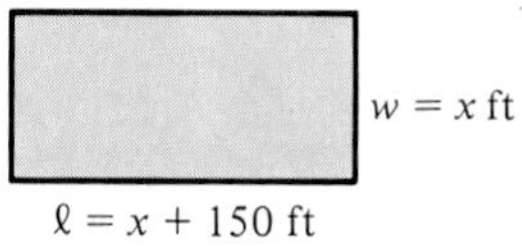

$w = x$ ft

$\ell = x + 150$ ft

FIGURE 2.1

$$P = 2l + 2w$$
$$2700 = 2(x + 150) + 2x$$
$$2700 = 2x + 300 + 2x$$
$$2700 = 4x + 300$$
$$2400 = 4x$$
$$600 = x$$
$$\text{width} = x = 600 \text{ feet}$$
$$\text{length} = x + 150 = 600 + 150 = 750 \text{ feet}$$

Check:

$$P = 2l + 2w$$
$$2700 = 2(750) + 2(600)$$
$$2700 = 1500 + 1200$$
$$2700 = 2700 \qquad \text{true}$$

Exercise Set 2.3

(a) Write an equation that can be used to solve the problem and (b) find the solution to the problem.

1. One number is 6 times another number. The sum of the two numbers is 56. Find the numbers.

2. Kathy is 15 years older than Dawn. The sum of their ages is 41. Find Kathy's and Dawn's ages.

3. The sum of two consecutive integers is 51. Find the two integers.

4. The sum of two consecutive even integers is 78. Find the two integers.

5. Twice a number decreased by 8 is 38. Find the number.

6. For two consecutive integers, the smaller plus 3 times the larger is 39. Find the integers.   9 + 10

7. One train travels 3 times as far as another. The sum of their distances is 48 miles. Find the distance traveled by each train.   12 & 36

8. The sum of three consecutive integers is 48. Find the three integers.   15 16 17

9. The sum of three consecutive even integers is 66. Find the three integers.   21 22 23

10. When $\frac{1}{3}$ of a number is added to 10, the sum is 15. Find the number.

11. When $\frac{3}{5}$ of a number is subtracted from 10, the difference is 4. Find the number.

12. The larger of two numbers is 5 times the smaller. Find the two numbers if twice the smaller equals 3 less than $\frac{1}{2}$ the larger.

13. The larger of two integers is two more than 4 times the smaller. Find the two numbers if 5 times the smaller number is 2 more than $\frac{1}{2}$ the larger number.

14. The larger of two integers is $\frac{5}{3}$ the smaller. If twice the smaller is subtracted from twice the larger, the difference is 12. Find the two numbers.

15. The sum of the angles of a triangle is 180°. Find the three angles if the two base angles are the same and the third angle is twice as large as the other two angles.

16. Find the three angles of a triangle if one angle is 20° greater than the smallest angle and the third angle is twice the smallest angle.

17. An *isosceles triangle* is a triangle with two sides of equal length. Find the three sides of the isosceles triangle if the larger side is 15 inches greater than each of the two smaller sides, and the sum of the lengths of the three sides is 45 inches.

18. A number increased by 8% is 54. Find the number.

19. The monthly rental fee for a telephone from the Southern Bell Telephone Company is $2.89. Radio Shack is selling telephones for $37.99. How long would it take for the monthly rental fee to equal the cost of a new phone?

20. It cost the Jacksons $10.50 a week to wash and dry their clothes at the corner laundry. If a washer and drier cost a total of $798, how many weeks will it take for the laundry cost to equal the cost of a washer and drier?

21. Ron Gigliotti buys a monthly bus pass, which entitles the owner to unlimited bus travel for $30 per month. Without the pass each bus ride costs $1.25. How many rides per month would Ron have to make so that it is less expensive to purchase the pass?

22. The cost of renting an automobile is $30 plus 18 cents a mile. Determine the maximum distance that Tony can drive if he has only $50.

23. Mountain Creek, Pennsylvania, is growing by 300 per year. If the present population is 5200, how long will it take for the population to reach 8800?

24. A city with a population of 15,000 is decreasing by 800 per year. In how many years will the population drop to 8000?

25. The cost of renting an automobile is $35 a day plus 20 cents per mile. How far can Kendra drive in 1 day if she has only $80?

26. Mrs. Hill used her own car for a 1-day 350-mile business trip. When an employee uses their own car for business the company reimburses the employee a fixed dollar amount plus 18 cents a mile. Mrs. Hill cannot remember the fixed amount but knows she was reimbursed a total of $83.00. Find the fixed amount the company reimburses their employees.

27. Bridget receives a flat weekly salary of $240 plus a 12% commission on the total dollar volume of all sales she makes. What must her dollar volume be in a week if she is to make a total weekly salary of $540?

28. The Computer Store has reduced the price of a computer by 15%. What is the original price of the computer if the sale price is $1275?

29. Collier County has a 4% sales tax. What is the maxi-

mum price of a car if the total cost, including tax, is to be $10,000?

30. J.R. is on a diet and can have only 400 calories for lunch. He orders a hot dog and French fries. If the hot dog and roll contain a total of 250 calories and each French fry contains 20 calories, how many French fries can he eat?

31. United Airlines wants the cost of a one-way ticket between Jacksonville, Florida, and Houston, Texas, to be exactly $499 including tax. What should be the price of the one-way ticket if there is a 7% sales tax on the cost of the ticket?

32. Debby Sunderland wishes to sell her paintings for $500, which includes a $6\frac{3}{4}\%$ sales tax. How much should she charge for each painting?

33. After Mrs. Englers is seated in a restaurant, she realizes that she only has $9.25. If she must pay 7% sales tax and wishes to leave a 15% tip, what is the maximum price of a lunch she can order?

34. Janet invested $4000 for one year in two savings accounts. She invested part of the money at 8% interest and the rest at 10%. If the total interest received is $370, how much was invested in each account?

35. The American Health Center offers two membership plans. One plan calls for a flat yearly payment of $510. A second plan calls for a $150 yearly fee plus a $6 payment for each day the center is used. How often, over a period of a year, would Ms. Smith have to use the cen-

ter so that the cost of the second plan equals the cost of the first plan?

36. The Midtown Tennis Club offers two payment plans for its members. Plan 1 is a monthly fee of $20 plus $8 per hour of court rental time. Plan 2 is no monthly fee, but court times costs $16.25 per hour. How many hours would Mrs. Levin have to play per month so that plan 1 becomes advantageous?

37. The length of a rectangle is to be 2 feet greater than twice its width. Find the dimensions of the rectangle if the perimeter is to be 40 feet. Use $P = 2l + 2w$.

38. Mrs. Gonzalez is planning to build a sandbox for her children. She wants its length to be 3 feet more than its width. Find the length and width of the sandbox if only 22 feet of lumber are available to form the frame.

39. The width of a rectangle is 1 meter more than $\frac{1}{2}$ its length. Find the length and width of the rectangle if its perimeter is 20 meters.

40. Nguyen, a landscape architect, wishes to fence in two equal areas as illustrated in the figure. If both areas are squares and the total amount of fencing used is 91 meters, find the dimensions of each square.

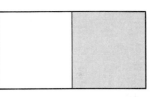

41. Taryn wishes to build a bookcase with four shelves. The height of the bookcase should be 3 feet greater than the width. If only 30 feet of wood are available to build the bookcase, what should the dimensions of the bookcase be?

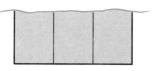

42. Mike wishes to fence in three rectangular areas along a river bank as illustrated in the following figure. Each rectangle is to have the same dimensions, and the length of each rectangle is to be 1 meter greater than its width. Find the length and width of each rectangle if the total amount of fencing used is 81 meters.

43. The first week of a going-out-of-business sale, Sam's General Store reduces all prices by 10%. The second week of the sale, Sam's reduces all items by 5 addi-

See Exercise 35.

tional dollars. If Debby bought a calculator for $49 during the second week of the sale, find the original price of the calculator.

44. In 1992 the property tax of a community is increased by 6% over the 1991 tax. An additional surcharge of $200 is also added for a special project. If the Peterson's 1992 tax totals $2108, find their property tax for the year 1991.

45. At a liquidation sale, Quality Photo Company reduces the price of all cameras by 1/4, and then takes an additional $10 off. If Mark purchases a Minolta camera for $290 during this sale, find the original price of the camera.

46. J.P. Richardson sells each of his paintings for $50. The gallery where he displays his work charges him $810 a month plus a 10% commission on sales. How many paintings must J.P. sell in a month to break even?

47. A farm is divided into three regions. The area of one region is twice as large as the area of the smallest region, and the area of the third region is 4 acres less than three times the area of the smallest region. If the total acreage of the farm is 512 acres, find the area of each of the three regions.

48. A 17 foot by 10 foot rectangular garden is divided into four parts. Two parts have the same area, the third part

is twice as large as each of the first two parts, and the last part has an area 20 square feet greater than each of the first two parts. Find the area of each of the four parts of the garden.

49. To find the average of a set of test grades, we divide the sum of the test grades by the number of test grades. On her first four algebra tests, Paula scored 87, 93, 97, and 96. (a) Write an equation that can be used to determine the score she needs to obtain on her fifth test to have a 90 average. (b) Explain how you determined your equation. (c) Solve the equation and determine the score.

50. Philip's scores on five physics hourly exams are 70, 83, 97, 84, and 74. If the final exam will count twice as much as each hourly exam, (a) what grade does Philip need on the final exam to have an 80 average? (b) If the highest possible score on the final exam is 100 points, is it possible for Philip to obtain a 90 average? Explain.

51. (a) Make up your own realistic word problem involving money. Represent this word problem as an equation.
    (b) Solve the equation and find the answer to the problem.

52. (a) Make up your own realistic word problem involving percents. Represent this word problem as an equation.
    (b) Solve the equation and find the answer to the problem.

Cumulative Review Exercises

[1.7] **53.** Evaluate $\dfrac{[2[(5-3)-4]]^2 \div (-8)}{-|8-5|-4^2}$.

[2.1] *Solve the following equations:*

54. $\frac{1}{2}x = 3(x-2)$ **55.** $\frac{1}{5}x + \frac{2}{3} = \frac{5}{4}x$

[2.2] **56.** Solve the equation $4x - 6y = 9$ for y.

JUST FOR FUN

1. On Monday Linda purchased shares in a money market fund. On Tuesday the value of the shares go up 5%, and on Wednesday the value falls 5%. How much did Linda pay for the shares on Monday if she sold them on Thursday for $59.85?

2. The Jerry Smith Auto Rental Agency charges $28 per day plus 15 cents a mile. If Denise rented a Chevrolet

Lumina for 3 days and the total bill was $121.68, including a 4% sales tax, how many miles did she drive?

3. Pick any number, multiply it by 2, add 33, subtract 13, divide by 2, and subtract the number you started with. You should end with the number 10. Show that this procedure will result in the answer 10 for any number, n, selected.

2.4

Rate and Mixture Problems

▶ 1 Solve rate (or motion) application problems containing only one rate.

▶ 2 Solve rate problems containing two rates.

▶ 3 Solve mixture application problems.

We have placed rate and mixture problems in the same section because rate problems involving two rates and mixture problems are solved using similar procedures.

Rate Problem

▶ **1**   A formula with many useful applications is

$$\text{Amount} = \text{rate} \cdot \text{time}$$

The "amount" in this formula can be a measure of many different quantities, including distance or length, area, volume, or number of items produced.

When applying this formula, we must make sure that the units are consistent. For example, if the rate is given in feet per *second*, the time must be given in *seconds*. If the rate is given as gallons per *hour*, the time must be given in *hours*. If the rate is given as items produced per *minute*, the time must be given in *minutes*. Problems that can be solved using this formula are called rate (or motion) problems because they involve motion, at a constant rate, for a certain period of time.

A nurse giving a patient an intravenous injection may use this formula to determine the drip rate of the fluid being injected. A company drilling for oil or water may use this formula to determine the amount of time needed to reach its goal. This formula may also be used when determining how fast a train or plane must travel to be at a certain point at a certain time.

EXAMPLE 1   An arborist is trying to save a white birch tree from an insect called borers. The treatment is to inject 200 cubic centimeters of a Lindane solution into the base of the tree over an 8-hour period. At what rate will the solution be injected into the tree?

Solution: We are given the volume and the time and are asked to find the rate.

$$\text{Volume} = \text{rate} \cdot \text{time}$$
$$200 = r \cdot 8$$
$$200 = 8r$$
$$r = 25 \text{ cc/hr}$$

Therefore, the rate of flow of the Lindane solution is 25 cc/hr.   ■

The rate formula is often used to calculate distance. For this use the word "amount" is replaced with the word "distance." This specific formula is called the *distance formula*.

Distance Formula

$$\text{Distance} = \text{rate} \cdot \text{time}$$

EXAMPLE 2   A conveyer belt at the Wonder Bread Baking Company is transporting a continuous length of dough at a rate of 1.5 feet per second. A cutting blade is activated at regular intervals to cut the dough into 10.8-inch lengths that will eventually become loaves of bread. At what time intervals should the cutting blade be activated?

Solution: Since the rate is given in feet per second and the dough's length is given in inches, one of these quantities must be changed. Since 1 foot equals 12 inches,

$$1.5 \text{ feet per second} = (1.5)(12) = 18 \text{ inches per second}$$
$$\text{Distance} = \text{rate} \cdot \text{time}$$
$$10.8 = 18 \cdot t$$
$$t = \frac{10.8}{18} = 0.6 \text{ second}$$

The blade should cut at 0.6-second intervals.   ■

Rate Problems with Two Rates

▶**2**   Sometimes when a problem has two different rates it is helpful to put the information in tabular form to help analyze the problem.

EXAMPLE 3   Two trains leave San Jose at the same time, traveling in the same direction on parallel tracks. One train travels at 80 miles per hour; the other travels at 60 miles per hour. In how many hours will they be 144 miles apart? (See Fig. 2.2.)

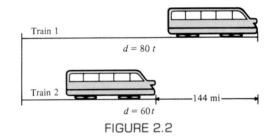

FIGURE 2.2

Solution: To solve this problem, we will use the distance formula.

Let t = time when the trains are 144 miles apart

Train	Rate	Time	Distance
1	80	t	$80t$
2	60	t	$60t$

The difference in their distances is 144 miles. Thus

$$\text{Distance train 1} - \text{distance train 2} = 144$$
$$80t - 60t = 144$$
$$20t = 144$$
$$t = 7.2$$

In 7.2 hours the trains will be 144 miles apart.   ■

EXAMPLE 4   Mrs. Sanders and her daughter Teresa jog regularly. Mrs. Sanders jogs at 5 miles per hour, Teresa at 4 miles per hour. Teresa begins jogging at noon. Mrs. Sanders begins at 12:30 P.M. and travels the same straight path.

(a) Determine at what time the mother and daughter will meet.
(b) How far from the starting point will they be when they meet?

Solution: (a) Since Mrs. Sanders is the faster jogger, she will cover the same distance in a shorter time period. When they meet they will both have traveled the same distance.

$$\text{Let } t = \text{time Teresa is jogging}$$

$$\text{then } t - \frac{1}{2} = \text{time Mrs. Sanders is jogging}$$

Jogger	Rate	Time	Distance
Teresa	4	t	$4t$
Mrs. Sanders	5	$t - \frac{1}{2}$	$5(t - \frac{1}{2})$

When they meet they will both have covered the same distance from the starting point.

$$\text{Teresa's distance} = \text{Mrs. Sander's distance}$$

$$4t = 5\left(t - \frac{1}{2}\right)$$

$$4t = 5t - \frac{5}{2}$$

$$-t = \frac{-5}{2}$$

$$t = \frac{5}{2}$$

They will meet $2\frac{1}{2}$ hours after Teresa begins jogging, or at 2:30 P.M.

(b) The distance can be found using either Mrs. Sanders's, or Teresa's rate. We will use Teresa's.

$$d = rt$$

$$= 4\left(\frac{5}{2}\right) = \frac{20}{2} = 10 \text{ miles}$$

They will meet 10 miles from the starting point. ∎

In Example 4, would the answer have changed if we let t represent the time Mrs. Sanders is jogging rather than the time Teresa is jogging? Try it and see.

EXAMPLE 5 A coffee bean grinding machine grinds coffee beans and then packs and seals the ground coffee in containers. When the machine runs at the normal rate, the machine packs 400 containers per hour. However, the rate of the machine can be increased to pack 600 containers per hour to meet a higher demand. At 9 A.M. on Tuesday the machine is turned on and begins running at the regular rate of 400 containers per hour. Later in the day the rate of the machine is increased to pack 600 containers per hour. At 5 P.M. the machine is turned off. If the machine produced the same number of containers running at the slower rate as at the faster rate, find (a) how long the machine was on at the slower rate, and (b) the total number of containers of coffee produced.

Solution: (a) We are told that the amounts produced at the two different rates are the same. We use this fact to help set up our equation. The machine was in operation from 9 A.M. to 5 P.M., a total of 8 hours.

$$\text{Let } t = \text{time machine is on at the slower rate}$$

$$\text{then } 8 - t = \text{time machine is on at the faster rate}$$

	Rate	Time	Amount
Slower rate	400	t	$400t$
Faster rate	600	$8 - t$	$600(8 - t)$

Since the amount produced at each speed is the same,

$$\text{amount at slower rate} = \text{amount at faster rate}$$
$$400t = 600(8 - t)$$
$$400t = 4800 - 600t$$
$$1000t = 4800$$
$$t = 4.8$$

Thus the machine was on for 4.8 hours at 400 units per hour and $8 - t$ or $8 - 4.8 = 3.2$ hours at hours at 600 units per hour.

(b) The total number of containers of coffee produced can be found by adding the two amounts.

$$\begin{aligned}
\text{total amount produced} &= 400t + 600(8 - t) \\
&= 400(4.8) + 600(3.2) \\
&= 1920 + 1920 \\
&= 3840
\end{aligned}$$

The answer to part (b) could have also been found by computing the number of containers of coffee produced at the initial rate of 400 units per hour and then doubling the value, since the same number of containers was produced at each rate. ■

EXAMPLE 6   On a 176-mile trip to their beach front cottage, the Lucianos drove at a steady speed for the first $\frac{1}{2}$ hour. During the last 3 hours of their trip, they increased their speed by 12 miles per hour. Find the Lucianos' speed during the first $\frac{1}{2}$ hour of their trip.

Solution:

$$\text{Let } r = \text{speed (or rate) during first part of their trip}$$

$$\text{then } r + 12 = \text{speed during second part of the trip}$$

	Rate	Time	Distance
First part	r	$\frac{1}{2}$	$\frac{1}{2}r$
Second part	$r + 12$	3	$3(r + 12)$

Since the total distance traveled is 176 miles,

$$\text{distance first } \tfrac{1}{2} \text{ hour } + \text{ distance last 3 hours } = 176 \text{ miles}$$
$$\tfrac{1}{2}r + 3(r + 12) = 176$$
$$2[\tfrac{1}{2}r + 3(r + 12)] = 2(176)$$
$$2(\tfrac{1}{2}r) + (2)(3)(r + 12) = 2(176)$$
$$r + 6(r + 12) = 352$$
$$r + 6r + 72 = 352$$
$$7r + 72 = 352$$
$$7r = 280$$
$$r = 40$$

Thus they traveled at 40 miles per hour during the first part of their trip. They traveled $r + 12$ or 52 miles per hour during the last part of their trip. ■

Mixture Problems

▶ **3** Any problem where two or more quantities are combined to produce a different quantity or where a single quantity is separated into two or more different quantities may be considered a **mixture problem.** As we did when working with rate problems containing two different rates, we will use tables to help organize the information.

Our first mixture problem example uses the **simple interest formula.**

$$\text{Interest} = \text{principal} \cdot \text{rate} \cdot \text{time}$$
$$\text{or} \quad I = prt$$

This formula is used to calculate the interest earned in a savings account that gives simple interest or the interest you must pay on a simple interest loan. For example, if $2000 is placed in a savings account giving 6% simple interest for a period of a year, the interest earned is found as follows:

$$I = prt$$
$$= 2000(0.06)(1) = 120$$

Thus, after 1 year $120 interest is earned.

EXAMPLE 7 Mr. and Mrs. McKane invested $8000 for 1 year, part at 7% and part at $5\tfrac{1}{4}\%$ simple interest. If they earned $458.50 total interest, how much was invested at each rate?

Solution: Let $x =$ amount invested at 7% simple interest.

If x is the amount invested at 7% simple interest, then the amount remaining, $8000 - x$, is invested at $5\tfrac{1}{4}\%$ simple interest. We will construct a table to help us visualize the solution. To find the interest earned in each account, we use the simple interest formula: $I = prt$.

Account	Principal	Rate	Time	Interest
7%	x	0.07	1	$0.07x$
$5\tfrac{1}{4}\%$	$8000 - x$	0.0525	1	$0.0525(8000 - x)$

Since the total interest from both accounts is $458.50, we write

$$\text{interest from 7\% account} + \text{interest from } 5\tfrac{1}{4}\% \text{ account} = \text{total interest}$$
$$0.07x + 0.0525(8000 - x) = 458.50$$

Now solve the equation.

$$0.07x + 420 - 0.0525x = 458.50$$
$$0.0175x + 420 = 458.50$$
$$0.0175x = 38.50$$
$$x = \frac{38.50}{0.0175} = 2200$$

Therefore, $2200 was invested at 7% and $8000 - x$ or $8000 - 2200 = \$5800$ was invested at $5\frac{1}{4}\%$ simple interest.   ∎

EXAMPLE 8   Marni's hot dog stand in Chicago sells hot dogs for $1.25 each and potato knishes for $1.50 each. If the sales for the day total $377 and a total of 278 items were sold, how many of each item were sold?

Solution:
$$\text{Let } x = \text{number of hot dogs sold}$$
$$\text{then } 278 - x = \text{number of knishes sold}$$

Item	Cost of Item	Number of items	Total Sales
Hot dogs	1.25	x	$1.25x$
Knishes	1.50	$278 - x$	$1.50(278 - x)$

$$\text{total sales of hot dogs} + \text{total sales of knishes} = \text{total sales}$$
$$1.25x + 1.50(278 - x) = 377$$
$$1.25x + 417 - 1.50x = 377$$
$$-0.25x + 417 = 377$$
$$-0.25x = -40$$
$$x = \frac{-40}{-0.25} = 160$$

Therefore, 160 hot dogs and $278 - 160$ or 118 knishes were sold.   ∎

In Example 8 you could have multiplied both sides of the equation by 100 to eliminate the decimal numbers, and then solved the equation.

EXAMPLE 9   Ali, a pharmacist, has both 6% and 15% phenobarbitol solutions. He receives a prescription for 0.5 liter of an 8% phenobarbitol solution. How much of each solution must he mix to fill the prescription?

Solution:
$$\text{Let } x = \text{number of liters of 6\% solution}$$
$$\text{then } 0.5 - x = \text{number of liters of 15\% solution}$$

The amount of phenobarbitol in a solution is found by multiplying the percent strength of phenobarbitol in the solution by the volume of the solution (see Fig. 2.3). We will draw a sketch of the problem and then construct a table.

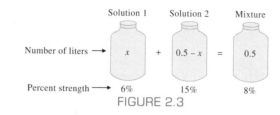

FIGURE 2.3

Solution	Strength of Solution	Number of Liters	Amount of Phenobarbitol
1	0.06	x	$0.06x$
2	0.15	$0.5 - x$	$0.15(0.5 - x)$
mixture	0.08	0.5	$0.08(0.5)$

$$\begin{pmatrix} \text{Amount of} \\ \text{phenobarbitol} \\ \text{in 6\% solution} \end{pmatrix} + \begin{pmatrix} \text{amount of} \\ \text{phenobarbitol} \\ \text{in 15\% solution} \end{pmatrix} = \begin{pmatrix} \text{amount of phenobarbitol} \\ \text{in mixture} \end{pmatrix}$$

$$0.06x + 0.15(0.5 - x) = 0.08(0.5)$$
$$0.06x + 0.075 - 0.15x = 0.04$$
$$0.075 - 0.09x = 0.04$$
$$-0.09x = 0.04 - 0.075$$
$$-0.09x = -0.035$$
$$x = \frac{-0.035}{-0.09} = 0.39 \quad \text{(to nearest hundredth)}$$

Ali must mix 0.39 liter of the 6% solution and $0.5 - x$ or $0.5 - 0.39 = 0.11$ liter of the 15% solution to make 0.5 liter of an 8% solution. ∎

Exercise Set 2.4

Round answers in this exercise set to the nearest hundredth.

1. A children's ride consists of a stagecoach being pulled by bulls. If the stagecoach covers a distance of 1.2 miles in 0.3 hours, find the average speed of the stagecoach.

See Exercise 1.

2. A patient is to receive 1200 cubic centimeters of intravenous fluid over a period of 3 hours. What should be the intravenous flow rate?

3. The $10\frac{1}{2}$ mile mule trip down to Phantom Ranch at the bottom of the Grand Canyon takes $5\frac{1}{2}$ hours. The return trip covers a distance of 8 miles and takes $4\frac{1}{2}$ hours. Find (a) the average speed going down the canyon; (b) the average speed coming up the canyon.

4. A laser can cut through a steel door at the rate of $\frac{1}{5}$ centimeter per minute. How thick is the door if it requires 32 minutes to cut through the steel door?

5. At the dolomite quarry in Corydon, Indiana, the conveyer belt transports crushed dolomite ore to a dump truck at an average of 1800 pounds per minute. How long will it take to fill a dump truck that can hold a maximum of 29,500 pounds of crushed ore?

6. Annette speed walks at a rate of 120 yards per minute. How long will it take her to speed walk 1350 feet?

7. A piece of coral is growing in the ocean at a rate of 3 pounds per year. How long will it take for the coral to gain 24 ounces?

8. Under certain conditions, sound travels at 1080 feet per second. If a loud explosion occurs $2\frac{1}{2}$ miles away from you, how long will it take before you hear the noise? There are 5280 feet in a mile.

9. Each year Americans use 18 billion disposable diapers. If you were to lay the disposable diapers end to end, in one day the distance covered by the diapers would be 9167.1 miles.

   (a) Determine the number of days it would take for disposable diapers placed end to end to reach from the earth to the moon, a distance of approximately 239,000 miles.

   (b) If the diapers were laid end to end for a year how far would they reach?

10. (a) A videocassette tape is 246 meters in length. Some videocassette recorders have three tape speeds: SP for standard play, LP for long play, and SLP for super long play (some recorders refer to SLP as EP for extended play). If the tape runs for 2 hours at SP or 4 hours at LP or 6 hours at SLP, find the rate of speed of the tape at all three speeds.

   (b) If a new 8-hour tape is to be developed for the SLP rate, find the length of the new tape.

11. The moving walkway, referred to as the "travelator," at the United Airlines Terminal at Chicago's O'Hare International Airport is like a moving conveyer belt that people walk on to increase their speed relative to the ground. If the length of one piece of the moving walkway is 275 feet, and the rate of the moving walkway is 120 feet per minute, (a) how long will it take a briefcase

See Exercise 11.

placed on one end of the moving walkway to reach the other end? (b) How long will it take a person walking at 150 feet per minute along side of the moving walkway to walk from one end of the moving walkway to the other end? (c) How long will it take a person walking at 150 feet per minute on the moving walkway to walk from one end to the other end? (d) If a person walks at 150 feet per minute, how much time will they save by walking on the moving walkway instead of walking along side the walkway?

12. Under certain conditions, sound travels at approximately 1080 feet per second. If Marsala yells down a canyon and hears her echo in 1.7 seconds, approximately how deep is the canyon?

Write an equation that can be used to solve the rate problem. Solve the equation and answer the question asked.

13. Two planes leave Cleveland at the same time. One plane flies east at 550 miles per hour. The other flies west at 650 miles per hour. In how many hours will they be 3000 miles apart?

14. Two hot air balloons leave Albuquerque, New Mexico, going in the same direction. One balloon travels at 16 miles per hour, and the second balloon travels at 14 miles per hour. In how many hours will they be 8 miles apart?

15. Two families on vacation leave San Diego and travel in opposite directions. After 3 hours the two cars are 330 miles apart. If one car travels at 60 miles per hour, find the speed of the other car.

16. Two runners enter the same marathon. If the faster runner's speed is 9.2 miles per hour and the two runners are 4.8 miles apart after 3 hours, find the slower runner's speed.

17. Two trains 520 miles apart travel toward each other on different tracks. One travels at 60 miles per hour, the other at 70 miles per hour. When will they pass each other?

18. Two cars leave from the same point at the same time, one traveling east and the other traveling west. If the car traveling west is moving 10 miles per hour faster than the car traveling east, and the two cars are 342 miles apart after 3 hours, find the speed of each car.

19. A passenger train leaves the Baltimore Depot 1.2 hours after a freight train leaves. The passenger train is traveling 20 miles per hour faster than the freight train. If the passenger train overtakes the freight train in 3 hours, find the speed of each train.

20. Two molding machines are turned on at 9 A.M. The older molding machine can produce 40 plastic buckets in 1 hour. The newer machine can produce 50 buckets in 1 hour. How long will it take the two machines to produce a total of 540 buckets?

21. A jogger and a cyclist head for the same destination at 8 A.M. from the same point. The average speed of the cyclist is four times the speed of the jogger. In 2 hours the cyclist is 18 miles ahead of the jogger. (a) At what rate did the cyclist ride? (b) How far did the cyclist ride?

22. Two hoses are being used to fill a pool. The hose with the larger diameter supplies twice as much water as the hose with the smaller diameter. The smaller hose is on for 1.5 hours before the hose with the larger diameter is turned on. If the total volume of water in the pool is 2100 gallons 3 hours after the larger-diameter hose is turned on, find the rate of flow from each hose.

23. Dr. McDonald hikes down to the bottom of Rim Rock Canyon, camps the night, and returns the next day. Her hiking speed down averages 2.6 miles per hour and her return trip averages 1.2 miles per hour. If she spent a total of 16 hours hiking find (a) how long it took her to reach the bottom of the canyon, and (b) the total distance traveled.

24. A Dodge has traveled 50 miles when a Chevy begins traveling in the same direction. If the Chevy travels at 65 miles per hour and the Dodge travels at 55 miles per hour, how long will it take for the Chevy to catch up to the Dodge?

25. Two machines are packing boxes with spaghetti. The smaller machine can package 400 boxes per hour and the larger machine can package 600 boxes per hour. If the larger machine is on for 2 hours before the smaller machine is turned on, how long will it take after the smaller machine is turned on for a total of 15,000 boxes of spaghetti to be boxed?

26. Mary began driving to an out of town college at a speed of 45 miles per hour. One-half hour after she left, Mary's parents realized Mary had forgotten to take her wallet. Her parents then tried to catch up to Mary in their car. If the parents traveled at 65 miles per hour, how long did it take them to overtake Mary?

Write an equation that can be used to solve the mixture problem. Solve the equation and answer the question asked.

27. Mr. Templeton invests $11,000 for one year, part at 9% and part at 10% simple interest. How much money was invested at each interest rate if the total interest earned from both investments is $1050? Use interest = principal · rate · time.

28. Ms. Feldman invested $10,000 for one year, part at 7% and part at $6\frac{1}{4}$%. If she earned a total interest of $656.50, how much was invested at each rate?

29. General Motors stock is selling at $45 per share and Reebok stock is selling at $14 per share. Bob Davis has a maximum of $9000 to invest. He wishes to purchase five times as many share of Reebok as of General Motors.

(a) How many shares of each will he purchase? Stocks can be purchased only in whole shares.

(b) How much money will be left?

30. The admission at an ice hockey game is $8.50 for adults and $6.50 for children. A total of 650 tickets were sold. How many tickets were sold to children and how many to adults if a total of $5045 was collected?

31. Victor has a total of 33 dimes and quarters. The total value of the coins is $4.50. How many dimes and how many quarters does he have?

32. Jim Kelly sells almonds for $6.00 per pound, and walnuts for $5.20 per pound. How many pounds of each should he mix to produce a 30-pound mixture that costs $5.50 per pound?

33. How many pounds of coffee costing $6.20 per pound must Larry mix with 18 pounds of coffee costing $5.80 per pound to produce a mixture that costs $6.10 per pound?

34. How many ounces of water should a chemist mix with 16 ounces of a 25% sulfuric acid solution to reduce it to a 10% solution?

35. How many ounces of pure vinegar should a cook add to 40 ounces of a 10% vinegar solution to make it a 25% vinegar solution?

36. Fifty pounds of a cement–sand mixture is 40% sand. How many pounds of sand must be added for the resulting mixture to be 60% sand?

37. How many liters of a 20% alcohol solution and how many liters of a 50% alcohol solution must be mixed to get 12 liters of a 30% alcohol solution?

38. Two acid solutions are available to a chemist. One is a 20% sulfuric acid solution, but the label that indicates the strength of the other sulfuric acid solution is missing. Two hundred milliliters of the 20% acid solution and 100 milliliters of the solution with the unknown strength are mixed together. Upon analysis, the mixture was found to have a 25% sulfuric acid concentration. Find the strength of the solution with the missing label.

39. The Bryerman Nursery sells in bulk two types of grass seeds. The lower-quality seeds have a germination rate of 76%, but the germination rate of the higher-quality seeds is not known. Twelve pounds of the higher-quality seeds are mixed with 16 pounds of the lower-quality seed. If a later analysis of the mixture finds that the mixture's germination rate was 82%, what is the germination rate of the higher-quality seed?

40. The Agway nursery is selling in bulk two types of sunflower seed for bird feeding. The striped sunflower seeds cost $1.20 per pound, while the all black sunflower seeds cost $1.60 per pound. How many pounds of each should Mr. Wicker mix to get a 20-pound mixture that sells for $30?

41. Some states allow a husband and wife to file individual state tax returns (on a single form) even though they file a joint federal return. It is usually to the taxpayer's advantage to do this when both the husband and wife work. The smallest amount of tax owed (or the largest refund) will occur when the husband's and wife's taxable incomes are the same.

Mr. Clar's 1992 taxable income was $28,200 and Mrs. Clar's income for that year was $32,450. The Clars' total tax deduction for the year was $6400. This deduction can be divided between Mr. and Mrs. Clar in any way they wish. How should the $6400 be divided between them to result in each person's having the same taxable income and therefore the greatest tax return?

42. A certain type of engine uses a fuel mixture of 15 parts of gasoline to 1 part of oil. How much pure gasoline must be mixed with a gasoline–oil mixture, which is 75% gasoline, to make 8 quarts of the mixture to run the engine?

Write an equation that can be used to solve the rate or mixture problem. Solve the equation and answer the question asked.

43. Two hikers visiting Yellowstone National Park start at the same point and hike in opposite directions around a thermal spring. The distance around the thermal spring is 8.2 miles. One hiker walks 0.4 miles per hour faster than the other. How fast does each hiker walk if they meet in 2 hours?

44. The Rappaports and Calters leave their homes at 8 A.M. planning to meet for a picnic at a point between them. If the Rappaports travel at 60 miles per hour and the Calters travel at 50 miles per hour, and they live 330 miles apart, at what time will they meet?

45. The Bernhams decide to invest $6000 in two stocks, Sears Department Stores and U.S. Steel. They wish to purchase three times as many shares of Sears as of U.S. Steel.

   (a) If Sears is selling at $34 a share and U.S. Steel at $23, how many shares of each will be purchased?

   (b) How much money will be left?

46. Barbara Anders invested $8000 for one year, part at 6% and part at 10% simple interest. How much was invested in each account if the same amount of interest was received from each account?

47. One sump pump can remove 10 gallons of water a minute. A larger sump pump can remove 20 gallons of water a minute. How long will it take the two pumps working together to empty a 15,000-gallon swimming pool?

48. A jetliner flew from Chicago to Los Angeles at an average speed of 500 miles per hour. Then it continued on over the Pacific Ocean to Hawaii at an average speed of 550 miles per hour. If the entire trip covered 5200 miles and the part over the ocean took twice as long as the part over land, what was the time of the entire trip?

49. Hal invested $4000 for one year in two savings accounts giving simple interest. He invested 2500 at 9% interest and the rest at a different interest rate. If the total interest received is $315, what was the rate on the second account?

50. How many quarts of pure antifreeze should Mr. Alberts add to 10 quarts of a 20% antifreeze solution to make a 50% antifreeze solution?

51. A small plane flew a round trip from Orno, Maine, to Tallahassee, Florida. The average flying speed down (with the tail wind) was 300 miles per hour. The average flying speed on the return trip (with a head wind) was 220 miles per hour. If the total flying time was 11.2 hours

   (a) how long did it take the plane to fly from Orno to Tallahassee?

   (b) Find the distance between the two airports.

52. A movie ticket for an adult costs $6.00 and a child's ticket costs $4.50. A total of 172 tickets are sold for a given show. If $894 is collected for the show, how many adults and how many children attended the show?

53. Diedre holds two part-time jobs. One job pays $6.00 per hour and the other pays $6.50 per hour. Last week she earned a total of $114 and worked for a total of 18 hours. How many hours did she work at each job?

54. Dan has a total of 12 bills in his wallet. Some are $5 bills and the rest are $10 bills. The total value of the

12 bills is $115. How many $5 bills and how many $10 bills does he have?

55. Annette Leopard and her crew sailed out to Goat Island at an average speed of 20 knots (a knot is a speed of 1 nautical mile per hour. There are about 6072.12 feet in a nautical mile). On her return trip she averaged 12 knots. The total time she spent sailing was 8 hours.

   (a) How long did it take for Annette and her crew to reach Goat Island?

   (b) What was the total distance traveled?

56. On a 100-mile trip to their cottage, the Ghents traveled at a steady speed for the first hour. The speed during the second hour of their trip was 16 miles per hour slower than the speed of the first hour. Find their speed during their first hour.

57. A pint of coffee (16 ounces) containing 3% caffeine is mixed with a half-gallon of coffee (64 ounces) containing 7% caffeine. What percent of caffeine will the mixture contain?

58. Judy has 60 ounces of water whose temperature is 92°C. How much water with a temperature of 20°C must she mix in with all the 92°C water to get a mixture whose temperature is 50°C? Neglect heat loss from the water to the surrounding air.

59. Philip has an 80% methyl alcohol solution. He wishes to make a gallon of windshield washer solution by mixing his methyl alcohol solution with water. If 128 ounces, or a gallon, of windshield washer fluid should contain 6% methyl alcohol, how much of the 80% methyl alcohol solution and how much water must be mixed?

60. Cheryl is making a meatloaf by combining chopped sirloin with veal. The sirloin contains 1.2 grams of fat per ounce, and the veal contains 0.3 gram of fat per ounce. If Cheryl wants her 64-ounce mixture to have only 0.8 gram of fat per ounce, find how much sirloin and how much veal she must use?

61. Sundance dairy has 400 quarts of whole milk containing 5% butterfat. How many quarts of low-fat milk containing 1.5% butterfat should be added to produce milk containing 2% butterfat?

62. Mike can ride his bike to work in $\frac{3}{4}$ hour. If he takes his car to work, the trip takes $\frac{1}{6}$ hour. If Mike drives his car an average of 14 miles per hour faster than he rides his bike, determine the distance he travels to work.

63. A machine that folds and seals milk cartons can produce 50 milk cartons per minute. A new machine can produce 70 milk cartons per minute. The older machine has made 200 milk cartons when the newer machine is turned on. If both machines then continue working, how long will it take after the new machine is turned on for the new machine to produce the same total number of milk cartons as the older machine?

64. The salinity (salt content) of the Atlantic Ocean averages 37 parts per thousand. If 64 ounces of water are collected and placed in the sun, how many ounces of pure water would need to evaporate to raise the salinity to 45 parts per thousand? Note that only the pure water is evaporated (the salt is left behind).

65. Mike's tractor/lawn mower has no speedometer. Explain how Mike can determine the speed of his tractor without the use of a speedometer.

66. **(a)** Make up your own realistic rate problem that can be represented as an equation.

   (b) Write the equation that represents your problem.

   (c) Solve the equation, and then find the answer to your problem.

67. **(a)** Make up your own realistic mixture problem that can be represented as an equation.

   (b) Write the equation that represents your problem.

   (c) Solve the equation, and then find the answer to your problem.

68. Two rockets are launched from Cape Canaveral. The first rocket, launched at noon, will travel at 8000 miles per hour. The second rocket will be launched some time later and travel at 9500 miles per hour. When should the second rocket be launched if the rockets are to meet at a distance of 38,000 miles from earth?

   (a) Explain how to find the solution to this problem.

   (b) Find the solution to the problem.

Cumulative Review Exercises

Solve the equation.

[2.1] **69.** $0.6x + 0.22 = 0.4(x - 2.3)$

70. $\frac{2}{9}x + 3 = x + \frac{1}{5}$

[2.2] **71.** Solve the equation $\frac{3}{5}(x - 2) = \frac{2}{7}(2x + 3y)$ for y.

[2.3] **72.** Hertz Automobile Rental Agency charges a daily fee of $30 plus 14 cents a mile. National Automobile Rental Agency charges a daily fee of $16 plus 24 cents a mile for the same car. What distance would you have to drive in 1 day to make the cost of renting from Hertz equal to the cost of renting from National?

JUST FOR FUN

1. Two cars labeled *A* and *B* are in a 500-lap race. Each lap is 1 mile. The lead car, *A*, is averaging 125 miles per hour when it reaches the halfway point. Car *B* is exactly 6.2 laps behind.

   (a) Find the average speed of car *B*.

   (b) At that instant how far behind, in seconds, is car *B* from car *A*?

2. Radar and sonar determine the distance of an object by emitting radio waves that travel at the speed of light, approximately 1000 feet per microsecond (a millionth of a second) in air. The device also determines the time it takes for its signal to travel to the object and return from the object.

If radar determines that it takes 0.6 second to receive the echo of its signal, how far is the object from the radar device?

3. Lester knits at a rate of eight stitches per minute. He is planning to knit an afghan 4 feet by 6 feet. How long will it take Lester to knit the afghan if the instructions on the skein of wool indicate that four stitches equal 1 inch and six rows equal 1 inch?

4. The radiator of an automobile has a capacity of 16 quarts. It is presently filled with a 20% antifreeze solution. How many quarts must Jorge drain and replace with pure antifreeze to make the radiator contain a 50% antifreeze solution?

2.5

Solving Linear Inequalities

▶ **1** Solve inequalities.

▶ **2** Graph solutions on the number line, and express solutions in interval notation and as solution sets.

▶ **3** Solve "and" type compound inequalities.

▶ **4** Solve continued inequalities.

▶ **5** Solve "or" type compound inequalities.

▶ **1** The inequality symbols are as follows:*

Inequality Symbols	
$>$	is greater than
$\geq$	is greater than or equal to
$<$	is less than
$\leq$	is less than or equal to

A mathematical expression containing one or more of these symbols is called an **inequality.** The direction of the inequality symbol is sometimes called the **sense** of the inequality.

Examples of inequalities in one variable are

$$2x + 3 \leq 5, \qquad 4x > 3x - 5, \qquad -3 \leq -x + 5, \qquad 2x + 3 \geq 0$$

To solve an inequality, we must isolate the variable on one side of the inequality symbol. To isolate the variable, we use the same basic techniques used in solving equations.

*$\neq$, not equal to, is also an inequality. $\neq$ means $<$ or $>$. Thus $2 \neq 3$ means $2 < 3$ or $2 > 3$.

Properties Used to Solve Inequalities

1. If $a > b$, then $a + c > b + c$.
2. If $a > b$, then $a - c > b - c$.
3. If $a > b$, and $c > 0$, then $ac > bc$.
4. If $a > b$ and $c > 0$, then $\dfrac{a}{c} > \dfrac{b}{c}$.
5. If $a > b$ and $c < 0$, then $ac < bc$.
6. If $a > b$ and $c < 0$, then $\dfrac{a}{c} < \dfrac{b}{c}$.

The first two properties state that the same number can be added to or subtracted from both sides of an inequality. The third and fourth properties state that both sides of an inequality can be multiplied or divided by any positive real number. The last two properties indicate that **when both sides of an inequality are multiplied or divided by a negative number, the sense of the inequality changes.**

Example of Multiplication by a Negative Number

Multiply both sides of the inequality by -1 and change the sense of the inequality.

$$4 > -2$$
$$-1(4) < -1(-2)$$
$$-4 < 2$$

Example of Division by a Negative Number

$$10 \geq -4$$
$$\frac{10}{-2} \leq \frac{-4}{-2}$$
$$-5 \leq 2$$

Divide both sides of the inequality by -2 and change the sense of the inequality.

COMMON STUDENT ERROR

Students often forget to change the sense (the direction) of the inequality when multiplying or dividing both sides of the inequality by a negative number.

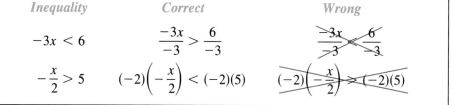

Inequality	*Correct*	*Wrong*
$-3x < 6$	$\dfrac{-3x}{-3} > \dfrac{6}{-3}$	$\dfrac{-3x}{-3} < \dfrac{6}{-3}$
$-\dfrac{x}{2} > 5$	$(-2)\left(-\dfrac{x}{2}\right) < (-2)(5)$	$(-2)\left(-\dfrac{x}{2}\right) > (-2)(5)$

EXAMPLE 1   Solve the inequality $2x + 6 < 12$.

Solution:

$$2x + 6 < 12$$
$$2x + 6 - 6 < 12 - 6$$
$$2x < 6$$
$$\frac{2x}{2} < \frac{6}{2}$$
$$x < 3$$

Note that the solution set is $\{x \mid x < 3\}$. Any number less than 3 will satisfy the inequality. ∎

▶ **2**   The solution set to an inequality in one variable can be graphed on the number line or written in interval notation.

Solution of Inequality	Solution Indicated on Number Line	Solution Represented in Interval Notation
$x > a$	○───→ a	(a, ∞)
$x \geq a$	●───→ a	$[a, \infty)$
$x < a$	←───○ a	$(-\infty, a)$
$x \leq a$	←───● a	$(-\infty, a]$
$a < x < b$	○───○ a b	(a, b)
$a \leq x \leq b$	●───● a b	$[a, b]$
$a < x \leq b$	○───● a b	$(a, b]$
$a \leq x < b$	●───○ a b	$[a, b)$

Note that a shaded circle on the number line indicates that the end point is part of the solution, and an unshaded circle indicates that the end point is not part of the solution. In interval notation brackets, [ ], are used to indicate that the end points are part of the solution and parentheses, ( ), indicate that the end points are not part of the solution. The symbol ∞ is read "infinity"; it indicates that the solution set continues indefinitely.

Solution of Inequality	Solution Illustrated on Number Line	Solution Represented in Interval Notation
$x \geq 5$	●───→ 5	$[5, \infty)$
$x < 3$	←───○ 3	$(-\infty, 3)$
$2 < x \leq 6$	○───● 2 6	$(2, 6]$
$-6 \leq x \leq -1$	●───● −6 −1	$[-6, -1]$

EXAMPLE 2   Solve the following inequality and give the answer both on the number line and in interval notation.

$$3(x - 2) \leq 5x + 8$$

Solution:

$$3(x - 2) \leq 5x + 8$$
$$3x - 6 \leq 5x + 8$$
$$3x - 5x - 6 \leq 5x - 5x + 8$$
$$-2x - 6 \leq 8$$
$$-2x - 6 + 6 \leq 8 + 6$$
$$-2x \leq 14$$
$$\frac{-2x}{-2} \geq \frac{14}{-2}$$
$$x \geq -7$$

Number Line   *Interval Notation*

                                    $[-7, \infty)$

                 -7

Note that the solution set is $\{x \mid x \geq -7\}$.   ■

In Example 2 we illustrated the answer on the number line, in interval notation, and as a solution set. Your instructor may indicate which way he or she prefers the answer be given.

EXAMPLE 3   Solve the equality $\frac{1}{2}(4x + 14) \geq 5x + 4 - 3x - 10$.

Solution:
$$\frac{1}{2}(4x + 14) \geq 5x + 4 - 3x - 10$$

$$\frac{1}{2}(4x + 14) \geq 2x - 6$$

$$\left(\frac{1}{2}\right)(4x) + \left(\frac{1}{2}\right)(14) \geq 2x - 6$$

$$2x + 7 \geq 2x - 6$$

$$2x - 2x + 7 \geq 2x - 2x - 6$$

$$7 \geq -6$$

Since 7 is always greater than or equal to -6, the solution set is the set of all real numbers, $\mathbb{R}$. The solution set can also be indicated on the number line or given in interval notation.

                        or $(-\infty, \infty)$ ■

If the solution to Example 3 were $7 \leq -6$, the answer would have been no solution, since 7 is never less than or equal to -6. When an inequality has no solution, its solution set is the empty or null set, $\emptyset$ or $\{\ \}$.

EXAMPLE 4   Solve the inequality $\dfrac{4 - 2y}{3} \geq \dfrac{2y}{4} - 3$.

Solution: Multiply both sides of the inequality by the least common denominator, 12.
$$\overset{4}{\cancel{12}}\left(\frac{4 - 2y}{\underset{1}{\cancel{3}}}\right) \geq 12\left(\frac{2y}{4} - 3\right)$$

$$4(4) + 4(-2y) \geq \overset{3}{\cancel{12}}\left(\frac{2y}{\underset{1}{\cancel{4}}}\right) + 12(-3)$$

$$16 - 8y \geq 6y - 36$$

$$16 \geq 14y - 36$$

$$52 \geq 14y$$

$$\frac{52}{14} \geq y$$

$$\frac{26}{7} \geq y \ \text{ or } \ y \leq \frac{26}{7}$$

              $\frac{26}{7}$

In interval notation the answer is $(-\infty, \frac{26}{7}]$. The solution set is $\{y \mid y \leq \frac{26}{7}\}$.   ■

HELPFUL HINT	Note that in Example 4 we indicated that $\frac{26}{7} \geq y$ can be written $y \leq \frac{26}{7}$. Generally, when writing a solution to an inequality we write the variable on the left.

For example,

$$a < x \quad \text{means} \quad x > a \qquad \text{(inequality symbol points to } a \text{ in both cases)}$$
$$a > x \quad \text{means} \quad x < a \qquad \text{(inequality symbol points to } x \text{ in both cases)}$$
$$-6 < x \quad \text{means} \quad x > -6 \qquad \text{(inequality symbol points to } -6 \text{ in both cases)}$$
$$-3 > x \quad \text{means} \quad x < -3 \qquad \text{(inequality symbol points to } x \text{ in both cases)}$$

EXAMPLE 5   A small single-engine airplane can carry a maximum weight of 1500 pounds when its gas tank is full. Nancy Johnson, the pilot, has to transport boxes weighing 80 pounds.

(a) Write an inequality that can be used to determine the maximum number of boxes that Nancy can safely place on her plane if she weighs 125 pounds.
(b) Find the number of boxes that Nancy can transport.

Solution: (a) Let n = number of boxes.

$$\text{Nancy's weight} + \text{weight of } n \text{ boxes} \leq 1500$$
$$125 \quad + \quad 80n \quad\quad \leq 1500$$

(b) $125 + 80n \leq 1500$
$$80n \leq 1375$$
$$n \leq 17.2$$

Therefore, Nancy can transport up to 17 boxes per trip.   ∎

EXAMPLE 6   A taxi's fare is \$1.75 for the first half-mile and \$1.10 for each additional half-mile. Any additional part of a half-mile will be rounded up to the next half-mile.

(a) Write an inequality that can be used to determine the maximum distance that Karen can travel if she has only \$12.35.
(b) Find the maximum distance that Karen can travel.

Solution: (a)   Let x = number of half-miles after the first
then $1.10x$ = cost of traveling x additional half-miles

$$\text{Cost of first half-mile} + \text{cost of additional half-miles} \leq \text{total cost}$$
$$1.75 \quad + \quad 1.10x \quad\quad \leq \quad 12.35$$

(b) $1.75 + 1.10x \leq 12.35$
$$1.10x \leq 10.60$$
$$x \leq \frac{10.60}{1.10}$$
$$x \leq 9.64$$

Karen can travel a distance less than or equal to 9 half-miles after the first half-mile, for a total of 10 half-miles, or 5 miles. If Karen travels for 10 half-miles after the first, she will owe $1.75 + 1.10(10) = 12.75$, which is more than she has.   ∎

EXAMPLE 7 For a business to realize a profit, the revenue (or income), R, must be greater than the cost, C. That is, a profit will be obtained only when $R > C$ (the company breaks even when $R = C$). A company that produces playing cards has a weekly cost equation of $C = 1525 + 1.7x$ and a weekly revenue equation of $R = 4.2x$, where x is the number of decks of playing cards produced and sold in a week. How many decks of cards must be produced and sold in a week for the company to make a profit?

Solution: The company will make a profit when $R > C$ or

$$4.2x > 1525 + 1.7x$$
$$2.5x > 1525$$
$$x > \frac{1525}{2.5}$$
$$x > 610$$

The company will make a profit when more than 610 decks are produced and sold in a week. ∎

Compound Inequalities

▶**3** A **compound inequality** is formed by joining two inequalities with the word *and* or *or*.

Examples of Compound Inequalities

$$3 < x \quad \text{and} \quad x < 5$$
$$x + 4 > 3 \quad \text{or} \quad 2x - 3 < 6$$
$$4x - 6 \geq -3 \quad \text{and} \quad x - 6 < 5$$

The solution of a compound inequality using the word *and* is all the numbers that make *both* parts of the inequality true. Consider

$$3 < x \quad \text{and} \quad x < 5$$

What are the numbers that satisfy both inequalities? The numbers that satisfy both may be easier to see if we graph the solution to each inequality on a number line (see Fig. 2.4). Note that the numbers that satisfy both inequalities are the numbers between 3 and 5. The solution set is $\{x \mid 3 < x < 5\}$.

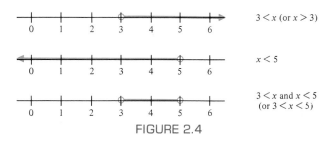

FIGURE 2.4

Recall from Chapter 1 that the intersection of two sets is the set of elements common to both sets. *To find the solution set of an inequality containing the word* **and** *take the* **intersection** *of the solution sets of the two inequalities.*

EXAMPLE 8 Solve $x + 2 \leq 5$ and $2x - 4 > -2$.

Solution: Begin by solving each inequality separately.

$$x + 2 \leq 5 \qquad 2x - 4 > -2$$
$$x \leq 3 \qquad 2x > 2$$
$$x > 1$$

Now take the intersection of the sets $\{x \mid x \leq 3\}$ and $\{x \mid x > 1\}$. When we find $\{x \mid x \leq 3\} \cap \{x \mid x > 1\}$, we are finding the values of x common to both sets. Figure 2.5 illustrates that the solution set is $\{x \mid 1 < x \leq 3\}$. In interval notation, the answer is $(1, 3]$.

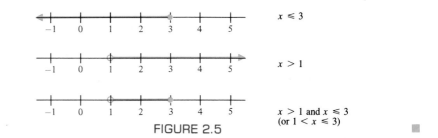

FIGURE 2.5

▶ **4** Sometimes a compound inequality using the word *and* as the connecting word can be written in a shorter form. For example, $3 < x$ and $x < 5$ can be written as $3 < x < 5$. The word *and* does not appear when the inequality is written in this form, but it is implied. Inequalities written in the form $a < x < b$ are called **continued inequalities.** The compound inequality $1 < x + 5$ and $x + 5 \leq 7$ can be written $1 < x + 5 \leq 7$.

EXAMPLE 9   Solve $1 < x + 5 \leq 7$.

Solution: $1 < x + 5 \leq 7$ means $1 < x + 5$ and $x + 5 \leq 7$. Solve each inequality separately.

$$1 < x + 5 \quad \text{and} \quad x + 5 \leq 7$$
$$-4 < x \qquad\qquad x \leq 2$$

Remember that $-4 < x$ means $x > -4$. Figure 2.6 illustrates that the solution set is $\{x \mid -4 < x \leq 2\}$. In interval notation, the answer is $(-4, 2]$.

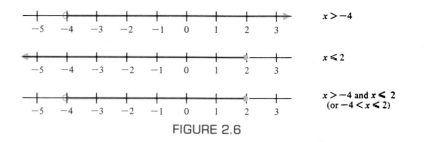

FIGURE 2.6

The inequality in Example 9, $1 < x + 5 \leq 7$, could have been solved in another way. We can still use the properties discussed earlier to solve continued inequalities. However, when working with such inequalities, whatever we do to one part we must do to all three parts. In Example 9, we could have subtracted 5 from all three parts to isolate the variable and obtain the answer.

$$1 < x + 5 \leq 7$$
$$1 - 5 < x + 5 - 5 \leq 7 - 5$$
$$-4 < x \leq 2$$

Note that this answer is the same as the answer obtained in Example 9.

EXAMPLE 10   Solve the inequality.

$$-3 \leq 2x - 7 < 8$$

Solution: We wish to isolate the variable x. We begin by adding 7 to all three parts of the inequality.

$$-3 + 7 \leq 2x - 7 + 7 < 8 + 7$$
$$4 \leq 2x < 15$$

Now divide all three parts of the inequality by 2.

$$\frac{4}{2} \leq \frac{2x}{2} < \frac{15}{2}$$

$$2 \leq x < \frac{15}{2}$$

or $\left[2, \dfrac{15}{2} \right)$

The solution set is $\left\{ x \mid 2 \leq x < \dfrac{15}{2} \right\}$.   ■

EXAMPLE 11   Solve the inequality.

$$-2 < \frac{4 - 3x}{5} < 8$$

Solution: Multiply all three parts by 5 to eliminate the denominator.

$$-2\,(5) < 5\left(\frac{4 - 3x}{5} \right) < 8\,(5)$$
$$-10 < 4 - 3x < 40$$
$$-10 - 4 < 4 - 4 - 3x < 40 - 4$$
$$-14 < -3x < 36$$

At this point we divide all three parts of the inequality by -3. Remember that when we multiply or divide an inequality by a negative number the sense of the inequality changes.

$$\frac{-14}{-3} > \frac{-3x}{-3} > \frac{36}{-3}$$

$$\frac{14}{3} > x > -12$$

Although $\frac{14}{3} > x > -12$ is correct, we generally write continued inequalities with the lesser value on the left. We will therefore rewrite the answer as

$$-12 < x < \frac{14}{3}$$

The answer may also be illustrated on the number line, written in interval notation, or written as a solution set.

or $(-12, \frac{14}{3})$

The solution set is $\{x \mid -12 < x < \frac{14}{3}\}$. ∎

HELPFUL HINT

You must be very careful when writing the solution to a continued inequality. In Example 11 we can change the answer from

$$\frac{14}{3} > x > -12 \quad \text{to} \quad -12 < x < \frac{14}{3}$$

This is correct since both say that x is greater than -12 and less than $\frac{14}{3}$. Notice that the inequality symbol in both cases is pointing to the smaller number.

In Example 11, had we written the answer $\frac{14}{3} < x < -12$, we would have given the **wrong** solution. Remember that the inequality $\frac{14}{3} < x < -12$ means that $\frac{14}{3} < x$ and $x < -12$. There is no number that is both greater than $\frac{14}{3}$ and less than -12. Also, by examining the inequality $\frac{14}{3} < x < -12$, it appears as if we are saying that -12 is a greater number than $\frac{14}{3}$, which is obviously *wrong*.

It would also be **wrong** to write the answer

$$-12 < x > \frac{14}{3} \quad \text{or} \quad \frac{14}{3} < x > -12$$

EXAMPLE 12 An average greater than or equal to 80 and less than 90 will result in a final grade of B in a course. Steven received grades of 85, 90, 68, and 70 on his first four exams in the course. Within what range of grades will Steven's fifth (and last) exam result in his receiving a final grade of B in the course?

Solution: Let x = Steven's last exam grade.

$$80 \le \text{average of five exams} < 90$$

$$80 \le \frac{85 + 90 + 68 + 70 + x}{5} < 90$$

$$80 \le \frac{313 + x}{5} < 90$$

$$400 \le 313 + x < 450$$

$$400 - 313 \le x < 450 - 313$$

$$87 \le x < 137$$

Steven would need a minimum grade of 87 to obtain a final grade of B. If the highest grade he could receive on the test is 100, it is impossible for him to obtain a final grade of A (90 average or higher). ∎

▶ **5**   The solution to a compound inequality using the word *or* is all the numbers that make *either* of the inequalities a true statement. Consider the compound inequality

$$x > 3 \quad \text{or} \quad x < 5$$

What are the numbers that satisfy the inequality? Let us graph the solution to each inequality on the number line (see Fig 2.7). Note that every real number satisfies at least one of the two inequalities. The solution set to the compound inequality is all real numbers, $\mathbb{R}$.

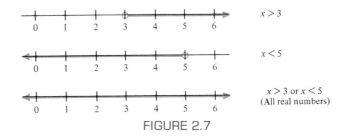

FIGURE 2.7

Recall from Chapter 1 that the *union* of two sets is the set of elements that belong to *either* of the sets. *To find the solution set of an inequality containing the word* **or,** *take the* **union** *of the solution sets of the two inequalities that comprise the compound inequality.*

EXAMPLE 13   Solve $x + 3 \le -1$ or $-4x + 3 < -5$.

Solution: Solve each inequality separately.

$$x + 3 \le -1 \quad \text{or} \quad -4x + 3 < -5$$
$$x \le -4 \quad \text{or} \quad -4x < -8$$
$$x > 2$$

Now graph each solution on number lines and then find the union (Fig. 2.8). The union is $x \le -4$ or $x > 2$.

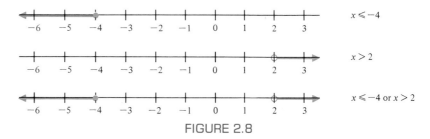

FIGURE 2.8

The solution set is $\{x \mid x \le -4\} \cup \{x \mid x > 2\}$. The union of these two sets can be written as $\{x \mid x \le -4 \text{ or } x > 2\}$. Thus the solution set is $\{x \mid x \le -4 \text{ or } x > 2\}$. In interval notation, the answer is $(-\infty, -4] \cup (2, \infty)$. ■

We often encounter inequalities in our daily lives. For example, on a highway there may be a minimum speed of 30 mph and a maximum speed of 55 mph. A restaurant may have a sign stating that maximum capacity is 300 people, and the minimum takeoff speed of an airplane may be 125 miles per hour.

HELPFUL HINT

There are various ways to write the solution to an inequality problem. Be sure to indicate the solution to an inequality problem in the form requested by your professor. Examples of the different forms follow.

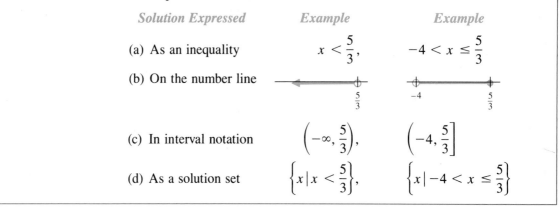

Solution Expressed	Example	Example
(a) As an inequality	$x < \dfrac{5}{3}$,	$-4 < x \le \dfrac{5}{3}$
(b) On the number line		
(c) In interval notation	$\left(-\infty, \dfrac{5}{3}\right)$,	$\left(-4, \dfrac{5}{3}\right]$
(d) As a solution set	$\left\{ x \mid x < \dfrac{5}{3} \right\}$,	$\left\{ x \mid -4 < x \le \dfrac{5}{3} \right\}$

Exercise Set 2.5

Express each inequality (a) using the number line, (b) in interval notation, and (c) as a solution set (set builder notation).

1. $x < -3$

2. $x > \dfrac{5}{2}$

3. $x \ge 5$

4. $-2 < x < 5$

5. $2 \le x < \dfrac{12}{5}$

6. $x \ge -\dfrac{6}{5}$

7. $-6 < x \le -4$

8. $-3 \le x \le 8$

9. $-4 \le x \le 5$

10. $x > -\dfrac{1}{2}$

Solve the inequality and graph the solution on the number line.

11. $x + 3 < 8$

12. $x + 5 \ge -5$

13. $2x > 12$

14. $3x \le -5$

15. $2x + 3 > 4$

16. $3 - x < -4$

17. $4x + 5 > 3x$

18. $x + 4 < 2x - 3$

19. $4x + 3 \le -2x + 9$

20. $4(x - 2) \le 4x - 8$

21. $14 > 3a - 6$

22. $4b - 6 \ge 2b + 12$

23. $-(x - 3) + 4 \le -2x + 5$

24. $\dfrac{y}{3} + \dfrac{2}{5} \le 4$

25. $2y - 6y + 10 \le 2(-2y + 3)$

Solve the inequality and give the solution in interval notation.

26. $\dfrac{c + 3}{2} + 5 > c + 2$

27. $(w - 5) \le \dfrac{3}{4}(2w + 6)$

28. $4 - 3x < 7 + 2x + 4$

29. $4 + \dfrac{3x}{2} < 6$

30. $\dfrac{3y - 6}{2} > \dfrac{2y + 5}{6}$

31. $\dfrac{5 - 6y}{3} \le 1 - 2y$

32. $\dfrac{3(x - 2)}{5} > \dfrac{5(2 - x)}{3}$

33. $x + 1 < 3(x + 2) - 2x$

34. $\dfrac{1}{2}\left(\dfrac{3}{5}y + 4\right) \le \dfrac{1}{3}(y - 6)$

Solve the inequality by the method illustrated in Examples 10 and 11. Give the solution in interval notation.

35. $4 < x + 3 < 9$

36. $-2 \le x - 5 < 7$

37. $-3 < 5x \le 8$

38. $-2 < -4x < 8$

39. $4 \le 2x - 3 < 7$

40. $-12 < 3x - 5 \le -4$

41. $\dfrac{1}{2} < 3x + 4 < 6$

42. $-7 < \dfrac{4 - 3x}{2} < 9$

43. $-6 < \dfrac{-2x - 3}{4} \le 8$

Solve the inequality and indicate the solution set.

44. $4 < \dfrac{4x - 3}{2} \le 12$

45. $-12 \le \dfrac{4 - 3x}{-5} < 2$

46. $\dfrac{3}{5} < \dfrac{-x - 5}{3} < 6$

47. $6 \le -3(2x - 4) < 12$

48. $-7 < \dfrac{4 - 2x}{3} < \dfrac{1}{3}$

49. $-15 < \dfrac{3(x - 2)}{5} \le 0$

50. $0 < \dfrac{2(x - 3)}{5} \le 12$

51. $1 < \dfrac{4 - 6x}{2} < 5$

52. $\dfrac{1}{8} \le 4 - 2(x + 3) \le 5$

Solve the inequality and indicate the solution set.

53. $x < 4$ and $x > 2$

54. $x < 4$ or $x > 2$

55. $x < 2$ and $x > 4$

56. $x < 2$ or $x > 4$

57. $x + 2 < 3$ and $x + 1 > -2$

58. $2x - 3 \le 5$ or $2x - 8 \ge 4$

59. $5x - 3 \le 7$ or $-x + 3 < -5$

60. $-2x - 3 < 2$ and $x + 6 > 4$

Solve the inequality and give the solution in interval notation.

61. $3x - 6 \le 4$ or $2x - 3 < 5$

62. $-x + 6 > -3$ or $4x - 2 < 12$

63. $4x + 5 \ge 5$ and $3x - 4 \le 2$

64. $x - 3 > -5$ and $-2x - 4 > -2$

65. $5x - 3 > 10$ and $4 - 3x < -2$

66. $x - 4 > 4$ or $3x - 5 \ge 1$

67. $4 - x < -2$ or $3x - 1 < -1$

68. $-x + 3 < 0$ or $2x - 5 \ge 3$

For each exercise, set up an inequality that can be used to solve the problem. Solve the problem and find the desired value.

69. Cal, a janitor, must move a large shipment of books from the first floor to the fifth floor. The sign on the elevator reads "maximum weight 900 pounds." If each box of books weigh 80 pounds, find the maximum number of boxes that Cal can place on the elevator.

70. If the janitor in Exercise 69, weighing 170 pounds, must ride up with the boxes, find the maximum number of boxes of books that can be placed in the elevator.

71. A telephone operator informs a customer in a phone booth that the charge for calling Denver, Colorado, is $4.25 for the first 3 minutes and 48 cents for each additional minute. Any additional part of a minute will be rounded up to the nearest minute. Find the maximum time the customer can talk if he has only $9.50.

72. A downtown parking garage in Austin charges $0.75 for the first hour and $0.50 for each additional hour. What is the maximum length of time you can park in the garage if you wish to pay no more than $3.75?

73. Miriam Davidson is considering writing and publishing her own book. She estimates her revenue equation to be $R = 6.42x$. Her cost equation is estimated to be $C = 10,025 + 1.09x$, where x is the number of books she sells. Find the minimum number of books she must sell to make a profit. See Example 7.

74. Peter Collinge is considering opening a dry-cleaning store. He estimates his cost equation to be $C = 8000 + 0.08x$ and his revenue equation to be $R = 1.85x$, where x is the number of garments dry cleaned in a year. Find the minimum number of garments that must be dry cleaned in a year for Peter to make a profit.

75. A nonprofit organization can purchase a $60 bulk-mailing permit and then send bulk mail at a rate of 8.4 cents per piece. Without the permit each piece of bulk mail would cost 16.7 cents. Find the minimum number of pieces of bulk mail that would have to be mailed for it to be financially worthwhile for an organization to purchase the bulk-mailing permit.

76. The cost for mailing a package first class is 29 cents for the first ounce and 23 cents for each additional ounce. What is the maximum weight of a package that can be mailed first class for $5.00?

77. To receive an A in a course you must obtain an average of 90 or higher on five exams. If Ray's first four exam grades are 90, 87, 96, and 95, what is the minimum grade Ray can receive on the fifth exam to get an A in the course?

78. To pass a course you need an average grade of 60 or more. If Maria's grades are 65, 72, 90, 47, and 62, find the minimum grade Maria can get on her sixth and last exam and pass the course.

79. For air to be considered "clean," the average of three pollutants must be less than 3.2 parts per million. If the first two pollutants are 2.7 and 3.42 ppm, what values of the third pollutant will result in clean air?

80. Ms. Mahoney's grades on her first four exams are 87, 92, 70, and 75. An average greater than or equal to 80 and less than 90 will result in a final grade of B. What range of grades on Ms. Mahoney's fifth and last exam will result in a final grade of B? Assume a maximum grade of 100.

81. The water acidity in a pool is considered normal when the average pH reading of three daily measurements is between 7.2 and 7.8. If the first two pH readings are 7.48 and 7.85, find the range of pH values for the third reading that will result in the acidity level being normal.

82. (a) Explain the step by step procedure to use to solve the inequality $a < bx + c < d$ for x.

   (b) Solve the inequality for x and write the answer in interval notation (assume $b > 0$).

83. (a) Explain how to solve the inequality $x + 5 < 3x - 8 \le 2(x + 7)$.

   (b) Solve the inequality and give the answer in interval notation.

Cumulative Review Exercises

[1.2] **84.** $A = \{1, 4, 6, 7, 9\}$, $B = \{1, 3, 4, 5, 6\}$. Find (a) $A \cup B$; (b) $A \cap B$.

85. $A = \{-3, 4, 5/2, \sqrt{7}, 0, -29/80\}$. List the elements that are (a) counting numbers, (b) whole numbers, (c) rational numbers, (d) real numbers.

[1.3] *Name the properties illustrated.*

86. $(3x + 6) + 4y = 3x + (6 + 4y)$

87. $3x + y = y + 3x$

[2.2] **88.** Solve the formula $R = L + (V - D)r$ for V.

JUST FOR FUN

Russell's first five exams were 82, 90, 74, 76, and 68. The final exam for the course is to count one-third in computing the final average. A final average greater than or equal to 80 and less than 90 will result in a final grade of B. What range of final-exam grades will result in Russell receiving a final grade of B in the course? Assume a maximum grade of 100 is possible.

2.6

Solving Equations and Inequalities Containing Absolute Value

▶ **1** Solve equations containing absolute value.

▶ **2** Solve inequalities of the form $|x| < a$.

▶ **3** Solve inequalities of the form $|x| > a$.

▶ **4** Solve inequalities of the form $|x| > a$ or $|x| < a$ when $a < o$.

Equations Containing
Absolute Value

FIGURE 2.9

▶ **1** In Section 1.4 we introduced the concept of absolute value. We stated that the absolute value of a number may be considered the distance (without sign) from the number 0 on the number line. The absolute value of 3, written $|3|$, is 3 since it is 3 units from 0 on the number line. Similarly, the absolute value of -3, written $|-3|$, is also 3 since it is 3 units from 0 on the number line (see Fig. 2.9).

Consider the equation $|x| = 3$; what values of x make this equation true? We know that $|3| = 3$ and $|-3| = 3$. The solutions to $|x| = 3$ are 3 and -3. When solving the equation $|x| = 3$, we are finding the values that are 3 units from 0 on the number line. When solving an equation of the form $|x| = a$, $a \geq 0$, we are finding the values that are a units from 0 on the number line.

Equations of the Form $|x| = a$

If $|x| = a$ and $a > 0$, then $x = a$ or $x = -a$.

EXAMPLE 1   Solve the equation $|x| = 4$.

Solution: Using the given rule, we get $x = 4$ or $x = -4$. The solution set is $\{-4, 4\}$.   ■

EXAMPLE 2   Solve the equation $|x| = 0$.

Solution: The only real number whose absolute value equals 0 is 0. Thus $|x| = 0$ has the solution set $\{0\}$.   ■

EXAMPLE 3   Solve the equation $|x| = -2$.

Solution: The absolute value of a number is never negative, so there are no solutions to this equation. The solution set is $\varnothing$.   ■

EXAMPLE 4   Solve the equation $|2w - 1| = 5$.

Solution: If we consider $2w - 1$ to be x, then $2w - 1$ must be 5 units from 0 on the number line. Thus the quantity $2w - 1$ must be equal to 5 or -5.

$$2w - 1 = 5 \quad \text{or} \quad 2w - 1 = -5$$
$$2w = 6 \qquad\qquad 2w = -4$$
$$w = 3 \qquad\qquad w = -2$$

Check:

$$w = 3, \quad |2(3) - 1| = 5 \qquad\qquad w = -2, \quad |2(-2) - 1| = 5$$
$$|6 - 1| = 5 \qquad\qquad\qquad |-4 - 1| = 5$$
$$|5| = 5 \qquad\qquad\qquad\qquad |-5| = 5$$
$$5 = 5 \quad \text{true} \qquad\qquad\qquad 5 = 5 \quad \text{true}$$

The solution set is $\{-2, 3\}$.   ■

EXAMPLE 5 Solve the equation $\left|\frac{2}{3}z - 6\right| + 4 = 6$.

Solution: The absolute value must be isolated on one side of the equation before the two cases can be written. Therefore, we begin by subtracting 4 from both sides of the equation to get the absolute value alone on one side of the equation.

$$\left|\frac{2}{3}z - 6\right| + 4 = 6$$

$$\left|\frac{2}{3}z - 6\right| = 2$$

Now we proceed as before.

$$\frac{2}{3}z - 6 = 2 \quad \text{or} \quad \frac{2}{3}z - 6 = -2$$

$$\frac{2}{3}z = 8 \qquad\qquad \frac{2}{3}z = 4$$

$$2z = 24 \qquad\qquad 2z = 12$$

$$z = 12 \qquad\qquad z = 6$$

The solution set is $\{6, 12\}$.

Inequalities Containing Absolute Value

▶**2** We just showed how to solve equations of the form $|x| = a$. Now you will learn how to solve inequalities containing absolute values. We will work first with inequalities of the form $|x| < a$.

Earlier we stated that the solution set to $|x| = 3$ is the values that were exactly 3 units from 0 on the number line. The solution set to $|x| = 3$ is $\{-3, 3\}$. Similarly, we can state that $|x| < 3$ is the set of values that are less than 3 units from the number 0 on the number line. This includes all real numbers between -3 and 3 (see Fig. 2.10). The solution set of $|x| < 3$ is $\{x \,|\, -3 < x < 3\}$.

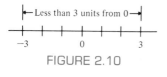

FIGURE 2.10

When we are asked to find the solution set to an inequality of the form $|x| < a$, we are finding the set of values that are less than a units from 0 on the number line. The solution set to $|x| \leq a$ is the set of values that are less than *or equal to* a units from 0 on the number line.

We can use the same reasoning process to solve more complicated problems, as shown in Example 6.

EXAMPLE 6 Solve the inequality $|2x - 3| < 5$.

Solution: The solution to this inequality will be the set of values such that the distance between $2x - 3$ and 0 on the number line will be less than 5 units (see Fig. 2.11). Using Fig. 2.11, we can see that $-5 < 2x - 3 < 5$.

Solving we get

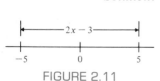

FIGURE 2.11

$$-5 < 2x - 3 < 5$$

$$-2 < 2x < 8$$

$$-1 < x < 4$$

The solution set is $\{x \,|\, -1 < x < 4\}$. When x is any number between -1 and 4, $2x - 3$ will be a number less than 5 units from 0 on the number line (or a number between -5 and 5).

Using the same reasoning process, we can see that to solve inequalities of the form $|x| < a$ we use the following procedure:

Inequalities of the Form $|x| < a$

If $|x| < a$ and $a > 0$, then $-a < x < a$.

EXAMPLE 7   Solve the inequality $|3x - 4| \leq 5$ and graph the solution on the number line.

Solution: Since this inequality is of the form $|x| \leq a$, we write

$$-5 \leq 3x - 4 \leq 5$$
$$-1 \leq 3x \leq 9$$
$$-\frac{1}{3} \leq x \leq 3$$

EXAMPLE 8   Solve the inequality $|4 - x| + 1 < 3$ and graph the solution on the number line.

Solution: First isolate the absolute value by subtracting 1 from both sides of the inequality. Then solve as in the previous examples.

$$|4 - x| + 1 < 3$$
$$|4 - x| < 2$$
$$-2 < 4 - x < 2$$
$$-6 < -x < -2$$
$$-1(-6) > -1(-x) > -1(-2)$$
$$6 > x > 2$$
$$\text{or} \quad 2 < x < 6$$

The solution set is $\{x \,|\, 2 < x < 6\}$. The answer in interval notation is $(2,6)$.

▶ **3**   Now we look at inequalities of the form $|x| > a$. Consider $|x| > 3$. This inequality represents the set of values that are greater than 3 units from 0 on the number line. The solution to $|x| > 3$ is $x > 3$ or $x < -3$ (see Fig. 2.12).

FIGURE 2.12

Similarly, $|x| > a$ is the set of values that are greater than a units from 0 on the number line.

EXAMPLE 9   Solve the inequality $|2x - 3| > 5$ and graph the solution on the number line.

Solution: The solution to $|2x - 3| > 5$ is the set of values such that the distance between $2x - 3$ and 0 on the number line will be greater than 5. The quantity $2x - 3$ must either be less than -5 or greater than 5 (see Fig. 2.13).

FIGURE 2.13

Since $2x - 3$ must be either less than -5 or greater than 5, we set up and solve the following compound inequality

$$2x - 3 < -5 \quad \text{or} \quad 2x - 3 > 5$$
$$2x < -2 \qquad\qquad 2x > 8$$
$$x < -1 \qquad\qquad x > 4$$

The solution set to $|2x - 3| > 5$ is $\{x \mid x < -1 \text{ or } x > 4\}$. When x is any number less than -1 or greater than 4, $2x - 3$ will be greater than 5 units from 0 on the number line (or a number less than -5 or greater than 5). ■

Using the same reasoning process, we can see that to solve inequalities of the form $|x| > a$ we use the following procedure.

Inequalities of the Form $|x| > a$

If $|x| > a$ and $a > 0$, then $x < -a$ or $x > a$.

EXAMPLE 10   Solve the inequality $|2x - 5| \ge 3$ and graph the solution on the number line.

Solution: Since this inequality is of the form $|x| \ge a$, we use the procedure given above.

$$2x - 5 \le -3 \quad \text{or} \quad 2x - 5 \ge 3$$
$$2x \le 2 \qquad\qquad 2x \ge 8$$
$$x \le 1 \qquad\qquad x \ge 4$$

The solution set is $\{x \mid x \le 1 \text{ or } x \ge 4\}$. In interval notation, the solution is $(-\infty, 1] \cup [4, \infty)$. ■

EXAMPLE 11   Solve the inequality $\left| \dfrac{3x - 4}{2} \right| \ge \dfrac{5}{12}$.

Solution: Since this inequality is of the form $|x| \ge a$, we write

$$\frac{3x - 4}{2} \le -\frac{5}{12} \quad \text{or} \quad \frac{3x - 4}{2} \ge \frac{5}{12}$$

Now multiply both sides of each inequality by the least common denominator, 12. Then solve each inequality.

$$\overset{6}{\cancel{12}}\left(\frac{3x - 4}{\underset{1}{\cancel{2}}}\right) \le \frac{-5}{\cancel{12}} \cdot \cancel{12} \quad \text{or} \quad \overset{6}{\cancel{12}}\left(\frac{3x - 4}{\underset{1}{\cancel{2}}}\right) \ge \frac{5}{\cancel{12}} \cdot \cancel{12}$$

$$6(3x - 4) \le -5 \qquad\qquad 6(3x - 4) \ge 5$$
$$18x - 24 \le -5 \qquad\qquad 18x - 24 \ge 5$$
$$18x \le 19 \qquad\qquad\qquad 18x \ge 29$$
$$x \le \frac{19}{18} \qquad\qquad\qquad x \ge \frac{29}{18}$$

$$\frac{19}{18} \qquad\qquad \frac{29}{18}$$

Summary of Procedures for Solving Equations and Inequalities Containing Absolute Value

For $a > 0$,

If $|x| = a$,   then $x = a$ or $x = -a$.

If $|x| < a$,   then $-a < x < a$.

If $|x| > a$,   then $x < -a$ or $x > a$.

▶ **4**   Note that in the summary box we specified that $a > 0$. Now let us consider what happens in an absolute value inequality when $a < 0$. Consider the inequality $|x| < -3$. Since $|x|$ will always have a value greater than or equal to 0 for any real number x, this inequality can never be true. Thus there is no value of x that makes this a true statement and the solution is the empty set, $\varnothing$. Whenever we have an absolute value inequality of this type, the solution will be the empty set.

EXAMPLE 12   Solve the inequality $|x - 4| - 3 < -5$.

Solution: Begin by adding 3 to both sides of the inequality.

$$|x - 4| - 3 < -5$$
$$|x - 4| < -2$$

Since $|x - 4|$ will always be greater than or equal to 0 for any real number x, this inequality can never be true. Thus the solution is the empty set, $\varnothing$. ■

Now consider the inequality $|x| > -3$. Since $|x|$ will always have a value greater than or equal to 0 for any real number x, this inequality will always be true. Since every value of x will make this inequality a true statement, the answer is all real numbers, $\mathbb{R}$. Whenever we have an absolute value inequality of this type, the solution will be all real numbers, $\mathbb{R}$.

EXAMPLE 13   Solve the inequality $|2x + 3| + 4 \geq -7$.

Solution: Begin by subtracting 4 from both sides of the inequality.

$$|2x + 3| + 4 \geq -7$$
$$|2x + 3| \geq -11$$

Since $|2x + 3|$ will always be greater than or equal to 0 for any real number x, this inequality is true for all real numbers. Thus the solution set is all real numbers, $\mathbb{R}$.   ■

Exercise Set 2.6

Find the solution set for each equation.

1. $|x| = 5$
4. $|x| = 0$

2. $|y| = 7$
5. $|x| = -2$

3. $|x| = 12$
6. $|x + 1| = 5$

7. $|x + 5| = 7$

8. $|3 + y| = \dfrac{3}{5}$

9. $|2w + 4| = 6$

10. $|3x - 4| = 0$

11. $|5 - 3x| = \dfrac{1}{2}$

12. $|3(y + 4)| = 12$

13. $|4(x - 2)| = 18$

14. $\left|\dfrac{x - 3}{4}\right| = 5$

15. $\left|\dfrac{3z + 5}{6}\right| - 3 = 6$

16. $\left|\dfrac{x - 3}{4}\right| + 4 = 4$

17. $\left|\dfrac{5x - 3}{2}\right| + 2 = 6$

18. $\left|\dfrac{2x + 3}{2}\right| + 1 = 4$

Find the solution set for each inequality.

19. $|y| \leq 5$
21. $|x - 7| \leq 9$
23. $|3z - 5| \leq 5$
25. $|2x + 3| - 5 \leq 10$

27. $|x - 5| \leq \dfrac{1}{2}$

29. $|2x - 6| + 5 \leq 2$

31. $\left|5 - \dfrac{3x}{4}\right| < 8$

20. $|x| \leq 9$
22. $|7 - x| < 5$
24. $|x - 3| - 2 < 3$
26. $|4 - 3x| - 4 < 11$

28. $|2x - 3| < -4$

30. $\left|\dfrac{2x - 1}{3}\right| \leq \dfrac{5}{3}$

32. $\left|\dfrac{x - 3}{2}\right| - 4 \leq -2$

Find the solution set for each inequality.

33. $|x| > 3$
35. $|x + 4| > 5$
37. $|3x + 1| > 4$

39. $\left|\dfrac{6 + 2z}{3}\right| > 2$

41. $|4x - 3| + 2 > 7$

43. $\left|\dfrac{2x - 4}{3}\right| > -5$

34. $|y| \geq 5$
36. $|5 - x| \geq 3$
38. $|4 - 3y| \geq 8$

40. $\left|\dfrac{5 - 3w}{4}\right| \geq 10$

42. $|2x - 1| - 4 \geq 8$

44. $\left|\dfrac{2x - 3}{4}\right| - 1 > 3$

45. $\left| \dfrac{x}{2} + 4 \right| \geq 5$ $\{x \mid x \leq -18 \text{ or } x \geq 2\}$

46. $\left| 4 - \dfrac{3x}{5} \right| \geq 9$ $\{z \mid z \leq -20 \text{ or } z \geq 2\}$

Find the solution set for each equation or inequality.

47. $|w| = 7$

48. $|x - 3| = 5$ $\{z \mid z \leq -20 \text{ or } z \geq 2\}$

49. $|x - 3| < 5$

50. $|z| \geq 2$

51. $|x + 5| > 9$

52. $|3x - 4| \leq -6$

53. $|2y + 4| < 1$ $\left\{ -\dfrac{11}{4}, \dfrac{7}{4} \right\}$

54. $|2x - 5| + 3 \leq 10$

55. $|4x + 2| = 9$

56. $|2x - 4| + 2 = 10$

57. $|5 + 2x| \geq 3$

58. $|4 - x| = 5$ $\left\{x \mid x \leq -\dfrac{2}{3} \text{ or } x \geq \dfrac{4}{3}\right\}$

59. $|4 + 3x| \leq 9$

60. $|x - 3| + 5 \geq 3$

61. $|3x - 5| + 4 = 2$

62. $|4 - 2x| - 5 = 5$

63. $\left| \dfrac{3x - 2}{4} \right| - 5 = 1$ $\{w \mid -16 < w < 8\}$

64. $\left| \dfrac{4c - 4}{5} \right| \leq 8$

65. $\left| \dfrac{w + 4}{3} \right| < 4$

66. $\left| \dfrac{3x + 4}{5} \right| > \dfrac{7}{5}$

67. $\left| \dfrac{3x - 2}{4} \right| + 5 \geq 5$

68. $\left| \dfrac{2x - 4}{5} \right| = 12$

69. $|2 - 3x| - 4 \geq -2$

70. $\left| \dfrac{3 - 2x}{4} \right| \geq 5$

71. $\left| 2 \left(\dfrac{3 - x}{5} \right) \right| < \dfrac{9}{5}$

72. $\left| 3 \left(x + \dfrac{1}{2} \right) \right| > 5$

73. (a) Explain how to find the solution to the equation $|ax + b| = c$ (assume $c > 0$ and $a \neq 0$).
   (b) Solve this equation for x.

74. (a) Explain how to find the solution to the inequality $|ax + b| < c$ (assume $a > 0$ and $c > 0$).
   (b) Solve this inequality for x.

75. (a) Explain how to find the solution to the inequality $|ax + b| > c$ (assume $a > 0$ and $c > 0$).
   (b) Solve this inequality for x.

76. (a) What is the first step in solving the inequality $-2|3x - 5| \leq -6$?
   (b) Solve the inequality and give the solution in interval notation.

Cumulative Review Exercises

Evaluate each of the following:

[1.7] **77.** $\dfrac{1}{3} + \dfrac{1}{4} \div \dfrac{2}{5} \left(\dfrac{1}{3} \right)^2$

   78. $4(x + 3y) - 5xy$ when $x = 1, y = 3$

[2.4] **79.** Raul swims across a lake averaging 2 miles an hour. Then he turns around and swims back across the lake, averaging 1.6 miles per hour. If the total time of his swimming was 1.5 hours, what was the width of the lake?

[2.5] **80.** Find the solution set to the inequality $3(x - 2) - 4(x - 3) > 2$.

JUST FOR FUN

1. Find all values of x such that $|x - 3| = |3 - x|$.

2. Find all values of x and y such that $|x - y| = |y - x|$.

3. Solve $|x + 6| = |2x - 3|$.

4. Solve $|x - 3| = |x + 5|$.

SUMMARY

GLOSSARY

Coefficient (or numerical coefficient) *(45):* The numerical part of a term.

Compound inequality *(87):* Two inequalities joined with the word *and* or *or*.

Conditional equation *(51):* An equation true only under specific conditions.

Constant *(45):* a term that consists of only a number.

Continued inequality *(45):* An inequality of the form $a < x < b$.

Degree of a term *(45):* The sum of the exponents on the variables in a term.

Equation *(46):* A mathematical statement of equality.

Equivalent equations *(46):* Equations with the same solution set.

Formula *(54):* An equation used to represent a scientific or real-life principle in mathematical terms.

Identity *(51):* An equation true for all real numbers.

Inconsistent equation *(52):* An equation that has no solution.

Inequality *(82):* A mathematical expression containing one or more inequality symbols.

Least common denominator *(49):* The smallest number divisible by a given set of numbers.

Like terms *(45):* Terms that have the same variables with the same exponents.

Linear equation *(46):* The standard form of a linear equation in one variable is $ax + b = c$, $a \neq 0$. A linear equation is also called a first-degree equation.

Sense of an inequality *(82):* The direction of the inequality symbol.

Simplify an expression *(45):* Combine like terms in the expression.

Solution of an equation *(46):* The number or numbers that make the equation true.

Solution set of an equation *(46):* The set of real numbers that make the equation true.

Subscript *(55):* Numbers or letters to the right of and below a variable.

Terms *(44):* The parts added or subtracted in an algebraic expression.

Unlike terms *(45):* Terms that are not "like" terms.

IMPORTANT FACTS

Properties of Equality

Reflexive property: $a = a$

Symmetric property: If $a = b$, then $b = a$.

Transitive property: If $a = b$ and $b = c$, then $a = c$.

Addition property: If $a = b$, then $a + c = b + c$.

Multiplication property: If $a = b$, then $ac = bc$.

Proportions

If $\dfrac{a}{b} = \dfrac{c}{d}$, then $ad = bc$

Distance Formula

distance = rate × time

Properties Used to Solve Inequalities

1. If $a > b$, then $a + c > b + c$.
2. If $a > b$, then $a - c > b - c$.
3. If $a > b$ and $c > 0$, then $ac > bc$.
4. If $a > b$ and $c > 0$, then $\dfrac{a}{c} > \dfrac{b}{c}$.
5. If $a > b$ and $c < 0$, then $ac < bc$.
6. If $a > b$ and $c < 0$, then $\dfrac{a}{c} < \dfrac{b}{c}$.

Absolute Value, for $a > 0$

If $|x| = a$, then $x = a$ or $x = -a$.
If $|x| < a$, then $-a < x < a$.
If $|x| > a$, then $x < -a$ or $x > a$.

Review Exercises

[2.1] State the degree of the term.

1. $15x^4y^6$ **2.** $6x$ **3.** $-4xyz^5$

Simplify the expression. If an expression cannot be simplified, so state.

4. $x^2 + 3x + 6$ **5.** $x^2 + 2xy + 6x^2 - 4$

6. $3(x + 4) - 3x - 4$ **7.** $2[-(x - y) + 3x] - 5y + 6$

Solve each equation. If an equation has no solution, so state.

8. $\dfrac{x-4}{5} = 9 - x$

9. $3(x+2) - 6 = 4(x-5)$

10. $3 + \dfrac{x}{2} = \dfrac{5}{6}$

11. $-6 - 2x = \dfrac{1}{2}(4x+12)$

12. $2\left(\dfrac{x}{2} - 4\right) = 3\left(x + \dfrac{1}{3}\right)$

13. $3x - 4 = 6x + 4 - 3x$

14. $3[2x - (x+4)] = -3$

[2.2] *Evaluate the formula for the values given.*

15. $P = \dfrac{nRT}{V}$: $n = 10$, $R = 100$, $T = 4$, $V = 20$

17. $h = \dfrac{1}{2}at^2 + v_0t + h_0$: $a = -32$, $v_0 = 60$; $h_0 = 120$, $t = 2$

16. $x = \dfrac{-b + \sqrt{b^2 - 4ac}}{2a}$: $a = 8$, $b = 10$, $c = -3$

18. $z = \dfrac{\bar{x} - \mu}{\dfrac{\sigma}{\sqrt{n}}}$: $\bar{x} = 60$, $\mu = 80$, $\sigma = 5$, $n = 25$

Solve for the variable indicated.

19. $A = lw$ for l

20. $A = \pi r^2 h$ for h

21. $P = 2l + 2w$ for w

22. $d = rt$ for r

23. $y = mx + b$ for m

24. $2x - 3y = 5$, for y

25. $P_1V_1 = P_2V_2$ for V_2

26. $S = \dfrac{3a + b}{2}$ for a

27. $K = 2(d + l)$ for l

28. $I = p + prt$ for t

29. $A = \dfrac{1}{2}h(b_1 + b_2)$, for b_1

30. $w = V_0t - 2l$ for t

[2.3] *Write an equation that can be used to solve the problem. Solve the problem and check your answer.*

31. The sum of a number and 4 times the number is 80. Find the number.

32. Paul is 4 years older than his sister. The sum of their ages is 36. Find the ages of Paul and his sister.

33. Four times a number increased by 12 is 32. Find the number.

34. One-fourth of a number plus 6 is 11. Find the number.

35. A number decreased by 60% is 20. Find the number.

36. The sum of two consecutive odd integers is 28. Find the integers.

37. A number decreased by 10% is 180. Find the number.

38. The larger of two numbers is 1 less than twice the smaller. When the smaller is subtracted from the larger the difference is 9. Find the two numbers.

[2.4] *Solve the following rate and mixture problems.*

39. Tanya is a quality control inspector at the Eastman Kodak Company. In a typical 8-hour work day, she inspects 245 rolls of film. What is Tanya's hourly inspection rate?

40. The Sampsons invest $10,000 in two accounts. One account pays 8% simple interest and the other account pays 5% simple interest. If the total interest for the year is $680, how much money was invested in each account?

41. Two trains leave Portland at the same time traveling in opposite directions. One train travels at 60 miles per hour and the other at 90 miles per hour. In how many hours will they be 400 miles apart?

42. Space Shuttle 2 takes off 0.5 hour after Shuttle 1 takes off. If Shuttle 2 travels 300 miles per hour faster than Shuttle 1 and overtakes Shuttle 1 exactly 5 hours after Shuttle 2 takes off, find (a) the speed of Shuttle 1, and (b) the distance from earth when Shuttle 2 overtakes Shuttle 1.

43. Mr. Tomlins, the owner of a gourmet coffee shop, has two coffees, one selling for $6.00 per pound and the other for $6.80 per pound. How many pounds of each type of coffee should he mix to make 40 pounds of coffee to sell for for $6.50 per pound?

[2.3–2.4] *Solve the following word problems.*

44. A blouse has been reduced by 12%. The sale price is $22. Find the original price.

45. Nicolle jogged for a distance and then turned around and walked back to her starting point. While jogging she averaged 7.2 miles per hour, and while walking she averaged 2.4 miles per hour. If the total time spent jogging and walking was 4 hours, find (a) how long she jogged and (b) the total distance she traveled.

46. Find the three angles of a triangle if one angle measures 25° greater than the smallest angle and the other angle measures 5° less than twice the smallest angle.

47. Two hoses are being used to fill a swimming pool. The hose with the larger diameter supplies 1.5 times as much water as the hose with the smaller diameter. The larger hose is on for 2 hours before the smaller hose is turned on. If 5 hours after the larger hose is turned on there are 3150 gallons of water in the pool, find the volume flow from each hose.

48. The sum of two consecutive integers is 49. Find the integers.

49. A clothier has two blue dye solutions, both made from the same dye. One solution is 6% blue dye and the other is 20% blue dye. How many ounces of the 20% solution must be mixed with 10 ounces of the 6% solution to result in the mixture being a 12% blue dye solution?

50. Ken invests $12,000 in two savings accounts. One account is paying 10% simple interest and the other account is paying 6% simple interest. If the same interest is obtained from both accounts, how much was invested at each rate?

51. The West Ridge Fitness Center has two membership plans. The first plan is a flat $40 per month fee plus $1.00 per visit. The second plan is $25 per month plus a $4.00 per visit charge. How many visits would Mike have to make per month to make it advantageous for him to select the first plan?

52. Two trains leave Tucson at the same time, along parallel tracks, traveling in opposite directions. The faster train travels 10 miles per hour faster than the slower train. Find the speed of the *faster* train if the trains are 510 miles apart after 3 hours.

[2.5] *Solve the inequality. Graph the solution on the real number line.*

53. $x - 3 \geq 4$

54. $2 - x \leq 5$

55. $2x + 4 > 9$

56. $16 \leq 4x - 5$

57. $\dfrac{4x + 3}{5} > -3$

58. $2(x - 3) > 3x + 4$

59. $-4(x - 2) \leq 6x + 4$

60. $\dfrac{x}{4} \geq 5 - 2x$

Write an inequality that can be used to solve the problem. Solve the inequality and answer the question.

61. A small airplane has a maximum load of 1525 pounds if it is to take off safely. If the passengers weigh 468 pounds, how many 80-pound boxes can be safely transported on the plane?

62. Jack, a telephone operator, informs a customer in a phone booth that the charge for calling Omaha, Nebraska, is $4.50 for the first 3 minutes and 95 cents each additional minute and any part thereof. How long can the customer talk if he has $8.65?

63. A fitness center guarantees that you will lose a minimum of 3 pounds the first week and $1\frac{1}{2}$ pounds each additional week. Find the maximum amount of time needed to lose 27 pounds.

Solve the inequality. Indicate the solution in interval notation.

64. $1 < x - 4 < 7$

65. $2 \leq x + 5 < 8$

66. $3 < 2x - 4 < 8$

67. $-12 < 6 - 3x < -2$

68. $-1 \leq \dfrac{2x - 3}{4} < 5$

69. $-8 < \dfrac{4 - 2x}{3} < 0$

70. Manuel's first four exam grades are 94, 73, 72, and 80. If a final average greater than or equal to 80 and less than 90 is needed to receive a final grade of B in the course, what range of grades on the fifth and last exam will result in Manuel receiving a B in the course? Assume a maximum grade of 100.

Find the solution set to each compound inequality.

71. $x < 3$ and $2x - 4 > -10$

72. $2x - 1 > 5$ or $3x - 2 \leq 7$

73. $3x + 5 > 2$ or $6 - x < 1$

74. $4x - 3 \leq 7$ and $2x - 1 \geq 3$

75. $4x - 5 < 11$ and $-3x - 4 \geq 8$

76. $\dfrac{5x - 3}{2} > 7$ or $\dfrac{2x - 1}{3} \leq -3$

[2.6] *Find the solution set to each of the following*

77. $|x| = 4$

78. $|x| < 3$

79. $|x| \geq 4$

80. $|x - 4| = 9$

81. $|x - 2| \geq 5$

82. $|4 - 2x| = 5$

83. $|3 - 2x| < 7$

84. $\left|\dfrac{2x - 3}{5}\right| = 1$

85. $\left|\dfrac{x - 4}{3}\right| < 6$

86. $\left|\dfrac{4 - x}{3}\right| \geq \dfrac{1}{2}$

87. $|4(2 - x)| > 5$

88. $|2x - 3| + 4 \geq -10$

[2.5–2.6] *Solve the inequality and give the answer in interval notation.*

89. $\dfrac{5x - 3}{2} > 6$

90. $\dfrac{x}{3} \leq 2x - 5$

91. $|x + 6| < -1$

92. $\left|\dfrac{x - 3}{4}\right| \leq 5$

93. $-6 \leq \dfrac{3 - 2x}{4} < 5$

94. $x \leq 4$ and $4x - 6 \geq -14$

95. $|3(x + 2)| \leq \dfrac{9}{2}$

96. $\dfrac{2x - 8}{4} > 5$ or $\dfrac{3x - 2}{5} \leq -7$

97. $|x + 3| - 2 < 7$

98. $|4 - 3x| \geq 5$

99. $\left|\dfrac{x - 4}{2}\right| - 3 > 5$

100. $\dfrac{3}{5} < \dfrac{2x - 4}{3} \leq \dfrac{9}{4}$

Practice Test

1. State the degree of the term $-6xy^2z^3$.

2. Solve the equation $3(x - 2) = 4(4 - x) + 5$.

3. Solve the equation $\dfrac{3}{5} - \dfrac{x}{2} = 4$.

4. Solve the equation $\dfrac{3x}{4} - 1 = 5 + \dfrac{2x - 1}{3}$.

5. Find the value of S_n for the given values.

$$S_n = \dfrac{a_1(1 - r^n)}{1 - r}, \qquad a_1 = 3, r = \dfrac{1}{3}, n = 3$$

6. Solve for b in the equation $c = \dfrac{a - 3b}{2}$.

7. Solve $A = \frac{1}{2}h(b_1 + b_2)$ for b_2

For each problem, write an equation that can be used to solve the problem. Solve the equation and answer the question asked.

8. The sum of two consecutive integers is 47. Find the two integers.

9. The cost of renting an automobile is \$35 a day and 15 cents a mile. How far can Valerie drive in 1 day on \$65?

10. Two joggers start at the same point at the same time and jog in opposite directions. Homer jogs at 4 miles per hour, while Frances jogs at $5\frac{1}{4}$ miles per hour. How far apart will they be in $1\frac{1}{4}$ hours?

11. How many liters of 12% salt solution must be added to 10 liters of 25% salt solution to get a 20% salt solution?

12. Millie Johnson has \$12,000 to invest. She places part of her money in a savings account paying 8% simple interest and the balance in a savings account paying 7% simple interest. If the total interest from the two accounts

at the end of one year is \$910, find the amount placed in each account.

13. Solve the following inequality and graph the solution on the number line.

$$\dfrac{6 - 2x}{5} \geq -12$$

14. Solve the inequality and write the solution in interval notation.

$$-4 < \dfrac{x + 4}{2} < 8$$

15. Find the solution set to the equation.

$$|x - 4| = 5$$

Find the solution set to the inequalities.

16. $|2x - 3| + 1 > 6$

17. $\left|\dfrac{2x - 3}{4}\right| \leq \dfrac{1}{2}$

Cumulative Review Test

1. Set $A = \{1, 4, 6, 7, 9, 12\}$. Set $B = \{2, 3, 4, 5, 6, 9, 10, 12\}$. Find (a) $A \cup B$; (b) $A \cap B$

2. Name the indicated properties:
   (a) $4x + y = y + 4x$ (b) $(2x)y = 2(xy)$
   (c) $2(x + 3) = 2x + 6$

Evaluate each expression.

4. $4 - |-3| - (6 + |-3|)^2$
5. $-4^2 + (-3)^2 - 2^3 + (-2)^0$
6. $x^3 - xy + y^2$ when $x = -3$ and $y = -2$

Solve the following equations.

8. $3x - 4 = -2(x - 3) - 9$

10. $\dfrac{x}{4} - 5 = 3x - \dfrac{1}{3}$

12. Explain the difference between a conditional linear equation, an identity, and an inconsistent linear equation, and give an example of each.

13. Evaluate the formula $x = \dfrac{-b + \sqrt{b^2 - 4ac}}{2a}$ for $a = 3$, $b = -8$, and $c = -3$.

Find the solution set for each of the following.

16. $|4z + 8| = 12$
17. $|2x - 4| - 6 \geq 18$
18. The Computer Tutor has reduced the price of a computer by 20%. Find the original price of the computer if the sale price is $1800.
19. Two cars leave Caldwell, New Jersey at the same time traveling in opposite directions. The car traveling west is

3. Insert $<$, $>$, or $=$ in the shaded area to make the statement true: $-|-3|$ ▨ $|-5|$.

7. $\dfrac{8 - \sqrt[3]{27} \cdot 3 \div 9}{|-5| - (5 - (12 \div 4))^2}$

9. $1.2(x - 3) = 2.4x - 4.98$

11. $\dfrac{\frac{1}{4}x + 2}{3} = \dfrac{x - 4}{4}$

14. Solve the formula $I = p + prt$ for t.
15. Solve the inequality and give the answer (a) on the number line, (b) as a solution set, and (c) in interval notation.
$$-4 < \frac{5x - 2}{3} < 2$$

moving 10 miles per hour faster than the car traveling east. If the two cars are 270 miles apart after 3 hours, find the speed of each car.

20. Mr. Kane has a 20% saltwater solution and a 50% saltwater solution. How much of each solution should he mix to get 2 liters of a 30% saltwater solution?

3

Graphs and Functions

See Section 3.3, Just for Fun Exercise 1.

The Cartesian Coordinate System, Distance, and Midpoint Formulas

▶**1**   Plot points in the Cartesian coordinate system.

▶**2**   Find the distance between two points.

▶**3**   Find the midpoint of a line segment.

In this chapter we discuss procedures for drawing graphs. A graph is a picture that shows the relationship between two or more variables in an equation. Many algebraic relationships are easier to understand if we can see a visual picture of them.

▶**1**   Before learning how to construct a graph, you must know the **Cartesian (or rectangular) coordinate system.**

The Cartesian coordinate system, named after the French mathematician and philosopher René Descarte (1596–1650), consists of two axes (or number lines) in a plane drawn perpendicular to each other (see Fig. 3.1). Note how the two axes yield four **quadrants,** labeled I, II, III, and IV.

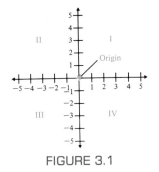

FIGURE 3.1

The horizontal axis is called the *x* **axis.** The vertical axis is called the *y* **axis.** The point of intersection of the two axes is called the **origin.** Starting from the origin and moving to the right, the numbers increase; moving to the left, the numbers decrease. Starting from the origin and moving up, the numbers increase; moving down, the numbers decrease.

To graph a point, it is necessary to know both its *x* coordinate and *y* coordinate. An **ordered pair** (x, y) is used to give the two coordinates of a point. If, for example, the *x* coordinate of a point is 3 and the *y* coordinate is 5, the ordered pair representing the point is (3, 5). Note that the *x* coordinate is always the first coordinate listed in the ordered pair. The point representing the ordered pair (3, 5) is plotted in Fig. 3.2.

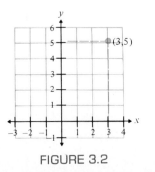

FIGURE 3.2

EXAMPLE 1 Plot each of the following points on the same set of axes.

(a) $A(4, 2)$ (b) $B(0, -3)$ (c) $C(-3, 1)$ (d) $D(4, 0)$

Solution: See Fig. 3.3.

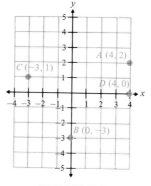

FIGURE 3.3

Notice that when the x coordinate is 0, as in part (b), the point is on the y axis. When the y coordinate is 0, as in part (d), the point is on the x axis.

EXAMPLE 2 List the ordered pair for each of the points shown in Figure 3.4.

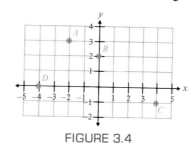

FIGURE 3.4

Solution: Remember to give the x value first in the ordered pair.

Point	Ordered Pair
A	$(-2, 3)$
B	$(0, 2)$
C	$(4, -1)$
D	$(-4, 0)$

▶ 2 Now we will see how to find the distance between any two points in a plane. After this, we will show how to find the midpoint of a given line segment. You need these two concepts to understand conic sections (Chapter 10).

Distance Between Two Points

To find the distance, d, between two points, we use the distance formula.

Distance Formula

The distance, d, between any two points (x_1, y_1) and (x_2, y_2) can be found by the distance formula

$$d = \sqrt{(x_2 - x_1)^2 + (y_2 - y_1)^2}$$

The distance between any two points will always be a positive number. Can you explain why? When finding the distance, it makes no difference which point we designate as point 1 (x_1, y_1) or point 2 (x_2, y_2). Note that the square of any real number will always be greater than or equal to zero. For example, $(5 - 2)^2 = (2 - 5)^2 = 9$.

EXAMPLE 3   Determine the distance between the points $(-1, 7)$ and $(-4, 3)$.

Solution: First plot the points (Fig. 3.5). Call $(-1, 7)$ point 2 and $(-4, 3)$ point 1. Thus (x_2, y_2) represents $(-1, 7)$ and (x_1, y_1) represents $(-4, 3)$. Now use the distance formula to find the distance, d.

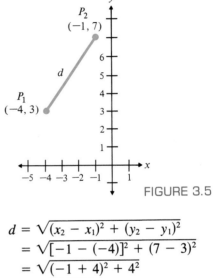

FIGURE 3.5

$$d = \sqrt{(x_2 - x_1)^2 + (y_2 - y_1)^2}$$
$$= \sqrt{[-1 - (-4)]^2 + (7 - 3)^2}$$
$$= \sqrt{(-1 + 4)^2 + 4^2}$$
$$= \sqrt{3^2 + 4^2}$$
$$= \sqrt{9 + 16}$$
$$= \sqrt{25}$$
$$= 5$$

Thus the distance between the points $(-1, 7)$ and $(-4, 3)$ is 5 units.   ■

If in Example 3 we had selected $(-4, 3)$ for point 2 and $(-1, 7)$ for point 1, our results would not have changed.

$$d = \sqrt{(x_2 - x_1)^2 + (y_2 - y_1)^2}$$
$$= \sqrt{[-4 - (-1)]^2 + (3 - 7)^2}$$
$$= \sqrt{(-4 + 1)^2 + (-4)^2}$$
$$= \sqrt{(-3)^2 + (-4)^2}$$
$$= \sqrt{9 + 16}$$
$$= \sqrt{25}$$
$$= 5$$

When using the distance formula, do not expect your distance to always come out as a rational number. If your answer is an irrational number, such as $\sqrt{187}$, you could use a calculator to obtain an approximate decimal answer. On a calculator with a square root key, we can determine that $\sqrt{187} \approx 13.674794$. Rounding this approximation to the nearest hundredth, we find that $\sqrt{187} \approx 13.67$. Appendix C gives square roots of integers from 1 to 100 rounded to the nearest hundredth.

HELPFUL HINT   Students will sometimes begin finding the distance correctly using the distance formula but will forget to take the square root of the sum $(x_2 - x_1)^2 + (y_2 - y_1)^2$ to obtain the correct answer.

Midpoint of a Line Segment

▶**3**   It is often necessary to find the midpoint of a line segment between two given points. To do this, we use the midpoint formula.

Midpoint Formula

Given any two points (x_1, y_1) and (x_2, y_2), the point halfway between the given points can be found by the midpoint formula:

$$\text{Midpoint} = \left(\frac{x_1 + x_2}{2}, \frac{y_1 + y_2}{2}\right)$$

EXAMPLE 4   Determine the midpoint of the line segment between the points $(-3, 7)$ and $(4, 2)$.

Solution: It makes no difference which points we label (x_1, y_1) and (x_2, y_2). Let us replace (x_1, y_1) with $(-3, 7)$ and (x_2, y_2) with $(4, 2)$ (see Fig. 3.6).

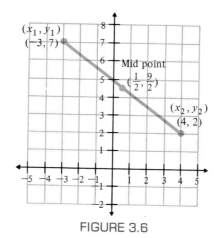

FIGURE 3.6

$$\text{Midpoint} = \left(\frac{x_1 + x_2}{2}, \frac{y_1 + y_2}{2}\right)$$

$$= \left(\frac{-3 + 4}{2}, \frac{7 + 2}{2}\right)$$

$$= \left(\frac{1}{2}, \frac{9}{2}\right)$$

The point $\left(\frac{1}{2}, \frac{9}{2}\right)$ is halfway between the points $(-3, 7)$ and $(4, 2)$.   ∎

Exercise Set 3.1

1. List the ordered pairs corresponding to the following points.

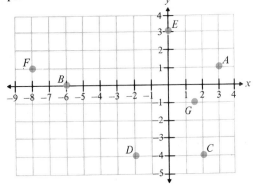

2. List the ordered pairs corresponding to the following points.

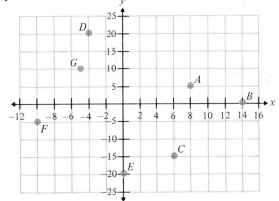

3. Graph the following points on the same set of axes.

(a) (4, 2)   (b) (−6, 2)

(c) (0, −1)   (d) (−2, 0)

4. Graph the following points on the same set of axes.

(a) (−4, −2)   (b) (3, 2)

(c) (2, −3)   (d) (−3, 3)

Determine the distance between the points. If a calculator with a square root key is available, round your answer to the nearest hundredth.

5. (2, −2) and (2, −5) $= 3$

7. (−4, 3) and (5, 3)

9. (1, 4) and (−3, 1)

11. (−3, −5) and (6, −2)

13. (0, 6) and (5, −1)

15. (−1.6, 3.5) and (−4.3, −1.7) $= 5.86$

17. $\left(\dfrac{3}{4}, 2\right)$ and $\left(-\dfrac{1}{2}, 6\right)$

6. (−5, 5) and (−5, 1)

8. (−1, −1) and (3, 2)

10. (−1, −4) and (4, 8) $= 13$

12. (5, 3) and (−5, −3)

14. (4.2, − 3.6) and (−2.6, 2.3)

16. $(3, -1)$ and $\left(\dfrac{1}{2}, 4\right)$

18. (4, 0) and $\left(-\dfrac{3}{5}, -4\right)$

Determine the midpoint of the line segment between the points.

19. (5, 2) and (−1, 4)

21. (−5, 3) and (5, −3)

23. (−2, −8) and (−6, −2)

25. (1, −6) and (−8, −4)

27. (−9.62, 12.58) and (3.52, 6.57)

29. $\left(\dfrac{5}{2}, 3\right)$ and $\left(2, \dfrac{9}{2}\right)$

20. (1, 4) and (2, 6)

22. (0, 8) and (4, −6)

24. (4, 7) and (1, −3)

26. (15.3, −6.2) and (8.2, −12.4)

28. $\left(3, \dfrac{1}{2}\right)$ and (2, −4)

30. $\left(-\dfrac{5}{2}, -\dfrac{11}{2}\right)$ and $\left(-\dfrac{7}{2}, \dfrac{3}{2}\right)$

Find the perimeter of the triangle determined by the indicated set of points. If a calculator with a square root key is available, give the answer rounded to the nearest hundredth. If a calculator with a square root key is not available leave your answer as a sum of square roots.

31. $A(7, 7), B(7, 1), C(-1, 1)$

32. $A(4, 3), B(-2, 3), C(5, 0)$

33. $A(0, -4), B(2, 3), C(4, 6)$

34. $A(-2, -1), B(6, -8), C(3, 5)$

35. When the distance between two different points is found using the distance formula, why must the distance always be a positive number?

Cumulative Review Exercises

[2.2] **36.** Evaluate $\dfrac{-b + \sqrt{b^2 - 4ac}}{2a}$ for $a = 2,\ b = 7,$ and $c = -15$.

[2.3] **37.** Hertz Automobile Rental Agency charges a daily fee of $30 plus 14 cents a mile. National Automobile Rental Agency charges a daily fee of $16 plus 24 cents a mile for the same car. What distance would you have to drive in 1 day to make the cost of renting from Hertz equal to the cost of renting from National?

[2.5] **38.** Solve the inequality $-4 \le \dfrac{4 - 3x}{2} < 5$. Write the solution in set builder notation.

[2.6] **39.** Find the solution set for the inequality $|3x + 2| > 5$.

3.2

Graphing Linear Equations

▶ **1** Write a linear equation in standard form.

▶ **2** Know what a graph represents.

▶ **3** Graph linear equations by plotting points.

▶ **4** Graph linear equations using intercepts.

▶ **5** Graph equations of the form $x = a$ and $y = a$.

▶ **6** Apply graphing to practical problems.

▶ **1** All the equations that we graph in this chapter will be straight lines. These equations are called linear equations. A **linear equation** is an equation whose graph will be a straight line. Linear equations are also called **first-degree equations** since the degree of their highest-powered term is the first degree.

A linear equation may be written in a number of different forms. One is standard form.

Standard Form of a Linear Equation

$$ax + by = c$$

where a, b, and c are real numbers, and a and b are not both 0.

Examples of Linear Equations in Standard Form

$$2x + 3y = 4$$
$$-x + 5y = -2$$

▶ **2** Consider the linear equation in two variables, $y = x + 1$. What is the solution? Since the equation contains two variables, its solutions must contain two numbers, one for each variable. One set of numbers that satisfies this equation is $x = 1$ and $y = 2$. To see that this is true, we substitute both values into the equation at the same time and see that the equation checks.

$$y = x + 1$$
$$2 = 1 + 1$$
$$2 = 2 \quad \text{true}$$

One solution to the equation $y = x + 1$ is the ordered pair $(1, 2)$. However, the equation $y = x + 1$ has many other solutions. If you check the ordered pairs $(2, 3)$, $(3, 4)$, $(-1, 0)$, $(\frac{1}{2}, \frac{3}{2})$, you will see that they are all solutions to the equation $y = x + 1$. How many possible solutions does the equation $y = x + 1$ have? The equation $y = x + 1$ has an unlimited or *infinite number* of possible solutions. Since it is not possible to list all the specific solutions to the equation, we illustrate them with a graph. **A graph of an equation is an illustration of the set of points that satisfy the equation.**

Graphing Equations by Plotting Points

▶ **3** All linear equations will be straight lines when graphed. Since only two points are needed to draw a straight line, when graphing linear equations we only need to find and plot two ordered pairs that satisfy the equation; but it is always a good idea to use a third ordered pair as a check. If the three points are not in a straight line, you have made a mistake. A set of points in a straight line is said to be **collinear.**

One method of finding ordered pairs that satisfy an equation is to solve the equation for y. Then substitute values for x and find the corresponding values of y.

EXAMPLE 1 Graph the equation $y = 3x + 6$.

Solution: This equation is already solved for y. We will find three ordered pairs that satisfy the equation by arbitrarily selecting three values for x, substituting them in the equation, and finding the corresponding values for y. In this equation we let x have values of 0, 2 and -3.

$$y = 3x + 6$$

x Value	y Value	x	y
$x = 0$	$y = 3(0) + 6 = 6$	0	6
$x = 2$	$y = 3(2) + 6 = 12$	2	12
$x = -3$	$y = 3(-3) + 6 = -3$	-3	-3

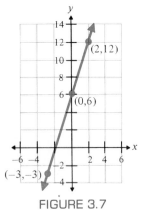

FIGURE 3.7

Now plot the three ordered pairs on the same set of axes (Fig. 3.7). Since the three points are collinear, everything appears correct. Connect the three points with a straight line. Place arrows at the ends of the line to show that the line continues infinitely in both directions. ■

To plot the equation $y = 3x + 6$, we used the three values $x = 0$, $x = 2$, and $x = -3$. We could have picked three entirely different values and obtained exactly the same graph. When selecting values to substitute for x, use whatever values make the equation easy to evaluate.

The graph in Example 1 represents the set of all ordered pairs that satisfies the equation $y = 3x + 6$. If we select any point on this line, the ordered pair representing that point will be a solution to the equation $y = 3x + 6$. Similarly, any solution to the equation will be represented by a point on the line.

EXAMPLE 2   Graph the equation $-2x + 3y = -6$.

Solution: We will first solve the equation for y. This will make it easier to select values to substitute for x that can be quickly evaluated.

$$-2x + 3y = -6$$
$$3y = 2x - 6$$
$$y = \frac{2x - 6}{3}$$
$$y = \frac{2}{3}x - \frac{6}{3}$$
$$y = \frac{2}{3}x - 2$$

Now we will select values for x that make $2x/3$ integral values. $2x/3$ will have integral values when x is a multiple of 3. We will therefore select $x = 0$, 3, and 6.

$$y = \frac{2}{3}x - 2$$

x Value	y Value	x	y
$x = 0$	$y = \frac{2}{3}(0) - 2 = -2$	0	-2
$x = 3$	$y = \frac{2}{3}(3) - 2 = 0$	3	0
$x = 6$	$y = \frac{2}{3}(6) - 2 = 2$	6	2

Now plot the points and draw the graph (Fig. 3.8).

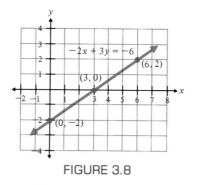

FIGURE 3.8

EXAMPLE 3   Graph the equation $\frac{5}{6}x - \frac{1}{2}y = \frac{5}{2}$.

Solution: We begin by multiplying both sides of the equation by the least common denominator, 6, to eliminate fractions. Then we proceed to solve the equation for y.

$$\frac{5}{6}x - \frac{1}{2}y = \frac{5}{2}$$

$$6\left(\frac{5}{6}x - \frac{1}{2}y\right) = 6 \cdot \frac{5}{2}$$

$$6\left(\frac{5}{6}x\right) - 6\left(\frac{1}{2}y\right) = 15$$

$$5x - 3y = 15$$

$$-3y = -5x + 15$$

$$y = \frac{-5x + 15}{-3}$$

$$y = \frac{5}{3}x - 5$$

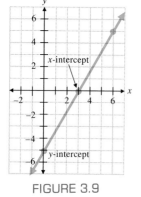

FIGURE 3.9

Now select values for x and solve for y in the equation $y = \frac{5}{3}x - 5$.

$$y = \frac{5}{3}x - 5$$

x Value	y Value	x	y
0	$y = \frac{5}{3}(0) - 5 = -5$	0	-5
3	$y = \frac{5}{3}(3) - 5 = 0$	3	0
6	$y = \frac{5}{3}(6) - 5 = 5$	6	5

Plot the points and draw the straight line (Fig. 3.9). ◼

Graphing Equations Using Intercepts

▶**4**   Let's examine two points on the graph shown in Fig. 3.9. Note that the graph crosses the x axis at the point $(3, 0)$. Therefore, 3 is called the **x intercept.** The graph crosses the y axis at the point $(0, -5)$. Therefore, -5 is called the **y intercept.**

HELPFUL HINT

Recall that the general form of an ordered pair is (x, y). When the y coordinate in the ordered pair is 0, then the x coordinate will be the x intercept. When the x coordinate of the ordered pair is 0, then the y coordinate will be the y intercept.

It is often convenient to graph linear equations by finding their x and y intercepts.

---

x and y Intercepts

To find the y intercept, set $x = 0$ and solve for y.
To find the x intercept, set $y = 0$ and solve for x.

EXAMPLE 4   Graph the equation $3y = 6x + 12$ by plotting the x and y intercepts.

Solution: To find the y intercept (that is, where the graph crosses the y axis), set $x = 0$ and solve for y.

$$3y = 6x + 12$$
$$3y = 6(0) + 12$$
$$3y = 0 + 12$$
$$3y = 12$$
$$y = \frac{12}{3} = 4$$

The graph crosses the y axis at $y = 4$. The ordered pair representing the y intercept is $(0, 4)$.

To find the x intercept (where the graph crosses the x axis), set $y = 0$ and solve for x.

$$3y = 6x + 12$$
$$3(0) = 6x + 12$$
$$0 = 6x + 12$$
$$-12 = 6x$$
$$-\frac{12}{6} = x$$
$$-2 = x$$

The graph crosses the x axis at $x = -2$. The ordered pair representing the x intercept is $(-2, 0)$. Now plot the intercepts and draw the graph (Fig. 3.10). ■

FIGURE 3.10

When graphing equations using just intercepts, you must be particularly careful. Since you are plotting only two points, you have no checkpoint. If one of your intercepts is wrong, your graph will be wrong. When graphing by plotting intercepts, you may wish to plot a third point as a checkpoint.

Now return to Example 3 and graph that equation by finding and plotting the x and y intercepts.

▶ **5**   Examples 5 and 6 illustrate how equations of the form $x = a$ and $y = a$, where a is a constant, are graphed.

EXAMPLE 5   Graph the equation $y = 3$.

Solution: This equation can be written as $y = 3 + 0x$. Thus, for any value of x selected, y will be 3. The graph of $y = 3$ is illustrated in Fig. 3.11.

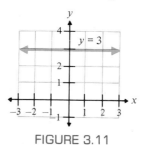

FIGURE 3.11

The graph of any equation of the form $y = a$ will always be a horizontal line for any real number a.

EXAMPLE 6   Graph the equation $x = -2$.

Solution: This equation can be written as $x = -2 + 0y$. Thus, for every value of y selected, x will have a value of -2 (see Fig. 3.12).

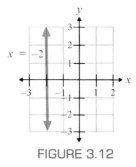

FIGURE 3.12 ∎

The graph of any equation of the form $x = a$ will always be a vertical line for any real number a.

Applications of Graphing

▶ **6**   Graphs are often used to show the relationship between variables. The axes of a graph do not have to be labeled x and y; they can be any designated variables. Consider the following example.

EXAMPLE 7   The yearly profit, p, of a tire store can be estimated by the formula $p = 20x - 30,000$ where x is the number of tires sold per year.

(a) Draw a graph of profits versus tires sold for up to 6000 tires.
(b) Estimate the number of tires that must be sold for the company to break even.
(c) Estimate the number of tires sold if the company has a $40,000 profit.

Solution: (a) The minimum number of tires that can be sold is 0. Therefore, negative values do not have to be indicated on the horizontal axis. We will arbitrarily select 3 values for x and find the corresponding values of p. The graph is illustrated in Fig. 3.13.

x	p
0	$-30,000$
2000	$10,000$
5000	$70,000$

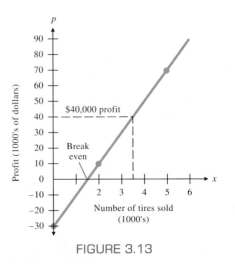

FIGURE 3.13

(b) To break even, approximately 1500 tires must be sold.

(c) To make a $40,000 profit, approximately 3500 tires must be sold.

Sometimes it is difficult to read an exact answer from a graph. To determine the exact number of tires needed to break even, substitute 0 for p in the equation $p = 20x - 30,000$ and solve for x. To determine the exact number of tires needed to obtain a $40,000 profit, substitute 40,000 for p and solve the equation for x. ■

EXAMPLE 8   Mr. Jordan is a part owner in a newly formed toy company. His monthly salary consists of $200 plus 10% of the company's net revenue for that month.

(a) Write an equation expressing his monthly salary, s, in terms of the company's net revenue, r.

(b) Draw a graph of his monthly salary for net revenue up to $20,000.

(c) If the company's net revenue for the month of April is $15,000, what will Mr. Jordan's monthly salary be?

Solution: (a) His salary consists of $200 plus 10% of the net revenues, r. Ten percent of r is $0.10r$. Thus the equation is

$$s = 200 + 0.10r$$

(b) Select values for r, find the corresponding values of s, and then draw the graph. We can select values for r that are between 0 and $20,000 (see Fig. 3.14).

r	s
0	200
10,000	1200
20,000	2200

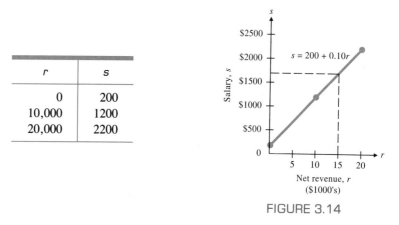

FIGURE 3.14

(c) When the net revenue is $15,000, Mr. Jordan's salary is about $1700. ■

Exercise Set 3.2

Graph each equation by solving the equation for y, selecting three arbitrary values for x, and finding the corresponding values of y.

1. $y = 4$ **2.** $x = 6$ **3.** $x = -2$ **4.** $y = 5$

5. $y = 4x - 2$

6. $y = -x + 3$

7. $y = x + 2$

8. $y = x - 4$

9. $y = -\dfrac{1}{2}x + 5$

10. $2y = 2x + 4$

11. $6x - 2y = 4$

12. $4x - y = 5$

13. $5x - 2y = 8$

14. $-2x + 4y = 8$

15. $6x + 5y = 30$

16. $-2x - 3y = 6$

17. $-6x - y = -7$

18. $8y - 16x = 24$

19. $y = 20x + 40$

20. $2y - 50 = 100x$

21. $-2x + 5y = 15$

22. $5x - 2y = -7$

23. $-4x - 3y = -12$

24. $4x = 5y + 2$

25. $y = \dfrac{2}{3}x$

26. $y = -\dfrac{3}{5}x$

27. $y = \dfrac{1}{2}x + 4$

28. $y = -\dfrac{2}{5}x + 2$

29. $x - \dfrac{2}{3}y = -2$

30. $\dfrac{4}{3}x - 2y = \dfrac{10}{3}$

31. $\dfrac{2}{3}x + \dfrac{1}{2}y = \dfrac{3}{2}$

32. $3x + \dfrac{3}{2}y = \dfrac{5}{2}$

Graph each equation using x and y intercepts.

33. $y = 8x + 4$ **34.** $y = -2x + 6$ **35.** $y = 2x + 3$ **36.** $y = -3x + 8$

37. $y = -6x + 5$ **38.** $y = 4x - 8$ **39.** $4y + 3x = 12$ **40.** $-2x + 3y = 10$

41. $4x = 3y - 9$ **42.** $7x + 14y = 21$ **43.** $\frac{1}{2}x + 2y = 4$ **44.** $30x + 25y = 50$

45. $6x - 12y = 24$ **46.** $25x + 50y = 100$ **47.** $-16y = 4x + 96$ **48.** $\frac{1}{3}x - 2y = 6$

49. $30y + x = 45$ **50.** $120x - 360y = 720$ **51.** $40x + 6y = 40$ **52.** $20x - 240 = -60y$

53. $\frac{1}{3}x + \frac{1}{4}y = 12$ **54.** $-\frac{1}{2}y - \frac{1}{3}x = -1$ **55.** $\frac{1}{2}y = \frac{3}{8}x - \frac{3}{4}$ **56.** $\frac{1}{6}x + \frac{1}{2}y = -1$

57. Using the formula distance = rate · time, $d = rt$, draw a graph of distance versus time for a constant rate of 50 miles per hour.

58. Using the simple interest formula interest = principal · rate · time, $i = prt$, draw a graph of interest versus time for a principal of $1000 and a rate of 8%.

59. The profit of a company that produces bicycles can be approximated by the formula $P = 60x - 80,000$, where x is the number of bicycles produced and sold.

(a) Draw a graph of profit versus the number of bicycles sold (for up to 5000 bicycles).

(b) Estimate the number of bicycles that must be sold for the company to break even.

(c) Estimate the number of bicycles that must be sold for the company to make $150,000 profit.

60. The auto rental fee from an auto rental agency is $40 a day plus 12 cents a mile.

(a) Write an equation expressing rental fee, F, in terms of miles, m.

(b) Draw a graph illustrating the rental fee versus the mileage for up to 200 miles.

(c) Estimate the rental fee for 1 day if Mary drives 60 miles.

(d) Estimate the number of miles Mary has driven if the rental fee is $60.

61. The weekly cost of operating a taxi is $50 plus 12 cents per mile.

(a) Write an equation expressing weekly cost, c, in terms of miles, m.

(b) Draw a graph illustrating weekly cost versus the number of miles, up to 200, driven per week.

(c) How many miles would Jack have to drive for the weekly cost to be $70?

(d) If the weekly cost is $60, how many miles did Jack drive?

62. Ellen Branston's weekly salary is $200 plus 15% commission on her weekly sales.

(a) Write an equation expressing Ellen's weekly salary, s, in terms of her weekly sales, x.

(b) Draw a graph of Ellen's weekly salary versus her weekly sales, for up to $5000 in sales.

(c) What is Ellen's weekly salary if her sales were $4000?

(d) If her salary for the week is $400, what are her weekly sales?

63. Ms. Tocci, a real estate agent, makes $150 per week plus a 1% sales commission on each property she sells.

See Exercise 63

(a) Write an equation expressing her weekly salary, s, in terms of sales, x.

(b) Draw a graph of her salary versus her weekly sales, for sales up to $100,000.

(c) If she sells one house per week for $80,000, what will be her weekly salary?

64. What does a graph of an equation represent?

65. Explain the procedure to follow to graph linear equations by plotting points.

66. Explain the procedure to follow to graph linear equations using the intercepts.

Cumulative Review Exercises

[2.1] **67.** Solve the equation $3x - 2 = \frac{1}{3}(3x - 3)$.

[2.2] **68.** Solve the folowung formula for p_2.

$$E = a_1 p_1 + a_2 p_2 + a_3 p_3$$

[2.5] **69.** Solve the inequality $\frac{3}{5}(x - 3) > \frac{1}{4}(3 - x)$ and indicate the solution (a) on the number line, (b) in interval notation, and (c) in set builder notation.

[2.6] **70.** Solve the equation $\left| \dfrac{x - 4}{3} \right| + 2 = 4$.

JUST FOR FUN

1. Graph $y = |x|$.

2. Graph $y = |x + 1|$.

3. Graph $y = |x - 2|$.

4. Graph $y = |x| - 2$.

5. Graph $y = \begin{cases} x + 3, x > 4 \\ 3x - 5, x \le 4 \end{cases}$

6. Graph $y = \begin{cases} 2x - 3, x \ge 2 \\ -3x + 7, x < 2 \end{cases}$

3.3

Slope of a Line

▶ **1**   Find the slope of a line.

▶ **2**   Determine when two lines are parallel or perpendicular.

▶ **1**   The **slope of a line** is the ratio of the vertical change to the horizontal change between any selected points on the line. As an example, consider the two points (3, 6) and (1, 2) on the line in Fig. 3.15(a). If we draw a line parallel to the x axis through the point (1, 2) and a line parallel to the y axis through the point (3, 6), the two lines intersect at (3, 2) (see Fig. 3.15b).

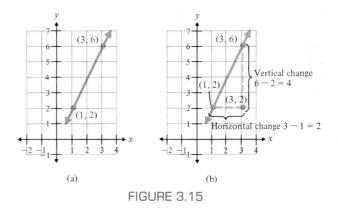

(a)   (b)

FIGURE 3.15

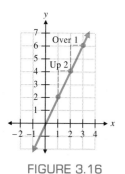

FIGURE 3.16

From Fig. 3.15b we can determine the slope of the line. The vertical change (along the y axis) is $6 - 2$, or 4 units. The horizontal change (along the x axis) is $3 - 1$, or 2 units.

$$\text{Slope} = \frac{\text{vertical change}}{\text{horizontal change}} = \frac{4}{2} = 2$$

Thus the slope of the line through the points (3, 6) and (1, 2) is 2. By examining the line connecting these two points, we can see that for each two units the graph moves up the y axis it moves 1 unit to the right on the x axis (see Fig. 3.16).

Let's now determine the procedure to find the slope of a line passing through the two points (x_1, y_1) and (x_2, y_2). Consider Fig. 3.17. The vertical change can be found by subtracting y_1 from y_2. The horizontal change can be found by subtracting x_1 from x_2.

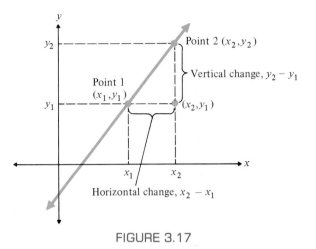

FIGURE 3.17

Slope

If $x_1 \neq x_2$, the slope of the line through the distinct points (x_1, y_1) and (x_2, y_2) is

$$\text{Slope} = \frac{\text{change in } y \text{ (vertical change)}}{\text{change in } x \text{ (horizontal change)}} = \frac{y_2 - y_1}{x_2 - x_1}$$

It makes no difference which two points on the line are selected when finding the slope of a line. It also makes no difference which point you label (x_1, y_1) or (x_2, y_2). The letter m is used to represent the slope of a line. The Greek capital letter delta, Δ, is used to represent the words "the change in." Thus the slope is sometimes indicated as

$$m = \frac{\Delta y}{\Delta x} = \frac{y_2 - y_1}{x_2 - x_1}$$

EXAMPLE 1　Find the slope of the line in Fig. 3.18.

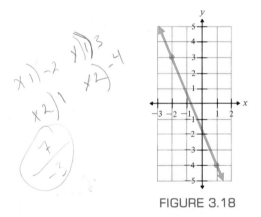

FIGURE 3.18

Solution:　Two points on the line are $(-2, 3)$ and $(1, -4)$. Let $(x_2, y_2) = (-2, 3)$ and $(x_1, y_1) = (1, -4)$. Then

$$m = \frac{y_2 - y_1}{x_2 - x_1} = \frac{3 - (-4)}{-2 - 1}$$

$$= \frac{3 + 4}{-3}$$

$$= -\frac{7}{3}$$

The slope of the line is $-\frac{7}{3}$. Note that if we had let $(x_1, y_1) = (-2, 3)$ and $(x_2, y_2) = (1, -4)$, the slope would remain the same. Try and see.   ■

A line that rises going from left to right (Fig. 3.19a) has a **positive slope.** A line that neither rises nor falls going from left to right (Fig. 3.19b) has **zero slope.** And a line that falls going from left to right (Fig. 3.19c) has a **negative slope.**

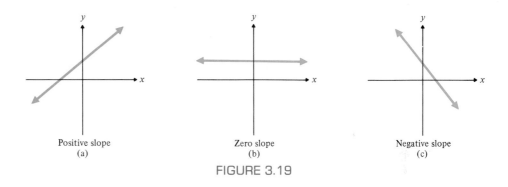

Positive slope
(a)

Zero slope
(b)

Negative slope
(c)

FIGURE 3.19

Consider the graph of $x = 3$ (Fig. 3.20). What is its slope? The graph is a vertical line and goes through the points $(3, 2)$ and $(3, 5)$. Let the point $(3, 5)$ represent (x_2, y_2) and let $(3, 2)$ represent (x_1, y_1). Then the slope of the line is

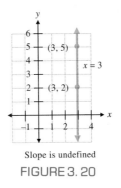

Slope is undefined

FIGURE 3. 20

$$m = \frac{y_2 - y_1}{x_2 - x_1} = \frac{5 - 2}{3 - 3} = \frac{3}{0}$$

Since it is meaningless to divide by 0, we say that the slope of this line is undefined. **The slope of any vertical line is undefined.**

COMMON STUDENT ERROR

When students are asked to give the slope of a horizontal or a vertical line, they often answer incorrectly. When asked for the slope of a *horizontal line*, your response should be "*the slope is 0.*" If you give your answer as "no slope," your teacher may well mark it wrong for these words may have various interpretations. When asked for the slope of a *vertical line*, your answer should be "*the slope is undefined.*" Again, if you use the words "no slope," this may be interpreted differently by your teacher and marked wrong.

Parallel and Perpendicular Lines

▶**2** Two lines in the same plane are **parallel** when they do not intersect no matter how far they are extended. Figure 3.21 illustrates two parallel lines, l_1 and l_2. For two lines not to intersect, they must rise or fall at the same rate. That is, their slopes must be the same. **Two distinct lines are parallel if their slopes are the same, and two distinct lines with the same slope are parallel lines.** If line l_1 has slope m_1 and line l_2 has slope m_2, and if $m_1 = m_2$, then lines l_1 and l_2 must be parallel lines.

Parallel lines

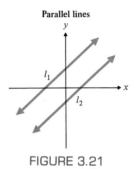

FIGURE 3.21

EXAMPLE 2   Two points on l_1 are $(1, 6)$ and $(-1, 2)$. Two points on l_2 are $(2, 3)$ and $(-1, -3)$. Determine if l_1 and l_2 are parallel lines.

Solution: First determine the slope of l_1.

$$m_1 = \frac{6 - 2}{1 - (-1)} = \frac{4}{2} = 2$$

Now determine the slope of l_2.

$$m_2 = \frac{3 - (-3)}{2 - (-1)} = \frac{6}{3} = 2$$

Since l_1 and l_2 have the same slope, 2, the two lines are parallel.   ■

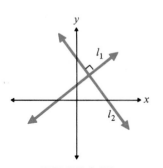

FIGURE 3.22

Two lines that cross at right angles (90° angles) are said to be **perpendicular lines.** Two perpendicular lines are illustrated in Fig. 3.22. The square where the two lines meet is used to indicate that the two lines meet at a right angle.

Two lines will be perpendicular to each other when their slopes are negative reciprocals. For any slope a, the negative reciprocal is $\dfrac{-1}{a}$. For example, for a slope of 2, its negative reciprocal is $\dfrac{-1}{2}$ or $-\dfrac{1}{2}$. For a slope of $-\dfrac{1}{3}$, its negative reciprocal is $\dfrac{-1}{-1/3} = (-1)(-3) = 3$.

Slope	Negative Reciprocal of Slope	Product
2	$-\dfrac{1}{2}$	$2\left(-\dfrac{1}{2}\right) = -1$
$-\dfrac{1}{3}$	3	$-\dfrac{1}{3}(3) = -1$
$-\dfrac{2}{5}$	$\dfrac{5}{2}$	$-\dfrac{2}{5}\left(\dfrac{5}{2}\right) = -1$

Notice that the product of a number and its negative reciprocal equals -1. If l_1 has slope m_1 and l_2 has slope m_2, and if $m_1 m_2 = -1$, then l_1 and l_2 must be perpendicular lines.

EXAMPLE 3 Two points on l_1 are $(6, 3)$ and $(2, -3)$. Two points on l_2 are $(0, 2)$ and $(6, -2)$. Determine if l_1 and l_2 are perpendicular lines.

Solution: First determine the slope of l_1.

$$m_1 = \frac{3 - (-3)}{6 - 2} = \frac{6}{4} = \frac{3}{2}$$

Now determine the slope of l_2.

$$m_2 = \frac{2 - (-2)}{0 - 6} = \frac{4}{-6} = \frac{-2}{3}$$

Finally, determine if $m_1 m_2 = -1$. If so, the lines are perpendicular.

$$m_1 m_2 = \frac{3}{2}\left(-\frac{2}{3}\right) = -1$$

Since the product of the slopes equals -1, the lines are perpendicular. Note that each slope is the negative reciprocal of the other. ■

Note: Any horizontal line is perpendicular to any vertical line, although the negative reciprocal test cannot be applied.

Exercise Set 3.3

Find the slope of the line through the given points. If the slope of the line is undefined, so state.

1. $(1, 5)$ and $(2, -3)$ 　　　　　　　　　　　　　　**2.** $(3, 1)$ and $(5, 4)$
3. $(5, 2)$ and $(1, 4)$ 　　　　　　　　　　　　　　**4.** $(5, 1)$ and $(2, 4)$
5. $(-1, 4)$ and $(0, 3)$ 　　　　　　　　　　　　　**6.** $(2, 3)$ and $(-2, 3)$
7. $(4, 2)$ and $(4, -1)$ 　　　　　　　　　　　　　**8.** $(6, -2)$ and $(-1, -2)$
9. $(-3, 4)$ and $(-1, -6)$ 　　　　　　　　　　　**10.** $(4, -3)$ and $(3, -4)$

11. $(2, 5)$ and $(-1, 5)$

12. $(-2, 3)$ and $(7, -3)$

13. $(2, -4)$ and $(-5, -3)$

14. $(-4, 0)$ and $(0, -6)$

15. $(2, 0)$ and $(-4, -2)$

16. $(-6, 2)$ and $(4, -3)$

Find the slope of the line in each of the given figures. If the slope of the line is undefined, so state.

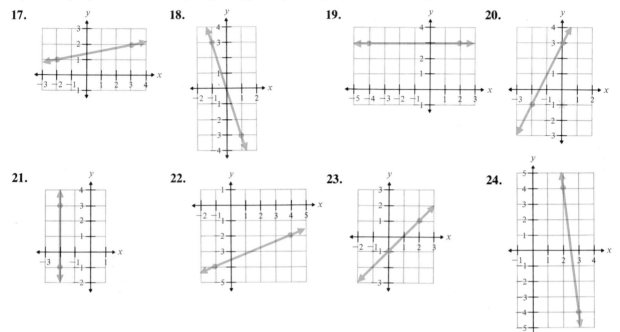

17.

18.

19.

20.

21.

22.

23.

24.

Two points on l_1 and two points on l_2 are given. Determine if l_1 is parallel to l_2, l_1 is perpendicular to l_2, or neither.

25. l_1: $(0, 4)$ and $(2, 8)$; l_2: $(0, -1)$ and $(3, 5)$

26. l_1: $(-1, 0)$ and $(2, 3)$; l_2: $(3, 2)$ and $(4, -1)$

27. l_1: $(3, 2)$ and $(-1, -2)$; l_2: $(2, 0)$ and $(3, -1)$

28. l_1: $(2, 3)$ and $(3, 2)$; l_2: $(1, 4)$ and $(4, 1)$

29. l_1: $(3, 4)$ and $(-2, 3)$; l_2: $(0, -3)$ and $(2, -1)$

30. l_1: $(6, 2)$ and $(3, 1)$; l_2: $(-6, -2)$ and $(-3, -1)$

31. l_1: $(0, 2)$ and $(6, -2)$; l_2: $(4, 0)$ and $(6, 3)$

32. l_1: $(-1, 3)$ and $(4, 2)$; l_2: $(1, -3)$ and $(4, 2)$

33. l_1: $(1, 5)$ and $(-2, -1)$; l_2: $(1, -2)$ and $(3, 2)$

34. l_1: $(1, -6)$ and $(0, -4)$; l_2: $(0, -6)$ and $(2, -2)$

Solve for the given variable if the line through the two given points is to have the given slope.

35. $(6, a)$ and $(3, 4)$, $m = 1$

36. $(1, 0)$ and $(4, x)$, $m = 3$

37. $(5, b)$ and $(2, -4)$, $m = 2$

38. $(6, 1)$ and $(4, d)$, $m = 3$

39. $(2, -3)$ and $(3, c)$, $m = -1$

40. $(y, -1)$ and $(3, 2)$, $m = -3$

41. $(x, 2)$ and $(3, -4)$, $m = 2$

42. $(-2, -3)$ and $(x, 4)$, $m = \dfrac{1}{2}$

43. $(3, 5)$ and $(x, 3)$, $m = \dfrac{2}{3}$

44. $(-4, -1)$ and $(y, 2)$, $m = \dfrac{-3}{5}$

45. Explain how to find the slope of a given line.

46. Explain what it means when the slope of a line is positive.

47. Explain what it means when the slope of a line is negative.

48. What is the slope of a horizontal line? Explain why this is so.

49. Explain why the slope of a vertical line is undefined.

50. When finding the slope of a line, how does the slope change if we interchange (x_1, y_1) and (x_2, y_2)? Explain your answer.

Cumulative Review Exercises

[2.3] **51.** The sum of three consecutive odd integers is 27. Find the three integers.

(a) Explain the procedure to solve an inequality of each form given on the right (assume b > 0).

(b) Solve each inequality for x.

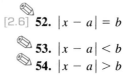

[2.6] **52.** $|x - a| = b$

53. $|x - a| < b$

54. $|x - a| > b$

JUST FOR FUN

1. The chapter opening photo on page 107 is the Castle at Chichén Itza, Mexico. Each side of the castle has a stairway consisting of 91 steps. The steps of the castle are quite narrow and steep, which makes them hard to climb. The average height of the steps is 14.2 inches, and the average width is 6.4 inches.

(a) Find the total vertical distance in inches of the 91 steps.

(b) Find the total horizontal distance in inches of the 91 steps.

(c) If a straight line were to be drawn connecting the tips of the steps, what would be the absolute value of the slope of this line?

2. A **tangent line** is a straight line that touches a curve at a single point (the tangent line may cross the curve at a different point if extended). The figure at the right, top, shows three tangent lines to the curve at points a, b, and c.

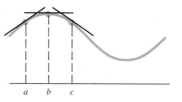

Note that the tangent line at point a has a positive slope, the tangent line at point b has a slope of 0, and the tangent line at point c has a negative slope. Now consider the curve below.

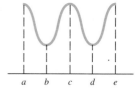

Assume that tangent lines are drawn at all points on the curve except at end points a and e. Where on the curve would the tangent lines have a positive slope, a slope of 0, a negative slope?

3.4

Slope—Intercept and Point—Slope Forms of a Linear Equation

▶ **1** Write linear equations in slope—intercept form.

▶ **2** Graph equations using the slope and y intercept.

▶ **3** Use slope to recognize parallel and perpendicular lines.

▶ **4** Write linear equations in point—slope form.

▶ **1** A linear equation written in the form $y = mx + b$ is said to be in **slope—intercept form.**

Slope—Intercept Form of a Linear Equation

$$y = mx + b$$

where **m is the slope** of the line and **b is the y intercept** of the line.

Examples of Equations in Slope—Intercept Form

$$y = 3x - 6, \qquad y = \frac{1}{2}x + \frac{3}{2}$$

This form is called the slope–intercept form because the *m* represents the slope of the graph and the *b* represents the *y* intercept.

slope ⌝   ⌐ *y* intercept

$$y = mx + b$$

Equation	Slope	y Intercept
$y = 3x - 6$	3	-6
$y = \dfrac{1}{2}x + \dfrac{3}{2}$	$\dfrac{1}{2}$	$\dfrac{3}{2}$

> To write an equation in slope–intercept form, solve the equation for *y*.

EXAMPLE 1 Write the equation $-3x + 4y = 8$ in slope–intercept form. State the slope and *y* intercept.

Solution: Solve for *y*.

$$-3x + 4y = 8$$
$$4y = 3x + 8$$
$$y = \frac{3x + 8}{4}$$
$$y = \frac{3}{4}x + \frac{8}{4}$$
$$y = \frac{3}{4}x + 2$$

The slope is $\frac{3}{4}$; the *y* intercept is 2.   ■

EXAMPLE 2 Write the equation of the line illustrated in Fig. 3.23.

Solution: If we can determine the line's slope and its *y* intercept, we can write the equation in slope–intercept form. By looking at the figure, we can determine that the *y* intercept is -5.

Notice that *y* changes 3 units for each unit change of *x*. Also note that the slope is negative since the line falls as it moves to the right. Therefore, the slope of the line is -3. We could also find the slope by taking two points on the line and finding $\Delta y / \Delta x$ for the two points selected.

Since the slope is -3 and the *y* intercept is -5, the equation of the line is $y = -3x - 5$.   ■

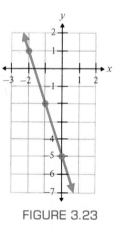

FIGURE 3.23

▶ **2** One reason for studying the slope–intercept form of a line is that the information obtained from an equation in this form can be useful in drawing the graph of a linear equation. Once we know the *y* intercept and the slope of a line we can (1) mark the *y* intercept on the set of axes, and (2) use the slope to get a second (and third) point on the graph. This procedure is illustrated in Example 3.

EXAMPLE 3   Graph $2y + 4x = 6$ using the y intercept and slope.

Solution: Begin by solving for y to get the equation in slope–intercept form.

$$2y + 4x = 6$$
$$2y = -4x + 6$$
$$y = -2x + 3$$

With the equation in this form, we see that the slope is -2 and the y intercept is 3. Now mark the y intercept, 3, on the set of axes (Fig. 3.24). Then use the slope to obtain a second point. The slope is negative; therefore, the graph must fall as it goes from left to right. Since the slope is -2, the vertical change to the horizontal change must be in the ratio of 2 to 1 (remember 2 means $\frac{2}{1}$). Thus, if we start at $y = 3$ and move down 2 units and to the right 1 unit, we will obtain a second point on the graph.

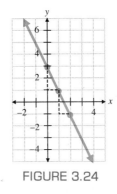

FIGURE 3.24

Continue this process of moving 2 units down and 1 unit to the right to get a third point. Now draw a line through the three points to get the graph. Note that we selected to move down and to the right to get the second and third points. We could have also selected to move up and to the left to get the second and third points. ■

If you were asked to graph $y = \frac{4}{3}x - 3$ using the y intercept and slope, how would you do it? To begin, you should mark your first point at -3 on the y axis. Then you could obtain your second point by moving up 4 units and to the right 3 units.

▶ **3**   In Section 3.3 we introduced parallel and perpendicular lines. Now we will explore those concepts further.

EXAMPLE 4   (a) Determine if the following lines are parallel.

$$2x - y = -4$$
$$2y = 4x - 2$$

(b) Graph both equations on the same set of axes.

Solution: (a) Recall from Section 3.3 that two distinct lines are parallel when they have the same slope. To compare the slopes of the two lines, write each in slope–intercept form by solving each equation for y.

$$2x - y = -4 \qquad\qquad 2y = 4x - 2$$
$$-y = -2x - 4 \qquad y = \frac{4x - 2}{2}$$
$$y = 2x + 4 \qquad\qquad y = 2x - 1$$

Since both lines have the same slope, 2, they are parallel.

(b) Both lines are graphed in Fig. 3.25.

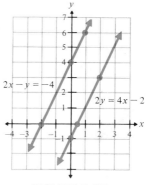

FIGURE 3.25

EXAMPLE 5   (a) Determine if the equations $2x + 3y = 6$ and $3x - 2y = 12$ will be perpendicular lines when graphed.

(b) Graph both equations on the same set of axes.

Solution: (a) To determine if the equations will be perpendicular lines when graphed, we must find and compare their individual slopes. Recall that two lines are perpendicular when their slopes are negative reciprocals of each other. If the product of their slopes equals -1, the lines will be perpendicular. To find the slopes, we will write the equations in slope–intercept form by solving the equations for y.

$$2x + 3y = 6 \qquad\qquad 3x - 2y = 12$$
$$3y = -2x + 6 \qquad\qquad -2y = -3x + 12$$
$$y = \frac{-2x + 6}{3} \qquad\qquad y = \frac{-3x + 12}{-2}$$
$$y = \frac{-2}{3}x + 2 \qquad\qquad y = \frac{3}{2}x - 6$$

Since $\frac{3}{2}$ is the negative reciprocal of $-\frac{2}{3}$, $\frac{3}{2}(-\frac{2}{3}) = -1$, the lines will be perpendicular when graphed.

(b) Both lines are graphed in Fig. 3.26.

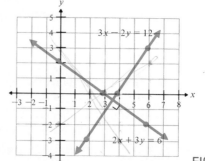

FIGURE 3.26

EXAMPLE 6   Determine the equation of the line that is parallel to the line $2x + 4y = 8$ and has a y intercept of 5.

Solution: If we know the slope of a line and its y intercept, we can use the slope–intercept form, $y = mx + b$, to write the equation. The slope of the given equation can be found by solving for y.

$$2x + 4y = 8$$
$$4y = -2x + 8$$
$$y = \frac{-2x + 8}{4}$$
$$y = -\frac{1}{2}x + 2$$

Two lines are parallel when they have the same slope. Therefore, the slope of the line parallel to the given line must be $-\frac{1}{2}$. Since its slope is $-\frac{1}{2}$ and its y intercept is 5, its equation must be

$$y = -\frac{1}{2}x + 5$$

EXAMPLE 7   Determine the equation of the line that is perpendicular to the line $2x + 4y = 8$ and has a y intercept of 5.

Solution: Two lines are perpendicular when their slopes are negative reciprocals of each other. From Example 6 we know that the slope of the given line is $-\frac{1}{2}$. Therefore, the slope of a line perpendicular to the given line must be $(-1) \div (-\frac{1}{2})$ or 2. The line perpendicular to the given line has a y intercept of 5. Thus the equation we are seeking is

$$y = 2x + 5$$

Point–Slope Form of a Linear Equation

▶ **4**   When the slope of a line and a point on the line are known, we can use the **point–slope form** to determine the equation of the line. The point–slope form can be developed by beginning with the slope between any two points (x, y) and (x_1, y_1) on a line.

$$m = \frac{y - y_1}{x - x_1} \quad \text{or} \quad \frac{m}{1} = \frac{y - y_1}{x - x_1}$$

now cross multiply to obtain

$$m(x - x_1) = y - y_1 \quad \text{or} \quad y - y_1 = m(x - x_1)$$

Point–Slope Form of a Linear Equation

$$y - y_1 = m(x - x_1)$$

where **m is the slope** of the line and **(x_1, y_1) is a point on the line.**

EXAMPLE 8   Write in slope–intercept form the equation of the line that goes through the point $(2, 3)$ and has a slope of 4.

Solution: Since we are given the slope of the line and a point on the line, we can write the equation in point–slope form. We can then solve the equation for y to write the equation in slope–intercept form. The slope, m, is 4. The point on the line is (2, 3); call this (x_1, y_1). Substitute 4 for m, 2 for x_1, and 3 for y_1 in the point–slope form of a line.

$$y - y_1 = m(x - x_1)$$
$$y - 3 = 4(x - 2) \qquad \text{Point–slope form}$$
$$y - 3 = 4x - 8$$
$$y = 4x - 5 \qquad \text{Slope–intercept form}$$

The graph of $y = 4x - 5$ has a slope of 4 and passes through the point (2, 3).   ∎

EXAMPLE 9   Determine, in slope–intercept form, the equation of the line through the points $(-1, -3)$ and $(4, 2)$.

Solution: When we are given two points on a line, we can find the slope of the line. We can then use the slope and one of the given points to find the equation of the line in point–slope form. Any equation in point–slope form can be changed to slope–intercept form by solving the equation for y. We must first find the slope between the two points. To determine the slope, let's designate $(-1, -3)$ as (x_1, y_1) and $(4, 2)$ as (x_2, y_2).

$$m = \frac{y_2 - y_1}{x_2 - x_1} = \frac{2 - (-3)}{4 - (-1)} = \frac{2 + 3}{4 + 1} = \frac{5}{5} = 1$$

We can now use the point–slope form with either point (one at a time) to determine the equation of the line. This example will be worked out using both points to show that the solutions obtained are identical.

Use point $(-1, -3)$ as (x_1, y_1):

$$y - y_1 = m(x - x_1)$$
$$y - (-3) = 1[x - (-1)]$$
$$y + 3 = x + 1$$
$$y = x - 2 \qquad \text{Slope–intercept form}$$

Use point $(4, 2)$ as (x_1, y_1):

$$y - y_1 = m(x - x_1)$$
$$y - 2 = 1(x - 4)$$
$$y - 2 = x - 4$$
$$y = x - 2 \qquad \text{Slope–intercept form}$$

The solutions are identical.   ∎

EXAMPLE 10   Determine, in standard form, the equation of the line that is parallel to $5x - 3y = 12$ and passes through the point $(-4, 6)$.

Solution: First find the slope of the given line.

$$5x - 3y = 12$$
$$-3y = -5x + 12$$
$$y = \frac{-5x + 12}{-3}$$
$$y = \frac{5}{3}x - 4$$

Since the slope of the given line is $\frac{5}{3}$, the slope of any line parallel to it must also be $\frac{5}{3}$. We now know the slope of the line, $\frac{5}{3}$, and a point on the line, $(-4, 6)$. We can therefore use the point–slope form to determine the equation.

$$y - y_1 = m(x - x_1)$$
$$y - 6 = \frac{5}{3}[x - (-4)]$$

We write the equation in standard form.

$$y - 6 = \frac{5}{3}(x + 4)$$
$$3(y - 6) = 5(x + 4)$$
$$3y - 18 = 5x + 20$$
$$-5x + 3y - 18 = 20$$
$$-5x + 3y = 38 \qquad \text{Standard form}$$

Note that $5x - 3y = -38$ is also an acceptable answer.   ■

EXAMPLE 11   Determine in slope–intercept form the equation of the line that is perpendicular to $5y = -10x + 7$ and passes through the point $(4, \frac{1}{3})$.

Solution: First determine the slope of the given line by solving the equation for y.

$$5y = -10x + 7$$
$$y = \frac{-10x + 7}{5}$$
$$y = -2x + \frac{7}{5}$$

Since the slope of the given line is -2, the slope of the line perpendicular to it must be $\frac{1}{2}$. The line we are seeking passes through the point $(4, \frac{1}{3})$. Using the point–slope form, we obtain

$$y - y_1 = m(x - x_1)$$
$$y - \frac{1}{3} = \frac{1}{2}(x - 4)$$

Multiply both sides of the equation by the least common denominator, 6, to eliminate fractions.

$$6\left(y - \frac{1}{3}\right) = 6\left[\frac{1}{2}(x - 4)\right]$$
$$6y - 2 = 3(x - 4)$$
$$6y - 2 = 3x - 12$$

Now write the equation in slope–intercept form.

$$6y = 3x - 10$$

$$y = \frac{3x - 10}{6}$$

$$y = \frac{3x}{6} - \frac{10}{6}$$

$$y = \frac{1}{2}x - \frac{5}{3} \qquad \text{Slope-intercept form}$$

HELPFUL HINT

We have discussed three forms of a linear equation:

Standard form:	$ax + by = c$
Slope–intercept form:	$y = mx + b$
Point–slope form:	$y - y_1 = m(x - x_1)$

Consider the equation $2y = 3x + 4$. We can write this equation in all three forms as follows:

Standard form: $\qquad -3x + 2y = 4$ (or $3x - 2y = -4$)

Slope–intercept form: $\quad y = \dfrac{3}{2}x + 2$
(solve for y)

Point–slope form: $\qquad y - 2 = \dfrac{3}{2}x$ or $y - 2 = \dfrac{3}{2}(x - 0)$

Note that to go from slope–intercept form to point–slope form we subtracted 2 from both sides of the equation. Also note that the point represented in point–slope form is $(0, 2)$, or the y intercept.

Exercise Set 3.4

Write an equation of the given line.

1.

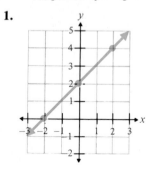

2.

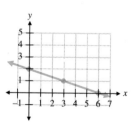

3.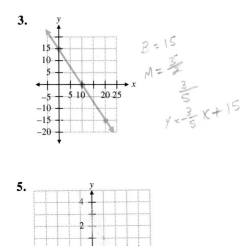

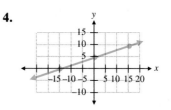
$B = 15$
$M = \frac{3}{2}$
$\frac{3}{5}$
$Y = -\frac{3}{5} X + 15$

4.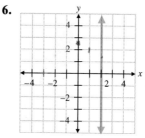

5.

6.

Draw the graph of each line using the information given.

7. y intercept of 3, slope of -2

8. y intercept of -1, slope of $3/4$

9. y intercept of $1/2$, slope of 3

10. y intercept of $5/2$, slope of $-1/2$

11. x intercept -3, slope is undefined

12. y intercept 2, slope of 0

Write each equation in slope intercept form (if not given in that form). Then use the y intercept and slope to draw the graph of the linear equation.

13. $y = -x + 2$

14. $2x + y = 6$

15. $20x - 30y = 60$

16. $5y = 2x - 5$

17. $-50x + 20y = 40$

18. $60x = -30y + 60$

Determine if the two given lines are parallel, perpendicular, or neither.

19. $y = 2x - 4$
$y = 2x + 3$

20. $2x + 3y = 6$
$y = -\frac{2}{3}x + 5$

21. $4x + 2y = 8$
$8x = 4 - 4y$

22. $3x - 5y = 10$
$3y + 5x = 5$

23. $2x + 5y = 10$
$-x + 3y = 9$

24. $6x + 2y = 8$
$4x - 9 = -y$

25. $y = \frac{1}{2}x - 6$
$-3y = 6x + 9$

26. $2y - 6 = -5x$
$y = -\frac{5}{2}x - 2$

27. $y = 2x - 6$
$x = -2y - 4$

28. $3x + 4y = 12$
$4x + 3y = 12$

29. $2x + y - 6 = 0$
$6x + 3y = 12$

30. $x - 3y = -9$
$y = 3x + 6$

31. $-6x + 4y = 6$
$3y = -2x + 12$

32. $-4x + 6y = 12$
$2x - 3y = 6$

Use the point–slope form to find the equation of a line with the properties given. Then write the equation in slope–intercept form.

33. Slope = 4, through (2, 3)

34. Slope = −2, through (−4, 5)

35. Slope = −1, through (6, 0)

36. Slope = $\frac{1}{2}$, through (−1, −5)

37. Slope = −$\frac{2}{3}$, through (−1, −2)

38. Slope = $\frac{3}{5}$, through (4, −2)

39. Through (4, 6) and (−2, 1)

40. Through (−4, −2) and (−2, 1)

41. Through (6, 3) and (5, 2)

42. Through (−4, 6) and (4, −6)

43. Through (1, 0) and (−2, 4)

44. Through (10, 3) and (0, −2)

Find the equation of a line with the properties given. Write the equation in the form indicated.

45. Through (1, 4) parallel to $y = 2x + 4$ (slope–intercept form)

46. Through (3, −2) parallel to $y = −3x + 6$ (standard form)

47. Through ($\frac{1}{2}$, 3) parallel to $2x + 3y − 9 = 0$ (standard form)

48. through ($\frac{1}{5}$, −$\frac{2}{3}$) parallel to $−3x = 2y + 6$ (slope–intercept form)

49. Through (−4, $\frac{3}{4}$) parallel to $y = \frac{2}{3}x − 5$ (slope–intercept form)

50. Through (2, 3) perpendicular to $y = 2x − 3$ (slope–intercept form)

51. Through (−2, 4) perpendicular to $4x − 2y = 8$ (standard form)

52. Through ($\frac{1}{2}$, −3) perpendicular to $5x = −2y + 3$ (standard form)

53. Through (−$\frac{2}{3}$, −4) perpendicular to $\frac{1}{2}x = y − 6$ (slope–intercept form)

54. With x intercept 2 and y intercept 3 (standard form)

55. With x intercept $\frac{1}{2}$ and y intercept −$\frac{1}{4}$ (standard form)

56. Through (2, 5) and parallel to the line with x intercept 1 and y intercept 3 (slope–intercept form)

57. Through (4, −2) and parallel to the line with x intercept −3 and y intercept 2 (slope–intercept form)

58. Through (−3, 4) and perpendicular to the line with x intercept 2 and y intercept 2 (standard form)

59. Through (6, 2) and perpendicular to the line with x intercept 2 and y intercept −3 (slope–intercept form)

60. Through the point (2, 1) parallel to the line through the points (3, 5) and (−2, 3) (slope–intercept form)

61. Through the point (6, −2) perpendicular to the line through the points (−2, $\frac{1}{2}$) and (4, 3) (standard form)

62. In this chapter we have discussed three forms of an equation. Name the three forms, and give one equation illustrating each form.

63. Consider the equation $y − x = 2$. Write the equation in:

(a) Standard form

(b) Slope–intercept form

(c) Point–slope form

64. Consider the equation $4 = 2y + 4x$. Write the equation in:

(a) Standard form

(b) Slope–intercept form

(c) Point–slope form

65. Explain how to determine without graphing the equations if two linear equations represent (a) parallel lines (b) perpendicular lines.

66. In Section 3.3 we explained how to graph linear equations by plotting points and by using the intercepts. In this section we discussed drawing graphs of linear equations using the slope and y intercept.

(a) Explain how to graph a linear equation by plotting points, by using the intercepts, and by using the slope and y intercept.

(b) Graph the equation $4x + 6y = 12$ using each of the three methods.

Cumulative Review Exercises

[2.5] **67.** What must you do when multiplying or dividing both sides of an inequality by a negative number?

[3.1] **68.** Determine the distance and midpoint between the points (4, −6) and (−5, 3).

[3.2] **69. (a)** Explain how to find the x and y intercepts of a linear equation.

(b) Graph $20y = 120 − 40x$ using the x and y intercepts.

70. The rental cost for Jalopy Car Rental Agency is $15 a day plus 10 cents a mile.

(a) Write an equation expressing the rental cost, C, in terms of miles, m.

(b) Draw a graph of the rental cost versus the mileage for up to 200 miles.

3.5

Relations and Functions

▶ **1** Identify relations.

▶ **2** Find the domain and range of a relation.

▶ **3** Identify functions.

▶ **4** Use function notation.

▶ **5** Graph a linear function.

▶ **1** If you plan to take additional mathematics courses, an understanding of relations and functions will be very helpful. In this section you are introduced to these important concepts. The function concept is discussed and expanded further in later sections of the text.

A **relation** is any set of ordered pairs. A relation may be indicated by (1) a set of ordered pairs, (2) a table of values, (3) a graph, (4) a rule, or (5) an equation. For example, each of the following indicates a relation.

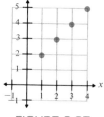

FIGURE 3.27

1. Set of ordered pairs {(1, 2), (2, 3), (3, 4), (4, 5)}

2. Table of values

x	1	2	3	4
y	2	3	4	5

3. Graph (see Figure 3.27)

4. Rule: For each integer from 1 to 4 inclusive, add 1 to obtain its corresponding value.

5. Equation: $y = x + 1$, for $1 \le x \le 4$, $x \in N$

Note that these five examples all indicate the same relation.

TABLE 3.1

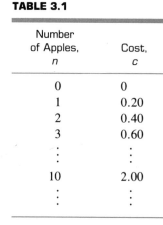

Number of Apples, n	Cost, c
0	0
1	0.20
2	0.40
3	0.60
⋮	⋮
10	2.00
⋮	⋮

Consider the following. An apple costs 20 cents. What will be your cost if you purchase 0 apples, 1 apple, 2 apples, and so on? We can indicate this using Table 3.1.

In general, we can see that the cost for purchasing n apples will be 20 cents times the number of apples, or $0.20n$. We can represent the cost of purchasing n apples, where n is a whole number, by the equation $c = 0.20n$. In the equation $c = 0.20n$ the cost, c, depends on the number of apples, n; thus we call c the *dependent variable* and n the *independent variable*.

▶ **2** In any relation the set of values that can be used for the independent variable is called its **domain.** The set of values that represent the dependent variable is called its **range.**

Consider the equation for the cost of apples, $c = 0.20n$. What is its domain and what is its range? The domain of this relation is the set of "input values" that can be used to represent n, the number of apples. Since we cannot purchase a fractional part of an apple or a negative amount of apples, the domain is the set of numbers $\{0, 1, 2, 3, \ldots\}$. Note that the values on the left side of Table 3.1 are the elements that make up the domain. When the values in the domain, $0, 1, 2, 3, \ldots$, are substituted for n in the formula $c = 0.20n$, the values we get out are $0.00, 0.20, 0.40, 0.60, \ldots$. These values appear on the right side of Table 3.1. The range is the set of these "output values." The range is $\{0.00, 0.20, 0.40, 0.60, \ldots\}$.

If we list the table of values in Table 3.1 as a set of ordered pairs, we get $\{(0, 0.00), (1, 0.20), (2, 0.40), (3, 0.60), \ldots\}$. Note that the *domain is the set of first coordinates in the set of ordered pairs,* and the *range is the set of second coordinates in the set of ordered pairs.*

When a graph is given, its domain and range can be determined by observation, as illustrated in Example 1.

EXAMPLE 1   State the domain and range of the relation shown in Fig. 3.28.

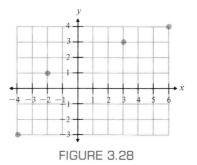

FIGURE 3.28

Solution: The domain is the set of x values (first coordinate in the set of ordered pairs).

Domain: $\{-4, -2, 3, 6\}$

The range is the set of y values (second coordinate in the set of ordered pairs).

Range: $\{-3, 1, 3, 4\}$

The numbers in the domain and range were listed from smallest to largest; however, you may list the numbers in any order.

EXAMPLE 2   State the domain and range of the relation shown in Fig. 3.29.

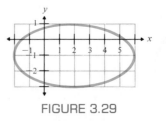

FIGURE 3.29

Solution: The domain is the set of x values. All values of x between -2 and 6 inclusive are indicated on the graph. We can indicate this using set builder notation.

Domain: $\{x \mid -2 \le x \le 6\}$

The range is the set of y values. All values of y between -3 and 1 inclusive are indicated on the graph.

Range: $\{y \mid -3 \le y \le 1\}$

EXAMPLE 3 Determine the domain and range of the relation shown in Fig. 3.30.

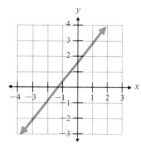

FIGURE 3.30

Solution: Since the line is extended indefinitely, every value of x will be included in the domain. The domain is the set of real numbers.

Domain: $\mathbb{R}$

The range will also be the set of real numbers since all values of y are included on the graph.

Range: $\mathbb{R}$

EXAMPLE 4 Determine the domain the range in each of the following relations.

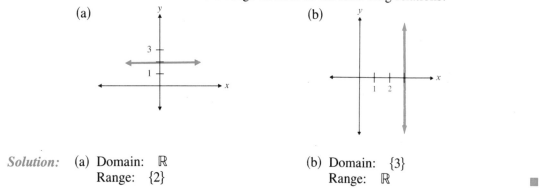

Solution: (a) Domain: $\mathbb{R}$
Range: $\{2\}$

(b) Domain: $\{3\}$
Range: $\mathbb{R}$

▶ **3** A *function* is a special type of relation. For a relation to be a function, each first coordinate in the set of ordered pairs must have a unique second coordinate. Is the set of ordered pairs $\{(1, 4), (2, 3), (3, 5), (-1, 3), (0, 6)\}$ a function? Do any of the ordered pairs have the same first coordinate and a different second coordinate? *Since no two ordered pairs have the same first coordinate, the set of ordered pairs is a function.* Note that the second coordinate in the ordered pairs may repeat.

Function

A **function** is a relation in which no two ordered pairs have the same first coordinate and a different second coordinate.

Now consider a second set of ordered pairs: $\{(-1, 3), (4, 2), (3, 1), (2, 6), (3, 5)\}$. Is this set of ordered pairs a function? Since two ordered pairs, $(3, 1)$ and $(3, 5)$, have the same first coordinate, this set of ordered pairs is not a function.

A function may also be defined as a relation in which each element of the domain corresponds to one and only one element of the range. In other words, each x-value must correspond to a unique y-value.

Let us graph each set of ordered pairs and observe them visually (Fig. 3.31). Note in part (a) that each x value has a unique y value. If a vertical line is drawn through any point, no other points are intersected. This relation is therefore a function. In part (b), if we draw a vertical line through the point $(3, 1)$, it will intersect the point $(3, 5)$. Thus each x value does not have a unique y value and the set of points is not a function.

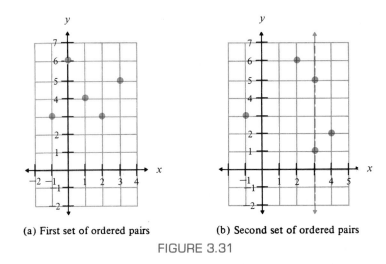

(a) First set of ordered pairs   (b) Second set of ordered pairs

FIGURE 3.31

To determine if a graph is a function, we can use the **vertical line test.** If a vertical line can be drawn through any part of the graph and the line intersects another part of the graph, the graph is not a function. If a vertical line cannot be drawn to intersect the graph at more than one point, the graph is a function.

EXAMPLE 5   Determine by using the vertical line test whether or not the graphs of the following are functions.

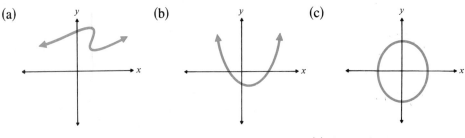

(a)   (b)   (c)

Solution: (a) Not a function   (b) Is a function   (c) Not a function

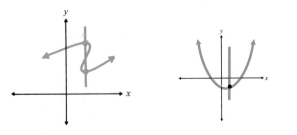

EXAMPLE 6   State the range and domain of the function illustrated in Fig. 3.32.

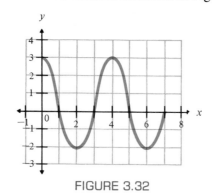

FIGURE 3.32

Solution: The domain is the set of x values, 0 through 7. Thus the domain is $\{x \mid 0 \leq x \leq 7\}$. The range is the set of y values, -2 through 3. The range is therefore $\{y \mid -2 \leq y \leq 3\}$. ■

Function Notation

▶ **4**   Consider the equation $y = 3x + 2$. By examining its graph (Fig. 3.33), we can see that it is a function.

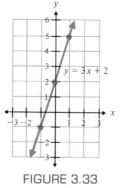

FIGURE 3.33

Since the value of y depends on the value of x and the equation is a function, we say that **y is a function of x.** The notation $y = f(x)$ is used to show that y is a function of the variable x. For this example, we can write

$$y = f(x) = 3x + 2$$

The notation $f(x)$ is read "f of x" and *does not mean f times x*. Other letters may be used to indicate functions. For example $g(x)$ and $h(x)$ also represent functions of x.

To evaluate a function for a specific value of x, substitute that value for x. For example,

$$f(x) = 3x + 2$$
$$f(1) = 3(1) + 2 = 5$$
$$f(-4) = 3(-4) + 2 = -10$$
$$f(0) = 3(0) + 2 = 2$$

The notation $f(1)$ is read "f at 1," $f(-4)$ is read "f at -4," and so on.

▶ **5**   All equations of the form $f(x) = ax + b$ will be **linear functions.** That is, they will be functions whose graphs are straight lines. We graph linear functions the same way we graph linear equations. We select values for the independent variable (often x) and find the corresponding values of the dependent variable (often y).

EXAMPLE 7   Graph $f(x) = 2x - 3$.

Solution: Remember that $f(x)$ is the same as y. We can make a chart of values by substituting values of x and finding the corresponding values of y [or of $f(x)$].

$$y = f(x) = 2x - 3$$

		x	y
$x = 0$	$y = f(0) = 2(0) - 3 = -3$	0	-3
$x = 2$	$y = f(2) = 2(2) - 3 = 1$	2	1
$x = 3$	$y = f(3) = 2(3) - 3 = 3$	3	3

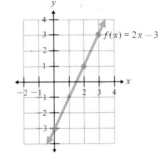

If we had wished, the vertical axis could have been labeled $f(x)$ instead of y.   ◾

An important part of the study of economics is the graphing of linear functions. The following example is a typical problem that might be found in an economics textbook.

EXAMPLE 8   It has been shown that the number of video tapes rented per week at a specific store is a function of the price of the video tape. The equation approximating the number of weekly rentals is

$$f(p) = -105p + 485, \qquad \$0.50 \le p \le \$4.50$$

where $f(p)$ is the number of rentals and p is the price per rental. (a) Construct a graph showing the relationship between the price of the tape and the number of rentals. (b) Estimate the number of weekly rentals if the rental cost is \$3.50.

Solution: (a) Construct a table of values and then graph the functions.

p	$f(p)$
1.00	380
2.00	275
3.00	170

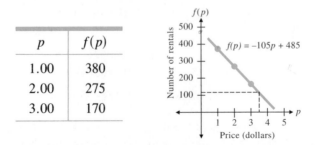

(b) If the rental cost is \$3.50, then approximately 120 tapes will be rented.   ◾

You should understand that all linear equations, other than equations of the form $x = a$, will be functions. The graph of $x = -1$ is given in Fig. 3.34. Note that it is not a function since it does not pass the vertical line test. Equations of the form $y = a$ are functions since they do pass the vertical line test. The graph of $y = 1$, given in Fig. 3.35, is a function.

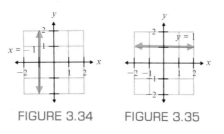

FIGURE 3.34   FIGURE 3.35

Exercise Set 3.5

Determine which of the relations are also functions. Give the range and domain of each relation or function.

1. {(1, 4), (2, 2), (3, 5), (4, 3), (5, 1)}

2. {(1, 1), (4, 4), (3, 3), (2, 2), (4, 1)}

3. {(3, −1), (5, 0), (1, 2), (4, 4), (2, 2), (7, 5)}

4. $\left\{(-1, 1), (0, -3), (3, 4), (4, 5), \left(-2, \dfrac{1}{2}\right)\right\}$

5. {(5, 0), (3, −4), (2, −1), (5, 2), (1, 1)}

6. {(6, 3), (−3, 4), (0, 3), (5, 2), (3, 5), (2, 5)}

7. $\left\{\left(\dfrac{1}{2}, \dfrac{2}{3}\right), (3, 0), (2, -1), (5, -3), (-2, 2), (0, 5)\right\}$

8. $\left\{\left(\dfrac{1}{5}, 2\right), \left(2, \dfrac{1}{2}\right), \left(\dfrac{2}{3}, 0\right), (-3, 2), (-3, -3), (5, 1)\right\}$

9. {(6, 0), (2, −3), (1, 5), (1, 0), (1, 2)}

10. {(3, −3), (3, −7), (3, −9), (3, 5)}

11. {(0, 3), (1, 3), (2, 2), (1, −1), (2, −7)}

12. {(3, 5), (2, 5), (1, 5), (0, 5), (−1, 5)}

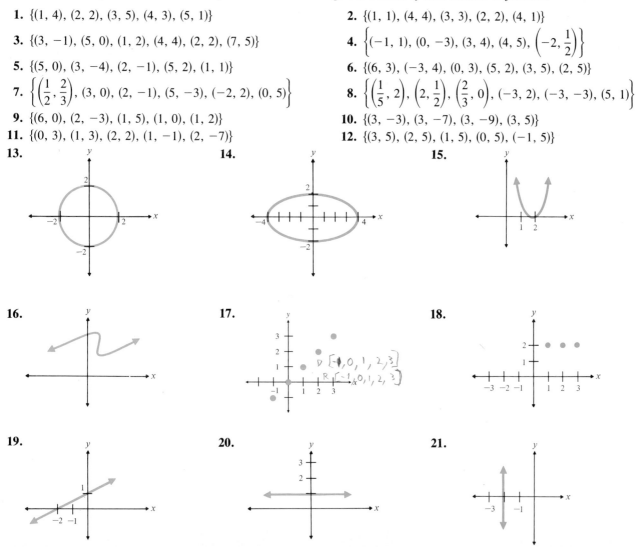

13.

14.

15.

16.

17.

18.

19.

20.

21.

22.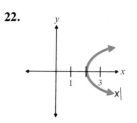

23.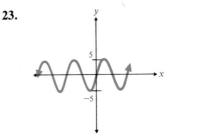

24.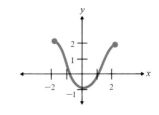

Evaluate the functions at the values indicated.

25. $f(x) = 2x + 7$; find   **(a)** $f(3)$ **(b)** $f(-2)$

26. $f(x) = -3x + 4$; find   **(a)** $f(0)$ **(b)** $f(1)$

27. $f(x) = 5x - 6$; find   **(a)** $f(2)$ **(b)** $f(3)$

28. $f(x) = 2x + 6$; find   **(a)** $f(-3)$ **(b)** $f(\tfrac{1}{2})$

29. $f(x) = 3 - 2x$; find   **(a)** $f(2)$ **(b)** $f(\tfrac{1}{2})$

30. $f(x) = \dfrac{1}{2}x - 4$; find   **(a)** $f(10)$ **(b)** $f(-4)$

31. $f(x) = \dfrac{2}{3}x - 3$; find   **(a)** $f(3)$ **(b)** $f(-12)$

32. $f(x) = -\dfrac{3}{4}x + \dfrac{1}{2}$; find   **(a)** $f(2)$ **(b)** $f(0)$

Graph each of the following functions.

33. $f(x) = 3x + 1$

3 10
0 1
-3 -8

34. $f(x) = -x + 3$

35. $f(x) = 2x - 1$

2 3
0 -1
-2 -5

36. $f(x) = 4x - 2$

37. $f(x) = -x - 2$

38. $f(x) = 3x - 5$

39. $f(x) = \dfrac{1}{2}x + 3$

1 3.5
0 3
-1 2.5

40. $f(x) = \dfrac{1}{3}x + 1$

$\dfrac{9}{3}$ 2
0 1
$\dfrac{-3}{6} = \dfrac{1}{3}$

41. $g(x) = 2x - 6$

42. $g(x) = 6x - 2$

43. The price of commodities, like soybeans, is determined by **supply and demand.** If too many soybeans are produced, then the supply will be greater than the demand, and the price will drop. If not enough soybeans are produced, the demand will be greater than the supply, and the price of soybeans will rise. Thus the price of soybeans is a function of the number of bushels of soybeans produced. The price of a bushel of soybeans can be estimated by the formula

$$f(Q) = -0.00004Q + 4.25,$$
$$10,000 \le Q \le 60,000$$

In this formula, $f(Q)$ is the price of a bushel of soybeans and Q is the annual number of bushels of soybeans produced.

(a) Construct a graph showing the relationship between the number of bushels of soybeans produced and the price of a bushel of soybeans.

(b) Estimate the cost of a bushel of soybeans if 40,000 bushels of soybeans are produced in a given year.

44. The average annual household expenditure is a function of the average annual household income. The average expenditure can be estimated by the function

$$f(i) = 0.6i + 5000, \qquad \$3500 \le i \le \$50,000.$$

In this formula, $f(i)$ is the average household expenditure and i is the average household income.

(a) Construct a graph showing the relationship between average household income and the average household expenditure.

(b) Estimate the average household expenditure for a family whose average household income is $30,000.

See Exercise 43.

Answer the following questions.

45. What is a relation?

46. What is a function?

47. Are all relations also functions? Explain.

48. Are all functions also relations? Explain.

49. Explain how the vertical line test is used to determine if a relation is a function.

50. What is the domain of a relation?

51. What is the range of a relation?

52. What are the range and domain of the function $f(x) = 3x - 2$? Explain your answer.

53. What are the range and domain of a function of the form $f(x) = ax + b$, $a \ne 0$. Explain your answer.

Cumulative Review Exercises

[1.7] **54.** Evaluate $\dfrac{-6^2 - 16 \div 2 \div |-4|}{5 - 3 \cdot 2 - 4 \div 2^2}$.

Solve the equations.

[2.1] **55.** $\dfrac{3}{4}x + \dfrac{1}{5} = \dfrac{2}{3}(x - 2)$

56. $2.6x - (-1.4x + 3.4) = 6.2$

[2.4] **57.** Two trains leave Chicago, Illinois, traveling in the same direction along parallel tracks. The first train leaves 3 hours before the second, and its speed is 15 miles per hour faster than the second. Find the speed of each train if they are 270 miles apart 3 hours after the second train leaves Chicago.

3.6

Graphing Linear Inequalities

▸ **1**   Graph linear inequalities.

A linear inequality results when the equal sign in a linear equation is replaced with an inequality sign. Examples of linear inequalities in two variables are:

$$2x + 3y > 2 \qquad 3y < 4x - 6$$
$$-x - 2y \le 3 \qquad 5x \ge 2y - 3$$

To Graph a Linear Inequality

---

1. Replace the inequality symbol with an equal sign.
2. Draw the graph of the equation in step 1. If the original inequality contains a $\ge$ or $\le$ symbol, draw the graph using a solid line. If the original inequality contains a $>$ or $<$ symbol, draw the graph using a dashed line.
3. Select any point not on the line and determine if this point is a solution to the original inequality. If the point selected is a solution, shade the region on the side of the line containing this point. If the selected point does not satisfy the inequality, shade the region on the side of the line not containing this point.

EXAMPLE 1   Graph the inequality $y < 2x - 4$.

Solution: Graph the equation $y = 2x - 4$. Since the original inequality contains a less-than sign, $<$, use a dashed line when drawing the graph (see Fig. 3.36). The dashed line indicates that the points on this line are not solutions to the inequality $y < 2x - 4$. Select a point not on the line and determine if this point satisfies the inequality. Often the easiest point to use is the origin, $(0, 0)$.

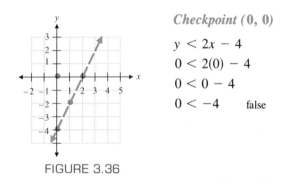

Checkpoint $(0, 0)$

$$y < 2x - 4$$
$$0 < 2(0) - 4$$
$$0 < 0 - 4$$
$$0 < -4 \qquad \text{false}$$

FIGURE 3.36

Since 0 is not less than -4, $0 \nless -4$, the point $(0, 0)$ does not satisfy the inequality. The solution will be all the points on the opposite side of the line from the point $(0, 0)$. Shade in this region (Fig. 3.37). Every point in the shaded area satisfies the given inequality. Let's check a few selected points: A, B, and C.

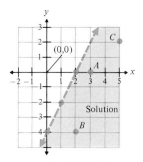

FIGURE 3.37

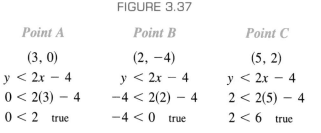

Point A	Point B	Point C
(3, 0)	(2, −4)	(5, 2)
$y < 2x - 4$	$y < 2x - 4$	$y < 2x - 4$
$0 < 2(3) - 4$	$-4 < 2(2) - 4$	$2 < 2(5) - 4$
$0 < 2$ true	$-4 < 0$ true	$2 < 6$ true

EXAMPLE 2 Graph the inequality $y \geq -\frac{1}{2}x$.

Solution: Graph the equation $y = -\frac{1}{2}x$. Since the inequality is $\geq$, we use a solid line to indicate that the points on the line are solutions to the inequality (Fig. 3.38). Since the point (0, 0) is on the line, we cannot select that point to find the solution. Let's arbitrarily select the point (3, 1).

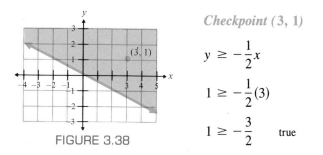

FIGURE 3.38

Checkpoint (3, 1)

$$y \geq -\frac{1}{2}x$$

$$1 \geq -\frac{1}{2}(3)$$

$$1 \geq -\frac{3}{2} \quad \text{true}$$

Since the point (3, 1) satisfies the inequality, every point on the same side of the line as (3, 1) will also satisfy the inequality $y \geq -\frac{1}{2}x$. Shade this region as indicated. Every point in the shaded region as well as every point on the line satisfies the inequality.

EXAMPLE 3 Graph the inequality $3x - 2y < -6$.

Solution: Graph the equation $3x - 2y = 6$. Since the inequality is $<$, we use a dashed line when drawing the graph (see Fig. 3.39). Substituting the checkpoint (0, 0) into the inequality results in a false statement.

Checkpoint (0, 0)

$$3x - 2y < -6$$

$$3(0) - 2(0) < -6$$

$$0 < -6 \quad \text{false}$$

FIGURE 3.39

The solution is therefore that part of the plane that does not contain the origin.

Exercise Set 3.6

Graph each inequality.

1. $x > 3$

2. $y < -2$

3. $x \geq \dfrac{5}{2}$

4. $y < x$

5. $y \geq 2x$

6. $y > -2x$

7. $y < 2x + 1$

8. $y \geq 3x - 1$

9. $y < -3x + 4$

10. $y \geq 2x + 4$

11. $y \geq \dfrac{1}{2}x - 4$

12. $y < 3x + 5$

13. $y \leq \dfrac{1}{3}x + 6$

14. $y > 6x + 1$

15. $y \leq -3x + 5$

16. $y \leq \dfrac{2}{3}x + 3$

17. $y > 5x - 4$

18. $y > \dfrac{2}{3}x - 1$

19. $2x + y < 4$

20. $3x - 4y \leq 12$

21. $2x \leq 5y + 10$

22. $\frac{5}{6}x - \frac{1}{6}y \leq \frac{2}{3}$ 　　　　　**23.** $-\frac{1}{6}x - \frac{1}{3}y > \frac{2}{3}$ 　　　　　**24.** $-\frac{5}{12}x + \frac{1}{4}y \geq \frac{3}{4}$

25. When graphing an inequality containing $\geq$ or $\leq$, why will points on the line be solutions to the inequality?

26. When graphing an inequality containing $>$ or $<$, why will points on the line not be solutions to the inequality?

Cumulative Review Exercises

[2.1] **27.** Solve the proportion $\frac{\frac{2}{3}(x-4)}{5} = \frac{x+8}{6}$.

[2.2] **28.** If $C = \bar{x} + Z\frac{\sigma}{\sqrt{n}}$, find C when $\bar{x} = 80$, $Z = 1.96$, $\sigma = 3$, and $n = 25$.

[2.3] **29.** El Gigundo Department Store is going out of business. The first week all items are being reduced by 10%. The second week all items are being reduced by an additional \$2. If during the second week Sean purchases a Motley Crüe CD for \$12.15, find the original cost of the CD.

[3.4] **30.** Write an equation of the line that passes through the point $(6, -2)$ and is perpendicular to the line $2x - y = 4$.

JUST FOR FUN

1. Graph $y < |x|$.
2. Graph $y \geq |x + 4|$.

SUMMARY

GLOSSARY

Cartesian coordinate system (or rectangular coordinate system) *(108):* Two number lines drawn perpendicular to each other, creating four quadrants.

Collinear points *(114):* Points in a straight line.

Domain *(139):* The set of first coordinates in a set of ordered pairs.

Function *(141):* A relation in which no two ordered pairs have the same first coordinate and a different second coordinate.

Graph *(114):* An illustration of the set of points that satisfy an equation.

Linear function *(143):* Equations of the form $f(x) = ax + b$ are linear functions.

Negative reciprocals *(126):* Two real numbers whose product is -1.

Negative slope *(125):* A line has a negative slope when it falls as it moves from left to right.

Ordered pair *(108):* The x and y coordinates of a point listed in parentheses, x first.

Origin *(108):* The point of intersection of the x and y axes.

Parallel lines *(126):* Two lines in the same plane that do not intersect no matter how far they are extended. Two lines are parallel when they have the same slope.

Perpendicular lines *(126):* Two lines that cross at right angles. Two lines are perpendicular when their slopes are negative reciprocals.

Positive slope *(125):* A line has a positive slope when it rises as it moves from left to right.

Range *(139):* The set of second coordinates in a set of ordered pairs.

Relation *(139):* Any set of ordered pairs.

Slope of a line *(123):* The ratio of the vertical change to the horizontal change between any two points on a line.

x Axis *(108):* The horizontal axis in the Cartesian coordinate system.

x Intercept *(116):* The value of x where a graph crosses the x axis.

y Axis *(108):* The vertical axis in the Cartesian coordinate system.

y Intercept *(116):* The value of y where a graph crosses the y axis.

IMPORTANT FACTS

Distance formula $d = \sqrt{(x_2 - x_1)^2 + (y_2 - y_1)^2}$

Midpoint formula $\left(\dfrac{x_1 + x_2}{2}, \dfrac{y_1 + y_2}{2} \right)$

Slope of a line, m $m = \dfrac{\Delta y}{\Delta x} = \dfrac{y_2 - y_1}{x_2 - x_1}$

Standard form of a linear equation $ax + by = c$

Slope–intercept form of a linear equation
$y = mx + b$

Point–slope form of a linear equation
$y - y_1 = m(x - x_1)$

To find the x intercept, set $y = 0$ and solve the equation for x.

To find the y intercept, set $x = 0$ and solve the equation for y.

To write an equation in slope–intercept form, solve the equation for y.

Review Exercises

[3.1]

1. Plot the ordered pairs on the same set of axes.

   (a) $A(5, 3)$ (b) $B(0, 6)$ (c) $C\left(5, \dfrac{1}{2}\right)$

   (d) $D(-4, 3)$ (e) $E(-6, -1)$ (f) $F(-2, 0)$

Find the length and the midpoint of the line segment between the two given points.

2. $(0, 0), (3, -4)$ 3. $(6, 2), (2, -1)$ 4. $(-2, -3), (3, 9)$
5. $(-4, 3), (-2, 5)$ 6. $(3, 4), (5, 4)$ 7. $(-3, 5), (-3, -8)$

[3.2] *Graph the equation by the method of your choice.*

8. $y = 4$ 9. $x = -2$ 10. $y = 4x$

11. $y = -3x + 4$ 12. $y = -\dfrac{1}{2}x + 2$ 13. $2x - 3y = 12$

14. $2y = 3x - 6$

15. $5x - 2y = 10$

16. $3x = 6y + 9$

17. $25x - 50y = 200$

18. $3x - 2y = 150$

19. $\frac{2}{3}x = \frac{1}{4}y + 20$

[3.3–3.4] *Determine the slope and y intercept of the equation.*

20. $y = -x + 5$

21. $y = -4x + \frac{1}{2}$

22. $3x + 6y = 9$

23. $3x + 5y = 12$

24. $9x + 7y = 15$

25. $36x - 72y = 144$

26. $x = -2$

27. $y = 6$

28. $y = -4x$

Determine the slope of the line through the two given points.

29. $(4, 6), (5, -1)$

30. $(-2, 3), (4, 1)$

31. $(-3, 5), (0, 6)$

32. $(-4, -2), (-3, 5)$

Two points on l_1 and two points on l_2 are given. Determine if l_1 is parallel to l_2, l_1 is perpendicular to l_2, or neither.

33. l_1: $(4, 3)$ and $(0, -3)$; l_2: $(1, -1)$ and $(2, -2)$

34. l_1: $(3, 2)$ and $(2, 3)$; l_2: $(4, 1)$ and $(1, 4)$

35. l_1: $(4, 0)$ and $(1, 3)$; l_2: $(5, 2)$ and $(6, 3)$

36. l_1: $(-3, 5)$ and $(2, 3)$; l_2: $(-4, -2)$ and $(-1, 2)$

Solve for the given variable if the line through the two given points is to have the given slope.

37. $(5, a)$ and $(4, 2)$; $m = 1$

38. $(3, 0)$ and $(5, y)$; $m = 3$

39. $(-2, -1)$ and $(4, y)$; $m = -6$

40. $(x, 2)$ and $(5, -2)$; $m = 2$

Write the equation of the line.

41.   **42.**   **43.**

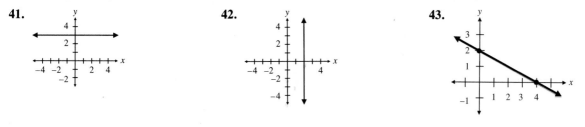

Determine if the two lines are parallel, perpendicular, or neither.

44. $y = 3x - 6$
$6y = 18x + 6$

45. $2x - 3y = 9$
$-3x - 2y = 6$

46. $y = \frac{4}{9}x + 5$
$4x = 9y + 9$

47. $4x = 6y + 3$
$-2x = -3y + 10$

48. $y = \frac{2}{5}x - 5$
$5x + 2y + 10 = 0$

49. $4x - 2y = 10$
$-2x + 4y = -8$

Find the equation of the line with the properties given. Write the answer in slope–intercept form.

50. Slope $= 2$, through $(3, 4)$

51. Slope $= -\frac{2}{3}$, through $(3, 2)$

52. Through $(4, 3)$ and $(2, 1)$

53. Through $(-2, 3)$ and $(0, -4)$

54. Through $(-6, 2)$ parallel to $y = 3x - 4$

55. Through $(4, -2)$ parallel to $2x - 5y = 6$

56. Through $(-3, 1)$ perpendicular to $y = \frac{3}{5}x + 5$

57. Through $(4, 2)$ perpendicular to $4x - 2y = 8$

(c) Estimate the number of bagels sold if the company has a \$20,000 profit.

58. The yearly profit of a bagel company can be estimated by the formula $p = 0.1x - 5000$, where x is the number of bagels sold per year.

(a) Draw a graph of profits versus bagels sold for up to 250,000 bagels.

(b) Estimate the number of bagels that must be sold for the company to break even.

59. Draw a graph illustrating the interest on a \$12,000 loan for a 1-year period for various interest rates up to 20%. Use interest $=$ principal $\cdot$ rate $\cdot$ time.

[3.5] *Give the range and domain of each relation.*

60. $\{(3, 4), (-2, 5), (0, -1), (6, 9)\}$

61. $\{(\frac{1}{2}, 2), (4, -6), (5, 3), (2, -1)\}$

62.
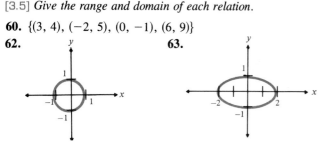

63.

64.
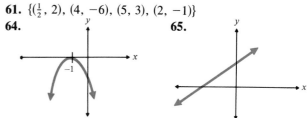

65.

Determine which of the following relations are functions.

66.
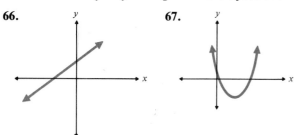

67.

68.
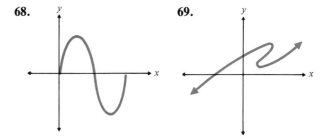

69.

70. $\{(0, 4), (5, 6), (-2, 4), (1, 3)\}$

71. $\{(6, -2), (-3, 5), (5, 2), (-1, 4)\}$

72. $\{(3, 2), (4, 2), (5, 4), (6, 6)\}$

73. $\{(1, 4), (2, 4), (3, 4), (4, 4)\}$

74. $\{(-3, 2), (4, 3), (3, 2), (4, 6)\}$

75. $\{(0, 4), (1, 3), (5, 2), (1, -1)\}$

Evaluate the functions at the values indicated.

76. $f(x) = 3x + 4$; find **(a)** $f(2)$ **(b)** $f(-3)$

77. $f(x) = -2x + 5$; find **(a)** $f(-1)$ **(b)** $f(\frac{1}{2})$

78. $f(x) = -\frac{1}{2}x + 4$; find **(a)** $f(4)$ **(b)** $f(3)$

79. $f(x) = 6 - 2x$; find **(a)** $f(0)$ **(b)** $f(-5)$

Graph each of the following functions.

80. $f(x) = 2x + 4$ **81.** $f(x) = 3x - 5$ **82.** $f(x) = -4x + 2$ **83.** $g(x) = \dfrac{1}{2}x + 2$

84. The profit or loss for a publishing company on a particular textbook can be estimated by using the function $f(n) = 24n - 200,000$, where $f(n)$ is the profit or loss, and n is the number of copies of the book sold.

    (a) Construct a graph showing the relationship between the number of books sold and the profit or loss for the publishing company for up to 15,000 books.

    (b) Estimate the number of books that must be sold for the company to break even.

    (c) Estimate the number of books that must be sold for the company's profit to be $100,000.

[3.6] *Graph the given inequality.*

85. $y \geq -3$ **86.** $x < 4$ **87.** $y < 3x$ **88.** $y > 2x + 1$

89. $y \leq 4x - 3$ **90.** $y \geq 6x + 5$ **91.** $y < -x + 4$ **92.** $y \leq \dfrac{1}{3}x - 2$

Practice Test

1. Find the length and midpoint of the line segment between the points $(1, 3)$ and $(-2, -1)$.

2. Find the slope and y intercept of $4x - 9y = 15$.

3. Write the equation of the following graph in slope–intercept form.

4. Write the equation of the line with a slope of 4 passing through the point $(-1, 3)$. Write the equation in slope–intercept form.

5. Write the equation of the line (in slope–intercept form) passing through the points $(3, -1)$ and $(-4, 2)$.

6. Write the equation of the line (in slope–intercept form) passing through the point $(-1, 4)$ perpendicular to $2x + 3y = 6$.

7. Graph $y = 2x - 2$.

8. Graph $2x + 3y = 10$.

9. The monthly profit (or loss) of Belushi's Video Store can be estimated by the function $f(c) = 2.5c - 800$, where c is the number of tapes rented each month, and $f(c)$ is the monthly profit or loss.

(a) Construct a graph showing the relationship between the number of tapes rented each month, for up to 2000 tapes, and the monthly profit or loss.

(b) Estimate the number of tapes rented if the monthly profit is $4000.

(c) Estimate the number of tapes that need to be rented each month for the company to break even.

10. State the domain and range of the following relation:

$$\{(4, 0), (2, -3), (\tfrac{1}{2}, 2), (6, 9)\}$$

11. Determine which, if any, of the following are functions.

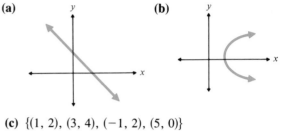

(a)   **(b)**

(c) $\{(1, 2), (3, 4), (-1, 2), (5, 0)\}$

12. $f(x) = 3x - 5$; find **(a)** $f(2)$ **(b)** $f(-4)$

13. Graph $f(x) = \dfrac{2}{3}x + 1$.

14. Graph $y \geq -3x + 5$.

15. Graph $y \leq 4x - 2$.

CHAPTER

4

Systems of Linear Equations and Inequalities

See Section 4.3, Example 6.

4.1

Solving Systems of Linear Equations

▶**1**   Solve a system of linear equations graphically.

▶**2**   Solve a system of linear equations by substitution.

▶**3**   Solve a system of linear equations by the addition method.

It is often necessary to find a common solution to two or more linear equations. We refer to the equations in this type of problem as **simultaneous linear equations** or as a **system of linear equations.**

$$\left.\begin{array}{l} (1)\ y = x + 5 \\ (2)\ y = 2x + 4 \end{array}\right\} \quad \text{system of linear equations}$$

A **solution to a system of equations** is an ordered pair or pairs that satisfy *all* equations in the system. The only solution to the system above is (1, 6).

Check in Equation (1)	*Check in Equation (2)*
(1, 6)	(1, 6)
$y = x + 5$	$y = 2x + 4$
$6 = 1 + 5$	$6 = 2(1) + 4$
$6 = 6$ true	$6 = 6$ true

The ordered pair (1, 6) satisfies *both* equations and is the solution to the system of equations.

A system of equations may consist of more than two equations. If a system consists of three equations in three variables, namely x, y, and z, the solution will be an *ordered triple* of the form (x, y, z). If the ordered triple (x, y, z) is a solution to the system, it must satisfy all three equations in the system. A system with three equations and three unknowns is referred to as a *third-order system*. Systems of equations may have more than three equations and three variables, but we will not discuss them in this book.

Graphing Method

▶**1**   To solve a system of linear equations in two variables graphically, graph all equations in the system on the same set of axes. The solution to the system will be the ordered pair (or pairs) common to all the lines, or the point of intersection of all lines in the system.

When two lines are graphed, three situations are possible, as illustrated below in Fig. 4.1. In Fig. 4.1a, lines 1 and 2 intersect at exactly one point. This system of equations has *exactly one solution*. This is an example of a **consistent** system of equations. A consistent system of equations is a system of equations that has a solution.

Lines 1 and 2 of Fig. 4.1b are different but parallel lines. The lines do not intersect, and this system of equations has *no solution*. This is an example of an **inconsistent** system of equations. An inconsistent system of equations is a system of equations that has no solution.

In Fig. 4.1c, lines 1 and 2 are actually the same line. In this case, every point on the line satisfies both equations and is a solution to the system of equations. This system has *an infinite number of solutions*. This is an example of a **dependent** system of equations. A dependent system of linear equations is a system of equations where both equations represent the same line. *Note that a dependent system is also a consistent system since it has a solution.*

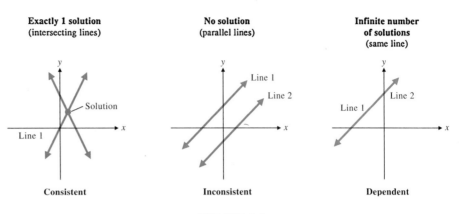

FIGURE 4.1

We can determine if a system of linear equations is consistent, inconsistent, or dependent by writing each equation in slope–intercept form and comparing the slopes and y intercepts. Note that if the slopes of the lines are different (Fig. 4.1a), the system is consistent. If the slopes are the same but the y intercepts different (Fig. 4.1b), the system is inconsistent, and if both the slopes and the y intercepts are the same (Fig. 4.1c), the system is dependent.

EXAMPLE 1   Without graphing the equations, determine if the following system of equations is consistent, inconsistent, or dependent.

$$2x + y = 3$$
$$4x + 2y = 12$$

Solution: Write each equation in slope–intercept form.

$$
\begin{array}{ll}
2x + y = 3 & 4x + 2y = 12 \\
y = -2x + 3 & 2y = -4x + 12 \\
& y = -2x + 6
\end{array}
$$

Since both equations have the same slope, -2, and different y intercepts, the lines are parallel lines. Therefore, the system is inconsistent and has no solution.   ∎

EXAMPLE 2   Solve the following system of equations graphically.

$$y = x + 2$$
$$y = -x + 4$$

Solution: Graph both equations on the same set of axes (Fig. 4.2).

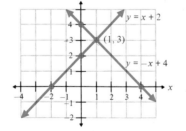

FIGURE 4.2

The solution is the point of intersection of the two lines, $(1, 3)$.   ∎

Substitution Method ▸**2** Often, an exact solution to a system of equations may be difficult to find on a graph. When an exact answer is necessary, the system should be solved algebraically, either by substitution or by addition (elimination) of equations.

To Solve a System of Equations by Substitution

1. Solve for a variable in either equation. (If possible, solve for a variable with a numerical coefficient of 1 to avoid working with fractions.)
2. Substitute the expression found for the variable in step 1 into the other equation.
3. Solve the equation determined in step 2 to find the value of one variable.
4. Substitute the value found in step 3 into the equation from step 1. Solve the equation to find the remaining variable.
5. Check your solution in all equations in the system.

EXAMPLE 3 Solve the system of equations by substitution.

$$y = 2x + 5$$
$$y = -4x + 2$$

Solution: Since both equations are already solved for y, we can substitute $2x + 5$ for y in the second equation and then solve for the remaining variable x.

$$2x + 5 = -4x + 2$$
$$6x + 5 = 2$$
$$6x = -3$$
$$x = -\frac{1}{2}$$

Now find y by substituting $-\frac{1}{2}$ for x in either of the original equations. We will use the first equation.

$$y = 2x + 5$$
$$= 2\left(-\frac{1}{2}\right) + 5$$
$$= -1 + 5$$
$$= 4$$

A check will show that the solution is $(-\frac{1}{2}, 4)$. ∎

EXAMPLE 4 Solve the system of equations by substitution.

$$2x + y = 11$$
$$x + 3y = 18$$

Solution: Begin by solving for one of the variables in either of the equations. You may solve for either of the variables; however, if you solve for a variable with a numerical coefficient of 1, you may avoid working with fractions. In this equation the y term in $2x + y = 11$ and the x term in $x + 3y = 18$ both have a numerical coefficient of 1.

Let's solve for y in $2x + y = 11$.

$$2x + y = 11$$
$$y = -2x + 11$$

Next, substitute $-2x + 11$ for y in the *other equation*, $x + 3y = 18$, and solve for the remaining variable, x.

$$x + 3y = 18$$
$$x + 3(-2x + 11) = 18$$
$$x - 6x + 33 = 18$$
$$-5x + 33 = 18$$
$$-5x = -15$$
$$x = 3$$

Finally, substitute $x = 3$ in the equation $y = -2x + 11$ and solve for y.

$$y = -2x + 11$$
$$y = -2(3) + 11 = 5$$

The solution is the ordered pair $(3, 5)$. ∎

If, when solving a system of equations by either substitution or the addition method, you arrive at an equation that is false, such as $5 = 6$ or $0 = 3$, the system is inconsistent and has no solution. If you obtain an equation that is true, such as $6 = 6$ or $0 = 0$, the system is dependent and has an infinite number of solutions.

HELPFUL HINT

You may successfully solve for one of the variables and forget to solve for the other. Remember that a solution must contain a numerical value for each variable in the system.

Addition Method

▶ **3** A third and often the easiest method of solving a system of equations is the addition or elimination method. The object of this process is to obtain two equations whose sum will be an equation containing only one variable. Keep in mind that your immediate goal is to obtain one equation containing only one unknown.

EXAMPLE 5 Solve the following system of equations using the addition method.

$$x + y = 6$$
$$2x - y = 3$$

Solution: Note that one equation contains a $+y$ and the other contains a $-y$. By adding the equations, we can eliminate the variable y and obtain one equation containing only one unknown, x.

$$
\begin{array}{r}
x + y = 6 \\
\underline{2x - y = 3} \\
3x \quad\quad = 9
\end{array}
$$

Now solve for the remaining variable, x.

$$\frac{3x}{3} = \frac{9}{3}$$
$$x = 3$$

Finally, solve for y by inserting $x = 3$ in either of the original equations.

$$x + y = 6$$
$$3 + y = 6$$
$$y = 3$$

The solution is (3, 3).   ∎

To Solve a System of Equations by the Addition (or Elimination) Method

1. If necessary, rewrite each equation so that the terms containing variables appear on the left side of the equal sign and any constants appear on the right side of the equal sign.
2. If necessary, multiply one or both equations by a constant(s) so that when the equations are added, the resulting sum will contain only one variable.
3. Add the respective sides of the equations. This will result in a single equation containing only one variable.
4. Solve for the variable in the equation in step 3.
5. Substitute the value found in step 4 into either of the original equations. Solve that equation to find the value of the remaining variable.
6. Check your solution in all equations in the system.

In step 2 of the procedure, we indicate that it may be necessary to multiply both sides of an equation by a constant. In this text we will use brackets, [ ], to indicate multiplication of *an entire equation* by a real number. Thus 4[ ] means to multiply the entire equation within the brackets by 4, and in general a[ ] means to multiply the entire equation within the brackets by the real number a. For example, $3[2x + 4 = -5]$ will give $6x + 12 = -15$ and $-2[5x - 6 = -8]$ will give $-10x + 12 = 16$.

EXAMPLE 6   Solve the following system of equations using the addition method.

$$2x + \ y = 11$$
$$x + 3y = 18$$

Solution: The object of the addition process is to obtain two equations whose sum will be an equation containing only one variable. To eliminate the variable x, we multiply the second equation by -2 and add the two equations.

$$2x + \ y = 11 \quad \text{gives} \quad 2x + \ y = \ \ 11$$
$$-2[x + 3y = 18] \qquad\qquad -2x - 6y = -36$$

Now add:

$$2x + y = 11$$
$$-2x - 6y = -36$$
$$\overline{-5y = -25}$$
$$y = 5$$

Solve for x.

$$2x + y = 11$$
$$2x + 5 = 11$$
$$2x = 6$$
$$x = 3$$

The solution is $(3, 5)$. Note that we could have first eliminated the variable y by multiplying the first equation by -3 and then adding. ■

EXAMPLE 7   Solve the following system of equations using the addition method.

$$2x + 3y = 6$$
$$5x - 4y = -8$$

Solution: The x variable can be eliminated by multiplying the first equation by -5 and the second by 2 and then adding the equations.

$$-5[2x + 3y = 6] \quad \text{gives} \quad -10x - 15y = -30$$
$$2[5x - 4y = -8] \qquad\qquad 10x - 8y = -16$$

$$-10x - 15y = -30$$
$$\underline{10x - 8y = -16}$$
$$-23y = -46$$
$$y = 2$$

The same value could be obtained for y by multiplying the first equation by 5 and the second by -2 and then adding. Try it now and see.

Solve for x.

$$2x + 3y = 6$$
$$2x + 3(2) = 6$$
$$2x + 6 = 6$$
$$2x = 0$$
$$x = 0$$

The solution to this system is $(0, 2)$. ■

EXAMPLE 8   Solve the following system of equations using the addition method.

$$2x + 3y = 7$$
$$5x - 7y = -3$$

Solution: We can eliminate the variable x by multiplying the first equation by -5 and the second by 2.

$$-5[2x + 3y = \quad 7] \quad \text{gives} \quad -10x - 15y = -35$$
$$2[5x - 7y = -3] \qquad\qquad 10x - 14y = -\ 6$$

$$-10x - 15y = -35$$
$$\underline{10x - 14y = -\ 6}$$
$$-29y = -41$$
$$y = \frac{41}{29}$$

We can now find x by substituting $y = \frac{41}{29}$ into one of the original equations and solving for x. If you try this, you will see that, although it can be done, it gets pretty messy. An easier method that can be used to solve for x is to go back to the original equations and eliminate the variable y.

$$7[2x + 3y = \quad 7] \quad \text{gives} \quad 14x + 21y = \quad 49$$
$$3[5x - 7y = -3] \qquad\qquad 15x - 21y = -9$$

$$14x + 21y = \quad 49$$
$$\underline{15x - 21y = -9}$$
$$29x \qquad\quad = 40$$
$$x = \frac{40}{29}$$

The solution is $\left(\dfrac{40}{29}, \dfrac{41}{29}\right)$. ■

EXAMPLE 9 Solve the following system of equations using the addition method.
$$0.5x + \quad 0.75y = -0.25$$
$$-0.25x - 0.125y = -0.625$$

Solution: When you have a system of equations containing decimals, your first step might be to multiply each equation in the system by the power of 10 that will eliminate the decimals from the equation. Then proceed as in all other examples. Since the first equation has decimals to the hundredth place, multiply each term in the equation by 100 to eliminate the decimals. Since the second equation has decimals to the thousandth place, multiply each term in the equation by 1000 to eliminate the decimals.

$$100\,(0.50x) + 100\,(0.75y) = 100\,(-0.25) \quad \text{or} \quad 50x + \quad 75y = -\ 25$$
$$1000\,(-0.25x) - 1000\,(0.125y) = 1000\,(-0.625) \quad \text{or} \quad -250x - 125y = -625$$

Now solve the resulting system of equations.
$$50x + 75y = -\ 25$$
$$-250x - 125y = -625$$

$$5[50x + 75y = -\ 25] \quad \text{gives} \quad 250x + 375y = -125$$
$$-250x - 125y = -625 \qquad\qquad \underline{-250x - 125y = -625}$$
$$250y = -750$$
$$y = -3$$

Now solve for x.

$$50x + 75y = -25$$
$$50x + 75(-3) = -25$$
$$50x - 225 = -25$$
$$50x = 200$$
$$x = 4$$

Thus the solution is $(4, -3)$. ■

EXAMPLE 10 Solve the system of equations using the addition method.

$$x + \frac{4}{3}y = 2$$

$$y = -\frac{2}{3}x + \frac{5}{2}$$

Solution: First, eliminate the fractions in the equations by multiplying each equation by its least common denominator. To eliminate fractions, we multiply the first equation by 3 and the second by 6.

$$x + \frac{4}{3}y = 2 \qquad\qquad y = -\frac{2}{3}x + \frac{5}{2}$$

$$3\left(x + \frac{4}{3}y\right) = 3 \cdot 2 \qquad 6 \cdot y = 6\left(-\frac{2}{3}x + \frac{5}{2}\right)$$

$$3x + 4y = 6 \qquad\qquad 6y = -4x + 15$$

The new system of equations is

$$3x + 4y = 6 \qquad \text{or} \qquad 3x + 4y = 6$$
$$6y = -4x + 15 \qquad\qquad 4x + 6y = 15$$

Now solve the system. We will solve for y by eliminating the terms containing x.

$$\begin{array}{ll} 4[3x + 4y = 6] & \text{gives} \\ -3[4x + 6y = 15] & \end{array} \qquad \begin{array}{r} 12x + 16y = 24 \\ -12x - 18y = -45 \\ \hline -2y = -21 \\ y = \dfrac{21}{2} \end{array}$$

Now solve for x by eliminating the y terms.

$$\begin{array}{ll} 3[3x + 4y = 6] & \text{gives} \\ -2[4x + 6y = 15] & \end{array} \qquad \begin{array}{r} 9x + 12y = 18 \\ -8x - 12y = -30 \\ \hline x = -12 \end{array}$$

We could also have found x by substituting $y = \frac{21}{2}$ in either of the original equations and then solving for x. The solution to the system is $(-12, \frac{21}{2})$. ■

EXAMPLE 11 Solve the following system of equations using the addition method.

$$2x + y = 3$$
$$4x + 2y = 12$$

Solution:

$$-2[2x + y = 3] \quad \text{gives} \quad -4x - 2y = -6$$
$$4x + 2y = 12 \qquad\qquad\qquad 4x + 2y = 12$$

$$\begin{array}{r} -4x - 2y = -6 \\ 4x + 2y = 12 \\ \hline 0 = 6 \end{array} \quad \text{false}$$

Since $0 = 6$ is a false statement, this system has no solution. The system is inconsistent and the lines will be parallel when graphed.   ∎

EXAMPLE 12   Solve the following system of equations using the addition method.

$$x - \frac{1}{2}y = 2$$
$$y = 2x - 4$$

Solution: First, align the x and y terms on the left side of the equation.

$$x - \frac{1}{2}y = 2$$
$$2x - y = 4$$

Now proceed as in previous examples.

$$-2\left[x - \frac{1}{2}y = 2\right] \quad \text{gives} \quad -2x + y = -4$$
$$2x - y = 4 \qquad\qquad\qquad 2x - y = 4$$

$$\begin{array}{r} -2x + y = -4 \\ 2x - y = 4 \\ \hline 0 = 0 \end{array} \quad \text{true}$$

Since $0 = 0$ is a true statement, the system is dependent and has an infinite number of solutions. Both equations represent the same line.   ∎

We have illustrated three methods that can be used to solve a system of linear equations: graphing, substitution, and the addition method. When you are given a system of equations, which method should you use to solve the system? When you need an exact solution, graphing should not be used. Of the two algebraic methods, the addition method may be the easiest to use if there are no numerical coefficients of 1 in the system. If one or more of the equations has a coefficient of 1, you may wish to use either method. We will present a fourth method, using determinants, in the optional Section 4.4.

Exercise Set 4.1

Determine which, if any, of the ordered pairs or ordered triples satisfy the system of linear equations.

1. $y = -6x$
   $y = -2x + 8$
   (a) $(0, 0)$ **(b)** $(-4, 16)$ **(c)** $(-2, 12)$

2. $x + 2y = 4$
   $y = 3x + 3$
   (a) $(0, 2)$ **(b)** $(-2, 3)$ **(c)** $(4, 15)$

3. $y = 2x + 4$
    $y = 2x - 1$
  (a) $(0, 4)$ **(b)** $(3, 10)$ **(c)** $(-2, 0)$

4. $2x - 3y = 6$
    $y = \frac{2}{3}x - 2$
  (a) $(3, 0)$ **(b)** $(3, -2)$ **(c)** $(1, -\frac{4}{3})$

5. $0.5y = -0.5x + 2$
    $2y = -2x + 8$
  (a) $(2, 5)$ **(b)** $(1, 3)$ **(c)** $(5, -1)$

6. $3x - 4y = 8$
    $2y = \frac{3}{2}x - 4$
  (a) $(1, -6)$ **(b)** $(-\frac{1}{3}, -\frac{9}{4})$ **(c)** $(0, -2)$

7. $2x + 3y = 6$
    $-2x + 5 = y$
  (a) $(\frac{1}{2}, \frac{5}{3})$ **(b)** $(2, 1)$ **(c)** $(\frac{9}{4}, \frac{1}{2})$

8. $x + 2y - z = -5$
    $2x - y + 2z = 8$
    $3x + 3y + 4z = 5$
  (a) $(1, 3, -2)$ **(b)** $(1, -2, 2)$ **(c)** $(0, 8, -2)$

9. $4x + y - 3z = 1$
    $2x - 2y + 6z = 11$
    $-6x + 3y + 12z = -4$
  (a) $(2, -1, -2)$ **(b)** $(\frac{1}{2}, 2, 1)$ **(c)** $(\frac{1}{2}, -3, \frac{2}{3})$

10. $2x - 3y + z = 1$
    $x + 2y + z = -1$
    $3x - y + 3z = 4$
  (a) $(1, 1, 4)$ **(b)** $(-3, -1, 4)$ **(c)** $(0, 3, 10)$

Write each equation in slope–intercept form. Without graphing the equations, state whether the system of equations is consistent, inconsistent, or dependent. Also indicate whether the system has exactly one solution, no solution, or an infinite number of solutions.

11. $2y = -x + 5$
    $x - 2y = 1$

12. $2x + y = 6$
    $2x - y = 6$

13. $3y = 2x + 3$
    $y = \frac{2}{3}x - 2$

14. $y = \frac{1}{2}x + 4$
    $2y = x + 8$

15. $2x - 3y = 4$
    $3x - 2y = -2$

16. $x + 2y = 6$
    $2x + y = 4$

17. $2x = 3y + 4$
    $6x - 9y = 12$

18. $x - y = 3$
    $2x - 2y = -2$

19. $y = \frac{3}{2}x + \frac{1}{2}$
    $3x - 2y = -\frac{1}{2}$

20. $x - y = 3$
    $\frac{1}{2}x - 2y = -6$

In Exercises 21 through 30, determine the solution to the system of equations graphically.

21. $y = x + 4$
    $y = -x + 2$

22. $y = 2x + 4$
    $y = -3x - 6$

23. $y = 2x - 1$
    $2y = 4x + 6$

24. $y = -2x - 1$
    $x + 2y = 4$

25. $2x + 3y = 6$
    $4x = -6y + 12$

26. $x + y = 1$
    $3x - y = -5$

27. $x + 3y = 4$
    $x = 1$

28. $2x - 5y = 10$
    $y = \frac{2}{5}x - 2$

29. $y = -5x + 5$
    $y = 2x - 2$

30. $2x - y = -4$
    $2y = 4x - 6$

Find the solution to each system of equations by substitution.

31. $x + 2y = 9$
    $x = 2y + 1$

32. $y = x + 2$
    $2y = -x - 2$

33. $x + y = 6$
    $x = y$

34. $2x + y = 3$
    $2y = 6 - 4x$

35. $2x + y = 3$
    $2x + y + 5 = 0$

36. $y = 2x + 4$
    $y = -\frac{3}{4}$

37. $x = \frac{1}{2}$
    $x + \frac{1}{3}y + 6 = 0$

38. $y = \frac{1}{3}x - 2$
    $x - 3y = 6$

39. $x - \frac{1}{2}y = 2$
    $y = 2x - 4$

40. $2x + 3y = 7$
    $6x - y = 1$

41. $3x + y = -1$
    $y = 3x + 5$

42. $y = -2x + 5$
    $x + 3y = 0$

43. $y = 2x - 13$
    $-4x - 7 = 9y$

44. $x = y + 4$
    $3x + 7y = -18$

45. $5x - 2y = -7$
    $5 = y - 3x$

46. $5x - 4y = -7$
    $x - \frac{3}{5}y = -2$

47. $x = 3y + 5$
    $y = \frac{2}{3}x + \frac{1}{2}$

48. $x + 2y = 4$
    $x + \frac{1}{2}y = 4$

49. $\frac{1}{2}x - \frac{1}{3}y = 2$
    $\frac{1}{4}x + \frac{2}{3}y = 6$

50. $\frac{1}{2}x + \frac{1}{4}y = 13$
    $\frac{1}{5}x + \frac{1}{8}y = 5$

Solve each system of equations using the addition method.

51. $x + y = -2$
    $x - y = 4$

52. $x - y = 12$
    $x + y = 2$

53. $-x + y = 5$
    $x + 2y = 1$

54. $x + y = 0$
    $-x + y = -2$

55. $3x + 2y = 15$
$x - 2y = -7$

56. $3x + 3y = 18$
$4x - y = 4$

57. $3x + y = 6$
$-6x - 2y = 10$

58. $2x + y = 14$
$-3x + y = -2$

59. $2x + y = 6$
$3x - 2y = 16$

60. $4x - 3y = 8$
$-2x + 5y = 14$

61. $2x - 5y = 13$
$5x + 3y = 17$

62. $4x = 2y + 6$
$y = 2x - 3$

63. $3y = 2x + 4$
$3y = 2x + 4$

64. $5x + 4y = 10$
$-3x - 5y = 7$

65. $4x - 3y = 8$
$-3x + 4y = 9$

66. $2x - y = 8$
$3x + y = 6$

67. $3x + 4y = 2$
$2x = -5y - 1$

68. $2x = 5y + 13$
$5x = -3y + 17$

69. $2y = -5x - 3$
$4x - 7y = 3$

70. $3x - 4y = 5$
$2x = 5y - 3$

71. $4x + 5y = 3$
$2x - 3y = 4$

72. $2x + 3y = 5$
$-3x - 4y = -2$

73. $0.2x + 0.5y = 1.6$
$-0.3x + 0.4y = -0.1$

74. $0.15x - 0.40y = 0.65$
$0.60x + 0.25y = -1.1$

75. $2.1x - 0.6y = 8.40$
$-1.5x - 0.3y = -6.00$

76. $-0.25x + 0.10y = 1.05$
$-0.40x - 0.625y = -0.675$

77. $2x - \dfrac{1}{3}y = 6$
$5x - y = 4$

78. $\dfrac{x}{2} - \dfrac{y}{3} = 1$
$\dfrac{x}{4} - \dfrac{y}{9} = \dfrac{2}{3}$

79. $\dfrac{x}{3} = 4 - \dfrac{y}{4}$
$3x = 4y$

80. $\dfrac{x}{4} - 3 = \dfrac{y}{6}$
$y = \dfrac{x}{2} + 2$

81. $\dfrac{x}{5} + \dfrac{y}{2} = 4$
$\dfrac{2x}{3} - y = \dfrac{8}{3}$

82. $\dfrac{2x}{3} - 4 = \dfrac{y}{2}$
$x - 3y = \dfrac{1}{3}$

83. $2(2x + y) = -2x + y + 8$
$3(x - y) = -(y + 1)$

84. $3(x - y) = 5(x + y) - 8$
$2x + 4(x + 3y) = 5(y - 2)$

85. Explain how you can determine, without graphing or solving, whether a system of two linear equations is consistent, inconsistent, or dependent.

86. When solving a system of equations by addition or substitution, how will you know if the system is inconsistent?

87. When solving a system of equations by addition or substitution, how will you know if the system is dependent?

Cumulative Review Exercises

[1.2] **88.** Explain the difference between a rational number and an irrational number.

[1.4] **89.** (a) Are all rational numbers real numbers?
(b) Are all irrational numbers real numbers?

[2.2] **90.** Find all numbers such that $|x - 4| = |4 - x|$.

[2.2] **91.** Evaluate $A = p\left(1 + \dfrac{r}{n}\right)^t$, when $p = 500$, $r = 0.08$, $n = 2$, and $t = 1$.

[3.5] **92.** Do the set of ordered pairs that follow represent a function? Explain your answer.
$\{(-3, 4), (7, 2), (-4, 5), (5, 0), (-3, 2)\}$

JUST FOR FUN

Solve each of the following using the addition method.

1. $\dfrac{x + 2}{2} - \dfrac{y + 4}{3} = 4$
$\dfrac{x + y}{2} = \dfrac{1}{2} + \dfrac{x - y}{3}$

2. $\dfrac{5x}{2} + 3y = \dfrac{9}{2} + y$
$\dfrac{1}{4}x - \dfrac{1}{2}y = 6x + 12$

Solve the system of equations.

3. $\dfrac{3}{a} + \dfrac{4}{b} = -1$
$\dfrac{1}{a} + \dfrac{6}{b} = 2$

4. $\dfrac{6}{x} + \dfrac{1}{y} = -1$
$\dfrac{3}{x} - \dfrac{2}{y} = -3$

Solve the following system of equations, where a and b represent any nonzero constants by (a) the substitution method and (b) the addition method.

5. $4ax + 3y = 19$
     $-ax + y = 4$

6. $ax = 2 - by$
     $-ax + 2by - 1 = 0$

4.2

Third-order Systems of Linear Equations

▶ **1** Solve third-order systems of equations.

▶ **1** A third-order system consists of three equations with three unknowns. The solution of a third-order system will be an ordered triple. A fourth-order system is one that consists of four equations and four unknowns. The procedures discussed in this section can be expanded to solve fourth- and higher-order systems.

Three methods that can be used to solve a third-order system are substitution, the addition method, and determinants. Determinants will be discussed in Section 4.4.

EXAMPLE 1   Solve the following system of equations by substitution.

$$x = 4$$
$$2x + y = 20$$
$$-x + 4y + 2z = 24$$

Solution: Substitute 4 for x in the equation $2x + y = 20$, and solve for y.

$$2x + y = 20$$
$$2(4) + y = 20$$
$$8 + y = 20$$
$$y = 12$$

Now substitute $x = 4$ and $y = 12$ in the last equation and solve for z.

$$-x + 4y + 2z = 24$$
$$-(4) + 4(12) + 2z = 24$$
$$-4 + 48 + 2z = 24$$
$$44 + 2z = 24$$
$$2z = -20$$
$$z = -10$$

Check: The solution must be checked in all three original equations.

$x = 4$	$2x + y = 20$	$-x + 4y + 2z = 24$
$4 = 4$ true	$2(4) + 12 = 20$	$-(4) + 4(12) + 2(-10) = 24$
	$20 = 20$ true	$24 = 24$ true

The solution is the ordered triple $(4, 12, -10)$. Remember that the ordered triple lists the x value first, the y value second, and the z value third.   ▪

Not every third-order system can be solved by substitution. When a third-order system cannot be solved using substitution, we can find the solution by the addition method as illustrated in Example 2.

EXAMPLE 2   Solve the system of equations.

(1) $3x + 2y + z = 4$
(2) $2x - 3y + 2z = -7$
(3) $x + 4y - z = 10$

Solution: For the sake of clarity the three equations have been labeled (1), (2), and (3). To solve this system of equations, we must first obtain two equations containing the same two variables. This is done by selecting two equations and using the addition method to eliminate one of the variables. For example, by adding equations (1) and (3) the variable z will be eliminated. Next we use a different pair of equations [either (1) and (2) or (2) and (3)] and use the addition method to eliminate the same variable that was eliminated previously. If we multiply equation (1) by -2 and add it to equation (2), the variable z will again be eliminated. We will then have two equations containing only two unknowns. Let us now work this example and then discuss it further.

$$
\begin{array}{rl}
(1) & 3x + 2y + z = 4 \\
(3) & x + 4y - z = 10 \\
\hline
(4) & 4x + 6y \phantom{{}- z} = 14
\end{array}
$$

Now select a different set of equations and eliminate the variable z.

$$
\begin{array}{rl}
(1) & -2[3x + 2y + z = 4] \quad \text{gives} \quad -6x - 4y - 2z = -8 \\
(2) & 2x - 3y + 2z = -7 \qquad\qquad\qquad 2x - 3y + 2z = -7 \\
& \hline
& \qquad\qquad\qquad\qquad (5) \quad -4x - 7y = -15
\end{array}
$$

We now have a system consisting of two equations with two unknowns.

$$
\begin{array}{rl}
(4) & 4x + 6y = 14 \\
(5) & -4x - 7y = -15
\end{array}
$$

Next we solve for one of the variables using a method presented earlier. If we add the two equations, the variable x will be eliminated.

$$
\begin{array}{r}
4x + 6y = 14 \\
-4x - 7y = -15 \\
\hline
-y = -1 \\
y = 1
\end{array}
$$

Next we substitute $y = 1$ in either one of the two equations containing only two variables [(4) or (5)] and solve for x.

$$
\begin{array}{rl}
(4) & 4x + 6y = 14 \\
& 4x + 6(1) = 14 \\
& 4x + 6 = 14 \\
& 4x = 8 \\
& x = 2
\end{array}
$$

Finally, substitute $x = 2$, $y = 1$ in any of the original equations and solve for z.

$$(1) \qquad 3x + 2y + z = \ \ 4$$
$$3(2) + 2(1) + z = \ \ 4$$
$$6 + 2 + z = \ \ 4$$
$$8 + z = \ \ 4$$
$$z = -4$$

The solution is the ordered triple $(2, 1, -4)$. ■

We selected first to eliminate the variable z by using equations (1) and (3) and then equations (1) and (2). We could have elected to eliminate either variable x or variable y first. For example, we could have eliminated variable x by multiplying equation (3) by -2 and then adding it to equation (2). We could also eliminate variable x by multiplying equation (3) by -3 and then adding it to equation (1). Solve the system above by first eliminating the variable x.

EXAMPLE 3 Solve the system of equations.

$$2x - 3y + 2z = -1$$
$$x + 2y \qquad = 14$$
$$x \qquad - 3z = -5$$

Solution: We will select to eliminate variable y from the first two equations.

$$2[2x - 3y + 2z = -1] \quad \text{gives} \quad 4x - 6y + 4z = -2$$
$$3[\ x + 2y \qquad = 14] \qquad \qquad \underline{3x + 6y \qquad = 42}$$
$$7x \qquad + 4z = \ 40$$

We now have two equations containing only the variables x and z.

$$7x + 4z = \ 40$$
$$x - 3z = -5$$

Let's now eliminate the variable x.

$$7x + 4z = \ 40 \quad \text{gives} \quad 7x + \ \ 4z = 40$$
$$-7[x - 3z = -5] \qquad \qquad \underline{-7x + 21z = 35}$$
$$25z = 75$$
$$z = \ \ 3$$

Now we solve for x by using one of the equations containing the variables x and z.

$$x - 3z = -5$$
$$x - 3(3) = -5$$
$$x - 9 = -5$$
$$x = \ \ 4$$

Finally, solve for y using any one of the original equations that contains a y.

$$x + 2y = 14$$
$$4 + 2y = 14$$
$$2y = 10$$
$$y = \ \ 5$$

The solution is (4, 5, 3).

Check:

$2x - 3y + 2z = -1$	$x + 2y = 14$	$x - 3z = -5$
$2(4) - 3(5) + 2(3) = -1$	$4 + 2(5) = 14$	$4 - 3(3) = -5$
$8 - 15 + 6 = -1$	$4 + 10 = 14$	$4 - \quad 9 = -5$
$-1 = -1$	$14 = 14$	$-5 = -5$
true	true	true

Geometrical Interpretation of Three Variables and Three Unknowns

When we have a system of linear equations in two variables, we can find its solution graphically using the Cartesian coordinate system. A linear equation in three variables, x, y, and z, can be graphed on a coordinate system with three axes drawn perpendicular to each other (see Fig. 4.3).

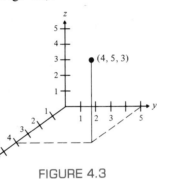

FIGURE 4.3

A point plotted in this type of three-dimensional system would appear to be a point in space. If we were to graph an equation such as $x + 2y + 3z = 4$, we would find that its graph would be a plane, not a line. In Example 3 we indicated the solution to be the ordered triple (4, 5, 3). This means that the three planes, one from each of the three given equations, all intersect at the point (4, 5, 3) in this three-dimensional system.

Exercise Set 4.2

Solve by substitution.

1. $x = 1$
$2x + y = 4$
$-3x - y + 4z = 15$

2. $2x + 3y = 9$
$4x - 6z = 12$
$y = 5$

3. $5x - 6z = -17$
$3x - 4y + 5z = -1$
$2z = -6$

4. $2x - 5y = 12$
$-3y = -9$
$2x - 3y + 4z = 8$

5. $x + 2y = 6$
$3y = 9$
$x + 2z = 12$

6. $x - y + 5z = -4$
$3x - 2z = 6$
$4z = 2$

Solve using the addition method.

7. $x + y - z = -3$
$x \quad + z = 2$
$2x - y + 2z = 3$

8. $x - 2y \quad = 2$
$2x + 3y \quad = 11$
$-y + 4z = 7$

9. $x \quad - 2z = -5$
$-y + 3z = 3$
$-2x \quad + z = 4$

10. $\begin{aligned} x - 3y \quad\quad &= 13 \\ 2y + z &= 1 \\ y - 2z &= 11 \end{aligned}$

11. $\begin{aligned} x + y + z &= 4 \\ x - 2y - z &= 1 \\ 2x - y - 2z &= -1 \end{aligned}$

12. $\begin{aligned} x - 2y + 3z &= -7 \\ 2x - y - z &= 7 \\ -x + 3y + 2z &= -8 \end{aligned}$

13. $\begin{aligned} 2x - 2y + 3z &= 5 \\ 2x + y - 2z &= -1 \\ 4x - y - 3z &= 0 \end{aligned}$

14. $\begin{aligned} 2x - y - z &= 4 \\ 4x - 3y - 2z &= -2 \\ 8x - 2y - 3z &= 3 \end{aligned}$

15. $\begin{aligned} x + 2y - 3z &= 5 \\ x + y + z &= 0 \\ 3x + 4y + 2z &= -1 \end{aligned}$

16. $\begin{aligned} x + 2y + 2z &= 1 \\ 2x - y + z &= 3 \\ 4x + y + 2z &= 0 \end{aligned}$

17. $\begin{aligned} 2x + 2y - z &= 2 \\ 3x + 4y + z &= -4 \\ 5x - 2y - 3z &= 5 \end{aligned}$

18. $\begin{aligned} x - 3y + 7z &= 13 \\ x + y + z &= 1 \\ x - 2y + 3z &= 4 \end{aligned}$

19. $\begin{aligned} -\frac{1}{4}x + \frac{1}{2}y - \frac{1}{2}z &= -2 \\ \frac{1}{2}x + \frac{1}{3}y - \frac{1}{4}z &= 2 \\ \frac{1}{2}x - \frac{1}{2}y + \frac{1}{4}z &= 1 \end{aligned}$

20. $\begin{aligned} \frac{2}{3}x + y - \frac{1}{3}z &= \frac{1}{3} \\ \frac{1}{2}x + y + z &= \frac{5}{2} \\ \frac{1}{4}x - \frac{1}{4}y + \frac{1}{4}z &= \frac{3}{2} \end{aligned}$

21. $\begin{aligned} x - \frac{2}{3}y - \frac{2z}{3} &= -2 \\ \frac{2x}{3} + y - \frac{2z}{3} &= \frac{1}{3} \\ -\frac{x}{4} + y - \frac{z}{4} &= \frac{3}{4} \end{aligned}$

22. What does the graph of an equation in two variables, such as $x + 2y = 3$, represent? What does a graph of an equation in three variables, such as $x + 2y + 3z = 6$, represent?

An equation in three variables, x, y and z, represents a plane. Consider a system of equations consisting of three equations in three variables. Answer the following questions.

23. If the three planes are parallel to one another as illustrated in the figure, how many points will be common to all three planes? Is the system consistent or inconsistent? Explain your answer.

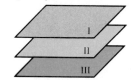

25. If the three planes are as illustrated in the figure, how many points will be common to all three planes? Is the system dependent? Explain your answer.

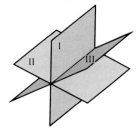

24. If two of the planes are parallel to each other and the third plane intersects each of the other two planes, how many points will be common to all three planes? Is the system consistent or inconsistent? Explain your answer.

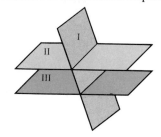

26. If the three planes are as illustrated as in the figure, how many points will be common to all three planes? Is the system consistent or inconsistent? Explain your answer.

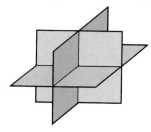

*When solving an **inconsistent system** of three equations in three unknowns, you will eventually arrive at a false statement such as $0 = 4$ or $0 = -9$. When solving a **dependent system** of three equations in three unknowns, you will eventually arrive at the true statement $0 = 0$. Determine whether the following systems are inconsistent, dependent, or neither.*

27. $\begin{aligned} 3x - 4y + z &= 4 \\ x + 2y + z &= 4 \\ -6x + 8y - 2z &= -8 \end{aligned}$

28. $\begin{aligned} 2x - 4y + 6z &= 8 \\ -x + 2y - 3z &= 6 \\ 3x + 4y + 5z &= 8 \end{aligned}$

29. $\begin{aligned} x + 3y + 2z &= 6 \\ x - 2y - 2 &= 8 \\ -3x - 9y - 6z &= -4 \end{aligned}$

30. $\begin{aligned} 2x - 2y + 4z &= 2 \\ -3x + y\quad\;\; &= -9 \\ 2x - y + z &= 5 \end{aligned}$

Cumulative Review Exercises

(2.4) **31.** Phillipa and her son Cameron go cross-country skiing. Phillipa averages 5 miles per hour, and Cameron averages 3 miles per hour. If Cameron begins $\frac{1}{6}$ hour before his mother, (a) determine how long after Cameron starts skiing, his mother will catch up with him. (b) How far from the starting point will they be when they meet?

[2.6] *Determine the solution set for each of the following:*

32. $\left| 4 - \dfrac{2x}{3} \right| > 5$

33. $\left| \dfrac{3x - 4}{2} \right| - 1 < 5$

[3.5] **34.** Graph the function $f(x) = \frac{1}{2}x - 3$.

JUST FOR FUN

1. Find the solution to the fourth-order system.

$$\begin{aligned} 3a + 2b - c &= 0 \\ 2a + 2c + d &= 5 \\ a + 2b - d &= -2 \\ 2a - b + c + d &= 2 \end{aligned}$$

4.3

Applications of Systems of Linear Equations

▶**1** Use systems of equations to solve practical application problems.

▶**1** Many of the applications solved in earlier chapters using only one variable can now be solved using two variables. The following example illustrates how Example 5 of Section 2.3 can be solved using two variables.

EXAMPLE 1   The sum of two numbers is 47. Find the two numbers if one number is 2 more than 4 times the other number.

Solution: Let x = smaller number
y = larger number

Statement	*Equation*
The sum of two numbers is 47	$x + y = 47$
One number is 2 more than 4 times the other number	$y = 4x + 2$

System of Equations

$$x + y = 47$$
$$y = 4x + 2$$

We will solve this system using substitution. Substitute $4x + 2$ in place of y in the first equation.

$$x + y = 47$$
$$x + (4x + 2) = 47$$
$$5x + 2 = 47$$
$$5x = 45$$

Smaller number: $x = 9$

Larger number: $y = 4x + 2$
$$= 4(9) + 2 = 38$$

The two numbers are 9 and 38. This answer checks with the solution obtained in Section 2.3. ∎

EXAMPLE 2 A plane travels 600 miles per hour with the wind and 450 miles per hour against the wind. Find the speed of the wind and the speed of the plane in still air.

Solution: Let x = speed of plane in still air
y = speed of wind

Speed of plane going with wind: $x + y = 600$

Speed of plane going against wind: $x - y = 450$

$$
\begin{aligned}
x + y &= 600 \\
x - y &= 450 \\
\hline
2x &= 1050 \\
x &= 525
\end{aligned}
$$

The plane's speed is 525 miles per hour in still air.

$$x + y = 600$$
$$525 + y = 600$$
$$y = 75$$

The wind's speed is 75 miles per hour. ∎

EXAMPLE 3 Mr. Duncan, a stationery salesman, receives a weekly salary plus a percentage of the sales he makes. One week, on sales of $3000, his total take-home pay was $560. The next week, on sales of $4000, his total take-home pay was $680. Find his weekly salary and his commission rate.

Solution: Let x = his weekly salary,
y = percentage of sales

The system of equations then becomes

$$x + 3000y = 560$$
$$x + 4000y = 680$$

$$
\begin{array}{ll}
x + 3000y = 560 \quad \text{gives} & x + 3000y = 560 \\
-1[x + 4000y = 680] & \underline{-x - 4000y = -680} \\
& -1000y = -120
\end{array}
$$

$$y = \frac{-120}{-1000} = 0.12$$

His commission is therefore 12% of sales. Now let us find his weekly salary.

$$x + 3000y = 560$$
$$x + 3000(0.12) = 560$$
$$x + 360 = 560$$
$$x = 200$$

His weekly salary is therefore $200.

■

EXAMPLE 4   Two large spools of electrical cable are connected as illustrated in Fig. 4.4. The point of connection between the two cables is kept stationary, and each spool of cable is to be rolled out by a team of three men. The lighter spool on the left is laid at a rate of 30 feet per minute faster than the heavier spool on the right. If a total of 1960 feet of cable is laid in 28 minutes, find the rate at which the cable is laid for each spool.

FIGURE 4.4

Solution: We let x represent the rate at which the cable on the right is laid and y the rate at which the cable on the left is laid. Both cables were laid for 28 minutes. We construct a table to aid us in finding the solution to the problem. Remember from Section 2.4 that distance = rate · time.

Cable	Rate	Time	Distance
On right	x	28	$28x$
On left	y	28	$28y$

Statement	*Equation*
The rate for laying the cable on the left is 30 feet per minute faster than the rate for the cable on the right.	$y = x + 30$
Total length (or distance) of cable laid is 1960 feet.	$28x + 28y = 1960$

System of Equations

$$y = x + 30$$
$$28x + 28y = 1960$$

We will solve this system by substitution. We substitute $x + 30$ for y in the equation $28x + 28y = 1960$.

$$28x + 28y = 1960$$
$$28x + 28(x + 30) = 1960$$
$$28x + 28x + 840 = 1960$$
$$56x + 840 = 1960$$
$$56x = 1120$$
$$x = 20$$

Thus the cable on the right is laid at a rate of 20 feet per minute. The cable on the left is laid 30 feet per minute faster. Thus, the cable on the left is laid at a rate of 50 feet per minute. ■

Now let us work some mixture problems using two variables.

EXAMPLE 5   A hot dog stand in Chicago sells hot dogs for $1.50 each and cans of soda for 75 cents each. If the total sales for the day equal $304.50 and a total of 278 items were sold, how many of each item were sold?

Solution: Let x = number of hot dogs sold
y = number of cans of soda sold

Item	Cost of Item	Number of Items	Dollar Sales
Hot dogs	1.50	x	$1.50x$
Cans of soda	0.75	y	$0.75y$

Since a total of 278 items were sold,

$$x + y = 278$$

Also,

dollar sales of hot dogs + dollar sales of cans of soda = total sales
$$1.50x + 0.75y = 304.50$$

The system of equations is therefore

$$x + y = 278$$
$$1.50x + 0.75y = 304.50$$

Let us solve this system using substitution, since $x + y = 278$, $y = 278 - x$. Now substitute.

$$1.50x + 0.75(278 - x) = 304.50$$
$$1.50x + 208.50 - 0.75x = 304.50$$
$$0.75x + 208.50 = 304.50$$
$$0.75x = 96$$
$$x = \frac{96}{0.75} = 128$$

Since $y = 278 - x$, the number of cans of soda purchased is $278 - 128$ or 150. Therefore, 128 hot dogs and 150 cans of soda were sold. ■

EXAMPLE 6   Martina, a chemist, wishes to mix a 15% sodium–iodine solution with a 40% sodium–iodine solution to get 6 liters of a 25% sodium–iodine solution. How many liters of the 15% solution and the 40% solution will she need to mix?

Solution: Let x = number of liters of the 15% solution
y = number of liters of the 40% solution

We will draw a sketch (Fig. 4.5) and then set up a table to help analyze the problem.

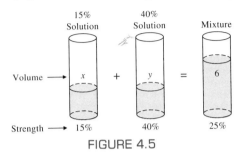

FIGURE 4.5

The amount of sodium–iodine in a solution is found by multiplying the percent strength of sodium–iodine in the solution by the volume of the solution (see Fig. 4.5).

Solution	Strength of Solution	Number of Liters	Amount of Sodium–Iodine
15% Solution	0.15	x	$0.15x$
40% Solution	0.40	y	$0.40y$
Mixture	0.25	6	0.25(6)

Since the sum of the volumes of the 15% solution and the 40% solution is 6 liters, our first equation is

$$x + y = 6$$

The second equation comes from the fact that

$$\left(\begin{array}{c} \text{amount of} \\ \text{sodium–iodine} \\ \text{in 15\% solution} \end{array} \right) + \left(\begin{array}{c} \text{amount of} \\ \text{sodium–iodine} \\ \text{in 40\% solution} \end{array} \right) = \left(\begin{array}{c} \text{amount of} \\ \text{sodium–iodine} \\ \text{in mixture} \end{array} \right)$$

or $0.15x + 0.40y = 0.25(6)$

The system of equations is therefore,

$$x + y = 6$$
$$0.15x + 0.40y = 0.25(6)$$

Solving $x + y = 6$ for y gives $y = -x + 6$. Substituting $-x + 6$ for y in the second equation gives

$$0.15x + 0.40y = 0.25(6)$$
$$0.15x + 0.40(-x + 6) = 0.25(6)$$
$$0.15x - 0.40x + 2.4 = 1.5$$
$$-0.25x + 2.4 = 1.5$$
$$-0.25x = -0.9$$
$$x = \frac{-0.9}{-0.25} = 3.6$$

Therefore, 3.6 liters of the 15% solution will be used. Since the two solutions must total 6 liters, $6 - 3.6$ or 2.4 liters of the 40% solution must be used. ∎

In Example 6, the equation $0.15x + 0.40y = 0.25(6)$ could have been simplified by multiplying both sides of the equation by 100. This would give the equation $15x + 40y = 25(6)$ or $15x + 40y = 150$. Then the system of equations would be $x + y = 6$ and $15x + 40y = 150$. If you solve this system, you should obtain the same solution.

Now let us look at some applications of third-order systems.

EXAMPLE 7   Tiny Tots Toys must borrow $25,000 to pay for an expansion. They are not able to obtain a loan for the total amount from a single bank, so they take out three loans from three different banks. They borrowed some of the money at a bank that charged them 8% interest. At the second bank, they borrowed $2000 more than one-half the amount borrowed from the first bank. The interest rate at the second bank is 10%. The balance of the $25,000 is borrowed from a third bank where they paid 9% interest. The total annual interest Tiny Tots Toys pays for the three loans is $2220. How much did they borrow at each rate?

Solution: Let $x =$ amount borrowed at first bank
$y =$ amount borrowed at second bank
$z =$ amount borrowed at third bank
Since the total amount borrowed is $25,000 we know that

$$x + y + z = 25{,}000$$

At the second bank, Tiny Tots Toys borrowed $2000 more than one-half the money borrowed from the first bank. Therefore, our second equation is

$$y = \frac{1}{2}x + 2000$$

Our last equation comes from the fact that the total annual interest charged by the three banks is $2220. The interest at each bank is found by multiplying the interest rate by the amount borrowed.

$$0.08x + 0.10y + 0.09z = 2220$$

Thus our system of equations is

$$(1) \quad x + y + z = 25{,}000$$

$$(2) \quad y = \frac{1}{2}x + 2000$$

$$(3) \quad 0.08x + 0.10y + 0.09z = 2220$$

Both sides of equation (2) can be multiplied by 2 to remove fractions.

$$2(y) = 2(\tfrac{1}{2}x + 2000)$$
$$2y = x + 4000$$
$$\text{or} \qquad -x + 2y = 4000$$

The decimals in equation (3) can be removed by multiplying both sides of the equation by 100 to get

$$8x + 10y + 9z = 222{,}000$$

Our system of equations is therefore

$$(1) \quad x + y + z = 25{,}000$$
$$(2) \quad -x + 2y = 4000$$
$$(3) \quad 8x + 10y + 9z = 222{,}000$$

There are various ways of solving this system. Let us start by multiplying equation (1) in the simplified set of equations by -9 and adding it to equation (3) to eliminate the variable z.

$$-9[x + y + z = 25{,}000] \quad \text{gives} \qquad \begin{aligned} -9x - 9y - 9z &= -225{,}000 \\ 8x + 10y + 9z &= \underline{222{,}000} \end{aligned}$$

$$(4) \quad \begin{aligned} -x + y &= -3{,}000 \end{aligned}$$

Now multiply equation (2) by -1 and add the results to equation (4) to eliminate the variable x.

$$\begin{aligned} (4) \qquad -x + y &= -3000 \quad \text{gives} \\ (2) \quad -1[-x + 2y &= 4000] \end{aligned} \qquad \begin{aligned} -x + y &= -3000 \\ \underline{x - 2y} &= \underline{-4000} \\ -y &= -7000 \\ y &= 7000 \end{aligned}$$

Now that we know the value of y we can solve for x.

$$\begin{aligned} -x + 2y &= 4000 \\ -x + 2(7000) &= 4000 \\ -x + 14{,}000 &= 4000 \\ -x &= -10{,}000 \\ x &= 10{,}000 \end{aligned}$$

Finally, solve for z.

$$\begin{aligned} x + y + z &= 25{,}000 \\ 10{,}000 + 7000 + z &= 25{,}000 \\ 17{,}000 + z &= 25{,}000 \\ z &= 8000 \end{aligned}$$

Thus, Tiny Tot Toys borrowed $10,000 at 8%, $7000 at 10%, and $8000 at 9% interest. ∎

EXAMPLE 8   Hobson, Inc., has a small manufacturing plant that makes three types of inflatable boats: one-person, two-person, and four-person models. Each boat requires the service of three departments: cutting, assembly, and packaging. The cutting, assembly, and packaging departments are allowed to use a total of 380, 330, and 120 person-hours per week, respectively. The time requirements for each boat and department are specified in the following table. Determine how many of each type of boat Hobson's must produce each week for its plant to operate at full capacity.

	Time (hr)		
Department	One-person Boat	Two-person Boat	Four-person Boat
Cutting	0.6	1.0	1.5
Assembly	0.6	0.9	1.2
Packaging	0.2	0.3	0.5

Solution: Let x = number of one-person boats
y = number of two-person boats
z = number of four-person boats

Then the total number of cutting hours for the three types of boats must equal 380 person-hours.

$$0.6x + 1.0y + 1.5z = 380$$

The total number of assembly hours must equal 330 person-hours.

$$0.6x + 0.9y + 1.2z = 330$$

The total number of packaging hours must equal 120 person-hours.

$$0.2x + 0.3y + 0.5z = 120$$

System of Equations

$$0.6x + 1.0y + 1.5z = 380$$
$$0.6x + 0.9y + 1.2z = 330$$
$$0.2x + 0.3y + 0.5z = 120$$

Multiplying each equation in the system by 10 will eliminate the decimal numbers.

Simplified System of Equations

$$6x + 10y + 15z = 3800$$
$$6x + 9y + 12z = 3300$$
$$2x + 3y + 5z = 1200$$

Let's first eliminate the variable x.

$$
\begin{array}{lll}
6x + 10y + 15z = 3800 & \text{gives} & 6x + 10y + 15z = 3800 \\
-1[6x + 9y + 12z = 3300] & & \underline{-6x - 9y - 12z = -3300} \\
& & y + 3z = 500
\end{array}
$$

$$
\begin{array}{lll}
6x + 10y + 15z = 3800 & \text{gives} & 6x + 10y + 15z = 3800 \\
-3[2x + 3y + 5z = 1200] & & \underline{-6x - 9y - 15z = -3600} \\
& & y = 200
\end{array}
$$

Note that when we added the last two equations, both variables x and z were eliminated at the same time. Now we proceed to solve for z.

$$y + 3z = 500$$
$$200 + 3z = 500$$
$$3z = 300$$
$$z = 100$$

Finally, find x.

$$6x + 10y + 15z = 3800$$
$$6x + 10(200) + 15(100) = 3800$$
$$6x + 2000 + 1500 = 3800$$
$$6x + 3500 = 3800$$
$$6x = 300$$
$$x = 50$$

Thus Hobson's should produce 50 one-person boats, 200 two-person boats, and 100 four-person boats per week.

Exercise Set 4.3

In each exercise, (a) express the problem as a system of linear equations and (b) use the method of your choice to find the solution to the problem.

1. The sum of two numbers is 73. Find the numbers if one number is 15 less than three times the other number.

2. The sum of two consecutive odd integers is 76. Find the two numbers.

3. The difference of two numbers is 25. Find the two numbers if the larger is 1 less than 3 times the smaller.

4. Two angles are **complementary angles** when the sum of their measures is 90°. If the larger of two complementary angles is 15° more than 2 times the smaller angle, find the two angles.

Angles *A* and *B* are complementary angles.

5. Two angles are **supplementary angles** when the sum of their measures is 180°. Find the two supplementary angles if one angle is 28° less than 3 times the other.

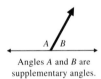
Angles *A* and *B* are supplementary angles.

6. A plane can travel 540 miles per hour with the wind and 490 miles per hour against the wind. Find the speed of the plane in still air and the speed of the wind.

7. A 50-foot length of rope is cut into two pieces. If one piece is 2 feet more than three times the other piece, find the length of the two pieces.

8. Steve Trinter, an electronics salesman, makes a weekly salary plus a commission on sales. One week his salary on sales of $4000 was $660. The next week his salary on sales of $6000 was $740. Find his weekly salary and his commission rate.

9. A car rental agency charges a daily fee plus a mileage fee. Mr. Dobson was charged $60 for 2 days and 100 miles and Mrs. Schwartz was charged $115 for 3 days and 400 miles. What is the agency's daily fee, and what is their mileage fee?

10. The total cost of printing a political leaflet consists of a fixed charge and an additional charge for each leaflet. If the total cost for 1000 leaflets is $550 and the total cost

See Exercise 13

for 2000 leaflets is $800, find the fixed charge and the charge for each leaflet.

11. Mr. Fiora, a druggist, needs 1000 milliliters of a 10% phenobarbital solution. He has only 5% and 25% phenobarbital solution available. How many milliliters of each solution should he mix to obtain the desired solution?

12. Mrs. Spinelli runs a grocery store. She wishes to mix 30 pounds of coffee to sell for a total cost of $170. To obtain the mixture, Mrs. Spinelli will mix coffee that sells for $5.20 per pound with coffee that sells for $6.30 per pound. How many pounds of each type of coffee should she use?

13. Mario owns a dairy. He has milk that is 5% butterfat and skim milk, without butterfat. How much 5% milk and how much skim milk should he mix to make 100 gallons of milk that is 3.5% butterfat?

14. Mr. and Mrs. McAdams invest a total of $8000 in two savings accounts. One account gives 10% interest and the other 8%. Find the amount placed in each account if they receive a total of $750 in interest after 1 year. Use interest = principal · rate · time.

15. Steve's recipe for Quiche Lorraine calls for 2 cups (16 ounces) of light cream that is 20% milk fat. It is often difficult to find light cream with 20% milk fat at the supermarket. What is commonly found is heavy cream, which is 36% milk fat, and half-and-half, which is 10.5% milk fat. How much of the heavy cream and how

much of the half-and-half should Steve mix to obtain the mixture necessary for the recipe?

16. Professor Collinge invested $30,000, part at 12% and part at 8%. If he had invested the entire amount at 9.5%, his total annual interest would be the same as the sum of the annual interest received from the two other accounts. How much was invested in each account?

17. Two cars start at the same point in Louisville and travel in opposite directions. One car travels 5 miles per hour faster than the other car. After 4 hours, the two cars are 420 miles apart. Find the speed of each car.

18. The Friendly Face Fruit Juice Company sells apple juice for 8.3 cents an ounce and raspberry juice for 9.3 cents an ounce. The company wishes to market and sell 8-ounce cans of apple–raspberry juice for 8.7 cents an ounce. How many ounces of each should be mixed?

19. Apple Computer stock is selling at $35 per share and Loews Theatre stock is selling at $20 per share. Geraldo has $6250 to invest. He wishes to purchase three times as many shares of Apple stock as of Loews. How many shares of each stock should he purchase?

20. A movie ticket for an adult costs $6.50 and a child's ticket costs $4.00. A total of 172 tickets are sold for a given show. If $1,013 is collected for the show, how many adults and how many children attended the show?

21. Safeway sells almonds for $6.00 per pound and walnuts for $5.40 per pound. How many pounds of each should they mix to produce a 30-pound mixture that costs $5.80 per pound?

22. Phong has a total of 12 bills in his wallet. Some are $5 bills and the rest are $10 bills. The total value of the bills is $115. How many $5 bills and how many $10 bills does he have?

23. A collection of dimes and quarters has a value of $3.55. If there are a total of 25 coins, how many dimes and quarters are there?

24. By traveling first at 40 mph and then at 50 mph, Jagat traveled 320 miles. Had he gone 10 mph faster for each period of time, he would have traveled 390 miles. How many hours did he travel at each rate?

25. Some states allow a husband and wife to file individual state tax returns (on a single form) even though they file a joint federal return. It is usually to the taxpayer's advantage to do this when both the husband and wife work. The smallest amount of tax owed (or the largest refund) will occur when the husband's and wife's taxable incomes are the same.

   Mr. Clar's 1988 taxable income was $26,200 and Mrs. Clar's income for that year was $22,450. The Clars' total tax deduction for the year was $12,400. This deduction can be divided between Mr. and Mrs. Clar in any way they wish. How should the $12,400 be divided between them to result in each person's having

the same taxable income and therefore the greatest tax return?

26. Two points on the line $y = ax + b$ are (3, 8) and $(-2, -17)$. Find the value of a and b.

27. Two points on the line $y = ax + b$ are $(-1, 7)$ and $(\frac{1}{2}, 4)$. Find the value of a and b.

28. Two photocopy machines are used to make large quantities of copies at Kinko's. The slower machine produces 75 copies per minute and the faster machine produces 120 copies per minute. The faster machine was in operation for 3 minutes when the slower machine was started. If they both continue copying together until they produced a total of 1335 copies, find the length of time both machines were in operation.

29. The sum of two times the first number and one-half the second number is 35. The sum of one-half the first number and one-third the second number is 15. Find the two numbers.

30. An automobile radiator has a capacity of 16 liters. How much pure antifreeze must be added to a mixture of water and antifreeze that is 18% antifreeze to make a mixture of 20% antifreeze that can be used to fill the radiator?

31. Animals in an experiment are to be kept on a strict diet. Each animal is to receive, among other things, 20 grams of protein and 6 grams of carbohydrates. The scientist only has two food mixes available of the following compositions.

	Protein (%)	Carbohydrate (%)
Mix A	10	6
Mix B	20	2

How many grams of each mix should be used to obtain the right diet for a single animal?

32. A company that makes children's wooden chairs makes two kinds of chairs. The basic model requires 1 hour to assemble and 0.5 hour to paint, and the deluxe model requires 3.2 hours to assemble but only 0.4 hour to paint. On a particular day the company has allocated 46.4 hours for assembling and 8.8 hours for painting. How many of each chair can be made?

33. By weight, one alloy of brass is 70% copper and 30% zinc. Another alloy of brass is 40% copper and 60% zinc. How many grams of each of these alloys need to be melted and combined to obtain 300 grams of a brass alloy that is 60% copper and 40% zinc?

34. A car travels 300 miles in the same amount of time that a truck takes to travel 240 miles. If the speed of the car is 10 miles per hour faster than the speed of the truck, find both speeds.

In each exercise (a) express the problem as a third-order system of linear equations and (b) solve the system of equations.

35. Three kinds of tickets are available for a Grateful Dead concert. The up-front floor tickets are the most expensive, the seats farther back on the floor are the second most expensive, and the balcony seats are the least expensive. The up-front floor seats are twice as expensive as the floor seats farther back. The balcony seats are $6 less than the floor seats in the back and $21 less than the up-front seats. Find the price of each seat.

36. The sum of twice the first number, three times the second number, and four times the third number is 40. The sum of the first and second numbers equals the third number. The third number is 2 more than twice the first number. Find the three numbers.

37. The sum of the measures of the angles of a triangle is $180°$. The smallest angle of a triangle is $\frac{2}{3}$ of the middle-sized angle. The largest angle is $30°$ less than 3 times the middle-sized angle. Find the measure of each angle.

38. Find a, b, and c so that the graph of the equation $y = ax^2 + bx + c$ passes through the points $(0, -3)$, $(2, 1)$, and $(-3, -24)$.

39. Find a, b, and c so that the graph of the equation $y = ax^2 + bx + c$ passes through the points $(2, 6)$, $(3, 17)$, and $(-1, -3)$.

40. Marion received a check for $10,000. She decided to divide the money (not equally) into three different investments. She placed part of her money in a savings account paying 7% interest. The second amount, which was twice the first amount, she placed into a certificate of deposit paying 9% interest. She placed the balance in a money market fund that yielded 10% interest. If Marion's total interest over the period of a year was $925.00, how much was placed in each account?

41. A 10% solution, a 12% solution, and a 20% solution of hydrogen peroxide are to be mixed to get 8 liters of a 13% solution. How many liters of each must be mixed if the volume of the 20% solution must be 2 liters less than the volume of the 10% solution?

42. An 8% solution, a 10% solution, and a 20% solution of sulfuric acid are to be mixed to get 100 milliliters of a 12% solution. If the *volume of acid* from the 8% solution is to equal half the *volume of acid* from the other two solutions, how much of each solution is needed?

43. Donaldson Furniture Company produces three types of rocking chairs: the children's model, the standard model, and the executive model. Each chair is made in three stages: cutting, construction, and finishing. The time needed for each stage of each chair is given in the following chart. During a specific week the company has available a maximum of 154 hours for cutting, 94 hours for construction, and 76 hours for finishing. Determine how many of each chair the company should make to be operating at full capacity.

	Children's	Standard	Executive
Cutting	5 hr	4 hr	7 hr
Construction	3 hr	2 hr	5 hr
Finishing	2 hr	2 hr	4 hr

44. By volume, one alloy is 60% copper, 30% zinc, and 10% nickel. A second alloy has percentages 50, 30, and 20, respectively, of the three metals. A third alloy is 30% copper and 70% nickel. How much of each alloy must be mixed so that 100 pounds of the resulting alloy is 40% copper, 15% zinc, and 45% nickel?

45. In the study of electronics it is necessary to analyze current flow through certain paths of a circuit. The study of three branches $(A, B, \text{and } C)$ of a circuit yields the following results:

$$I_A + I_B + I_C = 0$$
$$-8I_B + 10I_C = 0$$
$$4I_A - 8I_B = 6$$

where I_A, I_B, and I_C represent the current in branches A, B, and C, respectively. Determine the current in each branch of the circuit.

46. In the study of physics we often study the forces acting on an object. For three forces, F_1, F_2, and F_3, acting on a beam, the following equations were obtained.

$$3F_1 + F_2 - F_3 = 2$$
$$F_1 - 2F_2 + F_3 = 0$$
$$4F_1 - F_2 + F_3 = 3$$

Find the three forces.

Cumulative Review Exercises

[1.7] **47.** Evaluate $\frac{1}{2}x + \frac{2}{5}xy + \frac{1}{8}y$ when $x = -2$ and $y = 5$.

[3.1] **48.** Determine the length and midpoint of the line segment through the points $(6, -4)$ and $(2, -8)$.

[3.4] **49.** Write an equation of the line that passes through points $(6, -4)$ and $(2, -8)$.

[3.5] **50.** Explain how to determine if a graph is a function.

JUST FOR FUN

1. In an article published in the *Journal of Comparative Physiology and Psychology*, J. S. Brown discusses how we often approach a situation with mixed emotions. For example, when a person is asked to give a speech, he may be a little apprehensive about his ability to do a good job. At the same time, he would like the recognition that goes along with making the speech. J. S. Brown performed an experiment on trained rats. He placed their food in a metal box. He used that same box to administer small electrical shocks to the mice. Therefore, the rats "wished" to go into the box to receive food, yet did not "wish" to go into the box for fear of receiving a small shock. Using the appropriate apparatus, Brown arrived at the following relationships:

See Just for Fun, Exercise 1.

$$\text{pull (in grams) toward food} = -\frac{1}{5}d + 70, \qquad 30 < d < 172.5$$

$$\text{pull (in grams) away from shock} = -\frac{4}{3}d + 230, \qquad 30 < d < 172.5$$

where d is the distance in centimeters from the box (and food).

(a) Using the substitution method, find the distance at which the pull toward the food equals the pull away from the shock.

(b) If the rat is placed 100 cm from the box (or food), what will the rat do?

4.4

Solving Systems of Equations by Determinants (Optional)

▶ **1** Find the value of a second-order determinant.

▶ **2** Use Cramer's rule to solve second-order systems of equations.

▶ **3** Find the value of a third-order determinant.

▶ **4** Use Cramer's rule to solve third-order systems of equations.

Systems of linear equations can also be solved using determinants. Determinants are particularly useful when solving third- and higher-order systems of equations.

A **determinant** is a square array of numbers enclosed between two vertical bars. Examples of determinants are

$$\begin{vmatrix} 4 & -3 \\ 0 & 5 \end{vmatrix} \qquad \begin{vmatrix} 3 & 0 & 5 \\ 4 & -2 & 3 \\ 2 & \frac{1}{2} & -1 \end{vmatrix}$$

$$\text{(a)} \qquad\qquad\qquad \text{(b)}$$

The numbers that make up the array are called the **elements** of the determinant. The elements of determinant (a) are 4, -3, 0, and 5.

Determinant (a) is a **second-order determinant** since it has two rows and two columns of elements. Determinant (b) is a **third-order determinant.** Determinants can be of an order greater than 3.

The **principal diagonal** of a determinant is the line of elements from the upper-left corner to the lower-right corner. The **secondary diagonal** of a determinant is the line of elements from the lower-left corner to the upper-right corner.

$$\begin{vmatrix} a_1 & b_1 \\ a_2 & b_2 \end{vmatrix} \qquad \begin{vmatrix} a_1 & b_1 \\ a_2 & b_2 \end{vmatrix}$$

principal   secondary
diagonal   diagonal

▶ **1**   Every determinant represents a number. The **value of a second-order determinant** is the product of the elements in its principal diagonal minus the product of the elements in its secondary diagonal.

Value of a Second-order Determinant

$$\begin{vmatrix} a_1 & b_1 \\ a_2 & b_2 \end{vmatrix} = a_1 b_2 - a_2 b_1$$

EXAMPLE 1   Find the value of the determinant $\begin{vmatrix} 4 & 6 \\ -3 & 2 \end{vmatrix}$.

Solution: Here $a_1 = 4$, $a_2 = -3$, $b_1 = 6$, $b_2 = 2$.

$$\begin{vmatrix} 4 & 6 \\ -3 & 2 \end{vmatrix} = 4(2) - (-3)(6)$$

$$= 8 + 18$$

$$= 26$$

The determinant has a value of 26.   ■

EXAMPLE 2   Find the value of the determinant $\begin{vmatrix} -3 & 4 \\ 1 & 5 \end{vmatrix}$.

Solution:

$$\begin{vmatrix} -3 & 4 \\ 1 & 5 \end{vmatrix} = (-3)(5) - (1)(4)$$

$$= -15 - 4$$

$$= -19$$ ■

▶ **2**   Consider the system of equations

$$(1) \quad a_1 x + b_1 y = c_1$$
$$(2) \quad a_2 x + b_2 y = c_2$$

To eliminate the variable y, we can multiply both sides of equation (1) by b_2 and both sides of equation (2) by $-b_1$ and then add.

$$b_2[a_1 x + b_1 y = c_1] \quad \text{gives}$$
$$-b_1[a_2 x + b_2 y = c_2]$$

$$a_1 b_2 x + b_1 b_2 y = c_1 b_2$$
$$\underline{-a_2 b_1 x - b_1 b_2 y = -c_2 b_1}$$
$$(a_1 b_2 - a_2 b_1)x = c_1 b_2 - c_2 b_1$$

$$x = \frac{c_1 b_2 - c_2 b_1}{a_1 b_2 - a_2 b_1}$$

We can solve the system for y in a similar manner.

$$\begin{array}{l} -a_2[a_1x + b_1y = c_1] \\ a_1[a_2x + b_2y = c_2] \end{array} \quad \text{gives} \quad \begin{array}{r} -a_1a_2x - a_2b_1y = -a_2c_1 \\ \underline{a_1a_2x + a_1b_2y = a_1c_2} \\ (a_1b_2 - a_2b_1)y = a_1c_2 - a_2c_1 \\ y = \dfrac{a_1c_2 - a_2c_1}{a_1b_2 - a_2b_1} \end{array}$$

Both x and y have the same denominator, $a_1b_2 - a_2b_1$. Note that

$$\begin{vmatrix} a_1 & b_1 \\ a_2 & b_2 \end{vmatrix} = a_1b_2 - a_2b_1$$

The numerator of the expression used to find x is $c_1b_2 - c_2b_1$. Note that

$$\begin{vmatrix} c_1 & b_1 \\ c_2 & b_2 \end{vmatrix} = c_1b_2 - c_2b_1$$

The numerator of the expression used to find y is $a_1c_2 - a_2c_1$. Note that

$$\begin{vmatrix} a_1 & c_1 \\ a_2 & c_2 \end{vmatrix} = a_1c_2 - a_2c_1$$

Using the information above and making the appropriate substitutions, we can express both x and y as a quotient of two determinants.

$$x = \frac{c_1b_2 - c_2b_1}{a_1b_2 - a_2b_1} = \frac{\begin{vmatrix} c_1 & b_1 \\ c_2 & b_2 \end{vmatrix}}{\begin{vmatrix} a_1 & b_1 \\ a_2 & b_2 \end{vmatrix}}, \qquad y = \frac{a_1c_2 - a_2c_1}{a_1b_2 - a_2b_1} = \frac{\begin{vmatrix} a_1 & c_1 \\ a_2 & c_2 \end{vmatrix}}{\begin{vmatrix} a_1 & b_1 \\ a_2 & b_2 \end{vmatrix}}$$

A second-order system of equations can be solved by Cramer's rule, which uses 3 different determinants.

For discussion purposes, we will denote the three different determinants as

$$D = \begin{vmatrix} a_1 & b_1 \\ a_2 & b_2 \end{vmatrix} \qquad D_x = \begin{vmatrix} c_1 & b_1 \\ c_2 & b_2 \end{vmatrix} \qquad D_y = \begin{vmatrix} a_1 & c_1 \\ a_2 & c_2 \end{vmatrix}$$

determinant of numerical coefficients	a's in determinant D replaced by constants, c's	b's in determinant D replaced by constants, c's

---

Cramer's Rule

For a system of equations of the form

$$a_1x + b_1y = c_1$$
$$a_2x + b_2y = c_2$$

$$x = \frac{\begin{vmatrix} c_1 & b_1 \\ c_2 & b_2 \end{vmatrix}}{\begin{vmatrix} a_1 & b_1 \\ a_2 & b_2 \end{vmatrix}} = \frac{D_x}{D} \quad \text{and} \quad y = \frac{\begin{vmatrix} a_1 & c_1 \\ a_2 & c_2 \end{vmatrix}}{\begin{vmatrix} a_1 & b_1 \\ a_2 & b_2 \end{vmatrix}} = \frac{D_y}{D}, D \neq 0$$

EXAMPLE 3   Use determinants to evaluate the following system.

$$2x + y = 6$$
$$3x + y = 5$$

Solution: Both equations are given in the desired form, $ax + by = c$. We will refer to $2x + y = 6$ as equation 1 and $3x + y = 5$ as equation 2.

$$\begin{matrix} a_1 & b_1 & c_1 \\ \downarrow & \downarrow & \downarrow \\ 2x + & 1y = & 6 \\ 3x + & 1y = & 5 \\ \uparrow & \uparrow & \uparrow \\ a_2 & b_2 & c_2 \end{matrix}$$

We now find D, D_x, and D_y.

$$D = \begin{vmatrix} a_1 & b_1 \\ a_2 & b_2 \end{vmatrix} = \begin{vmatrix} 2 & 1 \\ 3 & 1 \end{vmatrix} = 2(1) - 3(1) = -1$$

$$D_x = \begin{vmatrix} c_1 & b_1 \\ c_2 & b_2 \end{vmatrix} = \begin{vmatrix} 6 & 1 \\ 5 & 1 \end{vmatrix} = 6(1) - 5(1) = 1$$

$$D_y = \begin{vmatrix} a_1 & c_1 \\ a_2 & c_2 \end{vmatrix} = \begin{vmatrix} 2 & 6 \\ 3 & 5 \end{vmatrix} = 2(5) - 3(6) = -8$$

$$x = \frac{D_x}{D} = \frac{1}{-1} = -1$$

$$y = \frac{D_y}{D} = \frac{-8}{-1} = 8$$

Thus the solution is $x = -1$, $y = 8$ or the ordered pair $(-1, 8)$.

Check:

$2x + y = 6$		$3x + y = 5$	
$2(-1) + 8 = 6$		$3(-1) + 8 = 5$	
$-2 + 8 = 6$		$-3 + 8 = 5$	
$6 = 6$	true	$5 = 5$	true

It makes no difference which equations you label 1 and 2 as long as you remain consistent. Rework the example letting $3x + y = 5$ represent equation 1 and $2x + y = 6$ equation 2.

EXAMPLE 4   Solve the following system using determinants.

$$2x - 4y = \quad 8$$
$$3x + 5y = -10$$

Solution:

$$a_1 = 2, \quad b_1 = -4, \quad c_1 = \quad 8$$
$$a_2 = 3, \quad b_2 = \quad 5, \quad c_2 = -10$$

$$D = \begin{vmatrix} a_1 & b_1 \\ a_2 & b_2 \end{vmatrix} = \begin{vmatrix} 2 & -4 \\ 3 & 5 \end{vmatrix} = 2(5) - 3(-4) = 22$$

$$D_x = \begin{vmatrix} c_1 & b_1 \\ c_2 & b_2 \end{vmatrix} = \begin{vmatrix} 8 & -4 \\ -10 & 5 \end{vmatrix} = 8(5) - (-10)(-4) = 0$$

$$D_y = \begin{vmatrix} a_1 & c_1 \\ a_2 & c_2 \end{vmatrix} = \begin{vmatrix} 2 & 8 \\ 3 & -10 \end{vmatrix} = 2(-10) - (3)(8) = -44$$

$$x = \frac{D_x}{D} = \frac{0}{22} = 0$$

$$y = \frac{D_y}{D} = \frac{-44}{22} = -2$$

Thus the solution is $(0, -2)$.   ■

When solving a system using determinants, if you obtain the quotient $0/0$ for any of the variables, the two equations represent the same line and there are an infinite number of solutions. If you obtain a quotient of the form $a/0$, $a \neq 0$, the two lines are parallel and there is no solution.

▶ **3**   A third-order determinant is evaluated as follows:

minor determinant of a_1 minor determinant of a_2 minor determinant of a_3

$$\begin{vmatrix} a_1 & b_1 & c_1 \\ a_2 & b_2 & c_2 \\ a_3 & b_3 & c_3 \end{vmatrix} = a_1 \begin{vmatrix} b_2 & c_2 \\ b_3 & c_3 \end{vmatrix} - a_2 \begin{vmatrix} b_1 & c_1 \\ b_3 & c_3 \end{vmatrix} + a_3 \begin{vmatrix} b_1 & c_1 \\ b_2 & c_2 \end{vmatrix}$$

This method of evaluating the determinant is called **expansion of the determinant by the minors of the first column.**

The determinant $\begin{vmatrix} b_2 & c_2 \\ b_3 & c_3 \end{vmatrix}$ is called the minor determinant of a_1. The minor determinant of a_1 is found by crossing out the elements in the same row and column in which the element a_1 appears.

$$\begin{vmatrix} a_1 & b_1 & c_1 \\ a_2 & b_2 & c_2 \\ a_3 & b_3 & c_3 \end{vmatrix}$$

The remaining elements form the minor determinant of a_1.

$$\begin{vmatrix} b_2 & c_2 \\ b_3 & c_3 \end{vmatrix}$$

The minor determinant of a_2 is found similarly.

$$\begin{vmatrix} a_1 & b_1 & c_1 \\ a_2 & b_2 & c_2 \\ a_3 & b_3 & c_3 \end{vmatrix} \qquad \text{minor determinant of } a_2$$

$$\begin{vmatrix} b_1 & c_1 \\ b_3 & c_3 \end{vmatrix}$$

The minor determinant of a_3 is found below.

$$\begin{vmatrix} a_1 & b_1 & c_1 \\ a_2 & b_2 & c_2 \\ a_3 & b_3 & c_3 \end{vmatrix} \qquad \text{minor determinant of } a_3 \qquad \begin{vmatrix} b_1 & c_1 \\ b_2 & c_2 \end{vmatrix}$$

EXAMPLE 5 Evaluate $\begin{vmatrix} 4 & -2 & 6 \\ 3 & 5 & 0 \\ 1 & -3 & -1 \end{vmatrix}$.

Solution:

$$\begin{vmatrix} 4 & -2 & 6 \\ 3 & 5 & 0 \\ 1 & -3 & -1 \end{vmatrix} = 4\begin{vmatrix} 5 & 0 \\ -3 & -1 \end{vmatrix} - 3\begin{vmatrix} -2 & 6 \\ -3 & -1 \end{vmatrix} + 1\begin{vmatrix} -2 & 6 \\ 5 & 0 \end{vmatrix}$$

$$= 4[5(-1) - (-3)0] - 3[(-2)(-1) - (-3)6] + 1[(-2)0 - 5(6)]$$
$$= 4(-5 + 0) - 3(2 + 18) + 1(0 - 30)$$
$$= 4(-5) - 3(20) + 1(-30)$$
$$= -20 - 60 - 30$$
$$= -110$$

The determinant has a value of -110. ∎

In Example 5 we found the solution using expansion of the determinant by the minor determinants of the first column. To evaluate a determinant, we can use expansion of the determinant by the minor determinants of any row or any column. To determine if the product of the element and its minor determinant is to be added or subtracted in obtaining the result, we use the following chart.

$$\begin{matrix} + & - & + \\ - & + & - \\ + & - & + \end{matrix}$$

If the element is in a position marked with a $+$, the product is to be added. If the element is in a position marked with a $-$, the product is to be subtracted.

EXAMPLE 6 Evaluate $\begin{vmatrix} 4 & -2 & 6 \\ 3 & 5 & 0 \\ 1 & -3 & -1 \end{vmatrix}$ using expansion of the determinant by the minors of the second row.

Solution: The second row of the chart is $-$, $+$, $-$. Thus the first product is to be subtracted, the second product added, and the third product subtracted.

$$\begin{vmatrix} 4 & -2 & 6 \\ 3 & 5 & 0 \\ 1 & -3 & -1 \end{vmatrix} = -3\underset{\underset{\text{subtract}}{\uparrow}}{\begin{vmatrix} -2 & 6 \\ -3 & -1 \end{vmatrix}} + 5\underset{\underset{\text{add}}{\uparrow}}{\begin{vmatrix} 4 & 6 \\ 1 & -1 \end{vmatrix}} - 0\underset{\underset{\text{subtract}}{\uparrow}}{\begin{vmatrix} 4 & -2 \\ 1 & -3 \end{vmatrix}}$$

$$= -3[(-2)(-1) - (-3)6] + 5[4(-1) - 1(6)] - 0[4(-3) - 1(-2)]$$
$$= -3(2 + 18) + 5(-4 - 6) - 0(-12 + 2)$$
$$= -3(20) + 5(-10) - 0$$
$$= -60 - 50$$
$$= -110$$ ∎

Note that the same answer was obtained by evaluating the determinant using the first column or the second row. When evaluating a determinant containing one or more 0's in a particular row or column, you may wish to evaluate the determinant by expansion of the minor determinants of that row or column.

▶ **4** Cramer's rule can be extended to third-order systems of equations as follows:

To evaluate the system

$$a_1x + b_1y + c_1z = d_1$$
$$a_2x + b_2y + c_2z = d_2$$
$$a_3x + b_3y + c_3z = d_3$$

with

$$D = \begin{vmatrix} a_1 & b_1 & c_1 \\ a_2 & b_2 & c_2 \\ a_3 & b_3 & c_3 \end{vmatrix} \qquad D_x = \begin{vmatrix} d_1 & b_1 & c_1 \\ d_2 & b_2 & c_2 \\ d_3 & b_3 & c_3 \end{vmatrix}$$

$$D_y = \begin{vmatrix} a_1 & d_1 & c_1 \\ a_2 & d_2 & c_2 \\ a_3 & d_3 & c_3 \end{vmatrix} \qquad D_z = \begin{vmatrix} a_1 & b_1 & d_1 \\ a_2 & b_2 & d_2 \\ a_3 & b_3 & d_3 \end{vmatrix}$$

then

$$x = \frac{D_x}{D}, \qquad y = \frac{D_y}{D}, \qquad z = \frac{D_z}{D}, \qquad D \neq 0$$

Note that the denominators of the expressions for x, y, and z are all the same determinant, D. Note that constants, the d's, replace the a's, the numerical coefficients of the x terms, in D_x. The d's replace the b's, the numerical coefficients of the y terms, in D_y. And the d's replace the c's, the numerical coefficients of the z terms in D_z.

EXAMPLE 7 Solve the following system of equations using determinants.

$$3x - 2y - z = -6$$
$$2x + 3y - 2z = 1$$
$$x - 4y + z = -3$$

Solution:
$$a_1 = 3 \qquad b_1 = -2 \qquad c_1 = -1 \qquad d_1 = -6$$
$$a_2 = 2 \qquad b_2 = 3 \qquad c_2 = -2 \qquad d_2 = 1$$
$$a_3 = 1 \qquad b_3 = -4 \qquad c_3 = 1 \qquad d_3 = -3$$

We will use expansion of the minor determinants by the first row to evaluate D.

$$D = \begin{vmatrix} 3 & -2 & -1 \\ 2 & 3 & -2 \\ 1 & -4 & 1 \end{vmatrix} = 3 \begin{vmatrix} 3 & -2 \\ -4 & 1 \end{vmatrix} - (-2) \begin{vmatrix} 2 & -2 \\ 1 & 1 \end{vmatrix} + (-1) \begin{vmatrix} 2 & 3 \\ 1 & -4 \end{vmatrix}$$

$$= 3(-5) + 2(4) - 1(-11)$$
$$= -15 + 8 + 11 = 4$$

We will evaluate D_x using expansion of the determinant by the minor determinants of the first column.

$$D_x = \begin{vmatrix} -6 & -2 & -1 \\ 1 & 3 & -2 \\ -3 & -4 & 1 \end{vmatrix} = (-6)\begin{vmatrix} 3 & -2 \\ -4 & 1 \end{vmatrix} - (1)\begin{vmatrix} -2 & -1 \\ -4 & 1 \end{vmatrix} + (-3)\begin{vmatrix} -2 & -1 \\ 3 & -2 \end{vmatrix}$$

$$= -6(-5) - 1(-6) - 3(7)$$

$$= 30 + 6 - 21 = 15$$

We will use expansion of the determinant by the minor determinants of the first row to evaluate D_y.

$$D_y = \begin{vmatrix} 3 & -6 & -1 \\ 2 & 1 & -2 \\ 1 & -3 & 1 \end{vmatrix} = 3\begin{vmatrix} 1 & -2 \\ -3 & 1 \end{vmatrix} - (-6)\begin{vmatrix} 2 & -2 \\ 1 & 1 \end{vmatrix} + (-1)\begin{vmatrix} 2 & 1 \\ 1 & -3 \end{vmatrix}$$

$$= 3(-5) + 6(4) - 1(-7)$$

$$= -15 + 24 + 7 = 16$$

We will evaluate D_z using expansion of the determinant by the minor determinants of the first row.

$$D_z = \begin{vmatrix} 3 & -2 & -6 \\ 2 & 3 & 1 \\ 1 & -4 & -3 \end{vmatrix} = 3\begin{vmatrix} 3 & 1 \\ -4 & -3 \end{vmatrix} - (-2)\begin{vmatrix} 2 & 1 \\ 1 & -3 \end{vmatrix} + (-6)\begin{vmatrix} 2 & 3 \\ 1 & -4 \end{vmatrix}$$

$$= 3(-5) + 2(-7) - 6(-11)$$

$$= -15 - 14 + 66 = 37$$

We found that $D = 4$, $D_x = 15$, $D_y = 16$, and $D_z = 37$.

$$x = \frac{D_x}{D} = \frac{15}{4}, \qquad y = \frac{D_y}{D} = \frac{16}{4} = 4, \qquad z = \frac{D_z}{D} = \frac{37}{4}$$

The solution to the system is $(\frac{15}{4}, 4, \frac{37}{4})$. Note the ordered triple lists x, y, and z in this order.   ■

When we are given a third-order system of equations in which one or more equations are missing a variable, we insert the variable with a coefficient of 0. This helps in aligning like terms. For example,

$$
\begin{aligned}
2x - 3y + 2z &= -1 \\
x + 2y &= 14 \\
x - 3z &= -5
\end{aligned}
\qquad \text{is written} \qquad
\begin{aligned}
2x - 3y + 2z &= -1 \\
x + 2y + 0z &= 14 \\
x + 0y - 3z &= -5
\end{aligned}
$$

when solving the system using determinants. Furthermore, it is very important to place the numbers in the correct column. In this example

$$D = \begin{vmatrix} 2 & -3 & 2 \\ 1 & 2 & 0 \\ 1 & 0 & -3 \end{vmatrix}, \qquad D_x = \begin{vmatrix} -1 & -3 & 2 \\ 14 & 2 & 0 \\ -5 & 0 & -3 \end{vmatrix}$$

$$D_y = \begin{vmatrix} 2 & -1 & 2 \\ 1 & 14 & 0 \\ 1 & -5 & -3 \end{vmatrix}, \qquad D_z = \begin{vmatrix} 2 & -3 & -1 \\ 1 & 2 & 14 \\ 1 & 0 & -5 \end{vmatrix}$$

When you are solving for a variable in a third-order system, if the denominator, D, has a value of 0 and any numerator (D_x, D_y, and D_z) does not have a value of zero, then the system is inconsistent and has no solution. If D, D_x, D_y, and D_z are all 0, then the system is dependent and there are infinitely many solutions.

Exercise Set 4.4

Solve the system of equations using determinants.

1. $x + 2y = 5$
$x - 2y = 1$

2. $3x - 2y = 4$
$3x + y = -2$

3. $x - 2y = -1$
$x + 3y = 9$

4. $3x - y = 3$
$4x - 3y = 14$

5. $3x + 4y = 8$
$2x - 3y = 9$

6. $6x + 3y = -4$
$9x + 5y = -6$

7. $2x = y + 5$
$6x + 2y = -5$

8. $x + 5y = 3$
$2x + 10y = 6$

9. $3x = -4y - 6$
$3y = -5x + 1$

10. $5x - 5y = 3$
$x - y = -2$

11. $6.3x - 4.5y = -9.9$
$-9.1x + 3.2y = -2.2$

12. $-1.1x + 8.3y = 36.5$
$3.5x + 1.6y = -4.1$

Solve the system of equations using determinants.

13. $x + y - z = -3$
$x + z = 2$
$2x - y + 2z = 3$

14. $2x - y + 3z = 0$
$x + 2y - z = 5$
$2y + z = 1$

15. $-x + y = 1$
$y - z = 2$
$x + z = -2$

16. $-x + 2y + 3z = -1$
$-3x - 3y + z = 0$
$2x + 3y + z = 2$

17. $2x + 2y + 2z = 0$
$-x - 3y + 7z = 15$
$3x + y + 4z = 21$

18. $x - 2y + 3z = 4$
$2x - y + z = -5$
$x + y - z = -2$

19. $x - y + 2z = 3$
$x - y + z = 1$
$2x + y + 2z = 2$

20. $2x + y - 2 = 0$
$3x + 2y + z = 3$
$x - 3y - 5z = 5$

21. $x + 2y + z = 1$
$x - y + z = 1$
$2x + y + 2z = 2$

22. $2x + y - 2z = -4$
$x + y + z = 1$
$x + y + 2z = 3$

23. $1.1x + 2.3y - 4.0z = -9.2$
$-2.3x + 4.6z = 6.9$
$-8.2y - 7.5z = -6.8$

24. $4.6y - 2.1z = 24.3$
$-5.6x + 1.8y = -5.8$
$2.8x - 4.7y - 3.1z = 7.0$

25. Describe a determinant, a second-order determinant, and a third-order determinant.

26. Given a second-order determinant of the form $\begin{vmatrix} a_1 & b_1 \\ a_2 & b_2 \end{vmatrix}$, how will the value of the determinant change if the a's are switched with the b's, $\begin{vmatrix} b_1 & a_1 \\ b_2 & a_2 \end{vmatrix}$? Explain your answer.

27. Given a second-order determinant of the form $\begin{vmatrix} a_1 & b_1 \\ a_2 & b_2 \end{vmatrix}$, how will the value of the determinant change if the a's are switched with each other and the b's are switched with each other, $\begin{vmatrix} a_2 & b_2 \\ a_1 & b_1 \end{vmatrix}$? Explain your answer.

Cumulative Review Exercises

[2.5] **28.** Solve the inequality $3(x - 2) < \dfrac{4}{5}(x - 4)$ and indicate the solution in interval notation.

Graph $3x + 4y = 8$.

[3.2] **29.** By plotting points.

30. Using the x and y intercepts.

[3.4] **31.** Using the slope and y intercept.

4.5

Solving Systems of Linear Inequalities

▶ **1** Solve systems of linear inequalities.

▶ **2** Solve linear programming problems.

▶ **3** Solve systems of linear inequalities containing absolute value.

▶ **1** In Section 3.6 we showed how to graph linear inequalities in two variables. In Section 4.1 we learned how to solve systems of equations graphically. In this section we show how to solve systems of linear inequalities graphically.

> **To Solve a System of Linear Inequalities**
>
> Graph each inequality on the same set of axes. The solution is the set of points that satisfies all the inequalities in the system.

EXAMPLE 1 Determine the solution to the system of inequalities.

$$x + y \leq 6$$
$$y > 2x - 3$$

Solution: First graph the inequality $x + y \leq 6$ (see Fig. 4.6). Now on the same set of axes graph inequality $y > 2x - 3$ (Fig. 4.7). The solution is the set of points common to both inequalities. It is the part of the graph that contains both shadings. The dashed line is not part of the solution, but the part of the solid line that satisfies both inequalities is.

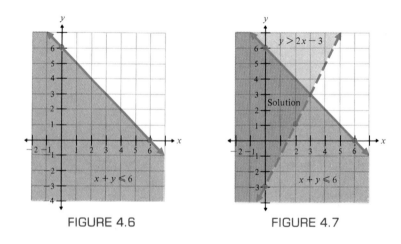

FIGURE 4.6   FIGURE 4.7   ■

EXAMPLE 2 Determine the solution to the system of inequalities.

$$2x + 3y \geq 4$$
$$2x - y > -6$$

Solution: Graph $2x + 3y \geq 4$ (see Fig. 4.8). Graph $2x - y > -6$ on the same set of axes (Fig. 4.9). The solution is the part of the graph with both shadings and the part of the solid line that satisfies both inequalities.

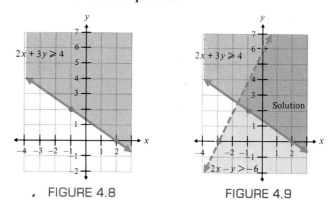

. FIGURE 4.8   FIGURE 4.9

EXAMPLE 3   Determine the solution to the system of inequalities.

$$y < 4$$
$$x > -2$$

Solution: The solution is illustrated in Fig. 4.10.

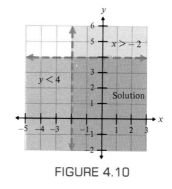

FIGURE 4.10

Linear Programming (Optional)   ▶**2**   There is a mathematical topic called **linear programming** for which you often have to graph more than two linear inequalities on the same set of axes. These inequalities are called **constraints.** The following two examples illustrate how to determine the solution to a set of more than two inequalities.

EXAMPLE 4   Determine the solution to the following set of inequalities.

$$x \geq 0$$
$$y \geq 0$$
$$2x + 3y \leq 12$$
$$2x + y \leq 8$$

Solution: The first two inequalities, $x \geq 0$ and $y \geq 0$, indicate that the solution must be in the first quadrant because that is the only quadrant where both x and y are positive. Figure 4.11 illustrates the graphs of the four inequalities.

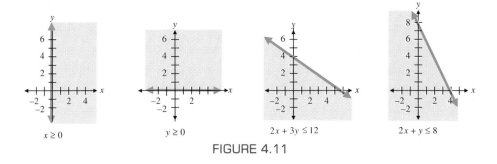

$x \geq 0$ $y \geq 0$ $2x + 3y \leq 12$ $2x + y \leq 8$

FIGURE 4.11

Figure 4.12 illustrates the graphs on the same set of axes and the solution set to the system of inequalities. Note that every point in the shaded area and every point on the lines that form the polygonal region is part of the answer.

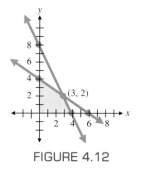

FIGURE 4.12

EXAMPLE 5 Determine the solution to the following set of inequalities.

$$x \geq 0$$
$$y \geq 0$$
$$x \leq 15$$
$$8x + 8y \leq 160$$
$$4x + 12y \leq 180$$

Solution: The first two inequalities indicate that the solution must be in the first quadrant. The third inequality indicates that x must be a value less than or equal to 15. Figure 4.13a indicates the graphs of the 3 bottom equations. Figure 4.13b indicates the solution to the system of inequalities.

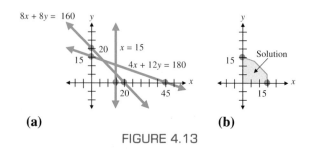

(a)   **(b)**

FIGURE 4.13

**Graphing
Inequalities with
Absolute Value
(Optional)**

▶ **3**   Now we will graph inequalities containing absolute value on the Cartesian coordinate system. Before we do some examples, let us recall the rules involving absolute value inequalities that we learned in Section 2.6. Recall that

> If $|x| < a$ and $a > 0$, then $-a < x < a$.
>
> If $|x| > a$ and $a > 0$, then $x < -a$ or $x > a$.

EXAMPLE 6   Graph $|x| < 3$ in the Cartesian coordinate system.

Solution: From the rules given above we know that $|x| < 3$ means $-3 < x < 3$. Now draw dashed vertical lines through -3 and 3 and shade the area between the two (Fig. 4.14).

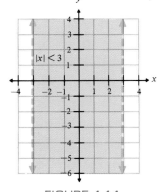

FIGURE 4.14

EXAMPLE 7   Graph $|y + 1| > 3$ in the Cartesian coordinate system.

Solution: From the rules given above we know that $|y + 1| > 3$ means $y + 1 > 3$ or $y + 1 < -3$. Solve each inequality.

$$y + 1 > 3 \quad \text{or} \quad y + 1 < -3$$
$$y > 2 \qquad\qquad y < -4$$

Now graph both inequalities and take the *union* of the two graphs. The solution is the shaded area in Fig. 4.15.

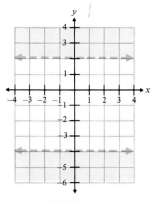

FIGURE 4.15

EXAMPLE 8   Graph the system of inequalities.

$$|x| < 3$$
$$|y + 1| > 3$$

Solution: Draw both inequalities on the same set of axes. Therefore, we combine the graph drawn in Example 6 with the graph drawn in Example 7 (see Fig. 4.16). The points common to both inequalities form the solution to the system.

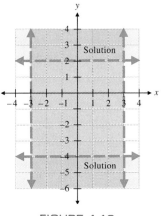

FIGURE 4.16

Exercise Set 4.5

Determine the solution to each system of inequalities.

1. $x - y > 2$
    $y < -2x + 3$

2. $y \geq 3x - 2$
    $y > -4x$

3. $y \leq x - 4$
    $y < -2x + 4$

4. $2x + 3y < 6$
    $4x - 2y \geq 8$

5. $y < x$
    $y \geq 3x + 2$

6. $-x + 3y \geq 6$
    $-2x - y > 4$

7. $4x - 2y < 6$
    $y \leq -x + 4$

8. $y \leq 3x + 4$
    $y > 2$

9. $-4x + 5y < 20$
    $x \geq -3$

10. $3x - 4y \leq 6$
    $y > -x + 4$

11. $x \leq 4$
    $y \geq -2$

12. $x \geq 0$
    $x - 3y < 6$

13. $5x + 2y > 10$
$\quad\;\; 3x - y > 3$

14. $3x + 2y > 8$
$\quad\;\; x - 5y < 5$

15. $-2x < y + 4$
$\quad\;\; 3x \geq y$

16. $y \leq 4x - 6$
$\quad\;\; 2x + 4y < 6$

17. $\frac{1}{2}x + 3y > 6$
$\quad\;\; y < 3x - 4$

18. $\frac{1}{2}x + \frac{1}{3}y \geq 2$
$\quad\;\; 2x - 3y \leq -6$

Determine the solution to each system of inequalities.

19.
$\quad\quad x \geq 0$
$\quad\quad y \geq 0$
$5x + 4y \leq 20$
$\;\; x + 2y \leq 6$

20.
$\quad\quad x \geq 0$
$\quad\quad y \geq 0$
$3x + 2y \leq 10$
$2x + 5y \leq 15$

21.
$\quad\quad x \geq 0$
$\quad\quad y \geq 0$
$\;\; x + y \leq 6$
$7x + 4y \leq 28$

22.
$\quad\quad x \geq 0$
$\quad\quad y \geq 0$
$8x + 3y \leq 24$
$2x + 3y \leq 12$

23.
$\quad\quad x \geq 0$
$\quad\quad y \geq 0$
$7x + 4y \leq 24$
$2x + 5y \leq 20$

24.
$\quad\quad x \geq 0$
$\quad\quad y \geq 0$
$5x + 4y \leq 16$
$\;\; x + 6y \leq 18$

25.
$\quad\quad x \geq 0$
$\quad\quad y \geq 0$
$\quad\quad x \leq 4$
$x + \;\; y \leq 6$
$x + 2y \leq 8$

26.
$\quad\quad x \geq 0$
$\quad\quad y \geq 0$
$\quad\quad x \leq 15$
$40x + 25y \leq 1000$
$\;\; 5x + 30y \leq 900$

27.
$\quad\quad x \geq 0$
$\quad\quad y \geq 0$
$\quad\quad x \leq 15$
$30x + 25y \leq 750$
$10x + 40y \leq 800$

Determine the solution to each of the following systems.

28. $|x| < 3$
$\quad\;\; y > x$

29. $|y| > 2$
$\quad\;\; y \leq x + 3$

30. $|x| > 1$
$\quad\;\; y \leq 3x + 2$

31. $|y| < 4$
$\quad\;\; y \geq -2x + 2$

32. $|x| \leq 3$
$\quad\;\; |y| > 2$

33. $|x| \geq 1$
$\quad\;\; |y| \geq 2$

34. $|x| < 2$
$\quad\;\; |y| \geq 3$

35. $|x + 2| < 3$
$\quad\;\; |y| > 4$

36. $|x - 3| \geq 2$
$\quad\;\; x + y < 5$

37. $|x - 2| > 1$
$\quad y > -2$

38. $|x - 3| \leq 4$
$\quad |y + 2| \leq 1$

39. $|x - 3| > 4$
$\quad |y + 1| \leq 3$

40. Is it possible for a system of linear inequalities to have no solution? Explain why this can or cannot occur. Make up an example to support your answer.

Cumulative Review Exercises

[2.2] **41.** A formula used when studying levers in physics is $f_1 d_1 + f_2 d_2 = f_3 d_3$. Solve this formula for f_2.

[3.5] *State the range and domain of the following functions:*

42. $\{(4, 3), (5, -2), (-1, 2), (0, -5)\}$

43. $f(x) = \frac{2}{3}x - 4$

44.

JUST FOR FUN

Determine the solution to each of the following systems.

1. $|2x - 3| - 1 > 3$
$\quad |y - 2| < 3$

2. $|x + \frac{3}{2}| > \frac{5}{2}$
$\quad |2y - \frac{1}{2}| \leq \frac{3}{2}$

3. $y < |x|$
$\quad y < 4$

4. $y \geq |x - 2|$
$\quad y \leq -|x - 2|$

SUMMARY

GLOSSARY

Consistent system of equations *(158):* A system of equations that has a solution.

Dependent system of equations *(158):* A system of equations that has an infinite number of solutions.

Determinant *(185):* A square array of numbers enclosed between two vertical bars. A determinant represents a number.

Elements of a determinant *(185):* The numbers that make up the array of numbers in a determinant.

Inconsistent system of equations *(158):* A system of equations that has no solution.

Second-order determinant *(185):* A determinant that has two rows and two columns of elements.

Solution to a system of linear equations *(158):* The ordered pair or pairs that satisfy all equations in the system.

System of linear equations *(158):* Two or more linear equations taken as a system.

System of linear inequalities *(194):* Two or more linear inequalities taken as a system.

Third-order determinant *(185):* A determinant that has three rows and three columns of elements.

Third-order system of linear equations *(169):* A system of linear equations consisting of three equations with three unknowns.

IMPORTANT FACTS

Value of a Second-order Determinant

$$\begin{vmatrix} a_1 & b_1 \\ a_2 & b_2 \end{vmatrix} = a_1 b_2 - a_2 b_1$$

Cramer's Rule For a system of equations of the form

$$a_1 x + b_1 y = c_1$$
$$a_2 x + b_2 y = c_2$$

$$x = \frac{\begin{vmatrix} c_1 & b_1 \\ c_2 & b_2 \end{vmatrix}}{\begin{vmatrix} a_1 & b_1 \\ a_2 & b_2 \end{vmatrix}} = \frac{D_x}{D} \quad \text{and} \quad y = \frac{\begin{vmatrix} a_1 & c_1 \\ a_2 & c_2 \end{vmatrix}}{\begin{vmatrix} a_1 & b_1 \\ a_2 & b_2 \end{vmatrix}} = \frac{D_y}{D}, \; D \neq 0$$

Value of a Third-order Determinant

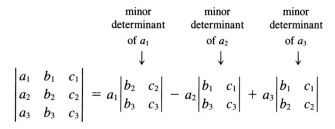

$$\begin{vmatrix} a_1 & b_1 & c_1 \\ a_2 & b_2 & c_2 \\ a_3 & b_3 & c_3 \end{vmatrix} = a_1 \begin{vmatrix} b_2 & c_2 \\ b_3 & c_3 \end{vmatrix} - a_2 \begin{vmatrix} b_1 & c_1 \\ b_3 & c_3 \end{vmatrix} + a_3 \begin{vmatrix} b_1 & c_1 \\ b_2 & c_2 \end{vmatrix}$$

Cramer's Rule For a system of equations of the form

$$a_1 x + b_1 y + c_1 z = d_1$$
$$a_2 x + b_2 y + c_2 z = d_2$$
$$a_3 x + b_3 y + c_3 z = d_3$$

$$x = \frac{\begin{vmatrix} d_1 & b_1 & c_1 \\ d_2 & b_2 & c_2 \\ d_3 & b_3 & c_3 \end{vmatrix}}{\begin{vmatrix} a_1 & b_1 & c_1 \\ a_2 & b_2 & c_2 \\ a_3 & b_3 & c_3 \end{vmatrix}} = \frac{D_x}{D}, \quad y = \frac{\begin{vmatrix} a_1 & d_1 & c_1 \\ a_2 & d_2 & c_2 \\ a_3 & d_3 & c_3 \end{vmatrix}}{\begin{vmatrix} a_1 & b_1 & c_1 \\ a_2 & b_2 & c_2 \\ a_3 & b_3 & c_3 \end{vmatrix}} = \frac{D_y}{D},$$

$$z = \frac{\begin{vmatrix} a_1 & b_1 & d_1 \\ a_2 & b_2 & d_2 \\ a_3 & b_3 & d_3 \end{vmatrix}}{\begin{vmatrix} a_1 & b_1 & c_1 \\ a_2 & b_2 & c_2 \\ a_3 & b_3 & c_3 \end{vmatrix}} = \frac{D_z}{D}, \; D \neq 0$$

Review Exercises

[4.1] *Write each equation in slope–intercept form. Without graphing or solving the system of equations, state whether the system of linear equations is consistent, inconsistent, or dependent. Also indicate whether the system has exactly one solution, no solution, or an infinite number of solutions.*

1. $x + 2y = 8$
$3x + 6y = 12$

2. $y = -3x - 6$
$2x + 3y = 8$

3. $y = \frac{1}{2}x + 4$
$x + 2y = 8$

4. $6x = 4y - 8$
$4x = 6y + 8$

Determine the solution to the system of equations graphically.

5. $y = x + 3$
$y = 2x + 5$

6. $x = -2$
$y = 3$

7. $2x + 2y = 8$
$2x - y = -4$

8. $2y = 2x - 6$
$\frac{1}{2}x - \frac{1}{2}y = \frac{3}{2}$

Find the solution to the system of equations by subsitution.

9. $y = 2x + 1$
$y = 3x - 2$

10. $y = -x + 5$
$y = 2x - 1$

11. $y = 2x - 8$
$2x - 5y = 0$

12. $-x + 3y = 9$
$x = -2y + 1$

13. $2x + y = 5$
$3x + 2y = 8$

14. $2x - y = 6$
$\frac{1}{2}x + 2y = 6$

15. $3x + y = 17$
$\frac{1}{2}x - \frac{3}{4}y = 1$

16. $x = -3y$
$\frac{1}{2}x + 2y = 3$

Find the solution to the system of equations using the addition method.

17. $x + y = 6$
$x - y = 10$

18. $x + 2y = -3$
$2x - 2y = 6$

19. $2x + 3y = 4$
$x + 2y = -6$

20. $0.6x + 0.5y = 2$
$0.25x - 0.2y = 1.65$

21. $4x - 3y = 8$
$2x + 5y = 8$

22. $-2x + 3y = 15$
$3x + 3y = 10$

23. $x + \frac{2}{5}y = \frac{9}{5}$
$x - \frac{3}{2}y = -2$

24. $2x + 2y = 8$
$y = 4x - 3$

25. $y = -\frac{3}{4}x + \frac{5}{2}$
$x + \frac{5}{4}y = \frac{7}{2}$

26. $2x - 5y = 12$
$x - \frac{4}{3}y = -2$

[4.2] *Determine the solution to the third-order system using substitution or the addition method.*

27. $x + 2y = 12$
$4x = 8$
$3x - 4y + 5z = 20$

28. $3x + 4y - 5z = 10$
$4x + 2z = 16$
$2z = -4$

29. $x + 5y + 5z = 6$
$3x + 3y - z = 10$
$x + 3y + 2z = 5$

30. $-x - y - z = -6$
$2x + 3y - z = 7$
$-3x + y + z = -6$

31. $3y - 2z = -4$
$3x - 5z = -7$
$2x + y = 6$

32. $3x + 2y - 5z = 19$
$2x - 3y + 3z = -15$
$5x - 4y - 2z = -2$

[4.3] *(a) Express the problem as a system of linear equations and (b) use the method of your choice to find the solution to the problem.*

33. The sum of two numbers is 48. Find the two numbers if the larger is 3 less than twice the smaller.

34. The difference of two numbers is 18. Find the two numbers if the larger is 4 times the smaller.

35. A plane can travel 600 miles per hour with the wind and 530 miles per hour against the wind. Find the speed of the wind and the speed of the plane in still air.

36. Curtis has a 30% acid solution and a 50% acid solution. How much of each must he mix to get 6 liters of a 40% acid solution?

37. The admission at an ice hockey game is $7.50 for adults and $5.50 for children. A total of 650 tickets were sold. How many tickets were sold to children and how many to adults if a total of $4395 was collected?

38. The sum of three numbers is 17. The first number is 1 more than the sum of the other two numbers, and the second number is three times the third number. Find the three numbers.

39. Mary has a total of $40,000 invested in three different savings accounts. She has some money invested in one account that gives 10% interest. The second account has $5000 less than the first account and gives 8% interest. The third account gives 6% interest. If the total annual interest that Mary receives is $3500, find the amount in each account.

[4.4] *Determine the solution to the system of equations using determinants.*

40. $5x + 6y = 14$
$x - 3y = 7$

41. $3x + 5y = -2$
$5x + 3y = 2$

42. $4x + 3y = 2$
$7x - 2y = -11$

43. $x + y + z = 8$
$x - y - z = 0$
$x + 2y + z = 9$

44. $x + 2y - 4z = 17$
$2x - y + z = -9$
$2x - y - 3z = -1$

45. $y + 3z = 4$
$-x - y + 2z = 0$
$x + 2y + z = 1$

[4.5] *Determine the solution to the system of inequalities.*

46. $-x + 3y > 6$
$2x - y \le 2$

47. $5x - 2y \le 10$
$3x + 2y > 6$

48. $y > 2x + 3$
$y < -x + 4$

49. $y \le -x + 4$
$2x + 4y > 6$

Determine the solution to the system of inequalities (optional).

50.
$$x \geq 0$$
$$y \geq 0$$
$$x + y \leq 6$$
$$4x + y \leq 8$$

51.
$$x \geq 0$$
$$y \geq 0$$
$$2x + y \leq 6$$
$$4x + 5y \leq 20$$

Determine the solution to the system of inequalities (optional).

52. $|x| \leq 3$
$|y| > 2$

53. $|x| > 4$
$|y - 2| \leq 3$

Practice Test

Write each equation in slope–intercept form. Then determine, without solving the system, whether the system of equations is consistent, inconsistent, or dependent. State whether the system has exactly one solution, no solution, or an infinite number of solutions.

1. $4x + 3y = -6$
$6y = 8x + 4$

2. $5x + 3y = 9$
$5x - 3y = 9$

Solve the system of equations graphically.

3. $y = 3x - 2$
$y = -2x + 8$

Solve the system of equations by substitution.

4. $y = 4x - 5$
$y = 2x + 7$

5. $3x + y = 8$
$x - y = 6$

Solve the system of equations using the addition method.

6. $2x + y = 5$
$x + 3y = -10$

7. $\frac{3}{2}x + y = 6$
$x - \frac{5}{2}y = -4$

(a) Express the problem as a system of linear equations and (b) use the method of your choice to find the solution to the problem.

8. Max has cashews that sell for \$7 a pound and peanuts that sell for \$5.50 a pound. How much of each must he mix to get 20 pounds of a mixture that sells for \$6.00 per pound?

Solve the system of equations.

9.
$$x = 2$$
$$2x + 3y = 10$$
$$-x + 3y - 2z = 10$$

10.
$$x + y + z = 2$$
$$-2x - y + z = 1$$
$$x - 2y - z = 1$$

Graph the system of inequalities and indicate its solution.

11. $y + 3x \leq 6$
$2x + y > 4$

12. $3x + 2y < 9$
$-2x + 5y \leq 10$

*13. $|x| > 3$
$|y| \leq 1$

* Optional

Cumulative Review Test

1. Evaluate $24 \div 4(2 - (5 - 2))^2 - 6$.

2. Consider the set of numbers

$$\{\tfrac{1}{2}, -4, 9, 0, \sqrt{3}, -4.63, 1\}.$$

List the elements of the set that are (a) natural numbers, (b) rational numbers, and (c) real numbers.

3. Write the numbers from smallest to largest

$$-1, |-4|, \tfrac{3}{4}, \tfrac{5}{8}, -|-8|, |-10|$$

Solve the following equations.

4. $-(3 - 2(x - 4)) = 3(x - 6)$

5. $\dfrac{1}{3}x = \dfrac{3}{5}x + 4$

6. $|4x - 3| + 2 = 10$

7. Solve the formula $R = 3(a + b)$ for b.

8. Find the solution set of the inequality

$$0 < \frac{3x - 2}{4} \le 8.$$

9. Find the length and the midpoint of the line segment through the points $(1, 5)$ and $(-3, 2)$.

Graph the following equations.

10. $-2x + 4y = 12$

11. $2y = 3x - 8$

12. Write in slope–intercept form the equation of the line that is parallel to the line $2x - 3y = 8$ and passes through the point $(2, 3)$.

13. Determine which of the following graphs on the right are functions. Explain how you determined your answer.

(a)   **(b)**   **(c)**

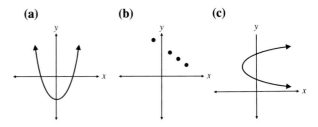

14. Graph the inequality $6x - 3y < 12$.

Solve the following systems of equations.

15. $3x + y = 6$
$y = 2x + 1$

16. $5x + 4y = 10$
$3x + 5y = -7$

17. $x - 2y = 0$
$2x + z = 7$
$y - 2z = -5$

18. If the largest angle of a triangle is nine times the measure of the smallest angle, and the middle-size angle is 70° greater than the measure of the smallest angle, find the measure of the three angles.

19. Dawn speed walks at 4 miles per hour and Judy jogs at 6 miles per hour. Dawn begins walking $\frac{1}{2}$ hour before Judy starts jogging. If Judy jogs on the same path that Dawn speed walks, how long after Judy begins jogging will she catch up to Dawn?

20. There are two different priced seats at a rock concert. The higher-priced seats sell for $20 and the less expensive seats sell for $16. If a total of 1000 tickets is sold and the total ticket sales are $18,400, how many of each type of seat is sold?

Polynomials

See Section 5.7, Exercise 31.

Exponents and Scientific Notation

▶ **1** Learn the meaning of exponents.

▶ **2** Learn the product rule for exponents.

▶ **3** Learn the quotient rule for exponents.

▶ **4** Learn the negative exponent rule.

▶ **5** Learn the zero exponent rule.

▶ **6** Write numbers in scientific notation.

▶ **7** Change numbers in scientific notation to numbers without exponents.

▶ **8** Use scientific notation in calculations.

▶ **1** Before we can discuss polynomials, we need to understand exponents. In this section we will review and then expand our knowledge of exponents. Recall from Section 1.6 that

$$2^3 = \underbrace{2 \cdot 2 \cdot 2}_{3 \text{ factors of } 2}$$

and

$$x^m = \underbrace{x \cdot x \cdot x \cdot x \cdot \cdots \cdot x}_{m \text{ factors of } x}$$

The quantity 2^3 is called an **exponential expression.** The 2 is the **base** and the 3 is the **exponent** of the expression.

▶ **2** Consider the multiplication $x^3 \cdot x^5$.

$$x^3 \cdot x^5 = (x \cdot x \cdot x) \cdot (x \cdot x \cdot x \cdot x \cdot x) = x^8$$

This problem could also be evaluated using the product rule for exponents.

Product Rule for Exponents*

If m and n are natural numbers and a is any real number, then

$$a^m \cdot a^n = a^{m+n}$$

To multiply expressions in exponential form, maintain the common base and add the exponents.

$$x^3 \cdot x^5 = x^{3+5} = x^8$$

EXAMPLE 1   Simplify each of the following.

(a) $3^2 \cdot 3^3$ (b) $x^3 \cdot x^9$ (c) $x \cdot x^6$

*The rules given in this section and the next section also apply for rational or fractional exponents. Rational exponents will be discussed in Section 8.2. We will review these rules again at that time.

Solution: (a) $3^2 \cdot 3^3 = 3^{2+3} = 3^5 = 243$
(b) $x^3 \cdot x^9 = x^{3+9} = x^{12}$
(c) $x \cdot x^6 = x^1 \cdot x^6 = x^{1+6} = x^7$ ∎

COMMON STUDENT ERROR

When multiplying expressions in exponential form the base remains the same. *Do not multiply out the common base.*

Correct	Wrong
$2^2 \cdot 2^3 = 2^5$	~~$2^2 \cdot 2^3 = 4^5$~~

When multiplying two expressions written in exponential form, *add the exponents. Do not multiply the exponents.*

Correct	Wrong
$2^2 \cdot 2^3 = 2^5$	~~$2^2 \cdot 2^3 = 2^6$~~

▶ **3** Consider the division $x^7 \div x^4$.

$$\frac{x^7}{x^4} = \frac{\overset{1}{\cancel{x}} \cdot \overset{1}{\cancel{x}} \cdot \overset{1}{\cancel{x}} \cdot \overset{1}{\cancel{x}} \cdot x \cdot x \cdot x}{\underset{1}{\cancel{x}} \cdot \underset{1}{\cancel{x}} \cdot \underset{1}{\cancel{x}} \cdot \underset{1}{\cancel{x}}} = x \cdot x \cdot x = x^3$$

This problem could also be solved using the quotient rule for exponents.

Quotient Rule for Exponents

If a is any nonzero real number and m and n are nonzero integers, then

$$\frac{a^m}{a^n} = a^{m-n}$$

To divide expressions in exponential form, maintain the common base and subtract the exponents.

$$\frac{x^7}{x^4} = x^{7-4} = x^3$$

EXAMPLE 2 Simplify each of the following.

(a) $\dfrac{5^4}{5^2}$ (b) $\dfrac{x^5}{x^2}$ (c) $\dfrac{y^2}{y^5}$

Solution: (a) $\dfrac{5^4}{5^2} = 5^{4-2} = 5^2 = 25$ (b) $\dfrac{x^5}{x^2} = x^{5-2} = x^3$ (c) $\dfrac{y^2}{y^5} = y^{2-5} = y^{-3}$ ∎

Notice in Example 2(c) that the answer contains a negative exponent. Let's do part (c) again by dividing out common factors.

$$\frac{y^2}{y^5} = \frac{\overset{1}{\cancel{y}} \cdot \overset{1}{\cancel{y}}}{\underset{1}{\cancel{y}} \cdot \underset{1}{\cancel{y}} \cdot y \cdot y \cdot y} = \frac{1}{y^3}$$

From this process and from Example 2(c), we can reason that $y^{-3} = 1/y^3$.

▶ **4**   This problem could also be done using the negative exponent rule.

Negative Exponent Rule

For any nonzero real number a and any whole number m,

$$a^{-m} = \frac{1}{a^m}, \qquad a \neq 0$$

EXAMPLE 3   Write each of the following without negative exponents.

(a) 2^{-3} (b) x^{-2} (c) $\dfrac{1}{x^{-3}}$ (d) $\dfrac{2}{y^{-5}}$

Solution: (a) $2^{-3} = \dfrac{1}{2^3} = \dfrac{1}{8}$ (b) $x^{-2} = \dfrac{1}{x^2}$

(c) $\dfrac{1}{x^{-3}} = \dfrac{1}{\dfrac{1}{x^3}} = \dfrac{1}{1} \cdot \dfrac{x^3}{1} = x^3$ (d) $\dfrac{2}{y^{-5}} = \dfrac{2}{\dfrac{1}{y^5}} = \dfrac{2}{1} \cdot \dfrac{y^5}{1} = 2y^5$

HELPFUL HINT

Notice that **a factor can be moved from a numerator to a denominator or from a denominator to a numerator simply by changing the** *sign of the exponent.*

Examples

$$\frac{1}{2^{-1}} = 2 \qquad x^{-3} = \frac{1}{x^3} \qquad 2^{-2} = \frac{1}{2^2}$$

$$6^{-1} = \frac{1}{6} \qquad \frac{1}{x^{-5}} = x^5 \qquad \frac{1}{3^{-2}} = 3^2$$

COMMON STUDENT ERROR

Correct	*Wrong*	*Wrong*
$\dfrac{1}{3^2} = 3^{-2}$	$\dfrac{1}{3^2} = 3^2$ ✗	$\dfrac{1}{3^2} = -3^2$ ✗
$\dfrac{1}{x^5} = x^{-5}$	$\dfrac{1}{x^5} = x^5$ ✗	$\dfrac{1}{x^5} = -x^5$ ✗

*An important concept that you must understand is that **an exponent applies only to the number or variable immediately preceding it unless parentheses are used.*** For example, when considering xy^2, only the y is squared. Similarly, in the expression xy^{-1} only the y is raised to the negative 1 power.

$$xy^{-1} = x \cdot y^{-1} = x \cdot \frac{1}{y} = \frac{x}{y}$$

Other Examples

$$2x^{-1} = 2 \cdot \frac{1}{x} = \frac{2}{x}$$

$$-x^{-2} = -1 \cdot x^{-2} = -1 \cdot \frac{1}{x^2} = \frac{-1}{x^2}$$

$$-5x^{-4} = -5 \cdot \frac{1}{x^4} = \frac{-5}{x^4}$$

COMMON STUDENT ERROR

Remember that the exponent refers only to the symbol or number directly preceding it. The correct procedure for evaluating the expression, -6^2, is

$$-6^2 = -(6)^2 = -(6)(6) = -36 \qquad \text{correct}$$

If the number -6 was to be squared, it would be written $(-6)^2$.

$$(-6)^2 = (-6)(-6) = 36 \qquad \text{correct}$$

Students often make the following error.

$$-6^2 = (-6)(-6) = 36 \qquad \text{wrong}$$

Generally, we do not leave exponential expressions with negative exponents. *When we indicate that an exponential expression is to be simplified, make sure that your answer is written without negative exponents.*

EXAMPLE 4   Simplify each of the following.

(a) $-3^2x^2y^{-3}$ (b) $(-4)^2x^{-2}yz^{-3}$ (c) $4^{-2}x^{-1}y^2$ (d) $\dfrac{3xz^2}{y^{-4}}$

Solution: (a) $-3^2x^2y^{-3} = -\dfrac{9x^2}{y^3}$ (b) $(-4)^2x^{-2}yz^{-3} = \dfrac{16y}{x^2z^3}$

(c) $4^{-2}x^{-1}y^2 = \dfrac{y^2}{4^2x^1} = \dfrac{y^2}{16x}$ (d) $\dfrac{3xz^2}{y^{-4}} = 3xy^4z^2$ ∎

▶ **5**   The last rule we will study in this section is the zero exponent rule. We introduce it now because it may be useful when we work scientific notation problems.

Consider the following: Any nonzero number divided by itself is 1. Therefore,

$$\frac{x^5}{x^5} = 1$$

By the quotient rule for exponents,

$$\frac{x^5}{x^5} = x^{5-5} = x^0$$

Since $x^0 = \dfrac{x^5}{x^5}$ and $\dfrac{x^5}{x^5} = 1$, by the transitive property of equality,

$$x^0 = 1$$

Here is the zero exponent rule.

Zero Exponent Rule

If a is any nonzero real number, then

$$a^0 = 1$$

The zero exponent rule illustrates that any nonzero real number with an exponent of 0 has a value of 1. We must specify that $a \neq 0$ because 0^0 is not a real number.

EXAMPLE 5 Simplify (assume that the base is not 0).

(a) x^0 (b) $3x^0$ (c) $(5x)^0$ (d) $-(a + b)^0$

Solution: (a) $x^0 = 1$ (b) $3x^0 = 3(1)$ (c) $(5x)^0 = 1$ (d) $-(a + b)^0 = -(1)$
                                              $= 3$ $= -1$ ∎

Now let's look at examples that combine a number of these properties.

EXAMPLE 6 Use the rules for exponents to simplify the expression.

(a) $4^2 \cdot 4^{-4}$ (b) $y^4 \cdot y^{-6}$ (c) $\dfrac{2^3}{2^{-2}}$ (d) $\dfrac{x^{-5}}{x^{-2}}$ (e) $z^3 \cdot z^{-3}$

Solution: (a) $4^2 \cdot 4^{-4} = 4^{2+(-4)} = 4^{-2} = \dfrac{1}{4^2} = \dfrac{1}{16}$

(b) $y^4 \cdot y^{-6} = y^{4+(-6)} = y^{-2} = \dfrac{1}{y^2}$

(c) $\dfrac{2^3}{2^{-2}} = 2^{3-(-2)} = 2^{3+2} = 2^5 = 32$

(d) $\dfrac{x^{-5}}{x^{-2}} = x^{-5-(-2)} = x^{-5+2} = x^{-3} = \dfrac{1}{x^3}$

(e) $z^3 \cdot z^{-3} = z^{3+(-3)} = z^0 = 1$ ∎

EXAMPLE 7 Simplify $\dfrac{(x^{-3})(2x^6)}{x^{-5}}$.

Solution:

$$\frac{(x^{-3})(2x^6)}{x^{-5}} = \frac{2x^{-3+6}}{x^{-5}}$$

$$= \frac{2x^3}{x^{-5}}$$

$$= 2x^8$$

The following Helpful Hint is very important. Read it carefully.

HELPFUL HINT _____

Notice in Example 7 that when we simplified $\dfrac{2x^3}{x^{-5}}$ our answer was $2x^8$. This answer can be obtained by rewriting x^{-5} in the denominator as an x^5 in the numerator

$$\frac{2x^3}{x^{-5}} = 2x^3 \cdot x^5 = 2x^8$$

When simplifying problems of this type, we can move a factor with a negative exponent from the numerator to the denominator, or from the denominator to the numerator, *by changing the sign of the exponent*.

If the same variable appears in both the numerator and denominator of an expression, we generally change the location of the variable with the *lesser* exponent.

Examples

$$\frac{x^{-5}}{x^{-3}} = \frac{1}{x^{-3} \cdot x^5} = \frac{1}{x^2} \qquad \frac{z^{-2}}{z^{-6}} = z^{-2} \cdot z^6 = z^4$$

$$\frac{y^6}{y^{-3}} = y^6 \cdot y^3 = y^9 \qquad \frac{w^{-5}}{w^4} = \frac{1}{w^4 \cdot w^5} = \frac{1}{w^9}$$

Now we will simplify an expression containing a number of variable factors.

$$\frac{x^{-6}y^{-3}z^{-4}}{x^{-4}y^{-7}z^2} = \frac{y^{-3}y^7}{x^{-4}x^6z^2z^4} = \frac{y^4}{x^2z^6}$$

EXAMPLE 8 Simplify $\left(\dfrac{4x^3y^{-2}}{3xy^5}\right)\left(\dfrac{6x^4y^3}{4x^{-2}y^2}\right)$.

Solution: Begin by simplifying each factor and writing each factor without negative exponents.

$$\left(\frac{4x^3y^{-2}}{3xy^5}\right)\left(\frac{6x^4y^3}{4x^{-2}y^2}\right) = \left(\frac{4x^2}{3y^7}\right)\left(\frac{3x^6y}{2}\right)$$

Now simplify further.

$$= \frac{4x^2}{3y^7} \cdot \frac{3x^6y}{2}$$

$$= \frac{2x^8}{y^6}$$

EXAMPLE 9   Simplify $\dfrac{(6x^{-3}y^5)(5x^{-7}y^{-2})}{20x^{-4}y^5}$.

Solution: We will begin by multiplying the factors in the numerator.

$$\frac{(6x^{-3}y^5)(5x^{-7}y^{-2})}{20x^{-4}y^5} = \frac{30x^{-10}y^3}{20x^{-4}y^5}$$

$$= \frac{3}{2x^6y^2}$$

EXAMPLE 10   Simplify (assume that all variables used as exponents are integers).

(a) $x^{3b} \cdot x^{4b+5}$ (b) $\dfrac{y^{3r-2}}{y^{2r+4}}$ (c) $\dfrac{(x^{2p+4})(x^{p-2})}{x^{3p+1}}$

Solution: (a) By the product rule for exponents,

$$x^{3b} \cdot x^{4b+5} = x^{3b+(4b+5)} = x^{7b+5}$$

(b) By the quotient rule for exponents,

$$\frac{y^{3r-2}}{y^{2r+4}} = y^{3r-2-(2r+4)} = y^{r-6}$$

(c) Using both the product and quotient rule, we get

$$\frac{(x^{2p+4})(x^{p-2})}{x^{3p+1}} = \frac{x^{(2p+4)+(p-2)}}{x^{3p+1}}$$

$$= \frac{x^{3p+2}}{x^{3p+1}}$$

$$= x^{3p+2-(3p+1)}$$

$$= x^{3p+2-3p-1}$$

$$= x^1 \quad \text{or} \quad x$$

COMMON STUDENT ERROR

Another common error made by students is to treat a term as a factor. Consider the expression $\dfrac{x^6 + y^{-7}}{y^4}$.

Correct

$$\frac{x^6 + y^{-7}}{y^4} = \frac{x^6}{y^4} + \frac{y^{-7}}{y^4} = \frac{x^6}{y^4} + \frac{1}{y^{11}}$$

Wrong

$$\frac{x^6 + y^{-7}}{y^4} = \frac{x^6}{y^4 \cdot y^7}$$

$$\frac{x^6 + y^{-7}}{y^4} = \frac{x^6}{y^4 + y^7}$$

Can you explain why the procedures on the right are wrong? What does $\dfrac{x^6 y^{-7}}{y^4}$ simplify to?

Scientific Notation

▶**6** When working with scientific problems, we often deal with very large and very small numbers. For example, the distance from the earth to the sun is about 93,000,000 miles. The wavelength of a yellow color of light is about 0.0000006 meter. Because it is difficult to work with many zeros, scientists often express such numbers with exponents. For example, the number 93,000,000 might be written 9.3×10^7 and the number 0.0000006 might be written 6.0×10^{-7}. Numbers such as 9.3×10^7 and 6.0×10^{-7} are in a form called **scientific notation.** Each number written in scientific notation is written with a number greater than or equal to 1 and less than 10 ($1 \le a < 10$) multiplied by some power of 10.

Examples of Numbers in Scientific Notation

$$3.2 \times 10^6$$
$$4.176 \times 10^3$$
$$2.64 \times 10^{-2}$$
$$1 \times 10^{-5}$$

Consider the number 32,400:

$$32,400 = 3.24 \times 10,000$$
$$= 3.24 \times 10^4$$

Note that $10,000 = 10^4$. Also note that there are four zeros in 10,000, the same number as the exponent in 10^4. The procedure for writing a number in scientific notation follows.

To Write a Number in Scientific Notation

1. Move the decimal in the original number to the right of the first non-zero digit. This will give you a number greater than or equal to 1 and less than 10.
2. Count the number of places you have moved the decimal to obtain the number in step 1. If the original number was 10 or greater, the count is to be considered positive. If the original number was less than 1, the count is to be considered negative.
3. Multiply the number obtained in step 1 by 10 raised to the count (power) found in step 2.

EXAMPLE 11 Write the following numbers using scientific notation.
(a) 68,900   (b) 0.000572   (c) 216,000   (d) 0.0074

Solution: (a) 68,900 means 68,900.

68,900. $= 6.89 \times 10^4$

The decimal point is moved 4 places. Since the original number is greater than 10, the exponent is positive.

(b) 0.000572 $= 5.72 \times 10^{-4}$

The decimal point is moved 4 places. Since the original number is less than 1, the exponent is negative.

(c) $216{,}000. = 2.16 \times 10^5$ (d) $0.0074 = 7.4 \times 10^{-3}$

5 places   3 places

▶ **7**

> **To Convert from a Number Given in Scientific Notation**
>
> 1. Observe the exponent on the power of 10.
> 2. (a) If the exponent is positive, move the decimal in the number to the right the same number of places as the exponent. It may be necessary to add zeros to the number. This will result in a number greater than or equal to 10.
>   (b) If the exponent is negative, move the decimal in the number to the left the same number of places as the exponent. It may be necessary to add zeros. This will result in a number less than 1.

EXAMPLE 12 Write each number without exponents.

(a) 2.1×10^4 (b) 8.73×10^{-3} (c) 1.45×10^8

Solution: (a) Moving the decimal four places to the right gives

$$2.1 \times 10^4 = 2.1 \times 10{,}000 = 21{,}000$$

(b) $8.73 \times 10^{-3} = 0.00873$ Move the decimal three places to the left.

(c) $1.45 \times 10^8 = 145{,}000{,}000$ Move the decimal eight places to the right.

▶ **8** We can use the rules of exponents discussed in this section when working with numbers written in scientific notation.

EXAMPLE 13 Simplify $(5.6 \times 10^6)(2 \times 10^{-4})$.

Solution: $(5.6 \times 10^6)(2 \times 10^{-4}) = (5.6 \times 2)(10^6 \times 10^{-4})$
$$= 11.2 \times 10^2$$
$$= 1120$$

EXAMPLE 14 Simplify $\dfrac{14.4 \times 10^{-5}}{3 \times 10^{-3}}$.

Solution: $\dfrac{14.4 \times 10^{-5}}{3 \times 10^{-3}} = \left(\dfrac{14.4}{3}\right)\left(\dfrac{10^{-5}}{10^{-3}}\right)$
$$= 4.8 \times 10^{-5-(-3)}$$
$$= 4.8 \times 10^{-5+3}$$
$$= 4.8 \times 10^{-2}$$
$$= 0.048$$

EXAMPLE 15 Multiply $(327{,}000)(0.00008)$.

Solution: Change each number to scientific notation form.

$$(327,000)(0.00008) = (3.27 \times 10^5)(8 \times 10^{-5})$$
$$= (3.27 \times 8)(10^5 \times 10^{-5})$$
$$= 26.16 \times 10^0$$
$$= 26.16 \times 1$$
$$= 26.16$$

Calculator Corner

Scientific Notation on Calculators

What will your calculator show when you multiply very large or very small numbers? The answer depends on whether your calculator has the ability to display an answer in scientific notation form. On calculators without this ability, you will probably get an error message since the answer will be too large or too small for the display.

Example: On a calculator without scientific notation

$\boxed{C}$ 8000000 $\boxed{\times}$ 600000 $\boxed{=}$ Error

If your calculator has the ability to give an answer in scientific notation form, you would get an answer given in scientific notation, as follows:

Example:

$\boxed{C}$ 8000000 $\boxed{\times}$ 600000 $\boxed{=}$ 4.8   12

The 4.8   12 means 4.8×10^{12}.

Example

$\boxed{C}$ .0000003 $\boxed{\times}$ .004 $\boxed{=}$ 1.2   −9

The 1.2   −9 means 1.2×10^{-9}.

Exercise Set 5.1

Simplify and write the answer without any negative exponents.

1. 3^{-2} **2.** 5^{-2} **3.** 1^{-2} **4.** x^{-4}

5. $5y^{-3}$ **6.** $\dfrac{1}{x^{-1}}$ **7.** $\dfrac{1}{x^{-4}}$ **8.** $\dfrac{3}{5y^{-2}}$

9. $\dfrac{2x}{y^{-3}}$ **10.** $\dfrac{6x^4}{y^{-1}}$ **11.** $\dfrac{5x^{-2}y^{-3}}{2z^{-1}}$ **12.** $\dfrac{4x^{-3}y}{z^4}$

13. $\dfrac{5x^{-2}y^{-3}}{z^{-4}}$ **14.** $\dfrac{10xy^5}{2z^{-3}}$ **15.** $\dfrac{4^{-1}x^{-1}}{y}$ **16.** $\dfrac{5^{-1}z}{x^{-1}y^{-1}}$

Evaluate. Assume all bases represented by variables are nonzero.

17. x^0 **18.** 3^0 **19.** $4x^0$ **20.** $5y^0$

21. $-(7x)^0$ **22.** $-2x^0$ **23.** $-3x^0$ **24.** $(x + y)^0$

25. $-(a + b)^0$ **26.** $3(a + b)^0$ **27.** $3x^0 + 4y^0$ **28.** $-4(x^0 - 3y^0)$

Simplify and write the answer without any negative exponents.

29. $6^3 \cdot 6^{-4}$

30. $7 \cdot 7^{-3}$

31. $x^2 \cdot x$

32. $x^2 \cdot x^4$

33. $x^6 \cdot x^{-2}$

34. $x^{-4} \cdot x^3$

35. $\dfrac{3^4}{3^2}$

36. $\dfrac{5^3}{5^5}$

37. $\dfrac{5^2}{5^{-2}}$

38. $\dfrac{7^{-5}}{7^{-3}}$

39. $\dfrac{x^{-9}}{x^2}$

40. $\dfrac{x^{-2}}{x}$

41. $\dfrac{x^0}{x^{-3}}$

42. $\dfrac{y^3}{2y^5}$

43. $\dfrac{3y^{-2}}{y^{-7}}$

44. $\dfrac{6x^4}{x^{-3}}$

45. $\dfrac{4x^{-3}}{x^{-1}}$

46. $\dfrac{x^{-3}}{x^{-5}}$

47. $2x^{-4} \cdot 6x^{-3}$

48. $(4x^2y^5)(2x^{-3}y^{-4})$

49. $(-3y^{-2})(-y^3)$

50. $(3x^4y^{-2})(2xy^{-3})$

51. $(2x^{-3}y^{-4})(6x^{-4}y^7)$

52. $(-2x^3y^4)(-x^{-3}y^5)$

53. $(5x^2y^{-2}z^4)(-2x^5y^2z)$

54. $(-3x^{-4}y^6z^{-4})(2x^3yz^3)$

55. $(2x^4y^7z^9)(4x^3y^{-5}z^{-12})$

56. $\dfrac{24x^3y^2}{8xy}$

57. $\dfrac{27x^5y^{-4}}{9x^3y^2}$

58. $\dfrac{6x^{-2}y^3}{2x^4y}$

59. $\dfrac{9xy^{-4}}{3x^{-2}y}$

60. $\dfrac{(x^{-2})(4x^2)}{x^3}$

61. $\dfrac{(2x^4)(6xy^3)}{4y^3}$

62. $\dfrac{(4xy^5)(5x^4y^{-3})}{2x^5y^9}$

63. $\dfrac{(-3x^{-1}y^{-2})(2x^4y^{-3})}{6xy^4}$

64. $\dfrac{(x^4y)(3x^4y^{-3}z)}{6x^8y^2z^4}$

65. $\dfrac{(4x^{-5}y^{-2})(3x^2y^{-5})}{24x^3y^{-4}}$

66. $\left(\dfrac{5x^2y^3}{4z^3}\right)\left(\dfrac{8xy^6}{2z^3}\right)$

67. $\left(\dfrac{2x^5}{y^{-3}}\right)\left(\dfrac{3xy^{-2}}{z^{-3}}\right)$

68. $\left(\dfrac{2x^{-2}y^{-3}}{z^3}\right)\left(\dfrac{x^3y^5}{z^{-2}}\right)$

69. $\left(\dfrac{3x^{-2}y^{-2}}{x^4y^{-5}}\right)\left(\dfrac{2x^3y^5}{9x^{-2}y^3}\right)$

70. $\left(\dfrac{4x^{-2}y^{-2}z^3}{5x^5y^2z^5}\right)\left(\dfrac{25xy^4z^{-2}}{16x^{-3}y^5z^{-4}}\right)$

71. $\left(\dfrac{x^4y^{-3}}{x^{-4}y^{-3}z}\right)\left(\dfrac{x^3y^4z}{x^5y^{-2}z^{-3}}\right)$

72. $\dfrac{(6x^4y^{-2}z^{-1})(5x^{-5}y^4)}{9x^{-2}y^6z^{-4}}$

Simplify each of the following. Assume that all variables represent integers.

73. $x^{4a} \cdot x^{3a+4}$

74. $y^{4r-2} \cdot y^{-2r+3}$

75. $w^{5b-2} \cdot w^{2b+3}$

76. $d^{5x+3} \cdot d^{-2x-3}$

77. $\dfrac{x^{2w+3}}{x^{w-4}}$

78. $\dfrac{y^{5m-1}}{y^{7m-1}}$

79. $\dfrac{(x^{3p+5})(x^{2p-3})}{x^{4p-1}}$

80. $\dfrac{(s^{2t-3})(s^{-t+5})}{s^{2t+4}}$

Express each number in scientific notation.

81. 3700

82. 3,610,000

83. 900

84. 0.00062

85. 0.047

86. 0.0000462

87. 19,000

88. 5,260,000,000

89. 0.00000186

90. 0.0003

91. 0.00000914

92. 37,000

Express each number without exponents.

93. 5.2×10^3

94. 1.63×10^{-4}

95. 4×10^7

96. 6.15×10^5

97. 2.13×10^{-5}

98. 9.64×10^{-7}

99. 3.12×10^{-1}

100. 4.6×10^1

101. 9×10^6

102. 7.3×10^4

103. 5.35×10^2

104. 1.04×10^{-2}

Express each solution without exponents.

105. $(5 \times 10^3)(3 \times 10^4)$

106. $(2.1 \times 10^1)(3 \times 10^{-4})$

107. $(1.6 \times 10^{-2})(4 \times 10^{-3})$

108. $\dfrac{6.4 \times 10^5}{2 \times 10^3}$

109. $\dfrac{8.4 \times 10^{-6}}{4 \times 10^{-4}}$

110. $\dfrac{25 \times 10^3}{5 \times 10^{-2}}$

111. $\dfrac{4 \times 10^5}{2 \times 10^4}$

112. $\dfrac{16{,}000}{0.008}$

113. $(700{,}000)(6{,}000{,}000)$

114. $(0.0006)(5{,}000{,}000)$

Express each solution using scientific notation.

115. $(0.003)(0.00015)$

116. $(230{,}000)(3000)$

117. $\dfrac{1{,}400{,}000}{700}$

118. $\dfrac{20{,}000}{0.0005}$

119. $\dfrac{0.0000426}{200}$

120. $\dfrac{(0.000012)(400{,}000)}{0.000006}$

121. The distance to the sun is 93,000,000 miles. If a spacecraft travels at a speed of 3100 miles per hour, how long will it take for it to reach the sun?

122. A computer can do one calculation in 0.0000004 second. How long would it take a computer to do a trillion (10^{12}) calculations?

123. If $x^{-1} = 5$, what is the value of x? Explain how you determined your answer.

124. If $x^{-1} = y^2$, what is x equal to? Explain how you determined your answer.

Cumulative Review Exercises

[1.6] **125.** Evaluate $\sqrt[3]{-125}$.

[2.1] **126.** Solve the equation
$-4.32 + 1.2(2x - 1.1) = 5.6x - 3.24.$

[2.3] **127.** What number when multiplied by 2 and divided by 5 gives 8?

[3.4] **128.** Write in slope–intercept form the equation for the line illustrated.

JUST FOR FUN

1. The Richter scale is used to measure the intensity of earthquakes. An earthquake that measures 1 on the Richter scale is barely detected by instruments. An earthquake that measures 2 on the Richter scale is 10 times as intense as an earthquake that measures 1. An earthquake that measures 3 on the Richter scale is 10 times as intense as one that measures 2, and $10 \cdot 10$ or 100 times as intense as one that measures 1, and so on.

(a) Use each Richter scale number as an exponent of a power of 10 and give its equivalent value. For example, 0 gives $10^0 = 1$, 1 gives $10^1 = 10$, 2 gives $10^2 = 100$, and so on.

(b) How many times more intense is an earthquake that measures 6 than an earthquake that measures 2 on the Richter scale?   10,000

(c) On October 17, 1989, an earthquake measuring 6.9 on the Richter scale, with its epicenter near Santa Cruz, California, did major damage to San Francisco, California, and the surrounding area. How many times more intense was the great San Francisco earthquake than the October, 1989, earthquake?

Richter scale

10
9   — Disastrous San Francisco earthquake in 1906 measured 8.3
8   — Great earthquake, tremendous damage
7   — Major earthquake, wide damage
6   — Moderately destructive
5
4   — Possible slight damage within a small area
3
2   — Barely perceptable even near epicenter
1   — Detectable only by instrument
0

2. A *light year* is the distance that light travels in one year.

(a) Find the number of miles in a light year if light travels at 1.86×10^5 miles *per second*.

(b) If the earth is 93,000,000 miles from the sun, how long does it take for light from the sun to reach the earth?

(c) Our Milky Way galaxy is about 6.25×10^{16} miles across. If a spaceship could travel at half the speed of light, how long would it take for the craft to travel from one end of the galaxy to the other?

5.2

More on Exponents

▶ **1**   Learn the three power rules.

▶ **1**   In Section 5.1 we introduced a number of properties of exponents. In this section we discuss the power rules of exponents. For the sake of clarity, we will refer to the three power rules we discuss as power rules 1, 2, and 3.

Consider the problem $(x^3)^2$.

$$(x^3)^2 = x^3 \cdot x^3 = x^{3+3} = x^6$$

This problem could also be evaluated using power rule 1.

Power Rule for Exponents

If a is a real number and m and n are integers, then

$$(a^m)^n = a^{m \cdot n} \qquad \text{Power rule 1}$$

To raise an expression in exponential form to a power, maintain the base and multiply the exponents.

$$(x^3)^2 = x^{3 \cdot 2} = x^6$$

EXAMPLE 1   Simplify each of the following.

(a) $(2^3)^2$ (b) $(x^3)^5$ (c) $(y^3)^{-5}$ (d) $(3^{-2})^3$

Solution: (a) $(2^3)^2 = 2^{3 \cdot 2} = 2^6 = 64$ (b) $(x^3)^5 = x^{3 \cdot 5} = x^{15}$

(c) $(y^3)^{-5} = y^{3(-5)} = y^{-15} = \dfrac{1}{y^{15}}$ (d) $(3^{-2})^3 = 3^{-2(3)} = 3^{-6} = \dfrac{1}{3^6}$ or $\dfrac{1}{729}$ ∎

HELPFUL HINT

Students often confuse the *product rule*

$$a^m \cdot a^n = a^{m+n}$$

with the *power rule*

$$(a^m)^n = a^{m \cdot n}$$

Note that, for example, $(x^3)^2 = x^6$, not x^5.

Two additional forms of the power rule for exponents follow.

Power Rules for Exponents

If a and b are real numbers and m is an integer, then

$$(ab)^m = a^m b^m \qquad \text{Power rule 2}$$

$$\left(\frac{a}{b}\right)^m = \frac{a^m}{b^m}, \ b \neq 0 \qquad \text{Power rule 3}$$

Note that when an expression within parentheses is raised to a power, each factor in the parentheses is raised to that power.

EXAMPLE 2   Simplify each of the following.

(a) $(4x^3y^{-2})^3$ (b) $(3xy^{-3})^{-2}$ (c) $\left(\frac{3}{x^{-2}}\right)^4$

Solution: (a)

$$(4x^3y^{-2})^3 = 4^3 x^9 y^{-6}$$

$$= 4^3 x^9 \cdot \frac{1}{y^6}$$

$$= \frac{64x^9}{y^6}$$

(b)

$$(3xy^{-3})^{-2} = 3^{-2} x^{-2} y^6$$

$$= \frac{1}{3^2} \cdot \frac{1}{x^2} \cdot y^6$$

$$= \frac{y^6}{9x^2}$$

(c)

$$\left(\frac{3}{x^{-2}}\right)^4 = \frac{3^4}{x^{-8}}$$

$$= \frac{81}{x^{-8}}$$

$$= 81x^8 \qquad \blacksquare$$

EXAMPLE 3   Simplify each of the following.

(a) $\left(\frac{2}{3}\right)^{-2}$ (b) $\left(\frac{x^2}{y}\right)^{-3}$

Solution: (a) $\left(\frac{2}{3}\right)^{-2} = \frac{2^{-2}}{3^{-2}}$ (b) $\left(\frac{x^2}{y}\right)^{-3} = \frac{x^{-6}}{y^{-3}}$

$$= \frac{1}{2^2} \cdot 3^2 \qquad\qquad = \frac{1}{x^6} \cdot y^3$$

$$= \frac{9}{4} \qquad\qquad\qquad = \frac{y^3}{x^6} \qquad \blacksquare$$

This problem could have also been worked by using the negative exponent rule first. For example, we could have started part (a) by writing $\left(\frac{2}{3}\right)^{-2} = \frac{1}{\left(\frac{2}{3}\right)^2}$. Work both parts of Example 3 now by first using the negative exponent rule.

Note that in Example 3 part (a) $\left(\frac{2}{3}\right)^{-2} = \frac{9}{4}$ or $\left(\frac{3}{2}\right)^2$. In part (b) we see that $\left(\frac{x^2}{y}\right)^{-3} = \frac{y^3}{x^6}$ or $\left(\frac{y}{x^2}\right)^3$. Using these examples as guides, can you guess what the answer to $\left(\dfrac{a}{b}\right)^{-4}$ will be without working the problem? If you answered $\left(\dfrac{b}{a}\right)^{4}$ or $\dfrac{b^4}{a^4}$ you are correct.

A rational expression raised to a negative exponent can be converted to a rational expression with a positive exponent by inverting the rational expression and changing the sign of the exponent.

For any a and b, $a \neq 0$, $b \neq 0$,
$$\left(\frac{a}{b}\right)^{-m} = \left(\frac{b}{a}\right)^{m}$$

Examples

$$\left(\frac{5}{9}\right)^{-3} = \left(\frac{9}{5}\right)^{3} \qquad \left(\frac{x^2}{y^3}\right)^{-4} = \left(\frac{y^3}{x^2}\right)^{4}$$

When you are given an expression raised to a negative exponent, you can use the property just given to rewrite the expression with a positive exponent.

EXAMPLE 4   Simplify each of the following.

(a) $\left(\dfrac{6x^2y^4}{2x^2y}\right)^{2}$ (b) $\left(\dfrac{3x^4y^{-2}}{6xy^3}\right)^{-3}$

Solution: Problems involving exponents can often be solved in more than one way. In general, it will be easier to simplify the expression within parentheses before using the power rule.

(a) $\left(\dfrac{6x^2y^4}{2x^2y}\right)^{2} = (3y^3)^2 = 9y^6$

(b) $\left(\dfrac{3x^4y^{-2}}{6xy^3}\right)^{-3} = \left(\dfrac{x^3}{2y^5}\right)^{-3} = \left(\dfrac{2y^5}{x^3}\right)^{3} = \dfrac{8y^{15}}{x^9}$ ■

EXAMPLE 5   Simplify $\left(\dfrac{6x^2y^{-2}}{5}\right)^{-3}\left(\dfrac{2xy^4}{3}\right)^{2}$.

Solution: $\left(\dfrac{6x^2y^{-2}}{5}\right)^{-3}\left(\dfrac{2xy^4}{3}\right)^{2} = \left(\dfrac{6x^2}{5y^2}\right)^{-3}\left(\dfrac{2xy^4}{3}\right)^{2}$

$$= \left(\frac{5y^2}{6x^2}\right)^{3}\left(\frac{2xy^4}{3}\right)^{2}$$

$$= \left(\frac{125y^6}{216x^6}\right)\left(\frac{4x^2y^8}{9}\right)$$

$$= \frac{125y^{14}}{486x^4}$$ ■

EXAMPLE 6 Simplify $\left(\dfrac{4x^{-2}y^5}{3x^5y^9}\right)^{-2}\left(\dfrac{2x^4y^{-7}}{5x^{-3}y^{-5}}\right)^3$

Solution: First, simplify the expressions within parentheses and then write them without negative exponents. Use the power rule to simplify further.

$$\left(\dfrac{4x^{-2}y^5}{3x^5y^9}\right)^{-2}\left(\dfrac{2x^4y^{-7}}{5x^{-3}y^{-5}}\right)^3 = \left(\dfrac{4}{3x^7y^4}\right)^{-2}\left(\dfrac{2x^7}{5y^2}\right)^3$$

$$= \left(\dfrac{3x^7y^4}{4}\right)^2\left(\dfrac{2x^7}{5y^2}\right)^3$$

$$= \left(\dfrac{9x^{14}y^8}{16}\right)\left(\dfrac{8x^{21}}{125y^6}\right)$$

$$= \dfrac{9x^{35}y^2}{250}$$

EXAMPLE 7 Simplify $\dfrac{(2p^{-3}q^4)^{-2}(3p^{-4}q^5)^3}{(4p^{-5}q^4)^{-3}}$.

Solution: First, use the power rule; then simplify further.

$$\dfrac{(2p^{-3}q^4)^{-2}(3p^{-4}q^5)^3}{(4p^{-5}q^4)^{-3}} = \dfrac{(2^{-2}p^6q^{-8})(3^3p^{-12}q^{15})}{4^{-3}p^{15}q^{-12}}$$

$$= \dfrac{2^{-2}3^3p^{-6}q^7}{4^{-3}p^{15}q^{-12}}$$

$$= \dfrac{3^3 \cdot 4^3 q^{19}}{2^2 p^{21}}$$

$$= \dfrac{27 \cdot 64 q^{19}}{4p^{21}}$$

$$= \dfrac{432 q^{19}}{p^{21}}$$

COMMON STUDENT ERROR

The power rule of exponents states that

$$(ab)^m = a^m b^m \qquad \text{correct}$$

An error commonly made by students throughout their mathematics career is to write

$$(a + b)^m = a^m + b^m \qquad \text{wrong}$$

Select some numbers for a, b, and m and show that $(a + b)^m \ne a^m + b^m$.

Summary of Rules of Exponents

For all real numbers a and b and all integers m and n:

Product rule	$a^m \cdot a^n = a^{m+n}$	
Quotient rule	$\dfrac{a^m}{a^n} = a^{m-n},$	$a \neq 0$
Negative exponent rule	$a^{-m} = \dfrac{1}{a^m},$	$a \neq 0$
Zero exponent rule	$a^0 = 1,$	$a \neq 0$
Power rules	$\begin{cases} (a^m)^n = a^{mn} \\ (ab)^m = a^m b^m \\ \left(\dfrac{a}{b}\right)^m = \dfrac{a^m}{b^m}, \end{cases}$	$b \neq 0$

Exercise Set 5.2

Simplify and write the answers without negative exponents.

1. $x^0 \cdot x^4$

2. $(2^2)^3$

3. $(3^2)^2$

4. $(3^2)^{-1}$

5. $(2^3)^{-2}$

6. $(x^3)^{-5}$

7. $(y^0)^3$

8. $(x^{-3})^{-2}$

9. $(-x)^2$

10. $(-x)^3$

11. $(-x)^{-3}$

12. $(-2x^{-2})^3$

13. $3(x^4)^{-2}$

14. $(2x^{-3})^2$

15. $(3x^2)^3$

16. $\left(\dfrac{3}{4}\right)^{-2}$

17. $\left(\dfrac{1}{2}\right)^{-3}$

18. $\left(\dfrac{2x}{3}\right)^{-2}$

19. $(-3x^2y)^4$

20. $-3(x^2y)^4$

21. $(4x^2y^{-2})^2$

22. $(5xy^3)^{-2}$

23. $(2x^3y)^{-3}$

24. $(3x^{-2}y)^{-2}$

25. $(-4x^{-4}y^5)^{-3}$

26. $3(x^2y)^{-4}$

27. $\left(\dfrac{6x}{y^2}\right)^2$

28. $\left(\dfrac{3x^2y^4}{z}\right)^3$

29. $\left(\dfrac{2x^4y^5}{x^2}\right)^3$

30. $\left(\dfrac{3x^5y^6}{6x^4y^7}\right)^3$

31. $\left(\dfrac{20x^3y^4}{4x^4y}\right)^3$

32. $\left(\dfrac{4xy}{y^3}\right)^{-3}$

33. $\left(\dfrac{3xy^5}{6xy^5}\right)^{-2}$

34. $\left(\dfrac{3x^{-2}}{y}\right)^{-2}$

35. $\left(\dfrac{4x^{-2}y}{x^{-5}}\right)^3$

36. $\left(\dfrac{4x^2y}{x^{-5}}\right)^{-3}$

37. $\left(\dfrac{6x^2y}{3xz}\right)^{-3}$

38. $\left(\dfrac{5xy}{z^{-2}}\right)^3$

39. $(2x^3y)(6x^4y^3)^2$

40. $(3x^4)^2(5x^3y^8)$

41. $(4x^2y^5)^2(6x^4y^3)$

42. $\dfrac{(2x^3y)^2}{z} \cdot \dfrac{xy^4}{z^3}$

43. $\left(\dfrac{3x^2y^{-2}}{y^3}\right)^2\left(\dfrac{xy^2}{3}\right)^{-2}$

44. $\left(\dfrac{x^2y}{3}\right)^2\left(\dfrac{9x^2y^{-4}}{x^4y^8}\right)$

45. $\left(\dfrac{3x^2y^5}{2z^{-1}}\right)\left(\dfrac{2z^3}{3xy^4}\right)^4$

46. $\dfrac{(2x^3y^2)^2(3xy)^{-1}}{(x^4y^3)^2}$

47. $\dfrac{(4x^2y^{-3})^2(xy^5)^{-3}}{(6x^4y^5)^3}$

48. $\left(\dfrac{x^6y^{-2}}{x^{-2}y^3}\right)^2\left(\dfrac{x^{-1}y^{-3}}{x^{-4}y^2}\right)^{-3}$

49. $\left(\dfrac{x^2y^{-3}z^4}{x^{-1}y^2z^3}\right)^{-1}\left(\dfrac{xy^2z}{x^{-3}y^{-7}z^3}\right)^3$

50. $\left(\dfrac{3x^{-4}y^{-2}}{6xy^{-4}z^2}\right)^2\left(\dfrac{4x^{-1}y^{-2}z^3}{2xy^2z^{-3}}\right)^{-2}$

51. $\left(\dfrac{6x^4y^{-6}z^4}{2xy^{-6}z^{-2}}\right)^{-2}\left(\dfrac{-x^4y^3}{2z^4}\right)^{-1}$

52. $\left(\dfrac{-x^3y^{-1}z^{-3}}{2xy^3z^{-4}}\right)^{-1}\left(\dfrac{2x^3y^6z^2}{4x^{-2}y^{-4}z}\right)^{-2}$

53. $\dfrac{(4x^{-1}y^{-2})^{-3}(xy^3)^2}{(3x^{-1}y^3)^2}$

54. $\dfrac{(3x^{-4}y^2)^3(4x^4y^3)^{-2}}{(2x^3y^5)^3}$

55. $\dfrac{(5y^3z^{-4})^{-1}(2y^{-3}z^2)^{-1}}{(3y^3z^{-2})^{-1}}$

56. $\dfrac{(4xyz^2)^3(2xy^2z^{-3})^2}{(3x^{-1}yz^2)^{-1}}$

Simplify each expression. Assume that all variables used as exponents are integers.

57. $x^{-m}(x^{3m+2})^2$

58. $y^{3b+2} \cdot (y^{2b+4})^2$

59. $(b^{5y-2})^y \cdot b^{5y}$

60. $\dfrac{x^{3a} \cdot (x^{2a})^3}{x^{4a-2}}$

61. $\dfrac{(m^{-5y+2})(m^{2y+3})}{m(m^{4y+1})}$

62. $\dfrac{(w^{2p+4})^2(w^{-p-8})}{w^3(w^{4p-3})}$

Indicate the exponents that when placed in the shaded area will result in the expression having a value of 1. Explain how you determined your answer.

63. $\left(\dfrac{x^2y^{-2}}{x^{-3}y^{-1}}\right)^2\left(\dfrac{x^{\blacksquare}y^3}{x^7y^{\blacksquare}}\right)$

64. $\left(\dfrac{x^{-2}y^3z}{x^4y^{-2}z^{-3}}\right)^3\left(\dfrac{x^2y^{-2}}{x^{\blacksquare}y^{\blacksquare}z^{\blacksquare}}\right)$

65. $\left(\dfrac{x^{-4}y^{-2}z^7}{x^{-2}y^5z^2}\right)^{-4}\left(\dfrac{x^{\blacksquare}y^5z^{-2}}{x^4y^{\blacksquare}z^{\blacksquare}}\right)$

66. On page 220 we state that $\left(\dfrac{a}{b}\right)^{-m} = \left(\dfrac{b}{a}\right)^m$. Starting with $\left(\dfrac{a}{b}\right)^{-m}$, use the negative exponent rule to show that this expression is equal to $\left(\dfrac{b}{a}\right)^m$.

Cumulative Review Exercises

[4.1] *Solve the system of equations by the method indicated.*

67. $x + 3y = 10$
$2x - 4y = -10$
substitution

68. $\frac{1}{2}x + \frac{3}{2}y = 7$
$x - 2y = -6$
addition

69. $x + y = 1$
$-3x + 2y = 12$
graphing

[4.3] **70.** Margarita wishes to mix a 40% saltwater solution with a 60% saltwater solution to get 20 liters of a 45% saltwater solution. How many liters each of the 40% solution and the 60% solution should she mix?

JUST FOR FUN

We will learn in Section 8.2 that the rules of exponents given in this and the previous section also apply when the exponents are rational numbers. Using this information and the rules of exponents evaluate each of the following.

1. $\left(\dfrac{x^{1/2}}{x^{-1}}\right)^{3/2}$

2. $\left(\dfrac{x^{5/8}}{x^{1/4}}\right)^3$

3. $\left(\dfrac{x^4}{x^{-1/2}}\right)^{-1}$

4. $\left(\dfrac{x^{1/2}y^{-3/2}}{x^5y^{5/3}}\right)^{2/3}$

5. $\left(\dfrac{x^{1/2}y^4}{x^{-3}y^{5/2}}\right)^2$

6. $\left(\dfrac{x^{9/5}}{y^{1/3}}\right)^{2/3}\left(\dfrac{x^{-1}y^{2/3}}{y^{-1/3}}\right)^2$

Addition and Subtraction of Polynomials

▶**1**   Identify polynomials.

▶**2**   Find the degree of a polynomial.

▶**3**   Add polynomials.

▶**4**   Subtract polynomials.

▶**1**   A **polynomial** is a finite sum of terms in which all variables have whole-number exponents and no variables appear in a denominator.

A polynomial with terms of the form ax^n is called a **polynomial in x.**

Examples of Polynomials in x	*Not Polynomials*
$5x$	$x^{1/2}$ (fractional exponent)
$6x^2 - \dfrac{1}{2}x + 4$	$2x^{-1}$ (negative exponent)

Polynomials can be in more than one variable, as follows.

Examples of Polynomials in x and y
$$3xy - 6x^2y$$
$$4x^2y - 3xy^2 + 5$$

A polynomial of only one term is called a **monomial.** A **binomial** is a two-term polynomial, and a **trinomial** is a three-term polynomial. Polynomials containing more than three terms are not given special names. The term "poly" is a prefix meaning many.

Examples of Monomials	*Examples of Binomials*	*Examples of Trinomials*
4	$x + 4$	$x^2 - 2x + 1$
$6x$	$x^2 - 6x$	$6x^2 + 3xy - 2y^2$
$\dfrac{1}{5}xyz^3$	$x^2y - y^2$	$\dfrac{1}{2}x + 3y - 6x^2y^2$

▶**2**   In Section 2.1 we stated that the **degree of a term** is the sum of the exponents on the variables in the term. Thus $3x^2y^3z$ is of degree 6 $(2 + 3 + 1 = 6)$.

The **degree of a polynomial** is the same as that of its highest-degree term.

Polynomial	*Degree of Polynomial*
$8x^3 + 2x^2 + 3x + 4$	third (x^3 is highest-degree term)
4	zero (4 or $4x^0$ is highest-degree term)
$6x^2 + 4x^2y^5 - 6$	seventh ($4x^2y^5$ is highest-degree term)

The polynomials $2x^3 + 4x^2 - 6x + 3$ and $4x^2 - 3xy + 5y^2$ are examples of polynomials in **descending order** of the variable x because the exponents on the variable x descend (or get lower) as the terms go from left to right. Polynomials are often written in descending order of a given variable.

EXAMPLE 1 Write each of the following polynomials in descending order of the variable x.

(a) $3x + 4x^2 - 6$ (b) $xy - 6x^2 + 3y^2$

Solution: (a) $3x + 4x^2 - 6 = 4x^2 + 3x - 6$
(b) $xy - 6x^2 + 3y^2 = -6x^2 + xy + 3y^2$ ∎

Addition of Polynomials

▶3

Adding Polynomials

To add polynomials, combine the like terms of the polynomials.

EXAMPLE 2 Simplify $(4x^2 - 6x + 3) + (2x^2 + 5x - 1)$.

Solution:
$(4x^2 - 6x + 3) + (2x^2 + 5x - 1)$
$= 4x^2 - 6x + 3 + 2x^2 + 5x - 1$ Remove the parentheses.
$= \underbrace{4x^2 + 2x^2}\ \underbrace{- 6x + 5x}\ \underbrace{+ 3 - 1}$ Rearrange terms.
$= \quad 6x^2 \qquad -x \qquad +2$ Combine like terms. ∎

EXAMPLE 3 Simplify $(3x^2y - 4xy + y) + (x^2y + 2xy + 3y - 2)$.

Solution:
$(3x^2y - 4xy + y) + (x^2y + 2xy + 3y - 2)$
$= 3x^2y - 4xy + y + x^2y + 2xy + 3y - 2$ Remove the parentheses.
$= \underbrace{3x^2y + x^2y}\ \underbrace{- 4xy + 2xy}\ \underbrace{+ y + 3y} - 2$ Rearrange terms.
$= \quad 4x^2y \qquad -2xy \qquad +4y \quad - 2$ Combine like terms. ∎

Subtraction of Polynomials

▶4

To Subtract Polynomials

1. Remove parentheses from the polynomials being subtracted, and change the sign of every term of the polynomial being subtracted.
2. Combine like terms.

EXAMPLE 4 Simplify $(3x^2 - 2x + 5) - (x^2 - 3x + 4)$.

Solution:
$(3x^2 - 2x + 5) - (x^2 - 3x + 4)$
$= 3x^2 - 2x + 5 - x^2 + 3x - 4$ Remove the parentheses (change the sign of each term being subtracted).
$= 3x^2 - x^2 - 2x + 3x + 5 - 4$ Rearrange terms.
$= 2x^2 + x + 1$ Combine like terms. ∎

EXAMPLE 5 Subtract $(-x^2 - 2x + 3)$ from $(x^3 + 4x + 6)$.

Solution:
$(x^3 + 4x + 6) - (-x^2 - 2x + 3)$
$= x^3 + 4x + 6 + x^2 + 2x - 3$ Remove the parentheses.
$= x^3 + x^2 + 4x + 2x + 6 - 3$ Rearrange terms.
$= x^3 + x^2 + 6x + 3$ Combine like terms. ∎

EXAMPLE 6   Simplify $x^2y - 4xy^2 + 5 - (2x^2y - 3y^2 + 4)$.

Solution: $x^2y - 4xy^2 + 5 - (2x^2y - 3y^2 + 4)$

$= x^2y - 4xy^2 + 5 - 2x^2y + 3y^2 - 4$ Remove the parentheses.

$= x^2y - 2x^2y - 4xy^2 + 3y^2 + 5 - 4$ Rearrange terms.

$= -x^2y - 4xy^2 + 3y^2 + 1$ Combine like terms.

Note that $-x^2y$ and $-4xy^2$ are not like terms since the variables have different exponents. Also, $-4xy^2$ and $3y^2$ are not like terms since the $3y^2$ does not contain the variable x. ∎

Exercise Set 5.3

Indicate those expressions that are polynomials. If the polynomial has a specific name, for example, "monomial," "binomial," and so on, give the name. If the expression is not a polynomial, state so.

1. $5y$ **2.** $5x^2 - 6x + 9$ **3.** -10 **4.** $5x^{-3}$

5. $4 - 5z$ **6.** $6xy + 3z$ **7.** $8x^2 - 2x + 8y^2$ **8.** $x - 3$

9. $3x^{1/2} + 2xy$ **10.** $-2x^2 + 5x^{-1}$ **11.** $2xy + 5y^2$ **12.** $4 - 3x$

Write the polynomial in descending order of the variable x. If the polynomial is already in descending order, state so. Give the degree of each polynomial.

13. $-8 - 4x - x^2$ **14.** $2x + 4 - x^2$

15. $6y^2 + 3xy + 10x^2$ **16.** $3x^3 - x + 4$

17. $5 + 2x^3 + x^2 - 3x$ **18.** $-4 + x - 3x^2 + 4x^3$

19. $-2x^4 + 5x^2 - 4$ **20.** $6x^3 + 4xy^2 - 5x^2y + 7y^3$

21. $5xy^2 + 3x^2y - 6 - 2x^3$ **22.** $4y^2 - 2xy - 3x^2 + 5x^4$

Simplify.

23. $(6x + 3) + (x - 5)$ **24.** $(2x - 3) + (3x - 4)$

25. $(3x - 4) - (2x + 2)$ **26.** $(6x + 3) - (4x - 2)$

27. $(-12x - 3) - (-5x - 7)$ **28.** $(12x - 3) - (-2x + 7)$

29. $(x^2 - 6x + 3) - (2x + 5)$ **30.** $(x - 4) - (3x^2 - 4x + 6)$

31. $(x + y - z) + (2x - y + 3)$ **32.** $(4y^2 + 6y - 3) - (2y^2 + 6)$

33. $(5x - 7) + (2x^2 + 3x + 12)$ **34.** $(-3x + 8) + (-2x^2 - 3x - 5)$

35. $(6y^2 - 6y + 4) - (-2y^2 - y + 7)$ **36.** $(x^2 - 6x + 7) + (-x^2 + 3x + 5)$

37. $(-2x^2 + 4x - 5) - (5x^2 + 3x + 7)$ **38.** $(5x^2 - x - 1) - (-3x^2 - 2x - 5)$

39. $(5x^2 - x + 12) - (x + 5)$ **40.** $(6x^2y - 3xy) - (4x^2y + 2xy)$

41. $(-3x^3 + 4x^2y + 3xy^2) + (2x^3 - x^2y + xy^2)$ **42.** $(9x^3 - 2x^2 + 4x - 7) + (2x^3 - 6x^2 - 4x + 3)$

43. $(-2xy^2 + 4) - (-7xy^2 + 12)$ **44.** $(x^2 + 6xy + y^2) - (3x^2 + 6xy + 3y^2)$

45. $(x^2 + xy - y^2) + (2x^2 - 3xy + y^2)$ **46.** $(x^2y + 6x^2 - 3xy^2) + (-x^2y - 12x^2 + 4xy^2)$

47. $(4x^3 - 6x^2 + 5x - 7) - (2x^2 + 6x - 3)$ **48.** $(4x^3 - 3x^2y + 4) - (2y^2 + 8)$

49. $(4x^2y + 2x - 3) + (3x^2y - 5x + 5)$ **50.** $(x^2y + x - y) + (2x^2y + 2x - 6y + 3)$

51. $(9x^3 - 4) - (x^2 + 5)$ **52.** $(9x^3 + 6x^2y + 3xy^2) - (-5x^2y + 5)$

53. Add $x^2 - 2x + 4$ and $3x + 12$.   **54.** Add $4x^2 - 6x + 5$ and $-2x - 8$.

55. Subtract $(4x - 6)$ from $(3x + 5)$.   **56.** Subtract $(-x^2 + 3x + 5)$ from $(4x^2 - 6x + 2)$.

57. Add $-2x^2 + 4x - 12$ and $-x^2 - 2x$.   **58.** Add $5x^2 + x + 9$ and $2x^2 - 12$.

59. Subtract $(5x^2 - 6)$ from $(2x^2 - 4x + 8)$.

60. Subtract $(x^2 - 6)$ from $(6x^2 - 5x + 3)$.

61. Add $3x^2 + 4x - 5$ and $4x^2 + 3x - 8$.

62. Add $-5x^2 - 3$ and $x^2 + 2x - 9$.

63. Subtract $(-6y^2 + 3y - 4)$ from $(9y^2 - 3y)$.

64. Subtract $(4x^2 + 7x - 9)$ from $(x^3 - 6x + 3)$.

65. Add $6x^2 + 3xy$ and $-2x^2 + 4xy + 3y$.

66. Add $4x^2 + 3y^2 + 4$ and $-3x^2 - 7 + y^2$.

67. Subtract $(-4x^2 + 6x)$ from $(x^3 - 6)$.

68. Subtract $(2x^2 - 6x + 4)$ from $(6x + 8)$.

69. Add $4x^2 + 3x + y^2$ and $4x - 3y - 5y^2$.

70. Add $x^2 + 3y$ and $y^2 + 5y + 7$.

71. Subtract $(5x^2y + 8)$ from $(-2x^2y + 6xy^2 + 8)$.

72. Subtract $(6x^2y + 3xy)$ from $(2x^2y + 12xy)$.

73. When one polynomial is subtracted from another polynomial, what happens to the signs of all the terms of the polynomial being subtracted?

Cumulative Review Exercises

[1.7] **74.** Evaluate $\dfrac{\left(\left|\frac{1}{2}\right| - \left|-\frac{1}{3}\right|\right)^2}{-\left|\frac{1}{3}\right| \cdot \left|-\frac{2}{5}\right|}$.

[2.5] **75.** Solve the inequality $-4 < \dfrac{6 - 3x}{2} \leq 5$ and give the answer in set builder notation.

[3.5] **76.** What is a function?

[4.2] **77.** Solve the system of equations:
$$-2x + 3y + 4z = 17$$
$$-5x - 3y + z = -1$$
$$-x - 2y + 3z = 18$$

5.4

Multiplication of Polynomials

▶ **1** Multiply a monomial by a monomial.

▶ **2** Multiply a polynomial by a monomial.

▶ **3** Multiply a binomial by a binomial.

▶ **4** Multiply a polynomial by a polynomial.

▶ **5** Square a binomial.

▶ **6** Find the product of the sum and difference of the same two terms.

Multiplying a Monomial by a Monomial

▶ **1** **When we multiply polynomials, each term of one polynomial must multiply each term of the other polynomial.** This results in monomials multiplying monomials. To multiply monomials, we use the rules of exponents presented earlier.

EXAMPLE 1   Multiply   (a) $(3x^2y)(4x^5y^3)$;   (b) $(-2a^4b^7)(-3a^8b^3c)$

Solution: (a) $(3x^2y)(4x^5y^3) = 3 \cdot 4 \cdot x^2 \cdot x^5 \cdot y \cdot y^3 = 12x^{2+5}y^{1+3} = 12x^7y^4$

(b) $(-2a^4b^7)(-3a^8b^3c) = (-2)(-3)a^4 \cdot a^8 \cdot b^7 \cdot b^3 \cdot c$
$$= 6a^{4+8}b^{7+3}c$$
$$= 6a^{12}b^{10}c$$

Multiplying a Monomial by a Polynomial

▶ **2** When multiplying two polynomials, where one of the polynomials is a monomial and the other is not, we can use the *expanded form of the distributive property* to perform the multiplication.

> **Distributive Property**
> $$a(b + c + d + \cdots + n) = ab + ac + ad + \cdots + an$$

EXAMPLE 2   Multiply $3x(4x + 5)$.

Solution: $3x(4x + 5) = (3x)(4x) + (3x)(5)$
$$= 12x^2 + 15x$$ ∎

EXAMPLE 3   Multiply $-4x^2(5x^3 - 3x + 5)$.

Solution: $-4x^2(5x^3 - 3x + 5) = (-4x^2)(5x^3) + (-4x^2)(-3x) + (-4x^2)(5)$
$$= -20x^5 + 12x^3 - 20x^2$$ ∎

EXAMPLE 4   Multiply $2xy(3x^2y + 6xy^2 + 4)$.

Solution: $2xy(3x^2y + 6xy^2 + 4) = (2xy)(3x^2y) + (2xy)(6xy^2) + (2xy)(4)$
$$= 6x^3y^2 + 12x^2y^3 + 8xy$$ ∎

Multiplying a Binomial by a Binomial

▶ **3**   Consider multiplying $(a + b)(c + d)$. Treating $(a + b)$ as a single term and using the distributive property, we get

$$(a + b)(c + d) = (a + b)\,c + (a + b)\,d$$
$$= ac + bc + ad + bd$$

When multiplying a binomial by a binomial, each term of the first binomial must be multiplied by each term of the second binomial and all the results added together. Binomials can be multiplied vertically as well as horizontally.

EXAMPLE 5   Multiply $(3x + 2)(x - 5)$.

Solution: List the binomials one beneath the other. It makes no difference which one is placed on top. Then multiply each term of the top binomial by each term of the bottom binomial as shown. Remember to align like terms so that like terms can be added.

$$
\begin{array}{r}
3x + 2 \\
x - 5 \\
\hline
\end{array}
$$

$-5\,(3x + 2) \longrightarrow -15x - 10$ Multiply the top expression by -5.

$x\,(3x + 2) \longrightarrow 3x^2 + 2x$ Multiply the top expression by x.

$$\overline{3x^2 - 13x - 10}$$ Add like terms in columns.   ∎

A convenient way to multiply binomials is called the FOIL method. To multiply two binomials using the FOIL method, list the binomials side by side. The word FOIL indicates that you multiply the **F**irst terms, **O**uter terms, **I**nner terms, and **L**ast terms of the two binomials. This procedure is illustrated in Example 6, where we multiply the same two binomials multiplied in Example 5.

EXAMPLE 6   Multiply $(3x + 2)(x - 5)$ using the FOIL method.

Solution:

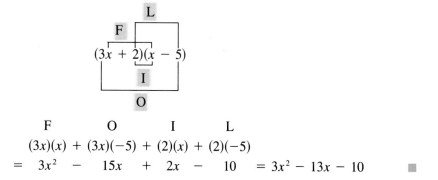

$$\begin{array}{cccc} \text{F} & \text{O} & \text{I} & \text{L} \\ (3x)(x) + (3x)(-5) + (2)(x) + (2)(-5) \end{array}$$
$$= \quad 3x^2 \quad - \quad 15x \quad + \quad 2x \quad - \quad 10 \quad = 3x^2 - 13x - 10 \quad \blacksquare$$

We performed the multiplications following the FOIL order. However, any order could be followed as long as each term of one binomial is multiplied by each term of the other binomial. We used FOIL rather than OILF or any other combination of letters because it is easier to remember.

EXAMPLE 7 Multiply $(3x^2 + 6)(x - 2y)$.

Solution: $(3x^2 + 6)(x - 2y)$

$$\begin{array}{cccc} \text{F} & \text{O} & \text{I} & \text{L} \\ (3x^2)(x) + (3x^2)(-2y) + (6)(x) + (6)(-2y) \end{array}$$
$$= 3x^3 - 6x^2y + 6x - 12y \quad \blacksquare$$

Multiplying a Polynomial by a Polynomial

▶**4** When multiplying a trinomial by a binomial or a trinomial by a trinomial, we generally multiply vertically. Remember that every term of the first polynomial must be multiplied by every term of the second polynomial. The procedure is illustrated in Examples 8 and 9.

EXAMPLE 8 Multiply $x^2 - 3x + 2$ by $2x^2 - 3$.

Solution: Place the longer polynomial on top; then multiply. Make sure you align like terms as you multiply so that the terms can be added

$$\begin{array}{r} x^2 - 3x + 2 \\ 2x^2 - 3 \end{array}$$

$-3\,(x^2 - 3x + 2) \longrightarrow \quad -3x^2 + 9x - 6$ Multiply top expression by -3.

$2x^2\,(x^2 - 3x + 2) \longrightarrow 2x^4 - 6x^3 + 4x^2$ Multiply top expression by $2x^2$.

$$\overline{\quad 2x^4 - 6x^3 + x^2 + 9x - 6\quad}$$ Add like terms in columns. ■

EXAMPLE 9 Multiply $3x^2 + 6xy - 5y^2$ by $x + 3y$.

Solution:

$$\begin{array}{r} 3x^2 + 6xy - 5y^2 \\ x + 3y \end{array}$$

$3y\,(3x^2 + 6xy - 5y^2) \longrightarrow \quad 9x^2y + 18xy^2 - 15y^3$ Multiply top expression by $3y$.

$x\,(3x^2 + 6xy - 5y^2) \longrightarrow 3x^3 + 6x^2y - 5xy^2$ Multiply top expression by x.

$$\overline{\quad 3x^3 + 15x^2y + 13xy^2 - 15y^3\quad}$$ Add like terms in columns. ■

Square of a Binomial ▶ **5**   Now we will study some special formulas. We will discuss formulas for finding the *square of a binomial* and for finding the *product of the sum and difference of the same two terms*. Example 10 will be used to introduce the formula for finding the square of a binomial.

EXAMPLE 10   Multiply $(x + 4)^2$.

Solution: $\begin{aligned}(x + 4)^2 &= (x + 4)(x + 4)\\ &= x^2 + 4x + 4x + 16\\ &= x^2 + 8x + 16\end{aligned}$ ■

Example 10 shows one method to square a binomial. We must often square a binomial, so we have special formulas for doing so.

Square of a Binomial

$$(a + b)^2 = a^2 + 2ab + b^2$$
$$(a - b)^2 = a^2 - 2ab + b^2$$

The square of a binomial is the sum of the square of the first term, twice the product of the two terms, and the square of the last term.

EXAMPLE 11   Expand $(3x + 5)^2$.

Solution: $\begin{aligned}(3x + 5)^2 &= (3x)^2 + 2(3x)(5) + (5)^2\\ &= 9x^2 + 30x + 25\end{aligned}$ ■

EXAMPLE 12   Expand $(2x - 3y)^2$.

Solution: $\begin{aligned}(2x - 3y)^2 &= (2x)^2 - 2(2x)(3y) + (3y)^2\\ &= 4x^2 - 12xy + 9y^2\end{aligned}$ ■

EXAMPLE 13   Expand $(4x^2 - 3y)^2$.

Solution: $\begin{aligned}(4x^2 - 3y)^2 &= (4x^2)^2 - 2(4x^2)(3y) + (3y)^2\\ &= 16x^4 - 24x^2y + 9y^2\end{aligned}$ ■

Examples 11 through 13 could also be done using the FOIL method.

COMMON STUDENT ERROR

Do not forget the middle term when squaring a binomial.

Correct	*Wrong*
$(x + 2)^2 = (x + 2)(x + 2)$	$(x + 2)^2 = x^2 + 4$
$\quad = x^2 + 4x + 4$	
$(x - 3)^2 = (x - 3)(x - 3)$	$(x - 3)^2 = x^2 + 9$
$\quad = x^2 - 6x + 9$	

EXAMPLE 14 Expand $[x + (y - 1)]^2$.

Solution: This problem looks more complicated than the previous examples, but it is worked the same way as the square of a binomial. Treat x as the first term and $(y - 1)$ as the second term.

$$[x + (y - 1)]^2 = (x)^2 + 2(x)(y - 1) + (y - 1)^2$$
$$= x^2 + (2x)(y - 1) + y^2 - 2y + 1$$
$$= x^2 + 2xy - 2x + y^2 - 2y + 1$$

None of the six terms are like terms, so no terms can be combined. Note that $(y - 1)^2$ is also the square of a binomial and was expanded as such. ∎

EXAMPLE 15 Use the FOIL method to multiply $(x + 6)(x - 6)$.

Solution: $(x + 6)(x - 6) = x^2 - 6x + 6x - 36 = x^2 - 36$ ∎

Difference of Two Squares

▶**6** Note in Example 15 that the outer and inner terms add to 0. By examining Example 15, we see that the product of the sum and difference of the same two terms is the difference of the squares of the two terms.

Product of Sum and Difference of the Same Two Terms

$$(a + b)(a - b) = a^2 - b^2$$

To multiply two binomials that differ only in the sign between their two terms, subtract the square of the second term from the square of the first term. Note that $a^2 - b^2$ represents a **difference of two squares.**

EXAMPLE 16 Multiply $(3x + 4)(3x - 4)$.

Solution: $(3x + 4)(3x - 4) = (3x)^2 - (4)^2 = 9x^2 - 16$ ∎

EXAMPLE 17 Multiply $(3x + 5y)(3x - 5y)$.

Solution: $(3x + 5y)(3x - 5y) = (3x)^2 - (5y)^2 = 9x^2 - 25y^2$ ∎

EXAMPLE 18 Multiply $(5x + y^3)(5x - y^3)$.

Solution: $(5x + y^3)(5x - y^3) = (5x)^2 - (y^3)^2 = 25x^2 - y^6$ ∎

EXAMPLE 19 Multiply $[4x + (3y + 2)][4x - (3y + 2)]$.

Solution: Treat $4x$ as the first term and $3y + 2$ as the second term. Then we have the sum and difference of the same two terms.

$$[4x + (3y + 2)][4x - (3y + 2)] = (4x)^2 - (3y + 2)^2$$
$$= 16x^2 - (9y^2 + 12y + 4)$$
$$= 16x^2 - 9y^2 - 12y - 4$$ ∎

Exercise Set 5.4

Multiply each of the following.

1. $(4xy)(6xy^4)$

3. $(\frac{5}{9}x^2y^5)(\frac{1}{5}x^5y^3z^2)$

5. $-2x(x^2 - 2x + 5)$

7. $5x^2(-4x^2 + 6x - 4)$

9. $-3x^2y(-2x^4y^2 + 3xy^3 + 4)$

11. $\frac{2}{3}yz(3x + 4y - 9y^2)$

2. $(-2xy^4)(3x^4y^6)$

4. $(\frac{2}{3}y^2z^4)(6x^5y^9z^{12})$

6. $-3x(-2x^2 + 5x - 6)$

8. $2y^3(3y^2 + 2y - 6)$

10. $3x^4(2xy^2 + 5x^7 - 6y)$

12. $\frac{1}{2}x^2y(4x^5y^2 + 3x - 6y^2)$

Multiply the binomials.

13. $(x - 4)(x + 5)$

15. $(3x + 1)(x - 3)$

17. $(x - y)(x + y)$

19. $(2x^2 - 3)(2x - 3)$

21. $(4 - x)(3 + 2x^2)$

23. $(\frac{1}{2}x + 2y)(2x - \frac{1}{3}y)$

25. $(4x^2 - 3y)(2y^2 - 3x)$

14. $(4x - 6)(3x - 5)$

16. $(6 - 2x)(5 + 3x)$

18. $(4 - 6z)(4 - 6z)$

20. $(6x - y)(2x + y)$

22. $(6y - 2x)(5x - 3)$

24. $(\frac{2}{5}x - \frac{1}{5}z)(\frac{1}{3}x + z)$

26. $(3xy^2 + y)(4x - 3xy)$

Multiply the polynomials.

27. $(x^2 - 3x + 2)(x - 4)$

29. $(y^2 - 3y + 4)(3 - 2y)$

31. $(7x - 3)(-2x^2 - 4x + 1)$

33. $(x^2y - 3xy^2)(x + 2y)$

35. $(a - 3b)(2a^2 - ab + 2b^2)$

37. $(x^3 - 2x^2 + 5x - 6)(2x^2 - 3x + 4)$

39. $(3x - 1)^3$

28. $(2x^2 - 3x + 4)(-3x + 4)$

30. $(5x^3 + 4x^2 - 6x + 2)(x + 5)$

32. $(x - 2)(4x^2 + 9x - 2)$

34. $(x^2 + y^2 + 2)(x + 2)$

36. $(3p + n)(p^2 - 3pn + n^2)$

38. $(2x^3 - 3x^2 + x)(-x^2 - 3x + 4)$

40. $(x - 2)^3$

Multiply using either the square of a binomial procedure or the product of the sum and difference of the same two terms procedure.

41. $(x + 4)(x - 4)$

43. $(2x - 1)(2x + 1)$

45. $(2x - 3y)^2$

47. $(2x + 5y)^2$

49. $(2y^2 - 5w)^2$

51. $(5m^2 + 2n)(5m^2 - 2n)$

53. $[y + (4 - 2x)]^2$

55. $[4 - (x - 3y)]^2$

57. $[(x + y) + 4]^2$

59. $[(x - 3y) - 5]^2$

42. $(x + 3)^2$

44. $(x + 2)(x + 2)$

46. $(4x - 2y)(4x + 2y)$

48. $(4x^2 + 3)^2$

50. $(3x^2 - 4y)(3x^2 + 4y)$

52. $[a + (b + 2)][a - (b + 2)]$

54. $[5x + (2y + 3)]^2$

56. $[w - (y + 2)]^2$

58. $[(x - 2y) - 3]^2$

60. $[(y + 2z) + 8]^2$

Multiply each of the following.

61. $(8r^5s^4)(-3rs^9)$

63. $3x(x^2 + 3x - 1)$

65. $(\frac{1}{3}x - \frac{2}{5}y)(\frac{1}{2}x - y)$

62. $(\frac{1}{5}a^5c^8)(\frac{2}{3}a^4b^5c^9)$

64. $-x(2x^2 - 6x + 5)$

66. $(\frac{1}{4}z + 3)(-4z + 5)$

67. $(3y + 4)(2y - 3)$

69. $-\frac{3}{5}x^2y\left(-\frac{2}{3}xy^4 + \frac{1}{9}xy + 3\right)$

71. $(2x - \frac{3}{4})(2x + \frac{3}{4})$

73. $(4x - 5y)^2$

75. $(x + 3)(2x^2 + 4x - 3)$

77. $(5x + 4)(x^2 - x + 4)$

79. $(2x - 3y)(3x^2 + 4xy - 2y^2)$

81. $(\frac{1}{2}r + \frac{1}{4}s)^2$

83. $\frac{2}{3}x^2y^4(\frac{3}{5}xy^3 - \frac{1}{4}x^4y + 2xy^3z^5)$

85. $[w + (3x + 4)][w - (3x + 4)]$

87. $(a + b)(a^2 - ab + b^2)$

89. $(a + 2b)(a^2 - 2ab + 4b^2)$

91. $(x + 3)^3$

93. $[(3m + 2) + n][(3m + 2) - n]$

95. $[(5x + 1) + 6y][(5x + 1) - 6y]$

68. $(3y + 4)(y + 1)$

70. $6xy(x^3 - 3x^2y + 4y^2)$

72. $(3m + 2n)^2$

74. $(2x^2 - 3y)(3x^2 + 2y)$

76. $(2x + 3)(4x^2 - 5x + 2)$

78. $(2x - 5)(3x^2 - 4x + 7)$

80. $(3x^2 + 6xy - 5y^2)(x - 3y)$

82. $(3w^2 + 4)(3w^2 - 4)$

84. $-\frac{3}{5}xy^3z^2(-xy^2z^5 - 5xy + \frac{1}{6}xz^7)$

86. $[3p + (2w - 3)][3p - (2w - 3)]$

88. $(x - 2y)(x^2 + 2xy + 3y^2)$

90. $(2m + n)(3m^2 - mn + 2n^2)$

92. $(2x + 3)^3$

94. $[3 + (x - y)][3 - (x - y)]$

96. $[x + (y + 3)]^2$

Cumulative Review Exercises

[1.7] **97.** Evaluate $\dfrac{3^3 - \sqrt[3]{-27} + 3^0}{3 - 3 \cdot 3 + 3 \div 3}$

[2.3] **98.** Ms. Jacobmeier invested a total of $10,000 in two savings accounts. One account earned 5% simple and the other account earned 6% simple interest annually. If the total interest earned from both accounts in one year is $560, how

much was invested in each account? Solve using only one variable.

[4.3] **99.** Solve Exercise 98 using two variables.

[3.4] **100.** Give the equation of the line that is parallel to the line $3x - 4y = 12$ and has a y intercept of -2.

JUST FOR FUN

Perform each of the following polynomial multiplications.

1. $[(y + 1) - (x + 2)]^2$.

2. Multiply $(x - 3y)^4$.

5.5

Division of Polynomials

▶**1** Divide a polynomial by a monomial.

▶**2** Divide a polynomial by a binomial.

Dividing a Polynomial by a Monomial

▶**1** In division of polynomials, division by zero is not permitted. When the denominator (or divisor) contains a variable, the variable cannot be a value that will result in the denominator being zero. (We discuss this concept further in Chapter 7.)

> **To divide a polynomial by a monomial,** divide each term of the polynomial by the monomial.

EXAMPLE 1   Divide $\dfrac{4x^2 - 8x}{2x}$.

Solution: $\dfrac{4x^2 - 8x}{2x} = \dfrac{4x^2}{2x} - \dfrac{8x}{2x} = 2x - 4$ ■

EXAMPLE 2   Divide $\dfrac{4x^3 - 6x^2 + 8x - 3}{2x}$.

Solution: $\dfrac{4x^3 - 6x^2 + 8x - 3}{2x} = \dfrac{4x^3}{2x} - \dfrac{6x^2}{2x} + \dfrac{8x}{2x} - \dfrac{3}{2x}$

$= 2x^2 - 3x + 4 - \dfrac{3}{2x}$ ■

EXAMPLE 3   Divide $\dfrac{4x^2y - 6x^4y^3 - 3x^5y^2 + 5x}{2xy^2}$.

Solution: $\dfrac{4x^2y - 6x^4y^3 - 3x^5y^2 + 5x}{2xy^2} = \dfrac{4x^2y}{2xy^2} - \dfrac{6x^4y^3}{2xy^2} - \dfrac{3x^5y^2}{2xy^2} + \dfrac{5x}{2xy^2}$

$= \dfrac{2x}{y} - 3x^3y - \dfrac{3x^4}{2} + \dfrac{5}{2y^2}$ ■

Dividing a Polynomial by a Binomial   ▶ **2**   We divide a polynomial by a binomial in much the same way as we perform long division.

EXAMPLE 4   Divide $\dfrac{x^2 + 7x + 10}{x + 2}$.

Solution: Rewrite the division problem as

$$x + 2 \overline{)x^2 + 7x + 10}$$

Divide x^2 (the first term in $x^2 + 7x + 10$) by x (the first term in $x + 2$).

$$\dfrac{x^2}{x} = x$$

Place the quotient, x, above the term containing x in the divisor.

$$x + 2 \overline{)x^2 + 7x + 10} \quad \overset{x}{}$$

Next, multiply the x by $x + 2$ as you would do in long division and place the product under their like terms.

$$\begin{array}{r} \text{times} \\ x \\ x + 2 \overline{)x^2 + 7x + 10} \\ x^2 + 2x \longleftarrow x(x+2) \\ \text{equals} \end{array}$$

Now subtract $x^2 + 2x$ from $x^2 + 7x$ by changing the signs of $x^2 + 2x$ and adding.

$$\begin{array}{r} x \\ x + 2 \overline{)x^2 + 7x + 10} \\ \underline{- x^2 - 2x} \\ 5x \end{array}$$

Now bring down the $+10$, the next term.

$$
\begin{array}{r}
x \\
x + 2 \overline{\smash{)}x^2 + 7x + 10} \\
\underline{x^2 + 2x } \\
5x + 10
\end{array}
$$

Determine the quotient of $5x$ divided by x.

$$
\frac{5x}{x} = +5
$$

Place the $+5$ above the constant in the dividend and multiply 5 by $x + 2$. Finally, finish the problem by subtracting.

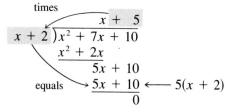

Thus $\dfrac{x^2 + 7x + 10}{x + 2} = x + 5$. There is no remainder.   ■

EXAMPLE 5   Divide $\dfrac{6x^2 - 5x + 5}{2x + 3}$.

Solution: In this problem we will mentally change the signs of the terms being subtracted and then add.

$$
\begin{array}{r}
3x - 7 \\
2x + 3 \overline{\smash{)}6x^2 - 5x + 5} \\
\underline{6x^2 + 9x} \longleftarrow 3x(2x + 3) \\
-14x + 5 \\
\underline{-14x - 21} \longleftarrow -7(2x + 3) \\
26 \longleftarrow \text{remainder}
\end{array}
$$

Thus $\dfrac{6x^2 - 5x + 5}{2x + 3} = 3x - 7 + \dfrac{26}{2x + 3}$.   ■

When dividing a polynomial by a binomial, the answer may be **checked** by multiplying the divisor by the quotient and then adding the remainder. You should end with the polynomial you began with. For example, to check Example 5 we do the following:

$$
\begin{aligned}
(2x + 3)(3x - 7) + 26 &= 6x^2 - 5x - 21 + 26 \\
&= 6x^2 - 5x + 5
\end{aligned}
$$

Since we got the polynomial we began with, our division is correct.

   When you are dividing a polynomial by a binomial, you should list both the polynomial and binomial in descending order. If a given powered term is missing, it is often helpful to include that term with a numerical coefficient of 0. For example, when dividing $(6x^2 + x^3 - 4) \div (x - 2)$, we rewrite the problem as $(x^3 + 6x^2 + 0x - 4) \div (x - 2)$ before beginning the division.

EXAMPLE 6   Divide $(3x^5 + 4x^2 - 12x - 17) \div (x^2 - 2)$.

Solution: Whenever a power of x is missing, we add that power of x with a coefficient of 0 to help align like terms.

$$
\begin{array}{r}
3x^3 \qquad\quad + 6x + 4 \\
x^2 + 0x - 2\overline{)3x^5 + 0x^4 + 0x^3 + 4x^2 - 12x - 17} \\
\underline{3x^5 + 0x^4 - 6x^3} \qquad\qquad\qquad \longleftarrow 3x^3(x^2 + 0x - 2) \\
6x^3 + 4x^2 - 12x \\
\underline{6x^3 + 0x^2 - 12x} \longleftarrow 6x(x^2 + 0x - 2) \\
4x^2 + 0x - 17 \\
\underline{4x^2 + 0x - 8} \longleftarrow 4(x^2 + 0x - 2) \\
-9
\end{array}
$$

To find the answer, we performed the divisions

$$\frac{3x^5}{x^2} = 3x^3, \qquad \frac{6x^3}{x^2} = 6x, \qquad \frac{4x^2}{x^2} = 4$$

The quotients $3x^3$, $6x$, and 4 were placed above their like terms in the dividend. The answer is $3x^3 + 6x + 4 - \dfrac{9}{x^2 - 2}$.   ∎

Exercise Set 5.5

Divide as indicated.

1. $\dfrac{6x + 8}{2}$

2. $\dfrac{3x + 6}{2}$

3. $\dfrac{4x^2 + 2x}{2x}$

4. $\dfrac{5y^3 + 6y^2 + 3y}{3y}$

5. $\dfrac{12x^2 - 4x - 8}{4}$

6. $\dfrac{15y^6 + 5y^2}{5y^4}$

7. $\dfrac{4x^5 - 6x^4 + 12x^3}{4x^2}$

8. $\dfrac{6x^2y - 9xy^2}{3xy}$

9. $\dfrac{4x^2y^2 - 8xy^3 + 3y^4}{2y^2}$

10. $\dfrac{15x^{12} - 5x^9 + 30x^6}{5x^6}$

11. $\dfrac{6x^2y - 12x^3y^2 + 9y^3}{2xy^2}$

12. $\dfrac{a^2b^2c - 6abc^2 + 5a^3b^5}{2abc^2}$

Divide as indicated.

13. $\dfrac{x^2 + 4x + 3}{x + 1}$

14. $\dfrac{x^2 + 7x + 10}{x + 5}$

15. $\dfrac{2x^2 + 13x + 15}{x + 5}$

16. $\dfrac{2x^2 + x - 10}{2x + 5}$

17. $\dfrac{6x^2 + x - 2}{2x - 1}$

18. $\dfrac{x^2 - 25}{x - 5}$

19. $\dfrac{4x^2 - 9}{2x - 3}$

20. $\dfrac{8x^2 + 6x - 27}{4x + 9}$

21. $\dfrac{x^3 + 3x^2 + 5x + 4}{x + 1}$

22. $\dfrac{4x^3 + 12x^2 + 7x - 3}{2x + 3}$

23. $\dfrac{9x^3 - 3x^2 - 3x + 4}{3x + 2}$

24. $\dfrac{2x^3 + 6x - 4}{x + 4}$

25. $\dfrac{4x^3 - 5x}{2x - 1}$

26. $\dfrac{3x^5 + 4x^2 - 12x - 8}{x^2 - 2}$

27. $\dfrac{4x^5 - 18x^3 + 8x^2 + 18x - 12}{2x^2 - 3}$

28. $\dfrac{3x^4 + 4x^3 - 32x^2 - 5x - 20}{3x^3 - 8x^2 - 5}$

Divide as indicated.

29. $\dfrac{6x^2 + 3x + 12}{2x}$

30. $\dfrac{6x^2 + 16x + 8}{3x + 2}$

31. $\dfrac{2x^2 + x - 10}{x - 2}$

32. $\dfrac{2x^3 - 3x^2 - 3x + 4}{x - 1}$

33. $\dfrac{12x^3 + 6x^2 + 3x + 9}{6x^2}$

34. $\dfrac{2x^2 + 7x - 15}{2x - 3}$

35. $\dfrac{-5x^3y^2 + 10xy - 6}{10x}$

36. $\dfrac{2x^2 + 13x + 15}{2x + 3}$

37. $\dfrac{9x^3 - x + 3}{3x - 2}$

38. $\dfrac{-x^3 - 6x^2 + 2x - 3}{x - 1}$

39. $\dfrac{3xyz + 6xyz^2 - 9x^3y^5z^7}{6xy}$

40. $\dfrac{6abc^3 - 5a^2b^3c^4 + 8ab^5c}{3ab^2c^3}$

41. $\dfrac{2x^4 - 8x^3 + 19x^2 - 33x + 15}{x^2 - x + 5}$

42. $\dfrac{3x^4 + 4x^3 - 32x^2 - 5x - 20}{x + 4}$

43. $\dfrac{2x^5 + 2x^4 - 3x^3 - 15x^2 + 18}{2x^2 - 3}$

44. **(a)** Explain how the answer may be checked when dividing a polynomial by a binomial.
   (b) Using your explanation in part (a) to check if the following division is correct.
   $$\frac{8x^2 + 2x - 15}{4x - 5} = 2x + 3$$

   (c) Check to see if the following division is correct.
   $$\frac{6x^2 - 23x + 14}{3x - 4} = 2x - 5 - \frac{6}{3x - 4}$$

Cumulative Review Exercises

[3.2, 3.4] **45.** Write (a) the standard form of a linear equation, (b) the slope–intercept form of a linear equation, and (c) the point–slope form of a linear equation.

[3.5] **46.** Is every function a relation? Is every relation a function? Explain the difference between a function and a relation.

47. If $f(x) = \frac{1}{2}x + \frac{3}{7}$, find $f(-\frac{2}{3})$.

[4.3] **48.** A postage meter at a local firm can stamp both first class postage, 29¢, and bulk postage, 19.8¢, on envelopes. If after 1 month the meter indicates that 550 envelopes were stamped and the total cost of the postage was $122.70, how many first-class and how many bulk-mail letters were stamped?

JUST FOR FUN

1. Divide $\dfrac{2x^3 - x^2y - 7xy^2 + 2y^3}{x - 2y}$.

2. Divide $\dfrac{x^3 + y^3}{x + y}$.

3. Divide $\dfrac{3x^3 - 5}{3x - 2}$. *Hint:* Answer contains fractions.

5.6

Synthetic Division (Optional)

▶ **1**   Divide polynomials by binomials using synthetic division.

▶ **1**   When a polynomial is divided by a binomial of the form $x - a$, the division process can be greatly shortened by a process called **synthetic division.** Consider the following examples. In the example on the right, we use only the numerical coefficients.

$$
\begin{array}{r}
2x^2 + 5x - 4 \\
x - 3\overline{)2x^3 - x^2 - 19x + 15} \\
\underline{2x^3 - 6x^2} \\
5x^2 - 19x \\
\underline{5x^2 - 15x} \\
-4x + 15 \\
\underline{-4x + 12} \\
3
\end{array}
\qquad
\begin{array}{r}
2 \quad +5 \quad -4 \\
1 - 3\overline{)2 \ -1 \ -19 \quad 15} \\
\underline{2 \ -6} \\
5 \ -19 \\
\underline{5 \ -15} \\
-4 \quad 15 \\
\underline{-4 \quad 12} \\
3
\end{array}
$$

Note that the variables do not play a role in determining the numerical coefficients of the quotient. This division problem can be done more quickly and easily using synthetic division, as outlined next.

1. Write the dividend in descending powers of x. Then list the numerical coefficients of each term in the dividend. If a given powered term is missing, place a 0 in the appropriate position to serve as a placeholder. In the preceding problem the numerical coefficients of the dividend are

$$2 \quad -1 \quad -19 \quad 15$$

2. When dividing by a binomial of the form $x - a$, place a to the left of the line of numbers from part 1. In this problem we are dividing by $x - 3$; thus $a = 3$. We write

$$3\rfloor \ 2 \ -1 \ -19 \ \ 15$$

3. Bring down the left number as follows:

$$
\begin{array}{r}
3\rfloor \ \ 2 \ \ -1 \ \ -19 \ \ \ 15 \\
\hline
2 \qquad\qquad\qquad
\end{array}
$$

4. Multiply the 3 by the number brought down, the 2, to get 6. Place 6 under the -1. Then add $-1 + 6$ to get 5.

$$
\begin{array}{r}
3\rfloor \ \ 2 \ \ -1 \ \ -19 \ \ \ 15 \\
6 \qquad\qquad\quad \\
\hline
2 \ \ \ \ 5 \qquad\qquad\quad
\end{array}
$$

5. Multiply the 3 by sum 5 to get 15. Place 15 under -19. Then add to get -4. Repeat this procedure as illustrated.

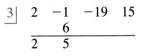

Note that the first three numerical values are identical to the numerical values obtained in the quotient when worked out by long division. The last digit, the 3, is identical to the remainder obtained by long division. The quotient must be one degree less than the dividend since we are dividing by $x - 3$. Since the original dividend was a third-degree polynomial, the quotient must be a second-degree polynomial.

The quotient is therefore $2x^2 + 5x - 4 + \dfrac{3}{x - 3}$.

EXAMPLE 1 Divide using synthetic division.

$$(6 - x^2 + x^3) \div (x + 2)$$

Solution: First, list the dividend in descending order of x.

$$(x^3 - x^2 + 6) \div (x + 2)$$

Since there is no x term, insert a 0 as a placeholder when listing the numerical coefficients.

$$
\begin{array}{r|rrrr}
-2 & 1 & -1 & 0 & 6 \\
   &   & -2 & 6 & -12 \\
\hline
   & 1 & -3 & 6 & -6
\end{array}
$$

Note that $x + 2 = x - (-2)$ and therefore $a = -2$.

Since the dividend is a third-degree equation, the quotient must be second degree. The quotient is $x^2 - 3x + 6 - \dfrac{6}{x + 2}$. ∎

EXAMPLE 2 Use synthetic division to divide.

$$(3x^4 + 11x^3 - 20x^2 + 7x + 35) \div (x + 5)$$

Solution:

$$
\begin{array}{r|rrrrr}
-5 & 3 & 11 & -20 & 7 & 35 \\
   &   & -15 & 20 & 0 & -35 \\
\hline
   & 3 & -4 & 0 & 7 & 0
\end{array}
$$

Since the dividend is of the fourth degree, the quotient must be of the third degree. The quotient is $3x^3 - 4x^2 + 0x + 7$ with no remainder. This can be simplified to $3x^3 - 4x^2 + 7$. ∎

EXAMPLE 3 Use synthetic division to divide.

$$(3x^3 - 6x^2 + 4x + 5) \div \left(x - \frac{1}{2} \right)$$

Solution:

$$
\begin{array}{r|rrrr}
\frac{1}{2} & 3 & -6 & 4 & 5 \\
   &   & \frac{3}{2} & -\frac{9}{4} & \frac{7}{8} \\
\hline
   & 3 & -\frac{9}{2} & \frac{7}{4} & \frac{47}{8}
\end{array}
$$

The solution is

$$3x^2 - \frac{9}{2}x + \frac{7}{4} + \frac{47}{8(x - \frac{1}{2})} \quad \text{or} \quad 3x^2 - 4.5x + 1.75 + \frac{5.875}{x - \frac{1}{2}}$$

∎

Exercise Set 5.6

Divide using synthetic division.

1. $(x^2 + x - 6) \div (x - 2)$

2. $(x^2 - 4x - 32) \div (x + 4)$

3. $(x^2 + 5x - 6) \div (x + 6)$

4. $(x^2 + 12x + 32) \div (x + 4)$

5. $(x^2 + 5x - 12) \div (x - 3)$

6. $(2x^2 - 9x + 15) \div (x - 6)$

7. $(3x^2 - 7x - 10) \div (x - 4)$

8. $(x^3 + 6x^2 + 4x - 7) \div (x + 5)$

9. $(4x^3 - 3x^2 + 2x) \div (x - 1)$

10. $(x^3 - 7x^2 - 13x + 5) \div (x - 2)$

11. $(3x^3 + 7x^2 - 4x + 12) \div (x + 3)$

12. $(3x^4 - 25x^2 - 20) \div (x - 3)$

13. $(5x^3 - 6x^2 + 3x - 6) \div (x + 1)$

14. $(y^4 - 1) \div (y - 1)$

15. $(x^4 + 16) \div (x + 4)$

16. $(2x^4 - x^2 + 5x - 12) \div (x - 3)$

17. $(y^5 + y^4 - 10) \div (y + 1)$

18. $(z^5 + 4z^4 - 10) \div (z + 1)$

19. $(3x^3 + 2x^2 - 4x + 1) \div \left(x - \dfrac{1}{3}\right)$

20. $(8x^3 - 6x^2 - 5x + 3) \div \left(x + \dfrac{3}{4}\right)$

21. $(2x^4 - x^3 + 2x^2 - 3x + 1) \div \left(x - \dfrac{1}{2}\right)$

22. $(9y^3 + 9y^2 - y + 2) \div \left(y + \dfrac{2}{3}\right)$

Cumulative Review Exercises

[2.5] **23.** Solve the inequality and graph the solution on the number line: $-1 < \dfrac{4(3x - 2)}{3} \le 5$

[3.1] **24.** Find the perimeter of the triangle determined by the following points. Write your answer as a sum of square roots: $A(3, 9)$; $B(-2, 6)$; $C(8, 4)$.

[3.2] **25.** Graph the equation $20x - 60y = 120$.

[5.2] **26.** Simplify $\dfrac{(4x^{-2}y^{-3})^{-2}(2xy^{-4})^3}{(3x^{-1}y^3)^2}$.

JUST FOR FUN

1. Divide $(0.2x^3 - 0.4x^2 + 0.32x - 0.64)$ by $(x - 0.4)$.

2. Synthetic division can be used to divide polynomials by binomials of the form $ax - b$, $a \ne 1$. To perform this division, divide $ax - b$ by a to obtain $x - \dfrac{b}{a}$. Then place b/a to the left of the numerical coefficients of the polynomial. Work the problem as explained previously.

After summing the numerical values below the line, divide all of them, except the remainder, by a. Write the quotient of the problem using these numbers.

(a) Use this procedure to divide $(9x^3 + 9x^2 + 5x + 12)$ by $(3x + 5)$.

(b) Explain why we do not divide the remainder by a.

5.7

Polynomial Functions

▶ **1**   Identify polynomial functions.

▶ **2**   Evaluate polynomial functions.

▶ **3**   Use polynomial functions in practical applications.

▶ **1**   The concept of function was first introduced in Section 3.5. In this section we expand on the concept of functions. Since functions are so important to mathematics and are a unifying concept, we also discuss them further in later chapters.

In Section 3.5 we discussed linear functions of the form $f(x) = ax + b$. Linear functions are a specific type of **polynomial function.** There are many other types of polynomial functions. The general form of a polynomial function is given below.

Polynomial Function

$$f(x) = a_nx^n + a_{n-1}x^{n-1} + a_{n-2}x^{n-2} + a_{n-3}x^{n-3} + \cdots + a_1x + a_0$$

where all exponents on x are whole numbers and $a_n, a_{n-1}, a_{n-2}, \ldots, a_1, a_0$ are all real numbers with $a_n \neq 0$.

Examples of Polynomial Functions

$$f(x) = 3x + 4 \qquad \text{(linear function, first degree)}$$

$$f(x) = 5x^2 - \frac{1}{2}x + 3 \qquad \text{(quadratic function, second degree)}$$

$$f(x) = 6x^3 - 4x \qquad \text{(cubic function, third degree)}$$

$$f(x) = \sqrt{2}x^4 - 6x \qquad \text{(fourth-degree function)}$$

Note that the right side of each of these functions is a polynomial since all exponents on x are whole numbers.

The graph of every equation of the form $y = a_nx^n + a_{n-1}x^{n-1} + a_{n-2}x^{n-2} + \cdots + a_1x + a_0$, where all exponents on x are whole numbers, will pass the vertical test. Every equation of this form is therefore a function. We will graph some polynomial functions in Section 5.8.

▶ **2** To evaluate a polynomial function for a specific value of the variable, substitute the value in the function whenever the variable appears.

EXAMPLE 1 $f(x) = 3x^3 - 6x^2 + 2x - 1$.

Evaluate: (a) $f(2)$ (b) $f(-2)$

Solution: (a) $f(x) = 3x^3 - 6x^2 + 2x - 1$ (b) $f(x) = 3x^3 - 6x^2 + 2x - 1$
$f(2) = 3(2)^3 - 6(2)^2 + 2(2) - 1$ $f(-2) = 3(-2)^3 - 6(-2)^2 + 2(-2) - 1$
$= 3(8) - 6(4) + 4 - 1$ $= 3(-8) - 6(4) - 4 - 1$
$= 24 - 24 + 4 - 1$ $= -24 - 24 - 4 - 1$
$= 3$ $= -53$ ∎

EXAMPLE 2 $f(x) = 3x + 2$

Find: (a) $f(1)$ (b) $f(a)$ (c) $f(a + b)$

Solution: (a) $f(x) = 3x + 2$
$f(1) = 3(1) + 2 = 3 + 2 = 5$
(b) How do we find $f(a)$? To find $f(1)$, we substituted 1 for each x in the function. Similarly, to find $f(a)$, we substitute a for each x in the function.

$$f(x) = 3x + 2$$
$$f(a) = 3a + 2$$

(c) To find $f(a + b)$, we substitute $(a + b)$ for each x in the function.

$$f(x) = 3x + 2$$
$$f(a + b) = 3(a + b) + 2$$
$$= 3a + 3b + 2 \qquad \blacksquare$$

EXAMPLE 3 $f(x) = x^2 + 2x - 3$

Find: (a) $f(3)$ (b) $f(a)$ (c) $f(a + b)$

Solution: (a) $f(x) = x^2 + 2x - 3$
$$f(3) = 3^2 + 2(3) - 3 = 9 + 6 - 3 = 12$$
(b) $f(x) = x^2 + 2x - 3$
$$f(a) = a^2 + 2a - 3$$
(c) $f(x) = x^2 + 2x - 3$
$$f(a + b) = (a + b)^2 + 2(a + b) - 3$$
$$= (a + b)(a + b) + 2(a + b) - 3$$
$$= a^2 + 2ab + b^2 + 2a + 2b - 3 \qquad \blacksquare$$

COMMON STUDENT ERROR

Students often make the incorrect assumptions that

$$f(a + b) = f(a) + f(b) \qquad \textbf{wrong}$$
$$f(a - b) = f(a) - f(b) \qquad \textbf{wrong}$$

To see why this is wrong, consider the function

$$f(x) = x^2 - 2x + 3$$
$$f(1) = 1^2 - 2(1) + 3 = 2$$
$$f(2) = 2^2 - 2(2) + 3 = 3$$
$$f(3) = 3^2 - 2(3) + 3 = 6$$

Note: $f(1 + 2) \neq f(1) + f(2)$
$$f(3) \neq f(1) + f(2)$$
$$6 \neq 2 + 3$$
$$6 \neq 5$$

Applications of Polynomial Functions

▶ **3**   Some applications of polynomials were discussed in Chapter 2. The examples discussed in that chapter were generally of the first degree. Now we examine additional applications of polynomial functions.

EXAMPLE 4   A polygon is a closed figure with straight line segments as sides. The number of different diagonals, d, in a polygon is a function of the number of sides n in the polygon.

$$d = f(n) = \frac{1}{2}n^2 - \frac{3}{2}n$$

(a) How many diagonals has a quadrilateral (four sides)?
(b) How many diagonals has an octagon (eight sides)?

Solution: (a) $n = 4$, $d = f(4) = \dfrac{1}{2}(4)^2 - \dfrac{3}{2}(4)$

$$= \frac{1}{2}(16) - 6$$

A quadrilateral has 2 diagonals.

$$= 8 - 6 = 2$$

(b) $n = 8$, $d = f(8) = \dfrac{1}{2}(8)^2 - \dfrac{3}{2}(8)$

$$= \frac{1}{2}(64) - 12$$

$$= 32 - 12 = 20$$ ∎

EXAMPLE 5 Neil Armstrong became the first person to walk on the moon on July 20, 1969. The velocity, v, of his spacecraft (the Eagle), in meters per second, was a function of time before touchdown, t.

$$v = f(t) = 3.2t + 0.45$$

The height, h, of the spacecraft above the moon's surface, in meters, was also a function of time before touchdown. Since we used $f(t)$ to represent velocity, we will use $g(t)$ to represent height.

$$h = g(t) = 1.6t^2 + 0.45t$$

What was the velocity of the spacecraft and distance from the surface of the moon at:

(a) 5 seconds from touchdown?
(b) 2 seconds from touchdown?
(c) touchdown?

Solution: (a) To find the velocity and height at 5 seconds from touchdown, substitute $t = 5$ into the appropriate formulas.

$$v = f(t) = 3.2t + 0.45$$
$$v = f(5) = 3.2(5) + 0.45$$
$$= 16.0 + 0.45 = 16.45 \text{ meters per second}$$
$$h = g(t) = 1.6t^2 + 0.45t$$
$$h = g(5) = 1.6(5)^2 + 0.45(5)$$
$$= 1.6(25) + 2.25$$
$$= 40 + 2.25 = 42.25 \text{ meters}$$

(b) At 2 seconds from touchdown,

$$v = 3.2(2) + 0.45 = 6.4 + 0.45 = 6.85 \text{ meters per second}$$
$$h = 1.6(2)^2 + 0.45(2) = 1.6(4) + 0.45(2)$$
$$= 6.4 + 0.9 = 7.3 \text{ meters}$$

(c) At touchdown, $t = 0$:

$$v = 3.2(0) + 0.45 = 0 + 0.45 = 0.45 \text{ meters per second}$$

Thus touchdown velocity was 0.45 meter per second.

$$h = 1.6(0^2) + 0.45(0) = 1.6(0) + 0 = 0 + 0 = 0 \text{ meters}$$

Thus at touchdown the Eagle was on the moon and the distance from the moon was 0. ∎

Exercise Set 5.7

1. $f(x) = 3x - 1$; find
   (a) $f(1)$ (b) $f(-3)$ (c) $f(a)$

2. $f(x) = 6x - 5$; find
   (a) $f(3)$ (b) $f(-4)$ (c) $f(b)$

3. $f(x) = 2x + 3$; find
   (a) $f(4)$ (b) $f(-2)$ (c) $f(a + b)$

4. $f(x) = -x + 4$; find
   (a) $f(-1)$ (b) $f(h)$ (c) $f(a + b)$

5. $f(x) = \frac{1}{2}x^2 - x + 4$; find
   (a) $f(2)$ (b) $f(3)$ (c) $f(-1)$

6. $f(x) = x^2 + 5x - 6$; find
   (a) $f(4)$ (b) $f(0.6)$ (c) $f(0)$

7. $f(x) = 2x^3 - x + 3$; find
   (a) $f(-1)$ (b) $f(2)$ (c) $f(\frac{2}{3})$

8. $f(x) = 0.4x^2 - 1.2x + 3.2$
   (a) $f(4)$ (b) $f(2)$ (c) $f(-3)$

9. $f(x) = -x^2 + 4x - 3$; find
   (a) $f(-4)$ (b) $f(0.3)$ (c) $f(c)$

10. $f(x) = x^2 + 2x + 5$; find
    (a) $f(4)$ (b) $f(b)$ (c) $f(b + 1)$

11. $f(x) = 3x^2 - 2x + 7$; find
    (a) $f(-2)$ (b) $f(\frac{1}{4})$ (c) $f(a + 3)$

12. $f(x) = x^2 - x + 1$; find
    (a) $f(4)$ (b) $f(a)$ (c) $f(a + 2)$

13. $f(x) = -x^2 - 2x + 5$; find
    (a) $f(-2)$ (b) $f(c)$ (c) $f(c + 2)$

14. $f(x) = -3x^2 + 4x - 6$; find
    (a) $f(-1)$ (b) $f(-x)$ (c) $f(x + 1)$

15. $f(x) = 2x^2 - 3x + 5$; find
    (a) $f(5)$ (b) $f(-x)$ (c) $f(2x)$

16. $f(x) = 4x^2 - 3x$; find
    (a) $f(-2)$ (b) $f(4)$ (c) $f(h + 2)$

17. $f(x) = x^2 + 3x - 4$; find
    (a) $f(h)$ (b) $f(h + 4)$ (c) $f(a + h)$

18. $f(x) = 2x^2 - x + 4$; find
    (a) $f(h)$ (b) $f(x - 1)$ (c) $f(a + h)$

19. $f(x) = 2x^2 - 3x + 1$; find
    (a) $f(h)$ (b) $f(x + 3)$ (c) $f(x + h)$

20. $f(x) = 3x^2 - 2x + 5$; find
    (a) $f(b)$ (b) $f(x + 2)$ (c) $f(a + b)$

Solve each word problem.

21. The sum s of the first n even counting numbers is given by the function $s = f(n) = n^2 + n$. Find the sum of:
    (a) The first 10 even numbers
    (b) The first 15 even numbers

22. Use the function given in Example 4, $d = f(n) = \frac{1}{2}n^2 - \frac{3}{2}n$, to find the number of diagonals in a figure with:
    (a) 10 sides
    (b) 6 sides

23. Use the functions given in Example 5, $v = f(t) = 3.2t + 0.45$ and $h = g(t) = 1.6t^2 + 0.45t$, to determine the velocity and height above the surface of the moon at:
    (a) 6 seconds from touchdown
    (b) 2.5 seconds from touchdown

24. The temperature, T, in degrees Celsius, in a sauna n minutes after being turned on is given by the function $T = f(n) = -0.03n^2 + 1.5n + 14$. Find the sauna's temperature after:
    (a) 3 minutes
    (b) 12 minutes

25. The stopping distance, d, in meters for a car traveling v kilometers per hour is given by the function $d = f(v) = 0.18v + 0.01v^2$. Find the stopping distance for speeds of:
    (a) 50 km/hr
    (b) 25 km/hr

26. The approximate number of accidents in one month, n, involving drivers x years of age can be approximated by the function $n = f(x) = 2x^2 - 150x + 4000$. Find the approximate number of accidents in one month that involved:
    (a) 18-year-olds
    (b) 25-year-olds

27. The number of centimeters that a specific spring will stretch, s, when a mass, m, in kilograms is attached to it is found by the function $s = f(m) = 3.4m - 0.3m^2$ (for $m \leq 6$ kg). How much will the spring stretch if the following masses are attached to it?
    (a) 2 kg
    (b) 4.2 kg

28. The profits earned, P, in millions of dollars, from constructing a building having x stories can be approximated by the function $P = f(x) = 0.02x^2 + 0.1x - 0.3$. Find the approximate profit earned from construction of an office building of:

(a) 3 stories

(b) 5 stories

29. The total number of oranges, N, in a square pyramid whose base is n by n oranges is given by the function

$$N = f(n) = \frac{1}{3}n^3 + \frac{1}{2}n^2 + \frac{1}{6}n$$

See Exercise 29.

Find the number of oranges if the base is:

(a) 6 by 6 oranges

(b) 8 by 8 oranges

30. If the cost of a ticket to a rock concert is increased by x dollars, the estimated increase in revenue, R, in thousands of dollars is given by the function $R = f(x) = 24 + 5x - x^2$, $x < 8$. Find the increase in revenue if the cost of the ticket is increased by:

(a) \$1

(b) \$4

31. A roller coaster at Knott's Berry Farm, California, has a 70-foot vertical drop. The speed of the last car t seconds after it starts its descent can be approximated, in feet per second, by the function $v = f(t) = 7.2t + 4.6$. The height of the last car from the bottom of the vertical drop, in feet, can be estimated by the function $h = g(t) = -0.8t^2 - 10t + 70$, $0 \le t \le 5$. Find the last car's speed, and the height from the bottom of the drop, after (a) 2 seconds, and (b) 5 seconds.

32. Is $y = 2x^2 + x^{-1} + 3$ a polynomial function? Explain your answer.

33. Is $y = 3x^2 + x^{1/2} + 5$ a polynomial function? Explain your answer.

34. Is $y = \sqrt{5}x^2 - \frac{3}{8}x + 9$ a polynomial function? Explain your answer.

Cumulative Review Exercises

[2.3] **35.** Bob is presently paying \$600 a month rent for his apartment. He is considering buying a house. Tim, a banker, informs Bob that at the present interest rate of $10\frac{1}{2}\%$ on a 30-year mortgage, he will pay a monthly mortgage payment of \$9.15 per thousand dollars of mortgage. How large a mortgage can Bob take out if his monthly mortgage payment is not to exceed his monthly rent?

[3.3] **36.** What is a graph?

37. What is the slope of a line?

[4.2] **38.** Solve the system of equations.

$$x + y + 3z = 9$$
$$x + y = 0$$
$$2y - z = 1$$

JUST FOR FUN

1. $f(x) = x^3 - 2x^2 + 6x + 3$; find $f(x + 3)$.

2. $f(x) = 2x^2 + 3x - 4$; find:

(a) $f(x + h)$

(b) $f(x + h) - f(x)$

(c) $\dfrac{f(x + h) - f(x)}{h}$

3. (a) Write a function in d that can be used to find the shaded area in the figure. Use 3.14 for π.

(b) Find the shaded area when $d = 4$ feet.

(c) Find the shaded area when $d = 6$ feet.

5.8

Graphing Polynomial Functions

▶ **1** Understand how to graph polynomial functions.

▶ **2** Graph quadratic functions.

▶ **3** Find the axis of symmetry and vertex of a parabola.

▶ **4** Graph cubic functions.

▶ **1** In Chapter 3 we graphed linear functions. Other polynomial functions can be graphed in much the same way. To graph polynomial functions, we can substitute values for x, find the corresponding values for y, and plot the points on the Cartesian coordinate system. When graphing linear functions, we had to plot only two points to draw the line. When plotting other polynomial functions, we must be sure to plot a sufficient number of points to get a true picture of the graph. The graphs of polynomial functions will be smooth curves and will pass the vertical line test for functions.

After plotting points, connect the points to get a smooth curve. When drawing the smooth curve, start with the point on the graph with the smallest x value and draw to the point having the next larger value of x. Continue this way through all of the points. Draw arrow tips on the ends of the graph to show that the graph continues in the same direction.

▶ **2** The first function we will graph in this section is a quadratic function.

Quadratic Function

Any function of the form

$$f(x) = ax^2 + bx + c, \qquad a \neq 0$$

where a, b, and c are real numbers, is a **quadratic function.**

In Section 3.5 we learned that $y = f(x)$. Therefore, any quadratic equation of the form $y = ax^2 + bx + c$, $a \neq 0$, will also be a quadratic function.

Every quadratic function will have the shape of a **parabola** (see Fig. 5.1) when graphed. When graphing a function of the form

$$f(x) = ax^2 + bx + c \ (\text{or } y = ax^2 + bx + c)$$

the sign of the numerical coefficient of the squared term, a, will determine whether the parabola will open upward or downward. When the squared term is positive, the parabola will open upward, as in Fig. 5.1a. When the squared term is negative, the parabola will open downward, as shown in Fig. 5.1b.

The **vertex** is the lowest point on a parabola that opens upward and the highest point on a parabola that opens downward.

$$y = ax^2 + bx + c$$
$$a > 0$$

$$y = ax^2 + bx + c$$
$$a < 0$$

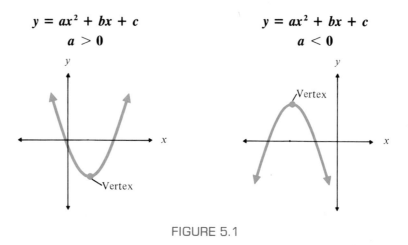

FIGURE 5.1

When graphing a quadratic equation, make sure you plot a sufficient number of points to show whether the parabola is opening upward or downward. We will graph a quadratic equation in Example 1.

EXAMPLE 1   Graph $f(x) = x^2 - 4x + 3$.

Solution: Since $f(x)$ is the same as y, graphing $f(x) = x^2 - 4x + 3$ is the same as graphing $y = x^2 - 4x + 3$. Make a table of values by substituting values for x and solving for $f(x)$ or y.

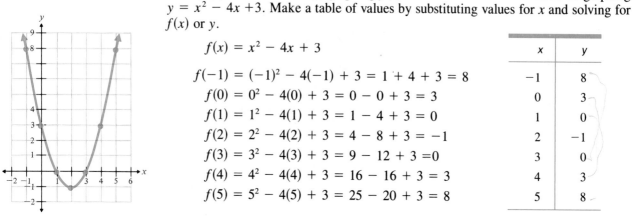

$$f(x) = x^2 - 4x + 3$$

	x	y
$f(-1) = (-1)^2 - 4(-1) + 3 = 1 + 4 + 3 = 8$	-1	8
$f(0) = 0^2 - 4(0) + 3 = 0 - 0 + 3 = 3$	0	3
$f(1) = 1^2 - 4(1) + 3 = 1 - 4 + 3 = 0$	1	0
$f(2) = 2^2 - 4(2) + 3 = 4 - 8 + 3 = -1$	2	-1
$f(3) = 3^2 - 4(3) + 3 = 9 - 12 + 3 = 0$	3	0
$f(4) = 4^2 - 4(4) + 3 = 16 - 16 + 3 = 3$	4	3
$f(5) = 5^2 - 4(5) + 3 = 25 - 20 + 3 = 8$	5	8

FIGURE 5.2

Now plot the points and connect them with a smooth curve (see Fig. 5.2).   ■

Notice that the graph in Fig. 5.2 is a function since it passes the vertical line test discussed in Section 3.5. The *domain* of this function, the set of values that can be used for x, is the set of real numbers, $\mathbb{R}$. The *range*, the corresponding set of values of y, is the set of real numbers greater than or equal to -1.

Domain: $\mathbb{R}$

Range: $\{y \mid y \geq -1\}$

▶**3**   When graphing quadratic functions, how do we decide what values to use for x? When the location of the vertex is unknown, this is a difficult question to answer. When the location of the vertex is known, it becomes more obvious which values to use.

Let us examine the parabola in Example 1 more closely (Fig. 5.3).

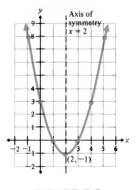

FIGURE 5.3

Notice that the parabola is *symmetric* about a vertical line through the vertex. This means that if we folded the page along this imaginary line, called the *axis of symmetry,* the right and left sides would coincide. Every parabola that opens upward or downward will have an axis of symmetry that will be a vertical line through its vertex. If we can determine the location of the axis of symmetry, we can use it as a guide in selecting values for x. When quadratic functions of the form $y = ax^2 + bx + c$ are graphed, the axis of symmetry of the parabola will be $x = -b/2a$. We will derive this formula in Section 9.4.

Axis of Symmetry

Given an equation of the form $y = ax^2 + bx + c$, its graph will be a parabola with axis of symmetry

$$x = \frac{-b}{2a}$$

Note in Figure 5.3 that the axis of symmetry is $x = 2$. We can find this by the axis of symmetry formula as follows:

$$f(x) = x^2 - 4x + 3$$
$$a = 1, \qquad b = -4, \qquad c = 3$$
$$x = \frac{-b}{2a} = \frac{-(-4)}{2(1)} = \frac{4}{2} = 2$$

The equation of the axis of symmetry is $x = 2$. The x coordinate of the vertex of the parabola is also at 2. The y coordinate of the vertex can now be found by substituting 2 for x in the function.

$$f(x) = x^2 - 4x + 3$$
$$f(2) = 2^2 - 4(2) + 3$$
$$= 4 - 8 + 3$$
$$= -1$$

Therefore when $x = 2$, $f(x)$ or $y = -1$. The coordinates of the vertex of the parabola are $(2, -1)$.

Now we will graph another quadratic function, this time making use of the axis of symmetry in selecting values for x.

EXAMPLE 2   Graph $f(x) = -x^2 + 2x + 3$.

Solution: Since the coefficient of the squared term, -1, is less than 0, this parabola will open downward. Now find the axis of symmetry.

$$x = \frac{-b}{2a} = \frac{-(2)}{2(-1)} = \frac{-2}{-2} = 1$$

The parabola will be symmetric about the line $x = 1$. Since the axis of symmetry is $x = 1$, we will choose values for x of $-2, -1, 0, 1, 2, 3, 4$. Note that we selected three values less than 1 and three values greater than 1. Now we find the corresponding values of y and draw the graph as shown in Fig. 5.4.

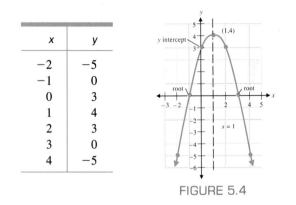

x	y
-2	-5
-1	0
0	3
1	4
2	3
3	0
4	-5

FIGURE 5.4

The parabola in Fig. 5.4 is a smooth curve. When graphing a parabola, or any polynomial function, it is often helpful to plot each point as it is determined. If the point does not appear to be part of the curve, check your calculations. Note in Fig. 5.4 that the domain is all real numbers, $\mathbb{R}$, and the range is the set of values less than or equal to 4, $\{y \mid y \leq 4\}$.

In Fig. 5.4, we see that the vertex is at the point $(1, 4)$, and the y intercept, where the graph crosses the y axis, is 3. The y intercept of any graph can be determined by substituting $x = 0$ into the equation and evaluating the equation. For the function in Example 2,

$$f(x) = -x^2 + 2x + 3$$
$$f(0) = -0^2 + 2(0) + 3 = 3$$

Can you explain why this procedure always gives the y intercept?

Also indicated in Fig. 5.4 are the x intercepts of the graphs. The x intercepts, where the graph crosses the x axis, are also called the **roots** of the equation. Therefore, the roots of $f(x) = -x^2 + 2x + 3$ are -1 and 3. In Chapter 9, we will discuss finding the roots of quadratic equations by using procedures other than graphing.

In this section we selected quadratic functions such that their parabolas will have vertices whose coordinates are integer values. In Section 9.4, we will discuss quadratic functions in more depth and consider parabolas whose vertices do not have x coordinates that are integer values.

▶ **4**   Now we will graph a cubic, or third degree, function by plotting points. Graphing third degree or higher functions by this method may not result in a totally accurate graph. To graph third degree or higher functions accurately requires a knowledge of calculus. However, this method will be sufficient for our needs.

EXAMPLE 3 Graph $f(x) = x^3 - 3x + 1$.

Solution: Select values for x and find the corresponding values of $f(x)$, or y. Plot the points and draw a smooth curve from point to point (see Fig. 5.5).

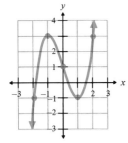

$f(x) = x^3 - 3x + 1$	x	y
$f(-3) = (-3)^3 - 3(-3) + 1 = -27 + 9 + 1 = -17$	-3	-17
$f(-2) = (-2)^3 - 3(-2) + 1 = -8 + 6 + 1 = -1$	-2	-1
$f(-1) = (-1)^3 - 3(-1) + 1 = -1 + 3 + 1 = 3$	-1	3
$f(0) = 0^3 - 3(0) + 1 = 0 - 0 + 1 = 1$	0	1
$f(1) = 1^3 - 3(1) + 1 = 1 - 3 + 1 = -1$	1	-1
$f(2) = 2^3 - 3(2) + 1 = 8 - 6 + 1 = 3$	2	3
$f(3) = 3^3 - 3(3) + 1 = 27 - 9 + 1 = 19$	3	19

FIGURE 5.5

Notice that the points $(-3, -17)$ and $(3, 19)$ listed in the table were not plotted on the graph in Fig. 5.5. Their y values are too small and too large, respectively. The arrows on the graph indicate that the graph continues in the same direction. The graph would pass through these two points if the vertical axis was extended. Unfortunately, there is no easy method to determine the vertices of the graph of a cubic equation. The graph in Fig. 5.5 is a function whose range and domain are both all real numbers, $\mathbb{R}$.

In Example 3, we graphed $f(x) = x^3 - 3x + 1$. If we multiply each term on the right sides of the function by -1, we obtain $f(x) = -x^3 + 3x - 1$. What will the graph of $f(x) = -x^3 + 3x - 1$ look like? We will graph this function in Example 4.

EXAMPLE 4 Graph $f(x) = -x^3 + 3x - 1$.

Solution:

$f(x) = -x^3 + 3x - 1$	x	y
$f(-3) = -(-3)^3 + 3(-3) - 1 = 17$	-3	17
$f(-2) = -(-2)^3 + 3(-2) - 1 = 1$	-2	1
$f(-1) = -(-1)^3 + 3(-1) - 1 = -3$	-1	-3
$f(0) = -(0)^3 + 3(0) - 1 = -1$	0	-1
$f(1) = -(1)^3 + 3(1) - 1 = 1$	1	1
$f(2) = -(1)^3 + 3(2) - 1 = -3$	2	-3
$f(3) = -(3)^3 + 3(3) - 1 = -19$	3	-19

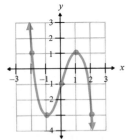

FIGURE 5.6

The graph is given in Fig. 5.6.

Notice for each point (x, y) on the graph of $f(x) = x^3 - 3x + 1$ in Fig. 5.5 the corresponding point on the graph of $f(x) = -x^3 + 3x - 1$ in Fig. 5.6 is $(x, -y)$. The graph in Fig. 5.5 is inverted to obtain the graph in Fig. 5.6.

When graphing quadratic functions, we stated that the sign of a, the coefficient of the squared term, determines whether the parabola opens upward or downward. When a is positive, the parabola opens upward, and when a is negative, the parabola opens downward.

Similarly, when graphing a cubic function, the sign of the coefficient of the cubed term determines whether the graph will eventually continue to increase or decrease as x increases. If the coefficient of the cubed term is positive, as in Example 3, the graph will eventually continue to increase, or rise, as x increases. If the coefficient of the cubed term is negative, as in Example 4, the graph will eventually continue to decrease, or fall, as x increases. Can you explain why this must happen?

Exercise Set 5.8

Indicate the axis of symmetry, the coordinates of the vertex, and whether the parabola opens up or down.

1. $y = x^2 + 2x - 7$ **2.** $y = x^2 + 4x - 9$

3. $y = -x^2 + 4x - 6$ **4.** $y = 3x^2 + 6x - 9$

5. $y = -3x^2 + 6x + 8$ **6.** $y = x^2 + 8x - 6$

7. $y = -4x^2 - 8x - 12$ **8.** $y = 2x^2 + 4x + 6$

9. $y = x^2 - x + 2$ **10.** $y = -x^2 + x + 8$

11. $y = 4x^2 + 12x - 5$ **12.** $y = -2x^2 - 6x - 5$

Graph each quadratic function and give its domain and range.

13. $y = x^2 - 1$ **14.** $y = x^2 + 4$ **15.** $y = -x^2 + 3$

16. $y = x^2 + 4x + 3$ **17.** $y = x^2 + 2x - 15$ **18.** $y = -x^2 + 10x - 21$

19. $y = -x^2 + 4x - 5$ **20.** $y = x^2 + 8x + 15$ **21.** $y = x^2 - 6x + 4$

22. $y = x^2 - 6x + 9$ **23.** $y = x^2 - 6x$ **24.** $f(x) = -x^2 - 4x + 4$

25. $f(x) = x^2 - 4x + 4$ **26.** $f(x) = x^2 - 2x + 1$ **27.** $f(x) = -x^2 + 4x - 8$

28. $f(x) = 2x^2 - 4x$

29. $f(x) = x^2 - 2x - 15$

30. $f(x) = -2x^2 - 8x + 4$

31. $f(x) = 3x^2 - 6x + 1$

32. $f(x) = 4x^2 + 8x + 4$

33. $f(x) = \frac{1}{2}x^2 - 2$

34. $f(x) = \frac{1}{2}x^2 + x - 4$

Graph each cubic function.

35. $y = x^3$

36. $y = x^3 + 1$

37. $y = x^3 + x$

38. $y = x^3 + 2x - 1$

39. $y = x^3 + x^2 - 3x - 1$

40. $f(x) = 2x^3 + x - 8$

41. $f(x) = -x^3 + 3x$

42. $f(x) = -x^3 + x - 6$

Graph each function.

43. $y = -x^2 - 2$

44. $y = x^2 + 2x + 3$

45. $y = x^2 + 6x - 2$

46. $f(x) = 2x^2 - 4x - 6$ **47.** $f(x) = x^3 - x^2 + 2x$ **48.** $f(x) = -2x^2 + 8x - 1$

49. $f(x) = -2x^3 + 6x^2 + 2x - 6$ **50.** $f(x) = -3x^2 - 6x + 5$ **51.** $y = \frac{1}{2}x^2 - 2x - 4$

52. $y = 2x^3 - 5x^2 + x - 3$

53. Explain how to determine if the graph of a quadratic function opens up or down.

54. What is the name given to the graph of a quadratic function?

55. Consider the function $f(x) = x^3$. Explain what happens to y as x increases? As x decreases?

56. Consider the function $f(x) = -x^3$. Explain what happens to y as x increases? As x decreases?

57. Consider the function $f(x) = x^4$. Explain what happens to y as x increases from -3 to 3?

58. Consider the function $f(x) = -x^4$. Explain what happens to y as x increases from -3 to 3?

59. (a) Make up your own quadratic function, and explain why it is a quadratic function. (b) Graph your function.

60. (a) Make up your own cubic function, and explain why it is a cubic function.

   (b) Graph your function.

Cumulative Review Exercises

Solve the system of equations $x - 4y = -16$
                                    $2x + 3y = -10$

[4.1] **61.** Using the addition method.

[4.4] **62.** Using determinants (optional topic).

[4.3] **63.** The sum of three numbers is 12. If the sum of the two smaller numbers equals the largest number, and the largest number is four less than twice the middle-size number, find the three numbers.

[5.4] **64.** Multiply $(3x^2 + 6x - 4)(5x - 3)$.

JUST FOR FUN

1. Graph $f(x) = x^4 - 3x^2 + 6$.

2. Graph $f(x) = x^4 - 2x^2 + 3x - 4$.

SUMMARY

GLOSSARY

Binomial *(224):* A two-term polynomial.
Degree of a polynomial *(224):* The same as the highest-degree term in the polynomial.

Descending order of the variable *(224):* Polynomial written so that the exponents on the variable decrease as terms go from left to right.

Monomial *(224):* A one-term polynomial.
Polynomial *(224):* A finite sum of terms in which all variables have whole-number exponents and no variables appear in a denominator.
Polynomial function *(241):* A function of the form $f(x) = a_n x^n + a_{n-1} x^{n-1} + a_{n-2} x^{n-2} + \cdots + a_1 x + a_0$
Quadratic function *(246):* A function of the form $f(x) = ax^2 + bx + c, a \neq 0.$

Scientific notation *(213):* Writing large and small numbers as a number greater than or equal to 1 and less than 10 multiplied by some power of 10.
Synthetic division *(238):* A shortened process of dividing a polynomial by a binomial of the form $x - a$.
Trinomial *(224):* A three-term polynomial.

IMPORTANT FACTS

Rules for Exponents

1. $a^m \cdot a^n = a^{m+n}$ product rule

2. $\dfrac{a^m}{a^n} = a^{m-n}, \quad a \neq 0$ quotient rule

3. $a^{-m} = \dfrac{1}{a^m}, \quad a \neq 0$ negative exponent rule

4. $a^0 = 1, \quad a \neq 0$ zero exponent rule
5. $(a^m)^n = a^{mn}$

$(ab)^m = a^m b^m$

$\left(\dfrac{a}{b}\right)^m = \dfrac{a^m}{b^m}, \quad b \neq 0$ power rules

FOIL Method to Multiply Two Binomials

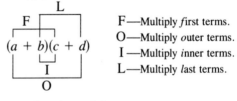

F—Multiply *first* terms.
O—Multiply *outer* terms.
I —Multiply *inner* terms.
L—Multiply *last* terms.

$$ac + ad + bc + bd$$

Special Product Formula

$$\left.\begin{array}{l}(a + b)^2 = a^2 + 2ab + b^2 \\ (a - b)^2 = a^2 - 2ab + b^2\end{array}\right\}$$ square of a binomial

$$(a + b)(a - b) = a^2 - b^2$$ product of sum and difference of the same two terms (or difference of two squares)

Review Exercises

[5.1–5.2] *Simplify and write the answer without negative exponents.*

1. $4^2 \cdot 4^1$

2. $x^3 \cdot x^5$

3. $y^5 \cdot y^2$

4. $\dfrac{3^4}{3^1}$

5. $\dfrac{x^6}{x^2}$

6. $\dfrac{y^{12}}{y^3}$

7. $\dfrac{y^5}{y^6}$

8. $\dfrac{x^4}{x^{-3}}$

9. $x^4 \cdot x^{-7}$

10. $3^{-2} \cdot 3^{-1}$

11. $2^{-3} \cdot 2^{-2}$

12. $3x^0$

13. $(3x^2)^2$

14. $\left(\dfrac{2}{3}\right)^{-1}$

15. $\left(\dfrac{3}{4}\right)^{-2}$

16. $\left(\dfrac{x}{y^2}\right)^{-1}$

17. $(6xy^2)(-2xy^4)$

18. $(7x^2y^5)(-3xy^4)$

19. $(4x^2y^{-3})(2x^{-4}y^2)$

20. $(2x^5y^{-4})(3x^2y)(5x^{-2}y^{-3})$

21. $\dfrac{6x^{-3}y^5}{2x^2y^{-2}}$

22. $\dfrac{12x^{-3}y^{-4}}{4x^{-2}y^5}$

23. $\dfrac{(5x^3y^2)(2xy^4z)}{20x^4y^{-2}z}$

24. $\left(\dfrac{3x^4y^2}{2x^{-2}y^3}\right)\left(\dfrac{4x^2y^{-3}}{6xy^4}\right)$

25. $\left(\dfrac{5x^2y}{x}\right)^3$

26. $\left(\dfrac{x^5y}{-3y^2}\right)^2$

27. $\left(\dfrac{2xy^3}{y^4}\right)^2$

28. $\left(\dfrac{x^2y}{x^{-1}y^{-3}}\right)^2$

29. $\left(\dfrac{-5x^{-2}y}{z^3}\right)^3$

30. $\left(\dfrac{6xy^3}{z^2}\right)^{-2}$

31. $\left(\dfrac{9x^{-2}y}{3xy}\right)^{-3}$

32. $(-2x^{-3}y^2)^{-4}$

33. $\left(\dfrac{5x^{-2}y^3}{xy^4}\right)^3$

34. $\left(\dfrac{2x^4y^6}{z^3}\right)^2(2x^2y^4)$

35. $\left(\dfrac{16x^4y^3z^{-2}}{4x^5y^2z^3}\right)^3$

36. $\left(\dfrac{2x^4y}{z^3}\right)^2\left(\dfrac{3x^{-2}y^4}{3xy^6}\right)$

37. $\left(\dfrac{3x^4y^{-2}}{6xy^{-3}}\right)^2\left(\dfrac{2x^{-1}y^5}{3x^4y^{-2}}\right)^{-3}$

38. $\left(\dfrac{x^{-2}y^{-2}z}{x^4y^{-4}z^3}\right)^{-1}\left(\dfrac{2x^2y^5}{4xy^{-2}z}\right)^3$

39. $\dfrac{(4x^{-2}y^3)^{-2}(x^4y^3)^4}{(2x^{-3}y^6)^2}$

40. $\dfrac{(5xy^{-3})^{-3}(xy^{-6})^2}{(x^{-3}y^5)^4}$

Express each number in scientific notation.

41. 0.0000742

42. $260{,}000$

43. $183{,}000$

44. 0.000001

Express each answer without exponents.

45. $(25 \times 10^{-3})(1.2 \times 10^6)$

46. $\dfrac{18 \times 10^3}{9 \times 10^5}$

47. $\dfrac{4{,}000{,}000}{0.02}$

48. $(0.004)(500{,}000)$

[5.3] *Indicate if the expression is a polynomial. If the expression is a polynomial, (a) give the special name of the polynomial if it has one, (b) write the polynomial in descending order of the variable x, and (c) give the degree of the polynomial.*

49. $5 - x$

50. $x^2 - 3 + 5x$

51. $x^2 - y^2 + xy$

52. $x^5y^3 - 6xy^3 + x^4y$

53. $-3 - 9x^2y + 6xy^3 + 2x^4$

54. $3x^2 + 6x^{-1} + 4$

[5.3–5.5] *Perform the indicated operation.*

55. $(4x + 3) + (6x - 8)$

56. $(5x^2 + 3x - 6) + (2x^2 - 7x - 9)$

57. $4x(x^2 + 2x + 3)$

58. $(x + 5)^2$

59. $\dfrac{15y^3 + 6y}{3y}$

60. $(3x^2 + 6x + 4) - (x^2 + 2x)$

61. $(2x + 3)(2x - 3)$

62. $(3xy + 1)(2x + 3y)$

63. $(6x^2 - 11x + 3) \div (3x - 1)$

64. $(x + y)(x - y)$

65. $(2x^3 - 4x^2 - 3x) - (4x^2 - 3x + 9)$

66. $-2xy^2(x^3 + x^2y^5 - 6y)$

67. $(3x - 2y)^2$

68. $(3x^2y + 6xy - 5y^2) - (4y^2 + 3xy)$

69. $(5xy - 6)(5xy + 6)$

70. $\dfrac{9xy - 6y^2 + 3y}{3y}$

71. $(2x - 5y^2)(2x + 5y^2)$

72. $(x^2 + x - 17) \div (x - 3)$

73. $\dfrac{4x^3y^2 + 8x^2y^3 + 12xy^4}{8xy^3}$

74. $[(x + 3y) + 2]^2$

75. $[(x + 3y) + 2][(x + 3y) - 2]$

76. $(-6xy + 6y^2 - 3x) - (y^2 + 3xy + 6x)$

77. $(3x^2 + 4x - 6)(2x - 3)$

78. $(4x^4 - 7x^2 - 5x + 4) \div (2x - 1)$

79. $(4x^3 + 6x - 5)(x + 3)$

80. $(4x^3 + 12x^2 + x - 10) \div (2x + 3)$

81. $(x^2y + 6xy + y^2)(x + y)$

[5.6] *Use synthetic division to obtain the quotient.*

82. $(3x^3 - 2x^2 + 10) \div (x - 3)$

83. $(2y^5 - 10y^3 + y - 1) \div (y + 1)$

84. $(x^5 - 20) \div (x - 2)$

85. $(2x^3 + x^2 + 5x - 3) \div \left(x - \dfrac{1}{2}\right)$

[5.7] *Evaluate each of the following functions at the indicated values.*

86. $f(x) = 3x^2 - 4x - 1$; find   **(a)** $f(2)$ **(b)** $f(-5)$

87. $f(x) = x^3 - 2x + 3$; find   **(a)** $f(\frac{1}{2})$ **(b)** $f(2)$

88. $f(x) = (x - 2)^2 + 5$ find   **(a)** $f(2)$ **(b)** $f(-3)$

89. $f(x) = x^2 + 2x - 1$; find   **(a)** $f(a)$ **(b)** $f(a + 2)$

90. $f(x) = x^2 - x + 3$; find   **(a)** $f(a)$ **(b)** $f(a + b)$

[5.8] *Graph each of the following functions and give the domain and range.*

91. $f(x) = x^2 - 4x + 4$

92. $y = x^2 - 1$

93. $f(x) = 2x^2 - 4x + 3$

Graph each of the following functions.

94. $f(x) = x^3$

95. $y = x^3 + 1$

96. $y = x^3 + 2x - 3$

97. The number of baskets of apples, N, that are produced by x trees in a small orchard ($x \le 100$) is given by the function $N = f(x) = 40x - 0.2x^2$. How many baskets of apples are produced by:

  (a) 20 trees?

  (b) 50 trees?

98. If a ball is dropped from the top of a 100-foot building, its height above the ground, h, at any time, t, can be found by the function $h = f(t) = -16t^2 + 100$, $t \le 2.5$. Find the height of the ball at:

  (a) 1 second

  (b) 2 seconds

See Exercise 97.

Practice Test

Simplify and write the answer without negative exponents.

1. $\left(\dfrac{3x^2y^3}{9x^5y^{-2}}\right)^2$

2. $\left(\dfrac{-3x^3y^{-2}}{y^5}\right)^2\left(\dfrac{x^3y^4}{x^{-2}y^5}\right)^{-3}$

3. (a) Give the specific name of the following polynomial.
   (b) Write the polynomial in descending powers of the variable x, and (c) state the degree of the polynomial.

$$-4x^2y^3 + 2x - 6x^4$$

Perform the operation indicated.

4. $(4x^3 - 3x - 4) - (2x^2 - 5x - 12)$

5. $(12x^6 - 6xy^2 + 15) \div 3x$

6. $(3x + y)(y - 2x)$

7. $(2x^2 + 3xy - 6y^2)(2x + y)$

8. $(2x^2 - 7x + 10) \div (2x + 3)$

9. $(6x^2y + 3y^2 + 5x) - (4x^2y + 2x - 4y^2)$

10. $(2x^3 - x^2 + 5x - 7) \div (x - 3)$

11. $3x^2y^4(-2x^5y^2 + 6x^2y^3 - 3x)$

12. $(2x + 3y)^2$

13. Use synthetic division to obtain the quotient.

$$(3x^4 - 12x^3 - 60x + 4) \div (x - 5)$$

14. $f(x) = x^3 - 2x + 3$; find $f(-3)$.

15. Graph $f(x) = x^2 - 4x + 2$ and give its domain and range.

16. A town presently has a population of 6000. The chamber of commerce estimates that the town's population, P, in n years can be approximated by the function

$$P = f(n) = 12n^2 + 10n + 6000, \qquad n \le 10$$

Find the town's expected population in 8 years.

CHAPTER

6

Factoring

See Section 6.5, Exercise 64.

Factoring a Monomial from a Polynomial and Factoring by Grouping

▶ **1** Find the greatest common factor.

▶ **2** Factor a monomial from a polynomial.

▶ **3** Factor by grouping.

Factoring is the opposite of the multiplication process. For example, in Chapter 5 we learned that

$$3x^2(6x + 3xy + 5x^3) = 18x^3 + 9x^3y + 15x^5$$

and

$$(6x + 3y)(2x - 5y) = 12x^2 - 24xy - 15y^2$$

In this chapter we learn how to determine the factors of a given expression. For example, we may show that

$$18x^3 + 9x^3y + 15x^5 = 3x^2(6x + 3xy + 5x^3)$$

and

$$12x^2 - 24xy - 15y^2 = (6x + 3y)(2x - 5y)$$

▶ **1** In earlier chapters we stated that if $a \cdot b = c$, then a and b are said to be **factors** of c. A given expression may have many factors. What are the integer factors of the number 12?

$$
\begin{array}{ll}
1 \cdot 12 = 12 & (-1)(-12) = 12 \\
2 \cdot 6 = 12 & (-2)(-6) = 12 \\
3 \cdot 4 = 12 & (-3)(-4) = 12
\end{array}
$$

Note that the factors of 12 are ± 1 (read "plus or minus 1"), ± 2, ± 3, ± 4, ± 6, ± 12. Generally, when asked to list the factors of a positive number, we list only the positive factors, although it should be understood the negatives of these factors are also factors. We would say the factors of 12 are 1, 2, 3, 4, 6, and 12.

What are the factors of $6x^3$?

$$
\begin{array}{ll}
\overset{\text{Factors}}{\overbrace{1 \cdot 6x^3}} = 6x^3 & \overset{\text{Factors}}{\overbrace{x \cdot 6x^2}} = 6x^3 \\
2 \cdot 3x^3 = 6x^3 & 2x \cdot 3x^2 = 6x^3 \\
3 \cdot 2x^3 = 6x^3 & 3x \cdot 2x^2 = 6x^3 \\
6 \cdot x^3 = 6x^3 & 6x \cdot x^2 = 6x^3
\end{array}
$$

Some factors of $6x^3$ are 1, 2, 3, 6, x, $2x$, $3x$, $6x$, x^2, $2x^2$, $3x^2$, $6x^2$, x^3, $2x^3$, $3x^3$, and $6x^3$. The opposite or negative of each of these factors is also a factor.

To factor a monomial from a polynomial, we must determine the greatest common factor (GCF) of each term in the polynomial. After the GCF is determined, we use the distributive property (in reverse) to factor the expression. The **greatest common factor** of two or more expressions is the greatest factor that divides (without remainder) each expression. Consider the three numbers 12, 18, and 24. The GCF of these three numbers is 6, since 6 is the greatest number that divides (is a factor of) each of these numbers. If you have forgotten how to find the GCF of a set of numbers, review an arithmetic or elementary algebra text before going any further.

The GCF of a collection of terms containing variables is easily found. Consider the terms x^3, x^4, x^5, and x^6. The GCF of these terms is x^3, since x^3 is the highest power of x that divides all four terms. Note that the GCF of a collection of terms will be the *lowest power of the common variable*.

EXAMPLE 1 Find the GCF of the following terms.

$$y^{12}, \quad y^4, \quad y^9, \quad y^7$$

Solution: Note that y^4 is the lowest power of y that appears in any of the four given terms. The GCF is therefore y^4. ∎

EXAMPLE 2 Find the GCF of the following terms.

$$x^3y^2, \quad xy^4, \quad x^4y^5$$

Solution: The lowest power of x that appears in any of three terms is x (or x^1). The lowest power of y that appears in any of the three terms is y^2. Thus the GCF of the three terms is xy^2. ∎

EXAMPLE 3 Find the GCF of the following terms.

$$6x^2y^3, \quad 9x^3y^4, \quad 24x^4$$

Solution: The GCF is $3x^2$. Since y does not appear in $24x^4$, it is not part of the GCF. ∎

EXAMPLE 4 Find the GCF of the following terms.

$$6(x - 3)^2, \quad 5(x - 3), \quad 18(x - 3)^4$$

Solution: The three numerical values have no common factor other than 1. The lowest power of $(x - 3)$ in any of the three terms is $(x - 3)$. Thus the GCF of the three terms is $(x - 3)$. ∎

Factoring
a Monomial
from a Polynomial

▶ **2**

> **To Factor a Monomial from a Polynomial**
>
> 1. Determine the greatest common factor of all terms in the polynomial.
> 2. Write each term as the product of the GCF and its other factor.
> 3. Use the distributive property to *factor out* the GCF.

When we factor a monomial from a polynomial, we are factoring out the greatest common factor. *The first step in any factoring problem will be to factor out the GCF.*

EXAMPLE 5 Factor $2x + 6$.

Solution: The GCF is 2.

$$2x + 6 = \boxed{2} \cdot x + \boxed{2} \cdot 3 \qquad \text{Write each term as a product of the GCF}$$
$$\text{and some other factor.}$$
$$= 2(x + 3) \qquad \text{Distributive property.}$$

Note that the GCF, 2, was *factored out* from each term in the polynomial. The $2x + 6$ in factored form is $2(x + 3)$. The factors of $2x + 6$ are 2 and $x + 3$. ∎

To check the factoring process, multiply the factors using the distributive property. The product should be the expression with which you began. For instance, in Example 5

Check: $2(x + 3) = 2 \cdot x + 2 \cdot 3 = 2x + 6$

EXAMPLE 6 Factor $15x^4 - 5x^3 + 20x^2$.

Solution: The GCF is $5x^2$.

$$15x^4 - 5x^3 + 20x^2 = \boxed{5x^2} \cdot 3x^2 - \boxed{5x^2} \cdot x + \boxed{5x^2} \cdot 4$$
$$= 5x^2(3x^2 - x + 4)$$
∎

EXAMPLE 7 Factor $20x^2y^3 + 6xy^4 - 12x^3y^5$.

Solution: The GCF is $2xy^3$.

$$20x^2y^3 + 6xy^4 - 12x^3y^5 = \boxed{2xy^3} \cdot 10x + \boxed{2xy^3} \cdot 3y - \boxed{2xy^3} \cdot 6x^2y^2$$
$$= 2xy^3(10x + 3y - 6x^2y^2)$$

Check: $2xy^3(10x + 3y - 6x^2y^2) = 20x^2y^3 + 6xy^4 - 12x^3y^5$ ∎

EXAMPLE 8 Factor $3x(5x - 2) + 4(5x - 2)$.

Solution: The GCF is $(5x - 2)$. Factoring out the GCF gives

$$3x\boxed{(5x - 2)} + 4\boxed{(5x - 2)} = (3x + 4)(5x - 2)$$

We could have also used the commutative property of multiplication to rewrite the expression as

$$\boxed{(5x - 2)}\,3x + \boxed{(5x - 2)}\,4 = (5x - 2)(3x + 4)$$

The answers $(3x + 4)(5x - 2)$ and $(5x - 2)(3x + 4)$ are equivalent answers and either may be given. ∎

EXAMPLE 9 Factor $9(2x - 5) + 6(2x - 5)^2$.

Solution: The GCF is $3(2x - 5)$.

$$9(2x - 5) + 6(2x - 5)^2 = \boxed{3(2x - 5)} \cdot 3 + \boxed{3(2x - 5)} \cdot 2(2x - 5)$$
$$= 3(2x - 5)[3 + 2(2x - 5)]$$

Now combine like terms:

$$= 3(2x - 5)[3 + 4x - 10]$$
$$= 3(2x - 5)(4x - 7)$$
∎

EXAMPLE 10   Factor $(2x - 5)(3x - 4) - (2x - 5)(x + 3)$.

Solution: Factor out the greatest common factor $(2x - 5)$ from both terms; then simplify.

$$(2x - 5)(3x - 4) - (2x - 5)(x + 3) = (2x - 5)[(3x - 4) - (x + 3)]$$
$$= (2x - 5)(3x - 4 - x - 3)$$
$$= (2x - 5)(2x - 7)   \blacksquare$$

Factoring by Grouping

▶ **3**   When a polynomial contains four *terms*, it may be possible to factor the polynomial by grouping. To factor by grouping, remove common factors from groups of terms. This procedure is illustrated in the following example. Factoring by grouping is important because you may use it to factor trinomials in Section 6.2.

EXAMPLE 11   Factor $ax + ay + bx + by$.

Solution: There is no factor (other than 1) common to all four terms. However, a is common to the first two terms and b is common to the last two terms. Factor a from the first two terms and b from the last two terms.

$$ax + ay + bx + by = a(x + y) + b(x + y)$$

Now $(x + y)$ is common to both terms. Factor out $(x + y)$.

$$a(x + y) + b(x + y) = (a + b)(x + y)$$

Thus $ax + ay + bx + by = (a + b)(x + y)$. $\blacksquare$

To Factor by Grouping

---

1. Arrange the four terms into two groups of two terms each. Each group of two terms must have a GCF.
2. Factor the GCF from each group of two terms.
3. If the two terms formed in step 2 have a GCF, factor out that GCF.

EXAMPLE 12   Factor $6x^2 + 9x + 8x + 12$ by grouping.

Solution: Factor a $3x$ from the first two terms and a 4 from the last two terms. Then factor the GCF, $2x + 3$, from the resulting two terms.

$$6x^2 + 9x + 8x + 12 = 3x(2x + 3) + 4(2x + 3)$$
$$= (3x + 4)(2x + 3)   \blacksquare$$

Factoring by grouping problems may be checked by multiplying the factors. Check the answer to Example 12 now.

EXAMPLE 13   Factor $ax - x + a - 1$ by grouping.

Solution: $ax - x + a - 1 = x(a - 1) + 1(a - 1)$
$$= (x + 1)(a - 1)$$

Note that $a - 1$ was expressed as $1(a - 1)$. $\blacksquare$

EXAMPLE 14 Factor $ax - x - a + 1$ by grouping.

Solution: When x is factored from the first two terms, we get

$$ax - x - a + 1 = x(a - 1) - a + 1$$

Now factor -1 from the last two terms to get a common factor of $a - 1$.

$$= x(a - 1) - 1(a - 1)$$
$$= (x - 1)(a - 1)$$

EXAMPLE 15 Factor $2x^2 + 4xy + 3xy + 6y^2$.

Solution: This problem contains two variables, x and y. The procedure to factor is basically the same. We will factor out a $2x$ from the first two terms and a $3y$ from the last two terms.

$$2x^2 + 4xy + 3xy + 6y^2 = 2x(x + 2y) + 3y(x + 2y)$$
$$= (2x + 3y)(x + 2y)$$

EXAMPLE 16 Factor $6r^4 - 9r^3s + 8rs - 12s^2$.

Solution: Factor a $3r^3$ from the first two terms and a $4s$ from the last two terms.

$$6r^4 - 9r^3s + 8rs - 12s^2 = 3r^3(2r - 3s) + 4s(2r - 3s)$$
$$= (3r^3 + 4s)(2r - 3s)$$

EXAMPLE 17 Factor $x^4 - 5x^3 + 2x^2 - 10x$.

Solution: The first step in *any* factoring problem is to determine if all the terms have a common factor. If so, begin by factoring out the common factor. In this example, x is common to all four terms. Begin the factoring process by factoring out an x. Then factor an x^2 from the first two terms in parentheses and a 2 from the last two terms.

$$x^4 - 5x^3 + 2x^2 - 10x = x(x^3 - 5x^2 + 2x - 10)$$
$$= x[x^2(x - 5) + 2(x - 5)]$$
$$= x[(x^2 + 2)(x - 5)]$$
$$= x(x^2 + 2)(x - 5)$$

Exercise Set 6.1

Factor out the greatest common factor. If an expression cannot be factored, so state.

1. $8n + 8$

2. $12x + 15$

3. $13x + 5$

4. $6x^2 + 3x - 9$

5. $16x^2 - 12x - 6$

6. $27y^3 - 9y^2 + 18y$

7. $7x^5 - 9x^4 + 3x^3$

8. $45y^{12} + 30y^{10}$

9. $24y^{15} - 9y^3 + 3y$

10. $38x^4 - 16x^5 - 9x^3$

11. $x + 3xy^2$

12. $2x^2y - 6x + 6x^3$

13. $6x + 5y + 5xy$

14. $3x^2y + 6x^2y^2 + 3xy$

15. $40x^2y^2 + 16xy^4 + 64xy^3$

16. $40x^2y^4z + 8x^6y^2z^2 - 4x^3y$

17. $36xy^2z^3 + 36x^3y^2z + 9x^2yz$

18. $19x^4y^{12}z^{13} - 8x^5y^3z^9$

19. $24x^6 + 8x^4 - 4x^3y$

20. $44x^5y + 11x^3y + 22x^2$

21. $52x^2y^2 + 16xy^3 + 26z$

22. $x(x + 2) + 3(x + 2)$

23. $5x(2x - 5) + 3(2x - 5)$

24. $4x(2x + 1)^2 + 1(2x + 1)$

25. $3x(4x - 5)^3 + 1(4x - 5)^2$

26. $4x(2x + 1) + 2x + 1$

27. $3x(2x + 5) - 6(2x + 5)^2$

28. $(x - 3)(x + 1) + (x - 3)(x + 2)$

29. $(3p - q)(2p - q) + (3p - q)(p - 2q)$

30. $(3r + 2)(3r - 1) - (3r + 2)(2r + 3)$

31. $(x - 2)(3x + 5) - (x - 2)(5x - 4)$

32. $6x^5(2x + 7) + 4x^3(2x + 7) - 2x^2(2x + 7)$

33. $6x^3(2x + 5) - 2x^2(2x + 5) - (2x + 5)$

34. $5a(3x - 2)^5 + 4(3x - 2)^4$

35. $4p(2r - 3)^7 - 3(2r - 3)^6$

36. $4a^2(5a - 3) + 2a(5a - 3) - 3(5a - 3)$

Factor by grouping.

37. $x^2 + 3x - 5x - 15$

38. $x^2 + 3x - 2x - 6$

39. $3x^2 + 9x + x + 3$

40. $x^2 + 4x + x + 4$

41. $4x^2 - 2x - 2x + 1$

42. $2x^2 + 6x - x - 3$

43. $8x^2 - 4x - 20x + 10$

44. $ax + ay + bx + by$

45. $2b + 2c + ab + ac$

46. $3ac + 3ad + 2bc + 2bd$

47. $35x^2 - 40xy + 21xy - 24y^2$

48. $15b^2 - 20bc - 18bc + 24c^2$

49. $x^3 - 3x^2 + 4x - 12$

50. $3x^2 - 18xy + 4xy - 24y^2$

51. $10x^2 - 12xy - 25xy + 30y^2$

52. $12x^2 - 9xy + 4xy - 3y^2$

53. $x^3 + 3x^2 - 2x - 6$

54. $2a^4 - 2a^3 - 5a^2 + 5a$

55. $2a^4b - 2ac^2 - 3a^3bc + 3c^3$

56. $8r^2 + 6rs - 12rs - 9s^2$

57. $3p^3 + 3pq^2 + 2p^2q + 2q^3$

58. $16r^3 - 4r^2s^2 - 4rs + s^3$

Factor completely.

59. $20p^3 - 18p^2 + 12p$

60. $6x^3 - 8x^2 - x$

61. $16xy^2z + 4x^3y - 8$

62. $80x^5y^3z^4 - 36x^2yz^3$

63. $5x^2 - 10x + 3x - 6$

64. $4x^2 + 6x - 6x - 9$

65. $14y^3z^5 - 28y^3z^6 - 9xy^2z^2$

66. $8x^2 - 20x - 4x + 10$

67. $7x^4y^9 - 21x^3y^7z^5 - 35y^8z^9$

68. $48x^2y + 16xy^2 + 33xy$

69. $15a^2 - 18ab - 20ab + 24b^2$

70. $7x(4x - 3) - 1(4x - 3)$

71. $3x(7x + 1) - 2(7x + 1)$

72. $3x(4x - 5) + 4x - 5$

73. $6x^2 - 9xy + 2xy - 3y^2$

74. $x^2 - 3xy + 2xy - 6y^2$

75. $5x(x + 3)^2 - 3(x + 3)$

76. $(4a - b)(2a + 3b) - (4a - b)(a - 5b)$

77. $(3c - d)(c + d) - (3c - d)(c - d)$

78. $2x^3y^3 + 6xyz^2 - 3x^2y^4 - 9y^2z^2$

79. $3x^5 - 15x^3 + 2x^3 - 10x$

80. $3x^2(b - 4) - 2x(b - 4) + 5(b - 4)$

✎ **81.** What is the first step in *any* factoring problem?

✎ **82.** What is the greatest common factor of the terms of an expression?

Cumulative Review Exercises

[2.3] **83.** The price of a shirt is increased by 10%. This price is then decreased by $10. If the final adjusted price of the shirt is $17.50, find the original price of the shirt.

[4.5] **84.** Find the solution to the system of inequalities.

$$y > -3x + 4$$
$$3x - 2y \le 6$$

Divide as indicated.

[5.5] **85.** $\dfrac{3x^2 - 6xy^2 + 12xy^3}{4x^2y^2}$

86. $\dfrac{6x^3 + 13x^2 - 6x - 18}{2x + 3}$

JUST FOR FUN

1. Factor $4x^2(x - 3)^3 - 6x(x - 3)^2 + 4(x - 3)$.
2. Factor
   (a) $(x + 1)^2 + (x + 1)$
   (b) $(x + 1)^3 + (x + 1)^2$
   (c) $(x + 1)^n + (x + 1)^{n-1}$
   (d) $(x + 1)^{n+1} + (x + 1)^n$
   *3. Factor $4x(x + 5)^{-2} + 2x(x + 5)^{-1}$.
   4. Factor $x(2x - 3)^{-1/2} + 6(2x - 3)^{1/2}$.

6.2

Factoring Trinomials

▶ **1** Factor trinomials of the form $x^2 + bx + c$.

▶ **2** Factor trinomials of the form $ax^2 + bx + c$, $a \ne 1$, using trial and error.

▶ **3** Factor trinomials of the form $ax^2 + bx + c$, $a \ne 1$, using grouping.

▶ **4** Factor trinomials that contain a common factor.

▶ **5** Factor using substitution.

▶ **1** In this section we learn how to factor trinomials of the form $ax^2 + bx + c$, $a \ne 0$.

Trinomials	*Coefficients*
$3x^2 + 2x - 5$	$a = 3, \quad b = 2, \quad c = -5$
$-\dfrac{1}{2}x^2 - 4x + 3$	$a = -\dfrac{1}{2}, \quad b = -4, \quad c = 3$

Factoring Trinomials of the form $x^2 + bx + c$

To Factor Trinomials of the Form $x^2 + bx + c$ (note $a = 1$)

1. Find two numbers (or factors) whose product is c and whose sum is b.
2. The factors of the trinomial will be of the form

$$(x + \;\;)(x + \;\;)$$

↑   ↑
one factor   other factor
determined   determined
in step 1   in step 1

If the numbers determined in step 1 are, for example, 3 and -5, the factors would be written $(x + 3)(x - 5)$. This procedure is illustrated in the following examples.

EXAMPLE 1   Factor $x^2 - x - 12$.

Solution: $a = 1, b = -1, c = -12$. We must determine two numbers whose product is c, -12, and whose sum is b, -1.

---

* Factoring problems with fractional and negative exponents are discussed in Section 8.2.

$$\begin{array}{cc} \textit{Factors of } -12 & \textit{Sum of Factors} \\ (1)(-12) & 1 + (-12) = -11 \\ (2)(-6) & 2 + (-6) = -4 \\ (3)(-4) & 3 + (-4) = -1 \\ (4)(-3) & 4 + (-3) = 1 \\ (6)(-2) & 6 + (-2) = 4 \\ (12)(-1) & 12 + (-1) = 11 \end{array}$$

The numbers we are seeking are 3 and -4.

$$x^2 - x - 12 = (x + 3)(x - 4)$$

one factor   other factor
of -12 of -12

Notice in Example 1 that we listed all the factors of -12. After the two factors are found whose product is c and whose sum is b, there is no need to go further in listing the factors. The factors were listed here to illustrate the fact that, for example, $(2)(-6)$ is a different set of factors than $(-2)(6)$. Note that as the positive factor increases the sum of the factors increases.

EXAMPLE 2   Factor $x^2 - 5x - 6$.

Solution: We must find two numbers whose product is -6 and whose sum is -5. The numbers are 1 and -6. *Note:* $(1)(-6) = -6$ and $1 + (-6) = -5$.

$$x^2 - 5x - 6 = (x + 1)(x - 6)$$

Since the factors may be placed in any order, $(x - 6)(x + 1)$ is also an acceptable answer.

HELPFUL HINT   **Checking Trinomial Factoring Problems**

Trinomial factoring problems can be checked by multiplying the factors obtained. If the factoring is correct, you will obtain the trinomial you started with. To check Example 2, we will multiply the factors using the FOIL method.

$$(x + 1)(x - 6) = x^2 - 6x + x - 6 = x^2 - 5x - 6$$

Since the product of the factors is the trinomial we began with, our factoring is correct. You should check all factoring problems.

The procedures used to factor trinomials of the form $x^2 + bx + c$ can be used on other types of trinomials, as illustrated in the following example.

EXAMPLE 3   Factor $x^2 + 2xy - 15y^2$.

Solution: We must find two numbers whose product is -15 and whose sum is 2. The two numbers are 5 and -3. Note $(5)(-3) = -15$ and $5 + (-3) = 2$. Since the last term of the trinomial contains a y^2, the second term of each factor must contain a y.

$$x^2 + 2xy - 15y^2 = (x + 5y)(x - 3y)$$

Check: $(x + 5y)(x - 3y) = x^2 - 3xy + 5xy - 15y^2$
$$= x^2 + 2xy - 15y^2$$ ■

If each term of a trinomial has a common factor, use the distributive property to remove the common factor before following the procedure outlined earlier.

EXAMPLE 4 Factor $3x^2 - 6x - 72$.

Solution: The factor 3 is common to all three terms of the trinomial. Factoring out the 3 gives
$$3x^2 - 6x - 72 = 3(x^2 - 2x - 24)$$

The 3 that was factored out is a part of the answer but plays no further part in the factoring process. Now continue to factor $x^2 - 2x - 24$ in the usual manner. We must find two numbers whose product is -24 and whose sum is -2. The numbers are -6 and 4.
$$3(x^2 - 2x - 24) = 3(x - 6)(x + 4)$$

Therefore, $3x^2 - 6x - 72 = 3(x - 6)(x + 4)$. ■

Factoring Trinomials of the Form $ax^2 + bx + c, a \neq 1$

▶**2** Now we will look at some examples of factoring trinomials of the form
$$ax^2 + bx + c, \qquad a \neq 1$$

Two methods of solving this type of trinomial will be illustrated. The first method, trial and error, involves trying various combinations until the correct combination is found. The second method makes use of factoring by grouping, a procedure that was presented in Section 5.1. You may use either method unless your instructor specifies that you should use a particular method.

Method 1: Trial and Error

Let us now discuss the trial and error method of factoring trinomials. As an aid in our explanation we will multiply $(2x + 3)(x + 1)$ using the FOIL method.

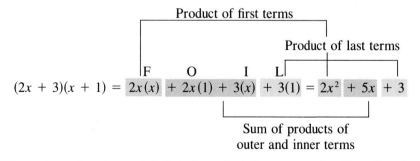

When given the trinomial $2x^2 + 5x + 3$ to factor, you should realize that the product of the first terms of the factors must be $2x^2$, the product of the last terms must be 3, and the sum of the products of the outer and inner terms must be $5x$.

To factor $2x^2 + 5x + 3$, we begin as shown here.

$$2x^2 + 5x + 3 = (2x \qquad)(x \qquad) \qquad \text{The product of first terms is } 2x^2.$$

Now fill in the second terms using positive integers whose product is 3. Only positive integers will be considered since the product of the last terms is positive, and the sum of the products of the outer and inner terms is also positive. The two possibilities are as follows:

$$(2x + 1)(x + 3)\Big\}$$
$$(2x + 3)(x + 1)\Big\}$$ The product of last terms is 3.

To determine which is the correct factoring process, find the sum of the products of the outer terms and inner terms. If either has a sum of $5x$, the middle term of the trinomial, then that factoring process is the correct one.

$(2x + 1)(x + 3) = 2x^2 + 6x + x + 3 = 2x^2 + 7x + 3$ Wrong middle term.
$(2x + 3)(x + 1) = 2x^2 + 2x + 3x + 3 = 2x^2 + 5x + 3$ Correct middle term.

Therefore, the factors of $2x^2 + 5x + 3$ are $2x + 3$ and $x + 1$. Thus

$$2x^2 + 5x + 3 = (2x + 3)(x + 1)$$

Note that if we had begun our factoring process by writing

$$2x^2 + 5x + 3 = (x \qquad)(2x \qquad)$$

we would have obtained the correct answer if we had continued with the procedure.

To Factor Trinomials of the Form $ax^2 + bx + c$, $a \neq 1$, Using Trial and Error

1. Write all pairs of factors of the coefficient of the squared term, a.
2. Write all pairs of factors of the constant, c.
3. Try various combinations of these factors until the correct middle term, bx, is found.

EXAMPLE 5   Factor $3x^2 - 13x + 10$.

Solution: The only factors of 3 are 1 and 3. Therefore, we write

$$3x^2 - 13x + 10 = (3x \qquad)(x \qquad)$$

The number 10 has both positive and negative factors. However, since the product of the last terms must be positive ($+10$), and the sum of the products of the outer and inner terms must be negative (-13), the two factors of 10 must both be negative. Why? The negative factors of 10 are $(-1)(-10)$ and $(-2)(-5)$. Below is a listing of the possible factors. We look for the factors that give us the correct middle term, $-13x$.

Possible Factors	Sum of Products of Outer and Inner Terms
$(3x - 1)(x - 10)$	$-31x$
$(3x - 10)(x - 1)$	$-13x$ ← Correct middle term.
$(3x - 2)(x - 5)$	$-17x$
$(3x - 5)(x - 2)$	$-11x$

Thus $3x^2 - 13x + 10 = (3x - 10)(x - 1)$.   ■

The following Helpful Hint is very important. Study it carefully.

HELPFUL HINT

Factoring by Trial and Error

When factoring a trinomial of the form $ax^2 + bx + c$, the sign of the constant, c, is very helpful in finding the solution. If $a > 0$, then:

1. When the constant, c, is positive, and the numerical coefficient of the x term, b, is positive, both numerical factors will be positive.

$$Example:\quad x^2 \;+\; 7x + 12 = (x \;+\; 3)(x \;+\; 4)$$
positive positive   positive positive

2. When c is positive and b is negative, both numerical factors will be negative.

$$Example:\quad x^2 \;-\; 5x + 6 = (x \;-\; 2)(x \;-\; 3)$$
negative positive   negative negative

In the first two cases the constant, c, was positive. *Whenever the constant is positive, the signs in both factors will be the same, either both positive or both negative.* When the constant is positive, both factors will contain the same sign as the sign of the x term.

3. When c is negative, one of the numerical factors will be positive and the other will be negative.

$$Example:\quad x^2 + x \;-\; 6 = (x \;+\; 3)(x \;-\; 2)$$
negative   positive negative

Here the constant is negative. *When the constant is negative, the factors will contain different signs. One factor will contain a positive sign and the other factor will contain a negative sign.*

EXAMPLE 6 Factor $8x^2 - 51x + 18$.

Solution: The factors of 8 will be either $4 \cdot 2$ or $8 \cdot 1$. When there is more than one set of factors for the first term, we generally try the medium-sized factors first. If this does not work, try the other factors. We will therefore try the $4 \cdot 2$ first.

Since the constant, 18, is positive and the x term, $-51x$, is negative the signs in the factors must both be negative. The possible factors of 18 are $(-18)(-1)$, $(-9)(-2)$, and $(-6)(-3)$. We need to find the set of factors whose sum of the products of the outer and inner terms is $-51x$.

Possible Factors	Sum of Products of Outer and Inner Terms
$(4x - 18)(2x - 1)$	$-40x$
$(4x - 9)(2x - 2)$	$-26x$
$(4x - 6)(2x - 3)$	$-24x$
$(4x - 3)(2x - 6)$	$-30x$
$(4x - 2)(2x - 9)$	$-40x$
$(4x - 1)(2x - 18)$	$-74x$

Since the 4 and 2 did not yield the correct middle term, we now try the 8 and 1.

$$(8x - 18)(x - 1) \qquad -26x$$
$$(8x - 9)(x - 2) \qquad -25x$$
$$(8x - 6)(x - 3) \qquad -30x$$
$$(8x - 3)(x - 6) \qquad \boxed{-51x}$$

The factors that give the correct term of $-51x$ are $(8x - 3)(x - 6)$. Note that once we find the factors we are seeking we can stop.

$$8x^2 - 51x + 18 = (8x - 3)(x - 6) \qquad \blacksquare$$

EXAMPLE 7   Factor $6x^2 - 11x - 10$.

Solution: The factors of 6 will be either $6 \cdot 1$ or $2 \cdot 3$. Therefore, the factors may be of the form $(6x \quad)(x \quad)$ or $(2x \quad)(3x \quad)$. We begin with the middle set of factors; thus we write

$$6x^2 - 11x - 10 = (2x \quad)(3x \quad)$$

The factors of -10 are $(-1)(10)$, $(1)(-10)$, $(-2)(5)$, and $(2)(-5)$. Since there are eight different factors of -10, there will be eight different pairs of possible factors to try. Can you list them? The correct factoring is

$$6x^2 - 11x - 10 = (2x - 5)(3x + 2) \qquad \blacksquare$$

Note in Example 7 that we were fortunate to find that the factors were of the form $(2x \quad)(3x \quad)$. If we had not been able to find the correct factors using these, we would have tried $(6x \quad)(x \quad)$.

Method 2:
Using Grouping

▶ **3**   Now we will discuss the grouping method of factoring trinomials of the form $ax^2 + bx + c$, $a \neq 1$.

To Factor Trinomials of the Form $ax^2 + bx + c$, $a \neq 1$, Using Grouping

---

1. Find two numbers whose product is $a \cdot c$ and whose sum is b.
2. Rewrite the bx term using the numbers found in step 1.
3. Factor by grouping.

EXAMPLE 8   Factor $2x^2 - 5x - 12$.

Solution: $a = 2$, $b = -5$, and $c = -12$. We must find two numbers whose product is $a \cdot c$, or $2(-12) = -24$, and whose sum is b, -5. The two numbers are -8 and 3. *Note:* $(-8)(3) = -24$ and $-8 + 3 = -5$. Now rewrite the bx term, $-5x$, using $-8x$ and $3x$.

$$\overbrace{}^{-5x}$$
$$2x^2 - 5x - 12 = 2x^2 - 8x + 3x - 12$$

Now factor by grouping as explained in Section 5.1. Factor out a $2x$ from the first two terms and a 3 from the last two terms.

$$= 2x(x - 4) + 3(x - 4)$$
$$= (2x + 3)(x - 4) \qquad \blacksquare$$

Note in Example 8 that we wrote $-5x$ as $-8x + 3x$. As we show below, the same answer would be obtained if we wrote $-5x$ as $3x - 8x$. Therefore, it makes no difference which factor is listed first when factoring by grouping.

$$2x^2 - 5x - 12 = 2x^2 \overbrace{+ 3x - 8x}^{-5x} - 12$$
$$= x(2x + 3) - 4(2x + 3)$$
$$= (x - 4)(2x + 3)$$

EXAMPLE 9 Factor $12x^2 - 19x + 5$.

Solution: We must find two numbers whose product is $(12)(5) = 60$ and whose sum is -19. Since the product of the numbers is positive and their sum is negative, the two numbers must both be negative. Why?

The two numbers are -15 and -4. *Note:* $(-15)(-4) = 60$ and $-15 + (-4) = -19$. Now rewrite the bx term, $-19x$, using $-15x$ and $-4x$. Then factor by grouping.

$$12x^2 - 19x + 5 = 12x^2 \overbrace{- 15x - 4x}^{-19x} + 5$$
$$= 3x(4x - 5) - 1(4x - 5)$$
$$= (3x - 1)(4x - 5) \qquad \blacksquare$$

Try Example 9 again, this time writing $-19x$ as $-4x - 15x$. If you do it correctly, you should get the same answer.

▶ **4** The first step when factoring any trinomial whose coefficient of the squared term is not 1 is to determine if all three terms have a common factor. If so, factor out that common factor. Then factor the remaining polynomial using either the trial and error method or the factoring by grouping method.

EXAMPLE 10 Factor $4x^2 + 10x - 6$.

Solution: By examining the three terms of the trinomial, we see that 2 is common to each term. We therefore begin by factoring out a 2.

$$4x^2 + 10x - 6 = 2(2x^2 + 5x - 3)$$

Now continue to factor $2x^2 + 5x - 3$ using either method 1 or 2.

$$4x^2 + 10x - 6 = 2(2x - 1)(x + 3) \qquad \blacksquare$$

It is important for you to realize that not every trinomial can be factored by the methods presented in this section. Consider the following example.

EXAMPLE 11 Factor $2x^2 + 6x + 5$.

Solution: If you try to factor this using either method 1 or 2, you will see it cannot be factored. Polynomials of this type are called **prime polynomials.** $\qquad \blacksquare$

HELPFUL HINT

We have introduced two different methods of factoring trinomials of the form $ax^2 + bx + c$, $a \neq 1$. Which method should you use? If your instructor tells you to use a specific method, use the method indicated. If you can use either method, you may wish to try factoring by the trial and error method first, especially if the number of possible factors of the constant is small. If you cannot quickly find the answer using trial and error, you may then wish to try the grouping method. After a little practice, you will be able to determine which method gives you the most success.

Factoring Using Substitution

▶ **5**   Sometimes a more complicated trinomial can be factored by substituting one variable for another. The next two examples illustrate **factoring using substitution.**

EXAMPLE 12   Factor $y^4 - y^2 - 6$.

Solution: If we can get this expression in the form $ax^2 + bx + c$, it will be easier to factor. Note that $(y^2)^2 = y^4$. If we substitute x for y^2, the trinomial becomes

$$y^4 - y^2 - 6 = (y^2)^2 - y^2 - 6$$
$$= x^2 - x - 6$$

Now proceed to factor $x^2 - x - 6$.

$$= (x + 2)(x - 3)$$

Finally, substitute y^2 in place of x to obtain

$$= (y^2 + 2)(y^2 - 3)$$

Thus $y^4 - y^2 - 6 = (y^2 + 2)(y^2 - 3)$. Note that x was substituted for y^2, and then y^2 was substituted back for x. ■

EXAMPLE 13   Factor $(x + 5)^2 + 3(x + 5) - 4$.

Solution: We will use basically the same procedure to factor this example as we used in Example 12. By substituting $a = x + 5$ in the equation, we obtain

$$(x + 5)^2 + 3(x + 5) - 4 = a^2 + 3a - 4$$

Now factor $a^2 + 3a - 4$.

$$= (a + 4)(a - 1)$$

Finally, replace a with $x + 5$ to obtain

$$= [(x + 5) + 4][(x + 5) - 1]$$
$$= (x + 9)(x + 4)$$

Note that a was substituted for $x + 5$, and then $x + 5$ was substituted back for a. ■

In Example 12 we used x in our substitution, whereas in Example 13 we used a. The letter selected is immaterial to the final answer.

Exercise Set 6.2

Factor each trinomial completely. If the trinomial cannot be factored, so state.

1. $x^2 + 7x + 6$
2. $x^2 - 6x + 10$
3. $p^2 - 3p - 10$
4. $y^2 - 12y + 11$
5. $w^2 - 7w + 9$
6. $x^2 - 16x + 64$
7. $x^2 - 34x + 64$
8. $x^2 - 11x - 30$
9. $a^2 - 18a + 45$
10. $x^2 - 11x + 10$
11. $y^2 - 9y + 15$
12. $p^2 - 17p - 60$
13. $x^2 - 4xy + 3y^2$
14. $x^2 - 6xy + 8y^2$
15. $z^2 - 7yz + 10y^2$
16. $x^2 - 12xy - 45y^2$
17. $5x^2 + 20x + 15$
18. $4x^2 + 12x - 16$
19. $x^3 - 3x^2 - 18x$
20. $x^3 + 11x^2 - 42x$
21. $x^3 - 5x^2 - 24x$
22. $5p^2 - 8p + 3$
23. $4w^2 + 13w + 3$
24. $3x^2 + 14x - 5$
25. $3x^2 - 11x - 6$
26. $2x^2 + 11x + 15$
27. $3w^2 - 2w - 8$
28. $5y^2 - 16y + 3$
29. $3y^2 - 2y - 5$
30. $3x^2 - 22xy + 7y^2$
31. $4x^2 + 4xy - 3y^2$
32. $6x^3 + 5x^2 - 4x$
33. $8x^2 + 2x - 20$
34. $12x^3 - 12x^2 - 45x$
35. $8x^2 - 8xy - 6y^2$
36. $18w^2 + 18wz - 8z^2$
37. $x^3y - 3x^2y - 18xy$
38. $a^3b - a^2b - 12ab$
39. $a^3b + 2a^2b - 35ab$
40. $3b^4c - 18b^3c^2 + 27b^2c^3$
41. $6p^3q^2 - 24p^2q^3 - 30pq^4$
42. $21x^2 + x - 2$
43. $35x^2 + 13x - 12$
44. $12a^2 - 11a - 5$
45. $8x^2 - 34x + 30$
46. $100b^2 - 90b + 20$

Factor each trinomial completely.

47. $x^4 + x^2 - 6$
48. $x^4 - 3x^2 - 10$
49. $x^4 + 5x^2 + 6$
50. $x^4 - 2x^2 - 15$
51. $6a^4 + 5a^2 - 25$
52. $(2x + 1)^2 + 2(2x + 1) - 15$
53. $4(x + 1)^2 + 8(x + 1) + 3$
54. $(2x + 3)^2 - (2x + 3) - 6$
55. $6(a + 2)^2 - 7(a + 2) - 5$
56. $6(p - 5)^2 + 11(p - 5) + 3$
57. $a^2b^2 - 8ab + 15$
58. $x^2y^2 + 10xy + 24$
59. $3x^2y^2 - 2xy - 5$
60. $3p^2q^2 + 11pq + 6$
61. $2a^2(5 - a) - 7a(5 - a) + 5(5 - a)$
62. $2y^2(y + 2) + 13y(y + 2) + 15(y + 2)$
63. $2x^2(x - 3) + 7x(x - 3) + 6(x - 3)$
64. $3x^2(x - 2) + 5x(x - 2) - 2(x - 2)$

Factor each trinomial completely.

65. $x^2 + 16x + 64$
66. $x^2 + 11xy + 24y^2$
67. $3y^2 - 33y + 54$
68. $x^3 - 10x^2 + 24x$
69. $y^4 - 7y^2 - 30$
70. $2x^2 + 5x + 3$
71. $3z^4 - 14z^2 - 5$
72. $5y^2 - 16y + 3$
73. $12x^2 + 16x - 3$
74. $6(r + 3)^2 + 13(r + 3) + 5$
75. $2x^2y^2 + 3xy - 9$
76. $7x^2 + 43x + 6$
77. $9x^2 - 15x - 36$
78. $12a^2 - 34ab + 24b^2$
79. $x^2(x + 3) + 3x(x + 3) + 2(x + 3)$
80. $8x^3y + 24x^2y - 32xy$
81. $5x^2 + 25xy + 20y^2$
82. $5a^3b^2 - 8a^2b^3 + 3ab^4$
83. $15a^2 + 16a - 15$
84. $12a^2 - 36a + 15$
85. $20y^2 + 13y - 15$
86. $x^2(x - 1) - x(x - 1) - 30(x - 1)$

87. When factoring any trinomial, what should the first step always be?

88. If the factors of a polynomial are $(2x + 3y)$ and $(x - 4y)$, find the polynomial. Explain how you determined your answer.

89. If the factors of a polynomial are 3, $(4x - 5)$, and $(2x - 3)$, find the polynomial. Explain how you determined your answer.

90. If we know that one factor of the polynomial $x^2 + 3x - 18$ is $x - 3$, how can we find the other factor? Find the other factor.

91. If we know that one factor of the polynomial $x^2 - xy - 6y^2$ is $x - 3y$, how can we find the other factor? Find the other factor.

92. (a) Explain in your own words the step-by-step procedure to follow to factor $6x^2 - x - 12$.

(b) Factor $6x^2 - x - 12$ using the procedure you explained in part (a).

93. (a) Explain in your own words the step by step procedure to follow to factor $8x^2 - 26x + 6$.

(b) Factor $8x^2 - 26x + 6$ using the procedure you explained in part (a).

Cumulative Review Exercises

[3.3] **94.** What is the slope of a horizontal line? Explain your answer.

95. What is the slope of a vertical line? Explain your answer.

[5.1] **96.** Write the answer in scientific notation form:
$$\frac{36,000,000}{0.0004}$$

[5.3] **97.** Simplify $2x^2y - 6xy^2 - (3x^2y + 2xy^2 - 6)$.

[5.4] **98.** Given $f(x) = x^2 - 4x + 6$, (a) indicate the axis of symmetry, and (b) graph the function.

JUST FOR FUN

1. Have you ever seen the "proof" that 1 is equal to 2? Here it is.

Let $a = b$	
$a^2 = b^2$	Square both sides of the equation.
$a^2 = b \cdot b$	
$a^2 = ab$	Substitute a for b.
$a^2 - b^2 = ab - b^2$	Subtract b^2 from both sides of the equation.
$(a + b)(a - b) = b(a - b)$	Factor both sides of the equation.
$\dfrac{(a + b)(a - b)}{a - b} = \dfrac{b(a - b)}{a - b}$	Divide both sides of the equation by $(a - b)$ and divide out common factors.
$a + b = b$	
$b + b = b$	Substitute b for a.
$2b = b$	Divide both sides of the equation by b.
$\dfrac{2b}{b} = \dfrac{b}{b}$	
$2 = 1$	

Obviously, $2 \neq 1$. Therefore, we must have made an error somewhere. Can you find it?

2. Factor completely $4a^{2n} - 4a^n - 15$.

3. Factor completely $12x^{2n}y^{2n} + 2x^ny^n - 2$.

6.3

Special Factoring Formulas

▶**1** Factor the difference of two squares.

▶**2** Factor perfect square trinomials.

▶**3** Factor the sum and difference of two cubes.

▶**1** In this section we present some special factoring formulas: factoring the difference of two squares, perfect square trinomials, and the sum and difference of two cubes. It will be to your advantage to memorize these formulas.

The expression $x^2 - 9$ is an example of the difference of two squares.

$$x^2 - 9 = (x)^2 - (3)^2$$

To factor the difference of two squares, it is convenient to use the difference-of-two-squares formula. This formula was first presented in Section 5.4.

Difference of Two Squares

$$a^2 - b^2 = (a + b)(a - b)$$

EXAMPLE 1   Factor each of the following differences of squares using the difference-of-two-squares formula.

(a) $x^2 - 16$ (b) $16x^2 - 9y^2$

Solution: Rewrite each expression as a difference of two squares.

(a) $x^2 - 16 = (x)^2 - (4)^2$ (b) $16x^2 - 9y^2 = (4x)^2 - (3y)^2$
$\qquad\qquad = (x + 4)(x - 4)$ $= (4x + 3y)(4x - 3y)$ ∎

EXAMPLE 2   Factor each of the following differences of squares.

(a) $x^6 - y^4$ (b) $z^4 - 81x^6$

Solution: Rewrite each expression as a difference of two squares and then use the difference-of-two-squares formula to factor the expression.

(a) $x^6 - y^4 = (x^3)^2 - (y^2)^2$ (b) $z^4 - 81x^6 = (z^2)^2 - (9x^3)^2$
$\qquad\qquad = (x^3 + y^2)(x^3 - y^2)$ $= (z^2 + 9x^3)(z^2 - 9x^3)$ ∎

EXAMPLE 3   Factor $4x^2 - 16y^2$ using the difference of two squares.

Solution: First remove the common factor, 4.

$$4x^2 - 16y^2 = 4(x^2 - 4y^2)$$

Now use the formula for the difference of two squares.

$$4(x^2 - 4y^2) = 4[(x)^2 - (2y)^2]$$
$$= 4(x + 2y)(x - 2y)$$ ∎

EXAMPLE 4   Factor $x^4 - 16y^4$.

Solution: $x^4 - 16y^4 = (x^2)^2 - (4y^2)^2$
$$= (x^2 + 4y^2)(x^2 - 4y^2)$$

Note that $(x^2 - 4y^2)$ is also a difference of two squares. We use the difference-of-two-squares formula a second time to obtain

$$= (x^2 + 4y^2)[(x)^2 - (2y)^2]$$
$$= (x^2 + 4y^2)(x + 2y)(x - 2y) \qquad \blacksquare$$

EXAMPLE 5   Factor $(x - 5)^2 - 9$ using the difference of two squares.

Solution: We can express $(x - 5)^2 - 9$ as a difference of two squares.

$$(x - 5)^2 - 9 = (x - 5)^2 - 3^2$$
$$= [(x - 5) + 3][(x - 5) - 3]$$
$$= (x - 2)(x - 8)$$

Thus $(x - 5)^2 - 9$ factors into $(x - 2)(x - 8)$. $\qquad \blacksquare$

Note: ***It is not possible to factor the sum of two squares of the form $a^2 + b^2$ over the set of real numbers.***

For example, it is not possible to factor $x^2 + 4$ since $x^2 + 4 = x^2 + 2^2$ and this is a sum of two squares.

▶**2**   In Section 5.4 we saw that

$$(a + b)^2 = a^2 + 2ab + b^2$$
$$(a - b)^2 = a^2 - 2ab + b^2$$

If we reverse the left and right sides of these two formulas, we obtain two special factoring formulas.

Perfect Square Trinomials

$$a^2 + 2ab + b^2 = (a + b)^2$$
$$a^2 - 2ab + b^2 = (a - b)^2$$

These two trinomials are called **perfect square trinomials** since each is the square of a binomial. *To be a perfect square trinomial, the first and last terms must be the squares of some expression and the middle term must be twice the product of the first and last terms.* When you are given a trinomial to factor, determine if it is a perfect square trinomial before you attempt to factor it by the procedures explained in Section 6.2. If it is a perfect square trinomial, you can factor it using the formulas given above.

Perfect Square Trinomials

$y^2 + 6y + 9$	or	$y^2 + 2(y)(3) + 3^2$
$9a^2b^2 - 24ab + 16$	or	$(3ab)^2 - 2(3ab)(4) + 4^2$
$(r + s)^2 + 6(r + s) + 9$	or	$(r + s)^2 + 2(r + s)(3) + 3^2$

Now let us factor some perfect square trinomials.

EXAMPLE 6   Factor $x^2 - 8x + 16$.

Solution: Since the first and last terms are squares, x^2, and 4^2, respectively, this trinomial might be a perfect square trinomial. To determine if it is, take twice the product of x and 4 to see if you obtain $8x$.

$$2(x)(4) = 8x$$

Since $8x$ is the middle term and since the sign of the middle term is negative, we factor as follows:

$$x^2 - 8x + 16 = (x - 4)^2 \qquad \blacksquare$$

EXAMPLE 7   Factor $9x^4 - 12x^2 + 4$.

Solution: The first term is a square $(3x^2)^2$, as is the last term 2^2. Since $2(3x^2)(2) = 12x^2$, we factor as follows:

$$9x^4 - 12x^2 + 4 = (3x^2 - 2)^2 \qquad \blacksquare$$

EXAMPLE 8   Factor $(a + b)^2 + 6(a + b) + 9$.

Solution: The first term $(a + b)^2$ is a square. The last term 9 is a square, 3^2. The middle term is $2(a + b)(3) = 6(a + b)$. Therefore, this is a perfect square trinomial. Thus

$$(a + b)^2 + 6(a + b) + 9 = [(a + b) + 3]^2 = (a + b + 3)^2 \qquad \blacksquare$$

EXAMPLE 9   Factor $x^2 - 6x + 9 - y^2$.

Solution: Since $x^2 - 6x + 9$ is a perfect square trinomial, we write

$$(x - 3)^2 - y^2$$

Now $(x - 3)^2 - y^2$ is a difference of two squares; therefore

$$(x - 3)^2 - y^2 = [(x - 3) + y][(x - 3) - y]$$
$$= (x - 3 + y)(x - 3 - y)$$

Thus $x^2 - 6x + 9 - y^2 = (x - 3 + y)(x - 3 - y)$. $\qquad \blacksquare$

Notice that the polynomial in Example 9 has four terms. In Section 6.1 we learned to factor polynomials with four terms using the grouping technique. If you study Example 9, you will see that no matter how you arrange the four terms they cannot be arranged so that the first two terms have a common factor and the last two terms have a common factor. When you come across a polynomial with four terms that you cannot factor by grouping, determine if you can rewrite three of the terms as the square of a binomial and then use the difference-of-two-squares formula to factor the expression.

EXAMPLE 10   Factor $4a^2 + 12ab + 9b^2 - 25$.

Solution: We first notice that this four-term polynomial cannot be factored using the factoring by grouping technique. We next look to see if three terms of the polynomial can be expressed as the square of a binomial. Since this can be done, we write the three terms as the square of a binomial. We complete our factoring using the difference-of-two-squares formula.

$$4a^2 + 12ab + 9b^2 - 25 = (2a + 3b)^2 - 5^2$$
$$= [(2a + 3b) + 5][(2a + 3b) - 5]$$
$$= (2a + 3b + 5)(2a + 3b - 5) \qquad \blacksquare$$

Sum and Difference of Two Cubes

▶ **3** Earlier in this section we discussed the difference of two squares. Now we will consider the sum and difference of two cubes. Consider the product of $(a + b)(a^2 - ab + b^2)$.

$$
\begin{array}{r}
a^2 - ab + b^2 \\
a + b \\
\hline
a^2 b - ab^2 + b^3 \\
a^3 - a^2 b + ab^2 \\
\hline
a^3 + b^3
\end{array}
$$

Thus $a^3 + b^3 = (a + b)(a^2 - ab + b^2)$. Using multiplication, we can also show that $a^3 - b^3 = (a - b)(a^2 + ab + b^2)$. The sum and the difference of two cubes are summarized in the following box.

Sum of Two Cubes

$$a^3 + b^3 = (a + b)(a^2 - ab + b^2)$$

Difference of Two Cubes

$$a^3 - b^3 = (a - b)(a^2 + ab + b^2)$$

EXAMPLE 11 Factor $x^3 + 27$.

Solution: Rewriting $x^3 + 27$ as a sum of two cubes gives $x^3 + 3^3$. Substitute x for a and 3 for b, and then factor using the sum-of-two cubes formula.

$$a^3 + b^3 = (a + b)(a^2 - ab + b^2)$$
$$x^3 + 3^3 = (x + 3)[x^2 - x(3) + 3^2]$$
$$= (x + 3)(x^2 - 3x + 9)$$

Thus $x^3 + 27 = (x + 3)(x^2 - 3x + 9)$. $\qquad \blacksquare$

EXAMPLE 12 Factor $27x^3 - 8y^6$.

Solution: We first determine that $27x^3$ and $8y^6$ have no common factors other than 1. Since we can express both $27x^3$ and $8y^6$ as cubes, we can factor the given expression using the difference-of-two-cubes formula.

$$27x^3 - 8y^6 = (3x)^3 - (2y^2)^3$$
$$= (3x - 2y^2)[(3x)^2 + (3x)(2y^2) + (2y^2)^2]$$
$$= (3x - 2y^2)(9x^2 + 6xy^2 + 4y^4)$$

Thus $27x^3 - 8y^6 = (3x - 2y^2)(9x^2 + 6xy^2 + 4y^4)$. $\qquad \blacksquare$

EXAMPLE 13 Factor $8y^3 - 64x^6$.

Solution: First factor the 8 common to both terms.

$$8y^3 - 64x^6 = 8(y^3 - 8x^6)$$

Next factor $y^3 - 8x^6$ by writing it as a difference of two cubes.

$$\begin{aligned} 8(y^3 - 8x^6) &= 8[(y)^3 - (2x^2)^3] \\ &= 8(y - 2x^2)[y^2 + y(2x^2) + (2x^2)^2] \\ &= 8(y - 2x^2)(y^2 + 2x^2y + 4x^4) \end{aligned}$$

Thus $8y^3 - 64x^6 = 8(y - 2x^2)(y^2 + 2x^2y + 4x^4)$. ∎

EXAMPLE 14 Factor $(x - 2)^3 + 64$.

Solution: Write $(x - 2)^3 + 64$ as a sum of two cubes, and then use the sum-of-two-cubes formula to factor.

$$\begin{aligned} (x - 2)^3 + (4)^3 &= [(x - 2) + 4][(x - 2)^2 - (x - 2)(4) + (4)^2] \\ &= (x - 2 + 4)(x^2 - 4x + 4 - 4x + 8 + 16) \\ &= (x + 2)(x^2 - 8x + 28) \end{aligned}$$

∎

COMMON STUDENT ERROR

Note that

$$(a + b)^2 = a^2 + 2ab + b^2$$
$$(a - b)^2 = a^2 - 2ab + b^2$$

Correct

$$a^3 + b^3 = (a + b)(a^2 - ab + b^2)$$
$$a^3 - b^3 = (a - b)(a^2 + ab + b^2)$$

not 2*ab*

Wrong

$$a^3 + b^3 = (a + b)(a^2 - 2ab + b^2)$$
$$a^3 - b^3 = (a - b)(a^2 + 2ab + b^2)$$

Exercise Set 6.3

Use the difference-of-two-squares formula or the perfect square trinomial formula to factor each expression. If the polynomial cannot be factored, so state.

1. $x^2 - 81$

2. $x^2 - 9$

3. $x^2 + 9$

4. $x^2 - 16$

5. $1 - 4x^2$

6. $1 - 9a^2$

7. $x^2 - 36y^2$

8. $25 - 16y^4$

9. $x^6 - 144y^4$

10. $81a^4 - 16b^2$

11. $x^6 - 4$

12. $x^2y^2 - 1$

13. $a^2b^2 - 49c^2$

14. $4a^2c^2 - 16x^2y^2$

15. $9x^2y^2 - 4x^2$

16. $25 - (x + y)^2$

17. $36 - (x - 6)^2$

18. $(2x + 3)^2 - 9$

19. $a^2 - (3b + 2)^2$

20. $a^2 + 2ab + b^2$

21. $x^2 + 10x + 25$

22. $25 - 10t + t^2$

23. $4 + 4x + x^2$

24. $y^2 - 8y + 16$

25. $4x^2 - 20xy + 25y^2$

26. $9y^2 + 6yz + z^2$

27. $9a^2 + 12a + 4$

28. $25a^2b^2 - 20ab + 4$

29. $w^4 + 16w^2 + 64$

30. $x^2 - 5xy + 25y^2$

31. $(x + y)^2 + 2(x + y) + 1$

32. $(x + 1)^2 + 6(x + 1) + 9$

33. $a^4 - 2a^2b^2 + b^4$

34. $(w - 3)^2 + 8(w - 3) + 16$

35. $x^2 + 6x + 9 - y^2$

36. $9 - (x^2 - 8x + 16)$

37. $25 - (x^2 + 4x + 4)$

38. $a^2 + 2ab + b^2 - 16c^2$

39. $9a^2 - 12ab + 4b^2 - 9$

40. $x^4 - 6x^2 + 9$

Factor using the sum- or difference-of-two-cubes formula.

41. $x^3 - 27$

42. $y^3 + 125$

43. $x^3 + y^3$

44. $a^3 - 125$

45. $x^3 - 8a^3$

46. $64 - a^3$

47. $y^3 + 1$

48. $w^3 - 216$

49. $27y^3 - 8x^3$

50. $5x^3 + 40y^3$

51. $24x^3 - 81y^3$

52. $y^6 + x^9$

53. $5x^3 - 625y^3$

54. $16y^6 - 250x^3$

55. $(x + 1)^3 + 1$

56. $(x - 3)^3 + 8$

57. $(x - y)^3 - 27$

58. $(2x + y)^3 - 64$

Factor using a special factoring formula.

59. $y^4 - 49x^2$

60. $a^4 - 4b^4$

61. $16y^2 - 81x^2$

62. $49 - 64x^2y^2$

63. $25x^4 - 81y^6$

64. $(x + y)^2 - 16$

65. $a^3 - 8$

66. $2a^2 - 24a + 72$

67. $x^3 - 64$

68. $9x^2y^2 + 24xy + 16$

69. $a^4 + 12a^2 + 36$

70. $27 - 8y^3$

71. $a^4 + 2a^2b^2 + b^4$

72. $8y^3 - 125x^6$

73. $x^2 - 2x + 1 - y^2$

74. $4r^2 + 4rs + s^2 - 9$

75. $(x + y)^3 + 1$

76. $9x^2 - 6xy + y^2 - 4$

77. Explain in your own words how to determine whether or not a trinomial is a perfect square trinomial.

78. Find two values of b that will make $4x^2 + bx + 9$ a perfect square trinomial. Explain how you determined your answer.

79. Find two values of c that will make $16x^2 + cx + 4$ a perfect square trinomial. Explain how you determined your answer.

80. Find the value of c that will make $25x^2 + 20x + c$ a perfect square trinomial. Explain how you determined your answer.

81. Find the value of d that will make $49x^2 - 42x + d$ a perfect square trinomial. Explain how you determined your answer.

82. Explain why a sum of two squares, $a^2 + b^2$, cannot be factored over the set of real numbers.

Cumulative Review Exercises

[1.2] **83.** Consider the set of elements

$$\{-2, \tfrac{5}{9}, -1.67, 0, \sqrt{3}, -\sqrt{6}, 3, 6\}.$$

List the elements that are:

(a) Counting numbers

(b) Rational numbers

(c) Irrational numbers

(d) Real numbers

Place the proper symbol, either $\in$ *or* $\subseteq$, *in the shaded area to make the statement true.*

84. $6 \blacksquare \{3, 4, 5, 6, 7\}$

85. $\{b\} \blacksquare \{a, b, c, d\}$

[2.2] **86.** Given the formula $z = \dfrac{p' - p}{\sqrt{\dfrac{pq}{n}}}$, find the value of

z when $p' = 0.4$, $p = 0.3$, $q = 0.7$, and $n = 4$.

[2.3] **87.** The length of a rectangular hallway is 2 feet greater than twice its width. Find the length and width of the hallway if its perimeter is 22 feet.

JUST FOR FUN

Factor each of the following.

1. $x^2 - 7$

2. $2x^2 - 15$

3. $(x - 8)^2 - (x - 5)^2$

4. $a^{2n} - 16a^n + 64$

6.4

A General Review of Factoring

▶ **1** Factor problems using a combination of factoring procedures.

▶ **1** We have presented a number of different factoring methods. Now we will combine problems and techniques from the previous sections.

A general procedure to factor any polynomial follows.

To Factor a Polynomial

1. Determine if the polynomial has a greatest common factor other than 1. If so, factor out the GCF from every term in the polynomial.
2. If the polynomial has two terms (or is a binomial), determine if it is a difference of two squares or a sum or difference of two cubes. If so, factor using the appropriate formula.
3. If the polynomial has three terms (or is a trinomial), determine if it is a perfect square trinomial. If so, factor accordingly. If it is not, then factor the trinomial using the method discussed in Section 6.2.
4. If the polynomial has more than three terms, try factoring by grouping. If that does not work, see if three of the terms are the square of a binomial.
5. As a final step, examine your factored polynomial to see if any factors listed have a common factor and can be factored further. If you find a common factor, factor it out at this point.

The following examples illustrate how to use the procedure.

EXAMPLE 1 Factor $3x^4 - 27x^2$.

Solution: First, determine if there is a greatest common factor other than 1. Since $3x^2$ is common to both terms, factor it out.

$$3x^4 - 27x^2 = 3x^2(x^2 - 9) = 3x^2(x + 3)(x - 3)$$

Note that $x^2 - 9$ is a difference of two squares.   ◼

EXAMPLE 2   Factor $3x^2y^2 - 24xy^2 + 48y^2$.

Solution: Begin by factoring the GCF, $3y^2$, from each term.

$$3x^2y^2 - 24xy^2 + 48y^2 = 3y^2(x^2 - 8x + 16) = 3y^2(x - 4)^2$$

Note that $x^2 - 8x + 16$ is a perfect square trinomial. If you did not recognize this, you would still obtain the correct answer by factoring the trinomial into $(x - 4)(x - 4)$.   ◼

EXAMPLE 3   Factor $24x^2 - 6xy + 16xy - 4y^2$.

Solution: Always begin by determining if the polynomial has a common factor. In this example the number 2 is common to all terms. Remove the common factor 2; then factor the remaining four-term polynomial by grouping.

$$\begin{aligned} 24x^2 - 6xy + 16xy - 4y^2 &= 2[12x^2 - 3xy + 8xy - 2y^2] \\ &= 2[3x(4x - y) + 2y(4x - y)] \\ &= 2(3x + 2y)(4x - y) \end{aligned}$$ ◼

EXAMPLE 4   Factor $10a^2b - 15ab + 20b$.

Solution: $$10a^2b - 15ab + 20b = 5b(2a^2 - 3a + 4).$$

Since $2a^2 - 3a + 4$ cannot be factored, we stop here.   ◼

EXAMPLE 5   Factor $2x^4y + 54xy$.

Solution: $$\begin{aligned} 2x^4y + 54xy &= 2xy(x^3 + 27) \\ &= 2xy(x + 3)(x^2 - 3x + 9). \end{aligned}$$

Note that $x^3 + 27$ is a sum of two cubes.   ◼

EXAMPLE 6   Factor $6x^2 - 3x + 6y^2 - 9$.

Solution: First, factor a 3 from all four terms.

$$6x^2 - 3x + 6y^2 - 9 = 3(2x^2 - x + 2y^2 - 3)$$

Now determine if the four terms within parentheses can be factored by grouping. Since these four terms cannot be factored by grouping, determine if three of the terms can be written as the square of a binomial. Since no matter how we rearrange the terms this cannot be done, we conclude that this expression cannot be factored further. Thus

$$6x^2 - 3x + 6y^2 - 9 = 3(2x^2 - x + 2y^2 - 3)$$ ◼

EXAMPLE 7   Factor $3x^2 - 18x + 27 - 3y^2$.

Solution: Factor a 3 from all four terms.

$$3x^2 - 18x + 27 - 3y^2 = 3(x^2 - 6x + 9 - y^2)$$

Now try factoring by grouping. Since the four terms within parentheses cannot be factored by grouping, determine if three of the terms can be written as the square of a binomial. Since this can be done, we express $x^2 - 6x + 9$ as $(x - 3)^2$ and then use the difference-of-two-squares formula. Thus

$$
\begin{aligned}
3x^2 - 18x + 27 - 3y^2 &= 3[(x - 3)^2 - y^2] \\
&= 3[(x - 3 + y)(x - 3 - y)] \\
&= 3(x - 3 + y)(x - 3 - y)
\end{aligned}
$$ ∎

Exercise Set 6.4

Factor each of the following completely.

1. $3x^2 + 3x - 36$
2. $2x^2 - 16x + 32$
3. $10s^2 + 19s - 15$
4. $6x^3y^2 + 10x^2y^3 + 8x^2y^2$
5. $8r^2 - 26r + 15$
6. $3x^3 - 12x^2 - 36x$
7. $2x^2 - 72$
8. $4x^2 - 4y^2$
9. $5x^5 - 45x$
10. $6x^2y^2z^2 - 24x^2y^2$
11. $3x^3 - 3x^2 - 12x^2 + 12x$
12. $2x^2y^2 + 6xy^2 - 10xy^2 - 30y^2$
13. $5x^4y^2 + 20x^3y^2 - 15x^3y^2 - 60x^2y^2$
14. $6x^2 - 15x - 9$
15. $x^4 - x^2y^2$
16. $4x^3 + 108$
17. $x^7y^2 - x^4y^2$
18. $x^4 - 16$
19. $x^5 - 16x$
20. $12x^2y^2 + 33xy^2 - 9y^2$
21. $4x^6 + 32y^3$
22. $12x^4 - 6x^3 - 6x^3 + 3x^2$
23. $2(a + b)^2 - 18$
24. $12x^3y^2 + 4x^2y^2 - 40xy^2$
25. $x^2 + 6xy + 9y^2$
26. $3x^2 - 30x + 75$
27. $(x + 2)^2 - 4$
28. $4y^4 - 36x^6$
29. $(2a + b)(2a - 3b) - (2a + b)(a - b)$
30. $pq + 6q + pr + 6r$
31. $(y + 3)^2 + 4(y + 3) + 4$
32. $b^4 + 2b^2 + 1$
33. $45a^4 - 30a^3 + 5a^2$
34. $(x + 1)^2 - (x + 1) - 6$
35. $x^3 + \dfrac{1}{27}$
36. $8y^3 - \dfrac{1}{8}$
37. $3x^3 + 2x^2 - 27x - 18$
38. $6y^3 + 14y^2 + 4y$
39. $a^3b - 16ab^3$
40. $x^6 + y^6$
41. $9 - (x^2 + 2xy + y^2)$
42. $x^2 - 2xy + y^2 - 25$
43. $24x^2 - 34x + 12$
44. $40x^2 + 52x - 12$
45. $7x^2 - 13x + 6$
46. $7(a - b)^2 + 4(a - b) - 3$
47. $x^4 - 81$
48. $(x + 2)^2 - 12(x + 2) + 36$
49. $5bc - 10cx - 6by + 12xy$
50. $16y^4 - 9y^2$
51. $3x^4 - x^2 - 4$
52. $x^2 + 16x + 64 - y^2$
53. $y^2 - (x^2 - 8x + 16)$
54. $4a^3 + 32$
55. $24ax + 18x + 36ay + 27y$
56. $2(y + 4)^2 + 5(y + 4) - 12$
57. $x^6 - 11x^3 + 30$
58. $a^2 + 12ab + 36b^2 - 16c^2$
59. $y - y^3$
60. $6x^4y + 15x^3y - 9x^2y$

61. $4x^2y^2 + 12xy + 9$ **62.** $x^4 - 2x^2y^2 + y^4$
63. $6r^2s^2 + rs - 1$ **64.** $4x^4 + 12x^2 + 9$

65. Explain the possible procedures that may be used to factor a polynomial of (a) two terms (b) three terms (c) four terms.

Cumulative Review Exercises

[2.1] *Solve the following equations.*   *Find the solution set for each inequality.*

   66. $4(x - 2) = 3(x - 4) - 4$ [2.5] **68.** $4(x - 3) < 6(x - 4)$.

   67. $-5(x - 2) + 3 = -5x - 6$ [2.6] **69.** $|2x - 3| > -4$.

6.5

Solving Equations Using Factoring

▸ **1** Know the standard form of a quadratic equation.

▸ **2** Use the zero-factor property to solve equations.

▸ **3** Use factoring to solve equations.

▸ **4** Use factoring to solve application problems.

▸ **1** Quadratic equations in two variables, x and y, were introduced and graphed in Section 5.8. In this section we explain how to solve quadratic equations in one variable using factoring. Every quadratic equation has a second-degree term as its highest term.

Examples of Quadratic Equations

$$3x^2 + 6x - 4 = 0$$
$$5x = 2x^2 - 4$$
$$(x + 4)(x - 3) = 0$$

Any quadratic equation can be written in standard form.

Standard Form of a Quadratic Equation

$$ax^2 + bx + c = 0, \qquad a \neq 0$$

where a, b, and c are real numbers.

Before going any further, convert each of the three quadratic equations given above to standard form, with $a > 0$.

▸ **2**   To solve equations using factoring, we make use of the **zero-factor property**.

Zero-factor Property

For all real numbers a and b, if $a \cdot b = 0$, then either $a = 0$ or $b = 0$, or both a and $b = 0$.

The zero-factor property indicates that, if the product of two factors equals zero, one (or both) of the factors must be zero.

EXAMPLE 1 Solve the equation $(x + 5)(x - 2) = 0$.

Solution: Since the product of the factors equals 0, according to the rule above, one or both factors must equal zero. Set each factor equal to 0 and solve each equation.

$$x + 5 = 0 \qquad \text{or} \qquad x - 2 = 0$$
$$x = -5 \qquad\qquad\qquad x = 2$$

Thus, if x is either -5 or 2, the product of the factors is 0.

Check:
$$x = -5 \qquad\qquad\qquad x = 2$$
$$(x + 5)(x - 2) = 0 \qquad (x + 5)(x - 2) = 0$$
$$(-5 + 5)(-5 - 2) = 0 \qquad (2 + 5)(2 - 2) = 0$$
$$0(-7) = 0 \qquad\qquad 7(0) = 0$$
$$0 = 0 \quad \text{true} \qquad\qquad 0 = 0 \quad \text{true} \qquad ■$$

▶ **3**

To Solve an Equation Using Factoring

1. Use the addition property to remove all terms from one side of the equation. This will result in one side of the equation being equal to 0.
2. Combine like terms in the equation and then factor.
3. Set each factor containing a variable equal to zero, solve the equations, and find the solutions.
4. Check the solutions in the original equation.

EXAMPLE 2 Solve the equation $2x^2 = 12x$.

Solution: First, make the right side of the equation equal to 0 by subtracting $12x$ from both sides of the equation. Then factor the left side of the equation.

$$2x^2 - 12x = 0$$
$$2x(x - 6) = 0$$

Now set each factor equal to zero.

$$2x = 0 \qquad \text{or} \qquad x - 6 = 0$$
$$x = 0 \qquad\qquad\qquad x = 6$$

The numbers 0 and 6 both satisfy the equation $2x^2 = 12x$. ■

COMMON STUDENT ERROR

The zero-factor property can be used only when one side of the equation is equal to 0.

Correct	*Wrong*
$(x - 4)(x + 3) = 0$	~~$(x - 4)(x + 3) = 2$~~
$x - 4 = 0$ or $x + 3 = 0$	~~$x - 4 = 2$~~ or ~~$x + 3 = 2$~~

Note that in the wrong process illustrated on the right the zero-factor property cannot be used since the right side of the equation is not equal to 0. Example 3 shows how to solve such problems correctly.

EXAMPLE 3   Solve the equation $(x - 1)(3x + 2) = 4x$.

Solution: Since the right side of the equation is not 0, we cannot use the zero-factor property. Begin by multiplying the factors on the left side of the equation. Then subtract $4x$ from both sides of the equation to obtain a 0 on the right side.

$$(x - 1)(3x + 2) = 4x$$
$$3x^2 - x - 2 = 4x$$
$$3x^2 - 5x - 2 = 0$$
$$(3x + 1)(x - 2) = 0$$

$$3x + 1 = 0 \qquad \text{or} \qquad x - 2 = 0$$
$$3x = -1 \qquad\qquad\qquad x = 2$$
$$x = -\frac{1}{3}$$

The solutions are $-\frac{1}{3}$ and 2. ∎

EXAMPLE 4   Solve the equation $3x^2 + 2x - 12 = -7x$.

Solution:

$$3x^2 + 2x - 12 = -7x \qquad\qquad \text{Add } 7x \text{ to both sides of equation.}$$
$$3x^2 + 9x - 12 = 0$$
$$3(x^2 + 3x - 4) = 0 \qquad\qquad \text{Factor out the 3.}$$
$$3(x + 4)(x - 1) = 0 \qquad\qquad \text{Factor the trinomial.}$$
$$x + 4 = 0 \qquad \text{or} \qquad x - 1 = 0 \qquad \text{Solve.}$$
$$x = -4 \qquad \text{or} \qquad x = 1$$

Since the 3 that was factored out is an expression not containing a variable, we do not have to set it equal to zero. Only the numbers -4 and 1 satisfy the equation $3x^2 + 2x - 12 = -7x$. ∎

EXAMPLE 5   Solve the equation $2x(x + 2) = x(x - 3) - 12$.

Solution:

$$2x(x + 2) = x(x - 3) - 12$$
$$2x^2 + 4x = x^2 - 3x - 12$$
$$x^2 + 7x + 12 = 0$$
$$(x + 4)(x + 3) = 0$$

$$x + 4 = -0 \qquad \text{or} \qquad x + 3 = 0$$
$$x = -4 \qquad \text{or} \qquad x = -3$$

∎

HELPFUL HINT	When solving an equation whose highest powered term has a *negative coefficient*, we generally make it positive by multiplying both sides of the equation by -1. This makes the factoring process easier.

For example, to solve the equation

$$-x^2 + 5x + 6 = 0$$
$$-1(-x^2 + 5x + 6) = -1 \cdot 0$$
$$x^2 - 5x - 6 = 0$$

Now solve the equation $x^2 - 5x - 6 = 0$ to obtain the solution.

$$(x - 6)(x + 1) = 0$$

$$x - 6 = 0 \qquad \text{or} \qquad x + 1 = 0$$
$$x = 6 \qquad\qquad\qquad x = -1$$

The numbers 6 and -1 both satisfy the original equation, $-x^2 + 5x + 6 = 0$.

The equations in Examples 1 through 5 were all quadratic equations that were placed in the form $ax^2 + bx + c = 0$ and solved by factoring. Other methods that can be used to solve quadratic equations of the above form include completing the square and the quadratic formula; we discuss these methods in Chapter 9.

The zero-factor property can be extended to three or more factors as illustrated in Example 6.

EXAMPLE 6   Solve the equation $2x^3 + 5x^2 - 3x = 0$.

Solution: First factor, then set each factor containing an x equal to 0.

$$2x^3 + 5x^2 - 3x = 0$$
$$x(2x^2 + 5x - 3) = 0$$
$$x(2x - 1)(x + 3) = 0$$

$$x = 0 \qquad \text{or} \qquad 2x - 1 = 0 \qquad \text{or} \qquad x + 3 = 0$$
$$2x = 1 \qquad\qquad\qquad x = -3$$
$$x = \frac{1}{2}$$

The numbers 0, $\frac{1}{2}$, and -3 all satisfy the given equation and are solutions to the equation. ∎

Note that the equation in Example 6 is not a quadratic equation because its highest-degree term is 3, not 2. This is an example of a *cubic*, or *third-degree, equation*.

▶**4**   Now let us look at some applications problems that use factoring in their solution.

EXAMPLE 7   The area of a triangle is 27 square inches. Find the base and height if its height is 3 inches less than twice its base.

Solution: Let x = base; then $2x - 3$ = height (Fig. 6.1).

FIGURE 6.1

$$\text{Area} = \frac{1}{2}(\text{base})(\text{height})$$

$$27 = \frac{1}{2}(x)(2x - 3)$$ Multiply both sides of the equation by 2 to remove fractions.

$$2(27) = 2\left[\frac{1}{2}(x)(2x - 3)\right]$$

$$54 = x(2x - 3)$$

$$54 = 2x^2 - 3x$$

or $2x^2 - 3x - 54 = 0$

$$(2x + 9)(x - 6) = 0$$

$2x + 9 = 0$ or $x - 6 = 0$

$2x = -9$ $x = 6$

$$x = -\frac{9}{2}$$

Since the dimensions of a geometric figure cannot be negative, we can eliminate $x = -\frac{9}{2}$ as an answer to our problem.

$$\text{Base} = x = 6 \text{ inches}$$

$$\text{Height} = 2x - 3 = 2(6) - 3 = 9 \text{ inches}$$ ■

EXAMPLE 8   A projectile on top of a building 384 feet high is fired upward with a velocity of 32 feet per second. The projectile's distance, s, above the ground at any time, t, is given by the formula $s = -16t^2 + 32t + 384$. Find the time that it takes for the object to strike the ground.

Solution   When the object strikes the ground, its distance from the ground is 0. Substituting $s = 0$ into the equation gives

$$0 = -16t^2 + 32t + 384$$

or $-16t^2 + 32t + 384 = 0$

$$-16(t^2 - 2t - 24) = 0$$

$$-16(t + 4)(t - 6) = 0$$

$t + 4 = 0$ or $t - 6 = 0$

$t = -4$ $t = 6$

Since it makes no sense to speak about a negative time, -4 is not a possible answer. The object will strike the ground in 6 seconds.   ■

Exercise Set 6.5

Solve each equation.

1. $x(x + 5) = 0$

2. $3x(x - 5) = 0$

3. $5x(x + 9) = 0$

4. $2(x + 3)(x - 5) = 0$

5. $(2x + 5)(x - 3)(3x + 6) = 0$

6. $x(2x + 3)(x - 5) = 0$

7. $4x - 12 = 0$

8. $9x - 27 = 0$

9. $-x^2 + 12x = 0$

10. $x^2 + 4x = 0$

11. $9x^2 = -18x$

12. $x^2 + 6x + 5 = 0$

13. $x^2 + x - 12 = 0$

14. $x(x + 6) = -9$

15. $x(x - 12) = -20$

16. $3y^2 - 2 = -y$

17. $-z^2 - 3z = -18$

18. $3x^2 = -21x - 18$

19. $3x^2 - 6x - 72 = 0$

20. $x^3 = 3x^2 + 18x$

21. $x^3 + 19x^2 = 42x$

22. $3x^2 - 9x - 30 = 0$

23. $2y^2 + 22y + 60 = 0$

24. $8x^2 + 14x - 15 = 0$

25. $-16x - 3 = -12x^2$

26. $-7x - 10 = -6x^2$

27. $-28x^2 + 15x - 2 = 0$

28. $-2y^2 + 24y - 22 = 0$

29. $3x^3 - 8x^2 - 3x = 0$

30. $z^3 + 16z^2 = -64z$

31. $3p^2 = 22p - 7$

32. $5w^2 - 16w = -3$

33. $3r^2 + r = 2$

34. $3x^2 = 7x + 20$

35. $4x^3 + 4x^2 - 48x = 0$

36. $x^2 - 25 = 0$

37. $6x^2 = 16x$

38. $4x^2 = 9$

39. $25x^3 - 16x = 0$

40. $2x^4 - 32x^2 = 0$

41. $(x + 4)^2 - 16 = 0$

42. $(2x + 5)^2 - 9 = 0$

43. $(x - 7)(x + 5) = -20$

44. $(x + 1)^2 = 3x + 7$

45. $6a^2 - 12 - 4a = 19a - 32$

46. $(x - 4)^2 - 4 = 0$

47. $(b - 1)(3b + 2) = 4b$

48. $2(a^2 - 3) - 3a = 2(a + 3)$

49. $2(x + 2)(x - 2) = (x - 2)(x + 3) - 2$

50. $2(a + 3)(a - 5) = 2(a - 1) + 8$

51. $2x^3 + 16x^2 + 30x = 0$

52. $18x^3 - 15x^2 = 12x$

For each exercise, write the problem as an equation. Solve the equation and answer the question.

53. The product of two consecutive positive integers is 72. Find the two integers.

54. The product of two consecutive positive even integers is 80. Find the two integers.

55. The product of two consecutive positive odd integers is 99. Find the two integers.

56. The product of two positive numbers is 108. Find the two numbers if one is 3 more than the other.

57. The product of two positive numbers is 35. Find the two numbers if one number is 3 less than twice the other number.

58. The product of two positive integers is 36. Find the two integers if one number is 4 times the other.

59. The area of a rectangle is 36 square feet. Find the length and width if the length is 4 times the width.

60. The area of a rectangle is 54 square inches. Find the length and width if the length is 3 inches less than twice the width.

61. The base of a triangle is 6 centimeters greater than its height. Find the base and height if the area is 80 square centimeters.

62. The height of a triangle is 1 centimeter less than twice its base. Find the base and height if the triangle's area is 33 square centimeters.

63. If the sides of a square are increased by 4 meters, the area becomes 121 square meters. Find the length of a side of the original square.

64. A model rocket is to be launched from a hill 80 feet above sea level. The launch site is next to the ocean (sea level), and the rocket will fall into the ocean. The rocket's distance, s, above sea level at any time, t, is found by the equation $s = -16t^2 + 64t + 80$. Find the time it takes for the rocket to strike the ocean.

65. A rubber ball is at the top of a 96-foot waterfall. The ball's distance from the pool of water at the bottom of the falls t seconds after it goes over the falls can be found by using the formula $d = -16t^2 + 96$. Find the ball's distance above the pool after (a) 1 second, (b) 1.5 seconds. (c) When will the ball hit the pool of water at the bottom of the falls?

See Exercise 65.

66. An egg is dropped from a helicopter that is 256 feet above the ground. The distance of the egg from the ground at any time t is given by the equation $s = -16t^2 + 256$. Find the time it takes for the egg to strike the ground.

67. A television is thrown downward from the top of a 640-foot-tall building with a velocity of 48 feet per second. The television's distance, s, from the ground at any time, t, is found by the equation $s = -16t^2 - 48t + 640$. Find the time it takes for the television to strike the ground.

68. (a) Explain in your own words how to solve an equation using factoring.

  (b) Solve the equation $-x - 20 = -12x^2$ using the procedure in part (a).

69. (a) Explain why the equation $(x + 3)(x + 4) = 2$ *cannot* be solved by writing $x + 3 = 2$ or $x + 4 = 2$.

  (b) Solve the equation $(x + 3)(x + 4) = 2$.

Cumulative Review Exercises

[2.3] **70.** Two distance runners, Carmen and Bob, run the same course with Bob running 1.2 miles per hours faster than Carmen. If Bob finishes in 4 hours and Carmen finishes in 5 hours, (a) what is the rate of each and (b) how long is the course?

[4.1] **71.** Solve the system of equations.

$$3x + 5y = 9$$
$$2x - y = 6$$

[4.5] **72.** Find the solution to the system of inequalities.

$$2y > 6x + 12$$
$$\tfrac{1}{2}y < \tfrac{3}{2}x + 2$$

[5.5] **73.** Divide $(6x^2 - x - 12) \div (2x - 3)$.

JUST FOR FUN

Write an equation that has the following values as its solution.

1. $2, -5$

2. $\tfrac{1}{2}, 3$

3. $-1, 0, 3$

Solve the following equations.

4. $x^6 - 7x^3 - 60 = 0$

5. $(x + 3)^2 + 2(x + 3) = 24$

6. $x^4 - 5x^2 + 4 = 0$

6.6

Using Factoring to Solve for a Variable in a Formula or Equation

▶ **1** Solve for a variable in a formula or equation.

▶ **1** In Section 2.2 we discussed solving for a variable in a formula, and in Section 6.1 we discussed factoring a monomial from a polynomial. Now we combine the information from both of these sections. To solve for a variable in a formula or equation, it is sometimes necessary to factor out the term we are solving for from all the terms that contain the variable. This procedure is illustrated in the following examples.

EXAMPLE 1 Solve the equation $x(y - 3) = 2y + 4$ for y.

Solution: To solve this equation for y, we need to isolate the terms containing y. We begin by using the distributive property.

$$x(y - 3) = 2y + 4$$
$$xy - 3x = 2y + 4$$

Now collect all terms containing the variable we are solving for, y, on one side of the equation and all terms not containing y on the other side of the equation. We will choose to collect the y terms on the left side of the equation and the terms not containing a y on the right side of the equation.

$xy \;\boxed{- 2y}\; - 3x = 2y \;\boxed{- 2y}\; + 4$	Subtract $2y$ from both sides of the equation.
$xy - 2y - 3x = 4$	
$xy - 2y - 3x \;\boxed{+ 3x}\; = \boxed{3x} + 4$	Add $3x$ to both sides of the equation.
$xy - 2y = 3x + 4$	Terms containing y are on one side of equation.

Now factor out the variable we are solving for, y, from all terms that contain the y, and then solve for y.

$y(x - 2) = 3x + 4$	Factor out y.
$\dfrac{y(x - 2)}{x - 2} = \dfrac{3x + 4}{x - 2}$	Divide both sides of the equation by $(x - 2)$.
$y = \dfrac{3x + 4}{x - 2}$	■

To Solve for a Variable in a Formula or Equation

1. If the formula contains a fraction or fractions, multiply all terms by the least common denominator to remove all fractions.
2. Use the distributive property, if necessary, to remove parentheses.
3. Collect all terms containing the variable that you are solving for on one side of the equation and all terms not containing the variable on the other side of the equation.
4. If there is more than one term containing the variable, factor out the variable you are solving for from each term containing the variable.
5. Isolate the variable you are solving for by dividing both sides of the equation by the factor the variable you are solving for is multiplied by.

We will give some examples of this procedure in this section. Additional examples will be given in Section 7.5.

EXAMPLE 2   A formula used when studying levers in physics is $d = \dfrac{fl}{f + w}$. Solve this formula for f.

Solution: We will illustrate the solution to this problem following the step-by-step procedure just given. We choose to collect the terms containing the variable f on the right side of the equation.

$$d = \frac{fl}{f + w}$$

Step (1): Multiply by the lcd, $f + w$.

$$d(f + w) = \frac{fl}{(f + w)}(f + w)$$

Step (2): Use the distributive property.

$$d(f + w) = fl$$

$$df + dw = fl$$

Step (3): Collect terms containing f on one side of equation, and terms not containing f on the other side of equation.

$$df - df + dw = fl - df$$

$$dw = fl - df$$

Step (4): Factor out the f from each term containing an f.

$$dw = f(l - d)$$

Step (5): Isolate the f by dividing both sides of the equation by $l - d$.

$$\frac{dw}{l - d} = \frac{f(l - d)}{l - d}$$

Thus $$f = \frac{dw}{l - d}$$ ∎

EXAMPLE 3   Solve the equation $2xy + 3yz = 6xz + 3$ for z.

Solution: First, we collect all terms containing the variable we are solving for, z, on one side of the equation and all terms not containing the variable z on the other side of the equation. The terms containing a z are $3yz$ and $6xz$. We will select to collect these terms on the left side of the equation and the terms that do not contain the z on the right side of the equation.

$$2xy + 3yz = 6xz + 3$$
$$2xy + 3yz - 6xz = 6xz - 6xz + 3$$
$$2xy + 3yz - 6xz = 3$$
$$2xy - 2xy + 3yz - 6xz = 3 - 2xy$$
$$3yz - 6xz = 3 - 2xy$$

The terms containing a z are now isolated on the left side of the equation. Factor out z, the variable we are solving for, from each term on the left side of the equation.

$$z(3y - 6x) = 3 - 2xy$$

Now solve for z by dividing both sides of the equation by $(3y - 6x)$.

$$\frac{z(3y - 6x)}{3y - 6x} = \frac{3 - 2xy}{3y - 6x}$$

$$z = \frac{3 - 2xy}{3y - 6x}$$ ∎

Consider the equation

$$\tfrac{1}{2}xy + \tfrac{3}{4}yz = \tfrac{3}{2}xz + \tfrac{3}{4}$$

How would you begin the solution? If you said multiply both sides of the equation by the least common denominator 4 to remove fractions, you answered correctly. After you multiply both sides of the equation by 4, you will end up with the equation given in Example 3. Try this now and see.

Exercise Set 6.6

Solve each equation for the indicated variable.

1. $3x + 2y = xy + 4$, for y

2. $x - 3xy = 6y$, for y

3. $2(x + y) = x(3 + y)$, for y

4. $-x(y + 5) = -3(x + y)$, for y

5. $y = \dfrac{x}{x - 1}$, for x

6. $y = \dfrac{x - 2}{x + 4}$, for x

7. $y = \dfrac{4 - x}{x - 2}$, for x

8. $x = \dfrac{y + 6}{y}$, for y

9. $3yz + 2 = 5x + z$, for z

10. $2(x + y) = 2 - 3xy$, for y

11. $2xyz + 3yz = -6xy$, for z

12. $\dfrac{1}{2}x + \dfrac{3}{4}xy = \dfrac{1}{3}(2x - y)$, for y

13. $\dfrac{1}{3}xy - 6y = 2y + 3$, for y

14. $-2(2xyz + 3) = x(2z + 3y)$, for z

15. $3rs - 2s = \dfrac{1}{2}(s + 2r)$, for r

16. $-rs + 2s = 4(r + rs)$, for r

17. $\dfrac{2}{3}x + ax = 2(x + 4) + 3$, for x

18. $\dfrac{1}{5}xy + 3y = -6x + 3$, for y

19. $\dfrac{3}{5}x - 2y = 6xy + \dfrac{4}{3}$, for y

20. $\dfrac{2}{3}(3xy - 6) = \dfrac{1}{2}(x + y)$, for y

Solve each formula for the indicated variable.

21. $I = P + Prt$, for P (banking)

22. $J_0 = I\omega - I\omega_0$ for I (physics)

23. $x_2 - x_1 = \dfrac{y_2 - y_1}{m}$, for m (mathematics)

24. $at_2 - at_1 + v_1 = v_2$, for a (physics)

25. $a_n = a_1 + nd - d$, for d (mathematics)

26. $R_1 + R_2 = \dfrac{R_1 R_2}{R_T}$ for R_T (electronics)

27. $Vr - R = O - Dr$ for r (economics)

28. $2P_1 - 2P_2 - P_1 P_c = P_2 P_c$ for P_c (economics)

29. $S_n - S_n r = a_1 - a_1 r^n$ for S_n (mathematics)

30. $2s - nf - nl = 0$ for n (mathematics)

31. $S_n - S_n r = a_1 - a_1 r^n$ for a_1 (mathematics)

32. $xm_1 + xm_2 + xm_3 = m_1 x_1 + m_2 x_2 + x_3 m_3$ for x (statistics)

33. $e = \dfrac{q_H + q_C}{q_H}$, for q_H (chemistry, physics)

34. $xm_1 + xm_2 + xm_3 = m_1 x_1 + m_2 x_2 + m_3 x_3$ for m_1 (statistics)

✎ **35.** Consider the equation $2xy - 3x = 4 - 5y$. If we solve this equation for y we get $y = \dfrac{3x + 4}{2x + 5}$. Now consider the equation $2ab - 3a = 4 - 5b$. Notice the similarity between this equation and the first equation. Without actually solving the equation, if this equation is solved for b, can you determine the answer? Give the solution and explain how you determined your answer.

Cumulative Review Exercises

[5.8] **36.** Graph the function
$$f(x) = x^3 - 2x - 4$$

Factor.

[6.1] **37.** $(4x - 3)(x + 2) - (x - 7)(x + 2)$.

[6.2] **38.** $6x^2 - 5x - 21$.

39. $2(3x - 2)^2 - 11(3x - 2) - 21$.

JUST FOR FUN

In more advanced mathematics courses you may need to solve an equation for y' (read y prime). When doing so, treat the y' as a different variable from y. Solve each of the following for y'.

1. $xy' + yy' = 1$

2. $xy - xy' = 3y' + 2$

3. $2xyy' - xy = x - 3y'$

SUMMARY

GLOSSARY

Factoring *(259):* The opposite of the multiplication process.

Greatest common factor *(259):* The greatest common factor of two or more expressions is the greatest factor that divides each expression.

Standard form of a quadratic equation *(284):* $ax^2 + bx + c = 0$, $a \neq 0$.

Zero-factor property *(285):* If $a \cdot b = 0$, then either $a = 0$ or $b = 0$ or both a and $b = 0$.

IMPORTANT FACTS

Special Factoring Formulas

$$a^2 - b^2 = (a + b)(a - b) \qquad \text{difference of two squares}$$

$$\left.\begin{array}{l} a^2 + 2ab + b^2 = (a + b)^2 \\ a^2 - 2ab + b^2 = (a - b)^2 \end{array}\right\} \qquad \text{perfect square trinomials}$$

$$a^3 + b^3 = (a + b)(a^2 - ab + b^2) \qquad \text{sum of two cubes}$$

$$a^3 - b^3 = (a - b)(a^2 + ab + b^2) \qquad \text{difference of two cubes}$$

Note: The sum of two squares $a^2 + b^2$ cannot be factored over the set of real numbers.

Review Exercises

[6.1] *Find the greatest common factor for each set of terms.*

1. $40x^2$, $36x^3$, $16x^5$

2. $12xy$, $36xy^2$, $18x^2y$

3. $15x^3y^2z^5$, $-6x^2y^3$, $30xy^4z$

4. $x(2x - 5)$, $3(2x - 5)^2$, $5(2x - 5)^3$

5. $x(x + 5)$, $x + 5$, $2(x + 5)^2$

6. $2x$, $(x - 2)$, $(x - 2)^2$

Factor. If an expression cannot be factored, so state.

7. $12x^2 + 4x - 8$

8. $60x^4 + 6x^9 - 18x^5y^2$

9. $24x^6 - 13y^5 + 6z$

10. $x(5x + 3) - 2(5x + 3)$

11. $2x(4x - 3) + 4x - 3$

12. $3x(x - 1)^2 - 2(x - 1)$

13. $4x(2x - 1) + 3(2x - 1)^2$

14. $12xy^4z^3 + 6x^2y^3z^2 - 15x^3y^2z^3$

Factor by grouping.

15. $x^2 + 3x + 2x + 6$

16. $x^2 - 5x + 3x - 15$

17. $x^2 - 7x + 7x - 49$

18. $3x^2 + x + 9x + 3$

19. $5x^2 + 20x - x - 4$

20. $5x^2 - xy + 20xy - 4y^2$

21. $12x^2 - 8xy + 15xy - 10y^2$

22. $(3x - y)(x + 2y) - (3x - y)(5x - 7y)$

23. $3a^4 - 12a^2b + 9a^2b - 36b^2$

[6.2] *Factor each trinomial.*

24. $x^2 + 8x + 15$

25. $x^2 - 8x + 15$

26. $x^2 - 6x - 27$

27. $x^2 - 12x - 45$

28. $x^2 - 5xy - 50y^2$

29. $x^2 - 15xy - 54y^2$

30. $2x^2 + 16x + 32$

31. $3x^2 - 18x + 27$

32. $x^3 - 3x^2 - 18x$

33. $8x^3 + 10x^2 - 25x$

34. $3x^2 + 13x + 4$

35. $4x^2 + 11xy - 3y^2$

36. $4x^3 - 9x^2 + 5x$

37. $12x^3 + 61x^2 + 5x$

38. $x^4 - 3x^2 - 10$

39. $x^4 - x^2 - 20$

40. $(x + 5)^2 + 10(x + 5) + 24$

41. $3(x + 2)^2 - 16(x + 2) - 12$

[6.3] *Use a special factoring formula to factor.*

42. $x^2 - 36$

43. $4x^2 - 16y^4$

44. $x^4 - 81$

45. $(x + 2)^2 - 9$

46. $(x - 3)^2 - 4$

47. $4x^2 - 12x + 9$

48. $9y^2 + 24y + 16$

49. $w^4 - 16w^2 + 64$

50. $a^2 + 6ab + 9b^2 - 4c^2$

51. $x^3 - 8$

52. $8x^3 + 27$

53. $27x^3 - 8y^3$

54. $27x^3 - 8y^6$

55. $8y^6 - 125x^3$

56. $(x + 1)^3 - 8$

[6.1–6.4] *Factor completely.*

57. $x^2y^2 - 2xy^2 - 15y^2$

58. $3x^3 - 18x^2 + 24x$

59. $3x^3y^4 + 18x^2y^4 - 6x^2y^4 - 36xy^4$

60. $3y^5 - 27y$

61. $2x^3y + 16y$

62. $5x^4y + 20x^3y + 20x^2y$

63. $6x^3 - 21x^2 - 12x$

64. $x^2 + 10x + 25 - y^2$

65. $3x^3 + 24y^3$

66. $x^2(x + 4) + 3x(x + 4) - 4(x + 4)$

67. $4(2x + 3)^2 - 12(2x + 3) + 5$

68. $4x^4 + 4x^2 - 3$

69. $(x - 1)x^2 - (x - 1)x - 2(x - 1)$

70. $30x^2 + 3x - 6$

71. $9ax - 3bx + 12ay - 4by$

72. $6p^2q^2 - 5pq - 6$

73. $9x^4 - 12x^2 + 4$

74. $4y^2 - (x^2 + 4x + 4)$

75. $6(2a + 3)^2 - 7(2a + 3) - 3$

76. $6x^4y^4 + 9x^3y^4 - 27x^2y^4$

77. $x^3 - \frac{8}{27}y^6$

[6.5] *Solve the equation.*

78. $(x - 5)(3x + 2) = 0$

79. $2x^2 = 3x$

80. $15x^2 + 20x = 0$

81. $x^2 - 2x - 24 = 0$

82. $x^2 + 8x + 15 = 0$

83. $x^2 = -2x + 8$

84. $3x^2 + 21x + 30 = 0$

85. $x^3 - 6x^2 + 8x = 0$

86. $12x^3 - 13x^2 - 4x = 0$

87. $8x^2 - 3 = -10x$

88. $4x^2 = 16$

89. $x(x + 3) = 2(x + 4) - 2$

Write the problem as an equation. Solve the equation and answer the question.

90. The product of two positive integers is 30. Find the integers if the larger is 4 less than twice the smaller.

91. The area of a rectangle is 63 square feet. Find the length and width of the rectangle if the length is 2 feet greater than the width.

92. The base of a triangle is 3 more than twice the height. Find the base and height if the area of the triangle is 22 square meters.

93. One square has a side 4 inches longer than the side of a second square. If the area of the larger square is 81 square inches, find the length of a side of each square.

94. A rocket is projected upward from the top of a 144-foot-tall building with a velocity of 128 feet per second. The rocket's distance from the ground, s, at any time, t, is given by the formula $s = -16t^2 + 128t + 144$. Find the time it takes for the rocket to strike the ground.

[6.6] *Solve each equation or formula for the indicated variable.*

95. $3x + 4y = xy + 6$, for y

96. $2xz - 3z = 6x$, for z

97. $2(3xyz - x) = x - 2z$, for z

98. $\frac{3}{5}(x - y) = \frac{1}{4}(xy - 2)$, for y

99. $x = \frac{y + 2}{y - 3}$, for y

100. $R = \frac{2l}{l + w}$, for l

101. $w - 6\beta = 3\alpha\beta + 2w$, for β

102. $\frac{3}{2}p = \frac{x_2 w_1 - x_1 w_1}{5}$, for w_1

Practice Test

Factor completely.

1. $4x^2y - 4x$

2. $3x^2 + 12x + 2x + 8$

3. $9x^3y^2 + 12x^2y^5 - 27xy^4$

4. $5(x - 2)^2 + 15(x - 2)$

5. $2x^2 + 4xy + 3xy + 6y^2$

6. $x^2 - 7xy + 12y^2$

7. $3x^3 - 6x^2 - 9x$

8. $6x^2 - 7x + 2$

9. $5x^2 + 17x + 6$

10. $81x^2 - 16y^4$

11. $27x^3y^6 - 8y^6$

12. $(x + 3)^2 + 2(x + 3) - 3$

13. $2x^4 + 5x^2 - 18$

Solve the equation.

14. $2(x - 5)(3x + 2) = 0$

15. $4x^2 - 18 = 21x$

16. $x^3 + 4x^2 - 5x = 0$

Solve for the indicated variable.

17. $3(x + 2y) = x(y - 5)$, for y

18. $\frac{2}{3}(w + p) = \frac{3}{4}(wp - 3)$, for p

19. The area of a triangle is 28 square meters. If the base of the triangle is 2 meters greater than 3 times the height, find the base and height of the triangle.

20. A baseball is projected upward from the top of a 448-foot-tall building with an initial velocity of 48 feet per second. The distance of the baseball, s, from the ground at any time, t, is given by the equation $s = -16t^2 + 48t + 448$. Find the time after which the baseball strikes the ground.

See Exercise 19.

Cumulative Review Test

1. Evaluate $\dfrac{\sqrt[3]{27} - \sqrt[3]{-8} + |-4|}{3^0 - 12 \div 3 \div 4 - 8}$.

2. Solve the equations $\frac{1}{3}(x - 6) = \frac{3}{4}(2x - 1)$.

3. Solve the formula $3P = \dfrac{2L - W}{4}$ for L.

4. Graph $4x - 3y = 9$.

5. Graph the inequality $2x - y \le 6$.

6. Indicate whether the following sets of ordered pairs are functions. Explain your answer.

  (a) $\{(0, 1), (3, -2), (-2, 6), (5, 6)\}$

  (b) $\{(1, 2), (3, 4), (5, 6), (1, 0)\}$

7. Solve the system of equations

$$3x - 2y = 8$$
$$2x - 5y = 10.$$

Simplify each of the following.

8. $\left(\dfrac{8x^{-2}y^3}{4xy^{-1}}\right)\left(\dfrac{2xy^5}{x^{-3}}\right)$

9. $\dfrac{(2p^4q^3)(3pq^4)^3}{(4p^{-2}q^3)^2}$

10. Simplify $3x^2 - 4x - 6 - (5x - 4x^2 - 6)$. Write the answer in descending powers of the variable.

11. Multiply $(x^2 - 3x - 6)(2x - 5)$.

12. Divide $\dfrac{9x^3y^5 - 8x^2y^4 - 12xy}{3x^2y}$.

13. If $f(x) = 3x^3 - 6x^2 - 4x + 3$, find $f(2)$.

14. Graph $f(x) = x^2 - 6x + 8$.

Factor.

15. $x^4 - 3x^3 + 2x^2 - 6x$

16. $12x^2y - 27xy + 6y$

17. $y^4 + 2y^2 - 24$

18. $8x^3 - 27y^6$

19. Kinko Copies charges 15 cents a page for making a master copy from material that must be hand fed into the copier. After the master copy is made, they can make additional copies from the master copy for 5 cents a page. John has a manuscript to be copied, but since the pages must be hand fed into the copier, John has a master copy made and six additional copies of the manuscript made from the master copy. If his total bill before tax is $279, how many pages are in the manuscript?

20. Santo's first four test grades are 68, 72, 90, and 86. What range of grades on his fifth test will result in an average greater than or equal to 70 and less than 80?

CHAPTER

7

Rational Expressions and Equations

See Section 7.6, Exercise 14.

Reducing Rational Expressions

▶ **1**   Find the domain of a rational expression.

▶ **2**   Reduce a rational expression to its lowest terms.

To successfully understand rational expressions, you must have a thorough understanding of the factoring techniques discussed in Chapter 6.

▶ **1**   A **rational expression** (also called an **algebraic fraction**) is an algebraic expression of the form p/q, where p and q are polynomials and $q \neq 0$. Examples of rational expressions are

$$\frac{2}{3}, \quad \frac{x + 3}{x}, \quad \frac{x^2 + 4x}{x - 3}, \quad \frac{x}{x^2 - 4}$$

Note that the denominator of an algebraic expression cannot equal 0 since division by 0 is not permitted. In the expression $(x + 3)/x$, x cannot have a value of 0, since the denominator would then equal 0. In $(x^2 + 4x)/(x - 3)$, x cannot have a value of 3 for that would result in the denominator having a value of 0. What values of x cannot be used in the expression $x/(x^2 - 4)$? If you answered 2 and -2, you answered correctly.

Whenever we list a rational expression containing a variable in the denominator, we always assume that the value or values of the variable that make the denominator 0 are excluded.

In Section 3.5 we discussed the domain of relations and functions. When discussing rational expressions, the **domain** will be the set of values that can be used to replace the variable. For example, in the expression $(x + 2)/(x - 3)$, the domain will be all real numbers except 3, $\{x \mid x \neq 3\}$. Note that 3 would make the denominator 0.

EXAMPLE 1   Find the domain of $\dfrac{x - 2}{3x - 8}$.

Solution: The domain will be all real numbers except those that make the denominator equal to 0. What value will make the denominator equal to 0? We can determine this by setting the denominator equal to 0 and solving the equation for x.

$$3x - 8 = 0$$
$$3x = 8$$
$$x = \frac{8}{3}$$

If x were $\frac{8}{3}$, the denominator would be 0. The domain is therefore all real numbers except $\frac{8}{3}$, $\{x \mid x \neq \frac{8}{3}\}$.   ■

EXAMPLE 2   Find the domain of $\dfrac{3}{y^2 - 2y - 15}$.

Solution: We must determine which values of y will make the denominator equal to 0. To do this, set the denominator equal to 0 and solve for y.

$$y^2 - 2y - 15 = 0$$
$$(y - 5)(y + 3) = 0$$
$$y - 5 = 0 \quad \text{or} \quad y + 3 = 0$$
$$y = 5 \qquad\qquad y = -3$$

Therefore, the domain will consist of all real numbers except -3 and 5.

$$\text{Domain:} \quad \{y \mid y \neq -3, y \neq 5\}$$

▶ **2** When we work problems containing rational expressions, we must make sure that we write the answer in the lowest terms. An algebraic fraction is **reduced to its lowest terms** when the numerator and denominator have no common factors other than 1. The fraction $\frac{6}{9}$ is not in reduced form because the 6 and 9 both contain the common factor of 3. When the 3 is factored out, the reduced fraction is $\frac{2}{3}$.

$$\frac{6}{9} = \frac{\overset{1}{\cancel{3}} \cdot 2}{\underset{1}{\cancel{3}} \cdot 3} = \frac{2}{3}$$

The rational expression $\dfrac{ab - b^2}{2b}$ is not in reduced form because both the numerator and denominator have a common factor of b. To reduce this expression, factor the b from each term in the numerator; then divide out the common factor b.

$$\frac{ab - b^2}{2b} = \frac{\cancel{b}(a - b)}{2\cancel{b}} = \frac{a - b}{2}$$

$\dfrac{ab - b^2}{2b}$ becomes $\dfrac{a - b}{2}$ when reduced to its lowest terms.

To Reduce Rational Expressions

1. Factor both numerator and denominator as completely as possible.
2. Divide both the numerator and the denominator by any common factors.

EXAMPLE 3 Reduce $\dfrac{x^2 + 2x - 3}{x + 3}$ to its lowest terms.

Solution: Factor the numerator; then divide out the common factor.

$$\frac{x^2 + 2x - 3}{x + 3} = \frac{\cancel{(x + 3)}(x - 1)}{\cancel{x + 3}} = x - 1$$

The rational expression reduces to $x - 1$.

When the terms in a numerator differ only in sign from the terms in a denominator, we can factor out -1 from either the numerator or denominator. **When -1 is factored from a polynomial, the sign of each term in the polynomial changes.** For example,

$$-2x + 3 = -1(2x - 3) = -(2x - 3)$$
$$6 - 5x = -1(-6 + 5x) = -(5x - 6)$$
$$-3x^2 + 5x - 6 = -1(3x^2 - 5x + 6) = -(3x^2 - 5x + 6)$$

EXAMPLE 4   Reduce $\dfrac{3x^2 + 19x - 14}{2 - 3x}$.

Solution: $\dfrac{3x^2 + 19x - 14}{2 - 3x} = \dfrac{(3x - 2)(x + 7)}{2 - 3x}$

$\qquad\qquad = \dfrac{(3x - 2)(x + 7)}{-(3x - 2)} = -(x + 7)$

Notice in Example 4 that $3x - 2$ appeared in the numerator and $2 - 3x$ appeared in the denominator. Since the expressions differ only in sign, we factored out -1 from $2 - 3x$. Expressions that differ only in sign are said to be *opposites*. Thus $3x - 2$ and $2 - 3x$ are opposites.

EXAMPLE 5   Reduce $\dfrac{5x^2y + 10xy^2 - 25x^2y^3}{5x^2y}$.

Solution: Factor the numerator. The greatest common factor of each term in the numerator is $5xy$. Then divide out the common factors.

$$\frac{5x^2y + 10xy^2 - 25x^2y^3}{5x^2y} = \frac{5xy(x + 2y - 5xy^2)}{5x^2y}$$

$$= \frac{x + 2y - 5xy}{x}$$

EXAMPLE 6   Reduce $\dfrac{x^2 - x - 12}{(x + 2)(x - 4) + x(x - 4)}$.

Solution: Factor the numerator and denominator. Note that each term in the denominator has a common factor of $x - 4$.

$$\frac{x^2 - x - 12}{(x + 2)(x - 4) + x(x - 4)} = \frac{(x + 3)(x - 4)}{(x - 4)[(x + 2) + x]}$$

$$= \frac{(x + 3)(x - 4)}{(x - 4)(2x + 2)}$$

$$= \frac{x + 3}{2x + 2}$$

COMMON STUDENT ERROR

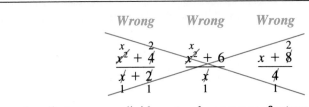

Remember that you can divide out only common **factors.** Only when expressions are **multiplied** can they be factors of the expression. None of the expressions above can be simplified from their original form.

EXAMPLE 7   Write each rational expression with the indicated denominator.

(a) $\dfrac{3}{xy^2}$, $x^5y^4z^6$ (b) $\dfrac{p-1}{p+3}$, $2p^2 + 5p - 3$

Solution: (a) To determine the expression that the numerator and denominator of the fraction $\dfrac{3}{xy^2}$ must be multiplied by to obtain a denominator of $x^5y^4z^6$, we can divide the desired denominator by the present denominator, as follows:

$$\frac{x^5y^4z^6}{xy^2} = x^4y^2z^6$$

Thus, both the numerator and denominator of $\dfrac{3}{xy^2}$ must be multiplied by $x^4y^2z^6$ to obtain an equivalent fraction with the denominator $x^5y^4z^6$.

$$\frac{3}{xy^2} \cdot \boxed{\frac{x^4y^2z^6}{x^4y^2z^6}} = \frac{3x^4y^2z^6}{x^5y^4z^6}$$

(b) To determine the expression that the numerator and denominator of the fraction $\dfrac{p-1}{p+3}$ must be multiplied by to obtain a denominator of $2p^2 + 5p - 3$, we can divide as follows:

$$\frac{2p^2 + 5p - 3}{p+3} = \frac{(2p-1)(p+3)}{(p+3)} = 2p - 1$$

Now multiply both the numerator and denominator of $\dfrac{p-1}{p+3}$ by $2p - 1$.

$$\frac{p-1}{p+3} \cdot \boxed{\frac{2p-1}{2p-1}} = \frac{2p^2 - 3p + 1}{2p^2 + 5p - 3}$$

Exercise Set 7.1

Determine the domain of each of the following.

1. $\dfrac{6}{x}$

2. $\dfrac{x}{x+5}$

3. $\dfrac{2x}{2x-6}$

4. $\dfrac{5}{x^2 - 9}$

5. $\dfrac{3}{x^2 + 4}$

6. $\dfrac{0}{-x - 2}$

7. $\dfrac{2}{(x - 2)^2}$

8. $\dfrac{-3}{x - \dfrac{1}{2}}$

9. $\dfrac{-2}{16 - r^2}$

10. $\dfrac{5}{(x + 3)(x - 2)}$

11. $\dfrac{3}{x^2 + 7x + 6}$

12. $\dfrac{x - 3}{x^2 + 4x - 21}$

13. $\dfrac{y - 1}{y - 1}$

14. $\dfrac{2x - 3}{3x - 7}$

15. $\dfrac{-2}{-8z + 15}$

16. $\dfrac{4x - 6}{x^2 + 6x + 9}$

17. $\dfrac{4x - 3}{x^2 - 16}$

18. $\dfrac{3a^2 - 6a + 4}{2a^2 + 3a - 2}$

19. $\dfrac{4 - 2x}{x^3 + 9x}$

20. $\dfrac{w^2 - 3w + 4}{3w^2 + 7w + 4}$

Write each expression in reduced form.

21. $\dfrac{x - xy}{x}$

22. $\dfrac{x^2 - 2x}{x}$

23. $\dfrac{5x^2 - 25x}{10}$

24. $\dfrac{3x + xy}{y + 3}$

25. $\dfrac{x^2 - 4}{x + 2}$

26. $\dfrac{x^2 - 4x^4}{x}$

27. $\dfrac{5x^2 - 10xy}{25x}$

28. $\dfrac{6x^2 + 12xy - 18y^2}{9}$

29. $\dfrac{4x^2 y + 12xy + 18x^3 y^3}{8xy^2}$

30. $\dfrac{4r - 2}{2 - 4r}$

31. $\dfrac{3 - 2x}{2x - 3}$

32. $\dfrac{3x - 5}{-9x + 15}$

33. $\dfrac{p + 4}{p^2 + 9p + 20}$

34. $\dfrac{4x^2 - 9}{2x^2 - x - 3}$

35. $\dfrac{x^2 - 2x - 24}{6 - x}$

36. $\dfrac{9x^2 - 64}{3x - 8}$

37. $\dfrac{4x^2 - 16x^4 + 6x^5 y}{8x^3 y}$

38. $\dfrac{x^2 + 5x + 6}{x^2 - 3x - 10}$

39. $\dfrac{y^2 - 10y + 24}{y^2 - 5y + 4}$

40. $\dfrac{9 - x}{-(x^2 - 8x - 9)}$

41. $\dfrac{x^5 + 4x^4 - 6x^3}{x^2 + 4x - 6}$

42. $\dfrac{x(x - 1) + x(x - 4)}{2x - 5}$

43. $\dfrac{(x + 1)(x - 3) + (x + 1)(x - 2)}{2(x + 1)}$

44. $\dfrac{(2x - 5)(x + 4) - (2x - 5)(x + 1)}{3(2x - 5)}$

45. $\dfrac{x^2 - 8x + 5x - 40}{x^2 - 2x + 5x - 10}$

46. $\dfrac{x^2 - 8x + 16}{x^2 + 3x - 4x - 12}$

47. $\dfrac{xy - yw + xz - zw}{xy + yw + xz + zw}$

48. $\dfrac{a^2 + 3a - ab - 3b}{a^2 - ab + 5a - 5b}$

49. $\dfrac{a^3 - b^3}{a^2 - b^2}$

50. $\dfrac{x^2 + 2x - 3}{x^3 + 27}$

51. $\dfrac{x^3 + 3x^2 - 4x - 12}{x^2 + 5x + 6}$

52. $\dfrac{25x^4 - 64}{15x^3 - 5x^2 + 24x - 8}$

Write each rational expression with the indicated denominator.

53. $\dfrac{5}{x}$, x^2

54. $\dfrac{3r}{r + 2}$, $5r + 10$

55. $\dfrac{w + 3}{w - 3}$, $w^2 - 9$

56. $\dfrac{5y}{y + 3}$, $y^2 - 3y - 18$

57. $\dfrac{1}{x^2 y^3}$, $2x^3 y^5 z^2$

58. $\dfrac{4}{p}$, $p^2(p - 2)$

59. $\dfrac{x + 1}{2(x - 4)}$, $8(x - 4)(x + 2)$

60. $\dfrac{p}{2p - 1}$, $6p^2 + p - 2$

61. $\dfrac{y}{3y + 4}$, $9y^2 + 15y + 4$

62. $\dfrac{z}{z + 2y}$, $z^3 + 8y^3$

In Exercises 63 through 66, (a) determine the polynomial to be placed in the shaded area that will result in a true statement; (b) explain how you determined your answer.

63. $\dfrac{\rule{1.5cm}{0.3cm}}{x^2 + 2x - 15} = \dfrac{1}{x - 3}$

64. $\dfrac{\rule{1.5cm}{0.3cm}}{3x + 4} = x - 3$

65. $\dfrac{y^2 - y - 20}{\rule{1.5cm}{0.3cm}} = \dfrac{y + 4}{y + 1}$

66. $\dfrac{\rule{1.5cm}{0.3cm}}{6p^2 + p - 15} = \dfrac{2p - 1}{2p - 3}$

67. Explain why $\dfrac{\sqrt{x}}{x + 1}$ is not a rational expression.

68. Explain why $\dfrac{2}{\sqrt{y} + 3}$ is not a rational expression.

69. Consider the expression $\dfrac{1}{x}$. What is the value of the fraction when **(a)** $x = 1$? **(b)** $x = 10$? **(c)** $x = 100$? **(d)** What happens to the value of the fraction as the denominator increases?

70. Consider the expression $\dfrac{1}{x}$. What is the value of the fraction when **(a)** $x = 1$? **(b)** $x = 0.5$ **(c)** $x = 0.1$? **(d)** What happens to the value of the fraction as the denominator approaches 0? (Assume $x > 0$.)

71. (a) Explain how to reduce a rational expression where the numerator and denominator differ only in sign.
(b) By the procedure explained in part (a), reduce the expression $\dfrac{3x^2 - 2x - 8}{-3x^2 + 2x + 8}$.

Cumulative Review Exercises

[2.2] **72.** Solve the formula $V = \frac{4}{3}\pi r^2 h$ for h.

[3.4] **73.** Find the slope and y intercept of the equation $3(y - 4) = -(x - 2)$.

[4.1] **74.** Solve the system of equations on the right.

$$3x + 5y = 12$$
$$2x - y = 8.$$

[5.1] **75.** Simplify $\dfrac{(4x^{-3}y^4)(2x^2y^{-1})}{12x^{-2}y^3}$.

JUST FOR FUN

Polynomial expressions are of the form $a_n x^n + a_{n-1}x^{n-1} + \cdots + a_1 x + a_0$ and **polynomial functions** are of the form $f(x) = a_n x^n + a_{n-1}x^{n-1} + \cdots + a_1 x + a_0$. There are also **rational expressions** of the form p/q, p and q polynomials, $q \neq 0$ and **rational functions** of the form $f(x) = \dfrac{p}{q}$, p and q polynomials, $q \neq 0$. The graph of every rational function will pass the vertical line test for functions. A discussion of rational functions is beyond the scope of this course, but just for fun you may wish to try to graph one.

1. Consider the rational function $f(x) = \dfrac{x^2 - 4}{x - 2}$.

(a) Determine its domain.
(b) Graph the function. (*Hint:* See what happens if you factor the numerator.)

2. Consider the rational function $y = \dfrac{x^2 - 4x + 3}{x - 1}$.

(a) Determine its domain.
(b) Graph the function.

7.2

Multiplication and Division of Rational Expressions

▶ **1** Multiply rational expressions.

▶ **2** Divide rational expressions.

▶ **1** To multiply fractions, we often divide out their common factors, then multiply the numerators together, and multiply the denominators together.

Multiplication of Rational Expressions

Multiplication

$$\frac{a}{b} \cdot \frac{c}{d} = \frac{a \cdot c}{b \cdot d}, \qquad b \neq 0, \quad d \neq 0$$

We follow the same basic procedure to multiply rational expressions as illustrated in Example 1.

EXAMPLE 1   Multiply $\dfrac{3x^2}{z^2} \cdot \dfrac{2z^5}{9x}$.

Solution: $\dfrac{\overset{1}{\cancel{3x^2}}^{x}}{\underset{1}{\cancel{z^2}}} \cdot \dfrac{\overset{z^3}{\cancel{2z^5}}}{\underset{3}{\cancel{9x}}_{1}} = \dfrac{x \cdot 2z^3}{1 \cdot 3} = \dfrac{2xz^3}{3}$ ∎

To Multiply Rational Expressions

1. Factor all numerators and denominators as far as possible.
2. Divide out the common factors.
3. Multiply numerators together and multiply denominators together.
4. Reduce the answer when possible.

EXAMPLE 2   Multiply $\dfrac{x-5}{4x} \cdot \dfrac{x^2 - 2x}{x^2 - 7x + 10}$.

Solution: $\dfrac{x-5}{4x} \cdot \dfrac{x^2 - 2x}{x^2 - 7x + 10} = \dfrac{\cancel{x-5}}{4\cancel{x}} \cdot \dfrac{\cancel{x}(x-2)}{(x-2)(x-5)} = \dfrac{1}{4}$ ∎

EXAMPLE 3   Multiply $\dfrac{2x-5}{x-4} \cdot \dfrac{x^2 - 8x + 16}{5 - 2x}$.

Solution: $\dfrac{2x-5}{x-4} \cdot \dfrac{x^2 - 8x + 16}{5 - 2x} = \dfrac{2x-5}{\cancel{x-4}} \cdot \dfrac{\cancel{(x-4)}(x-4)}{5 - 2x}$

$$= \dfrac{\cancel{2x-5}}{\cancel{x-4}} \cdot \dfrac{\cancel{(x-4)}(x-4)}{-1(\cancel{2x-5})}$$

$$= \dfrac{x-4}{-1} = -(x-4) \quad \text{or} \quad -x + 4$$ ∎

EXAMPLE 4   Multiply $\dfrac{x^2 - y^2}{x+y} \cdot \dfrac{x + 2y}{2x^2 - xy - y^2}$.

Solution: $\dfrac{x^2 - y^2}{x+y} \cdot \dfrac{x + 2y}{2x^2 - xy - y^2} = \dfrac{(x+y)(x-y)}{\cancel{x+y}} \cdot \dfrac{x + 2y}{(2x + y)(x-y)}$

$$= \dfrac{x + 2y}{2x + y}$$ ∎

EXAMPLE 5   Multiply $\dfrac{ab - ac + bd - cd}{ab + ac + bd + cd} \cdot \dfrac{b^2 + bc + bd + cd}{b^2 + bd - bc - cd}$.

Solution: Factor both numerators and denominators by grouping; then divide out common factors.

$$\dfrac{ab - ac + bd - cd}{ab + ac + bd + cd} \cdot \dfrac{b^2 + bc + bd + cd}{b^2 + bd - bc - cd} = \dfrac{a(b - c) + d(b - c)}{a(b + c) + d(b + c)} \cdot \dfrac{b(b + c) + d(b + c)}{b(b + d) - c(b + d)}$$

$$= \dfrac{\cancel{(a + d)}\cancel{(b - c)}}{\cancel{(a + d)}\cancel{(b + c)}} \cdot \dfrac{\cancel{(b + d)}\cancel{(b + c)}}{\cancel{(b - c)}\cancel{(b + d)}} = 1 \qquad ∎$$

Division of Rational Expressions

▶ **2**   To divide numerical fractions, we invert the divisor and proceed as in multiplication.

$$\dfrac{a}{b} \div \dfrac{c}{d} = \dfrac{a}{b} \cdot \dfrac{d}{c} = \dfrac{ad}{bc}, \qquad b \neq 0, \quad c \neq 0, \quad d \neq 0$$

We follow the same basic procedure to divide rational expressions, as illustrated in Example 6.

EXAMPLE 6   Divide $\dfrac{12x^4}{5y^3} \div \dfrac{3x^5}{10y}$.

Solution: $\dfrac{12x^4}{5y^3} \div \dfrac{3x^5}{10y} = \dfrac{\overset{4}{\cancel{12}}x^4}{\underset{1}{\cancel{5}}y^3} \cdot \dfrac{\overset{2}{\cancel{10}}y}{\underset{1}{\cancel{3}}x^5} = \dfrac{4 \cdot 2}{y^2 x} = \dfrac{8}{xy^2}$ ∎

To Divide Rational Expressions

Invert the divisor (the second or bottom fraction) and then multiply the resulting rational expressions.

EXAMPLE 7   Divide $\dfrac{x^2 - 9}{x + 4} \div \dfrac{x - 3}{x + 4}$.

Solution: $\dfrac{x^2 - 9}{x + 4} \cdot \dfrac{x + 4}{x - 3} = \dfrac{(x + 3)\cancel{(x - 3)}}{\cancel{x + 4}} \cdot \dfrac{\cancel{x + 4}}{\cancel{x - 3}} = x + 3$ ∎

EXAMPLE 8   Divide $\dfrac{12x^2 - 22x + 8}{3x} \div \dfrac{3x^2 + 2x - 8}{2x^2 + 4x}$.

Solution: $\dfrac{12x^2 - 22x + 8}{3x} \cdot \dfrac{2x^2 + 4x}{3x^2 + 2x - 8} = \dfrac{2(6x^2 - 11x + 4)}{3x} \cdot \dfrac{2x(x + 2)}{(3x - 4)(x + 2)}$

$$= \dfrac{2\cancel{(3x - 4)}(2x - 1)}{3\cancel{x}} \cdot \dfrac{2\cancel{x}\cancel{(x + 2)}}{\cancel{(3x - 4)}\cancel{(x + 2)}}$$

$$= \dfrac{4(2x - 1)}{3} \qquad ∎$$

EXAMPLE 9 Divide $\dfrac{x^4 - y^4}{x - y} \div \dfrac{x^2 + xy}{x^2 - 2xy + y^2}$.

Solution: $\dfrac{x^4 - y^4}{x - y} \cdot \dfrac{x^2 - 2xy + y^2}{x^2 + xy} = \dfrac{(x^2 + y^2)(x^2 - y^2)}{x - y} \cdot \dfrac{(x - y)(x - y)}{x(x + y)}$

$$= \dfrac{(x^2 + y^2)\cancel{(x + y)}\cancel{(x - y)}}{\cancel{x - y}} \cdot \dfrac{(x - y)(x - y)}{x\cancel{(x + y)}}$$

$$= \dfrac{(x^2 + y^2)(x - y)^2}{x}$$ ■

EXAMPLE 10 Perform the indicated operations.

$$\dfrac{x^4 - y^4}{x - y} \div \dfrac{x^2 + xy}{x^2 - 2xy + y^2} \cdot \dfrac{x^2}{x^2 - 2xy + y^2}$$

Solution: When a problem contains more than one operation, we follow the priority of operations given in Section 1.7. Since multiplication and division have the same priority, and there are no parentheses, we work from left to right. In Example 9 we found that

$$\dfrac{x^4 - y^4}{x - y} \div \dfrac{x^2 + xy}{x^2 + 2xy + y^2} = \dfrac{(x^2 + y^2)(x - y)^2}{x}$$

Using this result we can write:

$$\dfrac{x^4 - y^4}{x - y} \div \dfrac{x^2 + xy}{x^2 - 2xy + y} \cdot \dfrac{x^2}{x^2 - 2xy + y^2} = \dfrac{(x^2 + y^2)(x - y)^2}{x} \cdot \dfrac{x^2}{x^2 - 2xy + y^2}$$

$$= \dfrac{(x^2 + y^2)\cancel{(x - y)^2}}{x} \cdot \dfrac{x^2}{\cancel{(x - y)^2}}$$

$$= x(x^2 + y^2)$$ ■

Exercise Set 7.2

Multiply or divide as indicated. Write all answers in lowest terms.

1. $\dfrac{3x}{2y} \cdot \dfrac{y^3}{6}$

2. $\dfrac{16x^2}{y^4} \cdot \dfrac{5x^2}{4y^2}$

3. $\dfrac{9x^3}{4} \div \dfrac{3}{16y^2}$

4. $\dfrac{9y}{7z^2} \div \dfrac{3xy}{4z}$

5. $\dfrac{3y^3}{8x} \cdot \dfrac{9x^2}{y^3}$

6. $\dfrac{80m^4}{49x^5y^7} \cdot \dfrac{14x^{12}y^5}{25m^5}$

7. $\dfrac{12a^2}{4bc} \div \dfrac{3a^2}{bc}$

8. $\dfrac{-xy}{a} \div \dfrac{-2ax}{6y}$

9. $\dfrac{6x^5y^3}{5z^3} \cdot \dfrac{6x^4}{5yz^4}$

10. $(2x + 5) \cdot \dfrac{1}{4x + 10}$

11. $2xz \div \dfrac{4xy}{z}$

12. $\dfrac{1}{7x^2y} \div \dfrac{1}{21x^3y}$

13. $\dfrac{x - 3}{x + 5} \cdot \dfrac{2x^2 + 10x}{2x - 6}$

14. $\dfrac{3x - 2}{3x + 2} \cdot \dfrac{4x - 1}{1 - 4x}$

15. $\dfrac{4-x}{x-4} \cdot \dfrac{x-3}{3-x}$

16. $\dfrac{2a+2b}{3} \div \dfrac{a^2-b^2}{a-b}$

17. $\dfrac{x^2+7x+12}{x+4} \cdot \dfrac{1}{x+3}$

18. $\dfrac{x^2+3x-10}{2x} \cdot \dfrac{x^2-3x}{x^2-5x+6}$

19. $\dfrac{x^2+10x+21}{x+7} \div (x+3)$

20. $(x-3) \div \dfrac{x^2+3x-18}{x}$

21. $\dfrac{x^2-9x+14}{x^2-5x+6} \div \dfrac{x^2-5x-14}{x+2}$

22. $\dfrac{1}{x^2-17x+30} \div \dfrac{1}{x^2+7x-18}$

23. $\dfrac{6x^2-14x-12}{6x+4} \cdot \dfrac{x+3}{2x^2-2x-12}$

24. $\dfrac{a-b}{9a+9b} \div \dfrac{a^2-b^2}{a^2+2a+1}$

25. $\dfrac{(x+2)^2}{x-2} \div \dfrac{x^2-4}{2x-4}$

26. $\dfrac{2x+4y}{x^2+4xy+4y^2} \cdot \dfrac{x+2y}{2}$

27. $\dfrac{x^2-y^2}{8x^2-16xy+8y^2} \cdot \dfrac{4x-4y}{x+y}$

28. $\dfrac{6x^3-x^2-x}{2x^2+x-1} \cdot \dfrac{x^2-1}{x^3-2x^2+x}$

29. $\dfrac{x+2}{x^3-8} \cdot \dfrac{(x-2)^2}{x^2-4}$

30. $\dfrac{x^2+7x+10}{1-x} \div \dfrac{x^2+2x-15}{x-1}$

31. $\dfrac{x^2-y^2}{x^2-2xy+y^2} \div \dfrac{x+y}{x-y}$

32. $\dfrac{x^4-y^8}{x^2+y^4} \div \dfrac{x^2-y^4}{3x^2}$

33. $\dfrac{(x^2-y^2)^2}{(x^2-y^2)^3} \div \dfrac{x^2+y^2}{x^4-y^4}$

34. $\dfrac{2x^4+4x^2}{6x^2+14x+4} \div \dfrac{x^2+2}{3x^2+x}$

35. $\dfrac{8a^3-1}{4a^2+2a+1} \div \dfrac{a-1}{(a-1)^2}$

36. $\dfrac{27x^3+64}{3x+4} \div \dfrac{18x^2-24x+32}{x}$

37. $\dfrac{2x^3-7x^2+3x}{x^2+2x-3} \cdot \dfrac{x^2+3x}{(x-3)^2}$

38. $\dfrac{4x+y}{5x+2y} \cdot \dfrac{25x^2-5xy-6y^2}{20x^2-7xy-3y^2}$

39. $\dfrac{3r^2+17rs+10s^2}{6r^2+13rs-5s^2} \div \dfrac{6r^2+rs-2s^2}{6r^2-5rs+s^2}$

40. $\dfrac{x^2+2x-2x-4}{x^2+3x+4x+12} \cdot \dfrac{x^2+3x+3x+9}{x^2+3x-2x-6}$

41. $\dfrac{ac-ad+bc-bd}{ac+ad+bc+bd} \cdot \dfrac{pc+pd-qc-qd}{pc-pd+qc-qd}$

42. $\dfrac{a^2+2ab-ab-2b^2}{2a^2+2ad-2ab-2bd} \cdot \dfrac{a^2+ad-ac-cd}{a^2-ac+2ab-2bc}$

43. $\dfrac{2p^2+2pq-pq^2-q^3}{p^3+p^2+pq^2+q^2} \div \dfrac{p^3+p+p^2q+q}{p^3+p+p^2+1}$

44. $\dfrac{x^3-4x^2+x-4}{x^4-x^3+x^2-x} \cdot \dfrac{2x^3+2x^2+x+1}{2x^3-8x^2+x-4}$

45. $\dfrac{x^2+5x+6}{x^2-x-20} \cdot \dfrac{2x^2+6x-8}{x^2-9} \cdot \dfrac{x^2-3x}{x-1}$

46. $\dfrac{x^2-1}{x^2+x} \cdot \dfrac{2x+2}{1-x^2} \cdot \dfrac{x^2+x-2}{x^2-x}$

47. $\dfrac{x^3+64}{x-2} \cdot \dfrac{x^2-4}{x+4} \cdot \dfrac{x}{x+2}$

48. $\dfrac{2x^2-3x-14}{2x^2-9x+7} \div \dfrac{6x^2+x-15}{3x^2+2x-5} \cdot \dfrac{6x^2-7x-3}{2x^2-x-3}$

49. $\dfrac{a^2-b^2}{2a^2-3ab+b^2} \cdot \dfrac{2a^2-7ab+3b^2}{a^2+ab} \div \dfrac{ab-3b^2}{a^2+2ab+b^2}$

50. $\dfrac{10x^2-17x+3}{15x^2-8x+1} \div \dfrac{4x^2-12x+9}{3x^2+3xy-x-y} \cdot \dfrac{6x^2-11x+3}{2x^2-11x+12}$

51. $\dfrac{5x^2(x-1)-3x(x-1)-2(x-1)}{10x^2(x-1)+9x(x-1)+2(x-1)} \cdot \dfrac{2x+1}{x+3}$

52. $\dfrac{x^2(3x-y)-5x(3x-y)-24(3x-y)}{x^2(3x-y)-9x(3x-y)+8(3x-y)} \cdot \dfrac{x-1}{x+3}$

In Exercises 53 through 56, (a) determine the polynomial to be placed in the shaded area that will result in a true statement and (b) explain how you determined your answer.

53. $\dfrac{x^2-x-12}{x^2+2x-3} \cdot \dfrac{\rule{2cm}{0.4pt}}{x^2-2x-8} = 1$

54. $\dfrac{x^2-4}{(x+2)^2} \cdot \dfrac{2x^2+x-6}{\rule{2cm}{0.4pt}} = \dfrac{x-2}{2x+5}$

55. $\dfrac{x^2-9}{2x^2+3x-2} \div \dfrac{2x^2-9x+9}{\rule{2cm}{0.4pt}} = \dfrac{x+3}{2x-1}$

56. $\dfrac{4r^2-r-18}{\rule{2cm}{0.4pt}} \div \dfrac{4r^3-9r^2}{6r^2-9r+3} = \dfrac{3(r-1)}{r^2}$

Cumulative Review Exercises _____

[2.1] **57.** Solve the equation $\dfrac{4x + 2}{5} + 3 = 5$.

[2.5] **58.** Solve the inequality $-4 < 3x - 4 < 8$. Write the solution in interval notation.

[4.4] **59.** Solve the system of equations using determinants.

$$x + 2y = 4$$
$$2y = 6x + 6$$

[5.3] **60.** Simplify $3x^2y - 4xy + 2y^2 - (3xy + 6y^2 + 2x)$.

7.3

Addition and Subtraction of Rational Expressions

▸**1** Add or subtract rational expressions with a common denominator.

▸**2** Find the least common denominator (LCD).

▸**3** Add or subtract rational expressions with unlike denominators.

▸**1** Recall that when adding (or subtracting) two arithmetic fractions with a common denominator we add (or subtract) the numerators while keeping the common denominator.

Addition	Subtraction
$\dfrac{a}{c} + \dfrac{b}{c} = \dfrac{a + b}{c}, \quad c \neq 0$	$\dfrac{a}{c} - \dfrac{b}{c} = \dfrac{a - b}{c}, \quad c \neq 0$

To add or subtract rational expressions with a common denominator, we use the same principle as illustrated in Example 1.

EXAMPLE 1 Add $\dfrac{3}{x + 2} + \dfrac{x - 4}{x + 2}$.

Solution: Since the denominators are the same, we add the numerators and keep the common denominator.

$$\frac{3}{x + 2} + \frac{x - 4}{x + 2} = \frac{3 + (x - 4)}{x + 2}$$

$$= \frac{x - 1}{x + 2}$$

To Add or Subtract Expressions with a Common Denominator

1. Add or subtract the numerators.
2. Place the sum or difference of the numerators found in step 1 over the common denominator.
3. Reduce the fraction, if possible.

EXAMPLE 2   Add $\dfrac{x^2 + 3x - 2}{(x + 5)(x - 2)} + \dfrac{4x + 12}{(x + 5)(x - 2)}$.

Solution: $\dfrac{x^2 + 3x - 2}{(x + 5)(x - 2)} + \dfrac{4x + 12}{(x + 5)(x - 2)} = \dfrac{x^2 + 3x - 2 + (4x + 12)}{(x + 5)(x - 2)}$

$$= \dfrac{x^2 + 7x + 10}{(x + 5)(x - 2)}$$

$$= \dfrac{\cancel{(x + 5)}(x + 2)}{\cancel{(x + 5)}(x - 2)} = \dfrac{x + 2}{x - 2} \quad \blacksquare$$

When subtracting rational expressions, be sure to subtract the entire numerator of the fraction being subtracted. Study the common student error that follows very carefully.

COMMON STUDENT ERROR

The error presented here is often made by students. Study the information presented so that you will not make this error.

How do you simplify this problem?

$$\dfrac{4x}{x - 2} - \dfrac{2x + 1}{x - 2}$$

Correct

$$\dfrac{4x}{x - 2} - \dfrac{2x + 1}{x - 2} = \dfrac{4x - (2x + 1)}{x - 2}$$

$$= \dfrac{4x - 2x - 1}{x - 2}$$

$$= \dfrac{2x - 1}{x - 2}$$

Wrong

$$\cancel{\dfrac{4x}{x - 2} - \dfrac{2x + 1}{x - 2} = \dfrac{4x - 2x + 1}{x - 2}}$$

$$\cancel{= \dfrac{2x + 1}{x - 2}}$$

The procedure on the right is wrong because the *entire numerator*, $2x + 1$, must be subtracted from $4x$. In the procedure on the right, only $2x$ was subtracted. Note that **the sign of each term** (not just the first term) **in the numerator of the fraction being subtracted must change.**

EXAMPLE 3   Subtract $\dfrac{3x}{x - 6} - \dfrac{x^2 - 4x + 6}{x - 6}$.

Solution: $\dfrac{3x}{x - 6} - \dfrac{x^2 - 4x + 6}{x - 6} = \dfrac{3x - (x^2 - 4x + 6)}{x - 6}$

$$= \dfrac{3x - x^2 + 4x - 6}{x - 6}$$

$$= \dfrac{-x^2 + 7x - 6}{x - 6}$$

$$= \dfrac{-(x^2 - 7x + 6)}{x - 6}$$

$$= \dfrac{-(x - 6)(x - 1)}{x - 6}$$

$$= -(x - 1) \quad \blacksquare$$

▶**2** To add two numerical fractions with *unlike denominators,* we must first obtain a common denominator.

EXAMPLE 4 Add $\frac{3}{5} + \frac{4}{7}$.

Solution: The least common denominator (LCD) of 5 and 7 is 35. Thirty-five is the smallest number divisible by both 5 and 7. Now proceed to rewrite each fraction so that it has a denominator equal to the LCD.

$$\frac{3}{5} + \frac{4}{7} = \frac{3}{5} \cdot \boxed{\frac{7}{7}} + \frac{4}{7} \cdot \boxed{\frac{5}{5}}$$

$$= \frac{21}{35} + \frac{20}{35} = \frac{41}{35} \quad \text{or} \quad 1\frac{6}{35} \qquad \blacksquare$$

To add or subtract rational expressions with unlike denominators, we must first write each expression with a common denominator.

To Find the Least Common Denominator of Rational Expressions

1. Factor each denominator completely. Factors in any given denominator that occur more than once should be expressed as powers. For example, $(x + 5)(x + 5)$ should be expressed as $(x + 5)^2$.
2. List all different factors (other than 1) that appear in any of the denominators. When the same factor appears in more than one denominator, write the factor with the *highest* power that appears.
3. The least common denominator is the product of all the factors found in step 2.

EXAMPLE 5 Find the LCD.

$$\frac{3}{5x} - \frac{2}{x^2}$$

Solution: The factors that appear in the denominators are 5 and x. List each factor with its highest power. The LCD is the product of these factors.

$$\overset{\displaystyle \ulcorner \text{highest power of } x}{\text{LCD} = 5 \cdot x^2 = 5x^2} \qquad \blacksquare$$

EXAMPLE 6 Find the LCD.

$$\frac{1}{18x^3y} + \frac{5}{27x^2y^3}$$

Solution: The LCD of 18 and 27 is 54. The variable factors are x and y.

$$\text{LCD} = 54x^3y^3 \qquad \blacksquare$$

EXAMPLE 7 Find the LCD.

$$\frac{3}{x} - \frac{2y}{x + 5}$$

Solution: The factors that appear are x and $(x + 5)$. Note that the x in the second denominator, $x + 5$, is not a factor of that denominator since the operation is addition rather than multiplication.

$$\text{LCD} = x(x + 5)$$ ■

EXAMPLE 8   Find the LCD.

$$\frac{3}{2x^2 - 4x} + \frac{x^2}{x^2 - 4x + 4}$$

Solution: Factor both denominators.

$$\frac{3}{2x(x - 2)} + \frac{x^2}{(x - 2)^2}$$

The factors that appear are 2, x, and $x - 2$. List the highest powers of each of these factors that appear.

$$\text{LCD} = 2 \cdot x \cdot (x - 2)^2 = 2x(x - 2)^2$$ ■

EXAMPLE 9   Find the LCD.

$$\frac{5x}{x^2 - x - 12} - \frac{6x^2}{x^2 - 7x + 12}$$

Solution: Factor both denominators.

$$\frac{5x}{(x + 3)(x - 4)} - \frac{6x^2}{(x - 3)(x - 4)}$$
$$\text{LCD} = (x + 3)(x - 4)(x - 3)$$

Note that although $(x - 4)$ is a common factor of each denominator, the highest power of that factor that appears in either denominator is 1.   ■

▶ **3**   The method used to add or subtract rational expressions with unlike denominators is illustrated in Example 10.

EXAMPLE 10   Add $\dfrac{3}{x} + \dfrac{5}{y}$.

Solution: First, determine the LCD.

$$\text{LCD} = xy$$

Now write each fraction with the LCD. We do this by multiplying **both** numerator and denominator of each fraction by any factors needed to obtain the LCD.

In this problem the fraction on the left must be multiplied by y/y and the fraction on the right must be multiplied by x/x.

$$\frac{y}{y} \cdot \frac{3}{x} + \frac{5}{y} \cdot \frac{x}{x} = \frac{3y}{xy} + \frac{5x}{xy}$$

By multiplying both the numerator and denominator by the same factor, we are in effect multiplying by 1, which does not change the value of the fraction, only its appearance. Thus the new fraction is equivalent to the original fraction.

Now add the numerators while leaving the LCD alone.

$$\frac{3y}{xy} + \frac{5x}{xy} = \frac{3y + 5x}{xy} \quad \text{or} \quad \frac{5x + 3y}{xy}$$

Therefore, $\dfrac{3}{x} + \dfrac{5}{y} = \dfrac{5x + 3y}{xy}$.   ■

To Add or Subtract Two Rational Expressions with Unlike Denominators

1. Determine the LCD.
2. Rewrite each fraction as an equivalent fraction with the LCD. This is done by multiplying both the numerator and denominator of each fraction by any factors needed to obtain the LCD.
3. Leave the denominator in factored form, but multiply out the numerator.
4. Add or subtract the numerators while maintaining the LCD.
5. When possible, factor the remaining numerator and reduce fractions.

EXAMPLE 11   Add $\dfrac{5}{4x^2y} + \dfrac{3}{14xy^3}$.

Solution: The LCD is $28x^2y^3$. We must write each fraction with the denominator $28x^2y^3$. To do this, multiply the fraction on the left by $7y^2/7y^2$ and the fraction on the right by $2x/2x$.

$$\frac{7y^2}{7y^2} \cdot \frac{5}{4x^2y} + \frac{3}{14xy^3} \cdot \frac{2x}{2x} = \frac{35y^2}{28x^2y^3} + \frac{6x}{28x^2y^3}$$

$$= \frac{35y^2 + 6x}{28x^2y^3}$$ ■

EXAMPLE 12   Subtract $\dfrac{x + 2}{x - 4} - \dfrac{x + 3}{x + 4}$.

Solution: The LCD is $(x - 4)(x + 4)$.

$$\frac{x + 4}{x + 4} \cdot \frac{x + 2}{x - 4} - \frac{x + 3}{x + 4} \cdot \frac{x - 4}{x - 4} = \frac{(x + 4)(x + 2)}{(x + 4)(x - 4)} - \frac{(x + 3)(x - 4)}{(x + 4)(x - 4)}$$

Use the FOIL method to multiply each numerator.

$$= \frac{x^2 + 6x + 8}{(x + 4)(x - 4)} - \frac{x^2 - x - 12}{(x + 4)(x - 4)}$$

$$= \frac{x^2 + 6x + 8 - (x^2 - x - 12)}{(x + 4)(x - 4)}$$

$$= \frac{x^2 + 6x + 8 - x^2 + x + 12}{(x + 4)(x - 4)}$$

$$= \frac{7x + 20}{(x + 4)(x - 4)}$$ ■

EXAMPLE 13 Add $\dfrac{4}{x-3} + \dfrac{x+5}{3-x}$.

Solution: Note that each denominator is the opposite, or additive inverse, of the other. (The terms of one denominator differ only in sign from the terms of the other denominator.) When this special situation arises, we can multiply the numerator and denominator of either one of the fractions by -1 to obtain the LCD.

$$\frac{4}{x-3} + \frac{x+5}{3-x} = \frac{4}{x-3} + \boxed{\frac{-1}{-1}} \cdot \frac{(x+5)}{(3-x)}$$

$$= \frac{4}{x-3} + \frac{-x-5}{x-3}$$

$$= \frac{-x-1}{x-3} \qquad \blacksquare$$

EXAMPLE 14 Subtract $\dfrac{3x+4}{2x^2-5x-12} - \dfrac{2x-3}{5x^2-18x-8}$.

Solution: Factor the denominator of each expression.

$$\frac{3x+4}{(2x+3)(x-4)} - \frac{2x-3}{(5x+2)(x-4)}$$

The LCD is $(2x+3)(x-4)(5x+2)$.

$$\frac{3x+4}{(2x+3)(x-4)} - \frac{2x-3}{(5x+2)(x-4)} = \boxed{\frac{5x+2}{5x+2}} \cdot \frac{3x+4}{(2x+3)(x-4)} - \frac{2x-3}{(5x+2)(x-4)} \cdot \boxed{\frac{2x+3}{2x+3}}$$

$$= \frac{15x^2+26x+8}{(5x+2)(2x+3)(x-4)} - \frac{4x^2-9}{(5x+2)(2x+3)(x-4)}$$

$$= \frac{15x^2+26x+8-(4x^2-9)}{(5x+2)(2x+3)(x-4)}$$

$$= \frac{15x^2+26x+8-4x^2+9}{(5x+2)(2x+3)(x-4)}$$

$$= \frac{11x^2+26x+17}{(5x+2)(2x+3)(x-4)} \qquad \blacksquare$$

EXAMPLE 15 $\dfrac{x-1}{x-2} - \dfrac{x+1}{x+2} + \dfrac{x-6}{x^2-4}$.

Solution: First, factor x^2-4. The LCD of the three fractions is $(x+2)(x-2)$.

$$\frac{x-1}{x-2} - \frac{x+1}{x+2} + \frac{x-6}{x^2-4} = \frac{x-1}{x-2} - \frac{x+1}{x+2} + \frac{x-6}{(x+2)(x-2)}$$

$$= \frac{x+2}{x+2} \cdot \frac{x-1}{x-2} - \frac{x+1}{x+2} \cdot \frac{x-2}{x-2} + \frac{x-6}{(x+2)(x-2)}$$

$$= \frac{x^2+x-2}{(x+2)(x-2)} - \frac{x^2-x-2}{(x+2)(x-2)} + \frac{x-6}{(x+2)(x-2)}$$

$$= \frac{x^2+x-2-(x^2-x-2)+(x-6)}{(x+2)(x-2)}$$

$$= \frac{x^2+x-2-x^2+x+2+x-6}{(x+2)(x-2)}$$

$$= \frac{3x-6}{(x+2)(x-2)}$$

$$= \frac{3(x-2)}{(x+2)(x-2)} = \frac{3}{x+2}$$

In Exercise Set 7.3, Exercises 71 through 76 involve more than one operation. When working these exercises, work each part of the exercise separately, following the priority of operations given in Section 1.7. For example, if asked to simplify

$$\left(4 - \frac{1}{x-2}\right) \cdot \frac{x^2-4}{4x^2-5x-9}$$

you would begin by evaluating the expression within the parentheses. When you evaluate $\left(4 - \frac{1}{x-2}\right)$ you should obtain $\frac{4x-9}{x-2}$. Then you would evaluate

$$\frac{4x-9}{2} \cdot \frac{x^2-4}{4x^2-5x-9}$$

to get an answer of $\frac{x+2}{x-1}$. Work the entire problem now, starting from the beginning, to see if you obtain this answer.

Exercise Set 7.3

Add or subtract as indicated.

1. $\frac{2x-7}{3} - \frac{4}{3}$

2. $\frac{2x+3}{5} - \frac{x}{5}$

3. $\frac{x-4}{x} - \frac{x+4}{x}$

4. $\frac{-2x+6}{x^2+x-6} + \frac{3x-3}{x^2+x-6}$

5. $\frac{-t-4}{t^2-16} + \frac{2(t+4)}{t^2-16}$

6. $\frac{2x+4}{(x+2)(x-3)} - \frac{x+7}{(x+2)(x-3)}$

7. $\frac{4r+12}{3-r} - \frac{3r+15}{3-r}$

8. $\frac{x^2-2}{x^2+6x-7} - \frac{-4x+19}{x^2+6x-7}$

9. $\frac{-x^2}{x^2+5xy-14y^2} + \frac{x^2+xy+7y^2}{x^2+5xy-14y^2}$

10. $\frac{3r^2+15r}{r^3+2r^2-8r} + \frac{2r^2+5r}{r^3+2r^2-8r}$

11. $\dfrac{x^3 - 10x^2 + 35x}{x(x - 6)} - \dfrac{x^2 + 5x}{x(x - 6)}$

12. $\dfrac{3x^2 + 2x - 16}{3x^2 + 4x} - \dfrac{4x - 8}{3x^2 + 4x}$

13. $\dfrac{3x^2 - x}{2x^2 - x - 21} + \dfrac{3x - 8}{2x^2 - x - 21} - \dfrac{x^2 - x + 27}{2x^2 - x - 21}$

14. $\dfrac{2x^2 + 8x - 15}{2x^2 - 13x + 20} - \dfrac{2x + 10}{2x^2 - 13x + 20} - \dfrac{3x - 5}{2x^2 - 13x + 20}$

Find the least common denominator.

15. $\dfrac{5x}{x + 1} + \dfrac{6}{x + 2}$

16. $\dfrac{-4}{8x^2y^2} + \dfrac{7}{5x^4y^5}$

17. $\dfrac{x + 3}{16x^2y} - \dfrac{x^2}{3x^3}$

18. $\dfrac{9}{(x - 4)(x + 3)} - \dfrac{x + 8}{x - 4}$

19. $6z^2 + \dfrac{9z}{z - 3}$

20. $\dfrac{b^2 + 3}{18b} - \dfrac{b - 7}{12(b + 5)}$

21. $\dfrac{a - 2}{a^2 - 5a - 24} + \dfrac{3}{a^2 + 11a + 24}$

22. $\dfrac{6x + 5}{x^2 - 4} - \dfrac{3x}{x^2 - 5x - 14}$

23. $\dfrac{6}{x + 3} - \dfrac{x + 5}{x^2 - 4x + 3}$

24. $\dfrac{2a}{4a^2 + 7a + 3} - \dfrac{3}{2a^2 - a - 3}$

25. $\dfrac{3x - 5}{6x^2 + 13xy + 6y^2} + \dfrac{3}{3x^2 + 5xy + 2y^2}$

26. $\dfrac{6x}{(x - 2)(x + 3)} + \dfrac{2x + 4}{x^2 - 4} - \dfrac{2x}{2x + 4}$

27. $\dfrac{3}{x^2 + 3x - 4} - \dfrac{4}{4x^2 + 5x - 9} + \dfrac{x + 2}{4x^2 + 25x + 36}$

28. $\dfrac{x}{2x^2 - 7x + 3} + \dfrac{x - 3}{4x^2 + 4x - 3} - \dfrac{x^2 + 1}{2x^2 - 3x - 9}$

Add or subtract as indicated.

29. $\dfrac{4}{3x} + \dfrac{2}{x}$

30. $\dfrac{4}{3y} - \dfrac{1}{4y}$

31. $\dfrac{6}{x^2} + \dfrac{3}{2x}$

32. $3 + \dfrac{5}{x}$

33. $\dfrac{5}{6y} + \dfrac{3}{4y^2}$

34. $\dfrac{3x}{4y} + \dfrac{5}{6xy}$

35. $\dfrac{5}{12x^4y} - \dfrac{1}{5x^2y^3}$

36. $\dfrac{3}{4xy^3} + \dfrac{1}{6x^2y}$

37. $\dfrac{4x}{3xy} + 2$

38. $\dfrac{4}{x - 3} - \dfrac{2}{x}$

39. $\dfrac{5}{b - 2} + \dfrac{3x}{2 - b}$

40. $\dfrac{x}{x - y} - \dfrac{x}{y - x}$

41. $\dfrac{b}{a - b} + \dfrac{a + b}{b}$

42. $\dfrac{2}{x - 3} + \dfrac{4}{x - 1}$

43. $\dfrac{z + 5}{z - 5} - \dfrac{z - 5}{z + 5}$

44. $\dfrac{x + 7}{x + 3} - \dfrac{x - 3}{x + 7}$

45. $\dfrac{x}{x^2 - 9} - \dfrac{4(x - 3)}{x + 3}$

46. $\dfrac{4x}{x - 4} + \dfrac{x + 4}{x + 1}$

47. $\dfrac{2m + 1}{m - 5} - \dfrac{4}{m^2 - 3m - 10}$

48. $\dfrac{x}{x + 1} + \dfrac{1}{x^2 + 2x + 1}$

49. $\dfrac{-x^2 + 5x}{(x - 5)^2} + \dfrac{x + 1}{x - 5}$

50. $\dfrac{4}{(2x - 3)(x + 4)} - \dfrac{3}{(x + 4)(x - 4)}$

51. $\dfrac{x}{x^2 + 2x - 8} + \dfrac{x + 2}{x^2 - 3x + 2}$

52. $\dfrac{5x}{x^2 - 9x + 8} - \dfrac{3(x + 2)}{x^2 - 6x - 16}$

53. $5 - \dfrac{x - 1}{x^2 + 3x - 10}$

54. $\dfrac{3x}{2x - 3} + \dfrac{3x + 6}{2x^2 + x - 6}$

55. $\dfrac{3a - 4}{4a + 1} + \dfrac{3a + 6}{4a^2 + 9a + 2}$

56. $\dfrac{7}{3q^2 + q - 4} + \dfrac{9q + 2}{3q^2 - 2q - 8}$

57. $\dfrac{x + 3}{3x^2 + 6x - 9} + \dfrac{x - 3}{6x^2 - 15x + 9}$

58. $\dfrac{1}{x^2 - y^2} + \dfrac{4}{x^2 - 2xy - 3y^2}$

59. $\dfrac{x - y}{x^2 - 4xy + 4y^2} + \dfrac{x - 3y}{x^2 - 4y^2}$

60. $\dfrac{x + 2y}{x^2 - xy - 2y^2} - \dfrac{y}{x^2 - 3xy + 2y^2}$

61. $\dfrac{2x}{x - 3} - \dfrac{2x}{x + 3} + \dfrac{36}{x^2 - 9}$

62. $\dfrac{3}{p - 1} + \dfrac{4}{p + 1} + \dfrac{p + 2}{p^2 - 1}$

63. $\dfrac{x^2 + 2}{x^2 - x - 2} + \dfrac{1}{x + 1} - \dfrac{x}{x - 2}$

64. $\dfrac{2}{x^2 - 16} + \dfrac{x + 1}{x^2 + 8x + 16} + \dfrac{3}{x - 4}$

65. $\dfrac{3x + 2}{x - 5} + \dfrac{x}{3x + 4} - \dfrac{7x^2 + 24x + 28}{3x^2 - 11x - 20}$

66. $\dfrac{4}{3x - 2} - \dfrac{1}{x - 4} + 5$

67. $\dfrac{x}{x^2 - 10x + 24} - \dfrac{3}{x - 6} + 1$

68. $3 - \dfrac{4}{8r^2 + 2r - 15} + \dfrac{r + 2}{4r - 5}$

69. $\dfrac{3}{5x + 6} + \dfrac{x^2 - x}{5x^2 - 4x - 12} - \dfrac{4}{x - 2}$

70. $\dfrac{3}{x^2 - 13x + 36} + \dfrac{4}{2x^2 - 7x - 4} + \dfrac{1}{2x^2 - 17x - 9}$

Perform the indicated operations.

71. $\left(3 + \dfrac{1}{x + 3}\right)\left(\dfrac{x + 3}{x - 2}\right)$

72. $\left(\dfrac{3}{r + 1} - \dfrac{4}{r - 2}\right)\left(\dfrac{r - 2}{r + 10}\right)$

73. $\left(\dfrac{5}{a - 5} - \dfrac{2}{a + 3}\right) \div (3a + 25)$

74. $(x + 3)\left(\dfrac{x - 4}{x^2 + x - 6}\right) + \dfrac{2}{x + 1}$

75. $\left(\dfrac{x^2 + 4x - 5}{2x^2 + x - 3} \cdot \dfrac{2x + 3}{x + 1}\right) - \dfrac{2}{x + 2}$

76. $\left(\dfrac{x + 5}{x - 3} - x\right) \div \dfrac{1}{x - 3}$

In Exercises 77 and 78, (a) determine the polynomial to be placed in the shaded area that will result in a true statement; (b) explain how you determined your answer.

77. $\dfrac{2x^2 - 5x - 12}{x^2 + 3x - 10} + \dfrac{\rule{1.2cm}{0.3cm}}{x^2 + 3x - 10} = \dfrac{-3x^2 - 4x - 6}{x^2 + 3x - 10}$

79. $\dfrac{r^2 - 6}{r^2 - 5r + 6} - \dfrac{\rule{1.2cm}{0.3cm}}{r^2 - 5r + 6} = \dfrac{1}{r - 2}$

78. $\dfrac{5x^2 - 6}{x^2 - x - 1} - \dfrac{\rule{1.2cm}{0.3cm}}{x^2 - x - 1} = \dfrac{-2x^2 + 6x - 12}{x^2 - x - 1}$

Cumulative Review Exercises

[2.3] **80.** The price of a gray suit is increased by 20%. The suit price is then decreased by $20. If the suit sells for $196, find the original price of the suit.

[2.4] **81.** A bottling machine fills and caps bottles at a rate of 80 per minute for a certain time period. Then the machine is slowed down and fills and caps bottles at a rate of 60 per minute. If the sum of the two time periods was 14 minutes, and the number of bottles filled and capped at the higher rate in the first period was the same as the number filled and capped at the lower rate in the second period, determine (a) how long the machine was used at the faster rate, and (b) the total number of bottles filled and capped over the 14-minute period.

[5.5] **82.** Divide $\dfrac{9x^4y^6 - 3x^3y^2 + 5xy^5}{3xy^4}$.   **83.** Divide $\dfrac{6x^2 + 5x - 4}{3x + 4}$.

JUST FOR FUN

Add or subtract as indicated.

1. $\dfrac{3}{x^2 - x - 6} - \dfrac{2}{x^2 + x - 6}$ **2.** $\dfrac{3}{x^3 + 27} + \dfrac{4}{x^2 - 9}$

7.4

Complex Fractions

▶ **1** Recognize complex fractions.

▶ **2** Simplify complex fractions by multiplying by a common denominator.

▶ **3** Simplify complex fractions by simplifying the numerator and denominator.

▶ **1**   A **complex fraction** is one that has a fractional expression in its numerator or its denominator or both its numerator and denominator. Examples of complex fractions include

$$\frac{\dfrac{2}{3}}{5}, \quad \frac{\dfrac{x+1}{x}}{3x}, \quad \frac{\dfrac{x}{y}}{x+1}, \quad \frac{\dfrac{a+b}{a}}{\dfrac{a-b}{b}}, \quad \frac{3+\dfrac{1}{x}}{\dfrac{1}{x^2}+\dfrac{3}{x}}$$

The expression above the main fraction line is the numerator, and the expression below the main fraction line is the denominator of the complex fraction.

We will explain two methods that can be used to simplify complex fractions.

secondary fraction ⟶ $\dfrac{a+b}{a}$ ⟵ numerator of complex fraction

—————————— ⟵ main fraction line

secondary fraction ⟶ $\dfrac{a-b}{b}$ ⟵ denominator of complex fraction

Method 1

▶ **2**   The first method involves multiplying both the numerator and denominator of the complex fraction by a common denominator.

To Simplify a Complex Fraction (by Multiplying by a Common Denominator)

1. Find the least common denominator of each of the two secondary fractions.
2. Next find the LCD of the complex fraction. The LCD of the complex fraction will be the LCD of the two expressions found in step 1.
3. Multiply both secondary fractions by the LCD of the complex fraction found in step 2.
4. Simplify when possible.

EXAMPLE 1 Simplify $\dfrac{\dfrac{2}{3} + \dfrac{3}{4}}{\dfrac{3}{4} - \dfrac{1}{2}}$.

Solution: *Step 1:* The LCD of the numerator of the complex fraction is 12. The LCD of the denominator is 4.

Step 2: The LCD of the complex fraction is the LCD of 12 and 4, which is 12.

Step 3: Multiply both secondary fractions by 12.

$$\frac{12\left(\dfrac{2}{3} + \dfrac{3}{4}\right)}{12\left(\dfrac{3}{4} - \dfrac{1}{2}\right)} = \frac{12\left(\dfrac{2}{3}\right) + 12\left(\dfrac{3}{4}\right)}{12\left(\dfrac{3}{4}\right) - 12\left(\dfrac{1}{2}\right)}$$

Step 4: Simplify.

$$= \frac{\overset{4}{\cancel{12}}\left(\dfrac{2}{3}\right) + \overset{3}{\cancel{12}}\left(\dfrac{3}{4}\right)}{\overset{3}{\cancel{12}}\left(\dfrac{3}{4}\right) - \overset{6}{\cancel{12}}\left(\dfrac{1}{2}\right)}$$

$$= \frac{8 + 9}{9 - 6} = \frac{17}{3} \qquad \blacksquare$$

EXAMPLE 2 Simplify $\dfrac{\dfrac{2}{x^2} - \dfrac{3}{x}}{\dfrac{x}{5}}$.

Solution: The LCD of the numerator of the complex fraction is x^2. The LCD of the denominator is 5. Therefore, the LCD of the complex fraction is $5x^2$. Multiply the numerator and denominator by $5x^2$.

$$\frac{5x^2\left(\dfrac{2}{x^2} - \dfrac{3}{x}\right)}{5x^2\left(\dfrac{x}{5}\right)} = \frac{5x^2\left(\dfrac{2}{x^2}\right) - 5x^2\left(\dfrac{3}{x}\right)}{5x^2\left(\dfrac{x}{5}\right)}$$

$$= \frac{10 - 15x}{x^3}$$

$$= \frac{5(2 - 3x)}{x^3} \qquad \blacksquare$$

EXAMPLE 3 Simplify $\dfrac{a + \dfrac{1}{b}}{b + \dfrac{1}{a}}$.

Solution: Multiply the numerator and denominator of the complex fraction by its LCD, ab.

Chapter 7 Rational Expressions and Equations

$$\frac{ab\left(a + \dfrac{1}{b}\right)}{ab\left(b + \dfrac{1}{a}\right)} = \frac{a^2b + a}{ab^2 + b} = \frac{a(ab + 1)}{b(ab + 1)} = \frac{a}{b}$$

EXAMPLE 4   Simplify $\dfrac{a^{-1} + ab^{-2}}{ab^{-2} - a^{-2}b^{-1}}$.

Solution: Rewrite each expression without negative exponents.

$$\frac{\dfrac{1}{a} + \dfrac{a}{b^2}}{\dfrac{a}{b^2} - \dfrac{1}{a^2b}}$$

The LCD of the numerator is ab^2. The LCD of the denominator is a^2b^2. The LCD of the complex fraction is a^2b^2.

$$\frac{a^2b^2\left(\dfrac{1}{a} + \dfrac{a}{b^2}\right)}{a^2b^2\left(\dfrac{a}{b^2} - \dfrac{1}{a^2b}\right)} = \frac{ab^2 + a^3}{a^3 - b}$$

Method 2   ▶ **3**

To Simplify a Complex Fraction (by Simplifying Numerator and Denominator)

Complex fractions can also be simplified as follows:

1. Add or subtract each secondary fraction as indicated.
2. Invert and multiply the denominator of the complex fraction by the numerator of the complex fraction.
3. Simplify when possible.

Example 5 will show how Example 4 can be completed using the alternative procedure given above.

EXAMPLE 5   Simplify $\dfrac{a^{-1} + ab^{-2}}{ab^{-2} - a^{-2}b^{-1}}$.

Solution: $\dfrac{a^{-1} + ab^{-2}}{ab^{-2} - a^{-2}b^{-1}} = \dfrac{\dfrac{1}{a} + \dfrac{a}{b^2}}{\dfrac{a}{b^2} - \dfrac{1}{a^2b}}$

Now add the fractions in the numerator and subtract the fractions in the denominator. The lowest common denominator of the numerator of the complex fraction is ab^2. The LCD of the denominator of the complex fraction is a^2b^2.

$$\frac{\dfrac{b^2}{b^2}\cdot\dfrac{1}{a}+\dfrac{a}{b^2}\cdot\dfrac{a}{a}}{\dfrac{a^2}{a^2}\cdot\dfrac{a}{b^2}-\dfrac{1}{a^2b}\cdot\dfrac{b}{b}}=\frac{\dfrac{b^2}{ab^2}+\dfrac{a^2}{ab^2}}{\dfrac{a^3}{a^2b^2}-\dfrac{b}{a^2b^2}}=\frac{\dfrac{a^2+b^2}{ab^2}}{\dfrac{a^3-b}{a^2b^2}}$$

Now invert the denominator of the complex fraction and multiply it by the numerator.

$$\frac{a^2+b^2}{ab^{\cancel{2}}}\cdot\frac{a^{\cancel{2}}b^{\cancel{2}}}{a^3-b}=\frac{a(a^2+b^2)}{a^3-b}\quad\text{or}\quad\frac{a^3+ab^2}{a^3-b}$$

Note that we obtain the same answer by either method. ■

When doing the exercises, unless a particular method is specified, you may use either procedure.

Exercise Set 7.4

Simplify.

1. $\dfrac{1+\dfrac{3}{5}}{2+\dfrac{1}{5}}$

2. $\dfrac{1-\dfrac{9}{16}}{3+\dfrac{4}{5}}$

3. $\dfrac{2+\dfrac{3}{8}}{1+\dfrac{1}{3}}$

4. $\dfrac{\dfrac{3}{5}+\dfrac{2}{7}}{\dfrac{1}{5}+\dfrac{5}{6}}$

5. $\dfrac{\dfrac{4}{9}-\dfrac{3}{8}}{4-\dfrac{3}{5}}$

6. $\dfrac{1-\dfrac{x}{y}}{x}$

7. $\dfrac{\dfrac{x^2y}{4}}{\dfrac{2}{x}}$

8. $\dfrac{\dfrac{15a}{b^2}}{\dfrac{b^3}{5}}$

9. $\dfrac{\dfrac{8x^2y}{3z^3}}{\dfrac{4xy}{9z^5}}$

10. $\dfrac{\dfrac{36x^4}{5y^4z^5}}{\dfrac{9xy^2}{15z^5}}$

11. $\dfrac{x+\dfrac{1}{y}}{\dfrac{x}{y}}$

12. $\dfrac{x-\dfrac{x}{y}}{\dfrac{1+x}{y}}$

13. $\dfrac{\dfrac{9}{x}+\dfrac{3}{x^2}}{3+\dfrac{1}{x}}$

14. $\dfrac{\dfrac{2}{a}+\dfrac{1}{2a}}{a+\dfrac{a}{2}}$

15. $\dfrac{3-\dfrac{1}{y}}{2-\dfrac{1}{y}}$

16. $\dfrac{\dfrac{x}{x-y}}{\dfrac{x^2}{y}}$

17. $\dfrac{\dfrac{x}{y}-\dfrac{y}{x}}{\dfrac{x+y}{x}}$

18. $\dfrac{1}{\dfrac{1}{x}+y}$

19. $\dfrac{\dfrac{a^2}{b}-b}{\dfrac{b^2}{a}-a}$

20. $\dfrac{\dfrac{1}{x}+\dfrac{2}{x^2}}{2+\dfrac{1}{x^2}}$

21. $\dfrac{\dfrac{a}{b}-2}{\dfrac{-a}{b}+2}$

22. $\dfrac{\dfrac{x^2-y^2}{x}}{\dfrac{x+y}{x^3}}$

23. $\dfrac{\dfrac{4x+8}{3x^2}}{\dfrac{4x}{6}}$

24. $\dfrac{\dfrac{a}{a+1}-1}{\dfrac{2a+1}{a-1}}$

25. $\dfrac{\dfrac{x}{4}-\dfrac{1}{x}}{1+\dfrac{x+4}{x}}$

26. $\dfrac{1+\dfrac{x}{x+1}}{\dfrac{2x+1}{x-1}}$

27. $\dfrac{\dfrac{1}{x-1}+1}{\dfrac{1}{x+1}-1}$

28. $\dfrac{\dfrac{a+1}{a-1}+\dfrac{a-1}{a+1}}{\dfrac{a+1}{a-1}-\dfrac{a-1}{a+1}}$

29. $\dfrac{\dfrac{a-2}{a+2} - \dfrac{a+2}{a-2}}{\dfrac{a-2}{a+2} + \dfrac{a+2}{a-2}}$

30. $\dfrac{\dfrac{5}{5-x} + \dfrac{6}{x-5}}{\dfrac{3}{x} + \dfrac{2}{x-5}}$

Simplify.

31. $2a^{-2} + b$

32. $3a^{-2} + b^{-1}$

33. $(a^{-1} + b^{-1})^{-1}$

34. $\dfrac{a^{-1} + b^{-1}}{ab}$

35. $\dfrac{a^{-1} + b^{-1}}{\dfrac{1}{ab}}$

36. $\dfrac{a^{-1} + 1}{b^{-1} - 1}$

37. $\dfrac{\dfrac{a}{b} + a^{-1}}{\dfrac{b}{a} + a^{-1}}$

38. $\dfrac{a^{-1} + b^{-1}}{a^{-1}}$

39. $\dfrac{x^{-1} - y^{-1}}{x^{-1} + y^{-1}}$

40. $\dfrac{x^{-2} + \dfrac{1}{x}}{x^{-1} + x^{-2}}$

41. $\dfrac{a^{-1} + b^{-1}}{(a+b)^{-1}}$

42. $\dfrac{3a^{-1} - b^{-1}}{(a-b)^{-1}}$

43. $2x^{-1} - (3y)^{-1}$

44. $\dfrac{\dfrac{5}{x} + \dfrac{1}{y}}{(x-y)^{-1}}$

45. The efficiency of a jack, E, is given by the formula

$$E = \dfrac{\dfrac{1}{2}h}{h + \dfrac{1}{2}}$$

where h is determined by the pitch of the jack's thread.

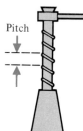

Pitch

Determine the efficiency of a jack whose value of h is:

(a) $\dfrac{2}{3}$ **(b)** $\dfrac{4}{5}$

46. If two resistors with resistances R_1 and R_2 are connected in parallel, their combined resistance, R_T, can be found from the formula on the right:

$$R_T = \dfrac{1}{\dfrac{1}{R_1} + \dfrac{1}{R_2}}$$

Simplify the right side of the formula.

47. If three resistors with resistances R_1, R_2, and R_3 are connected in parallel, their combined resistance can be found by the following formula:

$$R_T = \dfrac{1}{\dfrac{1}{R_1} + \dfrac{1}{R_2} + \dfrac{1}{R_3}}$$

Simplify the right side of this formula.

48. A formula used in the study of optics is

$$f = (p^{-1} + q^{-1})^{-1}$$

where p is the object's distance from a lens, q is the image distance from the lens, and f is the focal length of the lens. Express the right side of the formula without any negative exponents.

49. What is a complex fraction?

50. We have indicated two procedures for evaluating complex fractions. Which procedure do you prefer? Briefly explain why.

Cumulative Review Exercises

[1.7] **51.** Evaluate $\dfrac{\left|-\dfrac{3}{9}\right| - \left(-\dfrac{5}{9}\right) \cdot \left|-\dfrac{3}{8}\right|}{|-5 - (-3)|}$.

[2.6] **52.** Find the solution set to the inequality $\left|\dfrac{4 - 2x}{3}\right| \geq 3$.

[3.6] **53.** Graph the inequality $6y - 3x < 12$.

[5.6] **54.** Divide using synthetic division:
$$x^3 - 7x^2 - 13x + 9 \div (x - 2) \text{ (optional)}.$$

JUST FOR FUN

Simplify.

1. $\dfrac{1}{2a + \dfrac{1}{2a + \dfrac{1}{2a}}}$

2. $\dfrac{1}{x + \dfrac{1}{x + \dfrac{1}{x + 1}}}$

7.5

Solving Equations Containing Rational Expressions

▶ **1** Solve equations containing fractions by multiplying by the LCD.

▶ **2** Know when proposed solutions must be checked.

▶ **3** Solve proportions by cross multiplication.

▶ **4** Solve application problems using rational equations.

▶ **5** Solve for a variable in a formula containing fractions.

▶ **1** In Sections 7.1 through 7.4 we presented techniques to add, subtract, multiply, and divide rational expressions. In this section we present a method for solving equations containing fractions.

> **To Solve Equations Containing Fractions**
>
> 1. Determine the LCD of all fractions in the equation.
> 2. Multiply **both** sides of the equation by the LCD. This will result in every term in the equation being multiplied by the LCD.
> 3. Remove any parentheses and combine like terms on each side of the equation.
> 4. Solve the equation using the properties discussed in earlier sections.
> 5. Check the solution in the original equation.

In step 2, we multiply both sides of the equation by the LCD to eliminate fractions from the equation.

EXAMPLE 1 Solve $\dfrac{x}{3} + 2x = 7$.

Solution: $\boxed{3}\left(\dfrac{x}{3} + 2x\right) = 7 \cdot \boxed{3}$ Multiply both sides of the equation by the LCD, 3.

$\cancel{3}\left(\dfrac{x}{\cancel{3}}\right) + 3 \cdot 2x = 7 \cdot 3$ This has the effect of multiplying each term in the equation by the LCD.

$x + 6x = 21$

$7x = 21$

$x = 3$

Check: $x = 3$

$$\frac{x}{3} + 2x = 7$$

$$\frac{3}{3} + 2(3) = 7$$

$$1 + 6 = 7$$

$$7 = 7 \qquad \text{true} \qquad \blacksquare$$

EXAMPLE 2   Solve $\dfrac{3}{4} + \dfrac{5x}{9} = \dfrac{x}{6}$.

Solution: Multiply both sides of the equation by the LCD, 36.

$$36\left(\frac{3}{4} + \frac{5x}{9}\right) = \frac{x}{6} \cdot 36$$

$$\overset{9}{36}\left(\frac{3}{4}\right) + \overset{4}{36}\left(\frac{5x}{9}\right) = \frac{x}{6} \cdot \overset{6}{36}$$

$$27 + 20x = 6x$$

$$27 = -14x$$

$$x = \frac{27}{-14} \quad \text{or} \quad \frac{-27}{14} \qquad \blacksquare$$

▸**2**   **Whenever a variable appears in any denominator, you must check your proposed answer in the original equation. When checking, if a proposed answer makes any denominator equal to zero, that value is not a solution to the equation.** Such values are called **extraneous roots** or **extraneous solutions.** An extraneous root is a number obtained when solving an equation that is not a solution to the original equation.

EXAMPLE 3   Solve $3 - \dfrac{4}{x} = \dfrac{5}{2}$.

Solution: Multiply both sides of the equation by the LCD, $2x$.

$$2x\left(3 - \frac{4}{x}\right) = \left(\frac{5}{2}\right) \cdot 2x$$

$$2x(3) - 2x\left(\frac{4}{x}\right) = \left(\frac{5}{2}\right)2x$$

$$6x - 8 = 5x$$

$$x - 8 = 0$$

$$x = 8$$

Check: $x = 8$

$$3 - \frac{4}{x} = \frac{5}{2}$$

$$3 - \frac{4}{8} = \frac{5}{2}$$

$$3 - \frac{1}{2} = \frac{5}{2}$$

$$\frac{5}{2} = \frac{5}{2} \qquad \text{true}$$

Since 8 checks, it is the solution to the equation. $\blacksquare$

In this book, some checks will be omitted to conserve space. You should, however, check all answers when a variable appears in any denominator.

EXAMPLE 4 Solve $x + \dfrac{12}{x} = -7$.

Solution: $\boxed{x} \cdot \left(x + \dfrac{12}{x}\right) = -7 \cdot \boxed{x}$ *Multiply both sides of the equation by x.*

$$x(x) + x\left(\dfrac{12}{x}\right) = -7x$$

$$x^2 + 12 = -7x$$

$$x^2 + 7x + 12 = 0$$

$$(x + 3)(x + 4) = 0$$

$$x + 3 = 0 \quad \text{or} \quad x + 4 = 0$$

$$x = -3 \qquad\qquad x = -4$$

Checks of -3 and -4 will show that they are solutions to the equation. ■

EXAMPLE 5 Solve $\dfrac{2x}{x^2 - 4} + \dfrac{1}{x - 2} = \dfrac{2}{x + 2}$.

Solution: $\dfrac{2x}{(x + 2)(x - 2)} + \dfrac{1}{x - 2} = \dfrac{2}{x + 2}$

Multiply both sides of the equation by the LCD, $(x + 2)(x - 2)$.

$$(x + 2)(x - 2) \cdot \left[\dfrac{2x}{(x + 2)(x - 2)} + \dfrac{1}{x - 2}\right] = \dfrac{2}{x + 2} \cdot (x + 2)(x - 2)$$

$$(x+2)(x-2) \cdot \dfrac{2x}{(x+2)(x-2)} + (x + 2)(x-2) \cdot \dfrac{1}{x-2} = \dfrac{2}{x+2} \cdot (x+2)(x - 2)$$

$$2x + (x + 2) = 2(x - 2)$$

$$2x + x + 2 = 2x - 4$$

$$3x + 2 = 2x - 4$$

$$x + 2 = -4$$

$$x = -6$$

A check will show that -6 is the solution. ■

▶ 3 Proportions of the form $\dfrac{a}{b} = \dfrac{c}{d}$ were introduced in Section 2.1. Proportions are a specific type of fractional equation. Recall that to solve a proportion we often use *cross multiplication*.

$$\text{If } \dfrac{a}{b} = \dfrac{c}{d}, \text{ then } ad = bc, \, b \neq 0, \quad d \neq 0$$

When solving a proportion where the denominator of one or more of the ratios contains a variable, you must check to make sure that your answer is a true solution and not an extraneous solution.

Proportions are often used when working with similar figures. **Similar figures**

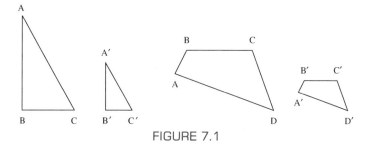

FIGURE 7.1

are figures whose corresponding angles are the same and whose corresponding sides are in proportion. Figure 7.1 illustrates two sets of similar figures.

In Fig. 7.1a, the ratio of the length of side AB to the length of side BC is the same as the ratio of the length of side $A'B'$ to the length of side $B'C'$. That is,

$$\frac{AB}{BC} = \frac{A'B'}{B'C'}$$

The unknown side of similar figures can often be found using proportions, as illustrated in Example 6.

EXAMPLE 6   Triangles ABC and $A'B'C'$ in Fig. 7.2 are similar figures. Find the length of sides AB and $B'C'$.

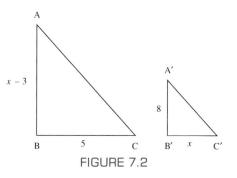

FIGURE 7.2

Solution: We can set up a proportion and then solve.

$$\frac{AB}{BC} = \frac{A'B'}{B'C'}$$

$$\frac{x-3}{5} = \frac{8}{x}$$

$$x(x-3) = 8 \cdot 5$$

$$x^2 - 3x = 40$$

$$x^2 - 3x - 40 = 0$$

$$(x-8)(x+5) = 0$$

$$x - 8 = 0 \quad \text{or} \quad x + 5 = 0$$

$$x = 8 \qquad\qquad x = -5$$

Since the length of the side of a triangle cannot be a negative number, -5 is not a possible answer. Substituting 8 for x, we see that the length of side $B'C'$ is 8 and the length of side AB is $8 - 3$ or 5.

Check:
$$\frac{AB}{BC} = \frac{A'B'}{B'C'}$$

$$\frac{5}{5} = \frac{8}{8}$$

$$1 = 1 \qquad \text{true}$$

Most proportions will be solved by using cross multiplication. Example 7 illustrates a proportion that can be more easily solved by multiplying both sides of the equation by the LCD, rather than by using cross multiplication.

EXAMPLE 7 Solve $\dfrac{x^2}{x-4} = \dfrac{16}{x-4}$.

Solution: This equation is a proportion, but if you try to solve it using cross multiplication, what will happen? You will end up with a third-degree or cubic equation. When this happens, try multiplying both sides of the equation by the LCD. We will multiply both sides of the equation by the LCD, $x - 4$.

$$(x-4) \cdot \frac{x^2}{x-4} = \frac{16}{x-4} \cdot (x-4)$$

$$x^2 = 16$$

$$x^2 - 16 = 0 \qquad \text{Factor the difference of squares.}$$

$$(x+4)(x-4) = 0$$

$$x + 4 = 0 \quad \text{or} \quad x - 4 = 0$$

$$x = -4 \qquad\qquad x = 4$$

Check:

$x = -4$	$x = 4$

$$\frac{x^2}{x-4} = \frac{16}{x-4} \qquad\qquad \frac{x^2}{x-4} = \frac{16}{x-4}$$

$$\frac{(-4)^2}{-4-4} = \frac{16}{-4-4} \qquad\qquad \frac{(4)^2}{4-4} = \frac{16}{4-4}$$

$$\frac{16}{-8} = \frac{16}{-8} \qquad\qquad \frac{16}{0} = \frac{16}{0} \qquad \text{no solution}$$

$$-2 = -2 \qquad \text{true}$$

Since 4 results in a denominator of 0, 4 is *not* a solution to the equation. The 4 is an extraneous root. The only solution to the equation is -4.

▶ **4** Now let us look at some applications of fractional equations.

EXAMPLE 8 A formula frequently used in optics is

$$\frac{1}{p} + \frac{1}{q} = \frac{1}{f}$$

where p represents the distance of the object from a mirror (or lens), q represents the distance of the image from the mirror (or lens), and f represents the focal length of the mirror (or lens). If a mirror has a focal length of 10 centimeters, how far from the mirror will the image appear when the object is 30 centimeters from the mirror?

Solution: The object distance, p, is 30 centimeters and the focal length, f, is 10 centimeters. We are asked to find the image distance, q.

$$\frac{1}{p} + \frac{1}{q} = \frac{1}{f}$$

$$\frac{1}{30} + \frac{1}{q} = \frac{1}{10}$$

Multiply both sides of the equation by the LCD, $30q$.

$$30q\left(\frac{1}{30} + \frac{1}{q}\right) = 30q\left(\frac{1}{10}\right)$$

$$30q\left(\frac{1}{30}\right) + 30q\left(\frac{1}{q}\right) = \overset{3}{30}q\left(\frac{1}{10}\right)$$

$$q + 30 = 3q$$

$$30 = 2q$$

$$15 = q$$

Thus the image will appear at a distance of 15 centimeters from the mirror.   ■

EXAMPLE 9   In electronics the total resistance, R_T, of resistors wired in a parallel circuit is determined by the formula

$$\frac{1}{R_T} = \frac{1}{R_1} + \frac{1}{R_2} + \frac{1}{R_3} + \cdots + \frac{1}{R_n}$$

where $R_1, R_2, R_3, \ldots, R_n$ are the resistances of the individual resistors (measured in ohms) in the circuit. Find the total resistance if two resistors, one of 200 ohms and the other of 300 ohms, are wired in a parallel circuit.

Solution: Since there are only two resistances, we use the formula

$$\frac{1}{R_T} = \frac{1}{R_1} + \frac{1}{R_2}$$

Let $R_1 = 200$ ohms and $R_2 = 300$ ohms; then

$$\frac{1}{R_T} = \frac{1}{200} + \frac{1}{300}$$

Multiply both sides of the equation by the LCD, $600R_T$.

$$600R_T \cdot \frac{1}{R_T} = 600R_T\left(\frac{1}{200} + \frac{1}{300}\right)$$

$$600R_T \cdot \frac{1}{R_T} = \overset{3}{600}R_T\left(\frac{1}{200}\right) + \overset{2}{600}R_T\left(\frac{1}{300}\right)$$

$$600 = 3R_T + 2R_T$$

$$600 = 5R_T$$

$$R_T = \frac{600}{5} = 120$$

Thus the total resistance of the parallel circuit is 120 ohms.   ■

EXAMPLE 10 If three identical resistors are to be wired in parallel, what should be the resistance of each resistor if the total resistance of the circuit is to be 300 ohms?

Solution: Let x = resistance of each resistor.

$$\frac{1}{R_T} = \frac{1}{R_1} + \frac{1}{R_2} + \frac{1}{R_3}$$

Since R_1, R_2, and R_3 are all the same value,

$$\frac{1}{300} = \frac{1}{x} + \frac{1}{x} + \frac{1}{x}$$

$$\frac{1}{300} = \frac{3}{x}$$

$$x = 900$$

Each of the three resistors should be 900 ohms. ■

▶ **5** Now we will do an example where we solve for a variable in a formula that contains fractions.

EXAMPLE 11 In Example 8 we used the formula $\dfrac{1}{p} + \dfrac{1}{q} = \dfrac{1}{f}$. Solve this formula for f.

Solution: Our goal is to isolate the variable f. We begin by multiplying both sides of the equation by the least common denominator, pqf, to eliminate fractions.

$$\frac{1}{p} + \frac{1}{q} = \frac{1}{f}$$

$$pqf\left(\frac{1}{p} + \frac{1}{q}\right) = pqf\left(\frac{1}{f}\right)$$

$$pqf\left(\frac{1}{p}\right) + pqf\left(\frac{1}{q}\right) = pqf\left(\frac{1}{f}\right)$$

$$qf + pf = pq$$

$$f(q + p) = pq$$

$$\frac{f(q + p)}{(q + p)} = \frac{pq}{q + p}$$

$$f = \frac{pq}{q + p} \quad \text{or,} \quad f = \frac{pq}{p + q}$$ ■

Exercise Set 7.5

Solve each equation, and then check your solution.

1. $\dfrac{2}{5} = \dfrac{x}{10}$

2. $\dfrac{3}{k} = \dfrac{9}{6}$

3. $\dfrac{x}{8} = \dfrac{-15}{4}$

4. $\dfrac{a}{25} = \dfrac{12}{10}$

5. $\dfrac{9}{3b} = \dfrac{-6}{2}$

6. $\dfrac{1}{4} = \dfrac{z + 1}{8}$

7. $\dfrac{4x + 5}{6} = \dfrac{7}{2}$

8. $\dfrac{a}{5} = \dfrac{a - 3}{2}$

(handwritten top margin:) $36x + 42 = 20x + 90$ $16x + 42 = 90$ $\frac{16x}{16} = \frac{48}{16}$ $x = 3$

(handwritten left:) $36x + 42$

9. $\dfrac{6x + 7}{10} = \dfrac{2x + 9}{6}$

10. $\dfrac{n}{10} = 9 - \dfrac{n}{5}$

11. $\dfrac{x}{3} - \dfrac{3x}{4} = \dfrac{1}{12}$

12. $\dfrac{2}{8} + \dfrac{3}{4} = \dfrac{w}{5}$

13. $\dfrac{3}{4} - x = 2x$

14. $\dfrac{2}{y} + \dfrac{1}{2} = \dfrac{5}{2y}$

15. $\dfrac{5}{3x} + \dfrac{3}{x} = 1$

16. $\dfrac{x}{4} - \dfrac{x}{6} = \dfrac{1}{4}$

17. $\dfrac{x - 1}{x - 5} = \dfrac{4}{x - 5}$

18. $\dfrac{2x + 3}{x + 1} = \dfrac{3}{2}$

19. $\dfrac{5y - 3}{7} = \dfrac{15y - 2}{28}$

20. $\dfrac{2}{x + 1} = \dfrac{1}{x - 2}$

21. $\dfrac{5}{-x - 6} = \dfrac{2}{x}$

22. $\dfrac{4}{y - 3} = \dfrac{6}{y + 3}$

23. $\dfrac{x - 2}{x + 4} = \dfrac{x + 1}{x + 10}$

24. $\dfrac{x - 3}{x + 1} = \dfrac{x - 6}{x + 5}$

25. $x - \dfrac{4}{3x} = -\dfrac{1}{3}$

26. $\dfrac{b}{2} - \dfrac{4}{b} = -\dfrac{7}{2}$

27. $\dfrac{2x - 1}{3} - \dfrac{3x}{4} = \dfrac{5}{6}$

28. $x + \dfrac{3}{x} = \dfrac{12}{x}$

29. $x + \dfrac{6}{x} = -5$

30. $\dfrac{15}{x} + \dfrac{9x - 7}{x + 2} = 9$

31. $\dfrac{3y - 2}{y + 1} = 4 - \dfrac{y + 2}{y - 1}$

32. $\dfrac{2b}{b + 1} = 2 - \dfrac{5}{2b}$

33. $\dfrac{1}{x + 3} + \dfrac{1}{x - 3} = \dfrac{-5}{x^2 - 9}$

34. $c - \dfrac{c}{3} + \dfrac{c}{5} = 26$

35. $\dfrac{2}{x - 3} - \dfrac{4}{x + 3} = \dfrac{8}{x^2 - 9}$

36. $\dfrac{x + 1}{x + 3} + \dfrac{x - 3}{x - 2} = \dfrac{2x^2 - 15}{x^2 + x - 6}$

37. $\dfrac{y}{2y + 2} + \dfrac{2y - 16}{4y + 4} = \dfrac{2y - 3}{y + 1}$

38. $\dfrac{3}{x + 3} + \dfrac{5}{x + 4} = \dfrac{12x + 19}{x^2 + 7x + 12}$

39. $\dfrac{1}{x + 2} + \dfrac{1}{x - 2} = \dfrac{4}{x^2 - 4}$

40. $\dfrac{4r - 1}{r^2 + 5r - 14} = \dfrac{1}{r - 2} - \dfrac{2}{r + 7}$

41. $\dfrac{5}{x^2 + 4x + 3} + \dfrac{2}{x^2 + x - 6} = \dfrac{3}{x^2 - x - 2}$

42. $\dfrac{2}{x^2 + 2x - 8} - \dfrac{1}{x^2 + 9x + 20} = \dfrac{4}{x^2 + 3x - 10}$

For each pair of similar figures, find the length of the two unknown sides (that is, those two sides indicated with the variable x).

43.

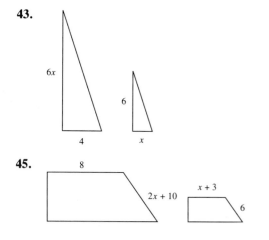

44.

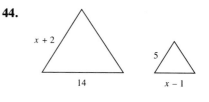

45.

46.

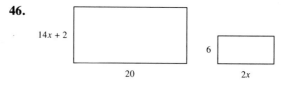

Solve the formula for the indicated variable.

47. $d = \dfrac{fl}{f + w}$ for w (physics)

48. $d = \dfrac{fl}{f + w}$ for f (physics)

49. $\dfrac{1}{p} + \dfrac{1}{q} = \dfrac{1}{f}$ for p (optics)

50. $\dfrac{1}{R_T} = \dfrac{1}{R_1} + \dfrac{1}{R_2}$ for R_T (electronics)

51. $\dfrac{1}{R_T} = \dfrac{1}{R_1} + \dfrac{1}{R_2}$ for R_1 (electronics)

52. $\dfrac{1}{p} + \dfrac{1}{q} = \dfrac{1}{f}$ for q (optics)

53. $z = \dfrac{\bar{x} - \mu}{\dfrac{\sigma}{\sqrt{n}}}$ for $\bar{x}$ (statistics)

54. $z = \dfrac{\bar{x} - \mu}{\dfrac{\sigma}{\sqrt{n}}}$ for μ (statistics)

55. $E = \dfrac{q}{\epsilon_0 A} - \dfrac{q'}{\epsilon_0 A}$ for q (physics)

56. $S_f - S_i = -\dfrac{Q}{T_1} + \dfrac{Q}{T_2}$ for Q (chemistry)

57. $\dfrac{1}{C_T} = \dfrac{1}{C_1} + \dfrac{1}{C_2} + \dfrac{1}{C_3}$ for C_T (electronics)

58. $\dfrac{n_1}{o} + \dfrac{n_2}{i} = \dfrac{n_1 - n_2}{r}$ for n_2 (optics)

Use the formula $\dfrac{1}{p} + \dfrac{1}{q} = \dfrac{1}{f}$ *for Exercises 59 through 64. See Example 8.*

59. Find the distance of the image from the mirror if the object is 12 inches from the mirror and the focal length is 6 inches.

60. If the object distance is 12 inches and the image distance is 20 inches, find the focal length of the mirror.

61. If the focal length of the mirror is 8 centimeters and the image distance from the mirror is 15 centimeters, find the object's distance from the mirror.

62. If the object distance is 15 centimeters and the image distance is 9 centimeters, find the focal length of the mirror.

63. A mirror has a focal length of 12 centimeters. Find the object's distance and image distance if the image distance is 3 times the object distance.

64. A mirror has a focal length of 2 inches. Find the object's distance and image distance if the object's distance is 3 inches more than the image distance.

Refer to Examples 9 and 10 for Exercises 65 through 68.

65. What is the total resistance in the circuit if resistors of 200 ohms and 700 ohms are connected in parallel?

66. What is the total resistance in the circuit if resistors of 500 ohms and 750 ohms are connected in parallel?

67. What is the total resistance in the circuit if resistors of 300 ohms, 500 ohms, and 3000 ohms are connected in parallel?

68. Three resistors of identical resistance are to be connected in parallel. What should be the resistance of each resistor if the circuit is to have a total resistance of 700 ohms?

69. What is an extraneous root?

70. Under what circumstances is it necessary to check your answers for extraneous roots?

71. Consider the equation $\dfrac{x}{4} - \dfrac{x}{3} = 2$ and the expression $\dfrac{x}{4} - \dfrac{x}{3} + 2$.

(a) What is the first step in solving the equation? Explain what effect the first step will have on the equation.

(b) Solve the equation.

(c) What is the first step in simplifying the expression? Explain what effect the first step has when simplifying the expression.

(d) Simplify the expression.

72. What are similar figures?

Cumulative Review Exercises

[3.5] **73.** $f(x) = \frac{1}{2}x^2 - 3x + 4$ find $f(5)$.

[4.5] **74.** Find the solution to the system of inequalities
$$y < 6x + 4$$
$$2y \geq -3x + 4$$

[5.8] **75.** Graph the function $f(x) = x^2 - 4x - 6$ and give the domain and range.

[6.3] **76.** Factor $8x^3 - 64y^6$

JUST FOR FUN

1. In Example 7 on page 58 we discussed the formula for converting a tax free investment, T_f, to a taxable investment, T_a. The formula we used was

$$T_a = \dfrac{T_f}{1 - f}$$

In the formula, f represents your federal tax bracket. Some investments, such as certain municipal bonds and municipal bond funds, are not only federally tax free but are also state and county or city tax free. When you wish to compare a taxable investment with an investment which is federal, state, and county tax free, you must replace *f* in the previous formula with your combined *true tax bracket,* which is $[f + (s + c)(1 - f)]$, where *s* is your state tax bracket, and *c* is your county or local tax bracket. If you are investing in a federal, state, and county tax free investment, the above formula then becomes

$$T_a = \frac{T_f}{1 - [f + (s + c)(1 - f)]}$$

Mr. Levy who lives in Detroit, Michigan, is burdened with a 4.6% state tax bracket, a 3% city tax bracket, and a 33% federal tax bracket. He is choosing between the Fidelity Michigan Triple *Tax Free* Money Market Portfolio yielding 6.01% and the Fidelity *Taxable* Cash Reserve Money Market Fund yielding 7.68%. With his tax brackets as given, (a) determine the taxable equivalent of the 6.01% tax free yield; (b) which investment should Mr. Levy make? Explain your answer.

2. The synodic period of Mercury is the time required for swiftly moving Mercury to gain one lap on Earth in their orbits around the sun. If the orbital periods (in Earth days) of the two planets are designated P_m and P_e, Mer-

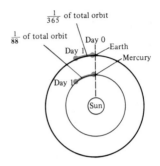

cury will be seen on the average to move $1/P_m$ of a revolution per day, while Earth moves $1/P_e$ of a revolution per day in pursuit. Mercury's daily gain on Earth is $(1/P_m - 1/P_e)$ of a revolution, so that the time for Mercury to gain one complete revolution on Earth, the synodic period *s*, may be found by the formula

$$\frac{1}{s} = \frac{1}{P_m} - \frac{1}{P_e}$$

If $P_e = 365$ days and P_m is 88 days, find the synodic period in units of terrestrial (Earth) days.

7.6

Applications of Rational Equations

▶ **1** Set up and solve work problems.

▶ **2** Set up and solve number problems.

▶ **3** Set up and solve motion problems.

Some applications of rational equations were illustrated in Section 7.5. In this section we examine some additional applications. We study work problems first.

Work Problems

▶ **1** Problems where two or more machines or people work together to complete a certain task are sometimes referred to as **work problems.** Work problems often involve equations containing fractions. Generally, work problems are based on the fact that the part of the work done by person 1 (or machine 1) plus the part of the work done by person 2 (or machine 2) is equal to the total amount of work done by both people (or both machines).

| Part of task done by first person or machine | + | Part of task done by second person or machine | = | 1 (one whole task completed) |

When we do work problems, we designate the total task completed as 1 (for 1 whole task completed). To determine the part of the task done by each person or machine we use the formula

Part of task completed = rate · time

This formula is very similar to the formula *amount = rate · time* that was discussed in Section 2.4.

Let us now discuss how to determine the rate. If, for example, John can do a particular task in 5 hours, then he could complete $\frac{1}{5}$ of the task in one hour. Thus his rate is $\frac{1}{5}$ of the task per hour. If Mary can do a job in 6 hours, then her rate is $\frac{1}{6}$ of the job per hour. Similarly, if Maria can do a job in x minutes, her rate is $\frac{1}{x}$ of the job per minute. *In general, if a person or machine can complete a task in x units of time, the rate is $\frac{1}{x}$.*

EXAMPLE 1 Bob can mow Mr. Richard's lawn in 3 hours. Steve can mow Mr. Richard's lawn in 4 hours. How long will it take to mow the lawn if both Bob and Steve work together?

Solution: Let x = time, in hours, for both boys together to mow the lawn. We construct a table to assist us in finding the part of the task completed by Bob and Steve.

	Rate of Work	Time Worked	Part of Task Completed
Bob	$\frac{1}{3}$	x	$\frac{x}{3}$
Steve	$\frac{1}{4}$	x	$\frac{x}{4}$

$$\left(\begin{array}{c} \text{part of lawn mowed} \\ \text{by Bob in } x \text{ hours} \end{array} \right) + \left(\begin{array}{c} \text{part of lawn mowed} \\ \text{by Steve in } x \text{ hours} \end{array} \right) = 1 \text{ (whole lawn mowed)}$$

$$\frac{x}{3} \qquad + \qquad \frac{x}{4} \qquad = \qquad 1$$

Multiply both sides of the equation by the LCD, 12; then solve for x.

$$12\left(\frac{x}{3} + \frac{x}{4} \right) = 12 \cdot 1$$

$$12\left(\frac{x}{3} \right) + 12\left(\frac{x}{4} \right) = 12$$

$$4x + 3x = 12$$

$$7x = 12$$

$$x = \frac{12}{7} \qquad 1.71 \text{ hours (to the nearest hundredth)}$$

The two boys together can mow the lawn in about 1.71 hours. Note that this is less time than it takes either boy to mow the lawn by himself (which is what we expect). ∎

EXAMPLE 2   A tank can be filled by one pipe in 4 hours and can be emptied by another pipe in 6 hours. If the valves to both pipes are open, how long will it take to fill the tank?

Solution: As one pipe is filling the tank, the other is emptying the tank. Thus the pipes are working against each other.

Let x = amount of time to fill the tank

	Rate of Work	Time Worked	Part of Tank Filled or Emptied
Pipe filling tank	$\dfrac{1}{4}$	x	$\dfrac{x}{4}$
Pipe emptying tank	$\dfrac{1}{6}$	x	$\dfrac{x}{6}$

Since the pipes are working against each other, we will *subtract* the part of the water being emptied from the part of the water filling the tank.

$$\left(\begin{array}{c}\text{part of water tank}\\ \text{filled in } x \text{ hours}\end{array}\right) - \left(\begin{array}{c}\text{part of water tank}\\ \text{emptied in } x \text{ hours}\end{array}\right) = 1 \text{ (whole tank filled)}$$

$$\frac{x}{4} \qquad - \qquad \frac{x}{6} \qquad = \qquad 1$$

$$12\left(\frac{x}{4} - \frac{x}{6}\right) = 12 \cdot 1$$

$$12\left(\frac{x}{4}\right) - 12\left(\frac{x}{6}\right) = 12$$

$$3x - 2x = 12$$

$$x = 12$$

The tank will be filled in 12 hours.   ■

EXAMPLE 3   Patty and Mike work in the toy department at Sears, where they assemble bicycles. When Patty and Mike work together, they can assemble a bicycle in 20 minutes. When Patty assembles a bike by herself, it takes her 36 minutes. How long would it take Mike to assemble the bike by himself?

Solution: Let x = amount of time for Mike to assemble the bike by himself.
    Let us make a chart of the information given. We know that when working together they can assemble the bike in 20 minutes. We use this information in the table that follows.

	Rate of work	Time	Part of Bicycle Completed
Patty	$\dfrac{1}{36}$	20	$\dfrac{20}{36}$
Mike	$\dfrac{1}{x}$	20	$\dfrac{20}{x}$

$$\left(\begin{array}{c}\text{part of bicycle}\\ \text{assembled by Patty}\end{array}\right) + \left(\begin{array}{c}\text{part of bicycle}\\ \text{assembled by Mike}\end{array}\right) = 1$$

$$\frac{20}{36} \quad + \quad \frac{20}{x} \quad = 1$$

$$36x\left(\frac{20}{36} + \frac{20}{x}\right) = 36x \cdot 1$$

$$36x\left(\frac{20}{36}\right) + 36x\left(\frac{20}{x}\right) = 36x$$

$$20x + 720 = 36x$$

$$720 = 16x$$

$$45 = x$$

Thus Mike can assemble a bike by himself in 45 minutes. ∎

Number Problems

▸**2** Now let us look at a problem where we must find a given number.

EXAMPLE 4 What number multiplied by the numerator and added to the denominator of the fraction $\frac{4}{7}$ makes the resulting fraction equal to $\frac{5}{3}$?

Solution: Let x = unknown number.

$$\frac{4x}{7+x} = \frac{5}{3} \qquad \text{Now cross multiply.}$$

$$3(4x) = 5(7+x)$$

$$12x = 35 + 5x$$

$$7x = 35$$

$$x = 5$$

The number is 5.

Check: $\dfrac{4\cdot 5}{7+5} = \dfrac{20}{12} = \dfrac{5}{3}.$ ∎

EXAMPLE 5 When the reciprocal of 3 times a number is subtracted from 1, the result is the reciprocal of twice the number. Find the number.

Solution: Let x = unknown number. Then $3x$ is 3 times the number, and $\dfrac{1}{3x}$ is the reciprocal of 3 times the number. Twice the number is $2x$, and $\dfrac{1}{2x}$ is the reciprocal of twice the number.

$$1 - \frac{1}{3x} = \frac{1}{2x} \qquad \text{The LCD is } 6x.$$

$$6x\left(1 - \frac{1}{3x}\right) = \frac{1}{2x} \cdot 6x$$

$$6x(1) - 6x\left(\frac{1}{3x}\right) = 6x\left(\frac{1}{2x}\right)$$

$$6x - 2 = 3$$

$$6x = 5$$

$$x = \frac{5}{6}$$

The number is $\frac{5}{6}$. ■

Rate Problems

▶ **3** The last type of problem we will look at is rate problems. Recall that we discussed rate problems earlier in Section 2.4. In that section we learned that distance = rate · time. Sometimes it is convenient to solve for the time when solving rate problems.

$$\text{Time} = \frac{\text{distance}}{\text{rate}}$$

EXAMPLE 6 Amy Schumacher can fly her plane 300 miles against the wind in the same time it takes her to fly 400 miles with the wind. If the wind blows at 20 miles per hour, find the speed of the plane in still air.

Solution: Let x = speed of plane in still air. Let us set up a table to help analyze the problem.

Plane	d	r	t
Against wind	300	$x - 20$	$\dfrac{300}{x - 20}$
With wind	400	$x + 20$	$\dfrac{400}{x + 20}$

Since the times are the same, we set up and solve the following equation:

$$\frac{300}{x - 20} = \frac{400}{x + 20}$$

$$300(x + 20) = 400(x - 20)$$

$$300x + 6000 = 400x - 8000$$

$$6000 = 100x - 8000$$

$$14{,}000 = 100x$$

$$140 = x$$

The speed of the plane in still air is 140 miles per hour. ■

EXAMPLE 7 Mary Kay rides her bike to and from her home to San Francisco City College. Going to school, she rides mostly downhill and averages 15 miles per hour. Coming home, mostly uphill, she averages only 6 miles per hour. If it takes her $\frac{1}{2}$ hour longer for her to get home than to ride to school, how far is the college from her home?

Solution: Let x = the distance from her home to the college. Note that in this problem the times are not equal. Her time returning is $\frac{1}{2}$ hour longer than going. Therefore, to make the times equal, we must add $\frac{1}{2}$ hour to her time going (or subtract $\frac{1}{2}$ hour from her time returning).

	d	r	t
Going	x	15	$\dfrac{x}{15}$
Returning	x	6	$\dfrac{x}{6}$

$$\text{Time going} + \frac{1}{2} \text{ hour} = \text{time returning}$$

$$\frac{x}{15} + \frac{1}{2} = \frac{x}{6}$$

$$30\left(\frac{x}{15}\right) + 30\left(\frac{1}{2}\right) = 30\left(\frac{x}{6}\right)$$

$$2x + 15 = 5x$$

$$15 = 3x$$

$$5 = x$$

Therefore, Mary Kay lives 5 miles from San Francisco City College. ■

EXAMPLE 8 The number 4 train in the New York City subway system goes from Woodlawn/Jerome Avenue in the Bronx to Flatbush Avenue/Brooklyn College in Brooklyn. The total one-way distance between these two stops is 24.2 miles. On this route, two tracks run parallel to each other, one for the local train and the other for the express train. The local train stops at every station (48 stops), while the express stops at only certain selected stations (33 stations). The local and express trains leave Woodlawn/Jerome Avenue at the same time. When the express reaches Flatbush Avenue/Brooklyn College, the local is at Wall Street, 7.8 miles from Flatbush. If the express averages 5.2 miles per hour faster than the local, find the speed of the two trains.

Solution: Let x = speed of local
then $x + 5.2$ = speed of express

In the same time period that the express reaches the end of the line, 24.2 miles, the local will have traveled $24.2 - 7.8 = 16.4$ miles.

Train	d	r	t
Local	16.4	x	$\dfrac{16.4}{x}$
Express	24.2	$x + 5.2$	$\dfrac{24.2}{x + 5.2}$

$$\frac{16.4}{x} = \frac{24.2}{x + 5.2}$$

$$16.4(x + 5.2) = 24.2x$$

$$16.4 + 85.28 = 24.2x$$

$$85.28 = 7.8$$

$$10.9 = x$$

The local averages 10.9 miles per hour and the express averages $10.9 + 5.2 = 16.1$ miles per hour.   ∎

EXAMPLE 9   Dawn, who lives in Buffalo, New York, travels to college in South Bend, Indiana. She travels on expressways through New York, Pennsylvania, Ohio, and part of Indiana. In New York and Pennsylvania, the speed limit on the expressway is 55 miles per hour, while in Ohio and Indiana the speed limit is 65 miles per hour. The total distance traveled by Dawn is 490 miles. If Dawn follows the speed limits, and the total trip takes 8 hours, how long did she drive at 55 miles per hour, and how long did she drive at 65 miles per hour?

Solution: Let x = number of miles driven at 55 mph
then $490 - x$ = number of miles driven at 65 mph

	Distance	Rate	Time
NY, PA	x	55	$\dfrac{x}{55}$
OH, IN	$490 - x$	65	$\dfrac{490 - x}{65}$

Since the total time is 8 hours, we write

$$\frac{x}{55} + \frac{490 - x}{65} = 8$$

The LCD of 55 and 65 is 715.

$$715\left(\frac{x}{55} + \frac{490 - x}{65}\right) = 715 \cdot 8$$

$$715\left(\frac{x}{55}\right) + 715\left(\frac{490 - x}{65}\right) = 5720$$

$$13x + 11(490 - x) = 5720$$

$$13x + 5390 - 11x = 5720$$

$$2x + 5390 = 5720$$

$$2x = 330$$

$$x = 165$$

Thus the time driven at 55 mph is $165 \div 55 = 3$ hours, and the time driven at 65 mph is $\dfrac{490 - 165}{65} = \dfrac{325}{65} = 5$ hours.   ∎

Notice that in Example 9 the answer to the problem was not the value obtained for *x*. The value obtained was a distance, and the question asked us to find the time. *When working word problems, you must read and work the problems very carefully and make sure you answer the original question that was asked.*

A number of examples discussed in this section can be solved in other ways. In this section our emphasis is on solving fractional equations, and so we worked these examples using fractions. It is important for you to realize that problems can often be solved in more than one way. Try solving Examples 6 and 9 in a different way now.

Exercise Set 7.6

Solve each problem.

1. At the Boeing Corporation it takes one computer 4 hours to print checks for its employees and a second computer 5 hours to complete the same job. How long will it take the two computers together to complete the job?

2. Ramon can mow a lawn on a rider lawn mower in 4 hours. Donna can mow the lawn in 6 hours with a push lawn mower. How long will it take them to mow the lawn together?

3. A $\frac{1}{2}$-inch-diameter hose can fill a swimming pool in 8 hours. A $\frac{4}{5}$-inch-diameter hose can fill the same pool in 5 hours. How long will it take to fill the pool when both hoses are used?

4. A conveyer belt operating at full speed can fill a tank with top soil in 3 hours. When a valve at the bottom of the tank is opened, the tank will empty in 4 hours. If the conveyer belt is operating at full speed and the valve at the bottom of the tank is open, how long will it take to fill the tank?

5. A factory making antifreeze has vats to hold the antifreeze. Each vat has an inlet valve and an outlet valve. The vat can be filled with antifreeze in 20 hours when the inlet valve is wide open and the outlet valve is closed. The vat can be emptied in 25 hours when the outlet valve is wide open and the inlet valve is closed. If a new vat is placed in operation and both the inlet valve and outlet valve are wide open, how long will it take to fill the vat?

6. When Mrs. Dellaquilla rides the power lawn mower and Mr. Dellaquilla uses the push lawn mower, they can mow their large lawn together in 2 hours. Mrs. Dellaquilla can mow the entire lawn by herself on the rider lawn mower in 3 hours. How long would it take Mr. Dellaquila to mow the entire lawn by himself using the push mower?

7. When two steam rollers are working together, they can roll a length of road in 10 hours. When the larger steam roller is working by itself, it can roll the length of road in 16 hours. How long would it take the smaller steam roller working by itself to roll the length of road?

8. Franki can plant a garden by herself in 4 hours. When her young son Garyn helps her, the total time it takes them to plant the garden is 3 hours. How long would it take Garyn to plant the garden himself?

9. When a professor and a graduate student work together, they can mark a set of mathematics final exams in 3.2 hours. When the professor marks the papers alone, it takes 5.7 hours. How long would it take the graduate student to mark the papers alone?

10. When only the cold water valve is opened, a washtub will fill in 8 minutes. When only the hot water valve is opened, the washtub will fill in 12 minutes. When the drain of the washtub is open, it will drain completely in 7 minutes. If both the hot and cold water valves are open and the drain is open, how long will it take for the washtub to fill?

11. A large tank is being used on the Donovan's farm to irrigate the crops. The tank has two inlet pipes and one outlet pipe. The two inlet pipes can fill the tank in 10 and 12 hours, respectively. The outlet pipe can empty the tank in 15 hours. If the tank is empty, how long would it take to fill the tank when all three valves are open?

12. A small township uses 3 pumps to remove water from flooded basements. One pump can remove all the water from a flooded basement in 6 hours. The second pump can remove the same amount of water in 5 hours, and the third pump requires only 4 hours to remove the water. If all 3 pumps work together to remove the water from the flooded basement, how long will it take to empty the basement?

13. It takes Fred twice as long as Nancy to knit an afghan. If together they knit an afghan in 12 hours, how long would it take Nancy to knit the afghan by herself?

14. A roofer requires 15 hours to put a new roof on a house. Anna, the roofer's apprentice, can reroof the house by herself in 20 hours. After working alone on a roof for 6 hours, the roofer leaves for another job. Anna takes over and completes the job. How long will it take Anna to complete the job?

15. Two pipes are used to fill an oil tanker. When the larger pipe is used alone, it takes 60 hours to fill the tanker. When the smaller pipe is used alone, it takes 80 hours to fill the tanker. If the large pipe begins filling the tanker, and after 20 hours the large pipe is closed down and the smaller pipe is opened, how much longer will it take to finish filling the tanker using only the smaller pipe?

16. What number multiplied by the numerator and added to the denominator of the fraction $\frac{4}{3}$ makes the resulting fraction $\frac{5}{2}$?

17. What number added to the numerator and multiplied by the denominator of the fraction $\frac{3}{2}$ makes the resulting fraction $\frac{1}{8}$?

18. One number is twice another. The sum of their reciprocals is $\frac{3}{4}$. Find the numbers.

19. The sum of the reciprocals of two consecutive integers is $\frac{11}{30}$. Find the two integers.

20. The sum of the reciprocals of two consecutive even integers is $\frac{5}{12}$. Find the two integers.

21. When a number is added to both the numerator and denominator of the fraction $\frac{5}{7}$, the resulting fraction is $\frac{4}{5}$. Find the number added.

22. When 3 is added to twice the reciprocal of a number, the sum is $\frac{31}{10}$. Find the number.

23. The reciprocal of 3 less than a certain number is twice the reciprocal of 6 less than twice the number. Find the number(s).

24. If 3 times a number is added to twice the reciprocal of the number, the answer is 5. Find the number(s).

25. If 3 times the reciprocal of a number is subtracted from twice the reciprocal of the square of the number, the difference is -1. Find the number(s).

26. A Greyhound Bus can travel 400 kilometers in the same time that an Amtrak train can travel 600 kilometers. If the speed of the train is 40 kilometers per hour greater than that of the bus, find the speeds of the bus and train.

27. The speed of a boat in still water is 20 miles per hour. It takes the same amount of time for the boat to travel 3 miles downstream (with the current) as it does to travel 2 miles upstream (against the current). Find the speed of the current.

28. The rate of a bicyclist is 8 miles per hour faster than that of a jogger. If the bicyclist travels 10 miles in the same amount of time that the jogger travels 5 miles, find the rate of the jogger.

29. Two cross-country skiers ski along the same path. One skier averages 8 miles per hour, while the other averages 6 miles per hour. If it takes the slower skier $\frac{1}{2}$ hour longer than the faster skier to reach the designated resting point, how far is the resting point from the starting point?

30. The current of a river is 3 miles per hour. It takes a motorboat a total of 3 hours to travel 12 miles upstream and return 12 miles downstream. What is the speed of the boat in still water?

31. Ray starts out on a boating trip at 8 A.M. Ray's boat can go 20 miles per hour in still water. (a) How far downstream can Ray go if the river's current is 5 miles per hour and he wishes to go down and back in 4 hours? (b) At what time must Ray turn back?

32. Chris Peckaitis drove from Santa Barbara, California, to Monterey, along scenic Route 1, a distance of 300 miles. For part of the trip she drove at a steady rate of 50 miles per hour, but in some of the more scenic areas she drove at a steady rate of 40 miles per hour. If the total time of the trip was 7.1 hours, how far did she travel at each speed?

33. A cyclist rode 30 miles before getting a flat tire, then the cyclist walked 2 miles to a service station. If the cycling rate was 4 times the walking rate and the total time of the outing was 6 hours, find the rate at which the cyclist was riding.

34. A train and car leave from a railroad station at the same time headed for the State Fair on the other side of the state. The car averages 50 miles per hour and the train averages 70 miles per hour. If the train arrives at the fair 2 hours ahead of the car, find the distance from the railroad station to the State Fair.

35. A train and a plane leave from Boston at the same time for a destination 900 miles away. If the speed of the plane is 5 times the speed of the train, and the plane arrives 12 hours before the train, find the speeds of the train and the plane.

36. Two brothers are long-distance swimmers. Jim averages 3.6 miles per hour and Pete averages 2.4 miles per hour. The brothers start swimming at the same time across Lake Mead to a point on the other side of the lake. If Jim arrives 0.2 hour ahead of Pete, find the distance they swam. (Of course each swimmer was accompanied by a boat for safety reasons.)

37. Two rockets are to be launched at the same time from NASA headquarters in Houston, Texas, and are to meet at a space station many miles from the earth. The first rocket is to travel at a speed of 20,000 miles per hour and the second rocket will travel at a speed of 18,000 miles per hour. If the first rocket is scheduled to reach the space station 0.6 hour before the second rocket, how far is the space station from NASA headquarters?

Cumulative Review Exercises

[5.3] **38.** Subtract $\frac{1}{2}x^2 - 3x^2 + 2xy - (\frac{3}{5}xy + 6y^2)$.

[5.4] **39.** Multiply $(4x^2 - 6x - 1)(3x - 4)$.

[5.5] **40.** Divide $(12x^2 + 7x + 12) \div (3x - 2)$.

[6.2] **41.** Factor $8x^2 + 26x + 15$.

7.7

Variation

▶ **1** Write an equation expressing direct variation.

▶ **2** Write an equation expressing inverse variation.

▶ **3** Write an equation expressing joint variation.

▶ **4** Write an equation containing a combination of variations.

In Sections 7.5 and 7.6 we saw many applications of equations containing fractions. In this section we see still more applications of rational equations.

Direct Variation

▶ **1** Many scientific formulas are expressed in terms of variations. A **variation** is an equation that relates one variable to one or more other variables using the operations of multiplication or division (or both operations). There are essentially three types of variation problems: direct, inverse, and joint variation.

In **direct variation** the two related variables will both increase together or both decrease together; that is, as one increases so does the other, and as one decreases so does the other.

Consider a car traveling at 30 miles an hour. The car travels 30 miles in 1 hour, 60 miles in 2 hours, and 90 miles in 3 hours. Notice that as the time increases, the distance traveled increases, and as the time decreases, the distance traveled decreases.

Also note that the ratio of distance to time is a constant (30) in each case:

$$\frac{\text{Distance}}{\text{Time}} = \frac{30}{1} = 30, \quad \frac{60}{2} = 30, \quad \frac{90}{3} = 30$$

The formula used to calculate distance traveled is

$$\text{Distance} = \text{speed} \cdot \text{time}$$

Since the speed has been specified as a constant, 30 miles per hour, the formula can be written

$$d = 30t$$

We say that distance varies directly as time or that distance is directly proportional to time.

The above equation is an example of a direct variation.

Direct Variation

The general form of a direct variation is

$$x = ky$$

In this formula, k is called the **constant of proportionality** or the variation constant.

EXAMPLE 1   The circumference of a circle, C, is directly proportional to (or varies directly as) its radius, r. Write the equation for the circumference of a circle if the constant of proportionality, k, is 2π.

Solution: $C = kr$ (C varies directly as r)
$C = 2\pi r$ (constant of proportionality is 2π)   ∎

EXAMPLE 2   The resistance (R) of a wire varies directly as its length (L).
(a) Write this variation as an equation.
(b) Find the resistance (measured in ohms) of a 20-foot length of wire assuming that the constant of proportionality for the wire is 0.007.

Solution: (a) $R = kL$
(b) $R = 0.007(20) = 0.14$
The resistance of the wire is 0.14 ohm.   ∎

In certain variation problems the constant of proportionality, k, may not be known. In such cases it can often be found by substituting given values in the variation and solving for k.

EXAMPLE 3   The gravitational force of attraction (F) between an object and the earth is directly proportional to the mass (m) of the object. If the force of attraction is 640 when the object's mass is 20, find the constant of proportionality.

Solution: $F = km$
$640 = k\,20$
$$\frac{640}{20} = \frac{20k}{20}$$
$32 = k$

Thus the constant of proportionality is 32.   ∎

EXAMPLE 4 x varies directly as the square of y. If x is 80 when y is 20, find x when y is 90.

Solution: Since the constant of proportionality is not given, we must first find k using the given information.

$$x = ky^2$$
$$80 = k(20)^2$$
$$80 = 400k$$
$$\frac{80}{400} = \frac{400k}{400}$$
$$0.2 = k$$

We now use $k = 0.2$ to find x when y is 90.

$$x = ky^2$$
$$x = 0.2(90)^2$$
$$x = 1620$$

Thus, when y equals 90, x equals 1620.

Inverse Variation

► **2** A second type of variation is **inverse variation.** When two quantities vary inversely, it means that as one quantity increases, the other quantity decreases, and vice versa.

To explain inverse variation, we use the formula, distance = speed · time. If we solve for time, we get time = distance/speed. Assume that the distance is fixed at 120 miles; then

$$\text{Time} = \frac{120}{\text{speed}}$$

Note that at a speed of 120 miles per hour it would take 1 hour to cover this distance. At a speed of 60 miles an hour, it would take 2 hours. At 30 miles an hour, it would take 4 hours. Note that as the speed decreases the time increases, and vice versa. Also note that the product of the speed and the time is a constant:

$$120 \cdot 1 = 120, \qquad 60 \cdot 2 = 120, \qquad 30 \cdot 4 = 120$$

The above equation can be written

$$t = \frac{120}{s}$$

This equation is an example of an inverse variation, where the time and speed are inversely proportional. The constant of proportionality is 120.

Inverse Variation

The general form of an inverse variation is

$$x = \frac{k}{y} \quad \text{or} \quad xy = k$$

Two quantities vary inversely, or are inversely proportional, when as one quantity increases the other quantity decreases. In the example just mentioned, as the speed increases the time decreases, and vice versa. Thus the speed and time are inversely proportional to each other.

EXAMPLE 5 The illuminance (I) of a light source varies inversely as the square of the distance (d) from the source. Assuming that the illuminance is 75 units at a distance of 6 meters, find the equation that expresses the relationship between the illuminance and the distance.

Solution: The general form of the equation is

$$I = \frac{k}{d^2} \qquad (\text{or} \qquad Id^2 = k)$$

To find k, we insert the given values for I and d.

$$75 = \frac{k}{6^2}$$

$$75 = \frac{k}{36}$$

$$(75)(36) = k$$

$$2700 = k$$

Thus the formula is $I = \dfrac{2700}{d^2}$. ■

EXAMPLE 6 x varies inversely as y. If $x = 8$ when $y = 15$, find x when $y = 18$.

Solution: First write the equation and solve for k.

$$x = \frac{k}{y}$$

$$8 = \frac{k}{15}$$

$$120 = k$$

Now substitute 120 for k in $x = \dfrac{k}{y}$ and find x when $y = 18$.

$$x = \frac{120}{y} = \frac{120}{18} = 6.7 \quad \text{(to the nearest tenth)}$$ ■

Joint Variation

▶ **3**   One quantity may vary directly as a product of two or more other quantities. This type of variation is called **joint variation.**

Joint Variation

The general form of a joint variation, where x varies directly as y and z, is

$$x = kyz$$

EXAMPLE 7   The area, A, of a triangle varies jointly as its base, b, and height, h. If the area of a triangle is 48 square inches when its base is 12 inches and its height is 8 inches, find the area of a triangle with a base of 15 inches and a height of 20 inches.

Solution: First write the joint variation; then solve for k.

$$A = kbh$$

$$48 = k(12)(8)$$

$$48 = k(96)$$

$$\frac{48}{96} = k$$

$$k = \frac{1}{2}$$

Now solve for the area of the given triangle.

$$A = kbh$$

$$= \frac{1}{2}(15)(20)$$

$$= 150 \text{ square inches} \qquad \blacksquare$$

Summary of Variations

Direct	Inverse	Joint
$x = ky$	$x = \dfrac{k}{y}$	$x = kyz$

Combined Variation

▶ **4** Often in real-life situations one variable varies as a combination of variables. The following examples illustrate the use of **combined variations.**

EXAMPLE 8 When studying the ability of a wire to stretch, or its elasticity, E, we learn that the elasticity of a wire is directly proportional to its length, L, and inversely proportional to its cross-sectional area, A. Express E in terms of L and A.

Solution: $E = \dfrac{kL}{A}$ $\qquad \blacksquare$

EXAMPLE 9 The electrostatic force, F, of attraction or repulsion between two electrical charges is jointly proportional to the two charges, q_1 and q_2, and inversely proportional to the square of the distance, d, between the two charges. Express F in terms of q_1, q_2, and d.

Solution: $F = \dfrac{kq_1q_2}{d^2}$ $\qquad \blacksquare$

EXAMPLE 10 A varies jointly as B and C and inversely as the square of D. If $A = 1$ when $B = 9$, $C = 4$, and $D = 6$, find A when $B = 8$, $C = 12$, and $D = 5$.

Solution: $A = \dfrac{kBC}{D^2}$

We must first find the constant of proportionality, k, by inserting the given values for A, B, C, and D and solving for k.

$$1 = \frac{k(9)(4)}{6^2}$$

$$1 = \frac{36k}{36}$$

$$1 = k$$

Thus the constant of proportionality equals 1. Now we find A for the given values of B, C, and D.

$$A = \frac{(1)(8)(12)}{(5)^2} = \frac{96}{25} = 3.84 \qquad \blacksquare$$

Exercise Set 7.7

Use your intuition to determine if the variation between the indicated quantiles is direct or inverse.

1. The speed and distance traveled by a car in a specified time period.
2. The distance between two cities on a map and the actual distance between the two cities.
3. The diameter of a hose and volume of water coming from the hose.
4. A given weight and the force needed to lift that weight.
5. The cubic-inch displacement in liters and the horsepower of the engine.
6. The volume of a balloon and its radius.
7. The light illuminating an object and the distance the light is from the object.
8. The length of a board and the force applied to the center needed to break the board.

9. The shutter opening of a camera and the amount of sunlight to reach the film.
10. A person's weight (due to the earth's gravity) and his distance from the earth.
11. The number of pages a person can read in a given period of time and his reading speed.
12. The time it takes an ice cube to melt in water and the temperature of the water.
13. The time needed to get proper exposure on a film and the aperture opening of the camera lens.
14. The time to reach a certain point for a plane flying with the wind and the speed of the wind.
15. The number of calories eaten and the amount of exercise required to burn off those calories.

For Exercises 16 through 33, (a) write the variation and (b) find the quantity indicated.

16. x varies directly as y. Find x when $y = 12$ and $k = 6$.
17. C varies directly as the square of Z. Find C when $Z = 9$ and $k = \frac{3}{4}$.
18. y varies directly as R. Find y when $R = 180$ and $k = 1.7$.
19. x varies inversely as y. Find x when $y = 25$ and $k = 5$.
20. R varies inversely as W. Find R when $W = 160$ and $k = 8$.
21. L varies inversely as the square of P. Find L when $P = 4$ and $k = 100$.
22. A varies directly as B and inversely as C. Find A when $B = 12$, $C = 4$, and $k = 3$.
23. A varies jointly as R_1 and R_2 and inversely as the square of L. Find A when $R_1 = 120$, $R_2 = 8$, $L = 5$, and $k = \frac{3}{2}$.
24. T varies directly as the square of D and inversely as F. Find T when $D = 8$, $F = 15$, and $k = 12$.
25. x varies directly as y. If x is 9 when y is 18, find x when y is 36.
26. Z varies directly as W. If Z is 7 when W is 28, find Z when W is 140.
27. y varies directly as the square of R. If y is 5 when $R = 5$, find y when R is 10.
28. S varies inversely as G. If S is 12 when G is 0.4, find S when G is 5.
29. C varies inversely as J. If C is 7 when J is 0.7, find C when J is 12.
30. x varies inversely as the square of P. If $x = 10$ when P is 6, find x when $P = 20$.
31. F varies jointly as M_1 and M_2 and inversely as d. If F is 20 when $M_1 = 5$, $M_2 = 10$, and $d = 0.2$, find F when $M_1 = 10$, $M_2 = 20$, and $d = 0.4$.

32. F varies jointly as q_1 and q_2 and inversely as the square of d. If F is 8 when $q_1 = 2$, $q_2 = 8$, and $d = 4$, find F when $q_1 = 28$, $q_2 = 12$, and $d = 2$.
33. S varies jointly as I and the square of T. If S is 8 when $I = 20$ and $T = 4$, find S when $I = 2$ and $T = 2$.
34. The volume of a gas, V, varies inversely as its pressure, P. If the volume, V, is 800 cc when the pressure is 200 millimeters (mm) of mercury, find the volume when the pressure is 25 mm of mercury.
35. The amount a spring will stretch, S, varies directly with the force (or weight), F, attached to the spring. If a spring stretches 1.4 inches when 20 pounds is attached, how far will it stretch when 10 pounds is attached?
36. The pressure P, on an object submerged in water varies directly with its depth, D. If the pressure at a depth of 50 feet is 21.6 pounds per square inch, find the pressure at a depth of 180 feet.
37. The intensity I, of light received at a source varies inversely as the square of the distance, d, from the source. If the light intensity is 20 footcandles at 15 feet, find the light intensity at 12 feet.
38. On earth the mass of an object varies directly with its weight. If an object with a weight of 256 pounds has a mass of 8 slugs, find the mass of an object weighing 120 pounds.
39. The weight, W, of an object in the earth's atmosphere varies inversely with the square of the distance, d, between the object and the center of the earth. A 140-pound person standing on earth is approximately 4000 miles from the earth's center. Find the weight (or gravitational force of attraction) of this person at a distance 100 miles from the earth's surface.

40. The wattage rating of an appliance, W, varies jointly as the square of the current, I, and the resistance, R. If the wattage is 1 watt when the current is 0.1 ampere and the resistance is 100 ohms, find the wattage when the current is 0.4 ampere and the resistance is 250 ohms.

41. The electrical resistance of a wire, R, varies directly as its length, L, and inversely as its cross-sectional area, A. If the resistance of a wire is 0.2 ohm when the length

is 200 feet and its cross-sectional area is 0.05 square inch, find the resistance of a wire whose length is 5000 feet with a cross-sectional area of 0.01 square inch.

42. Write a paragraph explaining the various types of variations. Include in your discussion the terms direct, inverse, joint, and combined variation. Give your own example of each type of variation.

Cumulative Review Exercises

[3.4] **43.** Solve for the variable d if the line through the two given points is to have the given slope: $(5, 1)$ and $(-4, d)$, $m = \frac{2}{3}$.

[4.3] **44.** Mr. Wilcox, a salesman, earns a base weekly salary plus a commission on his sales. The first week in February, on sales of $5000, his total in-

come was $550. The second week in February, on sales of $8000, his total income was $640. Find his weekly salary and his commission rate.

[4.5] **45.** Graph the inequality $|x - 2| < 4$ (optional).

[5.8] **46.** Graph $f(x) = -x^3 + 4x - 6$.

JUST FOR FUN

1. An article in the magazine *Outdoor and Travel Photography* states, "If a surface is illuminated by a point-source of light, the intensity of illumination produced is inversely proportional to the square of the distance separating them. In practical terms, this means that foreground objects will be grossly overexposed if your background subject is properly exposed with a flash. Thus, direct

flash will not offer pleasing results if there are any intervening objects between the foreground and the subject."

If the subject you are photographing is 4 feet from the flash, and the illumination on this subject is 1/16 of the light of the flash, what is the intensity of illumination on an intervening object that is 3 feet from the flash?

SUMMARY

GLOSSARY

Algebraic fraction (or rational expression) *(299):* An expression of the form p/q, where p and q are polynomials, $q \neq 0$.

Combined variation *(345):* A variation problem that involves two or more different types of variations.

Complex fraction *(318):* A fractional expression that has a fraction in its numerator or its denominator or both its numerator and denominator.

Constant of proportionality *(342):* The constant in a variation problem.

Domain of a rational expression *(299):* The set of values that can replace the variable in the expression.

Rational function *(304):* Functions of the form $f(x) = p/q$, where p and q are polynomials, $q \neq 0$.

Reduced to lowest terms *(300):* An algebraic fraction is reduced to its lowest terms when the numerator and denominator have no common factor other than 1.

Variation *(341):* An equation that relates one variable to one or more other variables using the operations of multiplication or division (or both operations).

IMPORTANT FACTS

Types of Variation

Direct	*Inverse*	*Joint*
$x = ky$	$x = \dfrac{k}{y}$	$x = kyz$

Review Exercises

[7.1] *Determine the domain of each of the following.*

1. $\dfrac{3}{x-4}$

2. $\dfrac{x}{x+1}$

3. $\dfrac{-2x}{x^2+5}$

4. $\dfrac{0}{(x+3)^2}$

5. $\dfrac{x+6}{x^2}$

6. $\dfrac{x^2-2}{x^2-3x-10}$

Write each expression in reduced form.

7. $\dfrac{x^2+xy}{x+y}$

8. $\dfrac{x^2-9}{x+3}$

9. $\dfrac{4-5x}{5x-4}$

10. $\dfrac{x^2+2x-3}{x^2+x-6}$

11. $\dfrac{2x^2-6x+5x-15}{2x^2+7x+5}$

12. $\dfrac{a^3-8}{a^2-4}$

[7.3] *Find the least common denominator.*

13. $\dfrac{6x}{x+1}-\dfrac{3}{x}$

14. $\dfrac{9x-3}{x+y}-\dfrac{4x+7}{x^2-y^2}$

15. $\dfrac{19x-5}{x^2+2x-35}+\dfrac{3x-2}{x^2+9x+14}$

16. $\dfrac{3}{(x+2)^2}-\dfrac{6(x+3)}{x^2-4}-\dfrac{4x}{x+1}$

[7.2–7.3] *Perform the indicated operation.*

17. $\dfrac{15x^2y^3}{3z}\cdot\dfrac{6z^3}{5xy^3}$

18. $\dfrac{1}{x-2}\cdot\dfrac{2-x}{2}$

19. $\dfrac{8xy^2}{z}\div\dfrac{x^4y^2}{4z^2}$

20. $\dfrac{4}{2x}+\dfrac{x}{x^2}$

21. $\dfrac{4x+4y}{x^2y}\cdot\dfrac{y^3}{8x}$

22. $\dfrac{4x^2-11x+4}{x-3}-\dfrac{x^2-4x+10}{x-3}$

23. $\dfrac{a-2}{a+3}\cdot\dfrac{a^2+4a+3}{a^2-a-2}$

24. $\dfrac{3x+3y}{x^2}\div\dfrac{x^2-y^2}{x^2}$

25. $\dfrac{6x^2-4x}{2x-3}-\dfrac{-3x+12}{2x-3}-\dfrac{2x+4}{2x-3}$

26. $\dfrac{5x}{3xy}-\dfrac{4}{x^2}$

27. $6+\dfrac{x}{x+2}$

28. $5-\dfrac{3}{x+3}$

29. $\dfrac{x^2-y^2}{x-y}\cdot\dfrac{x+y}{xy+x^2}$

30. $\dfrac{1}{a^2+8a+15}\div\dfrac{3}{a+5}$

31. $\dfrac{6x^2-4x}{2x-3}-\dfrac{-x+4}{2x-3}-\dfrac{6x^2+x-2}{2x-3}$

32. $\dfrac{a+c}{c}-\dfrac{a-c}{a}$

33. $\dfrac{4x^2+8x-5}{2x+5}\cdot\dfrac{x+1}{4x^2-4x+1}$

34. $(x+3)\div\dfrac{x^2-4x-21}{x-7}$

35. $\dfrac{x^2-3xy-10y^2}{6x}\div\dfrac{x+2y}{12x^2}$

36. $\dfrac{2}{3x}-\dfrac{3x}{3x-6}$

37. $\dfrac{x-4}{x-5}-\dfrac{3}{x+5}$

38. $\dfrac{4}{x+5}+\dfrac{6}{(x+5)^2}$

39. $\dfrac{x+3}{x^2-9}+\dfrac{2}{x+3}$

40. $\dfrac{1}{a-3}\cdot\dfrac{a^2-2a-3}{a^2+3a+2}$

41. $\dfrac{4x^2 - 16y^2}{9} \div \dfrac{(x + 2y)^2}{12}$

42. $\dfrac{4}{(x + 2)(x - 3)} - \dfrac{4}{(x - 2)(x + 2)}$

43. $\dfrac{2x^2 + 10x + 12}{(x + 2)^2} \cdot \dfrac{x + 2}{x^3 + 5x^2 + 6x}$

44. $\dfrac{x + 2}{x^2 - x - 6} + \dfrac{x - 3}{x^2 - 8x + 15}$

45. $\dfrac{x + 5}{x^2 - 15x + 50} - \dfrac{x - 2}{x^2 - 25}$

46. $\dfrac{y^4 - x^6}{x^3 - y^2} \div (y^2 - x^3)$

47. $\dfrac{1}{x + 3} - \dfrac{2}{x - 3} + \dfrac{6}{x^2 - 9}$

48. $\dfrac{x^3 + 27}{4x^2 - 4} \div \dfrac{x^2 - 3x + 9}{(x - 1)^2}$

49. $\dfrac{x - 4}{x - 5} - \dfrac{3}{x + 5} - \dfrac{10}{x^2 - 25}$

50. $\dfrac{x^2 - 8x + 16}{2x^2 - x - 6} \cdot \dfrac{2x^2 - 7x - 15}{x^2 - 2x - 24} \div \dfrac{x^2 - 9x + 20}{x^2 + 2x - 8}$

51. $\dfrac{x^2 - x - 56}{x^2 + 14x + 49} \cdot \dfrac{x^2 + 4x - 21}{x^2 - 9x + 8} + \dfrac{3}{x^2 + 8x - 9}$

[7.4] *Simplify the complex fraction.*

52. $\dfrac{4 - \dfrac{9}{16}}{1 + \dfrac{5}{8}}$

53. $\dfrac{\dfrac{15xy}{6z}}{\dfrac{3x}{z^2}}$

54. $\dfrac{\dfrac{36x^4y^2}{9xy^5}}{4z^2}$

55. $\dfrac{x + \dfrac{1}{y}}{y^2}$

56. $\dfrac{x - \dfrac{x}{y}}{\dfrac{1 + x}{y}}$

57. $\dfrac{\dfrac{4}{x} + \dfrac{2}{x^2}}{6 - \dfrac{1}{x}}$

58. $\dfrac{a^{-1} + 2}{a^{-1} + \dfrac{1}{a}}$

59. $\dfrac{x^{-2} + \dfrac{1}{x}}{\dfrac{1}{x^2} - \dfrac{1}{x}}$

[7.5] *Solve the equation.*

60. $\dfrac{3}{x} = \dfrac{8}{24}$

61. $\dfrac{4}{a} = \dfrac{16}{4}$

62. $\dfrac{x + 3}{5} = \dfrac{9}{5}$

63. $\dfrac{x}{6} = \dfrac{x - 4}{2}$

64. $\dfrac{3x + 4}{5} = \dfrac{2x - 8}{3}$

65. $\dfrac{x}{5} + \dfrac{x}{2} = 14$

66. $\dfrac{4}{x} - \dfrac{1}{6} = \dfrac{1}{x}$

67. $\dfrac{1}{x - 2} + \dfrac{1}{x + 2} = \dfrac{1}{x^2 - 4}$

68. $\dfrac{x - 3}{x - 2} + \dfrac{x + 1}{x + 3} = \dfrac{2x^2 + x + 1}{x^2 + x - 6}$

69. $\dfrac{x}{x^2 - 9} + \dfrac{2}{x + 3} = \dfrac{4}{x - 3}$

Solve for the variable indicated.

70. $S_n = \dfrac{a_1(r^n - 1)}{r - 1}$ for a_1

71. $\dfrac{V_1 P_1}{T_1} = \dfrac{V_2 P_2}{T_2}$, for P_2

72. $\dfrac{1}{f} = \dfrac{1}{p} + \dfrac{1}{q}$, for f

73. $\dfrac{1}{R_T} = \dfrac{1}{R_1} + \dfrac{1}{R_2}$, for R_2

Solve each problem.

74. Three resistors of 200, 400, and 1200 ohms, respectively, are wired in parallel. Find the total resistance of the circuit.

75. Two resistors are to be wired in parallel. One is to contain twice the resistance of the other. What should be the resistance of each resistor if the circuit's total resistance is to be 600 ohms?

76. What is the focal length of a mirror if the object distance is 12 centimeters and the image distance is 4 centimeters?

77. A mirror has a focal length of 10 centimeters. Find the object's distance from the lens if the image distance is twice the object's distance.

[7.6] *Solve each problem.*

78. It takes Dan 3 hours to mow Mr. Lee's lawn. It takes Kim 4 hours to mow the same lawn. How long will it take them working together to mow Mr. Lee's lawn?

79. Annette and Pete are both copy editors for a publishing company. Together they can edit a 500-page manuscript in 40 hours. If Annette by herself can edit the manuscript in 75 hours, how long will it take Pete to edit the manuscript by himself?

80. What number multiplied by the numerator and added to the denominator of the fraction $\frac{5}{8}$ makes the resulting value equal to 1?

81. When the reciprocal of twice a number is subtracted from 1, the result is the reciprocal of 3 times the number. Find the number.

82. Kit Waickman's motorboat can travel 15 miles per hour in still water. Traveling with the current of a river, the boat can travel 20 miles in the same time it takes to go 10 miles against the current. Find the rate of the current.

83. A small plane and a car start from the same location, at the same time, heading toward the same town 450 miles away. The speed of the plane is 3 times the speed of the car. The plane arrives at the town 6 hours ahead of the car. Find the speeds of the car and the plane.

[7.7] *Find the quantity indicated.*

84. A is directly proportional to B. If A is 120 when $B = 80$, find A when $B = 50$.

85. A is directly proportional to the square of C. If A is 5 when C is 5, find A when $C = 10$.

86. x is inversely proportional to y. If x is 20 when $y = 5$, find x when $y = 100$.

87. W is directly proportional to L and inversely proportional to A. If $W = 80$ when $L = 100$ and $A = 20$, find W when $L = 50$ and $A = 40$.

88. z is jointly proportional to x and y and inversely proportional to the square of r. If z is 12 when x is 20, $y = 8$, and $r = 8$, find z when $x = 10$, $y = 80$, and $r = 3$.

89. The scale of a map is 1 inch to 60 miles. How large a distance on the map represents 300 miles?

90. An electric company charges $0.162 per kilowatt-hour. What is the electric bill if 740 kilowatt-hours are used in a month?

91. The distance, d, an object drops in free fall is directly proportional to the square of the time, t. If an object falls 16 feet in 1 second, how far will an object fall in 5 seconds?

92. The area, A, of a circle varies directly with the square of its radius, r. If the area is 78.5 when the radius is 5, find the area when the radius is 8.

93. The time, t, it takes for an ice cube to melt is inversely proportional to the temperature of the water it is in. If it takes an ice cube 1.7 minutes to melt in 70°F water temperature, how long will it take the same-size ice cube to melt in 50°F water?

Practice Test

1. Find the domain of $\dfrac{x - 3}{x^2 - 3x - 28}$.

2. Reduce to lowest terms: $\dfrac{x^2 - 5x - 36}{9 - x}$.

Perform the operations indicated.

3. $\dfrac{6x^2y^4}{4z^2} \cdot \dfrac{8xz^3}{9y^4}$

4. $\dfrac{a^2 - 9a + 14}{a - 2} \cdot \dfrac{a^2 - 4a - 21}{(a - 7)^2}$

5. $\dfrac{x^2 - 9y^2}{3x + 6y} \div \dfrac{x + 3y}{x + 2y}$

6. $\dfrac{x^3 + y^3}{x + y} \div \dfrac{x^2 - xy + y^2}{x^2 + y^2}$

7. $\dfrac{5}{x} + \dfrac{3}{2x^2}$

8. $\dfrac{x - 5}{x^2 - 16} - \dfrac{x - 2}{x^2 + 2x - 8}$

9. $\dfrac{x + 1}{4x^2 - 4x + 1} + \dfrac{3}{2x^2 + 5x - 3}$

10. *Simplify* $\dfrac{\dfrac{1}{x} + \dfrac{1}{y}}{\dfrac{1}{x} - \dfrac{1}{y}}$.

11. *Simplify* $\dfrac{x + \dfrac{x}{y}}{x^{-1} + y^{-1}}$.

12. *Solve the equation:* $\dfrac{x}{3} - \dfrac{x}{4} = 5$.

13. *Solve the equation:*

$$\dfrac{x}{x - 8} + \dfrac{6}{x - 2} = \dfrac{x^2}{x^2 - 10x + 16}.$$

14. *P* varies directly as *Q* and inversely as *R*. If $P = 8$ when $Q = 4$ and $R = 10$, find *P* when $Q = 10$ and $R = 20$.

15. *W* varies jointly as *P* and *Q* and inversely as the square of *T*. If $W = 6$ when $P = 20$, $Q = 8$, and $T = 4$, find *W* if $P = 30$, $Q = 4$, and $T = 8$.

16. Kris can level a 1-acre field in 8 hours on his tractor. Heather can level a 1-acre field in 5 hours on her tractor. How long will it take them to level a 1-acre field if they work together?

8

Roots, Radicals, and Complex Numbers

See Section 8.7, Exercise 9.

8.1

Roots and Radicals

▸**1** Find principal square roots.

▸**2** Find cube and higher roots.

▸**3** Know when the root of a number is positive, negative, or not a real number.

▸**4** Evaluate radical expressions using absolute value.

In this chapter we expand on the concept of square root introduced in Chapter 1.

In the expression $\sqrt{x}$, the $\sqrt{}$ is called the **radical sign.** The number or expression within the radical sign is called the **radicand.**

$$\underset{\text{radicand}}{\overset{\text{radical sign}}{\sqrt{x}}}$$

The entire expression, including the radical sign and radicand, is called the **radical expression.** Another part of the radical expression is its index. The **index** tells the "root" of the expression. Square roots have an index of 2. The index of square roots is generally not written.

$$\sqrt{x} \quad \text{means} \quad \sqrt[2]{x}$$

▸**1** Every positive number has two square roots, a principal or positive square root and a negative square root. For any positive number x, the positive square root is written $\sqrt{x}$, and the negative square root is written $-\sqrt{x}$.

Number	Principal or Positive Square Root	Negative Square Root
25	$\sqrt{25}$	$-\sqrt{25}$
36	$\sqrt{36}$	$-\sqrt{36}$

> The **principal or positive square root** of a positive real number x, written $\sqrt{x}$, is that *positive* number whose square equals x.

Examples

$$\sqrt{25} = 5 \quad \text{since } 5^2 = 5 \cdot 5 = 25$$
$$\sqrt{36} = 6 \quad \text{since } 6^2 = 6 \cdot 6 = 36$$
$$\sqrt{\frac{4}{9}} = \frac{2}{3} \quad \text{since } \left(\frac{2}{3}\right)^2 = \left(\frac{2}{3}\right)\left(\frac{2}{3}\right) = \frac{4}{9}$$

Whenever we use the term square root in this book we will be referring to the principal or positive square root. Thus, if you are asked to find the value of

$\sqrt{25}$, your answer will be 5. Note that both $\sqrt{25}$ and $\sqrt{36}$ have square roots that are integers. Thus $\sqrt{25}$ and $\sqrt{36}$ are rational numbers. However, not every square root is a rational number. For example, $\sqrt{10}$ is not a rational number, but is an irrational number. There is no integer whose square is 10.

When we defined the principal square root, we indicated that we were taking the square root "of a positive real number x." Do you know why we had to specify that we were taking the square root of a *positive* real number? Consider the square root of -25, written $\sqrt{-25}$. What is its value? Is $\sqrt{-25}$ equal to 5? Is it equal to -5? The answer to both of these questions is no.

$$\sqrt{-25} \neq 5 \qquad \text{since } 5^2 = 25 \quad (\text{not } -25)$$
$$\sqrt{-25} \neq -5 \qquad \text{since } (-5)^2 = 25 \quad (\text{not } -25)$$

Since the square of any real number will always be greater than or equal to 0, there is no real number that will give -25 when squared. For this reason, $\sqrt{-25}$ is *not a real number*. We will discuss numbers like $\sqrt{-25}$ later in this chapter. Since the square of any real number cannot be negative, *the square root of a negative number will not be a real number*.

▶ **2**   Other types of radical expressions have different indexes. For example, $\sqrt[3]{x}$ is the third or cube root of x. The index of cube roots is 3. In the expression $\sqrt[5]{xy}$, read the fifth root of xy, the index is 5 and the radicand is xy.

Radical expressions that have indexes of 2, 4, 6, . . . or any even number are said to be **even roots**. Square roots are even roots since their index is 2. Radical expressions that have indexes of 3, 5, 7, . . . or any odd number are said to be **odd roots**.

Examples of Even Roots	*Examples of Odd Roots*
$\sqrt{9},\ \sqrt[4]{x},\ \sqrt[12]{9x^5}$	$\sqrt[3]{27},\ \sqrt[5]{x},\ \sqrt[17]{6x^4}$

Even Indexes

> The nth root of x, $\sqrt[n]{x}$, where n is an *even index* and x is a *positive* real number, is that *positive* number c such that $c^n = x$.

Examples

$$\sqrt{9} = 3 \qquad \text{since } 3^2 = 3 \cdot 3 = 9$$
$$\sqrt[4]{16} = 2 \qquad \text{since } 2^4 = 2 \cdot 2 \cdot 2 \cdot 2 = 16$$
$$\sqrt[4]{81} = 3 \qquad \text{since } 3^4 = 3 \cdot 3 \cdot 3 \cdot 3 = 81$$

Note that

$$-\sqrt{9} = -3$$
$$-\sqrt[4]{16} = -2$$
$$-\sqrt[4]{81} = -4$$

When considering radical expressions with even indexes, the radicand must be a positive value if the number is to be real. For example, is $\sqrt[4]{-16}$ equal to any real number? Is there a real number that when raised to the fourth power has a value of -16? Since any real number raised to an even power cannot be negative, there is no real number that equals $\sqrt[4]{-16}$. Thus, *when the index is even, the radicand must be nonnegative for the radical to be a real number*.

Odd Indexes

> The nth root of x, $\sqrt[n]{x}$, where n is an *odd index* and x is *any real number*, is that real number c such that $c^n = x$.

Examples

$$\sqrt[3]{8} = 2 \qquad \text{since } 2^3 = 2 \cdot 2 \cdot 2 = 8$$

$$\sqrt[3]{-8} = -2 \qquad \text{since } (-2)^3 = (-2)(-2)(-2) = -8$$

$$\sqrt[5]{243} = 3 \qquad \text{since } 3^5 = 3 \cdot 3 \cdot 3 \cdot 3 \cdot 3 = 243$$

$$\sqrt[5]{-243} = -3 \qquad \text{since } (-3)^5 = (-3)(-3)(-3)(-3)(-3) = -243$$

Note that

$$\sqrt[3]{8} = 2, \qquad \sqrt[3]{-8} = -2$$
$$-\sqrt[3]{8} = -2, \qquad -\sqrt[3]{-8} = -(-2) = 2$$

An odd root of a positive number is a positive number, and an odd root of a negative number is a negative number.

▶ **3**   Table 8.1 summarizes the information we have discussed.

TABLE 8.1

	n is even	n is odd
$x > 0$	$\sqrt[n]{x}$ is a positive real number	$\sqrt[n]{x}$ is a positive real number
$x < 0$	$\sqrt[n]{x}$ is not a real number	$\sqrt[n]{x}$ is a negative real number
$x = 0$	$\sqrt[n]{0} = 0$	$\sqrt[n]{0} = 0$

HELPFUL HINT

There is an important difference between $-\sqrt[4]{16}$ and $\sqrt[4]{-16}$. The number $-\sqrt[4]{16}$ is the negative of $\sqrt[4]{16}$. Since $\sqrt[4]{16} = 2$, $-\sqrt[4]{16} = -2$. However, $\sqrt[4]{-16}$ is not a real number since no real number when raised to the fourth power equals -16.

$$-\sqrt[4]{16} = -(\sqrt[4]{16}) = -2$$
$$\sqrt[4]{-16} \text{ is not a real number}$$

EXAMPLE 1   Indicate whether or not the radical expression is a real number. If the number is a real number find its value.

(a) $\sqrt[4]{-81}$ (b) $-\sqrt[4]{81}$ (c) $\sqrt[3]{-64}$ (d) $-\sqrt[3]{-64}$

Solution: (a) Not a real number. Even roots of negative numbers are not real numbers.

(b) Real number, $-\sqrt[4]{81} = -(\sqrt[4]{81}) = -(3) = -3$.

(c) Real number, $\sqrt[3]{-64} = -4$ since $(-4)^3 = -64$.

(d) Real number, $-\sqrt[3]{-64} = -(-4) = 4$ ■

▶ **4**   When $\sqrt{a^2}$ is evaluated for any nonzero, real number a, what will be the sign of the answer? Let us substitute in values for a, one positive and one negative, and then examine the results.

$$a = 2: \quad \sqrt{a^2} = \sqrt{2^2} = \sqrt{4} = 2$$
$$a = -2: \quad \sqrt{a^2} = \sqrt{(-2)^2} = \sqrt{4} = 2$$

By examining these examples, and other examples we can make up, we can reason that $\sqrt{a^2}$ *will always be a positive real number* for any nonzero real number a. Recall from Section 1.4 that the *absolute value* of any real number a, or $|a|$, is also a positive number for any nonzero number. We use these facts to reason that

$$\sqrt{a^2} = |a|, \qquad \text{for any real number } a$$

Examples

$$\sqrt{2^2} = |2| = 2 \qquad\qquad \sqrt{7^2} = |7| = 7$$
$$\sqrt{(-2)^2} = |-2| = 2 \qquad\qquad \sqrt{(-7)^2} = |-7| = 7$$

EXAMPLE 2   Use absolute value to evaluate each of the following.
(a) $\sqrt{5^2}$ (b) $\sqrt{(-5)^2}$ (c) $\sqrt{(-71)^2}$

Solution: (a) $\sqrt{5^2} = |5| = 5$
(b) $\sqrt{(-5)^2} = |-5| = 5$
(c) $\sqrt{(-71)^2} = |-71| = 71$ ■

You may be tempted to generalize that $\sqrt{a^2} = a$, *but this is only true when* $a \geq 0$. When $a < 0$, then $\sqrt{a^2} = -a$. This concept is illustrated using values of $a = 2$ and $a = -2$.

$a = 2$	$a = -2$
$\sqrt{a^2} = \sqrt{2^2} = \sqrt{4} = 2$	$\sqrt{a^2} = \sqrt{(-2)^2} = \sqrt{4} = 2$
Since $a = 2$, $\sqrt{a^2} = a$	Since $a = -2$, $\sqrt{a^2} = -a$
That is, $\sqrt{2^2} = 2$	That is, $\sqrt{(-2)^2} = -(-2) = 2$

If you do not know whether or not the a in the expression $\sqrt{a^2}$ is a number greater than or equal to zero, then you must write the answer to $\sqrt{a^2}$ as $|a|$.

To what is $\sqrt{(x + 1)^2}$ equal? Since we do not know whether $x + 1$ represents a positive or negative number, we must write the answer as $\sqrt{(x + 1)^2} = |x + 1|$. We will now show that $\sqrt{(x + 1)^2} = |x + 1|$ for values of $x = -3$ and $x = 5$.

$x = -3$	$x = 5$
$\sqrt{(x + 1)^2} = |x + 1|$	$\sqrt{(x + 1)^2} = |x + 1|$
$\sqrt{(-3 + 1)^2} = |-3 + 1|$	$\sqrt{(5 + 1)^2} = |5 + 1|$
$\sqrt{(-2)^2} = |-2|$	$\sqrt{6^2} = |6|$
$\sqrt{4} = 2$	$\sqrt{36} = 6$
$2 = 2$ true	$6 = 6$ true

Try a few other values for x and show that $\sqrt{(x + 1)^2} = |x + 1|$ for those values.

When we are asked to find the square root of an expression that is squared, and we do not know whether the expression represents a positive or a negative number, write the answer as the absolute value of the expression.

EXAMPLE 3 Write each of the following as an absolute value.

(a) $\sqrt{(x + 2)^2}$ (b) $\sqrt{(y - 7)^2}$ (c) $\sqrt{(x^2 - 5x + 6)^2}$

Solution: (a) $\sqrt{(x + 2)^2} = |x + 2|$

(b) $\sqrt{(y - 7)^2} = |y - 7|$

(c) $\sqrt{(x^2 - 5x + 6)^2} = |x^2 - 5x + 6|$

When we are asked to evaluate $\sqrt{a^2}$, and we do not know whether a represents a positive or negative number, we write the answer as $|a|$. However, if we are asked to evaluate $\sqrt{a^2}$ when we know that a represents a positive number, we can write the answer as a.

Exercise Set 8.1

Evaluate the radical expression if it is a real number. If the expression is not a real number, indicate so.

1. $\sqrt{25}$
2. $\sqrt[3]{27}$
3. $\sqrt[3]{-27}$
4. $\sqrt[5]{32}$
5. $\sqrt[3]{125}$
6. $\sqrt[4]{81}$
7. $\sqrt{-9}$
8. $\sqrt[6]{64}$
9. $\sqrt[3]{-8}$
10. $\sqrt[3]{216}$
11. $\sqrt{144}$
12. $\sqrt[4]{256}$
13. $\sqrt[5]{1}$
14. $\sqrt[3]{-125}$
15. $\sqrt[3]{343}$
16. $\sqrt[5]{-32}$
17. $\sqrt[4]{-16}$
18. $\sqrt[4]{16}$
19. $-\sqrt{-25}$
20. $\sqrt[3]{-64}$
21. $-\sqrt{36}$
22. $\sqrt{121}$
23. $\sqrt{\frac{1}{9}}$
24. $\sqrt{\frac{1}{4}}$
25. $\sqrt{-36}$
26. $\sqrt{\frac{4}{9}}$
27. $\sqrt[5]{-1}$
28. $-\sqrt{\frac{9}{16}}$

Use absolute value to evaluate the following.

29. $\sqrt{6^2}$
30. $\sqrt{(-6)^2}$
31. $\sqrt{(-1)^2}$
32. $\sqrt{(-17)^2}$
33. $\sqrt{(43)^2}$
34. $\sqrt{(-96)^2}$
35. $\sqrt{(147)^2}$
36. $\sqrt{(-147)^2}$
37. $\sqrt{(-83)^2}$
38. $\sqrt{(-89)^2}$
39. $\sqrt{(179)^2}$
40. $\sqrt{(213)^2}$

Write as an absolute value.

41. $\sqrt{(y - 8)^2}$
42. $\sqrt{(x - 7)^2}$
43. $\sqrt{(x - 3)^2}$
44. $\sqrt{(3x^2 - y)^2}$
45. $\sqrt{(3x + 5)^2}$
46. $\sqrt{(x^2 - 3x + 4)^2}$
47. $\sqrt{(6 - 3x)^2}$
48. $\sqrt{(4 - 5x^2)^2}$
49. $\sqrt{(y^2 - 4y + 3)^2}$
50. $\sqrt{(x^2 - 3x)^2}$
51. $\sqrt{(8a - b)^2}$
52. $\sqrt{(3w^4 - 4w)^2}$

53. (a) How many square roots does every positive real number have? Name them.

  (b) Find all square roots of the number 36.

  (c) When we refer to "the square root," which square root are we referring to?

  (d) Find the square root of 36.

54. (a) What are even roots? Give an example of an even root.

  (b) What are odd roots? Give an example of an odd root.

55. Explain why $\sqrt{-49}$ is not a real number.

56. Will a radical expression with an odd index and a real number as the radicand always be a real number? Explain your answer.

57. Will a radical expression with an even index and a real number as the radicand always be a real number? Explain your answer.

58. (a) To what is $\sqrt{a^2}$ equal?

  (b) To what is $\sqrt{a^2}$ equal if we know $a \geq 0$?

59. If $\sqrt[n]{x} = a$, then to what is x equal?

60. Select a value for x and show that $\sqrt{(2x - 1)^2} \neq 2x - 1$.

61. Select a value for x and show that $\sqrt{(5x - 3)^2} \neq 5x - 3$.

62. For what values of x will $\sqrt{(x - 1)^2} = x - 1$? Explain how you determined your answer.

63. For what values of x will $\sqrt{(x + 4)^2} = x + 4$? Explain how you determined your answer.

64. For what values of x will $\sqrt{(2x - 6)^2} = 2x - 6$? Explain how you determined your answer.

65. For what values of x will $\sqrt{(4x - 4)^2} = 4x - 4$? Explain how you determined your answer.

66. **(a)** For what values of a is $\sqrt{a^2} = |a|$? all real numbers
(b) For what values of a is $\sqrt{a^2} = a$?

67. Consider the expression $\sqrt[n]{x}$. Under what circumstances will this expression not be a real number?

68. Consider the expression $\sqrt[n]{x^n}$. Explain why this expression will be a real number for any real number x.

69. Consider the expression $\sqrt[n]{x^m}$. Under what circumstances will this expression not be a real number?

Cumulative Review Exercises

[6.1–6.5] *Factor*

70. $3y^2 - 18y + 27 - 3z^2$

71. $x^3 + \frac{1}{27}$

72. $(x + 2)^2 - (x + 2) - 12$

73. $2x^4 - 3x^3 - 6x^2 + 9x$

8.2

Rational Exponents

▶ **1** Change from a radical to an exponential expression.

▶ **2** Change from an exponential expression to a radical expression.

▶ **3** Use the rules of exponents with rational and negative exponents.

▶ **4** Factor expressions with rational exponents.

In this section we discuss changing radical expressions to exponential expressions, and vice versa. When you see a rational (or fractional) exponent, you should realize that the expression can be expressed as a radical expression.

In Section 8.1 we indicated that

$$\sqrt{a^2} = |a|$$

If we know that the a in the expression $\sqrt{a^2}$ represents a value greater than or equal to 0, then we can write

$$\sqrt{a^2} = a, \qquad a \geq 0$$

> **For the remainder of this chapter we make the assumption that all variables represent positive real numbers.**

We make this assumption so that we can write many answers without absolute value signs. With this assumption, when we evaluate a radical like $\sqrt{y^2}$, we can write the answer as y rather than $|y|$.

▶ **1** A radical expression of the form $\sqrt[n]{a}$ can be written as an exponential expression using the following rule.

> For any nonnegative number a, and n a positive integer
> $$\sqrt[n]{a} = a^{1/n}$$

Examples

$$\sqrt{6} = \sqrt[2]{6} = 6^{1/2} \qquad \sqrt[3]{x} = x^{1/3}$$
$$\sqrt{x} = \sqrt[2]{x} = x^{1/2} \qquad \sqrt[4]{y} = y^{1/4}$$
$$\sqrt[3]{9} = 9^{1/3} \qquad \sqrt[5]{3} = 3^{1/5}$$

We can expand the rule so that radicals of the form $\sqrt[n]{a^m}$ can be expressed as exponential expressions.

For any positive number a, and m and n integers, $n \geq 2$,

$$\sqrt[n]{a^m} = (\sqrt[n]{a})^m = \overset{\text{power}}{a^{m/n}} \underset{\text{index}}{}$$

This rule can be used to change an expression from radical form to exponential form, and vice versa. When changing a radical expression to exponential form, the *power* is placed in the *numerator*, and the *index or root* is placed in the *denominator* of the fractional exponent. Thus, for example, $\sqrt[3]{x^4}$ can be written $x^{4/3}$. Also $(\sqrt[5]{y})^2$ can be written $y^{2/5}$. Additional examples follow.

Examples

$$\sqrt{y^3} = y^{3/2} \qquad \sqrt[3]{z^2} = z^{2/3} \qquad \sqrt[5]{2^8} = 2^{8/5}$$
$$(\sqrt{z})^3 = z^{3/2} \qquad (\sqrt[4]{x})^3 = x^{3/4} \qquad (\sqrt[4]{6})^3 = 6^{3/4}$$

By the given rule, for nonnegative values of the variable we can write

$$\sqrt[3]{x^4} = (\sqrt[3]{x})^4, \qquad (\sqrt[5]{y})^2 = \sqrt[5]{y^2}$$

▶ **2** Exponential expressions with fractional exponents can be converted to radical expressions by following the reverse procedure. The *numerator* of the fractional exponent is the *power*, and the *denominator* of the fractional exponent is the *index or root* of the radical expression. Here are some examples.

Examples

$$x^{1/2} = \sqrt{x} \qquad\qquad 5^{1/3} = \sqrt[3]{5}$$
$$6^{2/3} = \sqrt[3]{6^2} \text{ or } (\sqrt[3]{6})^2 \qquad y^{3/10} = \sqrt[10]{y^3} \text{ or } (\sqrt[10]{y})^3$$
$$x^{9/5} = \sqrt[5]{x^9} \text{ or } (\sqrt[5]{x})^9 \qquad y^{10/3} = \sqrt[3]{y^{10}} \text{ or } (\sqrt[3]{y})^{10}$$

You may choose, for example, to write $6^{2/3}$ as either $\sqrt[3]{6^2}$ or $(\sqrt[3]{6})^2$.

When you rewrite a radical expression in exponential form, sometimes the radical expression can be simplified.

EXAMPLE 1 Write each of the following in exponential form and then simplify.
(a) $\sqrt[6]{(16)^3}$ (b) $(\sqrt[3]{x})^9$ (c) $\sqrt[12]{y^4}$ (d) $(\sqrt[10]{z})^5$

Solution: (a) $\sqrt[6]{(16)^3} = 16^{3/6} = 16^{1/2} = \sqrt{16} = 4$

(b) $(\sqrt[3]{x})^9 = x^{9/3} = x^3$

(c) $\sqrt[12]{y^4} = y^{4/12} = y^{1/3} = \sqrt[3]{y}$

(d) $(\sqrt[10]{z})^5 = z^{5/10} = z^{1/2} = \sqrt{z}$ ∎

Now consider $\sqrt[5]{x^5}$. When written in exponential form, this becomes $\sqrt[5]{x^5} = x^{5/5} = x^1 = x$. This leads to the following rule:

For any nonnegative number a,
$$\sqrt[n]{a^n} = (\sqrt[n]{a})^n = a^{n/n} = a$$

The rule indicates that, when the index and the exponent are the same in a radical expression, the expression simplifies to a (if a is assumed to be positive).

Examples

$$\sqrt{5^2} = 5 \qquad \sqrt[4]{y^4} = y$$
$$\sqrt{x^2} = x \qquad (\sqrt[5]{z})^5 = z$$

Remember, we are assuming that all the variables in this chapter represent positive numbers. This allows us to write the answers to $\sqrt{x^2}$ as x and $(\sqrt[5]{z})^5$ as z.

▶ **3**   In Sections 5.1 and 5.2 we introduced and discussed the rules of exponents. Let us review those rules now.

Rules of Exponents

For all real numbers a and b and all rational numbers m and n,

Product rule	$a^m \cdot a^n = a^{m+n}$
Quotient rule	$\dfrac{a^m}{a^n} = a^{m-n}, \qquad a \neq 0$
Negative exponent rule	$a^{-m} = \dfrac{1}{a^m}, \qquad a \neq 0$
Zero exponent rule	$a^0 = 1 \qquad a \neq 0$
Power rules	$\begin{cases} (a^m)^n = a^{mn} \\ (ab)^m = a^m b^m \\ \left(\dfrac{a}{b}\right)^m = \dfrac{a^m}{b^m}, \qquad b \neq 0 \end{cases}$

The rules of exponents apply for any real numbers a and b. When the exponents are rational numbers, the rules of exponents still apply. Using the rules of exponents, we will now work some problems in which the exponents are rational numbers.

EXAMPLE 2   Evaluate
(a) $8^{2/3}$;   (b) $8^{-2/3}$

Solution: (a) First rewrite the expression in radical form, and then evaluate.
$$8^{2/3} = (\sqrt[3]{8})^2 = 2^2 = 4$$

(b) Begin by using the negative exponent rule.

$$8^{-2/3} = \frac{1}{8^{2/3}} = \frac{1}{4}$$

Example 2(a) could also have been found as follows:

$$8^{2/3} = \sqrt[3]{8^2} = \sqrt[3]{64} = 4$$

However, it is generally easier to evaluate the root before the power.

EXAMPLE 3   Evaluate $\left(\dfrac{4}{25}\right)^{-1/2}$

Solution: Begin by using the negative exponent rule, and then write the expression in radical form.

$$\left(\frac{4}{25}\right)^{-1/2} = \frac{1}{\left(\dfrac{4}{25}\right)^{1/2}} = \frac{1}{\sqrt{\dfrac{4}{25}}} = \frac{1}{\dfrac{2}{5}} = \frac{5}{2}$$

Example 3 could also have been solved by using the fact that $\left(\dfrac{x}{y}\right)^{-m} = \left(\dfrac{y}{x}\right)^{m}$. This fact was discussed in Section 5.2. Using this information, the problem could be solved as follows:

$$\left(\frac{4}{25}\right)^{-1/2} = \left(\frac{25}{4}\right)^{1/2} = \sqrt{\frac{25}{4}} = \frac{5}{2}$$

EXAMPLE 4   Write each expression as a square root, and then simplify if possible:
(a) $-25^{1/2}$;   (b) $(-25)^{1/2}$

Solution: (a) Recall from earlier discussions that $-x^2$ means $-(x^2)$. The same principal applies here.

$$-25^{1/2} = -(25)^{1/2} = -\sqrt{25} = -5$$

(b) Since the negative sign is within the parentheses, we write $(-25)^{1/2} = \sqrt{-25}$, which is not a real number.

EXAMPLE 5   Simplify $x^{1/2} \cdot x^{2/3}$ and write the answer in exponential form.

Solution: Use the product rule, and then simplify the answer.

$$x^{1/2} \cdot x^{2/3} = x^{(1/2)+(2/3)} = x^{(3/6)+(4/6)} = x^{7/6}$$

EXAMPLE 6   Simplify $\dfrac{y^{3/5}}{y^{1/3}}$ and write the answer in exponential form.

Solution: Using the quotient rule, we write

$$\frac{y^{3/5}}{y^{1/3}} = y^{(3/5)-(1/3)} = y^{(9/15)-(5/15)} = y^{4/15}$$

EXAMPLE 7 Simplify $(x^{3/5})^{-1/4}$ and write the answer in radical form.

Solution: First, use the power rule; then use the negative exponent rule.

$$(x^{3/5})^{-1/4} = x^{(3/5)(-1/4)} = x^{-3/20} = \frac{1}{x^{3/20}} = \frac{1}{\sqrt[20]{x^3}} \quad \blacksquare$$

Note in Example 7 that we could have first applied the negative exponent rule and then applied the power rule. Try working Example 7 by first using the negative rule.

EXAMPLE 8 Simplify $\left(\dfrac{x^4y^{1/3}}{x^{1/2}y}\right)^2$ and write the answer in radical form.

Solution: We will begin this problem by using the power rule.

$$\left(\frac{x^4y^{1/3}}{x^{1/2}y}\right)^2 = \frac{x^8y^{2/3}}{xy^2}$$

Now we use the quotient rule.

$$\frac{x^8y^{2/3}}{xy^2} = x^{8-1}y^{(2/3)-2} = x^7y^{-4/3}$$

Now use the negative exponent rule to write the answer without negative exponents.

$$x^7y^{-4/3} = \frac{x^7}{y^{4/3}} = \frac{x^7}{\sqrt[3]{y^4}} \quad \blacksquare$$

Work Example 8 by first simplifying the expression within parentheses, and then using the power rule.

Factoring Expressions Containing Rational and Negative Exponents (Optional)

▶**4** One reason we introduce both rational and negative exponents is because you may need to use and understand them in a later mathematics course. In Section 6.1 we discussed factoring a monomial from a polynomial. Recall that the terms in polynomials must have nonnegative integer exponents. Now we will factor a common factor from expressions that are not polynomials. We begin by reviewing a factoring problem like those discussed in Section 6.1.

EXAMPLE 9 Factor $x^2 + x^7$.

Solution: Recall that we factor out any variables common to every term in the polynomial. In this problem, the variable x is common to each term. We factor out each variable with the *lowest* (or lesser) exponent. The lowest power of x is x^2.

$$x^2 + x^7 = x^2(1 + x^{\overset{\displaystyle\frown}{5}}) \qquad 7 - 2 = 5$$

Notice that the exponent on the x that is left in parentheses can be determined by *subtracting* the exponent on the x being factored out, the 2, from the exponent on the x it is being factored from, the 7 ($7 - 2 = 5$). The factoring can be checked using the laws of exponents. Notice that $x^2 \cdot x^5 = x^{2+5} = x^7$. $\quad \blacksquare$

In the previous example we used subtraction to determine the exponent on the variable left in parentheses after the common factor was factored out. The exponent was obtained by *subtracting the lesser exponent from the greater exponent*. We can use the same procedure when factoring out factors that have rational and negative exponents.

EXAMPLE 10   Factor $x^{1/5} + x^{6/5}$.

Solution: The lesser exponent is $\frac{1}{5}$. Thus we will factor out an $x^{1/5}$. When we factor out an $x^{1/5}$ from an $x^{6/5}$, what is left? We can determine the exponent on the remaining x by subtracting the smaller exponent (the exponent on the variable being factored out, the $\frac{1}{5}$) from the larger exponent (the exponent on the variable it is being factored from, the $\frac{6}{5}$).

$$x^{1/5} + x^{6/5} = x^{1/5}(1 + x^{\overset{\displaystyle \frac{6}{5} - \frac{1}{5} = \frac{5}{5} = 1}{1}})$$

Since it is not necessary to write an exponent of 1, the answer is

$$x^{1/5} + x^{6/5} = x^{1/5}(1 + x)$$

Check: We check the factoring process using both the distributive property and the laws of exponents.

$$
\begin{aligned}
x^{1/5}(1 + x) &= x^{1/5} \cdot 1 + x^{1/5} \cdot x \\
&= x^{1/5} + x^{(1/5)+1} \\
&= x^{1/5} + x^{(1/5)+(5/5)} \\
&= x^{1/5} + x^{6/5}
\end{aligned}
$$

Since we obtained the expression we started with, our factoring is correct. ■

Now let us look at some examples that contain negative exponents. We can subtract the exponent on the variable being factored out from the exponent on the variable it is being factored from to obtain the exponent on the remaining variable. As with the other examples, we subtract the lesser exponent from the greater exponent to obtain the remaining exponent.

EXAMPLE 11   Factor $y^{-3} + y^{-5}$, and then write the answer without negative exponents.

Solution: We must factor out the y with the lesser exponent. Since -5 is less than -3, we factor out a y^{-5}. Keep in mind that for an expression to be simplified the answer should be written without negative exponents.

$$y^{-3} + y^{-5} = y^{-5}(y^{\overset{\displaystyle -3 - (-5) = -3 + 5 = 2}{2}} + 1)$$

Thus,

$$y^{-3} + y^{-5} = y^{-5}(y^2 + 1) = \frac{y^2 + 1}{y^5}$$

Check of factoring: $y^{-5}(y^2 + 1) = y^{-5} \cdot y^2 + y^{-5} \cdot 1 = y^{-3} + y^{-5}$

Since our check resulted in the expression that we began with, our factoring is correct. ■

Now we will factor three-termed expressions containing rational exponents as if they were trinomials. To factor the following expressions, we will use the substitution procedure that was introduced in Section 6.2.

EXAMPLE 12   Factor $x^{2/3} - 7x^{1/3} + 10$.

Solution: We first notice that $x^{2/3}$ is the square of $x^{1/3}$, $(x^{1/3})^2 = x^{2/3}$. Therefore this expression can be changed to a trinomial using a substitution. We will substitute y for $x^{1/3}$ to obtain a trinomial in y.

$$x^{2/3} - 7x^{1/3} + 10 = (x^{1/3})^2 - 7x^{1/3} + 10$$
$$= y^2 - 7y + 10$$

We now factor the trinomial $y^2 - 7y + 10$ and then substitute $x^{1/3}$ back for each y.

$$y^2 - 7y + 10 = (y - 5)(y - 2)$$
$$= (x^{1/3} - 5)(x^{1/3} - 2)$$

Check: $(x^{1/3} - 5)(x^{1/3} - 2) = (x^{1/3})(x^{1/3}) + x^{1/3}(-2) + (-5)(x^{1/3}) + (-5)(-2)$
$$= x^{2/3} - 2x^{1/3} - 5x^{1/3} + 10$$
$$= x^{2/3} - 7x^{1/3} + 10$$

Since the check results in the original expression, our factoring is correct. Thus, $x^{2/3} - 7x^{1/3} + 10 = (x^{1/3} - 5)(x^{1/3} - 2)$.   ∎

EXAMPLE 13   Factor $6y^{2/5} + 5y^{1/5} - 6$.

Solution: We first notice that $y^{2/5}$ is the square of $y^{1/5}$, $(y^{1/5})^2 = y^{2/5}$. If we substitute z for $y^{1/5}$, we will obtain a trinomial in z.

$$6y^{2/5} + 5y^{1/5} - 6 = 6(y^{1/5})^2 + 5y^{1/5} - 6$$
$$= 6z^2 + 5z - 6$$

Now factor the trinomial and then substitute $y^{1/5}$ back for each z.

$$6z^2 + 5z - 6 = (3z - 2)(2z + 3)$$
$$= (3y^{1/5} - 2)(2y^{1/5} + 3)$$

Thus $6y^{2/5} + 5y^{1/5} - 6 = (3y^{1/5} - 2)(2y^{1/3} + 3)$.   ∎

In Example 12 we used the variable y in our substitution. In Example 13 we used the variable z. The final answer will be independent of the variable selected in the substitution, and the final answer will always be given with the variable in the original expression.

Exercise Set 8.2

In this exercise set assume all variables represent positive real numbers. Write each of the following in exponential form.

1. $\sqrt{x^3}$

2. $\sqrt{y^5}$

3. $\sqrt{4^5}$

4. $\sqrt[3]{z^2}$

5. $\sqrt[5]{x^4}$

6. $\sqrt[3]{z^5}$

7. $(\sqrt{x})^3$

8. $(\sqrt[3]{y})^2$

9. $(\sqrt[4]{5})^3$

10. $\sqrt[4]{x^7}$

11. $\sqrt[7]{y^2}$

12. $\sqrt[5]{x^9}$

Write each of the following in radical form.

13. $x^{1/2}$

14. $y^{2/3}$

15. $z^{3/2}$

16. $5^{1/2}$

17. $2^{1/4}$

18. $x^{5/8}$

19. $z^{9/4}$

20. $w^{3/5}$

21. $x^{4/5}$

22. $y^{17/4}$

23. $7^{1/3}$

24. $y^{1/5}$

Simplify each radical expression by changing the expression to exponential form. Write the answer in radical form.

25. $\sqrt{y^6}$

26. $\sqrt{x^{12}}$

27. $\sqrt{z^8}$

28. $\sqrt[3]{x^6}$

29. $\sqrt[3]{x^9}$

30. $\sqrt[6]{y^2}$

31. $\sqrt[10]{z^5}$

32. $\sqrt{2^4}$

33. $(\sqrt{5})^2$

34. $(\sqrt{6})^2$

35. $\sqrt[6]{y^6}$

36. $(\sqrt[4]{x})^4$

37. $(\sqrt[8]{x})^2$

38. $\sqrt[3]{4^6}$

39. $(\sqrt[3]{x})^{15}$

40. $(\sqrt[4]{y})^{40}$

41. $(\sqrt[10]{y})^5$

42. $\sqrt[4]{8^2}$

43. $\sqrt[18]{y^6}$

44. $(\sqrt[6]{8})^4$

Evaluate each of the following if possible. If the expression is not a real number, state so.

45. $4^{1/2}$

46. $8^{2/3}$

47. $27^{2/3}$

48. $-8^{1/3}$

49. $(-4)^{1/2}$

50. $\left(\dfrac{4}{9}\right)^{1/2}$

51. $\left(\dfrac{9}{25}\right)^{1/2}$

52. $\left(\dfrac{1}{8}\right)^{1/3}$

53. $-16^{1/2}$

54. $(-16)^{1/2}$

55. $-27^{1/3}$

56. $4^{-1/2}$

57. $27^{-1/3}$

58. $16^{-3/2}$

59. $4^{-3/2}$

60. $81^{-3/4}$

61. $\left(\dfrac{4}{49}\right)^{-1/2}$

62. $\left(\dfrac{25}{64}\right)^{-1/2}$

63. $\left(\dfrac{8}{27}\right)^{-2/3}$

64. $\left(\dfrac{81}{16}\right)^{-3/4}$

65. $25^{1/2} + 36^{1/2}$

66. $25^{-1/2} + 36^{-1/2}$

67. $8^{-1/3} + 9^{-1/2}$

68. $16^{-3/2} - 16^{-3/4}$

Simplify each of the following. Write the answer in exponential form.

69. $x^5 \cdot x^{1/2}$

70. $x^{1/3} \cdot x^{3/8}$

71. $\dfrac{x^{1/2}}{x^{1/3}}$

72. $(x^{2/3})^3$

73. $(x^{1/5})^{2/3}$

74. $x^{-3/5}$

75. $(x^{1/2})^{-2}$

76. $(z^{-1/4})^{-1/2}$

77. $(6^{-1/3})^0$

78. $\dfrac{x^4}{x^{-1/2}}$

79. $\dfrac{y^{-1/3}}{y^{-2}}$

80. $x^{-1/2} \cdot x^{-3/5}$

81. $x^{5/3} \cdot x^{-7/2}$

82. $(x^{-2/5})^{1/3}$

83. $\left(\dfrac{64}{x}\right)^{1/3}$

84. $\left(\dfrac{8}{y^4}\right)^{1/3}$

85. $\left(\dfrac{x^{3/7}}{x^{1/2}}\right)^2$

86. $\left(\dfrac{x^{-1/3}}{x^{-2}}\right)^{1/2}$

87. $\left(\dfrac{y^4}{y^{-2/5}}\right)^{-3}$

88. $\left(\dfrac{z^{1/2}y^3}{z^{1/2}}\right)^{1/2}$

89. $\dfrac{x^{3/4}y^{-2}}{x^{1/2}y^2}$

90. $\left(\dfrac{x^{1/2}y^3}{z^{-3}}\right)\left(\dfrac{xy^{2/3}}{z^2}\right)$

91. $\left(\dfrac{a^{1/2}b^{2/3}}{a^{-1/3}b^{3/5}}\right)^2$

92. $\left(\dfrac{x^{1/3}y^{2/3}}{z^5}\right)\left(\dfrac{x^2y}{z^6}\right)^{1/2}$

Factor each of the following. Write the answer without negative exponents. (Optional)

93. $x^{3/2} + x^{1/2}$

94. $x^{1/4} - x^{5/4}$

95. $y^{1/3} - y^{4/3}$

96. $x^{-1/2} + x^{1/2}$

97. $y^{-3/5} + y^{2/5}$

98. $x^2 - x^{-1}$

99. $y^{-1} - y$

100. $y^2 - y^{-2}$

101. $x^{-7} + x^{-5}$

102. $x^{-12} + x^{-7}$

103. $x^{-1/2} + x^{-5/2}$

104. $y^{-9/5} - y^{-4/5}$

105. $2x^{-4} - 6x^{-5}$ **106.** $3x^{1/2} - 6x^{-1/2}$ **107.** $5x - 10x^{-1}$

108. $4x^{-3} - 8x^{-4}$

Factor each of the following. (Optional)

109. $x^{2/3} + 2x^{1/3} - 3$ **110.** $x^{2/5} + 4x^{1/5} - 5$

111. $x + 6x^{1/2} + 9$ **112.** $x^{1/2} + x^{1/4} - 20$

113. $2x^{2/7} + x^{1/7} - 3$ **114.** $3x^{1/2} - 10x^{1/4} - 8$

115. $4x^{4/5} + 8x^{2/5} + 3$ **116.** $6x^{2/3} - 5x^{1/3} + 1$

117. $15x^{1/3} - 14x^{1/6} + 3$ **118.** $8x + 2x^{1/2} - 1$

122. By writing $\sqrt{\sqrt{x}}$ in exponential form, show that $\sqrt{\sqrt{x}} = \sqrt[4]{x}$, $x \geq 0$.

119. Under what specific conditions will $\sqrt[n]{a^n} = (\sqrt[n]{a})^n = a$?

123. Determine if $\sqrt[3]{\sqrt{x}} = \sqrt{\sqrt[3]{x}}$, $x \geq 0$. Explain how you determined your answer.

120. By selecting values for a and b, show that $(a^2 + b^2)^{1/2}$ *is not equal to* $a + b$.

121. By selecting values for a and b, show that $(a^{1/2} + b^{1/2})^2$ *is not equal to* $a + b$.

Cumulative Review Exercises

[3.5] **124.** Determine which of the following graphs are functions and which are relations.

(a)

(b)

(c)

[7.4] **125.** Simplify $\dfrac{a^{-2} + ab^{-1}}{ab^{-2} - a^{-2}b^{-1}}$.

[7.5] **126.** Solve the equation $\dfrac{3x - 2}{x + 4} = \dfrac{2x + 1}{3x - 2}$.

[7.6] **127.** Amy can fly her plane 500 miles against the wind in the same time it takes her to fly 560 miles with the wind. If the wind blows at 25 miles per hour, find the speed of the plane in still air.

JUST FOR FUN

Factor the following expressions. Write the answers without negative exponents.

1. $(6x - 5)^{-3} + (6x - 5)^{-2}$ **2.** $(2x + 3)^{-1/3} + (2x + 3)^{2/3}$

8.3

Multiplying and Simplifying Radicals

▶ **1** Know the product rule for radicals.

▶ **2** Simplify radicals whose radicands are natural numbers.

▶ **3** Simplify radicals whose radicands are variables.

▶ **4** Simplify radicals.

▶ **5** Simplify a product of two radicals.

▶ **1** We will first simplify radicals using the product rule.

Product Rule for Radicals

For nonnegative real numbers a and b,
$$\sqrt[n]{a} \cdot \sqrt[n]{b} = \sqrt[n]{ab}$$

Examples of the Product Rule

$$\sqrt{60} = \begin{cases} \sqrt{1} \cdot \sqrt{60} \\ \sqrt{2} \cdot \sqrt{30} \\ \sqrt{3} \cdot \sqrt{20} \\ \sqrt{4} \cdot \sqrt{15} \\ \sqrt{5} \cdot \sqrt{12} \\ \sqrt{6} \cdot \sqrt{10} \end{cases}$$

$\sqrt{60}$ can be factored into any of these forms

$$\sqrt[3]{60} = \begin{cases} \sqrt[3]{1} \cdot \sqrt[3]{60} \\ \sqrt[3]{2} \cdot \sqrt[3]{30} \\ \sqrt[3]{3} \cdot \sqrt[3]{20} \\ \sqrt[3]{4} \cdot \sqrt[3]{15} \\ \sqrt[3]{5} \cdot \sqrt[3]{12} \\ \sqrt[3]{6} \cdot \sqrt[3]{10} \end{cases}$$

$\sqrt[3]{60}$ can be factored into any of these forms

$$\sqrt{x^7} = \begin{cases} \sqrt{x} \cdot \sqrt{x^6} \\ \sqrt{x^2} \cdot \sqrt{x^5} \\ \sqrt{x^3} \cdot \sqrt{x^4} \end{cases}$$

$\sqrt{x^7}$ can be factored into any of these forms

$$\sqrt[3]{x^7} = \begin{cases} \sqrt[3]{x} \cdot \sqrt[3]{x^6} \\ \sqrt[3]{x^2} \cdot \sqrt[3]{x^5} \\ \sqrt[3]{x^3} \cdot \sqrt[3]{x^4} \end{cases}$$

$\sqrt[3]{x^7}$ can be factored into any of these forms

▶ **2** To help clarify our explanations, we will introduce **perfect power numbers**. A number is a **perfect square** if it is the square of a natural number. A number is a **perfect cube** if it is a cube of a natural number. A number is a **perfect fourth power number** if it is the fourth power of a natural number, and so on.

Some perfect square and perfect cube numbers follow.

Squares of natural numbers: $1^2, \; 2^2, \; 3^2, \; 4^2, \; 5^2, \; 6^2, \; 7^2, \; 8^2, \; 9^2, \ldots$

Perfect square numbers: $1, \; 4, \; 9, \; 16, \; 25, \; 36, \; 49, \; 64, \; 81, \ldots$

Cubes of natural numbers: $1^3, \; 2^3, \; 3^3, \; 4^3, \; 5^3, \; 6^3, \; 7^3, \; 8^3, \; 9^3, \ldots$

Perfect cube numbers: $1, \; 8, \; 27, \; 64, \; 125, \; 216, \; 343, \; 512, \; 729, \ldots$

Note that the square root of any perfect square number will be a whole number. For example,

$$\sqrt{36} = \sqrt{6^2} = 6^{2/2} = 6$$

Similarly, the cube root of any perfect cube number will be a whole number. For example,

$$\sqrt[3]{125} = \sqrt[3]{5^3} = 5^{3/3} = 5$$

The fourth root of any perfect fourth-power number will be a whole number, etc.

Now we will discuss how to simplify radicals with natural number radicands.

> **To Simplify Radicals With Natural Number Radicands**
>
> **1.** Write the radicand as the product of two numbers, one of which is the largest perfect power number for the given index.
> **2.** Use the product rule to write the expression as a product of roots.
> **3.** Find the roots of any perfect power numbers.

If we are simplifying a *square* root, we will write the radicand as the product of the largest *perfect square number* and another number. If we are simplifying a *cube* root, we will write the radicand as the product of the largest *perfect cube number* and another number, and so on.

EXAMPLE 1 Simplify $\sqrt{32}$.

Solution: Since we are evaluating a square root, we look for the largest perfect square that divides 32. The largest perfect square that divides, or is a factor of, 32 is 16.

$$\sqrt{32} = \sqrt{16 \cdot 2}$$
$$= \sqrt{16}\sqrt{2}$$
$$= 4\sqrt{2}$$
∎

In Example 1, if you first believed that 4 was the largest perfect square that divided 32, you could proceed as follows:

$$\sqrt{32} = \sqrt{4 \cdot 8}$$
$$= \sqrt{4}\sqrt{8}$$
$$= 2\sqrt{8} \qquad \text{4 is a perfect square that is a factor of 8.}$$
$$= 2\sqrt{4}\sqrt{2}$$
$$= 2 \cdot 2\sqrt{2}$$
$$= 4\sqrt{2}$$

Note that the final result is the same, but you must work harder to find the answer.

The chart on page 367 can be useful in determining the largest perfect square or perfect cube number that divides the radicand when simplifying square or cube root problems.

EXAMPLE 2 Simplify $\sqrt{60}$.

Solution: $\sqrt{60} = \sqrt{4}\sqrt{15} = 2\sqrt{15}$
∎

In Example 2, $\sqrt{15}$ can be factored into $\sqrt{5}\sqrt{3}$; however, since neither 5 nor 3 is a perfect square $\sqrt{15}$ cannot be simplified. The answer is $2\sqrt{15}$.

EXAMPLE 3 Simplify $\sqrt[3]{54}$.

Solution: The largest perfect cube number that is a factor of 54 is 27.
$$\sqrt[3]{54} = \sqrt[3]{27}\sqrt[3]{2} = 3\sqrt[3]{2}$$
∎

EXAMPLE 4 Simplify $\sqrt[3]{375}$.

Solution: $\sqrt[3]{375} = \sqrt[3]{125}\sqrt[3]{3} = 5\sqrt[3]{3}$
∎

▶**3**   Now we will discuss perfect powers of variables for a given index. The radicand x^n is a **perfect square** when n is a multiple of 2 (an even natural number). The radicand x^n is a **perfect cube** when n is a multiple of 3. In general, the radicand x^n is a **perfect power** when n is a **multiple of the index** of the radical (or when n is divisible by the index).

Following are some perfect square, perfect cube, and perfect fourth powers of the variable x.

$$\text{Perfect square powers of } x: \quad x^2, x^4, x^6, x^8, x^{10}, \ldots$$
$$\text{Perfect cube powers of } x: \quad x^3, x^6, x^9, x^{12}, x^{15}, \ldots$$
$$\text{Perfect fourth powers of } x: \quad x^4, x^8, x^{12}, x^{16}, x^{20}, \ldots$$
$$\text{Perfect powers of } x \text{ for index } n: \quad x^n, x^{2n}, x^{3n}, x^{4n}, x^{5n}, \ldots$$

HELPFUL HINT

A quick way of determining if a radicand x^n is a perfect power for a given index is to determine if the exponent n is divisible by the index of the radical. For example, consider $\sqrt[5]{x^{20}}$. Since the exponent, 20, is divisible by the index, 5, x^{20} is a perfect fifth power. Now consider $\sqrt[6]{x^{20}}$. Since the exponent, 20, is not divisible by the index, 6, x^{20} is not a perfect sixth power. Note, however, that x^{18} and x^{24} are both perfect sixth powers since 6 divides both 18 and 24.

A radical can often be simplified by writing the radical in exponential form, as illustrated in Example 5.

EXAMPLE 5   Simplify   (a) $\sqrt{x^4}$ (b) $\sqrt[3]{x^{12}}$ (c) $\sqrt[6]{y^{24}}$

Solution: (a) $\sqrt{x^4} = x^{4/2} = x^2$ (b) $\sqrt[3]{x^{12}} = x^{12/3} = x^4$ (c) $\sqrt[6]{y^{24}} = y^{24/6} = y^4$ ∎

To Simplify Radicals With Variable Radicands

1. Write each variable as the product of two factors, one of which is the largest perfect power of the variable for the given index.

2. Use the product rule to write the expression as a product of roots. Place all perfect powers under the same root.

3. Find the roots of any perfect powers.

EXAMPLE 6   Simplify   (a) $\sqrt{x^9}$ (b) $\sqrt[3]{x^{14}}$ (c) $\sqrt[5]{x^{23}}$

Solution: (a) The largest perfect square power less than or equal to x^9 is x^8.

$$\sqrt{x^9} = \sqrt{x^8 \cdot x} = \sqrt{x^8} \cdot \sqrt{x} = x^{8/2}\sqrt{x} = x^4\sqrt{x}$$

(b) The largest perfect cube power less than or equal to x^{14} is x^{12}.

$$\sqrt[3]{x^{14}} = \sqrt[3]{x^{12} \cdot x^2} = \sqrt[3]{x^{12}}\sqrt[3]{x^2} = x^{12/3}\sqrt[3]{x^2} = x^4\sqrt[3]{x^2}$$

(c) The largest perfect fifth power less than or equal to x^{23} is x^{20}.

$$\sqrt[5]{x^{23}} = \sqrt[5]{x^{20} \cdot x^3} = \sqrt[5]{x^{20}}\sqrt[5]{x^3} = x^{20/5}\sqrt[5]{x^3} = x^4\sqrt[5]{x^3}$$ ∎

EXAMPLE 7 Evaluate $\sqrt{x^{12}y^{17}}$.

Solution: x^{12} is a perfect square power. The highest perfect square power that is a factor of y^{17} is y^{16}. Write y^{17} as $y^{16} \cdot y^1$.

$$\sqrt{x^{12}y^{17}} = \sqrt{x^{12} \cdot y^{16} \cdot y} = \sqrt{x^{12}y^{16}} \sqrt{y}$$
$$= \sqrt{x^{12}} \sqrt{y^{16}} \sqrt{y}$$
$$= x^{12/2}y^{16/2} \sqrt{y}$$
$$= x^6 y^8 \sqrt{y} \qquad \blacksquare$$

Often the steps where we change the radical expression to exponential form is done mentally, and those steps are not illustrated. For instance, in Example 7, we might change $\sqrt{x^{12}y^{16}}$ to $x^6 y^8$ mentally, and not show the intermediate steps.

EXAMPLE 8 Simplify $\sqrt[4]{x^6 y^{23}}$.

Solution: We begin by finding the highest perfect fourth powers of x^6 and y^{23}. For an index of 4, the highest perfect power that is a factor of x^6 is x^4. The highest perfect power that is a factor of y^{23} is y^{20}.

$$\sqrt[4]{x^6 y^{23}} = \sqrt[4]{x^4 \cdot x^2 \cdot y^{20} \cdot y^3}$$
$$= \sqrt[4]{x^4 y^{20} \cdot x^2 y^3}$$
$$= \sqrt[4]{x^4 y^{20}} \sqrt[4]{x^2 y^3}$$
$$= xy^5 \sqrt[4]{x^2 y^3} \qquad \blacksquare$$

▶ **4** Now we give a general procedure for simplifying radicals.

To Simplify Radicals

1. If the radicand contains a numerical factor, write the number as a product of two numbers, one of which is the largest perfect power for the given index.
2. Write each variable factor as a product of two factors, one of which is the largest perfect power of the variable for the given index.
3. Use the product rule to write the expression as a product of roots. All the perfect powers (numbers and variables) should be placed under the same radical.
4. Simplify the root containing the perfect powers.

EXAMPLE 9 Simplify $\sqrt{80x^5 y^{12} z^3}$.

Solution: The highest perfect square that is a factor of 80 is 16. You should recognize that $80 = 16 \cdot 5$. The highest perfect square that is a factor of x^5 is x^4. Note that $x^5 = x^4 \cdot x$. The expression y^{12} is a perfect square. Finally, the highest perfect square that is a factor of z^3 is z^2. We know that $z^3 = z^2 \cdot z$. Place all the perfect squares under the same radical, and then simplify.

$$\sqrt{80x^5 y^{12} z^3} = \sqrt{16 \cdot 5 \cdot x^4 \cdot x \cdot y^{12} \cdot z^2 \cdot z}$$
$$= \sqrt{16x^4 y^{12} z^2 \cdot 5xz}$$
$$= \sqrt{16x^4 y^{12} z^2} \cdot \sqrt{5xz}$$
$$= 4x^2 y^6 z \sqrt{5xz} \qquad \blacksquare$$

EXAMPLE 10   Simplify $\sqrt[3]{54x^{17}y^{25}}$.

Solution: The highest perfect cube that is a factor of 54 is 27. The highest perfect cube that is a factor of x^{17} is x^{15}. The highest perfect cube that is a factor of y^{25} is y^{24}.

$$\sqrt[3]{54x^{17}y^{25}} = \sqrt[3]{27 \cdot 2 \cdot x^{15} \cdot x^2 \cdot y^{24} \cdot y}$$
$$= \sqrt[3]{27x^{15}y^{24} \cdot 2x^2y}$$
$$= \sqrt[3]{27x^{15}y^{24}} \cdot \sqrt[3]{2x^2y}$$
$$= 3x^5y^8\sqrt[3]{2x^2y}$$

HELPFUL HINT

In Example 8 we showed that

$$\sqrt[4]{x^6y^{23}} = xy^5\sqrt[4]{x^2y^3}$$

Note that the answer can be obtained by dividing the exponents on the variables in the radicand, 6 and 23, by the index, 4, and observing the quotients and remainders.

$$\underset{6 \div 4}{\text{quotient}} \quad \underset{23 \div 4}{\text{quotient}} \quad \underset{6 \div 4}{\text{remainder}} \quad \underset{23 \div 4}{\text{remainder}}$$

$$\sqrt[4]{x^6y^{23}} = x^1y^5\sqrt[4]{x^2y^3}$$

Can you explain why this procedure works? You may wish to use this procedure to work or check certain problems.

▶ **5**   When we are given the product of two or more radicals, we can use the product rule to multiply the radicals together and then simplify. Examples 11 and 12 illustrate this procedure.

EXAMPLE 11   Multiply and simplify.
(a) $\sqrt{2}\sqrt{8}$ (b) $\sqrt[3]{2x}\sqrt[3]{4x^2}$

Solution: (a) $\sqrt{2}\sqrt{8} = \sqrt{2 \cdot 8}$ (b) $\sqrt[3]{2x}\sqrt[3]{4x^2} = \sqrt[3]{2x \cdot 4x^2}$
$\qquad\qquad\qquad = \sqrt{16} = 4$ $\qquad\qquad\qquad\quad = \sqrt[3]{8x^3} = 2x$

EXAMPLE 12   Multiply and simplify.
(a) $\sqrt[4]{8x^3y}\sqrt[4]{8x^6y^2}$ (b) $\sqrt[3]{5xy^4}\sqrt[3]{50x^2y^{18}}$

Solution: (a) $\sqrt[4]{8x^3y}\sqrt[4]{8x^6y^2} = \sqrt[4]{8x^3y \cdot 8x^6y^2}$ (b) $\sqrt[3]{5xy^4}\sqrt[3]{50x^2y^{18}} = \sqrt[3]{5xy^4 \cdot 50x^2y^{18}}$
$\qquad\qquad\qquad\qquad\quad = \sqrt[4]{64x^9y^3}$ $\qquad\qquad\qquad\qquad\qquad = \sqrt[3]{250x^3y^{22}}$
$\qquad\qquad\qquad\qquad\quad = \sqrt[4]{16x^8}\sqrt[4]{4xy^3}$ $\qquad\qquad\qquad\qquad\qquad = \sqrt[3]{125x^3y^{21}}\sqrt[3]{2y}$
$\qquad\qquad\qquad\qquad\quad = 2x^2\sqrt[4]{4xy^3}$ $\qquad\qquad\qquad\qquad\qquad = 5xy^7\sqrt[3]{2y}$

Note that when simplifying a radical the radicand of your simplified answer should not have any variable with an exponent greater than or equal to the index.

EXAMPLE 13   Multiply and simplify $\sqrt{2x}(\sqrt{8x} - \sqrt{32})$.

Solution: Begin by using the distributive property.

$$\sqrt{2x}(\sqrt{8x} - \sqrt{32}) = (\sqrt{2x})(\sqrt{8x}) + (\sqrt{2x})(-\sqrt{32})$$
$$= \sqrt{16x^2} - \sqrt{64x}$$
$$= 4x^2 - \sqrt{64}\sqrt{x}$$
$$= 4x - 8\sqrt{x} \qquad \blacksquare$$

Note in Example 13 that the same answer could be obtained by first simplifying $\sqrt{8x}$ and $\sqrt{32}$ and then multiplying. You may wish to try this now.

EXAMPLE 14 Multiply and simplify $\sqrt[3]{3x^2y}(\sqrt[3]{9xy^5} + \sqrt[3]{18x^8y^{10}})$.

Solution: $\sqrt[3]{3x^2y}(\sqrt[3]{9xy^5} + \sqrt[3]{18x^8y^{10}}) = (\sqrt[3]{3x^2y})(\sqrt[3]{9xy^5}) + (\sqrt[3]{3x^2y})(\sqrt[3]{18x^8y^{10}})$
$$= \sqrt[3]{27x^3y^6} + \sqrt[3]{54x^{10}y^{11}}$$
$$= 3xy^2 + \sqrt[3]{27x^9y^9} \cdot \sqrt[3]{2xy^2}$$
$$= 3xy^2 + 3x^3y^3\sqrt[3]{2xy^2} \qquad \blacksquare$$

We will do additional multiplication of radical problems in Sections 8.4 and 8.5.

Exercise Set 8.3

Simplify each expression. In this exercise set, assume that all variables represent positive real numbers.

1. $\sqrt{50}$ 2. $\sqrt{40}$ 3. $\sqrt{32}$ 4. $\sqrt{72}$

5. $\sqrt[3]{16}$ 6. $\sqrt[3]{24}$ 7. $\sqrt[3]{54}$ 8. $\sqrt[4]{80}$

9. $\sqrt{x^3}$ 10. $\sqrt{y^5}$ 11. $\sqrt{x^{11}}$ 12. $\sqrt{a^{30}}$

13. $\sqrt{b^{27}}$ 14. $\sqrt[3]{y^7}$ 15. $\sqrt[4]{y^9}$ 16. $\sqrt[4]{b^{23}}$

17. $\sqrt{24x^3}$ 18. $\sqrt{20x^7}$ 19. $\sqrt[3]{24y^7}$ 20. $\sqrt[4]{16x^{10}}$

21. $\sqrt{x^3y^7}$ 22. $\sqrt{50xy^4}$ 23. $\sqrt[3]{81x^6y^8}$ 24. $\sqrt[3]{16x^3y^6}$

25. $\sqrt[3]{54x^{12}y^{13}}$ 26. $\sqrt[4]{x^9y^{12}z^{15}}$ 27. $\sqrt[5]{64x^{12}y^7}$ 28. $\sqrt[3]{18w^{12}v^9r^{31}}$

29. $\sqrt[3]{32c^4w^9z}$ 30. $\sqrt[4]{32x^8y^9z^{19}}$ 31. $\sqrt[3]{81x^7y^{21}z^{50}}$ 32. $\sqrt[3]{18x^4y^7z^{15}}$

Simplify each expression.

33. $\sqrt{5}\sqrt{5}$ 34. $\sqrt{60}\sqrt{5}$ 35. $\sqrt[3]{2}\sqrt[3]{4}$

36. $\sqrt[3]{2}\sqrt[3]{28}$ 37. $\sqrt[3]{3}\sqrt[3]{54}$ 38. $\sqrt{5x^2}\sqrt{8x^3}$

39. $\sqrt{15xy^4}\sqrt{6xy^3}$ 40. $(\sqrt{6xy^2})^2$ 41. $(\sqrt{4x^3y^2})^2$

42. $\sqrt{9x^3y^7}\sqrt{3xy^4}$ 43. $\sqrt[3]{5xy^2}\sqrt[3]{25x^4y^{12}}$ 44. $\sqrt[3]{9x^7y^{12}}\sqrt[3]{6x^4y}$

45. $(\sqrt[3]{2x^3y^4})^2$ 46. $(\sqrt[3]{5x^2y^6})^2$ 47. $\sqrt[4]{12xy^4}\sqrt[4]{2x^3y^9z^7}$

48. $\sqrt[4]{3x^9y^{12}}\sqrt[4]{54x^4y^7}$ 49. $\sqrt[5]{x^{24}y^{30}z^9}\sqrt[5]{x^{13}y^8z^7}$ 50. $\sqrt[4]{8x^4yz^3}\sqrt[4]{2x^2y^3z^7}$

Simplify each expression.

51. $\sqrt{2}(\sqrt{6} + \sqrt{2})$ 52. $\sqrt{5}(\sqrt{5} + \sqrt{3})$

53. $\sqrt{3}(\sqrt{12} - \sqrt{6})$ 54. $2(2\sqrt{8} - 3\sqrt{2})$

55. $\sqrt{2}(\sqrt{18} + \sqrt{8})$ 56. $\sqrt{2x}(\sqrt{8x} - \sqrt{32})$

57. $\sqrt{3y}(\sqrt{27y^2} - \sqrt{y})$ 58. $\sqrt[3]{x}(\sqrt[3]{x^2} + \sqrt[3]{x^5})$

59. $\sqrt[3]{2x^2y}(\sqrt[3]{4xy^5} + \sqrt[3]{12x^{10}y})$ 60. $\sqrt[4]{2x^3y^2}(\sqrt[4]{8x^5y^7} - \sqrt[4]{3x^5y^6})$

61. $2\sqrt[3]{x^4y^5}(\sqrt[3]{8x^{12}y^4} + \sqrt[3]{16xy^9})$ 62. $\sqrt[3]{4x^2y^6}(\sqrt[3]{9x^8y^5} - \sqrt[3]{7x^9y})$

63. $3\sqrt{2xy^4}(\sqrt{20x^4y^8} - 2\sqrt{6xy^9})$ 64. $\sqrt[5]{8x^4y^6}(\sqrt[5]{4x^6y^9} - \sqrt[5]{10xy^7})$

Simplify each expression.

65. $\sqrt{24}$

66. $\sqrt{200}$

67. $\sqrt[3]{32}$

68. $\sqrt[4]{162}$

69. $\sqrt[3]{x^5}$

70. $\sqrt[3]{y^{13}}$

71. $\sqrt{36x^5}$

72. $\sqrt[3]{80x^{11}}$

73. $\sqrt{x^5y^{12}}$

74. $\sqrt[3]{x^9y^{11}z}$

75. $\sqrt[4]{16ab^{17}c^9}$

76. $\sqrt[5]{32a^2b^5}$

77. $\sqrt{75}\sqrt{6}$

78. $\sqrt[4]{8}\sqrt[4]{10}$

79. $\sqrt{15x^2}\sqrt{6x^5}$

80. $\sqrt{14xy^2}\sqrt{3xy^3}$

81. $\sqrt{20xy^4}\sqrt{6x^5y^7}$

82. $\sqrt{6}(4 - \sqrt{2})$

83. $\sqrt{x}(\sqrt{x} + 3)$

84. $\sqrt{y}(\sqrt{y^3} - 2)$

85. $\sqrt[3]{4xy^2}\sqrt[3]{4xy^4}$

86. $(\sqrt[3]{4x^5y^2})^2$

87. $\sqrt[3]{y}(2\sqrt[3]{y} - \sqrt[3]{y^8})$

88. $\sqrt[3]{2x^9y^6z}\sqrt[3]{12xy^4z^3}$

89. $\sqrt[3]{3xy^2}(\sqrt[3]{4x^4y^3} - \sqrt[3]{8x^5y^4})$

90. $\sqrt[4]{4xy^2}(\sqrt[4]{2x^5y^6} + \sqrt[4]{5x^9y^2})$

91. We stated that for nonnegative real numbers a and b that $\sqrt[n]{a} \cdot \sqrt[n]{b} = \sqrt[n]{ab}$. Why is it necessary to specify that both a and b are nonnegative real numbers?

Cumulative Review Exercises

[1.2] **92.** What is a rational number?

93. What is a real number?

94. What is an irrational number?

[1.3] **95.** What is the definition of $|a|$?

[2.2] **96.** Solve the formula $E = \frac{1}{2}mv^2$ for m.

[2.5] **97.** Solve the inequality $-4 < 2x - 3 \le 5$ and indicate the solution (a) on the number line, (b) in internal notation, and (c) in set builder notation.

8.4

Dividing and Simplifying Radicals

▶ **1** Know the quotient rule for radicals.

▶ **2** Know when a radical is simplified.

▶ **3** Rationalize a denominator.

▶ **4** Rationalize a denominator using the conjugate of the denominator.

▶ **1** In mathematics we sometimes need to divide one radical expression by another. To divide radicals, or to simplify radicals containing fractions, we use the quotient rule for radicals.

Quotient Rule for Radicals

For nonnegative real numbers a and b,

$$\frac{\sqrt[n]{a}}{\sqrt[n]{b}} = \sqrt[n]{\frac{a}{b}} \qquad b \ne 0$$

Examples 1 through 3 illustrate how the quotient rule is used to simplify radical expressions.

EXAMPLE 1   Simplify each of the following.

(a) $\dfrac{\sqrt{75}}{\sqrt{3}}$ (b) $\dfrac{\sqrt[3]{24}}{\sqrt[3]{3}}$

Solution: (a) $\dfrac{\sqrt{75}}{\sqrt{3}} = \sqrt{\dfrac{75}{3}} = \sqrt{25} = 5$ (b) $\dfrac{\sqrt[3]{24}}{\sqrt[3]{3}} = \sqrt[3]{\dfrac{24}{3}} = \sqrt[3]{8} = 2$ ∎

EXAMPLE 2   Simplify each of the following.

(a) $\sqrt{\dfrac{9}{4}}$ (b) $\sqrt[3]{\dfrac{8}{27}}$

Solution: (a) $\sqrt{\dfrac{9}{4}} = \dfrac{\sqrt{9}}{\sqrt{4}} = \dfrac{3}{2}$ (b) $\sqrt[3]{\dfrac{8}{27}} = \dfrac{\sqrt[3]{8}}{\sqrt[3]{27}} = \dfrac{2}{3}$ ∎

EXAMPLE 3   Simplify each of the following.

(a) $\sqrt{\dfrac{16x^2}{8}}$ (b) $\sqrt{\dfrac{4x^5y^7}{16x^3y^{13}}}$ (c) $\sqrt[4]{\dfrac{15xy^5}{3x^9y}}$

Solution: (a) $\sqrt{\dfrac{16x^2}{8}} = \sqrt{2x^2} = \sqrt{2}\sqrt{x^2} = \sqrt{2}x$ or $x\sqrt{2}$

(b) $\sqrt{\dfrac{4x^5y^7}{16x^3y^{13}}} = \sqrt{\dfrac{x^2}{4y^6}} = \dfrac{\sqrt{x^2}}{\sqrt{4y^6}} = \dfrac{x}{2y^3}$

(c) $\sqrt[4]{\dfrac{15xy^5}{3x^9y}} = \sqrt[4]{\dfrac{5y^4}{x^8}} = \dfrac{\sqrt[4]{5y^4}}{\sqrt[4]{x^8}} = \dfrac{\sqrt[4]{y^4}\sqrt[4]{5}}{x^2} = \dfrac{y\sqrt[4]{5}}{x^2}$ ∎

▶ **2**   After you have simplified a radical expression, you should check it to make sure that it is simplified as far as possible. A radical is simplified as far as possible when the following three conditions are met.

A Radical Expression Is Simplified When the Following Are All True

1. There are no perfect powers that are factors of any radicand.
2. No radicand contains fractions.
3. There are no radicals in any denominator.

▶ **3**   When the denominator of a fraction contains a radical, we generally simplify the expression by **rationalizing the denominator.** To rationalize a denominator is to remove all radicals from the denominator. Denominators are rationalized because, without a calculator, it is often easier to evaluate a fraction with a whole-number denominator than one where the denominator contains a radical.

To rationalize a denominator, multiply both the numerator and the denominator of the fraction by the denominator, or by a radical that will result in the radicand in the denominator becoming a perfect power. The following examples illustrate the procedure to be used.

EXAMPLE 4 Simplify $\dfrac{1}{\sqrt{5}}$.

Solution: To simplify this expression, we must rationalize the denominator.

$$\frac{1}{\sqrt{5}} = \frac{1}{\sqrt{5}} \cdot \frac{\sqrt{5}}{\sqrt{5}} = \frac{\sqrt{5}}{\sqrt{25}} = \frac{\sqrt{5}}{5}$$

In Example 4, multiplying both the numerator and denominator by $\sqrt{5}$ is equivalent to multiplying the fraction by 1, which does not change the value of the original fraction.

EXAMPLE 5 Simplify each of the following.

(a) $\sqrt{\dfrac{2}{3}}$ (b) $\dfrac{x}{3\sqrt{2}}$ (c) $\sqrt{\dfrac{x}{y}}$

Solution: (a) $\sqrt{\dfrac{2}{3}} = \dfrac{\sqrt{2}}{\sqrt{3}} = \dfrac{\sqrt{2}}{\sqrt{3}} \cdot \dfrac{\sqrt{3}}{\sqrt{3}} = \dfrac{\sqrt{6}}{3}$

(b) $\dfrac{x}{3\sqrt{2}} = \dfrac{x}{3\sqrt{2}} \cdot \dfrac{\sqrt{2}}{\sqrt{2}} = \dfrac{x\sqrt{2}}{3 \cdot 2} = \dfrac{x\sqrt{2}}{6}$

(c) $\sqrt{\dfrac{x}{y}} = \dfrac{\sqrt{x}}{\sqrt{y}} = \dfrac{\sqrt{x}}{\sqrt{y}} \cdot \dfrac{\sqrt{y}}{\sqrt{y}} = \dfrac{\sqrt{xy}}{\sqrt{y^2}} = \dfrac{\sqrt{xy}}{y}$

EXAMPLE 6 Simplify $\sqrt[3]{\dfrac{3}{5}}$.

Solution: $\sqrt[3]{\dfrac{3}{5}} = \dfrac{\sqrt[3]{3}}{\sqrt[3]{5}}$. Since the denominator is a cube root, we must multiply the numerator and denominator by the cube root of an expression that will result in the product of the radicands in the denominator being a perfect cube. Multiply both the numerator and denominator by $\sqrt[3]{5^2}$.

$$\frac{\sqrt[3]{3}}{\sqrt[3]{5}} = \frac{\sqrt[3]{3}}{\sqrt[3]{5}} \cdot \frac{\sqrt[3]{5^2}}{\sqrt[3]{5^2}}$$

$$= \frac{\sqrt[3]{3} \cdot \sqrt[3]{5^2}}{\sqrt[3]{5^3}}$$

$$= \frac{\sqrt[3]{3}\sqrt[3]{25}}{5}$$

$$= \frac{\sqrt[3]{75}}{5}$$

Calculator Corner

Square roots can be found on a calculator that has a square root key, $\boxed{\sqrt{x}}$. To evaluate $\sqrt{\dfrac{2}{3}}$ on a calculator press the following keys

$$\boxed{c}\ 2\ \boxed{\div}\ 3\ \boxed{=}\ \boxed{\sqrt{x}}\ 0.8164965$$

Cube and higher roots can be found using a scientific calculator. The keys to press depend on your calculator. Generally, to find cube and higher roots, you will use either a $\boxed{\sqrt[x]{y}}$ or $\boxed{y^{1/x}}$ key. Recall that $\sqrt[3]{x} = x^{1/3}$. Some calculators use slight variations of these keys. If one of these keys is not available, often a cube or higher root may be found using the combination of $\boxed{\text{inv}}\ \boxed{y^x}$ keys. To find $\sqrt[3]{8}$ on a scientific calculator, you would press one of the following sequences of keys, depending on your calculator.

Evaluate $\sqrt[3]{8}$:

$$\boxed{c}\ 8\ \boxed{\sqrt[x]{y}}\ 3\ \boxed{=}\ 2$$
or $\quad \boxed{c}\ 8\ \boxed{y^{1/x}}\ 3\ \boxed{=}\ 2$
or $\quad \boxed{c}\ 8\ \boxed{\text{inv}}\ \boxed{y^x}\ 3\ \boxed{=}\ 2$

To evaluate $\sqrt[4]{\dfrac{5}{6}}$ on a calculator, you would press

or appropriate key or keys for your calculator
$$\downarrow$$
$$\boxed{c}\ 5\ \boxed{\div}\ 6\ \boxed{=}\ \boxed{\sqrt[x]{y}}\ 4\ \boxed{=}\ 0.9554427$$

In Example 6 we showed that $\sqrt[3]{\dfrac{3}{5}} = \dfrac{\sqrt[3]{75}}{5}$. Now use your calculator to show that $\sqrt[3]{\dfrac{3}{5}} = \dfrac{\sqrt[3]{75}}{5}$. What approximate decimal value is each radical expression equivalent to?

EXAMPLE 7 Simplify $\sqrt[3]{\dfrac{x}{2y^2}}$.

Solution: First use the quotient rule to rewrite the given cube root as the quotient of two cube roots.

$$\sqrt[3]{\frac{x}{2y^2}} = \frac{\sqrt[3]{x}}{\sqrt[3]{2y^2}}$$

Then multiply both the numerator and denominator by the cube root of an expression that will result in the product of the radicands in the denominator being a perfect cube.

The resulting exponents in the radicand in the denominator must be divisible by 3. Since the denominator is $\sqrt[3]{2y^2}$, we must multiply the expression by $\sqrt[3]{2^2y}$. Note that $2 \cdot 2^2 = 2^3$ and $y^2 \cdot y = y^3$. Multiply both the numerator and denominator by $\sqrt[3]{2^2y}$ and then simplify.

$$\frac{\sqrt[3]{x}}{\sqrt[3]{2y^2}} = \frac{\sqrt[3]{x}}{\sqrt[3]{2y^2}} \cdot \boxed{\frac{\sqrt[3]{2^2 y}}{\sqrt[3]{2^2 y}}}$$

$$= \frac{\sqrt[3]{x}\sqrt[3]{4y}}{\sqrt[3]{2^3 y^3}}$$

$$= \frac{\sqrt[3]{4xy}}{2y} \qquad\blacksquare$$

EXAMPLE 8 Simplify $\sqrt{\dfrac{12x^3 y^5}{5z}}$.

Solution: $\sqrt{\dfrac{12x^3 y^5}{5z}} = \dfrac{\sqrt{12x^3 y^5}}{\sqrt{5z}}$

Now simplify the numerator.

$$\frac{\sqrt{12x^3 y^5}}{\sqrt{5z}} = \frac{\sqrt{4x^2 y^4}\sqrt{3xy}}{\sqrt{5z}}$$

$$= \frac{2xy^2 \sqrt{3xy}}{\sqrt{5z}}$$

Now rationalize the denominator.

$$\frac{2xy^2 \sqrt{3xy}}{\sqrt{5z}} = \frac{2xy^2 \sqrt{3xy}}{\sqrt{5z}} \cdot \boxed{\frac{\sqrt{5z}}{\sqrt{5z}}}$$

$$= \frac{2xy^2 \sqrt{15xyz}}{\sqrt{5^2 z^2}}$$

$$= \frac{2xy^2 \sqrt{15xyz}}{5z} \qquad\blacksquare$$

▶**4** When the denominator of a rational expression is a binomial that contains a radical, we again rationalize the denominator. We do this by multiplying both the numerator and the denominator of the fraction by the **conjugate** of the denominator. The conjugate of a binomial is a binomial having the same two terms with the sign of the second term changed.

Binomial	*Its Conjugate*
$3 + \sqrt{2}$	$3 - \sqrt{2}$
$2\sqrt{3} - \sqrt{5}$	$2\sqrt{3} + \sqrt{5}$
$\sqrt{x} + \sqrt{y}$	$\sqrt{x} - \sqrt{y}$
$a + \sqrt{b}$	$a - \sqrt{b}$

When a binomial is multiplied by its conjugate, the outer and inner terms will sum to zero.

EXAMPLE 9 Multiply $(2 + \sqrt{3})(2 - \sqrt{3})$.

Solution: Multiply using the FOIL method.

$$\begin{array}{cccc} \text{F} & \text{O} & \text{I} & \text{L} \end{array}$$
$$2(2) + 2(-\sqrt{3}) + 2(\sqrt{3}) + \sqrt{3}(-\sqrt{3}) = 4 - 2\sqrt{3} + 2\sqrt{3} - \sqrt{9}$$
$$= 4 - \sqrt{9}$$
$$= 4 - 3 = 1 \qquad \blacksquare$$

Note in Example 9 that we would get the same results using the difference-of-two-squares formula, $(a + b)(a - b) = a^2 - b^2$.

$$(2 + \sqrt{3})(2 - \sqrt{3}) = 2^2 - (\sqrt{3})^2$$
$$= 4 - 3 = 1$$

EXAMPLE 10   Multiply $(\sqrt{3} - \sqrt{5})(\sqrt{3} + \sqrt{5})$.

Solution: $(\sqrt{3} - \sqrt{5})(\sqrt{3} + \sqrt{5}) = (\sqrt{3})^2 - (\sqrt{5})^2$
$$= 3 - 5 = -2 \qquad \blacksquare$$

EXAMPLE 11   Simplify $\dfrac{5}{2 + \sqrt{3}}$.

Solution: To simplify this expression, we must rationalize the denominator. We do this by multiplying both the numerator and denominator by $2 - \sqrt{3}$, which is the conjugate of $2 + \sqrt{3}$.

$$\frac{5}{2 + \sqrt{3}} \cdot \frac{2 - \sqrt{3}}{2 - \sqrt{3}} = \frac{5(2 - \sqrt{3})}{(2 + \sqrt{3})(2 - \sqrt{3})}$$
$$= \frac{5(2 - \sqrt{3})}{4 - 3}$$
$$= 5(2 - \sqrt{3}) \quad \text{or} \quad 10 - 5\sqrt{3} \qquad \blacksquare$$

EXAMPLE 12   Simplify $\dfrac{6}{\sqrt{5} - \sqrt{2}}$.

Solution: $\dfrac{6}{\sqrt{5} - \sqrt{2}} \cdot \dfrac{\sqrt{5} + \sqrt{2}}{\sqrt{5} + \sqrt{2}} = \dfrac{6(\sqrt{5} + \sqrt{2})}{5 - 2}$

$$= \frac{\overset{2}{\cancel{6}}(\sqrt{5} + \sqrt{2})}{\underset{1}{\cancel{3}}}$$
$$= 2(\sqrt{5} + \sqrt{2}) \quad \text{or} \quad 2\sqrt{5} + 2\sqrt{2} \qquad \blacksquare$$

EXAMPLE 13   Simplify $\dfrac{x - \sqrt{y}}{x + \sqrt{y}}$.

Solution: Multiply both the numerator and denominator of the fraction by the conjugate of the denominator, $x - \sqrt{y}$.

$$\frac{x - \sqrt{y}}{x + \sqrt{y}} \cdot \frac{x - \sqrt{y}}{x - \sqrt{y}} = \frac{x^2 - x\sqrt{y} - x\sqrt{y} + \sqrt{y^2}}{x^2 - y}$$

$$= \frac{x^2 - 2x\sqrt{y} + y}{x^2 - y}$$

Remember that you cannot divide out the x^2 terms or the y terms because they are not factors.   ■

COMMON STUDENT ERROR

The following simplifications are correct because the numbers and variables divided out are not within square roots.

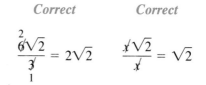

An expression within a square root cannot be divided by an expression not within the square root.

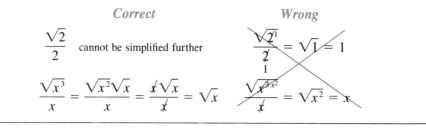

Exercise Set 8.4

In this exercise set assume that all variables represent positive real numbers.

Simplify each expression.

1. $\sqrt{\dfrac{27}{3}}$

2. $\sqrt{\dfrac{4}{25}}$

3. $\dfrac{\sqrt{3}}{\sqrt{27}}$

4. $\sqrt{\dfrac{16}{25}}$

5. $\sqrt[3]{\dfrac{2}{16}}$

6. $\dfrac{\sqrt[3]{108}}{\sqrt[3]{3}}$

7. $\dfrac{\sqrt{24}}{\sqrt{3}}$

8. $\sqrt[3]{\dfrac{x^3}{27}}$

9. $\sqrt{\dfrac{x^4}{25}}$

10. $\dfrac{\sqrt[3]{2x^6}}{\sqrt[3]{16x^3}}$

11. $\sqrt{\dfrac{16x^4}{4}}$

12. $\sqrt{\dfrac{27x^6}{3x^2}}$

13. $\sqrt{\dfrac{2x}{8x^5}}$

14. $\sqrt{\dfrac{25x^2y^5}{5x^4y}}$

15. $\sqrt{\dfrac{72x^2y^5}{8x^2y^7}}$

16. $\sqrt{\dfrac{x^4y^5}{4x^2y}}$

Simplify each expression.

17. $\dfrac{1}{\sqrt{3}}$

18. $\dfrac{3}{\sqrt{3}}$

19. $\dfrac{1}{\sqrt{2}}$

20. $\dfrac{2}{\sqrt{2}}$

21. $\dfrac{x}{\sqrt{5}}$ **22.** $\dfrac{2x}{\sqrt{6}}$ **23.** $\dfrac{x}{\sqrt{y}}$ **24.** $\sqrt{\dfrac{1}{3}}$

25. $\sqrt{\dfrac{x}{2}}$ **26.** $\sqrt{\dfrac{4}{5}}$ **27.** $\sqrt{\dfrac{5}{8}}$ **28.** $\sqrt{\dfrac{x}{2y^2}}$

29. $\dfrac{2\sqrt{3}}{\sqrt{5}}$ **30.** $\dfrac{2x}{\sqrt{18}}$ **31.** $\dfrac{2\sqrt{3}}{\sqrt{32}}$ **32.** $\sqrt{\dfrac{3x}{4y}}$

Simplify each expression.

33. $\dfrac{1}{\sqrt[3]{2}}$ **34.** $\dfrac{2}{\sqrt[3]{4}}$ **35.** $\dfrac{1}{\sqrt[3]{3}}$ **36.** $\dfrac{5}{\sqrt[3]{x}}$

37. $\sqrt[3]{\dfrac{5x}{y}}$ **38.** $\sqrt[3]{\dfrac{1}{4x}}$ **39.** $\sqrt[3]{\dfrac{5x}{4y^2}}$ **40.** $\dfrac{3}{\sqrt[4]{x}}$

41. $\dfrac{5x}{\sqrt[4]{2}}$ **42.** $\sqrt[4]{\dfrac{5}{3x^3}}$ **43.** $\sqrt[4]{\dfrac{2x}{4y^2}}$ **44.** $\sqrt[3]{\dfrac{3}{2y^2}}$

Simplify each expression.

45. $\sqrt{\dfrac{8x^5y}{2z}}$ **46.** $\sqrt{\dfrac{18x^4y^6}{3z}}$ **47.** $\sqrt{\dfrac{5xy^4}{2z}}$ **48.** $\sqrt{\dfrac{20y^4z^3}{3x}}$

49. $\sqrt{\dfrac{5xy^6}{6z}}$ **50.** $\sqrt{\dfrac{15x^5z^7}{2y}}$ **51.** $\sqrt{\dfrac{18x^4y^3}{2z}}$ **52.** $\sqrt{\dfrac{45y^{12}z^{10}}{2x}}$

53. $\sqrt[3]{\dfrac{15x^6y^7}{z^2}}$ **54.** $\sqrt[3]{\dfrac{8xy^2}{2z^2}}$ **55.** $\sqrt[3]{\dfrac{32x^4y^9}{4x^5}}$ **56.** $\sqrt[4]{\dfrac{5x^4y^5z}{2x^7}}$

Simplify each expression.

57. $(3 - \sqrt{3})(3 + \sqrt{3})$ **58.** $(4 + \sqrt{2})(4 - \sqrt{2})$

59. $(6 - \sqrt{5})(6 + \sqrt{5})$ **60.** $(\sqrt{8} - 3)(\sqrt{8} + 3)$

61. $(\sqrt{x} + 5)(\sqrt{x} - 5)$ **62.** $(\sqrt{6} + x)(\sqrt{6} - x)$

63. $(\sqrt{x} + y)(\sqrt{x} - y)$ **64.** $(\sqrt{x} + \sqrt{y})(\sqrt{x} - \sqrt{y})$

65. $(x + \sqrt{y})(x - \sqrt{y})$ **66.** $(\sqrt{7} + \sqrt{3})(\sqrt{7} - \sqrt{3})$

67. $(5 - \sqrt{y})(5 + \sqrt{y})$ **68.** $(\sqrt{3} - \sqrt{5})(\sqrt{3} + \sqrt{5})$

Simplify each expression.

69. $\dfrac{3}{1 + \sqrt{2}}$ **70.** $\dfrac{1}{2 + \sqrt{3}}$ **71.** $\dfrac{3}{\sqrt{6} - 5}$ **72.** $\dfrac{3}{\sqrt{2} + 5}$

73. $\dfrac{4}{\sqrt{2} - 7}$ **74.** $\dfrac{2}{\sqrt{2} + \sqrt{3}}$ **75.** $\dfrac{\sqrt{5}}{\sqrt{5} - \sqrt{6}}$ **76.** $\dfrac{8}{\sqrt{5} - \sqrt{8}}$

77. $\dfrac{1}{\sqrt{17} - \sqrt{8}}$ **78.** $\dfrac{2}{6 + \sqrt{x}}$ **79.** $\dfrac{5}{\sqrt{x} - 3}$ **80.** $\dfrac{5}{3 + \sqrt{x}}$

81. $\dfrac{4}{\sqrt{x} - y}$ **82.** $\dfrac{\sqrt{8x}}{x + \sqrt{y}}$ **83.** $\dfrac{\sqrt{2} - 1}{\sqrt{2} + 1}$ **84.** $\dfrac{\sqrt{x} - 2}{\sqrt{x} + 4}$

85. $\dfrac{\sqrt{x} - \sqrt{2y}}{\sqrt{x} - \sqrt{y}}$ **86.** $\dfrac{\sqrt{a^3} + \sqrt{a^7}}{\sqrt{a}}$ **87.** $\dfrac{2\sqrt{xy} - \sqrt{xy}}{\sqrt{x} + \sqrt{y}}$ **88.** $\dfrac{2}{\sqrt{x + 2} - 3}$

Simplify each expression.

89. $\sqrt{\dfrac{x}{9}}$

90. $\sqrt[4]{\dfrac{x^4}{16}}$

91. $\sqrt{\dfrac{2}{5}}$

92. $\sqrt{\dfrac{x}{y}}$

93. $(\sqrt{5} + \sqrt{6})(\sqrt{5} - \sqrt{6})$

94. $\sqrt[3]{\dfrac{1}{3}}$

95. $\sqrt{\dfrac{24x^3y^6}{5z}}$

96. $\dfrac{6}{4 - \sqrt{y}}$

97. $\sqrt{\dfrac{12xy^4}{2x^3y^4}}$

98. $\dfrac{4x}{\sqrt[3]{5y}}$

99. $(\sqrt{x} + 3)(\sqrt{x} - 3)$

100. $\dfrac{\sqrt{x}}{\sqrt{x} + \sqrt{y}}$

101. $\dfrac{7\sqrt{x}}{\sqrt{98}}$

102. $\sqrt{\dfrac{2xy^4}{18xy^2}}$

103. $\sqrt[4]{\dfrac{3}{2x}}$

104. $\sqrt{\dfrac{25x^2y^5}{3z}}$

105. $\sqrt[3]{\dfrac{32y^{12}z^{10}}{2x}}$

106. $\dfrac{\sqrt{3} + \sqrt{4}}{\sqrt{2} + \sqrt{3}}$

107. $\sqrt{\dfrac{2p}{q}}$

108. $\sqrt[4]{\dfrac{2}{9x}}$

109. $(\sqrt{y} - 3)(\sqrt{y} + 3)$

110. $(\sqrt{y} - x)(\sqrt{y} + x)$

111. $\sqrt[4]{\dfrac{2x^7y^{12}z^4}{3x^9}}$

112. $\dfrac{3}{\sqrt{y + 3} - \sqrt{y}}$

113. Consider the expression $\dfrac{1}{\sqrt{18}}$. Rationalize the denominator by:
(a) First simplifying $\sqrt{18}$ and then rationalizing.
(b) Multiplying both numerator and denominator by $\dfrac{\sqrt{2}}{\sqrt{2}}$.
(c) Multiplying both numerator and denominator by $\dfrac{\sqrt{18}}{\sqrt{18}}$.

114. Use a calculator or Appendix C to evaluate
(a) $\dfrac{\sqrt{2}}{2}$
(b) $\dfrac{1}{\sqrt{2}}$

115. Which is greater $\dfrac{2}{\sqrt{2}}$ or $\dfrac{3}{\sqrt{3}}$? Explain how you determined your answer.

116. Which is greater $\dfrac{\sqrt{3}}{2}$ or $\dfrac{2}{\sqrt{3}}$? Explain how you determined your answer.

117. Use a calculator to determine if $\sqrt[3]{\dfrac{2}{3}}$ is equal to $\dfrac{\sqrt[3]{18}}{3}$.

118. Use a calculator to determine if $\sqrt[4]{\dfrac{5}{9}}$ is equal to $\dfrac{\sqrt[4]{30}}{3}$.

119. What are the three conditions that need to be met for a radical expression to be simplified?

120. We stated that for nonnegative real numbers a and b, $b \neq 0$, $\dfrac{\sqrt[n]{a}}{\sqrt[n]{b}} = \sqrt[n]{\dfrac{a}{b}}$. Why is it necessary to specify that both a and b are nonnegative real numbers?

Cumulative Review Exercises

[2.4] **121.** Two cars leave from West Point at the same time traveling in opposite directions. One travels 10 miles per hour faster than the other. If the two cars are 270 miles apart after 3 hours, find the speed of each car.

[3.2–3.4] **122.** Explain how to graph a linear equation by (a) plotting points, (b) using the intercepts, and (c) using the slope and y intercepts. Plot $y = 3x - 4$ using each method.

[5.4] **123.** Multiply $(4x^2 - 3x - 2)(2x - 3)$.

[6.5] **124.** Solve the equation $(2x - 3)(x - 2) = 4x - 6$.

Addition and Subtraction of Radicals

▶**1** Add and subtract radicals.

▶**1** **Like radicals** are radicals having the same radicand and index. **Unlike radicals** are radicals differing in either the radicand or the index.

Examples of Like Radicals	*Examples of Unlike Radicals*	
$\sqrt{5}, 3\sqrt{5}$	$\sqrt{5}, \sqrt[3]{5}$	indexes differ
$5\sqrt{7}, -2\sqrt{7}$	$\sqrt{5}, \sqrt{7}$	radicands differ
$\sqrt{x}, 5\sqrt{x}$	$\sqrt{x}, \sqrt{2x}$	radicands differ
$\sqrt[3]{2x}, -4\sqrt[3]{2x}$	$\sqrt{x}, \sqrt[3]{x}$	indexes differ
$\sqrt[4]{xy}, -\sqrt[4]{xy}$	$\sqrt[3]{xy}, \sqrt[3]{x^2y}$	radicands differ

Like radicals are added and subtracted in much the same way that like terms are added or subtracted. To add or subtract like radicals, add or subtract their numerical coefficients and multiply this sum or difference by the like radical.

Examples of Adding Like Radicals

$$3\sqrt{5} + 2\sqrt{5} = (3 + 2)\sqrt{5} = 5\sqrt{5}$$
$$5\sqrt{x} - 7\sqrt{x} = (5 - 7)\sqrt{x} = -2\sqrt{x}$$
$$\sqrt[3]{4x} + 5\sqrt[3]{4x} = (1 + 5)\sqrt[3]{4x} = 6\sqrt[3]{4x}$$
$$4\sqrt{5x} - y\sqrt{5x} = (4 - y)\sqrt{5x}$$

EXAMPLE 1 Simplify each of the following.

(a) $6 + 4\sqrt{2} - \sqrt{2} + 3$ (b) $2\sqrt[3]{x} + 5x + 4\sqrt[3]{x} - 3$

Solution: (a) $6 + 4\sqrt{2} - \sqrt{2} + 3 = 3\sqrt{2} + 9$ (or $9 + 3\sqrt{2}$)

(b) $2\sqrt[3]{x} + 5x + 4\sqrt[3]{x} - 3 = 6\sqrt[3]{x} + 5x - 3$ ■

It is sometimes possible to convert unlike radicals into like radicals by simplifying one or more of the radicals.

EXAMPLE 2 Simplify $\sqrt{3} + \sqrt{27}$.

Solution: Since $\sqrt{3}$ and $\sqrt{27}$ are unlike radicals, they cannot be added in their present form. We can simplify $\sqrt{27}$ to obtain like radicals.

$$\sqrt{3} + \sqrt{27} = \sqrt{3} + \sqrt{9}\sqrt{3}$$
$$= \sqrt{3} + 3\sqrt{3}$$
$$= 4\sqrt{3}$$ ■

To Add or Subtract Radicals

---

1. Simplify each radical expression.
2. Combine like radicals (if there are any).

EXAMPLE 3   Simplify $4\sqrt{24} + \sqrt{54}$.

Solution:
$$4\sqrt{24} + \sqrt{54} = 4\sqrt{4} \cdot \sqrt{6} + \sqrt{9} \cdot \sqrt{6}$$
$$= 4 \cdot 2\sqrt{6} + 3\sqrt{6}$$
$$= 8\sqrt{6} + 3\sqrt{6}$$
$$= 11\sqrt{6}$$

EXAMPLE 4   Simplify $2\sqrt{45} - \sqrt{80} + \sqrt{20}$.

Solution:
$$2\sqrt{45} - \sqrt{80} + \sqrt{20} = 2\sqrt{9} \cdot \sqrt{5} - \sqrt{16} \cdot \sqrt{5} + \sqrt{4} \cdot \sqrt{5}$$
$$= 2 \cdot 3\sqrt{5} - 4\sqrt{5} + 2\sqrt{5}$$
$$= 6\sqrt{5} - 4\sqrt{5} + 2\sqrt{5}$$
$$= 4\sqrt{5}$$

EXAMPLE 5   Simplify $\sqrt[3]{27} + \sqrt[3]{81} - 4\sqrt[3]{3}$.

Solution:
$$\sqrt[3]{27} + \sqrt[3]{81} - 4\sqrt[3]{3} = 3 + \sqrt[3]{27}\sqrt[3]{3} - 4\sqrt[3]{3}$$
$$= 3 + 3\sqrt[3]{3} - 4\sqrt[3]{3}$$
$$= 3 - \sqrt[3]{3}$$

EXAMPLE 6   Simplify $\sqrt{x^2} - \sqrt{x^2 y} + x\sqrt{y}$.

Solution:
$$\sqrt{x^2} - \sqrt{x^2 y} + x\sqrt{y} = x - \sqrt{x^2}\sqrt{y} + x\sqrt{y}$$
$$= x - x\sqrt{y} + x\sqrt{y}$$
$$= x$$

EXAMPLE 7   Simplify $\sqrt[3]{x^{10}y^2} - \sqrt[3]{x^4 y^8}$.

Solution:
$$\sqrt[3]{x^{10}y^2} - \sqrt[3]{x^4 y^8} = \sqrt[3]{x^9} \cdot \sqrt[3]{xy^2} - \sqrt[3]{x^3 y^6} \cdot \sqrt[3]{xy^2}$$
$$= x^3\sqrt[3]{xy^2} - xy^2\sqrt[3]{xy^2}$$

Now factor out the common factor $\sqrt[3]{xy^2}$.
$$= (x^3 - xy^2)\sqrt[3]{xy^2}$$

EXAMPLE 8   Simplify $4\sqrt{2} - \dfrac{1}{\sqrt{8}} + \sqrt{32}$.

Solution:
$$4\sqrt{2} - \frac{1}{\sqrt{8}} + \sqrt{32} = 4\sqrt{2} - \frac{1}{\sqrt{8}} \cdot \frac{\sqrt{2}}{\sqrt{2}} + \sqrt{16}\sqrt{2}$$
$$= 4\sqrt{2} - \frac{\sqrt{2}}{\sqrt{16}} + 4\sqrt{2}$$
$$= 4\sqrt{2} - \frac{\sqrt{2}}{4} + 4\sqrt{2}$$
$$= \left(4 - \frac{1}{4} + 4\right)\sqrt{2}$$
$$= \frac{31\sqrt{2}}{4}$$

Now that we have discussed the addition and subtraction of radical expressions, we can do a few more multiplication problems involving radicals.

EXAMPLE 9   Simplify $(3\sqrt{6} - 4)(2 + 5\sqrt{6})$.

Solution: Use the FOIL method to multiply, and then combine like terms.

$$(3\sqrt{6})(2) + (3\sqrt{6})(5\sqrt{6}) + (-4)(2) + (-4)(5\sqrt{6}) = 6\sqrt{6} + 15\sqrt{36} - 8 - 20\sqrt{6}$$
$$= 6\sqrt{6} + 15(6) - 8 - 20\sqrt{6}$$
$$= 6\sqrt{6} + 90 - 8 - 20\sqrt{6}$$
$$= 82 - 14\sqrt{6} \qquad \blacksquare$$

EXAMPLE 10   Simplify $(3\sqrt{6} - \sqrt{3})^2$.

Solution: $(3\sqrt{6} - \sqrt{3})^2 = (3\sqrt{6} - \sqrt{3})(3\sqrt{6} - \sqrt{3})$

Now multiply the factors using the FOIL method.

$$(3\sqrt{6})(3\sqrt{6}) + (3\sqrt{6})(-\sqrt{3}) + (-\sqrt{3})(3\sqrt{6}) + (-\sqrt{3})(-\sqrt{3}) = 9(6) - 3\sqrt{18} - 3\sqrt{18} + 3$$
$$= 54 - 3\sqrt{18} - 3\sqrt{18} + 3$$
$$= 57 - 6\sqrt{18}$$
$$= 57 - 6\sqrt{9}\sqrt{2}$$
$$= 57 - 18\sqrt{2} \qquad \blacksquare$$

EXAMPLE 11   Simplify $(\sqrt[3]{x} - \sqrt[3]{2y^2})(\sqrt[3]{x^2} - \sqrt[3]{8y})$.

Solution: Multiply the factors using the FOIL method.

$$(\sqrt[3]{x})(\sqrt[3]{x^2}) + (\sqrt[3]{x})(-\sqrt[3]{8y}) + (-\sqrt[3]{2y^2})(\sqrt[3]{x^2}) + (-\sqrt[3]{2y^2})(-\sqrt[3]{8y}) = \sqrt[3]{x^3} - \sqrt[3]{8xy} - \sqrt[3]{2x^2y^2} + \sqrt[3]{16y^3}$$
$$= x - 2\sqrt[3]{xy} - \sqrt[3]{2x^2y^2} + 2y\sqrt[3]{2} \qquad \blacksquare$$

COMMON STUDENT ERROR

The product rule of radicals presented in Section 8.3 is

$$\sqrt[n]{a} \cdot \sqrt[n]{b} = \sqrt[n]{ab} \qquad Correct$$

The quotient rule of radicals presented in Section 8.4 is

$$\frac{\sqrt[n]{a}}{\sqrt[n]{b}} = \sqrt[n]{\frac{a}{b}} \qquad Correct$$

Students often incorrectly assume similar properties exist for addition and subtraction. They do not.

Wrong

$$\sqrt[n]{a} + \sqrt[n]{b} = \sqrt[n]{a + b}$$
$$\sqrt[n]{a} - \sqrt[n]{b} = \sqrt[n]{a - b}$$

To illustrate that $\sqrt[n]{a} + \sqrt[n]{b} \neq \sqrt[n]{a + b}$, let n be a square root (index 2), $a = 9$, $b = 16$.

$$\sqrt[n]{a} + \sqrt[n]{b} \neq \sqrt[n]{a + b}$$
$$\sqrt{9} + \sqrt{16} \neq \sqrt{9 + 16}$$
$$3 + 4 \neq \sqrt{25}$$
$$7 \neq 5$$

Exercise Set 8.5

In this exercise set assume all variables represent positive real numbers.

Simplify each expression.

1. $4\sqrt{3} - 2\sqrt{3}$
2. $6\sqrt[3]{7} - 8\sqrt[3]{7}$
3. $4\sqrt{10} + 6\sqrt{10} - \sqrt{10} + 2$
4. $2\sqrt{3} - 2\sqrt{3} - 4\sqrt{3} + 5$
5. $12\sqrt[3]{15} + 5\sqrt[3]{15} - 8\sqrt[3]{15}$
6. $4\sqrt{x} + \sqrt{x}$
7. $3\sqrt{y} - 6\sqrt{y}$
8. $3\sqrt{y} - \sqrt{y} + 3$
9. $3\sqrt{5} - \sqrt[3]{x} + 4\sqrt{5} + 3\sqrt[3]{x}$
10. $\sqrt{x} + \sqrt{y} + x + 3\sqrt{y}$
11. $5 + 4\sqrt[3]{x} - 8\sqrt[3]{x}$
12. $5\sqrt{x} + 4 + 3\sqrt{x} + 2x - \sqrt{x}$

Simplify each expression.

13. $\sqrt{8} - \sqrt{12}$
14. $\sqrt{75} + \sqrt{108}$
15. $-6\sqrt{75} + 4\sqrt{125}$
16. $3\sqrt{250} + 5\sqrt{160}$
17. $-4\sqrt{90} + 3\sqrt{40} + 2\sqrt{10}$
18. $8\sqrt{45} + 7\sqrt{20} + 2\sqrt{5}$
19. $4\sqrt{32} - \sqrt{18} + 2\sqrt{128}$
20. $5\sqrt{8} + 2\sqrt{50} - 3\sqrt{72}$
21. $2\sqrt{5x} - 3\sqrt{20x} - 4\sqrt{45x}$
22. $3\sqrt{27x^2} - 2\sqrt{108x^2} - \sqrt{48x^2}$
23. $3\sqrt{50x^2} - 3\sqrt{72x^2} - 8x\sqrt{18}$
24. $\sqrt[3]{54} - \sqrt[3]{16}$
25. $4\sqrt[3]{5} - 5\sqrt[3]{40}$
26. $\sqrt[3]{108} + 2\sqrt[3]{32}$
27. $2\sqrt[3]{16} + \sqrt[3]{54}$
28. $\sqrt[3]{27} - 5\sqrt[3]{8}$
29. $3\sqrt{45x^3} + \sqrt{5x}$
30. $2\sqrt[3]{x^4y^2} + 3x\sqrt[3]{xy^2}$
31. $2a\sqrt{20a^3b^2} + 2b\sqrt{45a^5}$
32. $x\sqrt[3]{x^2y} - \sqrt[3]{8x^5y}$
33. $3y\sqrt[4]{48x^5} - x\sqrt[4]{3x^5y^4}$
34. $\sqrt{4x^7y^5} + 3x^2\sqrt{x^3y^5} - 2xy\sqrt{x^5y^3}$
35. $x\sqrt[3]{27x^5y^2} - x^2\sqrt[3]{x^2y^2} + 2\sqrt[3]{x^8y^2}$
36. $2\sqrt[3]{x^7y^7} - 3x\sqrt[3]{x^4y^7}$
37. $\sqrt[3]{16x^9y^{10}} - 2x^2y\sqrt[3]{2x^3y^7}$
38. $x\sqrt[3]{x^7y^5} - xy^2\sqrt[3]{xy^2}$

Simplify each expression.

39. $\dfrac{1}{\sqrt{2}} + \dfrac{\sqrt{2}}{2}$
40. $\dfrac{1}{\sqrt{3}} + \dfrac{\sqrt{3}}{3}$
41. $\sqrt{3} - \dfrac{1}{\sqrt{3}}$
42. $\sqrt{6} - \sqrt{\dfrac{2}{3}}$
43. $\sqrt{\dfrac{1}{6}} + \sqrt{24}$
44. $3\sqrt{2} - \dfrac{2}{\sqrt{8}} + \sqrt{50}$
45. $\sqrt{\dfrac{1}{2}} + 3\sqrt{2} + \sqrt{18}$
46. $\dfrac{3}{\sqrt{18}} - 2\sqrt{18} + \sqrt{\dfrac{5}{8}}$
47. $4\sqrt{x} + \dfrac{1}{\sqrt{x}} + \sqrt{\dfrac{1}{x}}$
48. $\dfrac{1}{3} + \dfrac{1}{\sqrt{3}} + \sqrt{75}$
49. $\dfrac{1}{2}\sqrt{18} - \dfrac{3}{\sqrt{2}} - 3\sqrt{50}$
50. $\dfrac{\sqrt{3}}{3} + 2\sqrt{\dfrac{1}{3}} + \sqrt{12}$

Simplify each expression.

51. $(\sqrt{3} + 4)(\sqrt{3} + 5)$
52. $(\sqrt{3} + 1)(\sqrt{3} - 6)$
53. $(1 + \sqrt{5})(6 + \sqrt{5})$
54. $(3 - \sqrt{2})(4 - \sqrt{8})$
55. $(4 - \sqrt{2})(5 + \sqrt{2})$
56. $(5\sqrt{6} + 3)(4\sqrt{6} - 2)$
57. $(\sqrt{5} + \sqrt{3})(\sqrt{5} + \sqrt{3})$
58. $(4\sqrt{3} + \sqrt{2})(\sqrt{3} - \sqrt{2})$
59. $(\sqrt{2} - \sqrt{3})(\sqrt{3} + \sqrt{8})$
60. $(\sqrt{3} + 4)^2$
61. $(2 - \sqrt{3})^2$
62. $(2\sqrt{5} - 3)^2$
63. $(\sqrt{x} + \sqrt{3})(\sqrt{x} - \sqrt{12})$
64. $(\sqrt{y} + \sqrt{6z})(\sqrt{2z} - \sqrt{8y})$
65. $(2\sqrt{3x} - \sqrt{y})(3\sqrt{3x} + \sqrt{y})$
66. $(\sqrt[3]{9} + \sqrt[3]{2})(\sqrt[3]{3} + \sqrt[3]{4})$
67. $(\sqrt[3]{4} - \sqrt[3]{6})(\sqrt[3]{2} - \sqrt[3]{36})$
68. $(\sqrt[3]{4x} - \sqrt[3]{2y})(\sqrt[3]{4x} + \sqrt[3]{10})$

Simplify each expression.

69. $\sqrt{5} + 2\sqrt{5}$
70. $-2\sqrt{x} - 3\sqrt{x}$
71. $\sqrt{125} + \sqrt{20}$
72. $3\sqrt{7} + 2\sqrt{63} - 2\sqrt{28}$
73. $\dfrac{\sqrt{6}}{2} + \dfrac{1}{\sqrt{6}}$
74. $(\sqrt{5} + 2)(7 + \sqrt{5})$
75. $-\sqrt[4]{x} + 6\sqrt[4]{x} - 2\sqrt[4]{x}$
76. $2\sqrt[3]{81} + 4\sqrt[3]{24}$
77. $2 + 3\sqrt{y} - 6\sqrt{y} + 5$
78. $4\sqrt{3} - \dfrac{3}{\sqrt{3}} + 2\sqrt{18}$
79. $(3\sqrt{2} - 4)(\sqrt{2} + 5)$
80. $(\sqrt{5} + \sqrt{2})(\sqrt{2} + \sqrt{20})$

81. $4\sqrt{3x^3} - \sqrt{12x}$

82. $2b\sqrt[4]{a^4b} + ab\sqrt[4]{16b}$

83. $\dfrac{3}{\sqrt{y}} - \sqrt{\dfrac{9}{y}} + \sqrt{y}$

84. $(\sqrt{x} - \sqrt{5y})(\sqrt{x} - \sqrt{5y})$

85. $2\sqrt[3]{24a^3y^4} + 4a\sqrt[3]{81y^4}$

86. $(\sqrt[3]{x^2} - \sqrt[3]{y})(\sqrt[3]{x} - 2\sqrt[3]{y^2})$

87. $2x\sqrt[3]{xy} + 5y\sqrt[3]{x^4y^4}$

88. $(\sqrt[3]{a} + 5)(\sqrt[3]{a^2} - 3)$

89. $\dfrac{2}{\sqrt{50}} - 3\sqrt{50} - \dfrac{1}{\sqrt{8}}$

90. $\sqrt{48} + 2\sqrt{75} - 3\sqrt{27} - 5\sqrt{3}$

91. Use a calculator or Appendix C to estimate $\sqrt{3} + 3\sqrt{2}$.

92. Use a calculator or Appendix C to estimate $2\sqrt{3} + \sqrt{5}$.

93. Which is greater, $\dfrac{1}{\sqrt{3} + 2}$ or $2 + \sqrt{3}$? (Do not use a calculator or tables.) Explain how you determined your answer.

94. Which is greater, $\dfrac{1}{\sqrt{3}} + \sqrt{75}$ or $\dfrac{2}{\sqrt{12}} + \sqrt{48} + 2\sqrt{3}$? (Do not use a calculator or tables.) Explain how you determined your answer.

Cumulative Review Exercises

[5.1–5.2] **95.** Simplify $\dfrac{(2x^{-2}y^3)^2(x^{-1}y^{-3})}{(xy^2)^{-2}}$.

[6.5] **96.** Solve the equation $20x^2 + 3x - 9 = 0$.

[8.2] **97.** Simplify $\left(\dfrac{x^{3/4}y^{2/3}}{x^{1/2}y}\right)^2$

[8.3] **98.** Simplify $\sqrt[3]{3x^2y}\,(\sqrt[3]{9x^4y^3} - \sqrt[3]{x^{10}y^7})$.

JUST FOR FUN

1. *Simplify.* $\dfrac{1}{\sqrt[5]{3x^7y^9}} + \dfrac{2\sqrt[5]{81x^3y}}{3x^2y^2}$

2. $\dfrac{1}{\sqrt[4]{3x^5y^6z^{13}}}$

8.6

Solving Radical Equations

▶ **1** Solve radical equations containing one radical.

▶ **2** Solve radical equations containing two radicals.

▶ **3** Solve for a variable in a formula.

▶ **1** A **radical equation** is an equation that contains a variable in a radicand. Some examples of radical equations are

$$\sqrt{x} = 4, \qquad \sqrt[3]{y + 4} = 9, \qquad \sqrt{x - 2} = 4 + \sqrt{x + 8}$$

To Solve Radical Equations

1. Rewrite the equation so that one radical containing a variable is isolated by itself on one side of the equation.
2. Raise each side of the equation to a power equal to the index of the radical.
3. Collect and combine like terms.
4. If the remaining equation still contains a term with a variable in a radicand, repeat steps 1 through 3.
5. Solve the resulting equation for the unknown variable.
6. Check all solutions in the original equations for extraneous roots.

Recall from Section 7.5 that an extraneous root is a number obtained when solving an equation that is not a solution to the original equation.

The following examples illustrate the procedure for solving radical equations.

EXAMPLE 1   Solve the equation $\sqrt{x} = 6$.

Solution: The square root containing the variable is already by itself on one side of the equation. Square both sides of the equation.

$$\sqrt{x} = 6$$
$$(\sqrt{x})^2 = (6)^2$$
$$x = 36$$

Check: $\sqrt{x} = 6$
$$\sqrt{36} = 6$$
$$6 = 6 \qquad \text{true}$$

EXAMPLE 2   Solve the equation $\sqrt{x + 4} - 6 = 0$.

Solution: $\sqrt{x + 4} - 6 = 0$

$\qquad \sqrt{x + 4} = 6$ Isolate the radical containing the variable.

$\qquad (\sqrt{x + 4})^2 = 6^2$ Square both sides of the equation.

$\qquad\qquad x + 4 = 36$ Now solve for the variable.

$\qquad\qquad\qquad x = 32$

A check will show that 32 is the solution.

EXAMPLE 3   Solve the equation $\sqrt[3]{x} + 4 = 6$.

Solution: Since the 4 is outside the radical, we first subtract 4 from both sides of the equation to isolate the radical.

$$\sqrt[3]{x} + 4 = 6$$
$$\sqrt[3]{x} = 2$$

Now cube both sides of the equation.

$$(\sqrt[3]{x})^3 = 2^3$$
$$x = 8$$

A check will show that 8 is the solution.

EXAMPLE 4   Solve the equation $\sqrt{2x - 3} = x - 3$.

Solution: Square both sides of the equation.

$$(\sqrt{2x - 3})^2 = (x - 3)^2$$
$$2x - 3 = (x - 3)(x - 3)$$
$$2x - 3 = x^2 - 6x + 9$$
$$0 = x^2 - 8x + 12$$

Now factor.

$$x^2 - 8x + 12 = 0$$
$$(x - 6)(x - 2) = 0$$
$$x - 6 = 0 \quad \text{or} \quad x - 2 = 0$$
$$x = 6 \qquad\qquad x = 2$$

Check:

$x = 6$	$x = 2$
$\sqrt{2x - 3} = x - 3$	$\sqrt{2x - 3} = x - 3$
$\sqrt{2(6) - 3} = 6 - 3$	$\sqrt{2(2) - 3} = 2 - 3$
$\sqrt{9} = 3$	$\sqrt{1} = -1$
$3 = 3 \qquad$ true	$1 = -1 \qquad$ false

The 6 is a solution, but 2 is not a solution to the equation. The 2 is an extraneous root (or extraneous solution). Note that 2 satisfies the equation $(\sqrt{2x - 3})^2 = (x - 3)^2$, but not the original equation $\sqrt{2x - 3} = x - 3$. ∎

HELPFUL HINT

Don't forget to check your solutions in the original equation. Remember that when you raise both sides of an equation to a power you may introduce extraneous solutions.

Consider the equation $x = 2$. Note what happens when you square both sides of the equation.

$$x = 2$$
$$x^2 = 2^2$$
$$x^2 = 4$$

Note that the equation $x^2 = 4$ has two solutions, $+2$ and -2. Since the original equation $x = 2$ has only one solution, 2, we introduced the extraneous root -2.

EXAMPLE 5   Solve the equation $2x - 5\sqrt{x} - 3 = 0$.

Solution: First, write the equation with the square root containing the variable by itself on one side of the equation.

$$2x - 5\sqrt{x} - 3 = 0$$
$$-5\sqrt{x} = -2x + 3$$
$$\text{or} \quad 5\sqrt{x} = 2x - 3$$

Now square both sides of the equation.

$$(5\sqrt{x})^2 = (2x - 3)^2$$
$$25x = (2x - 3)(2x - 3)$$
$$25x = 4x^2 - 12x + 9$$
$$0 = 4x^2 - 37x + 9$$
$$0 = (4x - 1)(x - 9)$$

$$4x - 1 = 0 \quad \text{or} \quad x - 9 = 0$$
$$4x = 1 \qquad\qquad x = 9$$
$$x = \frac{1}{4}$$

Check:

$$x = \frac{1}{4}$$

$$2x - 5\sqrt{x} - 3 = 0$$

$$2\left(\frac{1}{4}\right) - 5\sqrt{\frac{1}{4}} - 3 = 0$$

$$\frac{1}{2} - 5\left(\frac{1}{2}\right) - 3 = 0$$

$$-5 = 0 \quad \text{false}$$

$$x = 9$$

$$2x - 5\sqrt{x} - 3 = 0$$

$$2(9) - 5\sqrt{9} - 3 = 0$$

$$18 - 5(3) - 3 = 0$$

$$18 - 15 - 3 = 0$$

$$0 = 0 \quad \text{true}$$

The solution is 9. Note that $\frac{1}{4}$ is an extraneous root and is not a solution. ■

▶ **2**   Now we will look at some equations that contain two radical expressions.

EXAMPLE 6   Solve the equation $3\sqrt{x - 1} = 2\sqrt{2x + 2}$.

Solution: Since the two radicals appear on different sides of the equation, we square both sides of the equation.

$$(3\sqrt{x - 1})^2 = (2\sqrt{2x + 2})^2$$

$$9(x - 1) = 4(2x + 2)$$

$$9x - 9 = 8x + 8$$

$$x - 9 = 8$$

$$x = 17$$

A check will show that 17 is the solution. ■

EXAMPLE 7   Solve the equation $3\sqrt[3]{x - 2} = \sqrt[3]{17x - 14}$.

Solution:
$$3\sqrt[3]{x - 2} = \sqrt[3]{17x - 14}$$

$$(3\sqrt[3]{x - 2})^3 = (\sqrt[3]{17x - 14})^3 \qquad \text{Cube both sides of the equation.}$$

$$27(x - 2) = 17x - 14$$

$$27x - 54 = 17x - 14$$

$$10x - 54 = -14$$

$$10x = 40$$

$$x = 4$$

A check will show that the solution is 4. ■

When a radical equation contains two radical terms and a third nonradical term, you will sometimes need to raise both sides of the equation to a given power twice to obtain the solution. Before you raise both sides of the equation to the given power, you need to isolate a radical term. This procedure is illustrated in Example 8.

EXAMPLE 8   Solve the equation $\sqrt{5x - 1} - \sqrt{3x - 2} = 1$.

Solution: We must isolate one variable term on one side of the equation. We will begin by adding $\sqrt{3x - 2}$ to both sides of the equation to isolate $\sqrt{5x - 1}$. Then we will square both sides of the equation and combine like terms.

$$\sqrt{5x - 1} - \sqrt{3x - 2} = 1$$
$$\sqrt{5x - 1} = 1 + \sqrt{3x - 2}$$
$$(\sqrt{5x - 1})^2 = (1 + \sqrt{3x - 2})^2$$
$$5x - 1 = (1 + \sqrt{3x - 2})(1 + \sqrt{3x - 2})$$
$$5x - 1 = 1 + \sqrt{3x - 2} + \sqrt{3x - 2} + (\sqrt{3x - 2})^2$$
$$5x - 1 = 1 + 2\sqrt{3x - 2} + 3x - 2$$
$$5x - 1 = 3x - 1 + 2\sqrt{3x - 2}$$

Now we isolate the remaining radical term. We then square both sides of the equation again and solve the resulting equation.

$$2x = 2\sqrt{3x - 2}$$
$$(2x)^2 = (2\sqrt{3x - 2})^2$$
$$4x^2 = 4(3x - 2)$$
$$4x^2 = 12x - 8$$
$$4x^2 - 12x + 8 = 0$$
$$4(x^2 - 3x + 2) = 0$$
$$4(x - 2)(x - 1) = 0$$
$$x - 2 = 0 \quad \text{or} \quad x - 1 = 0$$
$$x = 2 \qquad\qquad x = 1$$

A check will show that both 2 and 1 are solutions to the equation.   ■

You may be given a formula for which you are asked to solve for a variable which is a radicand. To do so, follow the same general procedure used to solve a radical equation. Begin by isolating the radical expression. Then raise both sides of the equation to the same power as the index of the radical. This procedure is illustrated in Example 9.

EXAMPLE 9   A formula in statistics for finding the maximum error of estimation is $E = Z\dfrac{\sigma}{\sqrt{n}}$. Solve this equation for n.

Solution: First multiply both sides of the equation by the least common denominator $\sqrt{n}$ to eliminate fractions. Then isolate $\sqrt{n}$. Finally, solve for n by squaring both sides of the equation.

$$E = Z\frac{\sigma}{\sqrt{n}}$$
$$\sqrt{n}(E) = \left(Z\frac{\sigma}{\sqrt{n}}\right)\sqrt{n}$$
$$\sqrt{n}(E) = Z\sigma$$
$$\sqrt{n} = \frac{Z\sigma}{E}$$
$$(\sqrt{n})^2 = \left(\frac{Z\sigma}{E}\right)^2$$
$$n = \left(\frac{Z\sigma}{E}\right)^2 \quad \text{or} \quad n = \frac{Z^2\sigma^2}{E^2}$$ ■

Exercise Set 8.6

Solve each equation and then check your solution(s). If the equation has no real solution, so state.

1. $\sqrt{x} = 5$

2. $\sqrt{x} = 9$

3. $\sqrt[3]{x} = 2$

4. $\sqrt[3]{x} = 4$

5. $\sqrt[4]{x} = 3$

6. $\sqrt{x - 3} + 5 = 6$

7. $-\sqrt{2x + 4} = -6$

8. $\sqrt{x} + 3 = 5$

9. $\sqrt[3]{2x + 11} = 3$

10. $\sqrt[3]{6x - 3} = 3$

11. $\sqrt[3]{3x + 4} = 7$

12. $2\sqrt{4x - 3} = 10$

13. $\sqrt{2x - 3} = 2\sqrt{3x - 2}$

14. $\sqrt{8x - 4} = \sqrt{7x + 2}$

15. $\sqrt{5x + 10} = -\sqrt{3x + 8}$

16. $\sqrt[4]{x + 8} = \sqrt[4]{2x}$

17. $\sqrt[3]{6x + 1} = \sqrt[3]{2x + 5}$

18. $\sqrt[3]{3x + 1} = 2$

19. $\sqrt{x^2 + 9x + 3} = -x$

20. $\sqrt{x^2 + 3x + 9} = x$

21. $\sqrt{m^2 + 4m - 20} = m$

22. $\sqrt{5a + 1} - 11 = 0$

23. $\sqrt{2y + 3} + y = 0$

24. $\sqrt{x^2 + 3} = x + 1$

25. $-\sqrt{x} = 2x - 1$

26. $\sqrt{3x + 4} = x - 2$

27. $\sqrt{x^2 + 8} = x + 2$

28. $\sqrt[3]{3x - 1} + 4 = 0$

29. $\sqrt[3]{x - 12} = \sqrt[3]{5x + 16}$

30. $\sqrt{6x - 1} = 3x$

31. $\sqrt{x + 7} = 2x - 1$

32. $\sqrt[3]{4x - 3} - 3 = 0$

Solve each equation. You will have to square both sides of the equation twice to eliminate all radicals (see Example 8).

33. $\sqrt{2a - 3} = \sqrt{2a} - 1$

34. $\sqrt{x + 2} = \sqrt{x + 16}$

35. $\sqrt{x + 1} = 2 - \sqrt{x}$

36. $\sqrt{x + 3} = \sqrt{x} - 3$

37. $\sqrt{x + 7} = 5 - \sqrt{x - 8}$

38. $\sqrt{y + 2} = 1 + \sqrt{y - 3}$

39. $\sqrt{b - 3} = 4 - \sqrt{b + 5}$

40. $\sqrt{4x - 3} = 2 + \sqrt{2x - 5}$

41. $\sqrt{2x + 8} - \sqrt{2x - 4} = 2$

42. $\sqrt{y + 1} = \sqrt{y + 2} - 1$

43. $\sqrt{2x + 4} - \sqrt{x + 3} - 1 = 0$

44. $2 + \sqrt{x + 8} = \sqrt{3x + 12}$

Solve each formula for the indicated variable.

45. $p = \sqrt{2v}$, for v

46. $l = \sqrt{4r}$, for r

47. $v = \sqrt{2gh}$, for g

48. $v = \sqrt{\dfrac{2E}{m}}$, for E

49. $v = \sqrt{\dfrac{FR}{m}}$ for F

50. $\omega = \sqrt{\dfrac{a_0}{x_0}}$, for x_0

51. $x = \sqrt{\dfrac{m}{k}} V_0$, for m

52. $T = 2\pi \sqrt{\dfrac{L}{32}}$, for L

53. Consider the equation $\sqrt{x + 3} = -\sqrt{2x - 1}$. Explain why this equation can have no real solution.

54. Consider the equation $-\sqrt{x^2} = \sqrt{(-x)^2}$. By studying the equation, can you determine its solution? Explain your answer.

55. Consider the equation $\sqrt[3]{x^2} = -\sqrt[3]{x^2}$. By studying the equation, can you determine its solution? Explain your answer.

56. Why is it necessary to check solutions to radical equations?

Cumulative Review Exercises

Perform the indicated operation.

[7.2] **57.** $\dfrac{4a^2 - 9b^2}{4a^2 + 12ab + 9b^2} \cdot \dfrac{6a^2b}{8a^2b^2 - 12ab^3}$.

58. $(t^2 - t - 12) \div \dfrac{t^2 - 9}{t^2 - 3t}$.

[7.3] **59.** $\dfrac{2}{x + 3} - \dfrac{1}{x - 3} + \dfrac{2x}{x^2 - 9}$.

[7.5] **60.** Solve the equation $2 + \dfrac{3x}{x - 1} = \dfrac{8}{x - 1}$.

JUST FOR FUN

Solve.

1. $\sqrt{4x + 1} - \sqrt{3x - 2} = \sqrt{x - 5}$

2. $\dfrac{x + \sqrt{x + 3}}{x - \sqrt{x + 3}} = 3$

3. $\sqrt{\sqrt{x + 25} - \sqrt{x}} = 5$

4. $\sqrt{\sqrt{x + 9} + \sqrt{x}} = 3$

Solve each of the following equations for n.

5. $z = \dfrac{\overline{x} - \mu}{\dfrac{\sigma}{\sqrt{n}}}$

6. $z = \dfrac{p' - p}{\sqrt{\dfrac{pq}{n}}}$

8.7

Applications of Radicals (Optional)

▶**1**   Learn some applications of radical equations.

▶**1**   Now we will look at a few of the many applications involving radicals.

EXAMPLE 1   A telephone pole is at a right, or 90°, angle with the ground, see Figure 8.1. The length of a wire, l, from any point on the pole, at height a, to any point on the ground at a distance of b from the pole's base can be found by the formula $l = \sqrt{a^2 + b^2}$.

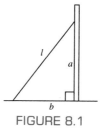

FIGURE 8.1

Find the length of wire that connects to the pole 40 feet above the ground, and which is anchored to the ground 20 feet from the base of the pole.

Solution: If we substitute 40 for a and 20 for b in the formula we get

$$l = \sqrt{a^2 + b^2}$$
$$= \sqrt{(40)^2 + (20)^2}$$
$$= \sqrt{1600 + 400}$$
$$= \sqrt{2000}$$
$$\approx 44.7$$

Thus the wire's length is about 44.7 feet. ■

The formula used in Example 1 is a special case of the Pythagorean Theorem that we will discuss in the next chapter. The above formula can be adapted to many similar situations involving right triangles.

EXAMPLE 2   During the sixteenth and seventeenth centuries, Galileo Galilei did numerous experiments with objects falling freely under the influence of gravity. He showed, for example, that an object dropped from, say, 10 feet hit the ground with a higher velocity than an object dropped from 5 feet. A formula that can be used to determine the velocity of an object after it has fallen a certain distance (and neglecting wind resistance) is

$$v = \sqrt{2gh}$$

where g is the acceleration due to gravity and h is the distance the object has fallen. On Earth the acceleration of gravity is approximately 32 feet per second squared.

(a) Find the velocity of an object after it has fallen 10 feet.
(b) Find the velocity of an object after it has fallen 100 feet.

Solution: (a) $v = \sqrt{2gh} = \sqrt{2(32)h} = \sqrt{64h}$.

At $h = 10$ feet, $v = \sqrt{64(10)} = \sqrt{640} \approx 25.3$ feet per second.
After an object has fallen 10 feet, its velocity is 25.3 feet per second.

(b) After falling 100 feet, $v = \sqrt{64(100)} = \sqrt{6400} = 80$ feet per second. ∎

EXAMPLE 3 The length of time it takes for a pendulum to make one complete swing back and forth is called the period of the pendulum. The period of a pendulum, T, can be found by the formula $T = 2\pi\sqrt{\dfrac{L}{32}}$, where L is the length of the pendulum in feet. Find the period of a pendulum if its length is 4 feet.

Solution: Substitute 4 for L and 3.14 for π in the formula. If you have a calculator that has a $\boxed{\pi}$ key, use that key to enter π.

$$T = 2\pi\sqrt{\frac{L}{32}}$$

$$= 2(3.14)\sqrt{\frac{4}{32}}$$

$$= 2(3.14)\sqrt{0.125}$$

$$\approx 2.22$$

Thus, the period is about 2.22 seconds. If you have a grandfather clock with a four-foot pendulum, it will take about 2.22 seconds for it to swing once back and forth. ∎

EXAMPLE 4 The area of a triangle is $A = \frac{1}{2}bh$. If the height is not known, but we know the length of each of the three sides, we can use *Hero's formula* to find the area.

$$A = \sqrt{S(S - a)(S - b)(S - c)}$$

where a, b, and c are the lengths of the three sides and

$$S = \frac{a + b + c}{2}$$

Use Hero's formula to find the area of a triangle with sides of 3 inches, 4 inches, and 5 inches.

Solution: The triangle is illustrated in Fig. 8.2. Let $a = 3$, $b = 4$, and $c = 5$. First, find the value of S.

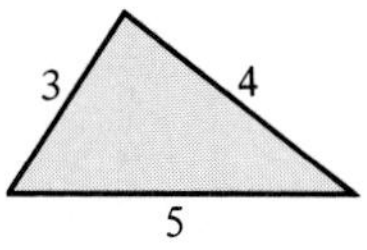

$$S = \frac{3 + 4 + 5}{2} = \frac{12}{2} = 6$$

FIGURE 8.2

Now find the area.

$$A = \sqrt{S(S - a)(S - b)(S - c)}$$
$$= \sqrt{6(6 - 3)(6 - 4)(6 - 5)}$$
$$= \sqrt{6(3)(2)(1)}$$
$$= \sqrt{36} = 6$$

The triangle has an area of 6 square inches.

Exercise Set 8.7

Use the formula given in Example 1 to find the length of side x.

1.   **2.**   **3.**   **4.**

Use the formula given in Example 1 to answer Exercises 5 through 9. Leave your answer in terms of a square root if a calculator with a square root key is not available for use.

5. How long a wire does a phone company worker need to reach from the top of a 4-meter telephone pole to a point 1.5 meters from the base of the pole?

6. Ms. Song Tran places an extension ladder against her house. The base of the ladder is 2 meters from the house and the ladder rests against the house 6 meters above the ground. How far is her ladder extended?

7. Find the length of the diagonal of a rectangle with a length of 12 inches and width of 5 inches.

8. A football field is 120 yards long from end zone to end zone. Find the length of the diagonal from one end zone to the other if the width of the field is 53.3 yards.

9. A regulation baseball diamond is a square with 90 feet between bases. How far is second base from home plate?

10. When you are given the area of a square, the length of a side can be found by the formula $s = \sqrt{A}$. Find the side of a square that has an area of 64 square inches.

11. Find the side of a square that has an area of 60 square meters.

12. When you are given the area of a circle, its radius can be found by the formula $r = \sqrt{A/\pi}$. Find the radius of a circle of an area of 20 square/inches.

13. Find the velocity of an object after it has fallen 80 feet. Use $v = \sqrt{2gh}$. Refer to Example 2.

14. Find the velocity of an object after it has fallen 50 feet.

15. Find the period of the pendulum if its length is 8 feet. Use $T = 2\pi \sqrt{L/32}$. Refer to Example 3.

16. Find the period of a 40-foot pendulum.

17. Find the area of a triangle if its three sides are

6 inches, 8 inches, and 10 inches. Use $A = \sqrt{S(S - a)(S - b)(S - c)}$. Refer to Example 4.

18. Find the area of a triangle if its three sides are 4 inches, 10 inches, and 12 inches.

19. For any planet, its "year" is the time it takes for the planet to revolve once around the sun. The number of Earth days in a given planet's year, N, is approximated by the formula

$$N = 0.2(\sqrt{R})^3$$

where R is the mean distance of the planet to the sun in millions of kilometers. Find the number of Earth days in the year of the planet Earth whose mean distance to the sun is 149.4 million kilometers.

20. Find the number of earth days in the year of the planet Mercury whose mean distance to the sun is 58 million kilometers.

21. When two forces, F_1 and F_2, pull at right angles to each other as illustrated below, the resultant, or the effective force, R, can be found by the formula $R = \sqrt{F_1^2 + F_2^2}$. Two cars are trying to pull a third out of the mud, as illustrated. If car A is exerting a force of 600 pounds and car B is exerting a force of 800 pounds, find the resulting force on the car stuck in the mud.

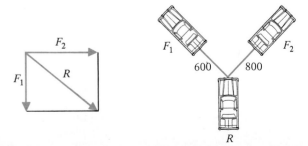

22. The escape velocity, in meters per second, or the velocity needed for a spacecraft to escape a planet's gravitational field, is found by the formula $v_e = \sqrt{2gR}$, where g is the force of gravity of the planet and R is the radius of the planet in meters. Find the escape velocity for Earth where $g = 9.75$ meters per second squared and $R = 6{,}370{,}000$ meters.

23. A formula used in the study of shallow-water wave motion is $c = \sqrt{gH}$, in which c is wave velocity, H is water depth, and g is the acceleration due to gravity. Find the wave velocity if the water's depth is 10 feet. (Use $g = 32$ ft/sec².)

24. The length of the diagonal of a rectangular solid is given by $d = \sqrt{a^2 + b^2 + c^2}$. Find the length of the

diagonal of a suitcase of length 37 inches, width 15 inches, and depth 9 inches.

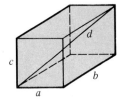

25. A formula we have already mentioned and will be discussing in more detail shortly is the quadratic formula

$$x = \frac{-b \pm \sqrt{b^2 - 4ac}}{2a}$$

(a) Find x when $a = 1$, $b = 0$, $c = -4$.
(b) Find x when $a = 1$, $b = 1$, $c = -12$.
(c) Find x when $a = 2$, $b = 5$, $c = -12$.
(d) Find x when $a = -1$, $b = 4$, $c = 5$.

Cumulative Review Exercises

[7.1] **26.** Reduce
$$\frac{(2x + 3)(3x - 4) - (2x + 3)(5x - 1)}{(2x + 3)}.$$

27. Find the domain of $\dfrac{x + 2}{3x^2 - 10x - 8}$.

[6.6] **28.** Solve the formula $P_1P_2 - P_1P_3 = P_2P_3$ for P_2

[7.5] **29.** Solve the equation $\dfrac{4b}{b + 1} = 4 - \dfrac{5}{b}$.

JUST FOR FUN

1. The force of gravity on the moon is $\frac{1}{6}$ of that on Earth. If an object falls off a rocket 100 feet above the surface of the moon, with what velocity will it strike the moon? Use $v = \sqrt{2gh}$. (See Example 2.)

2. A formula used to determine the frequency of a vibrating spring is $f = \dfrac{1}{2\pi}\sqrt{K/m}$, where f is the frequency of oscillation in cycles per second (also called hertz), K is the spring stiffness constant, and m is the mass of the spring.

Find the resulting frequency of a spring with a stiffness constant of 10^5 dynes/cm and a mass of 1000 grams.

8.8

Complex Numbers

▶ **1** Recognize an imaginary number.
▶ **2** Recognize a complex number.
▶ **3** Add and subtract complex numbers.
▶ **4** Multiply complex numbers.
▶ **5** Find the conjugate of a complex number.
▶ **6** Divide complex numbers.

► **1**   In Section 8.1 we stated that the square root of negative numbers, such as $\sqrt{-4}$, are not real numbers. Numbers like $\sqrt{-4}$ are called **imaginary numbers.** There is no real number that when multiplied by itself gives -4.

$$\sqrt{-4} \neq 2 \qquad \text{since} \qquad 2 \cdot 2 = 4$$
$$\sqrt{-4} \neq -2 \qquad \text{since} \qquad (-2)(-2) = 4$$

Numbers such as $\sqrt{-4}$ are called imaginary because when they were first introduced many mathematicians refused to believe that they existed. Although they do not belong to the set of real numbers, the imaginary numbers do exist and are very useful in mathematics.

Every imaginary number has a factor of $\sqrt{-1}$. For example,

$$\sqrt{-4} = \sqrt{4}\sqrt{-1}$$
$$\sqrt{-9} = \sqrt{9}\sqrt{-1}$$
$$\sqrt{-7} = \sqrt{7}\sqrt{-1}$$

The $\sqrt{-1}$, called the **imaginary unit,** is often denoted by the letter i.

$$i = \sqrt{-1}$$

We can therefore write

$$\sqrt{-4} = \sqrt{4}\sqrt{-1} = 2\sqrt{-1} = 2i$$
$$\sqrt{-9} = \sqrt{9}\sqrt{-1} = 3\sqrt{-1} = 3i$$
$$\sqrt{-7} = \sqrt{7}\sqrt{-1} = \sqrt{7}i \quad \text{or} \quad i\sqrt{7}$$

In this book we will generally write $i\sqrt{7}$ rather than $\sqrt{7}i$ to avoid confusion with $\sqrt{7i}$.

To help in writing square roots of negative numbers in terms of i, we give the following rule.

For any positive real number n,
$$\sqrt{-n} = i\sqrt{n}$$

Examples

$$\sqrt{-4} = i\sqrt{4} = 2i \qquad \sqrt{-3} = i\sqrt{3}$$
$$\sqrt{-25} = i\sqrt{25} = 5i \qquad \sqrt{-10} = i\sqrt{10}$$

Any number that can be expressed in the form bi, where b is any nonzero real number and $i = \sqrt{-1}$, is an **imaginary number.** For example, $3i$ and $i\sqrt{7}$ are imaginary numbers. Since the numbers $\sqrt{-4}$ and $\sqrt{-15}$ can also be placed in the bi form, these numbers are also imaginary numbers.

► **2**   Now we are prepared to discuss complex numbers.

> **Complex Number**
>
> Every number of the form
>
> $$a + bi$$
>
> where a and b are real numbers, is a complex number.

Every real number and every imaginary number is also a complex number. A complex number has two parts: a real part, a, and an imaginary part, b.

$$\underset{\text{real part}}{a} + \underset{\text{imaginary part}}{bi}$$

If $b = 0$, the complex number is a real number. If $a = 0$, the complex number is a *pure imaginary number*.

Examples of Complex Numbers

$3 + 4i$	$a = 3, b = 4$	
$5 - i\sqrt{3}$	$a = 5, b = -\sqrt{3}$	
5	$a = 5, b = 0$	(real number, $b = 0$)
$2i$	$a = 0, b = 2$	(imaginary number, $a = 0$)
$-i\sqrt{7}$	$a = 0, b = -\sqrt{7}$	(imaginary number, $a = 0$)

We stated that all real numbers and all imaginary numbers are also complex numbers. The relationship between the various sets of numbers is illustrated in Fig. 8.3.

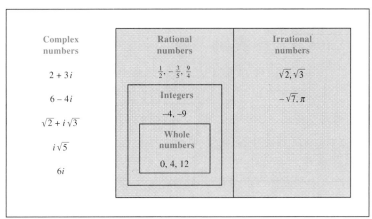

Real numbers are shaded

FIGURE 8.3

EXAMPLE 1 Write each of the following complex numbers in the form $a + bi$.

(a) $3 + \sqrt{-16}$ (b) $5 - \sqrt{-12}$ (c) 4 (d) $\sqrt{-18}$ (e) $6 + \sqrt{5}$

Solution: (a) $3 + \sqrt{-16} = 3 + \sqrt{16}\sqrt{-1}.$
$$= 3 + 4i$$

(b) $5 - \sqrt{-12} = 5 - \sqrt{12}\sqrt{-1}$
$$= 5 - \sqrt{4}\sqrt{3}\sqrt{-1}$$
$$= 5 - 2\sqrt{3}i \quad \text{or} \quad 5 - 2i\sqrt{3}$$

(c) $4 = 4 + 0i$

(d) $\sqrt{-18} = 0 + \sqrt{-18}$
$$= 0 + \sqrt{9}\sqrt{2}\sqrt{-1}$$
$$= 0 + 3\sqrt{2}i \quad \text{or} \quad 0 + 3i\sqrt{2}$$

(e) Both 6 and $\sqrt{5}$ are real numbers. $(6 + \sqrt{5}) + 0i.$ ■

Complex numbers can be added, subtracted, multiplied, and divided. To perform these operations, we make use of the definitions that $i = \sqrt{-1}$ and

$$i^2 = -1$$

▶ **3**   We will first explain how to add or subtract complex numbers. The procedures used to multiply and divide complex numbers will be explained shortly.

To Add (or Subtract) Complex Numbers

1. Change all imaginary numbers to *bi* form.
2. Add (or subtract) the real parts of the complex numbers.
3. Add (or subtract) the imaginary parts of the complex numbers.
4. Write the answer in the form $a + bi$.

EXAMPLE 2   Add $(4 + 13i) + (-6 - 8i)$.

Solution: $(4 + 13i) + (-6 - 8i) = 4 + 13i - 6 - 8i$
$$= 4 - 6 + 13i - 8i$$
$$= -2 + 5i$$ ■

EXAMPLE 3   Subtract $(-6 - \frac{1}{3}i) - (\frac{5}{2} - 4i)$.

Solution: $\left(-6 - \dfrac{1}{3}i\right) - \left(\dfrac{5}{2} - 4i\right) = -6 - \dfrac{1}{3}i - \dfrac{5}{2} + 4i$

$$= -6 - \dfrac{5}{2} - \dfrac{1}{3}i + 4i$$

$$= -\dfrac{12}{2} - \dfrac{5}{2} - \dfrac{1}{3}i + \dfrac{12}{3}i$$

$$= -\dfrac{17}{2} + \dfrac{11}{3}i$$ ■

EXAMPLE 4   Add $(6 - \sqrt{-8}) + (4 + \sqrt{-18})$.

Solution:
$$
\begin{aligned}
(6 - \sqrt{-8}) + (4 + \sqrt{-18}) &= (6 - \sqrt{8}\sqrt{-1}) + (4 + \sqrt{18}\sqrt{-1}) \\
&= (6 - \sqrt{4}\sqrt{2}\sqrt{-1}) + (4 + \sqrt{9}\sqrt{2}\sqrt{-1}) \\
&= (6 - 2i\sqrt{2}) + (4 + 3i\sqrt{2}) \\
&= 6 + 4 - 2i\sqrt{2} + 3i\sqrt{2} \\
&= 10 + i\sqrt{2}
\end{aligned}
$$ ∎

▶ **4**

> **To Multiply Complex Numbers**
>
> 1. Change all imaginary numbers to bi form.
> 2. Multiply the complex numbers as you would multiply polynomials.
> 3. Substitute -1 for each i^2.
> 4. Combine the real parts and the imaginary parts and then write the answer in $a + bi$ form.

EXAMPLE 5 Multiply $3i(5 - 2i)$.

Solution:
$$
\begin{aligned}
3i(5 - 2i) &= 3i(5) + (3i)(-2i) \\
&= 15i - 6i^2 \\
&= 15i - 6(-1) \\
&= 15i + 6 \quad \text{or} \quad 6 + 15i
\end{aligned}
$$ ∎

EXAMPLE 6 Multiply $\sqrt{-4}(\sqrt{-2} + 7)$.

Solution: First, write each imaginary number in bi form.
$$
\begin{aligned}
\sqrt{-4}(\sqrt{-2} + 7) &= 2i(i\sqrt{2} + 7) \\
&= (2i)(i\sqrt{2}) + (2i)(7) \\
&= 2i^2\sqrt{2} + 14i \\
&= 2(-1)\sqrt{2} + 14i \\
&= -2\sqrt{2} + 14i
\end{aligned}
$$ ∎

> **COMMON STUDENT ERROR**
>
> What is $\sqrt{-4} \cdot \sqrt{-2}$?
>
> | *Correct* | *Wrong* |
> |---|---|
> | $\begin{aligned}\sqrt{-4} \cdot \sqrt{-2} &= 2i \cdot i\sqrt{2} \\ &= 2i^2\sqrt{2} \\ &= 2(-1)\sqrt{2} \\ &= -2\sqrt{2}\end{aligned}$ | $\begin{aligned}\sqrt{-4} \cdot \sqrt{-2} &= \sqrt{8} \\ &= \sqrt{4} \cdot \sqrt{2} \\ &= 2\sqrt{2}\end{aligned}$ |
>
> Recall that $\sqrt{a} \cdot \sqrt{b} = \sqrt{ab}$ for nonnegative integers a and b.

EXAMPLE 7   Multiply $(3 + 5i)(2 - 3i)$.

Solution: We may begin by multiplying using the FOIL method.

$$(3)(2) + (3)(-3i) + (5i)(2) + (5i)(-3i) = 6 - 9i + 10i - 15i^2$$
$$= 6 - 9i + 10i - 15(-1)$$
$$= 6 - 9i + 10i + 15$$
$$= 21 + i$$

EXAMPLE 8   Multiply $(3 - \sqrt{-8})(\sqrt{-2} + 5)$.

Solution: $(3 - \sqrt{-8})(\sqrt{-2} + 5) = (3 - \sqrt{8}\sqrt{-1})(\sqrt{2}\sqrt{-1} + 5)$
$$= (3 - 2i\sqrt{2})(i\sqrt{2} + 5)$$

Now use the FOIL method to multiply.

$$= 3(i\sqrt{2}) + (3)(5) + (-2i\sqrt{2})(i\sqrt{2}) + (-2i\sqrt{2})(5)$$
$$= 3i\sqrt{2} + 15 - 2i^2(2) - 10i\sqrt{2}$$
$$= 3i\sqrt{2} + 15 - 2(-1)(2) - 10i\sqrt{2}$$
$$= 3i\sqrt{2} + 15 + 4 - 10i\sqrt{2}$$
$$= 19 - 7i\sqrt{2}$$

▶ **5**   The **conjugate** of a complex number $a + bi$ is $a - bi$. For example,

Complex Number	*Conjugate of Complex Number*
$3 + 4i$	$3 - 4i$
$1 - i\sqrt{3}$	$1 + i\sqrt{3}$
$2i$ (or $0 + 2i$)	$-2i$ (or $0 - 2i$)

When a complex number is multiplied by its conjugate using the FOIL method, the inner and outer parts will sum to zero. For example,

$$(5 + 2i)(5 - 2i) = 25 - 10i + 10i - 4i^2$$
$$= 25 - 4i^2$$
$$= 25 - 4(-1) = 25 + 4 = 29$$

▶ **6**

To Divide Complex Numbers

---

1. Change all imaginary numbers to *bi* form.
2. Write the division problem as a fraction.
3. Rationalize the denominator of the fraction by multiplying both the numerator and denominator of the fraction by the conjugate of the denominator.
4. Write the answer in $a + bi$ form.

EXAMPLE 9 Simplify $\dfrac{4 + i}{i}$.

Solution: Multiply both numerator and denominator by $-i$, the conjugate of i.

$$\frac{4 + i}{i} \cdot \frac{-i}{-i} = \frac{(4 + i)(-i)}{-i^2}$$

$$= \frac{-4i - i^2}{-i^2} \qquad \text{Now substitute } i^2 = -1.$$

$$= \frac{-4i - (-1)}{-(-1)}$$

$$= \frac{-4i + 1}{1} = 1 - 4i$$ ∎

EXAMPLE 10 Divide $\dfrac{6 - 5i}{2 - i}$.

Solution: Multiply both numerator and denominator by $2 + i$, the conjugate of $2 - i$.

$$\frac{6 - 5i}{2 - i} \cdot \frac{2 + i}{2 + i} = \frac{12 + 6i - 10i - 5i^2}{4 - i^2}$$

$$= \frac{12 - 4i - 5(-1)}{4 - (-1)}$$

$$= \frac{17 - 4i}{5} \quad \text{or} \quad \frac{17}{5} - \frac{4}{5}i$$ ∎

EXAMPLE 11 A concept needed for the study of electronics is *impedance*. Impedance affects the flow of current in a circuit. The impedance, Z, in a circuit is found by the formula $Z = \dfrac{V}{I}$, where V is voltage and I is current. Find Z when $V = 1.6 - 0.3i$ and $I = -0.2i$, where $i = \sqrt{-1}$.

Solution: $Z = \dfrac{V}{I} = \dfrac{1.6 - 0.3i}{-0.2i}$. Now multiply both the numerator and denominator by $0.2i$.

$$Z = \frac{1.6 - 0.3i}{-0.2i} \cdot \frac{0.2i}{0.2i} = \frac{0.32i - 0.06i^2}{-0.04i^2}$$

$$= \frac{0.32i + 0.06}{0.04}$$

$$= \frac{0.32i}{0.04} + \frac{0.06}{0.04}$$

$$= 8i + 1.5 \quad \text{or} \quad 1.5 + 8i$$ ∎

Most algebra books use i as the imaginary unit. However, most electronics books use j as the imaginary unit because i is often used to represent current.

Knowing that $i = \sqrt{-1}$ and $i^2 = -1$, we can find other powers of i. For example,

$$i^3 = i^2 \cdot i = -1 \cdot i = -i \qquad i^6 = i^4 \cdot i^2 = 1(-1) = -1$$
$$i^4 = i^2 \cdot i^2 = (-1)(-1) = 1 \qquad i^7 = i^4 \cdot i^3 = 1(-i) = -i$$
$$i^5 = i^4 \cdot i^1 = 1 \cdot i = i \qquad i^8 = i^4 \cdot i^4 = (1)(1) = 1$$

Note that powers of i rotate through the four numbers $i, -1, -i, 1$.

Exercise Set 8.8

Write each expression as a complex number in the form $a + bi$.

1. 3
2. $\sqrt{9}$
3. $3 + \sqrt{-4}$
4. $-\sqrt{5}$
5. $6 + \sqrt{3}$
6. $\sqrt{-8}$
7. $\sqrt{-25}$
8. $2 + \sqrt{-5}$
9. $4 + \sqrt{-12}$
10. $\sqrt{-4} + \sqrt{-16}$
11. $\sqrt{-25} - 2i$
12. $3 + \sqrt{-72}$
13. $9 - \sqrt{-9}$
14. $\sqrt{75} - \sqrt{-128}$
15. $2i - \sqrt{-80}$
16. $\sqrt{288} - \sqrt{-96}$

Add or subtract as indicated.

17. $(12 - 6i) + (3 + 2i)$
18. $(6 - 3i) - 2(2 - 4i)$
19. $(12 + \frac{5}{9}i) - (4 - \frac{3}{4}i)$
20. $(\frac{5}{8} + \sqrt{-4}) + (\frac{2}{3} + 7i)$
21. $(13 - \sqrt{-4}) - (-5 + \sqrt{-9})$
22. $(7 + \sqrt{5}) + (2\sqrt{5} + \sqrt{-5})$
23. $(\sqrt{3} + \sqrt{2}) + (3\sqrt{2} - \sqrt{-8})$
24. $(3 - \sqrt{-72}) + (4 - \sqrt{-32})$
25. $(19 + \sqrt{-147}) + (\sqrt{-75})$
26. $(13 + \sqrt{-108}) - (\sqrt{49} - \sqrt{-48})$
27. $(\sqrt{12} + \sqrt{-49}) - (\sqrt{49} - \sqrt{-12})$
28. $(\sqrt{20} - \sqrt{-12}) + (2\sqrt{5} + \sqrt{-75})$

Multiply as indicated.

29. $2(-3 - 2i)$
30. $-3(\sqrt{5} + 2i)$
31. $i(6 + i)$
32. $2i(2 - 5i)$
33. $-3.5i(6.4 - 1.8i)$
34. $\sqrt{-5}(2 + 3i)$
35. $\sqrt{-4}(\sqrt{3} + 2i)$
36. $\sqrt{-8}(\sqrt{2} - \sqrt{-2})$
37. $\sqrt{-6}(\sqrt{3} + \sqrt{-6})$
38. $-\sqrt{-2}(3 - \sqrt{-8})$
39. $(3 + 2i)(1 + i)$
40. $(3 - 4i)(6 + 5i)$
41. $(4 - 6i)(3 - i)$
42. $(3i + 4)(2i - 3)$
43. $(\frac{1}{4} + \sqrt{-3})(2 + \sqrt{3})$
44. $(2 - 3i)(4 + \sqrt{-4})$
45. $(5 - \sqrt{-8})(\frac{1}{4} + \sqrt{-2})$
46. $(\frac{3}{5} - \frac{1}{4}i)(\frac{2}{3} + \frac{2}{5}i)$

Divide as indicated.

47. $\dfrac{-5}{-3i}$
48. $\dfrac{2}{5i}$
49. $\dfrac{2 + 3i}{2i}$
50. $\dfrac{1 + i}{-3i}$
51. $\dfrac{2 + 5i}{5i}$
52. $\dfrac{6}{2 + i}$
53. $\dfrac{7}{7 - 2i}$
54. $\dfrac{4 + 2i}{1 + 3i}$
55. $\dfrac{6 - 3i}{4 + 2i}$
56. $\dfrac{4 - 3i}{4 + 3i}$
57. $\dfrac{4}{6 - \sqrt{-4}}$
58. $\dfrac{5}{3 + \sqrt{-5}}$
59. $\dfrac{\sqrt{6}}{\sqrt{3} - \sqrt{-9}}$
60. $\dfrac{\sqrt{2}}{5 + \sqrt{-12}}$
61. $\dfrac{\sqrt{10} + \sqrt{-3}}{5 - \sqrt{-20}}$
62. $\dfrac{12 - \sqrt{-12}}{\sqrt{3} + \sqrt{-5}}$

Perform the indicated operation.

63. $(4 - 2i) + (3 - 5i)$
64. $(\frac{1}{2} + i) - (\frac{3}{5} - \frac{2}{3}i)$
65. $(8 - \sqrt{-6}) - (2 - \sqrt{-24})$
66. $(\sqrt{8} - \sqrt{2}) - (\sqrt{12} - \sqrt{48})$
67. $5.2(4 - 3.2i)$
68. $-0.6i(3 + 5i)$

69. $\sqrt{-6}(\sqrt{3} - \sqrt{-10})$

70. $(5 + 2i)(3 - 5i)$

71. $(\sqrt{3} + 2i)(\sqrt{6} - \sqrt{-8})$

72. $\dfrac{6}{2i}$

73. $\dfrac{5 - 4i}{2i}$

74. $\dfrac{1}{2 - 3i}$

75. $\dfrac{4}{\sqrt{3} - \sqrt{-4}}$

76. $\dfrac{5 - 2i}{3 + 2i}$

77. $\left(5 - \dfrac{5}{9}i\right) - \left(2 - \dfrac{3}{5}i\right)$

78. $\dfrac{4}{7}\left(4 - \dfrac{3}{5}i\right)$

79. $\left(\dfrac{2}{3} - \dfrac{1}{5}i\right)\left(\dfrac{3}{5} - \dfrac{3}{4}i\right)$

80. $\sqrt{\dfrac{4}{9}}\left(\sqrt{\dfrac{25}{36}} - \sqrt{-\dfrac{4}{25}}\right)$

81. $\dfrac{5 - 3i}{-5i}$

82. $\dfrac{-1 - 2i}{2 + \sqrt{-5}}$

83. $(5.23 - 6.41i) - (8.56 - 4.5i) - 7.1i$

84. $(\sqrt{-6} + 3)(\sqrt{-15} + 5)$

85. Find the impedance, Z, using the formula $Z = \dfrac{V}{I}$ when $V = 1.8 + 0.5i$ and $I = 0.6i$. See Example 11.

86. Refer to Exercise 85. Find the impedance when $V = 2.4 - 0.6i$ and $I = -0.4i$

87. Under certain conditions, the total impedance, Z_T, of a circuit is given by the formula

$$Z_T = \dfrac{Z_1 Z_2}{Z_1 + Z_2}$$

Find Z_T when $Z_1 = 2 - i$ and $Z_2 = 4 + i$.

88. Explain why $i^{14} = -1$.

89. Explain why $i^{23} = -i$.

90. Determine whether i^{65} is equal to i, -1, $-i$, or 1. Show how you determined your answer.

91. Determine whether i^{-1} is equal to i, -1, $-i$, or 1. Show how you determined your answer.

Answer true or false.

92. Every real and every imaginary number is a complex number.

93. Every complex number is a real number.

94. The product of two pure imaginary numbers is always a real number.

95. The sum of two pure imaginary numbers is always an imaginary number.

96. The product of two complex numbers is always a real number.

97. The sum of two complex numbers is always a complex number.

98. List, if possible, a number that is *not* (a) a rational number, (b) an irrational number, (c) a real number, (d) an imaginary number, and (e) a complex number.

99. Write a paragraph or two explaining the relationship between the real numbers, imaginary numbers, and complex numbers. Include in your discussion how the various types of numbers relate to each other.

Cumulative Review Exercises

[4.3] **100.** Mr. Tomlins, a grocer, has two coffees, one selling for $5.00 per pound and the other for $5.80 per pound. How many pounds of each type of coffee should he mix to make 40 pounds of coffee to sell for $5.50 per pound?

[5.8] **101.** Find the axis of symmetry, the coordinates of the vertex, and whether the graph opens up or down. Then graph the function, $f(x) = x^2 - 2x + 1$.

Factor.

[6.2] **102.** $4x^4 + 12x^2 + 9$.

103. $15r^2s^2 + rs - 6$.

The Square Root Function (Optional)

▶ **1** Recognize square root functions.

▶ **2** Find the domain of square root functions.

▶ **3** Graph square root functions.

▶ **1** We have discussed functions in several previous chapters. Here we introduce a new type of function, the square root function.

Square Root Function

$$f(x) = \sqrt{x}, \qquad x \geq 0$$

for any algebraic expression x.

▶ **2** The radicand of the square root function must be greater than or equal to 0. If the radicand is less than 0, the square root becomes an imaginary number. Recall that the **domain** is the set of values that can be used for the independent variable, x. Thus the domain of $f(x) = \sqrt{x}$ (or $y = \sqrt{x}$) is all real numbers greater than or equal to 0.

EXAMPLE 1 Find the domain of $f(x) = \sqrt{x - 2}$.

Solution: We must find the values of x that will result in the radicand being nonnegative. When $x \geq 2$, the radicand will be greater than or equal to 0. The domain is the set of all real numbers greater than or equal to 2.

$$D = \{x \mid x \geq 2\}$$ ■

EXAMPLE 2 Find the domain of $f(x) = \sqrt{6 - \dfrac{3}{5}x}$.

Solution: We must find the set of values of x that makes the radicand greater than or equal to 0. To find these values, set the radicand greater than or equal to 0 and solve for x.

$$6 - \frac{3}{5}x \geq 0$$

$$6 - 6 - \frac{3}{5}x \geq 0 - 6 \qquad \text{Subtract 6 from both sides of the inequality.}$$

$$-\frac{3}{5}x \geq -6$$

$$\left(-\frac{5}{3}\right)\left(-\frac{3}{5}x\right) \leq -6\left(-\frac{5}{3}\right) \qquad \text{Multiply both sides of the inequality by } -\frac{5}{3} \text{ and change the sense of the inequality.}$$

$$x \leq 10$$

Thus the domain is the set of real numbers less than or equal to 10.

$$D = \{x \mid x \leq 10\}.$$ ■

▶ **3**   Now let us graph some square root functions.

EXAMPLE 3   Graph the function $y = \sqrt{x}$.

Solution: The domain of this function is all real numbers greater than or equal to 0. Thus, when we graph, we can only use values of x such that $x \geq 0$. We will substitute values for x that are perfect square numbers. The graph is illustrated in Fig. 8.4.

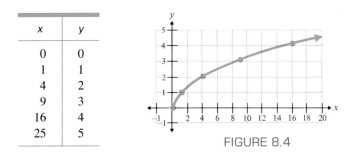

x	y
0	0
1	1
4	2
9	3
16	4
25	5

FIGURE 8.4

Note that the graph in Fig. 8.4 is a function since it passes the vertical line test. Also note that the range of $f(x)$, or the values of y, is the set of values greater than or equal to 0, $R:\{y \mid y \geq 0\}$.

EXAMPLE 4   Graph the function $y = -\sqrt{x + 3}$.

Solution: The domain is the set of values greater than or equal to -3. Therefore, when we substitute values for x, we will use values that are greater than or equal to -3. We will also select values that make the radicand a perfect square. The graph is shown in Fig. 8.5.

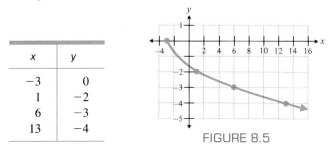

x	y
-3	0
1	-2
6	-3
13	-4

FIGURE 8.5

Note that the range of the graph in Fig. 8.5 is $\{y \mid y \leq 0\}$.

Exercise Set 8.9

Find the domain of the function.

1. $f(x) = \sqrt{x + 2}$

2. $f(x) = \sqrt{x - 2}$

3. $f(x) = \sqrt{4x}$

4. $f(x) = \sqrt{6 - x}$

5. $f(x) = \sqrt{4 - x}$

6. $f(x) = \sqrt{3x - 5}$

7. $f(x) = -\sqrt{x + 4}$

8. $f(x) = \sqrt{9 - 2x}$

9. $f(x) = -\sqrt{3x + 5}$

10. $f(x) = \sqrt{5x + 7}$ **11.** $f(x) = \sqrt{7x - \dfrac{1}{2}}$ **12.** $f(x) = \sqrt{\dfrac{2}{3}x - 4}$

13. $f(x) = \sqrt{\dfrac{5}{3} - 4x}$ **14.** $f(x) = \sqrt{6 - \dfrac{3}{2}x}$

Find the domain of the following functions. Graph the function and state its range.

15. $f(x) = \sqrt{x + 4}$ **16.** $f(x) = \sqrt{x - 4}$

17. $f(x) = \sqrt{x - 2}$ **18.** $f(x) = \sqrt{x + 2}$

19. $f(x) = -\sqrt{x + 3}$ **20.** $f(x) = -\sqrt{x - 3}$

21. $f(x) = \sqrt{3 - x}$ **22.** $f(x) = \sqrt{5 - x}$

23. $f(x) = -\sqrt{2 - x}$ **24.** $f(x) = \sqrt{2x + 5}$

25. $f(x) = \sqrt{2x}$ **26.** $f(x) = -\sqrt{2x}$

27. $f(x) = \sqrt{2x + 1}$ **28.** $f(x) = \sqrt{2x - 4}$

29. $f(x) = \sqrt{2x + 4}$ **30.** $f(x) = \sqrt{3x}$

31. Consider the square root function $y = \sqrt{x^2}$
    (a) What is its domain?
    (b) Graph the function.
    (c) What is its range?
32. Consider the square root function $y = \sqrt{x^2 + 1}$
    (a) What is its domain?
    (b) Graph the function.
    (c) What is its range?
33. Consider the square root function $y = \sqrt{x^2 - 4}$. Is 0 in the domain of this function? Explain your answer.

34. Consider the square root function $y = \sqrt{4 - x^2}$. Is 0 in the domain of this function? Explain your answer.
35. Consider the square root function $y = \sqrt{(x + 2)^2}$. What is the domain of this function? Explain your answer.
36. Explain why the range of a function of the form $y = \sqrt{x}$ must always be $\{y \mid y \geq 0\}$.
37. Consider the equation $y = \sqrt[3]{x}$. Is this a square root function? Explain. Graph the equation and determine if $y = \sqrt[3]{x}$ is a function. State the domain and range of the function or relation.

Cumulative Review Exercises

[4.3] 38. Lionel, a chemist, has 1 liter of a 20% sulfuric acid solution. How much of a 12% sulfuric acid solution must be mixed with the 1 liter of the 20% solution to make a 15% sulfuric acid solution?

[7.4] 39. Simplify $\dfrac{\dfrac{2}{a} + \dfrac{1}{2a}}{\dfrac{a}{2} - a}$.

[7.6] 40. When one printing machine runs alone, it can complete a specific printing order in 20 minutes. When a second and newer printing machine runs alone, it can complete the same job in 15 minutes. If the older printing machine runs for 8 minutes and is then turned off, and the newer machine starts running, how much longer will it take the newer printer to finish the job?

[8.6] 41. Solve the equation $\sqrt{x + 6} + \sqrt{x + 2} - 1 = 1$

JUST FOR FUN

For each square root function, (a) find the domain, (b) graph the function, and (c) find the range.

1. $f(x) = \sqrt{4 - x^2}$
2. $f(x) = \sqrt{x^2 - 4}$

3. Square root functions are one type of *radical functions*. Radical functions include square root functions, and cube and higher root functions. Graph $f(x) = \sqrt[3]{x} - 2$.

SUMMARY

GLOSSARY

Complex number *(397):* A number of the form $a + bi$. Every real number and every imaginary number are complex numbers.

Conjugate *(377):* The conjugate of $a + b$ is $a - b$.

Domain *(404):* The set of values that can be used for the independent variable.

Imaginary number *(396):* A number of the form bi, $b \neq 0$.

Imaginary unit *(396):* $\sqrt{-1}$. *Note:* $\sqrt{-1} = i$.

Index *(353):* The root of an expression.

Like radicals *(382):* Radicals that have the same index and radicand.

Perfect power *(367):* For $\sqrt[m]{x^n}$, x^n is a perfect power if n is a multiple of m.

Positive (or principal) square root *(353):* The principal square root of a positive real number x, written $\sqrt{x}$, is the positive number whose square equals x.

Radical equation *(386):* An equation that contains a variable in a radicand.

Radical expression *(353):* The radical sign and the radicand together are a radical expression.

Radical sign *(353):* $\sqrt{}$.

Rationalizing the denominator *(374):* To rationalize a denominator means to remove all radicals from the denominator.

Square root function *(404):* A function of the form $f(x) = \sqrt{x}$, $x \geq 0$.

IMPORTANT FACTS

$$\sqrt[n]{x} = a \text{ if } a^n = x$$
$$\sqrt{a^2} = |a|$$
$$\sqrt{a^2} = a, \qquad a \geq 0$$
$$\sqrt[n]{a^n} = a, \qquad a \geq 0,$$
$$\sqrt[n]{x} = x^{1/n}, \qquad x \geq 0$$
$$\sqrt[n]{x^m} = (\sqrt[n]{x})^m = x^{m/n}, \qquad x > 0$$
$$\sqrt[n]{a}\sqrt[n]{b} = \sqrt[n]{ab}, \qquad a \geq 0, b \geq 0$$
$$\frac{\sqrt[n]{a}}{\sqrt[n]{b}} = \sqrt[n]{\frac{a}{b}}, \qquad a \geq 0, b > 0$$

A Radical Is Simplified When the Following Are All True

1. There are no perfect powers in any radicand.
2. No radicand contains fractions.
3. There are no radicals in any denominator.

Review Exercises

[8.1] *Evaluate each expression.*

1. $\sqrt{9}$
2. $\sqrt{25}$
3. $\sqrt[3]{-8}$
4. $\sqrt[4]{256}$
5. $\sqrt[3]{27}$
6. $\sqrt[3]{-27}$
7. $\sqrt{6^2}$
8. $\sqrt{(19)^2}$

Use absolute value to evaluate the following.

9. $\sqrt{(-7)^2}$
10. $\sqrt{(-93)^2}$

Write as an absolute value.

11. $\sqrt{x^2}$
12. $\sqrt{(x-2)^2}$
13. $\sqrt{(x-y)^2}$
14. $\sqrt{(2x-3)^2}$

For the remainder of these review exercises, assume that all variables represent positive real numbers.

[8.2] *Write each of the following in exponential form.*

15. $\sqrt{x^5}$
16. $\sqrt[3]{x^5}$
17. $\sqrt[4]{y^3}$
18. $\sqrt[7]{5^2}$

Write each of the following in radical form.

19. $x^{1/2}$
20. $y^{3/5}$
21. $2^{1/4}$
22. $8^{3/4}$

Simplify each radical expression by changing the expression to exponential form. Write the answer in radical form.

23. $\sqrt{5^6}$
24. $\sqrt[3]{3^6}$
25. $\sqrt[6]{y^2}$
26. $\sqrt{x^{10}}$
27. $(\sqrt[3]{4})^6$
28. $\sqrt[5]{9^{10}}$
29. $\sqrt[20]{x^4}$
30. $\sqrt[3]{x^3}$

Evaluate each of the following if possible. If the expression is not a real number, state so.

31. $8^{1/3}$

32. $-25^{1/2}$

33. $(-25)^{1/2}$

34. $\left(\dfrac{8}{27}\right)^{-1/3}$

Simplify each of the following. Write the answer without negative exponents.

35. $x^{1/2} \cdot x^{2/5}$

36. $\left(\dfrac{64}{y^6}\right)^{1/3}$

37. $\left(\dfrac{y^{-3/5}}{y^{1/5}}\right)^{2/3}$

38. $\left(\dfrac{x^4 y^{-2}}{z^{1/2}}\right)^2 \left(\dfrac{xy^{1/2}}{z^2}\right)$

Factor each of the following. Write the answer without negative exponents. (optional)

39. $x^{1/5} + x^{6/5}$

40. $x^{-3} - x^{-4}$

41. $x^{-1/2} + x^{-2/3}$

42. $4x^{-6} + 6x^{-4}$

Factor each of the following. (optional)

43. $x^{1/2} - 5x^{1/4} + 4$

44. $x^{2/5} - 2x^{1/5} - 15$

45. $6x - 5x^{1/2} - 6$

46. $8x^{2/3} + 10x^{1/3} - 3$

[8.3–8.5] *Simplify each expression. Assume that all variables represent positive real numbers.*

47. $\sqrt{24}$

48. $\sqrt{80}$

49. $\sqrt[3]{16}$

50. $\sqrt[3]{54}$

51. $\sqrt{50x^3y^7}$

52. $\sqrt[3]{9x^6y^5}$

53. $\sqrt[4]{16x^9y^{12}}$

54. $\sqrt[3]{125x^7y^{10}}$

55. $\sqrt[5]{32x^{12}y^7z^{17}}$

56. $\sqrt{20}\sqrt{5}$

57. $\sqrt{5x}\sqrt{8x^5}$

58. $\sqrt[3]{2x^4y^5}\sqrt[3]{16x^4y^4}$

59. $(\sqrt[3]{5x^2y^3})^2$

60. $\sqrt[4]{8x^4y^7}\sqrt[4]{2x^5y^9}$

61. $\sqrt{3x}(\sqrt{12x} - \sqrt{20})$

62. $\sqrt[3]{y}(4\sqrt[3]{y^2} - \sqrt[3]{y^{10}})$

63. $\sqrt[3]{2x^2y^3}(\sqrt[3]{4x^4y^7} + \sqrt[3]{9xy^{12}})$

64. $\sqrt[4]{3x^3y^2}(\sqrt[4]{2x^5y^9} + \sqrt[4]{27x^9y^3})$

65. $\sqrt{\dfrac{1}{4}}$

66. $\sqrt{\dfrac{36}{25}}$

67. $\sqrt[3]{\dfrac{3}{81}}$

68. $\sqrt[3]{\dfrac{x^3}{8}}$

69. $\dfrac{\sqrt[3]{2x^9}}{\sqrt[3]{16x^6}}$

70. $\sqrt{\dfrac{32x^2y^5}{2x^4y}}$

71. $\sqrt[3]{\dfrac{54x^3y^6}{2y^3}}$

72. $\sqrt{\dfrac{75x^2y^5}{3x^4y^7}}$

73. $\dfrac{1}{\sqrt{2}}$

74. $\dfrac{1}{\sqrt{3}}$

75. $\dfrac{x}{\sqrt{7}}$

76. $\sqrt{\dfrac{2}{5}}$

77. $\sqrt{\dfrac{y}{x}}$

78. $\sqrt{\dfrac{3x}{5y}}$

79. $\dfrac{2}{\sqrt[3]{x}}$

80. $\sqrt[3]{\dfrac{3x}{5y}}$

81. $\sqrt{\dfrac{3x^2}{y}}$

82. $\sqrt{\dfrac{18x^4y^5}{3z}}$

83. $\sqrt{\dfrac{25x^2y^5}{3z}}$

84. $\sqrt{\dfrac{5x^3y^5}{2}}$

85. $\sqrt{\dfrac{27x^4z^5}{5y}}$

86. $\sqrt{\dfrac{20y^6z^9}{6x^3}}$

87. $\sqrt[3]{\dfrac{4x^5y^3}{x^6}}$

88. $\sqrt[3]{\dfrac{y^6}{2x^2}}$

89. $(3 - \sqrt{2})(3 + \sqrt{2})$

90. $(\sqrt{3} - \sqrt{5})(\sqrt{3} + \sqrt{5})$

91. $(\sqrt{x} + y)(\sqrt{x} - y)$

92. $(x - \sqrt{y})(x + \sqrt{y})$

93. $(\sqrt{3} + 5)^2$

94. $(\sqrt{5} - \sqrt{20})^2$

95. $(\sqrt{x} - \sqrt{3y})(\sqrt{x} + \sqrt{5y})$

96. $(2\sqrt{x} + 3\sqrt{y})(\sqrt{x} - \sqrt{y})$

97. $(\sqrt[3]{2x} - \sqrt[3]{3y})(\sqrt[3]{3x} - \sqrt[3]{2y})$

98. $\dfrac{5}{2 + \sqrt{5}}$

99. $\dfrac{3}{4 - \sqrt{2}}$

100. $\dfrac{x}{3 + \sqrt{x}}$

101. $\dfrac{\sqrt{x}}{\sqrt{x} + \sqrt{y}}$

102. $\dfrac{\sqrt{5} + \sqrt{2}}{\sqrt{6} - \sqrt{3}}$

103. $\dfrac{\sqrt{x} - 3}{\sqrt{x} + 3}$

104. $\dfrac{\sqrt{x} - 2\sqrt{y}}{\sqrt{x} - \sqrt{y}}$

105. $\dfrac{4}{\sqrt{y + 2} - 3}$

106. $2\sqrt{3} - \sqrt{3} + 5\sqrt{3}$

107. $\sqrt[3]{x} + 3\sqrt[3]{x} - 2\sqrt[3]{x}$

108. $\sqrt[3]{16} - \sqrt[3]{54}$

109. $\sqrt{3} + \sqrt{27} - \sqrt{192}$

110. $\sqrt[3]{16} - 5\sqrt[3]{54} + 2\sqrt[3]{64}$ **111.** $4\sqrt{2} - \dfrac{3}{\sqrt{32}} + \sqrt{50}$ **112.** $3\sqrt{x^5y^6} - \sqrt{16x^7y^8}$

113. $2\sqrt[3]{x^7y^8} - \sqrt[3]{x^4y^2} + 3\sqrt[3]{x^{10}y^2}$

[8.6] *Solve each equation; then check your solutions.*

114. $\sqrt[3]{x} = 4$ **115.** $\sqrt{3x + 4} = 5$ **116.** $2 + \sqrt[3]{x} = 4$

117. $\sqrt{x^2 + 2x - 4} = x$ **118.** $\sqrt[3]{x - 9} = \sqrt[3]{5x + 3}$ **119.** $\sqrt{x^2 + 5} = x + 1$

120. $\sqrt{x + 3} = \sqrt{3x + 9}$ **121.** $\sqrt{6x - 5} - \sqrt{2x + 6} - 1 = 0$

Solve each equation or formula for the variable indicated.

122. $V = \sqrt{\dfrac{2L}{w}}$, for L **123.** $r = \sqrt{\dfrac{A}{\pi}}$, for A

[8.7] *Solve each problem.*

124. How long a wire does a phone company need to reach the top of a 5 meter telephone pole from a point 2 meters from the base of the pole? Use $l = \sqrt{a^2 + b^2}$.

125. Use the formula $v = \sqrt{2gh}$ to find the velocity of an object after it has fallen 20 feet ($g = 32$).

126. Use the formula $T = 2\pi\sqrt{L/32}$ to find the period of a pendulum (T) if its length L is 64 feet. You may leave your answer in simplified radical form.

[8.8] *Write each expression as a complex number in the form $a + bi$.*

127. 5   **128.** -6 **129.** $2 - \sqrt{-4}$ **130.** $3 + \sqrt{-16}$

Perform the operations indicated.

131. $(3 + 2i) + (4 - i)$ **132.** $(4 - 6i) - (3 - 4i)$ **133.** $(5 + \sqrt{-9}) - (3 - \sqrt{-4})$

134. $(\sqrt{3} + \sqrt{-5}) + (2\sqrt{3} - \sqrt{-7})$ **135.** $4(3 + 2i)$ **136.** $-2(\sqrt{3} - i)$

137. $\sqrt{8}(\sqrt{-2} + 3)$ **138.** $\sqrt{-6}(\sqrt{6} + \sqrt{-6})$ **139.** $(4 + 3i)(2 - 3i)$

140. $(6 + \sqrt{-3})(4 - \sqrt{-15})$ **141.** $\dfrac{2}{3i}$ **142.** $\dfrac{-3}{-5i}$

143. $\dfrac{2 + \sqrt{3}}{2i}$ **144.** $\dfrac{5}{3 + 2i}$ **145.** $\dfrac{\sqrt{3}}{5 - \sqrt{-6}}$

146. $\dfrac{\sqrt{5} + 3i}{\sqrt{2} - \sqrt{-8}}$

[8.9] *State the domain of each of the following functions.*

147. $f(x) = \sqrt{x - 6}$ **148.** $f(x) = \sqrt{3x + 5}$

149. $f(x) = \sqrt{10 - 4x}$ **150.** $f(x) = \sqrt{\dfrac{2}{3}x - 4}$

State the domain of each function. Graph the function and state the range.

151. $f(x) = \sqrt{x}$ **152.** $f(x) = \sqrt{x + 2}$

153. $f(x) = \sqrt{5 - x}$ **154.** $f(x) = -\sqrt{x - 3}$

Practice Test

1. Use absolute value to evaluate $\sqrt{(-26)^2}$.

3. Simplify $\left(\dfrac{y^{2/3} \cdot y^{-1}}{y^{1/4}}\right)^2$.

2. Write as an absolute value $\sqrt{(3x - 4)^2}$.

*4. Factor $2x^{2/3} + x^{1/3} - 10$.

Simplify. Assume that all variables represent positive real numbers.

5. $\sqrt{50x^5y^8}$

6. $\sqrt[3]{4x^5y^2}\sqrt[3]{10x^6y^8}$

7. $\sqrt{\dfrac{2x^4y^5}{8z}}$

8. $\sqrt[3]{\dfrac{1}{x}}$

9. $\dfrac{\sqrt{2}}{2 + \sqrt{8}}$

10. $\sqrt{27} + 2\sqrt{3} - 5\sqrt{75}$

11. $\sqrt[3]{8x^3y^5} + 2\sqrt[3]{x^6y^8}$

12. $(\sqrt{5} - 3)(2 - \sqrt{8})$

Solve the equations.

13. $\sqrt{4x - 3} = 7$

14. $\sqrt{x^2 - x - 12} = x + 3$

15. $\sqrt{x - 15} = \sqrt{x} - 3$

16. Solve the formula $w = \dfrac{\sqrt{2gh}}{4}$ for h.

19. Divide $\dfrac{\sqrt{5}}{2 - \sqrt{-8}}$.

17. A ladder is placed against a house. If the base of the ladder is 5 feet from the house and the ladder rests on the house 12 feet above the ground, find the length of the ladder.

*20. **(a)** State the domain of $f(x) = \sqrt{x + 2}$.
(b) Graph the function and state the range.

18. Multiply $(6 - \sqrt{-4})(3 + \sqrt{-2})$.

Cumulative Review Test

1. Solve the equation $\frac{1}{5}(x - 3) = \frac{3}{4}(x + 3) - x$.

2. Graph $5x - 2y = 6$.

3. Define (a) a relation and (b) a function.

4. Solve the system of equations $x - 4y = 6$
$3x - y = 2$.

5. Simplify $\left(\dfrac{3x^2y^{-2}}{x^4y^{-5}}\right)\left(\dfrac{2xy^3}{x^2y^{-3}}\right)^{-1}$.

6. Multiply $(3x^2 - 4x - 6)(2x - 5)$.

7. Divide $\dfrac{3x^2 + 10x + 10}{x + 2}$.

8. Graph $f(x) = x^3 + 4x - 1$.

9. Factor $2x^2 - 12x + 18 - 2y^2$.

10. Find the domain of $\dfrac{x - 4}{5x - 3}$.

11. Reduce $\dfrac{(x + 2)(x - 4) + (x - 1)(x - 4)}{3(x - 4)}$.

12. Multiply $\dfrac{4x^2 + 8x + 3}{2x^2 - x - 1} \cdot \dfrac{x^2 - 1}{4x^2 + 12x + 9}$.

13. Subtract $\dfrac{x + 1}{x^2 + 2x - 3} - \dfrac{x}{2x^2 + 11x + 15}$.

14. Solve the equation $4 - \dfrac{5}{y} = \dfrac{4y}{y + 1}$.

15. Simplify $\left(\dfrac{x^2y^{1/2}}{x^{1/4}}\right)^2$

16. Simplify $\sqrt[3]{4x^{10}y^{20}} \cdot \sqrt[3]{4x^3y^9}$.

17. Solve the equation $\sqrt{2x^2 + 7} + 3 = 8$.

18. Divide $\dfrac{2}{3 + \sqrt{-6}}$.

19. Jim by himself can paint the living room in his house in 2 hours. Jim's son Mike can paint the same room by himself in 3 hours. How long will it take them to paint the room if they work together?

20. A wire goes from the top of a 30-foot telephone pole to a point on the ground 20 feet from the base of the pole. What is the length of the wire?

*From an optional section.

Quadratic Functions and the Algebra of Functions

See Section 9.4 Exercise 47.

9.1

Solving Quadratic Equations by Completing the Square

▶1 Know the square root property.

▶2 Use the square root property to solve for a variable in a formula.

▶3 Write perfect square trinomials.

▶4 Solve quadratic equations by completing the square.

In this section we introduce two important concepts, the square root property and the procedure for completing the square. We will introduce the square root property first, since we use it in our explanation of completing the square. The square root property is very important, and it will be used in several sections in this book.

In Section 6.5, we solved quadratic, or second degree, equations by factoring. Quadratic equations that cannot be solved by factoring can be solved using the completing-the-square procedure, or by using the quadratic formula which is presented in Section 9.2.

The Square Root Property

▶1 In Section 8.1 we stated that every positive number has two square roots. Thus far we have been using only the positive square root. In this section we use both the positive and negative square roots of a number.

The positive square root of 25 is 5.

$$\sqrt{25} = 5$$

The negative square root of 25 is -5.

$$-\sqrt{25} = -5$$

Notice that $5 \cdot 5 = 25$ and $(-5)(-5) = 25$. The two square roots of 25 are $+5$ and -5. A convenient way to indicate the two square roots of a number is to use the plus or minus symbol, $\pm$. For example, the square roots of 25 can be indicated ± 5, read "plus or minus 5." The equation $x^2 = 25$ has two solutions, the two roots of 25, ± 5. If you check each root, you will see that each value satisfies the equation. The square root property can be used to find the solutions to equations of the form $x^2 = a$.

Square Root Property

If $x^2 = a$, where a is a real number, then $x = \pm\sqrt{a}$.

EXAMPLE 1 Solve the equation $x^2 - 9 = 0$.

Solution: Add 9 to both sides of the equation to get the variable by itself on one side of the equation.

$$x^2 = 9$$

By the square root property,

$$x = \pm\sqrt{9}$$
$$x = \pm 3$$

Check in the original equation.

$$x = 3 \qquad\qquad\qquad x = -3$$
$$x^2 - 9 = 0 \qquad\qquad x^2 - 9 = 0$$
$$3^2 - 9 = 0 \qquad\qquad (-3)^2 - 9 = 0$$
$$0 = 0, \quad \text{true} \qquad\qquad 0 = 0, \quad \text{true}$$

In both cases the check is true, which means that both 3 and -3 are solutions to the original equation. ■

EXAMPLE 2   Solve the equation $x^2 + 5 = 65$.

Solution: Begin by subtracting 5 from both sides of the equation.

$$x^2 = 60$$
$$x = \pm\sqrt{60} = \pm\sqrt{4}\sqrt{15} = \pm 2\sqrt{15}$$

The solutions are $2\sqrt{15}$ and $-2\sqrt{15}$. ■

Not all quadratic equations have real solutions, as is illustrated in Example 3.

EXAMPLE 3   Solve the equation $x^2 + 7 = 0$.

Solution:
$$x^2 + 7 = 0$$
$$x^2 = -7$$
$$x = \pm\sqrt{-7} = \pm i\sqrt{7}$$

The solutions are $i\sqrt{7}$ and $-i\sqrt{7}$. ■

EXAMPLE 4   Solve the equation $(x - 4)^2 = 32$.

Solution: Begin by taking the square root of both sides of the equation.

$$(x - 4)^2 = 32$$
$$x - 4 = \pm\sqrt{32}$$
$$x = 4 \pm \sqrt{32}$$
$$x = 4 \pm \sqrt{16}\sqrt{2}$$
$$x = 4 \pm 4\sqrt{2}$$

The solutions are $4 + 4\sqrt{2}$ and $4 - 4\sqrt{2}$. ■

▶**2**   When solving for a squared variable in a *formula,* you may need to use the square root property to solve for the variable. However, *when you use the square root property in most formulas, you will use only the principal or positive root,* because you are generally solving for a quantity that cannot be negative.

Consider a right triangle, see Figure 9.1. The two smaller sides of a right triangle are called the **legs** and the side opposite the right angle is called the **hypotenuse.** The Pythagorean theorem expresses the relationship between the legs of the triangle and its hypotenuse.

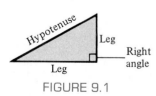

FIGURE 9.1

Pythagorean Theorem

The square of the hypotenuse of a right triangle is equal to the sum of the squares of the two legs. If a and b represent the legs and c represents the hypotenuse, then

$$a^2 + b^2 = c^2$$

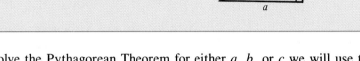

When we solve the Pythagorean Theorem for either a, b, or c we will use the square root property. However, when we use the square root property we use only the positive square root since a length will not be a negative number.

EXAMPLE 5   Find the hypotenuse of the right triangle whose legs are 3 feet and 4 feet.

Solution: It is often helpful when using the Pythagorean theorem to draw a picture of the problem before using the formula (Fig. 9.2). When drawing the picture, it makes no difference which leg is called a and which leg is called b.

$$a^2 + b^2 = c^2$$
$$3^2 + 4^2 = c^2$$
$$9 + 16 = c^2$$
$$25 = c^2$$
$$\sqrt{25} = c \qquad \text{Square root property using only the positive root.}$$
$$5 = c$$

FIGURE 9.2   The hypotenuse is 5 feet.   ■

Perfect Square Trinomials

▶ **3**   Now that we know the square root property we can focus our attention on the procedure for completing the square. To understand the procedure you need to know how to form perfect square trinomials.

A perfect square trinomial is a trinomial that can be expressed as the square of a binomial. Some examples follow.

Perfect Square Trinomials	*Factors*	*Square of a Binomial*
$x^2 + 8x + 16$	$= (x + 4)(x + 4)$	$= (x + 4)^2$
$x^2 - 8x + 16$	$= (x - 4)(x - 4)$	$= (x - 4)^2$
$x^2 + 10x + 25$	$= (x + 5)(x + 5)$	$= (x + 5)^2$
$x^2 - 10x + 25$	$= (x - 5)(x - 5)$	$= (x - 5)^2$

Notice that each of the squared terms in the trinomials given above has a numerical coefficient of 1. When the coefficient of the squared term is 1, there is an important relationship between the coefficient of the x term and the constant. In every perfect square trinomial the constant term is the square of one-half the coefficient of the x term.

Let us examine some perfect square trinomials whose coefficients of the squared term are 1.

$$x^2 + 8x + 16 = (x + 4)^2$$
$$[\tfrac{1}{2}(8)]^2 = (4)^2$$

$$x^2 - 10x + 25 = (x - 5)^2$$
$$[\tfrac{1}{2}(-10)]^2 = (-5)^2$$

Note that when such a perfect square trinomial is written as the square of a binomial the constant in the binomial is one-half the value of the coefficient of the x term in the perfect square trinomial. For example,

$$x^2 + 8x + 16 = (x + 4)^2$$
$$\tfrac{1}{2}(8)$$

$$x^2 - 10x + 25 = (x - 5)^2$$
$$\tfrac{1}{2}(-10)$$

Completing the Square

▶ **4**   Now we introduce the procedure for completing the square. To solve a quadratic equation by completing the square, we add (or subtract) a constant to (or from) both sides of the equation so that the remaining trinomial is a perfect square trinomial. Then we use the square root property to solve the resulting equation. We will now summarize the procedure.

To Solve a Quadratic Equation by Completing the Square

1. Use the multiplication (or division) property if necessary to make the numerical coefficient of the squared term equal to 1.
2. Rewrite the equation with the constant by itself on the right side of the equation.
3. Take one-half the numerical coefficient of the first-degree term, square it, and add this quantity to both sides of the equation.
4. Replace the trinomial with its equivalent squared binomial.
5. Take the square root of both sides of the equation.
6. Solve for the variable.
7. Check your answers in the original equation.

EXAMPLE 6   Solve the equation $x^2 + 6x + 5 = 0$ by completing the square.

Solution: Since the numerical coefficient of the squared term is 1, step 1 is not necessary.

Step 2:   Move the constant, 5, to the right side of the equation by subtracting 5 from both sides of the equation.

$$x^2 + 6x + 5 = 0$$
$$x^2 + 6x = -5$$

Step 3: Determine the square of one-half the numerical coefficient of the first-degree term.

$$\frac{1}{2}(6) = 3, \qquad 3^2 = 9$$

Add the number obtained to both sides of the equation.

$$x^2 + 6x \boxed{+ 9} = -5 \boxed{+ 9}$$
$$x^2 + 6x + 9 = 4$$

Step 4: By following this procedure we produce a perfect square trinomial on the left side of the equation. The expression $x^2 + 6x + 9$ is a perfect square trinomial that can be expressed as $(x + 3)^2$.

$$x^2 + 6x + 9 = 4$$

$$\begin{array}{l}\llcorner \frac{1}{2} \text{ the numerical coefficient of the} \\ \quad \text{first-degree term is } \frac{1}{2}(6) = +3\end{array}$$

$$(x \overset{\frown}{+} 3)^2 = 4$$

Step 5: Take the square root of both sides of the equation.

$$x + 3 = \pm\sqrt{4}$$
$$x + 3 = \pm 2$$

Step 6: Finally, solve for x by subtracting 3 from both sides of the equation.

$$x + 3 \boxed{- 3} = \boxed{-3} \pm 2$$
$$x = -3 \pm 2$$
$$x = -3 + 2 \quad \text{or} \quad x = -3 - 2$$
$$x = -1 \qquad\qquad x = -5$$

Step 7: Check both solutions in the original equation.

$x = -1$	$x = -5$
$x^2 + 6x + 5 = 0$	$x^2 + 6x + 5 = 0$
$(-1)^2 + 6(-1) + 5 = 0$	$(-5)^2 + 6(-5) + 5 = 0$
$1 - 6 + 5 = 0$	$25 - 30 + 5 = 0$
$0 = 0$ true	$0 = 0$ true

Since each number checks, both -1 and -5 are solutions to the original equation. ∎

EXAMPLE 7 Solve the equation $-x^2 = -3x - 18$ by completing the square.

Solution: The numerical coefficient of the squared term must be 1, not -1. Therefore, we begin by multiplying both sides of the equation by -1 to make the coefficient of the squared term equal to 1.

$$-x^2 = -3x - 18$$
$$-1(-x^2) = -1(-3x - 18)$$
$$x^2 = 3x + 18$$

Now move all terms except the constant to the left side of the equation.

$$x^2 - 3x = 18$$

Take half the numerical coefficient of the x term, square it, and add this product to both sides of the equation. Then write the left side of the equation as the square of a binomial.

$$\frac{1}{2}(-3) = -\frac{3}{2}, \qquad \left(-\frac{3}{2}\right)^2 = \frac{9}{4}$$

$$x^2 - 3x + \frac{9}{4} = 18 + \frac{9}{4}$$

$$\left(x - \frac{3}{2}\right)^2 = 18 + \frac{9}{4}$$

$$\left(x - \frac{3}{2}\right)^2 = \frac{72}{4} + \frac{9}{4}$$

$$\left(x - \frac{3}{2}\right)^2 = \frac{81}{4}$$

$$x - \frac{3}{2} = \pm\sqrt{\frac{81}{4}}$$

$$x - \frac{3}{2} = \pm\frac{9}{2}$$

$$x = \frac{3}{2} \pm \frac{9}{2}$$

$$x = \frac{3}{2} + \frac{9}{2} \qquad \text{or} \qquad x = \frac{3}{2} - \frac{9}{2}$$

$$x = \frac{12}{2} = 6 \qquad\qquad x = -\frac{6}{2} = -3 \qquad\blacksquare$$

In the following examples we will not illustrate some of the intermediate steps.

EXAMPLE 8 Solve the equation $x^2 - 6x + 17 = 0$.

Solution:

$$x^2 - 6x + 17 = 0$$
$$x^2 - 6x = -17$$
$$x^2 - 6x + 9 = -17 + 9$$
$$(x - 3)^2 = -8$$
$$x - 3 = \pm\sqrt{-8}$$
$$x - 3 = \pm 2i\sqrt{2}$$
$$x = 3 \pm 2i\sqrt{2}$$

The solutions are $3 + 2i\sqrt{2}$ and $3 - 2i\sqrt{2}$. Note that the solutions to the equation $x^2 - 6x + 17 = 0$ are not real. The solutions are complex numbers. $\blacksquare$

EXAMPLE 9 Solve the equation $-3m^2 + 6m + 24 = 0$ by completing the square.

Solution: To solve an equation by completing the square, the numerical coefficient of the squared term should be 1. Since the numerical coefficient of the squared term is -3, we multiply both sides of the equation by $-\frac{1}{3}$ to make the numerical coefficient of the squared term equal to 1.

$$-3m^2 + 6m + 24 = 0$$

$$-\frac{1}{3}(-3m^2 + 6m + 24) = -\frac{1}{3}(0)$$

$$m^2 - 2m - 8 = 0$$

Now proceed as before.

$$m^2 - 2m = 8$$

$$m^2 - 2m + 1 = 8 + 1$$

$$(m - 1)^2 = 9$$

$$m - 1 = \pm 3$$

$$m = 1 \pm 3$$

$$m = 1 + 3 \quad \text{or} \quad m = 1 - 3$$

$$m = 4 \qquad\qquad m = -2$$

It you were asked to solve the equation $-\frac{1}{4}x^2 + 2x - 8 = 0$ by completing the square, what would you do first? If you answered "multiply both sides of the equation by -4 to make the coefficient of the squared term 1," you answered correctly. To solve the equation $\frac{2}{3}x^2 + 2x - 8 = 0$, you could multiply both sides of the equation by $\frac{3}{2}$ to obtain a squared term with a coefficient of 1.

In examples 6 through 9 we solved quadratic equations by completing the square. Generally, quadratic equations that cannot be solved using factoring will be solved using the *quadratic formula* that will be presented in the next section. We introduced the completing the square procedure because we use the procedure to derive the quadratic formula in Section 9.2. We will also use the completing the square procedure quite a bit in Chapter 10.

Exercise Set 9.1

Use the square root property to solve each equation for the given variable.

1. $x^2 = 25$ **2.** $x^2 = 18$ **3.** $y^2 = 75$

4. $x^2 - 7 = 19$ **5.** $z^2 + 12 = 40$ **6.** $y^2 + 15 = 80$

7. $(x - 4)^2 = 16$ **8.** $(y - 3)^2 = 45$ **9.** $(z + \frac{1}{3})^2 = \frac{4}{9}$

10. $(x - 0.2)^2 = 0.64$ **11.** $(x + 1.8)^2 = 0.81$ **12.** $(x + \frac{1}{2})^2 = \frac{4}{9}$

13. $(2x - 5)^2 = 12$ **14.** $(4y + 1)^2 = 8$ **15.** $(2y + \frac{1}{2})^2 = \frac{4}{25}$

16. $(3x - \frac{1}{4})^2 = \frac{9}{25}$

Use the square root property to solve each of the formulas for the indicated variable. Use only the positive root in each case.

17. $A = s^2$, for s **18.** $r = \frac{1}{2}t^2$, for t **19.** $A = \pi r^2$, for r

20. $k = mv^2$, for v **21.** $F_x^2 + F_y^2 = F^2$, for F **22.** $a^2 + b^2 = c^2$, for b

23. $V^2 = V_x^2 + V_y^2$, for V_x **24.** $E = \frac{1}{9}rp^2$, for p **25.** $L = a^2 - b^2$, for b

26. $H = i^2R$, for i **27.** $v = m + nt^2$, for t **28.** $E = \frac{1}{2}mv^2$, for v

29. $d = b^2 - 4ac$, for b **30.** $V = \pi r^2 h$, for r **31.** $w = 3l + 2d^2$, for d

32. $A = P(1 + r)^2$, for r

Use the Pythagorean theorem to find the quantity indicated. You may leave your answer in square root form if a calculator with a square root key is not available and the number is not a perfect square.

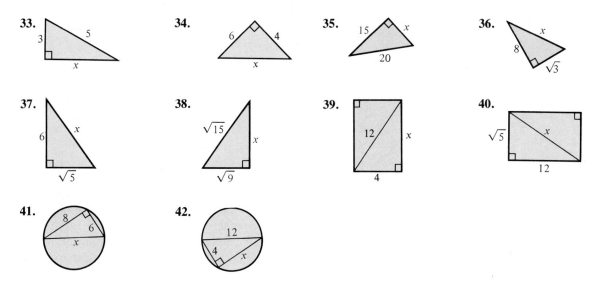

33.   **34.**   **35.**   **36.**

37.   **38.**   **39.**   **40.**

41.   **42.**

Solve each equation by completing the square.

43. $x^2 + 2x - 3 = 0$ **44.** $x^2 - 6x + 8 = 0$

45. $x^2 - 4x - 5 = 0$ **46.** $x^2 + 8x + 12 = 0$

47. $x^2 + 3x + 2 = 0$ **48.** $x^2 + 4x - 32 = 0$

49. $x^2 - 8x + 15 = 0$ **50.** $x^2 - 9x + 14 = 0$

51. $x^2 + 2x + 15 = 0$ **52.** $x^2 + 5x + 4 = 0$

53. $x^2 = -5x - 6$ **54.** $x^2 - 2x + 4 = 0$

55. $x^2 + 9x + 18 = 0$ **56.** $x^2 - 9x + 18 = 0$

57. $x^2 = 15x - 56$ **58.** $x^2 = 3x + 28$

59. $-4x = -x^2 + 12$ **60.** $x^2 + 3x + 6 = 0$

61. $\frac{1}{2}x^2 + x - 3 = 0$ **62.** $x^2 - 4x + 2 = 0$

63. $6x + 6 = -x^2$ **64.** $x^2 - x - 3 = 0$

65. $-x^2 + 5x = -8$ **66.** $\frac{1}{4}x^2 + \frac{3}{4}x - \frac{3}{2} = 0$

67. $-\frac{1}{4}x^2 - \frac{1}{2}x = 0$ **68.** $2x^2 - 6x = 0$

69. $12x^2 - 4x = 0$ **70.** $6x^2 = 9x$

71. $-\frac{1}{2}x^2 - x + \frac{3}{2} = 0$ **72.** $2x^2 + 2x - 24 = 0$

73. $2x^2 + 18x + 4 = 0$ **74.** $2x^2 = 8x + 90$

75. $3x^2 + 33x + 72 = 0$ **76.** $3x^2 + 2x - 1 = 0$

77. $\frac{2}{3}x^2 + \frac{4}{3}x + 1 = 0$ **78.** $3x^2 - 8x + 4 = 0$

79. $-3x^2 + 6x = 6$ **80.** $2x^2 - x = -5$

81. $\frac{5}{2}x^2 + \frac{3}{2}x - \frac{5}{4} = 0$ **82.** $\frac{3}{4}x^2 - \frac{1}{2}x - \frac{3}{20} = 0$

83. $3x^2 + \frac{1}{2}x = -4$ **84.** $2x^2 - \frac{1}{3}x = -2$

85. Solve by completing the square: $x^2 + 2ax + a^2 = k$.

86. Solve by completing the square: $(x \pm a)^2 - k = 0$.

87. The product of two consecutive positive odd integers is 63. Find the two odd integers.

88. The larger of two integers is 2 more than twice the smaller. Find the two numbers if their product is 12.

89. The length of a rectangle is 2 feet more than twice its width. Find the dimensions of the rectangle if the area is 60 square feet.

90. Find the length of the side of a square whose diagonal is 10 feet longer than the length of its side.

91. Find the length of the side of a square whose diagonal is 12 feet longer than the length of a side.

92. The length of a rectangle is 3 inches more than its width. If the length of the diagonal is 15 inches, find the dimensions of the rectangle.

93. What is the first step in the completing the square process?

94. Explain how to determine if a trinomial is a perfect square trinomial.

95. Write a paragraph explaining in your own words how to construct a perfect square trinomial.

Cumulative Review Exercises

[6.6] **96.** Solve the equation for z: $2xy - 3yz = -xy + z$.

[8.1] **97.** Express $\sqrt{(x^2 - 4x)^2}$ as an absolute value.

[8.2] **98.** Evaluate $25^{-1/2}$.

99. Simplify $\dfrac{x^{3/4} y^{1/2}}{x^{1/4} y^2}$

JUST FOR FUN

1. The surface area, S, and volume, V, of a right circular cylinder of radius r and height h are given by the formulas

$$S = 2\pi r^2 + 2\pi rh, \qquad V = \pi r^2 h$$

(a) Find the surface area of the cylinder if it has a height of 10 inches and a volume of 160 cubic inches.

(b) Find the radius if the height is 10 inches and the surface area is 160 square inches. (You may leave your answer in square root form.)

2. The distance formula, $d = \sqrt{(x_2 - x_1)^2 + (y_2 - y_1)^2}$, was discussed in Section 3.1. Derive the distance formula using the Pythagorean theorem and the square root property.

Write each equation in the form $a(x - h)^2 + b(y - k)^2 = c$ by completing the square twice: once for the x values and once for the y values. (We will be doing this type of problem in Chapter 10.)

3. $x^2 + 4x + y^2 - 6y = 3$

4. $4x^2 + 9y^2 - 48x + 72y + 144 = 0$

5. $x^2 - 4y^2 + 4x - 16y - 28 = 0$

9.2

Solving Quadratic Equations by the Quadratic Formula

▶**1** Derive the quadratic formula.

▶**2** Use the quadratic formula to solve equations.

▶**3** Use the discriminant to determine the number of solutions to a quadratic equation.

▶**4** Study applications of quadratic equations.

▶**1** The quadratic formula can be used to solve any quadratic equation. In fact, *it is the most useful and most versatile method of solving quadratic equations.* It is generally used in place of the completing the square method because of its efficiency.

The standard form of a quadratic equation is $ax^2 + bx + c = 0$, where a is the numerical coefficient of the squared term, b is the numerical coefficient of the first-degree term, and c is the constant.

Quadratic Equation in Standard Form	*Values of a, b, and c*
$x^2 - 3x + 4 = 0$	$a = 1, \quad b = -3, \quad c = 4$
$3x^2 - 4 = 0$	$a = 3, \quad b = 0, \quad c = -4$
$-5x^2 + 3x = 0$	$a = -5, \quad b = 3, \quad c = 0$

We can derive the quadratic formula by starting with a quadratic equation in standard form and completing the square as discussed in the preceding section.

$$ax^2 + bx + c = 0$$

$$ax^2 + bx = -c \qquad \text{Subtract } c \text{ from both sides of the equation.}$$

$$x^2 + \frac{b}{a}x = \frac{-c}{a} \qquad \text{Divide both sides of the equation by } a.$$

$$x^2 + \frac{b}{a}x + \frac{b^2}{4a^2} = \frac{-c}{a} + \frac{b^2}{4a^2} \qquad \begin{array}{l}\text{Take } 1/2 \text{ of } b/a \text{ to get } b/2a, \text{ square this}\\ \text{expression to get } b^2/4a^2, \text{ and add this}\\ \text{expression to both sides of the equation.}\end{array}$$

$$\left(x + \frac{b}{2a}\right)^2 = \frac{-c}{a} + \frac{b^2}{4a^2} \qquad \begin{array}{l}\text{Express the left side of the equation as the}\\ \text{square of a binomial.}\end{array}$$

$$\left(x + \frac{b}{2a}\right)^2 = \frac{4a}{4a} \cdot \frac{-c}{a} + \frac{b^2}{4a^2} \qquad \begin{array}{l}\text{Obtain a common denominator so that the}\\ \text{fractions can be added.}\end{array}$$

$$\left(x + \frac{b}{2a}\right)^2 = \frac{-4ac + b^2}{4a^2}$$

$$\left(x + \frac{b}{2a}\right)^2 = \frac{b^2 - 4ac}{4a^2}$$

$$x + \frac{b}{2a} = \pm\sqrt{\frac{b^2 - 4ac}{4a^2}} \qquad \begin{array}{l}\text{Take the square root of both sides of the}\\ \text{equation.}\end{array}$$

$$x + \frac{b}{2a} = \pm\frac{\sqrt{b^2 - 4ac}}{2a}$$

$$x = \frac{-b}{2a} \pm \frac{\sqrt{b^2 - 4ac}}{2a} \qquad \text{Subtract } b/2a \text{ from both sides of the equation.}$$

$$x = \frac{-b \pm \sqrt{b^2 - 4ac}}{2a} \qquad \text{The quadratic formula.}$$

▶ **2**

To Solve a Quadratic Equation by the Quadratic Formula

1. Write the quadratic equation in standard form, $ax^2 + bx + c = 0$, and determine the numerical values for a, b, and c.
2. Substitute the values for a, b, and c in the quadratic formula and then evaluate the formula to obtain the solution.

The Quadratic Formula

$$x = \frac{-b \pm \sqrt{b^2 - 4ac}}{2a}$$

EXAMPLE 1 Solve the equation $x^2 + 2x - 8 = 0$ using the quadratic formula.

Solution: $a = 1, b = 2, c = -8$.

$$x = \frac{-b \pm \sqrt{b^2 - 4ac}}{2a}$$

$$x = \frac{-(2) \pm \sqrt{(2)^2 - 4(1)(-8)}}{2(1)}$$

$$= \frac{-2 \pm \sqrt{4 + 32}}{2}$$

$$= \frac{-2 \pm \sqrt{36}}{2}$$

$$= \frac{-2 \pm 6}{2}$$

$$x = \frac{-2 + 6}{2} \quad \text{or} \quad x = \frac{-2 - 6}{2}$$

$$= \frac{4}{2} = 2 \qquad\qquad = \frac{-8}{2} = -4$$

Both 2 and -4 are solutions to the equation.

COMMON STUDENT ERROR

The entire numerator of the quadratic formula must be divided by $2a$.

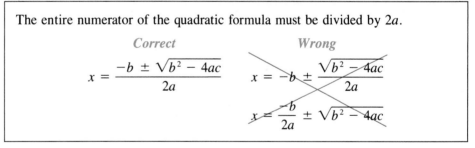

Correct

$$x = \frac{-b \pm \sqrt{b^2 - 4ac}}{2a}$$

Wrong

$$x = -b \pm \frac{\sqrt{b^2 - 4ac}}{2a}$$

$$x = \frac{-b}{2a} \pm \sqrt{b^2 - 4ac}$$

EXAMPLE 2 Solve the equation $2x^2 + 4x - 5 = 0$ using the quadratic formula.

Solution: $a = 2, b = 4, c = -5$.

$$x = \frac{-b \pm \sqrt{b^2 - 4ac}}{2a}$$

$$x = \frac{-4 \pm \sqrt{(4)^2 - 4(2)(-5)}}{2(2)} = \frac{-4 \pm \sqrt{56}}{4} = \frac{-4 \pm 2\sqrt{14}}{4}$$

Now factor out a 2 from both terms in the numerator; then divide out the common factor.

$$x = \frac{\overset{1}{2}(-2 \pm \sqrt{14})}{\underset{2}{\cancel{4}}} = \frac{-2 \pm \sqrt{14}}{2}$$

Thus, the solutions are $\dfrac{-2 + \sqrt{14}}{2}$ and $\dfrac{-2 - \sqrt{14}}{2}$.

COMMON STUDENT ERROR

Many students use the quadratic formula correctly until the last step, where they make an error. Below are illustrated both the correct and incorrect procedures for simplifying an answer.

When *both* terms in the numerator *and* the denominator have a common factor, that common factor may be divided out, as follows:

Correct

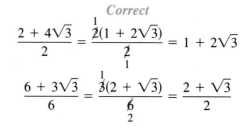

Below are some common errors. Study them carefully so you will not make them. Can you explain why each of the following procedures is wrong?

Wrong

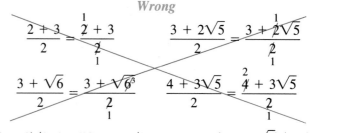

Note that $(2 + 3)/2$ simplifies to $5/2$. However, $(3 + 2\sqrt{5})/2$, $(3 + \sqrt{6})/2$, and $(4 + 3\sqrt{5})/2$ cannot be simplified any further.

EXAMPLE 3   Solve the quadratic equation $-2p^2 - 5p = 6$.

Solution: Do not let the change in variable worry you. The quadratic formula is used exactly the same way as when x is the variable.

$$-2p^2 - 5p - 6 = 0$$

$$a = -2, \qquad b = -5 \qquad c = -6$$

$$p = \frac{-b \pm \sqrt{b^2 - 4ac}}{2a}$$

$$= \frac{-(-5) \pm \sqrt{(-5)^2 - 4(-2)(-6)}}{2(-2)}$$

$$= \frac{5 \pm \sqrt{25 - 48}}{-4}$$

$$= \frac{5 \pm \sqrt{-23}}{-4}$$

$$= \frac{5 \pm i\sqrt{23}}{-4}$$

$$p = \frac{5 + i\sqrt{23}}{-4} \quad \text{or} \quad p = \frac{5 - i\sqrt{23}}{-4}$$

$$= \frac{-5 - i\sqrt{23}}{4} \qquad\qquad = \frac{-5 + i\sqrt{23}}{4}$$

The solutions are $\dfrac{-5 + i\sqrt{23}}{4}$ and $\dfrac{-5 - i\sqrt{23}}{4}$. Note that neither solution is a real solution. ■

EXAMPLE 4   Solve the equation $x^2 + \frac{2}{5}x - \frac{1}{3} = 0$ using the quadratic formula.

Solution: We could solve this equation using the quadratic formula with $a = 1$, $b = \frac{2}{5}$, and $c = -\frac{1}{3}$.

When a quadratic equation contains fractions, it is generally easier to begin by multiplying both sides of the equation by the least common denominator. In this example, the least common denominator is 15.

$$15\left(x^2 + \frac{2}{5}x - \frac{1}{3}\right) = 15(0)$$

$$15x^2 + 6x - 5 = 0$$

Now use the quadratic formula with $a = 15$, $b = 6$, $c = -5$.

$$x = \frac{-b \pm \sqrt{b^2 - 4ac}}{2a}$$

$$= \frac{-6 \pm \sqrt{6^2 - 4(15)(-5)}}{2(15)}$$

$$= \frac{-6 \pm \sqrt{336}}{30}$$

$$= \frac{-6 \pm \sqrt{16}\sqrt{21}}{30}$$

$$= \frac{-6 \pm 4\sqrt{21}}{30}$$

$$= \frac{\overset{1}{\cancel{2}}(-3 \pm 2\sqrt{21})}{\underset{15}{\cancel{30}}}$$

$$= \frac{-3 \pm 2\sqrt{21}}{15}$$

The solutions are $\dfrac{-3 + 2\sqrt{21}}{15}$ and $\dfrac{-3 - 2\sqrt{21}}{15}$. The same solution could be obtained using the fractional values stated earlier. ■

When you are given a quadratic equation where all the numerical coefficients have a common factor, you should factor the quadratic equation before using the quadratic formula. For example, consider the equation $15x^2 + 30x - 45 = 0$. In this equation, $a = 15$, $b = 30$, and $c = -45$. If we use the quadratic formula, we would eventually obtain $x = 1$ and $x = -3$ as solutions. By factoring the equation before using the formula, we get

$$15x^2 + 30x - 45 = 0$$

$$15(x^2 + 2x - 3) = 0$$

If we consider $x^2 + 2x - 3 = 0$, then $a = 1$, $b = 2$, and $c = -3$. If we use these new values of a, b, and c in the quadratic formula, we will obtain the identical solution, $x = 1$ and $x = -3$. However, the calculations with these smaller values of a, b, and c are greatly simplified. Solve both equations now using the quadratic formula to convince yourself. Why can this procedure always be used to obtain the correct answer?

▶ **3**   The expression under the square root sign in the quadratic formula is called the **discriminant.**

$$\underbrace{b^2 - 4ac}_{\text{discriminant}}$$

The discriminant gives the number and type of solutions to a quadratic equation.

Solutions of a Quadratic Equation

For a quadratic equation of the form $ax^2 + bx + c = 0$, $a \neq 0$:
If $b^2 - 4ac > 0$, then the quadratic equation has two distinct real solutions.
If $b^2 - 4ac = 0$, then the quadratic equation has a single real solution (also called a double root).
If $b^2 - 4ac < 0$, then the quadratic equation has no real solution.

EXAMPLE 5   (a) Find the discriminant of the equation $x^2 - 8x + 16 = 0$.
(b) Use the quadratic formula to find the solution.

Solution: (a) $a = 1$, $b = -8$, $c = 16$.

$$b^2 - 4ac = (-8)^2 - 4(1)(16)$$
$$= 64 - 64 = 0$$

Since the discriminant equals zero, there is a single real solution.

(b) $x = \dfrac{-b \pm \sqrt{b^2 - 4ac}}{2a}$

$= \dfrac{-(-8) \pm \sqrt{0}}{2(1)} = \dfrac{8 \pm 0}{2} = \dfrac{8}{2} = 4$

The only solution is 4.   ■

EXAMPLE 6   Without actually finding the solutions, determine if the following equations have two distinct real solutions, a single real solution, or no real solution.
(a) $2x^2 - 4x + 6 = 0$
(b) $x^2 - 5x - 8 = 0$
(c) $4x^2 - 12x = -9$

Solution: We use the discriminant of the quadratic formula to answer these questions.
(a) $b^2 - 4ac = (-4)^2 - 4(2)(6) = 16 - 48 = -32$
Since the discriminant is negative, this equation has no real solution.
(b) $b^2 - 4ac = (-5)^2 - 4(1)(-8) = 25 + 32 = 57$
Since the discriminant is positive, this equation has two distinct real solutions.

(c) First, rewrite $4x^2 - 12x = -9$ as $4x^2 - 12x + 9 = 0$.

$$b^2 - 4ac = (-12)^2 - 4(4)(9) = 144 - 144 = 0$$

Since the discriminant is zero, this equation has a single real solution.

▸**4** We have already seen many applications of quadratic equations. In Sections 5.7, 6.5, 7.5, 7.6, 8.6, 8.7, and 9.1, we solved application problems whose equations were quadratic. Any of those problems could have been solved using the quadratic formula rather than by factoring or completing the square. Now we will look at a few more of the many types of application problems that when expressed as equations will be quadratic equations.

An important formula in physics is $h = \frac{1}{2}at^2 + v_0t + h_0$. When an object is projected upward from an initial height of h_0, with an initial velocity of v_0, this formula can be used to find the height of the object above the ground, h, at any time t. The a in the formula is the acceleration of gravity. The acceleration of gravity will always be a negative number. Since the acceleration of earth's gravity is -32 ft/sec^2, we use -32 for a in the formula when discussing the Earth. This formula can also be used in discussing objects on the moon and other planets, but the value of a in the formula will need to change for each planetary body.

EXAMPLE 7 Jennifer is standing on top of a building and throws a ball upward from a height of 60 feet with an initial velocity of 30 feet per second. Use the formula $h = \frac{1}{2}at^2 + v_0t + h_0$ to determine:

(a) how long after the ball is thrown, to the nearest tenth of a second, will the ball be 25 feet from the ground?

(b) how long after the ball is thrown will the ball strike the ground?

Solution: (a) We will illustrate this problem with a diagram (see Figure 9.3). Here $a = -32$, $v_0 = 30$, and $h_0 = 60$. We are asked to find the time, t, it takes for the ball to reach a height, h, of 25 feet above the ground. We substitute these values into the formula and then solve for t.

$$h = \frac{1}{2}at^2 + v_0t + h_0$$

$$25 = \frac{1}{2}(-32)t^2 + 30t + 60$$

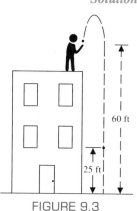

FIGURE 9.3

Now write the quadratic equation in standard form and solve for t by using the quadratic formula.

$$0 = -16t^2 + 30t + 35$$

$$\text{or} \quad -16t^2 + 30t + 35 = 0$$

$$a = -16, \qquad b = 30, \qquad c = 35$$

$$t = \frac{-b \pm \sqrt{b^2 - 4ac}}{2a} = \frac{-30 \pm \sqrt{(30)^2 - 4(-16)(35)}}{2(-16)} = \frac{-30 \pm \sqrt{3140}}{-32}$$

$$t = \frac{-30 + \sqrt{3140}}{-32} \quad \text{or} \quad t = \frac{-30 - \sqrt{3140}}{32}$$

$$\approx -0.8 \qquad\qquad\qquad \approx 2.7$$

Since time cannot be negative, the only acceptable answer is 2.7 seconds. Thus, about 2.7 seconds after the ball is thrown, it will be 25 feet above the ground.

(b) We wish to find the time at which the ball strikes the ground. When the ball strikes the ground, its distance above the ground is 0. Thus we substitute $h = 0$ into the formula and solve.

$$h = \frac{1}{2}at^2 + v_0t + h_0$$

$$0 = \frac{1}{2}(-32)t^2 + 30t + 60$$

$$0 = -16t^2 + 30t + 60$$

$$a = -16, \qquad b = 30, \qquad c = 60$$

$$t = \frac{-b \pm \sqrt{b^2 - 4ac}}{2a} = \frac{-30 \pm \sqrt{(30)^2 - 4(-16)(60)}}{2(-16)} = \frac{-30 \pm \sqrt{4740}}{-32}$$

$$t = \frac{-30 + \sqrt{4740}}{-32} \quad \text{or} \quad t = \frac{-30 - \sqrt{4740}}{-32}$$

$$t \approx 3.1 \qquad\qquad t \approx -1.2$$

Since time cannot be negative, the answer is 3.1 seconds. Thus the ball will strike the ground approximately 3.1 seconds after the ball is thrown.   ∎

In example 7a if we find the values of $(-30 \pm \sqrt{3140})/2$ using a calculator, we do not round off the value we get when we evaluate $\sqrt{3140}$. The only answer we round is the final answer.

If you check an answer that has been rounded back in the original equation, you may find that the two sides of the equation do not give exactly the same answers. This slight difference is due to the rounding of the final answer. When the check does not result in an exactly true match, we say there is a **round-off error.** In example 7a, if you substitute 2.7 for t in the equation $25 = \frac{1}{2}(-32)t^2 + 30t + 60$, you will see that the right side of the equation has a value of 24.36, which is close to the 25 on the left side of the equation. This small difference is due to round-off error.

In Example 7 we neglected air resistance. However, air resistance does not play a significant part in the accuracy of the answer until the object reaches a much greater velocity. Notice that we used a positive number to represent the initial velocity. If the object is projected upward, the initial velocity is considered positive, and if the object is projected downward, the initial velocity is considered negative. If an object is projected upward *from the ground,* then h_0 has a value of 0 and the formula simplifies to $h = \frac{1}{2}at^2 + v_0t$. We will discuss additional problems similar to this one in Section 9.4.

Now let us look at a rate problem. Rate problems were introduced and discussed in Section 2.4. They were also discussed in Sections 4.3 and 7.6.

EXAMPLE 8   In 5 hours the Dechs traveled 12 miles down river in their motorboat and then returned. If the rate of the river's current is 2 miles per hour, find the rate (or speed) of the motorboat in still water.

Solution: We are asked to find the rate of the boat in still water. Let $r =$ the rate of the boat in still water. We know that the total time of the trip is 5 hours. Thus the time down river plus the time up river must sum to 5 hours. Since distance = rate · time, we can find the time by dividing the distance by the rate.

	Distance	Rate	Time
Down river (with current)	12	$r + 2$	$\dfrac{12}{r + 2}$
Up river (against current)	12	$r - 2$	$\dfrac{12}{r - 2}$

Time down river + Time up river = Total time

$$\frac{12}{r + 2} + \frac{12}{r - 2} = 5$$

$$(r + 2)(r - 2)\left(\frac{12}{r + 2} + \frac{12}{r - 2}\right) = (r + 2)(r - 2)(5)$$ Multiply by the LCD to remove fractions.

$$(r + 2)(r - 2)\left(\frac{12}{r + 2}\right) + (r + 2)(r - 2)\left(\frac{12}{r - 2}\right) = (r + 2)(r - 2)(5)$$ Distributive property.

$$12(r - 2) + 12(r + 2) = 5(r^2 - 4)$$
$$12r - 24 + 12r + 24 = 5r^2 - 20$$
$$24r = 5r^2 - 20$$
$$\text{or } 5r^2 - 24r - 20 = 0$$

Using the quadratic formula with $a = 5$, $b = -24$, and $c = -20$, we obtain

$$r = \frac{24 \pm \sqrt{976}}{10}$$

$$r \approx 5.5 \quad \text{or} \quad r \approx -0.7$$

Since the rate will not be negative, the rate or speed of the boat in still water is about 5.5 miles per hour. ∎

Notice that in working with real-life situations most answers do not come out to be integral values.

Let us do one example involving a work problem. Work problems were discussed in Section 7.6. You may wish to review that section before reading the next example.

EXAMPLE 9 Two sump pumps are emptying a flooded basement. Together they can empty the basement in 6 hours. One of the pumps has a slightly larger horsepower and could do the job by itself in 2 hours less time than the other pump could if it were working alone. How long would it take each pump to complete the job alone?

Solution: Recall from Section 7.6 that the rate of work multiplied by the time worked gives the part of the task completed.

Let t = number of hours for slower pump to complete the job by itself
Then $t - 2$ = number of hours for the faster pump to complete the job by itself

	Rate of Work	Time Worked	Part of Task Completed
Slower pump	$\dfrac{1}{t}$	6	$\dfrac{6}{t}$
Faster pump	$\dfrac{1}{t-2}$	6	$\dfrac{6}{t-2}$

$$\left(\begin{array}{c}\text{Part of task}\\\text{by slower pump}\end{array}\right) + \left(\begin{array}{c}\text{part of task}\\\text{by faster pump}\end{array}\right) = 1$$

$$\frac{6}{t} + \frac{6}{t-2} = 1$$

$$t(t-2)\left(\frac{6}{t} + \frac{6}{t-2}\right) = t(t-2)(1) \qquad \begin{array}{l}\text{Multiply both sides of}\\\text{equation by the LCD}\\ t(t-2).\end{array}$$

$$t(t-2)\left(\frac{6}{t}\right) + t(t-2)\left(\frac{6}{t-2}\right) = t^2 - 2t \qquad \begin{array}{l}\text{Distributive}\\\text{property.}\end{array}$$

$$6(t-2) + 6t = t^2 - 2t$$

$$6t - 12 + 6t = t^2 - 2t$$

$$t^2 - 14t + 12 = 0$$

Using the quadratic formula, we obtain the following results:

$$t = \frac{14 \pm \sqrt{148}}{2}$$

$$t \approx 13.1 \quad \text{or} \quad t \approx 0.9$$

Both 13.1 and 0.9 satisfy the equation $\dfrac{6}{x} + \dfrac{6}{x-2} = 1$ (with some round-off error).
However, if we use 0.9 as an answer, then the faster pump could complete the task in a negative amount of time ($x - 2 = 0.9 - 2 = -1.1$ hours), which is not possible. Therefore, 0.9 hour is not a possible solution. The only possible answer is 13.1 hours. The slower pump takes approximately 13.1 hours by itself, and the faster pump takes approximately $13.1 - 2$ or 11.1 hours by itself to empty the basement.

Exercise Set 9.2

Determine whether each equation has two distinct real solutions, a single real solution, or no real solution.

1. $x^2 + 4x - 3 = 0$

2. $3x^2 + x + 3 = 0$

3. $2x^2 - 4x + 7 = 0$

4. $-2x^2 + x - 8 = 0$

5. $5x^2 + 3x - 7 = 0$

6. $2x^2 = 16x - 32$

7. $4x^2 - 24x = -36$

8. $5x^2 - 4x = 7$

9. $2x^2 - 8x + 5 = 0$

10. $3x^2 - 5x - 9 = 0$

11. $-3x^2 + 5x - 8 = 0$

12. $x^2 + 4x - 8 = 0$

13. $x^2 + 7x - 3 = 0$

14. $2x^2 - 6x + 9 = 0$

15. $4x^2 - 9 = 0$

16. $6x^2 - 5x = 0$

Use the quadratic formula to solve the equation.

17. $x^2 - 3x + 2 = 0$

18. $x^2 + 6x + 8 = 0$

19. $x^2 - 9x + 20 = 0$

20. $x^2 - 3x - 10 = 0$

21. $x^2 + 5x - 24 = 0$

22. $x^2 - 6x = -5$

23. $x^2 = 13x - 36$

24. $x^2 - 36 = 0$

25. $x^2 - 25 = 0$

26. $x^2 - 6x = 0$

27. $x^2 - 3x = 0$

28. $z^2 - 17z + 72 = 0$

29. $4p^2 - 7p + 13 = 0$

30. $2x^2 - 3x + 2 = 0$

31. $2y^2 - 7y + 4 = 0$

32. $2x^2 - 7x = -5$

33. $6x^2 = -x + 1$

34. $4r^2 + r - 3 = 0$

35. $2x^2 - 4x - 1 = 0$

36. $3w^2 - 4w + 5 = 0$

37. $4s^2 - 8s + 6 = 0$

38. $x^2 - 7x + 3 = 0$

39. $4x^2 = x + 5$

40. $x^2 - 2x - 1 = 0$

41. $6x^2 - 21x = 27$

42. $-x^2 + 2x + 15 = 0$

43. $-2x^2 + 11x - 15 = 0$

44. $-6x^2 + 5x - 9 = 0$

45. $-2x^2 - x - 3 = 0$

46. $9x^2 + 6x + 1 = 0$

47. $2x^2 + 6x = 0$

48. $3x^2 - 5x = 0$

49. $4x^2 - 7 = 0$

50. $-2x^2 = -10$

51. $\dfrac{1}{2}x^2 + 2x + \dfrac{2}{3} = 0$

52. $x^2 - \dfrac{x}{5} - \dfrac{1}{3} = 0$

53. $-x^2 + \dfrac{11}{3}x + \dfrac{10}{3} = 0$

54. $x^2 - \dfrac{7}{6}x + \dfrac{2}{3} = 0$

55. $0.1x^2 + 0.6x - 1.2 = 0$

56. $-2.3x^2 + 5.6x + 0.4 = 0$

57. $-1.62x^2 - 0.94x + 4.85 = 0$

58. $1.74x^2 - 2.04x + 6.2 = 0$

In exercises 59 through 76 use a calculator or Appendix C as needed to give the answer in decimal form.

59. Twice the square of a positive number increased by 3 times the number is 14. Find the number.

60. Three times the square of a positive number decreased by twice the number is 21. Find the number.

61. The length of a rectangular garden is 2 feet less than 3 times its width. Find the length and width if the area of the garden is 21 square feet.

62. Lora Moore wishes to form a rectangular region along a river bank by constructing fencing as illustrated in the diagram on the right. If she has only 400 feet of fencing and wishes to enclose an area of 15,000 square feet, find the dimensions of the rectangular region.

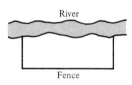

63. John Williams, a professional photographer, has a 6-inch by 8-inch photo. He wishes to reduce the photo by the same amount on each side so that the resulting photo will have half the area of the original photo. By how much will he have to reduce the length of each side?

In Exercises 64 through 67, use the equation $h = \frac{1}{2}at^2 + v_0t + h_0$. Refer to Example 7.

64. A horseshoe is projected upward from an initial height of 80 feet with an initial velocity of 60 feet per second. How long after the horseshoe is projected upward (a) will it be 20 feet from the ground, and (b) will it strike the ground?

65. A sandbag is thrown *downward* from a helicopter 120

feet above the ground with an initial velocity of 20 feet per second. How long after the sandbag is thrown will it strike the ground?

66. Jim is on the fourth floor of an 8-story building and Courtney is on the roof. Jim is 60 feet above the ground while Courtney is 120 feet above the ground. (a) If Jim throws a rock upward with an initial velocity of 100 feet per second and at the same time Courtney throws a rock upward at 60 feet per second, whose rock will hit the ground first? (b) Will the rocks ever be at the same distance above the ground? If so, at what times.

67. Gravity on the moon is about one-sixth of that on the Earth. Suppose Neil Armstrong is standing on a hill on the moon 60 feet above the ground below. If he jumps upward with a velocity of 40 feet per second, how long will it take for him to land on the ground below the hill?

68. Mr. Winkle jogs 6 miles, and then turns around and jogs back to his starting point. The first part of his jog he was going mostly uphill, so his speed was 2 miles per hour slower than his speed returning. If the total time he spent jogging was $1\frac{3}{4}$ hours, find his rate going and his rate returning.

69. A truck driver is transporting a heavy load from Detroit, Michigan, to Chicago, Illinois. On his return trip to Detroit, since his truck is lighter, he averages 10 miles per hour faster than his trip down. If the distance traveled each way is 300 miles and the total time spent driving is 11 hours, find his speed going and coming.

70. Two mechanics take 6 hours to rebuild an engine when they work together. If they worked alone, the more experienced mechanic could complete the job 1 hour faster than the less experienced mechanic. How long would it take each of them to rebuild the engine working alone?

71. A small electric heater requires 6 minutes longer to raise the temperature in the unheated garage to a comfortable level than does a larger electric heater. Together the two heaters can raise the garage temperature to a comfortable level in 42 minutes. How long would it take each heater by itself to raise the temperature in the garage to a comfortable level?

72. The temperature, T, in degrees Fahrenheit in a car's radiator during the first 4 minutes of driving is a function of time, t. The temperature can be found by the formula $T = 6.2t^2 + 12t + 32$, $0 \le t \le 4$. (a) What is the car's radiator temperature at the instant the car is turned on? (b) What is the car's radiator temperature after the car has been driven for 1 minute? (c) How long after the car has begun operating will the car's radiator temperature reach 120°F?

73. A video store's weekly profit, P, in thousands of dollars is a function of the rental price of the tapes, t. The profit equation is $P = 0.2t^2 + 1.5t - 1.2$, $0 \le t \le 5$.

(a) What is the store's weekly profit or loss if they charge $1 per tape? (b) What is the weekly profit if they charge $5 per tape? (c) At what tape rental price will their weekly profit be 1.6 thousand dollars?

74. The cost, C, in thousands of dollars of a ranch house in Roanoke, Virginia, is a function of the number of square feet of the house, s. The cost of a house can be approximated by the formula

$$C = -0.01s^2 + 80s + 20,000, \qquad 1200 \le s \le 4000$$

(a) Find the cost of a 1500-square-foot house.

(b) What square footage house can Mr. Dodge purchase if he has $150,000 to spend on a house?

75. At a local college, records have shown that the average person's grade point average, G, is a function of the number of hours they study and do homework per week, h. The grade point average can be estimated by the equation $G = 0.01h^2 + 0.2h + 1.2$, $0 \le h \le 8$. (a) What is the average GPA of the average student who studies for 0 hours a week? (b) What is the average GPA of a student who studies 4 hours per week? (c) To obtain a 3.2 GPA, how many hours per week would the average student need to study?

76. The roller coaster at Busch Gardens in Williamsburg, Virginia, has a number of steep drops. One of its drops has a vertical distance of 62 feet. The speed of the last car, in feet per second, at a time t seconds after it has begun its drop can be calculated by the formula $s = 6.74t + 2.3$, $0 \le t \le 4$. The height of the last car from the bottom of the drop at any time t seconds after it has begun its drop can be found by the formula $h = -3.3t^2 - 2.3t + 62$, $0 \le t \le 4$. (a) Find the time it takes for the last car to travel from the top of the drop to the bottom of the drop, and (b) find the speed of the last car when it reaches the bottom of the drop.

77. Consider the two equations $-6x^2 + \frac{1}{2}x - 5 = 0$ and $6x^2 - \frac{1}{2}x + 5 = 0$? Must the solution to these two equations be the same? Explain your answer.

78. (a) What is the discriminant of a quadratic equation of the form $ax^2 + bx + c = 0$? (b) Write a paragraph or two explaining the relationship between the value of the discriminant and the number of real solutions to a quadratic equation. In your paragraph, explain *why* the value of the discriminant determines the number of real solutions.

Cumulative Review Exercises

[8.3] **79.** Simplify $\sqrt[5]{64x^9y^{12}z^{20}}$.

80. Simplify $\sqrt[3]{4x^2y^8}(\sqrt[3]{2x^5y^4} + \sqrt[3]{6xy})$.

[8.4] **81.** Simplify $\dfrac{x + \sqrt{y}}{x - \sqrt{y}}$.

[8.6] **82.** Solve the equation $\sqrt{2x + 4} - 1 = \sqrt{x + 3}$.

JUST FOR FUN

Solve each of the following using the quadratic formula.

1. $x^2 - \sqrt{5}x - 10 = 0$.
2. $x^2 + 5\sqrt{6}x + 36 = 0$.
3. $a^4 - 5a^2 + 4 = 0$.
4. By increasing the speed of his jet by 200 miles per hour, Jack can decrease the time needed to cover 4000 miles by 1 hour. What is its speed?
5. A metal cube expands when heated. If each edge is increased 0.20 millimeter after being heated and the total

volume increases by 6 cubic millimeters, find the original length of a side of the cube.

Use the quadratic formula to solve the following equations for x.

6. $x^2 + x - y = 0$
7. $x^2 + 3yx + 4 = 0$
8. $yx^2 + 3x - 5 = 0$

9.3

Quadratic and Other Inequalities in One Variable

▶ 1 Solve quadratic inequalities using a sign graph.
▶ 2 Solve other types of inequalities.

In Section 2.5 we discussed linear inequalities in one variable. Now we will discuss quadratic inequalities in one variable. This Section will also serve to review both interval notation and set builder notation, which were introduced in Section 2.5.

When the equal sign in a quadratic equation of the form $ax^2 + bx + c = 0$ is replaced by an inequality sign, we get a quadratic inequality.

Examples of Quadratic Inequalities

$$x^2 + x - 12 \geq 0, \qquad 2x^2 - 9x - 5 < 0$$

The **solution to a quadratic inequality** is the set of all values that makes the inequality a true statement. For example, if we substitute 5 for x in $x^2 + x - 12 > 0$, we obtain

$$x^2 + x - 12 > 0$$
$$5^2 + 5 - 12 > 0$$
$$25 + 5 - 12 > 0$$
$$18 > 0 \qquad \text{true}$$

Since the inequality is true when x is 5, 5 satisfies the inequality. However, the solution is not only 5, for there are other values that satisfy (or are solutions to) the inequality. Does 4 satisfy the inequality? Does 2 satisfy the inequality?

▶ 1 A number of methods can be used to find the solution to quadratic inequalities. We will study only the method that uses a **sign graph**. Consider the inequality $x^2 + x - 12 > 0$. The first step in obtaining the solution is to factor the left side of the inequality

$$x^2 + x - 12 > 0$$
$$(x + 4)(x - 3) > 0$$

We must now determine the values of x that make the product of the factors $(x + 4)$ and $(x - 3)$ greater than 0, or positive. The product of two positive numbers is a positive number, and the product of two negative numbers is also a positive number. Therefore, we must find the values of x that make both factors positive numbers or both factors negative numbers. One way to do this is to use number lines. Consider the factor $x - 3$. For values of x less than 3, the value of $x - 3$ will be negative, and for values of x greater than 3, the value of $x - 3$ will be positive. This is illustrated in Fig. 9.4.

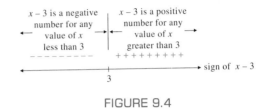

FIGURE 9.4

Now consider the factor $x + 4$. This factor will be negative when x is less than -4 and positive when x is greater than -4; see Fig. 9.5.

FIGURE 9.5

If we draw the two number lines together and draw a dashed vertical line through the values -4 and 3, we get the sign graph in Fig. 9.6. Note that when we place the two lines together the -4 is marked to the left of 3, since -4 is less than 3. Also note that the two vertical lines divide the sign graph into three vertical regions labeled a, b, and c.

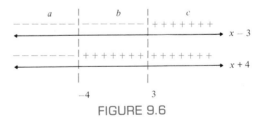

FIGURE 9.6

Next determine the sign of the product of the numbers in each region. If the two signs in a given vertical region are the same, $(-)(-)$ or $(+)(+)$, the product will be positive. If the signs in a given vertical region are different, $(+)(-)$ or $(-)(+)$, the product will be negative (see Fig. 9.7).

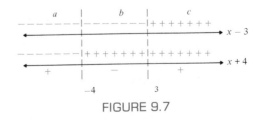

FIGURE 9.7

Since we are solving the inequality $x^2 + x - 12 > 0$, the solutions are the numbers in regions a and c. Any number less than -4 or greater than 3 is a solution to the inequality. As with linear inequalities the solution may be given in several different forms. The solution may be given as an illustration on a number line, in interval notation, or in set builder notation.

The following chart reviews the three methods of illustrating a solution to an inequality.

Solution to an Inequality	Illustrated on the Number Line	Indicated in Interval Notation	Indicated in Set Builder Form
$x < -4$ or $x > 3$	$-4 \quad 3$	$(-\infty, -4) \cup (3, \infty)$	$\{x \mid x < -4$ or $x > 3\}$
$x \leq -4$ or $x \geq 3$	$-4 \quad 3$	$(-\infty, -4] \cup [3, \infty)$	$\{x \mid x \leq -4$ or $x \geq 3\}$
$-4 < x < 3$	$-4 \quad 3$	$(-4, 3)$	$\{x \mid -4 < x < 3\}$
$-4 \leq x \leq 3$	$-4 \quad 3$	$[-4, 3]$	$\{x \mid -4 \leq x \leq 3\}$

Recall from Chapter 1 that the symbol $\cup$ represents the union of two sets.

EXAMPLE 1   Graph the solution to $x^2 + 9x \leq -18$ on the number line.

Solution: First, write the quadratic inequality in the form $x^2 + 9x + 18 \leq 0$; then factor.

$$x^2 + 9x + 18 \leq 0$$
$$(x + 6)(x + 3) \leq 0$$

Now draw a sign graph using one number line for each factor (see Fig. 9.8). Note that the -6 was placed to the left of -3 since $-6 < -3$. Since the inequality we are evaluating is $(x + 6)(x + 3) \leq 0$, the solution will be the numbers in region b, where the product of the factors is negative. The solution is illustrated in Fig. 9.9. Note the darkened dots at -6 and -3. They indicate that the -6 and -3 are a part of the solution.

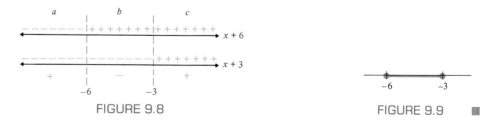

FIGURE 9.8   FIGURE 9.9   ■

EXAMPLE 2   Determine the solution to the inequality $x^2 - 10x + 25 \geq 0$. Indicate the solution (a) on the real number line, (b) in interval notation, and (c) in set builder form.

Solution:

$$x^2 - 10x + 25 \geq 0$$
$$(x - 5)(x - 5) \geq 0$$

or

$$(x - 5)^2 \geq 0$$

This is a special case in which both factors are the same. Since any real number squared will be greater than or equal to 0, the solution to this inequality is all real numbers, $\mathbb{R}$.

(b) $(-\infty, \infty)$ (c) $\{x \mid -\infty < x < \infty\}$ ∎

EXAMPLE 3   Determine the solution to the inequality $6x^2 + 7x - 3 \geq 0$. Express the solution in interval notation.

Solution:
$$6x^2 + 7x - 3 \geq 0$$
$$(2x + 3)(3x - 1) \geq 0$$

To find out where each factor equals 0, we can set each factor equal to 0 and solve.

$$2x + 3 = 0 \qquad 3x - 1 = 0$$
$$2x = -3 \qquad 3x = 1$$
$$x = -3/2 \qquad x = 1/3$$

Now construct a sign graph (Fig. 9.10). The solution is all values less than or equal to $-3/2$ or all values greater than or equal to $1/3$. In interval notation, the solution is $(-\infty, -\frac{3}{2}] \cup [\frac{1}{3}, \infty)$.

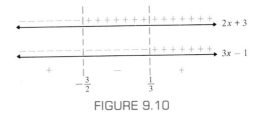

FIGURE 9.10   ∎

▶**2**   This procedure used to solve quadratic inequalities can be used to solve other types of inequalities, as illustrated in the following examples.

EXAMPLE 4   Solve the inequality $(x - 2)(x + 3)(x + 5) < 0$. Illustrate the solution on the number line and give the solution in set builder notation.

Solution: We will again use a sign graph. However, since there are three factors, we will need three number lines; see Fig. 9.11.

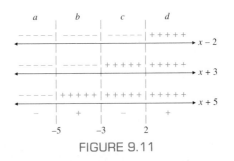

FIGURE 9.11

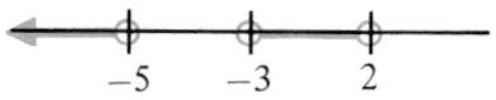

FIGURE 9.12

The products in regions a and c will be negative because the product of an odd number of negative factors is negative. The products in regions b and d will be positive because the product of an even number of negative factors is positive. The solution is all real numbers less than -5 or all real numbers between -3 and 2. The solution is illustrated in Fig. 9.12. The solution in set builder notation is $\{x \mid x < -5$ or $-3 < x < 2\}$. ∎

EXAMPLE 5 Solve the inequality $\dfrac{x + 3}{x - 4} \le 0$ and graph the solution on the number line.

Solution: We will use basically the same procedure as in previous examples. Recall that a quotient of two numbers with like signs is positive. Construct a sign graph (Fig. 9.13).

FIGURE 9.13

FIGURE 9.14

The solution is indicated on the number line in Fig. 9.14. Note that the darkened circle at the 3 indicates that the 3 is part of the solution. When $x = 3$, the quotient is equal to 0. However, since we cannot divide by 0, the 4 is not part of the solution. Thus we place an open dot at 4. ∎

EXAMPLE 6 Solve the inequality $\dfrac{6}{x - 2} > 4$ and give the solution in interval notation.

Solution: Your first reaction might be to multiply both sides of the inequality by $x - 2$. This is wrong. Do you know why? Remember that if we multiply both sides of an inequality by a negative number we must change the sense (or direction) of the equality symbol. Since we do not know the value of x, we do not know whether $x - 2$ represents a positive or negative number. To solve this inequality, subtract 4 from both sides of the inequality. Then rewrite 4 with a denominator of $x - 2$.

$$\frac{6}{x - 2} > 4$$

$$\frac{6}{x - 2} - 4 > 0$$

$$\frac{6}{x - 2} - \frac{4(x - 2)}{x - 2} > 0$$

$$\frac{6 - 4(x - 2)}{x - 2} > 0$$

$$\frac{6 - 4x + 8}{x - 2} > 0$$

$$\frac{-4x + 14}{x - 2} > 0$$

We can now multiply both sides of the inequality by -1 to make the x term in the numerator positive. Remember, when we multiply an inequality by a negative number, we must change the sense of the inequality.

$$-1\left(\frac{-4x + 14}{x - 2}\right) < -1(0)$$

$$\frac{4x - 14}{x - 2} < 0$$

Now determine the solution using a sign graph (Fig. 9.15).

FIGURE 9.15

The solution is the set of values between 2 and $\frac{7}{2}$. The solution in interval notation is $(2, \frac{7}{2})$.

EXAMPLE 7   Determine the solution to $\dfrac{(x - 3)(x + 4)}{x + 1} \geq 0$. Illustrate the solution on the number line and give the solution in interval notation.

Solution: When a problem involves both multiplication and division, any combination of an odd number of negative factors will result in the answer being negative. An even number of negative factors will result in a positive answer. Set up a sign graph with one number line for each factor as illustrated in Fig. 9.16.

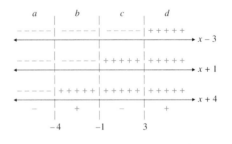

FIGURE 9.16

FIGURE 9.17

The answer is illustrated in Fig. 9.17. Notice -1 is not part of the solution since we cannot divide by 0. In interval notation the solution is $[-4, -1) \cup [3, \infty)$.

Exercise Set 9.3

Solve each inequality and graph the solution on the number line.

1. $x^2 - 3x - 10 \geq 0$ **2.** $x^2 + 8x + 7 < 0$ **3.** $x^2 + 4x > 0$ **4.** $x^2 - 5x \geq 0$

5. $x^2 - 16 < 0$ **6.** $x^2 - 25 \leq 0$ **7.** $x^2 + 9x + 20 \geq 0$ **8.** $x^2 - 13x + 42 > 0$

9. $x^2 - 6x \le 27$

10. $x^2 \ge -9x - 14$

11. $x^2 \le 2x + 35$

12. $x^2 \ge -4x$

13. $x^2 < 36$

14. $x^2 + 10x \le -21$

15. $2x^2 + 10x + 12 > 0$

16. $3x^2 - 21x - 24 \le 0$

17. $2x^2 - 7x - 15 \le 0$

18. $6x^2 > 5x + 6$

19. $4x^2 - 11x \le 20$

20. $5x^2 \le -21x - 4$

21. $2y^2 + y < 15$

22. $3a^2 - 4a + 1 > 0$

23. $3x^2 + x - 10 \ge 0$

24. $2y^2 - 5y + 2 \ge 0$

Solve each inequality and give the solution in interval notation.

25. $(x - 1)(x + 1)(x + 4) > 0$

26. $(x - 3)(x + 2)(x + 5) \le 0$

27. $(x - 4)(x - 1)(x + 3) \le 0$

28. $(2x - 4)(x + 3)(x + 6) > 0$

29. $x(x - 3)(2x + 6) \ge 0$

30. $(x - 3)(x + 4)(x - 2) \le 0$

31. $(2x + 6)(3x - 6)(x + 6) > 0$

32. $(2x - 1)(x + 5)(3x + 6) \ge 0$

33. $(x + 2)(x + 2)(x - 3) \ge 0$

34. $(x + 3)^2(x - 5) \le 0$

35. $(x + 3)(x + 3)(x - 4) < 0$

36. $x(x - 5)(x + 5) > 0$

Solve each inequality and give the solution in set builder notation.

37. $\dfrac{x + 3}{x - 1} > 0$

38. $\dfrac{x - 5}{x + 2} < 0$

39. $\dfrac{x - 4}{x - 1} \le 0$

40. $\dfrac{x + 6}{x + 2} \ge 0$

41. $\dfrac{2x - 4}{x - 1} < 0$

42. $\dfrac{3x + 6}{x + 4} \ge 0$

43. $\dfrac{3x + 6}{2x - 1} \ge 0$

44. $\dfrac{3x + 4}{2x - 1} < 0$

45. $\dfrac{x + 4}{x - 4} \le 0$

46. $\dfrac{x + 3}{x} \ge 0$

47. $\dfrac{4x - 2}{2x - 4} > 0$

48. $\dfrac{3x + 5}{x - 2} \le 0$

Solve each inequality and give the solution in both interval notation and set builder notation.

49. $\dfrac{(x + 2)(x - 4)}{x + 6} < 0$

50. $\dfrac{(x - 3)(x - 6)}{x + 4} \ge 0$

51. $\dfrac{(x - 6)(x - 1)}{x - 3} \ge 0$

52. $\dfrac{x + 6}{(x - 2)(x + 4)} > 0$

53. $\dfrac{x - 6}{(x + 4)(x - 1)} \le 0$

54. $\dfrac{x}{(x + 3)(x - 3)} \le 0$

55. $\dfrac{(x - 3)(2x + 5)}{(x - 6)} > 0$

56. $\dfrac{x(x - 3)}{2x + 6} < 0$

57. $\dfrac{(x + 2)(2x - 3)}{x} \ge 0$

58. $\dfrac{(x - 4)(3x - 2)}{x + 2} \le 0$

Solve each inequality. Graph the solution on the number line.

59. $\dfrac{2}{x - 3} \ge -1$

60. $\dfrac{3}{x - 1} \le -1$

61. $\dfrac{4}{x - 2} \ge 2$

62. $\dfrac{2}{2a - 1} > 2$

63. $\dfrac{2x - 5}{x - 4} \le 1$

64. $\dfrac{2x}{x + 1} > 1$

65. $\dfrac{w}{3w - 2} > -2$

66. $\dfrac{x - 1}{2x + 6} \le -3$

67. $\dfrac{x}{3x - 1} < -1$

68. $\dfrac{4x - 5}{x + 3} < 3$

69. $\dfrac{x + 8}{x + 2} > 1$

70. $\dfrac{4x + 2}{2x - 3} \ge 2$

71. What is the solution to the inequality $(x + 3)^2(x - 1)^2 \geq 0$? Explain your answer.

72. What is the solution to the inequality $x^2(x - 3)^2(x + 4)^2 \leq 0$? Explain your answer.

73. What is the solution to the inequality $x^2(x - 3)^2(x + 4)^2 < 0$? Explain your answer.

74. What is the solution to the inequality $\dfrac{x^2}{(x + 1)^2} \geq 0$? Explain your answer.

75. What is the solution to the inequality $\dfrac{x^2}{(x - 3)^2} > 0$? Explain your answer.

Cumulative Review Exercises

In Exercises 76 and 77 find the domain of the function.

[7.1] **76.** $f(x) = \dfrac{3}{x^2 - 4}$.

[8.9] **77.** $f(x) = \sqrt{4 - x}$.

[7.4] **78.** Simplify $\dfrac{ab^{-2} - a^{-1}b}{a^{-2} + ab^{-1}}$.

[8.8] **79.** Multiply $(\sqrt{-8} + \sqrt{2})(\sqrt{-2} - \sqrt{8})$..

JUST FOR FUN

Solve the following inequalities. Graph the solution on the number line.

1. $(x + 1)(x - 3)(x + 5)(x + 9) \geq 0$

2. $\dfrac{(x - 4)(x + 2)}{x(x + 6)} \geq 0$

3. $x^2 - x + 1 > 0$

9.4

Quadratic Functions

▶ **1** Identify quadratic functions.

▶ **2** Derive the equation for the axis of symmetry, and find the vertex of a parabola.

▶ **3** Find the roots of a quadratic equation.

▶ **4** Sketch graphs of quadratic equations.

▶ **5** Write a quadratic equation, given its roots.

▶ **6** Apply quadratic equations to some practical situations.

▶ **1**   We introduced quadratic functions in Section 5.8. In this section we study quadratic functions in more depth.

Quadratic Function
$$f(x) = ax^2 + bx + c, \qquad a \neq 0$$

Since y may replace $f(x)$, equations of the form $y = ax^2 + bx + c$, $a \neq 0$, are also quadratic functions. Recall from Section 5.8 that every quadratic equation of the form above, when graphed, will be a **parabola.** The graph of $y = ax^2 + bx + c$ will have one of the general forms indicated in Fig. 9.18. Note that both parabolas are functions since they pass the vertical line test.

When a quadratic equation is in the form given above, the sign of the numerical

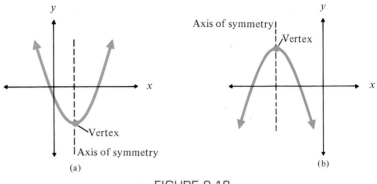

FIGURE 9.18

coefficient of the squared term, a, will determine whether the parabola will open upward (Fig. 9.18a) or downward (Fig. 9.18b). When the coefficient of the squared term is positive, the parabola will open upward, and when the coefficient of the squared term is negative, the parabola will open downward. The **vertex** is the lowest point on a parabola that opens upward and the highest point on a parabola that opens downward.

▶**2**   Graphs of quadratic equations of the form $y = ax^2 + bx + c$ will have **symmetry** about a line through the vertex. This means that if we fold the paper along this imaginary line, called the **axis of symmetry,** the two halves of the graph would coincide.

Axis of Symmetry of a Parabola

For an equation of the form $y = ax^2 + bx + c, a \neq 0$, the axis of symmetry of the parabola is

$$x = \frac{-b}{2a}$$

At this point we have the knowledge to see where this formula comes from. We will derive the formula $x = \dfrac{-b}{2a}$ and the coordinates of the vertex of a parabola by beginning with a quadratic equation in standard form and completing the square on the first two terms.

$$y = ax^2 + bx + c$$

$$y = a\left(x^2 + \frac{b}{a}x\right) + c$$

$$y = a\left[x^2 + \frac{b}{a}x + \left(\frac{b}{2a}\right)^2\right] + c - a\left(\frac{b}{2a}\right)^2$$

$$y = a\left(x + \frac{b}{2a}\right)^2 + c - a\left(\frac{b^2}{4a^2}\right)$$

$$y = a\left(x + \frac{b}{2a}\right)^2 + c\left(\frac{4a}{4a}\right) - \frac{b^2}{4a}$$

$$y = a\left(x + \frac{b}{2a}\right)^2 + \frac{4ac}{4a} - \frac{b^2}{4a}$$

$$y = a\left(x + \frac{b}{2a}\right)^2 + \frac{4ac - b^2}{4a}$$

Since a, b, and c are all constants, the expression $\dfrac{4ac - b^2}{4a}$ must be a constant. The expression $\left(x + \dfrac{b}{2a}\right)^2$ will always be a number greater than or equal to 0. The minimum value of $\left(x + \dfrac{b}{2a}\right)^2$ will be 0 when $x = \dfrac{-b}{2a}$. If a is positive, then $y = a\left(x + \dfrac{b}{2a}\right)^2 + \dfrac{4ac - b^2}{4a}$ will have a minimum of $\dfrac{4ac - b^2}{4a}$ when $x = \dfrac{-b}{2a}$. Thus the axis of symmetry, upon which the vertex lies, is $x = -\dfrac{b}{2a}$.

By observing the equation above, we can also see that the coordinates of the vertex of a parabola are as follows.

> **Coordinates of the Vertex of a Parabola**
>
> For an equation of the form $y = ax^2 + bx + c$, $a \ne 0$, the coordinates of the vertex of the parabola are
>
> $$\left(\dfrac{-b}{2a}, \dfrac{4ac - b^2}{4a}\right)$$

The y value of the vertex can also be found by evaluating the function at $\dfrac{-b}{2a}$. Thus the coordinates of a parabola may also be represented as

$$\left(\dfrac{-b}{2a}, f\left(\dfrac{-b}{2a}\right)\right)$$

Roots of an Equation

▶**3** When we discuss the equation $y = ax^2 + bx + c$ we often speak about the x intercepts of the graph. If we set y equal to 0 and solve the equation $ax^2 + bx + c = 0$, we are finding the *solutions* or the *roots* of the equation.

The x-intercepts of a graph are sometimes referred to as the **roots** of the equation because at all x-intercepts the value of y is 0; see Figure 9.19. Therefore, to find the roots of an equation algebraically (or the x-intercepts of the graph), set $y = 0$ and solve the resulting equation for x.

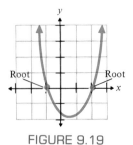

FIGURE 9.19

A graph of a quadratic equation of the form $y = ax^2 + bx + c$ will have two distinct x intercepts (Fig. 9.20a), a single x intercept (Fig. 9.20b), or no x intercept (Fig. 9.20c). In Section 9.2 we mentioned that, when the discriminant $b^2 - 4ac$ is greater than zero, there are two distinct real solutions; when equal to zero, there is a single real solution (also called a double root); and when less than zero, there is no real solution. This concept is illustrated in Fig. 9.20.

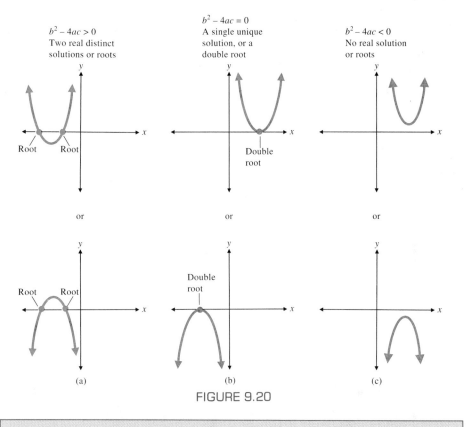

$b^2 - 4ac > 0$
Two real distinct
solutions or roots

$b^2 - 4ac = 0$
A single unique
solution, or a
double root

$b^2 - 4ac < 0$
No real solution
or roots

Root   Root

Double
root

or

or

or

Root   Root

Double
root

(a)

(b)

(c)

FIGURE 9.20

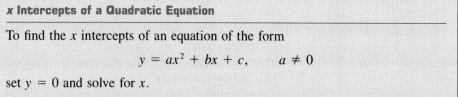

---

x Intercepts of a Quadratic Equation

To find the x intercepts of an equation of the form

$$y = ax^2 + bx + c, \qquad a \neq 0$$

set $y = 0$ and solve for x.

If we set $y = 0$ when finding the x intercepts, we get an equation of the form

$$ax^2 + bx + c = 0$$

We learned earlier in this chapter that equations of this form may be solved (1) by factoring, if the trinomial is factorable; (2) by the quadratic formula; or (3) by completing the square.

The x intercepts may also be found by graphing the equation and finding the point(s) of intersection of the graph with the x axis. However, this method may result in an inaccurate answer since you may only be able to estimate the points of intersection.

▶ **4**   We can sketch the graph of a quadratic equation by noticing whether the graph opens upward or downward, finding the vertex, finding the x intercepts, and finding the y intercept. Recall that, to find the y intercept, set $x = 0$ and solve for y. Example 1 illustrates how a quadratic function may be sketched.

EXAMPLE 1   Consider the equation $y = -x^2 + 8x - 12$.

(a) Determine whether the parabola opens upward or downward.

(b) Find the y intercept.

(c) Find the vertex.

(d) Find the x intercepts (if they exist).

(e) Sketch the graph.

Solution:　(a) Since a is -1, which is less than 0, the parabola opens downward.

(b) To find the y intercept, set $x = 0$ and solve for y.

$$y = -(0)^2 + 8(0) - 12 = -12$$

The y intercept is -12.

(c) $x = \dfrac{-b}{2a} = \dfrac{-8}{2(-1)} = 4$

$$y = \dfrac{4ac - b^2}{4a} = \dfrac{4(-1)(-12) - 8^2}{4(-1)} = \dfrac{48 - 64}{-4} = 4$$

The vertex is at $(4, 4)$. The y value of the vertex could also be found by substituting 4 for x in the original function and finding the corresponding value of y, 4.

(d) To find the x intercepts, we set $y = 0$.

$$0 = -x^2 + 8x - 12$$

or $\qquad -x^2 + 8x - 12 = 0$

We can multiply both sides of the equation by -1 and then factor.

$$-1(-x^2 + 8x - 12) = -1(0)$$
$$x^2 - 8x + 12 = 0$$
$$(x - 6)(x - 2) = 0$$

$$x - 6 = 0 \quad \text{or} \quad x - 2 = 0$$
$$x = 6 \quad \text{or} \qquad x = 2$$

Thus the x intercepts are 2 and 6. The roots could also be found by the quadratic formula (or by completing the square). Why not, at this time, find the roots by the quadratic formula?

(e) Now we use all this information to sketch the graph (Fig. 9.21). ■

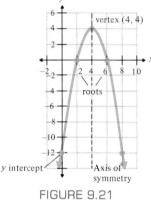

FIGURE 9.21

EXAMPLE 2　Consider the equation $y = 2x^2 + 3x + 4$.

(a) Determine whether the parabola opens upward or downward.

(b) Find the y intercept.

(c) Find the vertex.

(d) Find the x intercepts, if they exist.

(e) Sketch the graph.

Solution:　(a) Since a is 2, which is greater than 0, the parabola opens upward.

(b) $y = 2(0)^2 + 3(0) + 4 = 4$. The y intercept is 4.

(c) $x = -\dfrac{b}{2a} = \dfrac{-3}{2(2)} = \dfrac{-3}{4}$

$$y = \dfrac{4ac - b^2}{4a} = \dfrac{4(2)(4) - 3^2}{4(2)} = \dfrac{32 - 9}{8} = \dfrac{23}{8}$$

The vertex is $\left(-\frac{3}{4}, \frac{23}{8}\right)$.

(d) To find the x intercepts, set $y = 0$.

$$0 = 2x^2 + 3x + 4$$

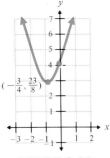

FIGURE 9.22

This trinomial cannot be factored. To determine if this equation has any real roots, we will determine the discriminant.

$$b^2 - 4ac = 3^2 - 4(2)(4) = 9 - 32 = -23$$

Since the discriminant is less than 0, this equation has no real roots. We should have expected this answer because the y value of the vertex is a positive number and therefore above the x axis. Since the parabola opens upward, it cannot intersect the x axis.

(e) The graph is sketched in Fig. 9.22. ∎

Note that when sketching by this procedure the exact curve of the graph may be slightly inaccurate, for we are not plotting point by point. However, for our needs a sketch is generally sufficient.

▶ **5** If we are given the roots of an equation, we can find the equation. This procedure is illustrated in Example 3.

EXAMPLE 3 Write an equation that has roots of -3 and 2.

Solution: If the roots are -3 and 2, the factors must be $(x + 3)$ and $(x - 2)$. Therefore, the equation is

$$0 = (x + 3)(x - 2)$$

or $\qquad 0 = x^2 + x - 6$ ∎

In Example 3 we gave the answer as $0 = (x + 3)(x - 2)$. Many other equations have roots -3 and 2. In fact, any equation of the form $0 = a(x + 3)(x - 2)$, for any real number a, will also have roots of -3 and 2. Can you explain why this is true?

Since the equation $0 = x^2 + x - 6$ has roots of -3 and 2, the graph of the function $y = x^2 + x - 6$ has x intercepts at -3 and 2.

▶ **6** Here are some applications of quadratic functions.

EXAMPLE 4 Consider the squares in Fig. 9.23. The number of complete squares below (or above) the diagonal of a square whose length and width are divided into N equal parts can be determined by the formula

$$C = \frac{1}{2}(N^2 - N)$$

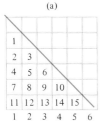

(a)

For example, in Fig. 9.23a, $N = 4$ and the number of complete squares below (or above) the diagonal is

$$C = \frac{1}{2}(N^2 - N) = \frac{1}{2}(4^2 - 4) = \frac{1}{2}(12) = 6$$

In Figure 9.23b, $N = 6$ and the number of complete squares below (or above) the diagonal is

$$C = \frac{1}{2}(N^2 - N) = \frac{1}{2}(6^2 - 6) = \frac{1}{2}(30) = 15$$

(a) Use the formula $C = \frac{1}{2}(N^2 - N)$ to determine the number of complete squares below the diagonal if the length and width of a square are divided into 12 equal parts.

(b) If there are 36 complete squares below the diagonal, determine the number of equal units in the length and width of the square.

(b)

FIGURE 9.23

Solution: (a) $N = 12$.

$$C = \frac{1}{2}(N^2 - N) = \frac{1}{2}(12^2 - 12) = \frac{1}{2}(132) = 66$$

(b) We are told that $C = 36$ and we must find N.

$$C = \frac{1}{2}(N^2 - N)$$

$$36 = \frac{1}{2}(N^2 - N)$$

$$72 = N^2 - N$$

or

$$N^2 - N - 72 = 0$$

$$(N - 9)(N + 8) = 0$$

$$N - 9 = 0 \quad \text{or} \quad N + 8 = 0$$

$$N = 9 \qquad\qquad N = -8$$

Since N must be a positive number, the only possible answer to our question is 9. Thus the length and width of the square are divided into nine equal parts. ■

EXAMPLE 5 When an object is projected in an upward direction from ground level, its distance (or height), d, from the ground at any time, t, is determined by the formula $d = v_0 t + \frac{1}{2}gt^2$, where v_0 is the initial velocity with which the object is projected and g is the force of gravity. The gravity of the Earth is -32 feet per second squared. If an object is projected upward with an initial velocity of 128 feet per second:

(a) Determine the height at $t = 1$ second.
(b) Sketch a graph of distance versus time.
(c) What is the maximum height the object will reach?
(d) At what time will the object reach its maximum height?
(e) At what time will the object strike the ground?

Solution: (a) The initial velocity, v_0, is 128 and gravity, g, is -32.

$$d = v_0 t + \frac{1}{2}gt^2$$

$$= 128t + \frac{1}{2}(-32)t^2$$

$$= 128t - 16t^2$$

When $t = 1$ second, the height is

$$d = 128(1) - 16(1)^2$$

$$= 128 - 16 = 112 \text{ feet}$$

(b) Find the vertex of $d = 128t - 16t^2$, which can be written $d = -16t^2 + 128t$.

$$t = \frac{-b}{2a} = \frac{-128}{-32} = 4$$

When $t = 4$, $d = -16(4)^2 + 128(4) = 256$. Therefore, the vertex is (4,256).

To find the roots set $d = 0$.

$$d = -16t^2 + 128t$$
$$0 = -16t^2 + 128t$$
$$0 = -16t(t - 8)$$
$$-16t = 0 \quad \text{or} \quad t - 8 = 0$$
$$t = 0 \qquad\qquad t = 8$$

Thus the roots are at 0 and 8. The graph is sketched in Fig. 9.24.

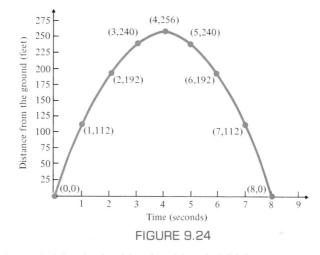

FIGURE 9.24

(c) The maximum height obtained by the object is 256 feet.

(d) By observation of the graph, we see that the object reaches its maximum height of 256 feet at $t = 4$ seconds.

(e) The object strikes the ground at $t = 8$ seconds. It takes 4 seconds for the object to reach its maximum height and 4 seconds to return to the ground. In part (b) we found the roots to be 0 and 8. Note that the object is on the ground at $t = 0$ seconds and $t = 8$ seconds. ■

Maximum/Minimum Problems

In Fig. 9.18 we can see that a parabola that opens upward has a minimum value and a parabola that opens downward has a maximum value. The **maximum or minimum value** of a parabola of the form $y = ax^2 + bx + c$ will be the y coordinate of the vertex. Therefore, the maximum or minimum value of a problem that can be represented as a quadratic equation will be

$$y = \frac{4ac - b^2}{4a}$$

The maximum or minimum value will occur at

$$x = \frac{-b}{2a}$$

EXAMPLE 6   When a cannon is fired at a certain angle, the distance of the shell above the ground, h, in meters, at time, t, in seconds, is given by the formula $h = -4.9t^2 + 24t + 5$. Find the maximum height attained by the shell.

Solution: Since this formula is a quadratic function, its graph will be a parabola. Since

$a = -4.9$, the parabola opens downward and the function has a maximum value. We can use $y = (4ac - b^2)/4a$ to find the maximum height.

$$y = \frac{4ac - b^2}{4a}$$

$$= \frac{4(-4.9)(5) - (24)^2}{4(-4.9)}$$

$$= \frac{-98 - 576}{-19.6}$$

$$= \frac{-674}{-19.6}$$

$$\approx 34.4 \text{ meters}$$

Thus the maximum height obtained by the shell is about 34.4 meters. The shell reaches this maximum height at

$$t = \frac{-b}{2a} = \frac{-24}{2(-4.9)} \approx 2.4 \text{ seconds}$$ ■

Exercise Set 9.4

(a) Determine whether the parabola opens upward or downward, (b) find the y intercept, (c) find the vertex, and (d) find the x intercepts (if they exist). You may need to use a calculator or Appendix C to find an approximate value for the x intercepts if they are irrational. (e) Sketch the graph.

1. $y = x^2 + 8x + 15$

2. $y = x^2 + 2x - 3$

3. $f(x) = x^2 - 6x + 4$

4. $y = -x^2 + 4x - 5$

5. $y = x^2 + 6x + 9$

6. $y = 2x^2 + 4x + 2$

7. $f(x) = x^2 - 4x + 4$

8. $y = -2x^2 + 4x - 8$

9. $y = 2x^2 - x - 6$

10. $y = -3x^2 + 6x - 9$

11. $y = 2x^2 - 3x - 9$

12. $f(x) = 3x^2 - 5x - 12$

13. $y = 3x^2 + 4x + 3$

14. $f(x) = -3x^2 - 2x - 6$

15. $f(x) = -2x^2 - 6x + 4$

16. $f(x) = 2x^2 + x - 6$

17. $y = x^2 + 4$

18. $y = -x^2 + 4$

19. $y = -9x^2 + 4$

20. $y = x^2 + 4x$

21. $y = -x^2 + 6x$

22. $f(x) = 3x^2 + 10x$ **23.** $f(x) = -5x^2 + 5$ **24.** $f(x) = 2x^2 - 6x + 4$

25. $y = 3x^2 + 4x - 6$ **26.** $f(x) = -x^2 + 3x - 5$ **27.** $y = -x^2 + 3x + 6$

28. $f(x) = -2x^2 - 6x + 5$ **29.** $f(x) = -4x^2 + 6x - 9$ **30.** $y = -2x^2 + 5x + 4$

Write an equation in two variables that has the x intercepts given.

31. 4, 6   **32.** -2, 5   **33.** 3, -4 **34.** 0, 4

35. 2, -3 **36.** -1, -6 **37.** 2, 2   **38.** $\frac{1}{2}$, 3

39. $-2, \frac{2}{3}$ **40.** $-\frac{3}{5}, \frac{2}{3}$ **41.** $-\frac{1}{2}, \frac{2}{3}$ **42.** $\frac{3}{5}, \frac{1}{4}$

43. Use the formula given in Example 4, $C = \frac{1}{2}(N^2 - N)$, to determine the number of whole squares below the diagonal if the length and width of a square are divided into eight equal parts.

44. The numbers 1 through 144 form a 12 by 12 square array of numbers. The number 1 is in the upper-left corner and the number 144 is in the lower-right corner. If a diagonal is drawn from number 1 to number 144, how many numbers lie below and above the diagonal?

45. Quentin has planted a 6-foot-tall willow tree. Its height, h, in feet, t years after being planted can be estimated by the formula $h = -0.3t^2 + 6.5t + 6$, $t \le 10$. Find the height of the tree after 8 years.

46. The number of centimeters that a specific spring will stretch, s, when a mass, m, in kilograms, is attached to it can be found by the formula $s = 3.7m - 0.5m^2$ (for $m \le 7$ kg). How far will the spring stretch if an object with a mass of 4 kilograms is attached?

47. A person of weight w on the end of a diving board causes it to dip d inches. The relationship between the weight on the board and the dip is $d = 0.00003w^2 + 0.05w$. How much will the board dip if a 200-pound person stands on the tip of the board?

48. The pressure in pounds per square inch within an aerosol spray can diminishes with the amount of time the spray is applied. The pressure, p, inside a spray can after t minutes of application is found by the formula $p = 120 - 4.3t^2$. Find the pressure within the can after 2 minutes.

49. An object is projected upward with an initial velocity of 192 feet per second. Use the formula given in Example 5,

$d = v_0t + \frac{1}{2}gt^2$, to answer the following:
(a) Find the object's distance from the ground in 3 seconds.
(b) Make a graph of distance versus time.
(c) What is the maximum height the object will reach?
(d) At what time will it reach its maximum height?
(e) At what time will the object strike the ground?

50. An object is projected upward with an initial velocity of 160 feet per second. Answer the questions asked in Exercise 49.

51. The Rochester Philharmonic is trying to determine the price to charge for concert tickets. If the price is too low, they will not make enough money to cover expenses, and if their price is too high, not enough people will wish to pay the price of a ticket. They estimate their total income, I, in hundreds of dollars, per show, can be approximated by the formula $I = -x^2 + 24x - 44$, $0 \le x \le 24$, where x is the cost of a ticket.
(a) Draw a graph of income versus the cost of a ticket.
(b) Determine the minimum cost of a ticket for the Philharmonic to break even.
(c) Determine the maximum cost of a ticket that the Philharmonic can charge and break even.
(d) How much should they charge if they are to receive the maximum income?
(e) Find the maximum income.

52. A company earns a weekly profit of P dollars according to the formula $P = -0.4x^2 + 80x - 200$. Find the number of items the company must sell each week to obtain the largest profit. Find the maximum profit.

53. Ramon throws a ball into the air with an initial velocity of 32 feet per second. The height of the ball at any time t is given by the formula $h = 32t - 16t^2$. At what time does the ball reach its maximum height? What is the maximum height?

54. When a baseball is thrown upward with a velocity of 64 feet per second from the top of a 160-foot-tall building, its distance from the ground, h, at any time, t, is $h = -16t^2 + 64t + 160$. Find the maximum height obtained by the baseball.

55. Consider a quadratic function of the form $y = ax^2 + bx + c$. To find the x intercepts of the graph, we set $y = 0$ and solve the resulting equation, $0 = ax^2 + bx + c$. Explain why (a) if the discriminant is negative, the graph of $y = ax^2 + bx + c$ has no x intercept, (b) if the discriminant is 0, the graph has one x inter-

cept, and (c) if discriminant is positive, the graph has two distinct x intercepts.

56. How can you determine whether a quadratic function will be a parabola that opens up or down when graphed?

57. By observing the value of the coefficient of the squared term in a quadratic function, and by determining coordinates of the vertex of the corresponding parabola, explain how you can determine the number of x intercepts the parabola has.

58. How will the graphs of $y = x^2 - 4x + 4$ and $y = -x^2 + 4x - 4$ compare? Explain your answer.

59. Consider the equations $y = x^2 - 8x + 12$ and $y = -x^2 + 8x - 12$. (a) Without graphing, can you explain how the graphs of the two equations compare? (b) Will the graphs have the same x intercepts? Explain your answer. (c) Will the graphs have the same vertex? Explain your answer. (d) Graph both equations on the same set of axes.

Cumulative Review Exercises

[4.1] **60.** Solve the system of equations algebraically.
$$2x + 3y = 8$$
$$\tfrac{1}{2}x - 2y = 4$$

[4.3] **61.** Kenji wishes to put part of his $15,000 in a money-market account earning 12% interest and the balance in a savings account earning 5.5% in-

terest. How much money should he invest in each account if he wishes the total interest for the year to be $900? Use $i = prt$.

[5.3] **62.** State the degree of the polynomial $6x^4 + 5x^3y^2 + 6y^4 - \tfrac{1}{2}x^3$.

[6.2] **63.** Factor $6x^2 + 17x - 45$.

JUST FOR FUN

1. What are the dimensions of the rectangle with the largest area that can be enclosed with 64 feet of fence?

2. A landscaper wishes to make a rectangular region along the bank of a river. She need only fence on three sides, as illustrated. What dimensions should she use if she wishes to maximize the rectangle's area and has only 100 feet of fencing?

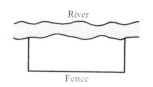

9.5

The Algebra of Functions

► **1** Find the sum, difference, product, and quotient of functions.

► **2** Find the composition of functions.

Now that we have introduced a wide variety of functions, including linear functions (Section 3.5), polynomial functions (Section 5.8), rational functions (Just For Fun, Section 7.1), the square root function (Section 8.9) and quadratic functions (Sections 5.8 and 9.4), we can discuss some operations performed on functions.

► **1** We can add, subtract, multiply, and divide functions. We can also perform another type of operation called finding the composite function. We will first discuss addition, subtraction, multiplication, and division of functions. Then we will discuss how to determine composite functions.

If $f(x)$ represents one function and $g(x)$ represents a second function, then the operations on functions may be performed as follows:

Sum of functions: $(f + g)(x) = f(x) + g(x)$

Difference of functions: $(f - g)(x) = f(x) - g(x)$

Product of functions: $(fg)(x) = f(x) \cdot g(x)$

Quotient of functions: $\left(\dfrac{f}{g}\right)(x) = \dfrac{f(x)}{g(x)};\quad g(x) \neq 0$

EXAMPLE 1 If $f(x) = x^2 + x - 6$ and $g(x) = x - 2$, find (a) $(f + g)(x)$, (b) $(f - g)(x)$, and (c) $(g - f)(x)$.

Solution: (a) $(f + g)(x) = f(x) + g(x) = (x^2 + x - 6) + (x - 2)$
$$= x^2 + x - 6 + x - 2$$
$$= x^2 + 2x - 8$$

(b) $(f - g)(x) = f(x) - g(x) = (x^2 + x - 6) - (x - 2)$
$$= x^2 + x - 6 - x + 2$$
$$= x^2 - 4$$

(c) $(g - f)(x) = g(x) - f(x) = (x - 2) - (x^2 + x - 6)$
$$= x - 2 - x^2 - x + 6$$
$$= -x^2 + 4$$

EXAMPLE 2 If $f(x) = x^2 + x - 6$ and $g(x) = x - 2$, find (a) $(f + g)(2)$ and (b) $f(2) + g(2)$.

Solution: (a) In part (a) of Example 1, we found that for these particular functions $(f + g)(x) = x^2 + 2x - 8$. Using this information, we can evaluate $(f + g)(2)$ by substituting 2 for each x in $(f + g)(x)$.

$$(f + g)(x) = x^2 + 2x - 8$$
$$(f + g)(2) = 2^2 + 2(2) - 8 = 4 + 4 - 8 = 0$$

Thus $(f + g)(2) = 0$.

(b) We will evaluate $f(2) + g(2)$ by substituting 2 for x in $f(x)$ and 2 for x in $g(x)$. We will then add the values to obtain the sum of the functions when x is 2.

$$f(x) + g(x) = (x^2 + x - 6) + (x - 2)$$
$$f(2) + g(2) = (2^2 + 2 - 6) + (2 - 2)$$
$$= (4 + 2 - 6) + (2 - 2) = 0 + 0 = 0$$

We see by comparing parts (a) and (b) that $(f + g)(2) = f(2) + g(2)$. ∎

EXAMPLE 3   If $f(x) = x^2 + x - 6$ and $g(x) = x - 2$, find (a) $(fg)(x)$ and (b) $\left(\dfrac{f}{g}\right)(x)$.

Solution: (a) $(fg)(x) = f(x)g(x) = (x^2 + x - 6)(x - 2)$
$$= x^3 - x^2 - 8x + 12$$

If you have forgotten how to multiply polynomials, see Section 5.4.

(b) $\left(\dfrac{f}{g}\right)(x) = \dfrac{f(x)}{g(x)} = \dfrac{x^2 + x - 6}{x - 2}$

$$= \dfrac{(x + 3)(x - 2)}{x - 2} = x + 3$$

In the original quotient, $\dfrac{x^2 + x - 6}{x - 2}$, x may not have a value of 2. Therefore, x cannot be 2 in the domain of the simplified quotient. We write the answer as

$$\left(\dfrac{f}{g}\right)(x) = x + 3, \quad x \neq 2$$ ∎

▶ **2**   Now let us consider another operation performed on functions, the **composition operation.** When we use the composition operation on two functions, the result is called the **composite function.** The composition operation is indicated by a small open circle, ∘, or by using parentheses and brackets. If $f(x)$ and $g(x)$ are two functions, then *the composite function of function f with function g* may be indicated by

$$(f \circ g)(x) \quad \text{or} \quad f[g(x)]$$

The composite function of function g with function f may be indicated by

$$(g \circ f)(x) \quad \text{or} \quad g[f(x)]$$

To explain how to determine a composite function, we will use the notation $f[g(x)]$ to represent the composite function of function f with function g.

Consider the functions $f(x) = x^2 - 2x + 3$ and $g(x) = x - 5$. How would you find $f(4)$? In Section 3.5 we indicated that to find $f(4)$ you substitute 4 for each x in $f(x)$.

$$f(x) = x^2 - 2x + 3$$
$$f(4) = 4^2 - 2(4) + 3 = 16 - 8 + 3 = 11$$

How would you find $f(a)$? To find $f(a)$, you substitute a for each x in $f(x)$.

$$f(x) = x^2 - 2x + 3$$
$$f(a) = a^2 - 2a + 3$$

Using the illustrations just presented, how do you think you would find $f[g(x)]$? If you answered "substitute function $g(x)$ for each x in function $f(x)$," you answered correctly.

$$f(x) = x^2 - 2x + 3, \qquad g(x) = x - 5$$
$$f[g(x)] = (x - 5)^2 - 2(x - 5) + 3$$
$$= (x - 5)(x - 5) - 2x + 10 + 3$$
$$= x^2 - 10x + 25 - 2x + 13$$
$$= x^2 - 12x + 38$$

Therefore, the composite function of function f with function g is $x^2 - 12x + 38$.

$$(f \circ g)(x) = f[g(x)] = x^2 - 12x + 38$$

How do you think we would determine $g[f(x)]$ or $(g \circ f)(x)$? If you answered, "substitute the function $f(x)$ for each x in function $g(x)$" you answered correctly.

$$g(x) = x - 5, \qquad f(x) = x^2 - 2x + 3$$
$$g[f(x)] = (x^2 - 2x + 3) - 5$$
$$= x^2 - 2x + 3 - 5$$
$$= x^2 - 2x - 2$$

Therefore, the composite function of function g with function f is $x^2 - 2x - 2$.

$$(g \circ f)(x) = g[f(x)] = x^2 - 2x - 2$$

By comparing the above illustrations we see that $f[g(x)] \neq g[f(x)]$.

EXAMPLE 4   Given $f(x) = x^2 + 4$ and $g(x) = \sqrt{x - 2}$, find (a) $(f \circ g)(x)$ and (b) $(g \circ f)(x)$.

Solution: (a) To find $(f \circ g)(x)$, we substitute $g(x)$, which is $\sqrt{x - 2}$, for each x in $f(x)$. You should realize that $\sqrt{x - 2}$ is a real number only when $x \geq 2$.

$$f(x) = x^2 + 4$$
$$(f \circ g)(x) = f[g(x)] = (\sqrt{x - 2})^2 + 4 = x - 2 + 4 = x + 2, \ x \geq 2.$$

Since values of $x < 2$ are not in the domain of $g(x)$, values of $x < 2$ are not in the domain of $(f \circ g)(x)$.

(b) To find $(g \circ f)(x)$, we substitute $f(x)$, which is $x^2 + 4$, for each x in $g(x)$.

$$g(x) = \sqrt{x - 2}$$
$$(g \circ f)(x) = g[f(x)] = \sqrt{(x^2 + 4) - 2} = \sqrt{x^2 + 2} \qquad \blacksquare$$

EXAMPLE 5   Given $f(x) = x - 1$ and $g(x) = x + 7$, find (a) $(f \circ g)(x)$ (b) $(f \circ g)(2)$, (c) $(g \circ f)(x)$, and (d) $(g \circ f)(2)$.

Solution: (a)
$$f(x) = x - 1$$
$$(f \circ g)(x) = f[g(x)] = (x + 7) - 1 = x + 6$$

(b) $(f \circ g)(2)$ can be found by substituting 2 for each x in $(f \circ g)(x)$.

$$(f \circ g)(x) = x + 6$$
$$(f \circ g)(2) = 2 + 6 = 8$$

(c)
$$g(x) = x + 7$$
$$(g \circ f)(x) = g[f(x)] = (x - 1) + 7 = x + 6$$

(d) Since $(g \circ f)(x) = x + 6$, $(g \circ f)(2) = 2 + 6 = 8$. $\qquad \blacksquare$

HELPFUL HINT

Do not confuse finding the product of two functions with finding a composite function.

Product of functions f and g: $(fg)(x)$ or $(f \cdot g)(x)$

Composite function of f with g: $(f \circ g)(x)$

When multiplying functions f and g, we can use a dot between the f and g. When finding the composite function of f with g, we use a small *open* circle.

Exercise Set 9.5

Given $f(x) = x^2 + 2x - 8$ and $g(x) = x - 2$, find the following:

1. (a) $(f + g)(x)$ (b) $(f + g)(2)$

2. (a) $(f - g)(x)$ (b) $(f - g)(4)$

3. (a) $(fg)(x)$ (b) $(fg)(-1)$

4. (a) $\left(\dfrac{f}{g}\right)(x)$ (b) $\left(\dfrac{f}{g}\right)(0)$

5. (a) $(f \circ g)(x)$ (b) $(f \circ g)(3)$

6. (a) $(g \circ f)(x)$ (b) $(g \circ f)(3)$

Given $f(x) = x^2 - 4$ and $g(x) = x + 2$, find the following:

7. (a) $(g + f)(x)$ (b) $(g + f)(-2)$

8. (a) $(g - f)(x)$ (b) $(g - f)(5)$

9. (a) $(gf)(x)$ (b) $(gf)(-4)$

10. (a) $\left(\dfrac{f}{g}\right)(x)$ (b) $\left(\dfrac{f}{g}\right)(4)$

11. (a) $(g \circ f)(x)$ (b) $(g \circ f)(4)$

12. (a) $(f \circ g)(x)$ (b) $(f \circ g)(4)$

Given $f(x) = x - 4$ and $g(x) = \sqrt{x + 6}$, $x \geq -6$, find the following:

13. (a) $(f + g)(x)$ (b) $(f + g)(3)$

14. (a) $(f - g)(x)$ (b) $(f - g)(3)$

15. (a) $(fg)(x)$ (b) $(fg)(10)$

16. (a) $\left(\dfrac{g}{f}\right)(x)$ (b) $\left(\dfrac{g}{f}\right)(10)$

17. (a) $(f \circ g)(x)$ (b) $(f \circ g)(7)$

18. (a) $(g \circ f)(x)$ (b) $(g \circ f)(8)$

19. Will $(f + g)(x) = (g + f)(x)$ for all values of x? Explain your answer and give an example to support your answer.

20. Will $(f - g)(x) = (g - f)(x)$ for all values of x? Explain your answer and give an example to support your answer.

21. Will $(fg)(x) = (gf)(x)$ for all values of x? Explain your answer and give an example to support your answer.

22. Will $\left(\dfrac{f}{g}\right)(x) = \left(\dfrac{g}{f}\right)(x)$ for all values of x? Explain your answer and give an example to support your answer.

23. Will $(f \circ g)(x) = (g \circ f)(x)$ for all values of x? Explain your answer and give an example to support your answer.

24. (a) Make up your own polynomial function, $f(x)$, and determine using your function if $f(x) + f(x) = 2 \cdot f(x)$ for your function. (b) Do you believe that $f(x) + f(x) = 2 \cdot f(x)$ for all functions? Explain your answer.

25. Consider the functions $f(x) = \sqrt{x + 5}$, $x \geq -5$, and $g(x) = x^2 - 5$, $x \geq 0$. (a) Show that for $x \geq 0$ $(f \circ g)(x) = (g \circ f)(x)$; (b) Explain why we need to stipulate that $x \geq 0$ for part (a) to be true.

Cumulative Review Exercises

[3.5] 26. Are all functions relations? Are all relations functions? Explain your answer.

[5.7] 27. Are all linear functions polynomial functions? Are all quadratic functions polynomial functions? Explain your answer.

28. Give an example of a polynomial function that is not a linear or quadratic function.

[7.1] 29. Read the Just For Fun at the end of Section 7.1. Are all polynomial functions rational functions? Are all rational functions polynomial functions? Explain your answer.

[8.9] 30. Are any polynomial functions square root functions? Are any square root functions polynomial functions? Explain your answer.

9.6

Inverse Functions

▶**1** Identify one-to-one functions.

▶**2** Find the inverse function of a set of ordered pairs.

▶**3** Find inverse functions.

▶**4** Show that $(f \circ f^{-1})(x) = (f^{-1} \circ f)(x) = x$.

▶**1** Inverse functions are an important concept that must be understood before we can discuss exponential and logarithmic functions in Chapter 11. However, before we discuss inverse functions we need to explain what is meant by *one-to-one functions*.

Consider the function $f(x) = x^2$ (see Fig. 9.25). Note that it is a function since it passes the vertical line test. For each value of x, there is a unique value of y. Does each value of y also have a unique value of x? The answer is no, as illustrated in Fig. 9.26. Note that for the indicated value of y there are two values of x, x_1 and x_2. If we limit the domain of $f(x) = x^2$ to values of x greater than or equal to 0, then each x value has a unique y value and each y value also has a unique x value (see Fig. 9.27).

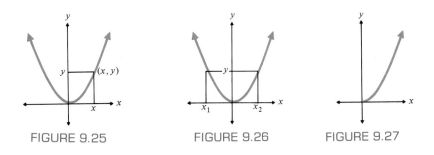

FIGURE 9.25   FIGURE 9.26   FIGURE 9.27

The function $f(x) = x^2$, $x \geq 0$, Fig. 9.27, is an example of a one-to-one function. A **one-to-one function** is a function where each value in the range has a unique value in the domain. Thus, if y is a one-to-one function of x, in addition to each x value having a unique y value (the definition of a function), each y value must also have a unique x value. For a function to be a one-to-one function, it must pass not only a **vertical line test** (the test to ensure that it is a function) but also a **horizontal line test** (to test the one-to-one criteria).

EXAMPLE 1   Determine which functions are one-to-one functions.

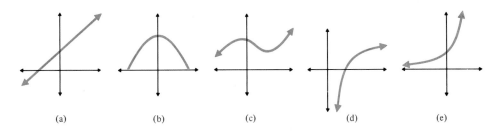

(a)   (b)   (c)   (d)   (e)

Solution: (a), (d), and (e) are one-to-one functions since they pass both the vertical line test and the horizontal line test. The graphs that follow show that (b) and (c) do not pass the horizontal line test and that each y does not have a unique x. The graphs in (b) and (c) are therefore not one-to-one functions.

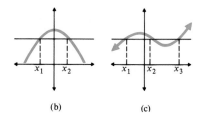

(b)   (c)

▶ **2**   Now that we have discussed one-to-one functions we can introduce inverse functions. If a function is one-to-one, its **inverse function** may be obtained by interchanging the first and second coordinates in each ordered pair of the function. Thus, for each ordered pair (x, y) in the function, the ordered pair (y, x) will be in the inverse function. For example,

Function: $\{(1, 4), (2, 0), (3, 7), (-2, 1), (-1, -5)\}$

Inverse function: $\{(4, 1), (0, 2), (7, 3), (1, -2), (-5, -1)\}$

Note that the domain of the function becomes the range of the inverse function, and the range of the function is the domain of the inverse function.

If we graph the points in the function and the points in the inverse function (Fig. 9.28), we see that the points are symmetric about the line $y = x$.

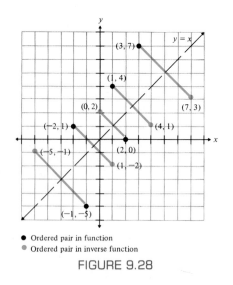

● Ordered pair in function
● Ordered pair in inverse function

FIGURE 9.28

If $f(x)$ is used to represent a function, the notation $f^{-1}(x)$ represents the inverse function of $f(x)$. Note the -1 in the notation is *not* an exponent.

> **Inverse Function**
>
> If $f(x)$ is a one-to-one function with ordered pairs of the form (x, y), then its inverse function, $f^{-1}(x)$, will be a one-to-one function with ordered pairs of the form (y, x).

When a function $f(x)$ and its inverse function $f^{-1}(x)$ are graphed on the same set of axes, $f(x)$ *and* $f^{-1}(x)$ *will be symmetric about the line* $y = x$.

▶ **3** When a one-to-one function is given in the form of an equation, its inverse function can be found by following this procedure:

> **To find the Inverse Function of a One-to-One Function of the Form y = f(x)**
>
> 1. Interchange the two variables x and y.
> 2. Solve the equation for y. The resulting equation will be the inverse function.

The following example will illustrate the procedure.

EXAMPLE 2 (a) Find the inverse function of $y = f(x) = 4x + 2$.
(b) On the same set of axes, graph both $f(x)$ and $f^{-1}(x)$.

Solution: (a) $y = 4x + 2$. First interchange x and y.   (b)

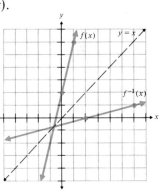

$$x = 4y + 2$$

Now solve for y.

$$x - 2 = 4y$$

$$\frac{x - 2}{4} = y$$

$$y = f^{-1}(x) = \frac{x - 2}{4}$$

Note the symmetry of $f(x)$ and $f^{-1}(x)$ about the line $y = x$. ■

EXAMPLE 3 (a) Find the inverse function of $y = f(x) = \dfrac{-3x - 2}{4}$.

(b) On the same set of axes, graph both $f(x)$ and $f^{-1}(x)$.

Solution: (a) $y = \dfrac{-3x - 2}{4}$. Interchange x and y; then solve for y.

$$x = \frac{-3y - 2}{4}$$

$$4x = -3y - 2$$

$$4x + 2 = -3y$$

$$\frac{4x + 2}{-3} = y$$

$$\text{or}\quad y = \frac{-4x - 2}{3}$$

Thus $f^{-1}(x) = \dfrac{-4x - 2}{3}$.

(b)

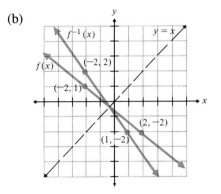

▶4 In Section 9.5 we discussed the composition of functions. If two functions $f(x)$ and $f^{-1}(x)$ are inverses of each other, then $(f \circ f^{-1})(x) = x$ and $(f^{-1} \circ f)(x) = x$.

EXAMPLE 4 In Example 3 we determined that $f(x) = \dfrac{-3x - 2}{4}$ and $f^{-1}(x) = \dfrac{-4x - 2}{3}$ are inverse functions. Show that $(f \circ f^{-1})(x) = x$ and $(f^{-1} \circ f)(x) = x$.

Solution: (a) To determine $(f \circ f^{-1})(x)$, we substitute $f^{-1}(x)$ for each x in $f(x)$.

$$f(x) = \frac{-3x - 2}{4}$$

$$(f \circ f^{-1})(x) = \frac{-3\left(\dfrac{-4x - 2}{3}\right) - 2}{4}$$

$$= \frac{-(-4x - 2) - 2}{4}$$

$$= \frac{4x + 2 - 2}{4} = \frac{4x}{4} = x$$

To determine $(f^{-1} \circ f)(x)$, substitute $f(x)$ for each x in $f^{-1}(x)$.

$$f^{-1}(x) = \frac{-4x - 2}{3}$$

$$(f^{-1} \circ f)(x) = \frac{-4\left(\dfrac{-3x - 2}{4}\right) - 2}{3}$$

$$= \frac{-(-3x - 2) - 2}{3}$$

$$= \frac{3x + 2 - 2}{3} = \frac{3x}{3} = x$$

Thus $(f \circ f^{-1})(x) = (f^{-1} \circ f)(x) = x$.

Exercise Set 9.6

In Exercises 1 through 10, determine whether each function is a one-to-one function.

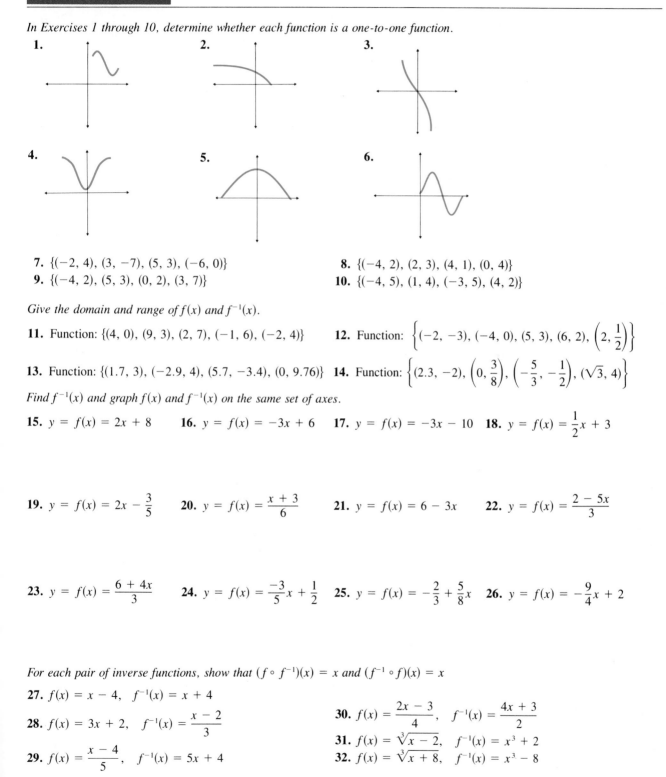

1.

2.

3.

4.

5.

6.

7. $\{(-2, 4), (3, -7), (5, 3), (-6, 0)\}$

8. $\{(-4, 2), (2, 3), (4, 1), (0, 4)\}$

9. $\{(-4, 2), (5, 3), (0, 2), (3, 7)\}$

10. $\{(-4, 5), (1, 4), (-3, 5), (4, 2)\}$

Give the domain and range of $f(x)$ and $f^{-1}(x)$.

11. Function: $\{(4, 0), (9, 3), (2, 7), (-1, 6), (-2, 4)\}$

12. Function: $\left\{(-2, -3), (-4, 0), (5, 3), (6, 2), \left(2, \frac{1}{2}\right)\right\}$

13. Function: $\{(1.7, 3), (-2.9, 4), (5.7, -3.4), (0, 9.76)\}$

14. Function: $\left\{(2.3, -2), \left(0, \frac{3}{8}\right), \left(-\frac{5}{3}, -\frac{1}{2}\right), (\sqrt{3}, 4)\right\}$

Find $f^{-1}(x)$ and graph $f(x)$ and $f^{-1}(x)$ on the same set of axes.

15. $y = f(x) = 2x + 8$

16. $y = f(x) = -3x + 6$

17. $y = f(x) = -3x - 10$

18. $y = f(x) = \frac{1}{2}x + 3$

19. $y = f(x) = 2x - \frac{3}{5}$

20. $y = f(x) = \frac{x + 3}{6}$

21. $y = f(x) = 6 - 3x$

22. $y = f(x) = \frac{2 - 5x}{3}$

23. $y = f(x) = \frac{6 + 4x}{3}$

24. $y = f(x) = \frac{-3}{5}x + \frac{1}{2}$

25. $y = f(x) = -\frac{2}{3} + \frac{5}{8}x$

26. $y = f(x) = -\frac{9}{4}x + 2$

For each pair of inverse functions, show that $(f \circ f^{-1})(x) = x$ and $(f^{-1} \circ f)(x) = x$

27. $f(x) = x - 4, \quad f^{-1}(x) = x + 4$

28. $f(x) = 3x + 2, \quad f^{-1}(x) = \frac{x - 2}{3}$

29. $f(x) = \frac{x - 4}{5}, \quad f^{-1}(x) = 5x + 4$

30. $f(x) = \frac{2x - 3}{4}, \quad f^{-1}(x) = \frac{4x + 3}{2}$

31. $f(x) = \sqrt[3]{x - 2}, \quad f^{-1}(x) = x^3 + 2$

32. $f(x) = \sqrt[3]{x + 8}, \quad f^{-1}(x) = x^3 - 8$

33. What are one-to-one functions?

34. What kind of functions can have inverse functions?

35. What is the relationship between the domain and range of a function and the domain and range of its inverse function?

36. Write a paragraph explaining why, if $f(x)$ and $g(x)$ are inverse functions, then $(f \circ g)(x) = x$ and $(g \circ f)(x) = x$.

Cumulative Review Exercises

[4.2] **37.** Solve the system of equations.

$$2x + 3y - 4z = 18$$
$$x - y - z = 3$$
$$x - 2y - 2z = 2$$

[5.6] **38.** Divide using synthetic division.

$$(x^3 + 6x^2 + 6x - 8) \div (x + 2)$$

[8.4] **39.** Simplify $\sqrt{\dfrac{24x^3y^2}{3xy^3}}$.

[9.1] **40.** Solve the equation $x^2 + 2x - 6 = 0$ by completing the square.

JUST FOR FUN

Find $f^{-1}(x)$ and graph $f(x)$ and $f^{-1}(x)$ on the same set of axes. Use the fact that the domain of $f(x)$ is the range of $f^{-1}(x)$ and the range of $f(x)$ is the domain of $f^{-1}(x)$ to graph $f^{-1}(x)$.

1. $y = \sqrt{x^2 - 9}, x \geq 3$

2. $y = \sqrt{x^2 - 9}, x \leq -3$

SUMMARY

GLOSSARY

Axis of symmetry *(441):* Imaginary line about which a parabola is symmetric.

Discriminant *(426):* For a quadratic equation of the form $ax^2 + bx + c = 0$, the discriminant is $b^2 - 4ac$.

Inverse function: *(456):* If $f(x)$ is a one-to-one function with ordered pairs of the form (x, y), then its inverse function, $f^{-1}(x)$, will be a one-to-one function with ordered pairs of the form (y, x).

One-to-one function: *(455):* A function where each value in the range has a unique value in the domain.

Parabola: *(440):* The shape of the graph of a quadratic function.

Perfect square trinomial: *(415):* A trinomial that can be expressed as the square of a binomial.

Quadratic function: *(440):* Functions of the form $f(x) = ax^2 + bx + c, a \neq 0$.

Root of a quadratic equation: *(442):* The solution to a quadratic equation of the form $ax^2 + bx + c = 0$, $a \neq 0$.

Solution to a quadratic equation: *(442):* The value or values of the variable that make the equation a true statement.

Standard form of a quadratic equation: *(421):* $ax^2 + bx + c = 0, a \neq 0$.

Vertex: (441): The lowest point on a parabola that opens upward or the highest point on a parabola that opens downward.

IMPORTANT FACTS

Pythagorean theorem $\quad a^2 + b^2 = c^2$

Quadratic formula $\quad x = \dfrac{-b \pm \sqrt{b^2 - 4ac}}{2a}$

Discriminant $\quad b^2 - 4ac$

If $b^2 - 4ac > 0$, the quadratic equation has two distinct real solutions.
If $b^2 - 4ac = 0$, the quadratic equation has one real solution.
If $b^2 - 4ac < 0$, the quadratic equation has no real solution.

Vertex at $\left(\dfrac{-b}{2a}, \dfrac{4ac - b^2}{4a} \right)$ for an equation of the form $y = ax^2 + bx + c$.

Algebra of Functions

Sum of functions: $\quad (f + g)(x) = f(x) + g(x)$
Difference of functions: $\quad (f - g)(x) = f(x) - g(x)$
Product of functions: $\quad (fg)(x) = f(x) \cdot g(x)$
Quotient of functions: $\quad \left(\dfrac{f}{g} \right)(x) = \dfrac{f(x)}{g(x)}, \; g(x) \neq 0$
Composite function of f with g: $\quad (f \circ g)(x) = f[g(x)]$
Composite function of g with f: $\quad (g \circ f)(x) = g[f(x)]$

If $f(x)$ and $g(x)$ are inverse functions then $(f \circ g)(x) = (g \circ f)(x) = x$.

Review Exercises

[9.1] *Use the square root property to solve each equation for x.*

1. $(x - 4)^2 = 20$

2. $(3x - 4)^2 = 60$

3. $(x - \frac{2}{3})^2 = \frac{1}{9}$

4. $(2x - \frac{1}{2})^2 = 4$

Use the square root property to solve each of the formulas for the indicated variable. Use only the positive square root.

5. $F_T^2 = F_a^2 + F_b^2$, for F_b

6. $L = 3r + 2s^2$, for s

Use the Pythagorean Theorem to solve for x.

7.

8.

Solve each equation by completing the square.

9. $x^2 - 10x + 16 = 0$ **10.** $x^2 - 8x + 15 = 0$ **11.** $x^2 - 14x + 13 = 0$

12. $x^2 + x - 6 = 0$ **13.** $x^2 - 3x - 54 = 0$ **14.** $x^2 = -5x + 6$

15. $x^2 + 2x - 5 = 0$ **16.** $x^2 - 3x + 8 = 0$ **17.** $2x^2 - 8x = -64$

18. $2x^2 - 4x = 30$ **19.** $4x^2 - 2x + 12 = 0$ **20.** $-x^2 - 6x + 10 = 0$

[9.2] *Determine whether the equation has two distinct real solutions, a single real solution, or no real solution.*

21. $3x^2 - 4x - 20 = 0$ **22.** $-3x^2 + 4x = 9$ **23.** $2x^2 + 6x + 7 = 0$

24. $x^2 - x + 8 = 0$ **25.** $x^2 - 12x = -36$ **26.** $3x^2 - 4x + 5 = 0$

27. $-3x^2 - 4x + 8 = 0$ **28.** $x^2 - 9x + 6 = 0$

Solve by the quadratic formula.

29. $x^2 - 9x + 14 = 0$ **30.** $x^2 + 7x - 30 = 0$ **31.** $x^2 = 7x - 10$

32. $5x^2 - 7x = 6$ 2,   **33.** $x^2 - 18 = 7x$ **34.** $x^2 - x + 30 = 0$

35. $6x^2 + x - 15 = 0$ **36.** $2x^2 + 4x - 3 = 0$ **37.** $-2x^2 + 3x + 6 = 0$

38. $x^2 - 6x + 7 = 0$ **39.** $3x^2 - 4x + 6 = 0$ **40.** $3x^2 - 6x - 8 = 0$

41. $2x^2 + 3x = 0$ **42.** $2x^2 - 5x = 0$

[9.1–9.2] *Find the solution to the quadratic equation by the method of your choice.*

43. $x^2 - 11x + 24 = 0$ **44.** $x^2 - 16x + 63 = 0$ **45.** $x^2 = -3x + 40$

46. $x^2 + 6x = 27$ **47.** $x^2 - 4x - 60 = 0$ **48.** $x^2 - x + 42 = 0$

49. $x^2 + 11x + 12 = 0$ **50.** $x^2 = 25$ **51.** $x^2 + 6x = 0$

52. $2x^2 + 5x = 3$ **53.** $3x^2 = 9x - 10$ **54.** $6x^2 + 5x = 6$

55. $x^2 + 3x - 6 = 0$ **56.** $3x^2 - 11x + 10 = 0$ **57.** $-3x^2 - 5x + 8 = 0$

58. $-2x^2 + 6x = -9$ **59.** $2x^2 - 5x = 0$ **60.** $3x^2 + 5x = 0$ 0,

61. $x^2 + \dfrac{5x}{4} = \dfrac{3}{8}$ **62.** $x^2 = \dfrac{5}{6}x + \dfrac{25}{6}$

[9.3] *Graph the solution to each inequality on the number line.*

63. $x^2 + 6x + 5 \geq 0$ **64.** $x^2 + 2x - 15 \leq 0$

65. $x^2 \leq 11x - 30$ **66.** $2x^2 + 6x > 0$

67. $3x^2 + 8x > 16$ **68.** $4x^2 - 16 \leq 0$

69. $5x^2 - 25 > 0$ **70.** $9x^2 > 25$

Solve each inequality and give the solution in both interval notation and set builder notation.

71. $\dfrac{x + 2}{x - 3} > 0$ **72.** $\dfrac{x - 5}{x + 2} \leq 0$

73. $\dfrac{2x - 4}{x + 1} \geq 0$ **74.** $\dfrac{3x + 5}{x - 6} < 0$

75. $(x + 3)(x + 1)(x - 2) > 0$ **76.** $x(x - 3)(x - 5) \leq 0$

77. $(x + 4)(x - 1)(x - 3) \geq 0$ **78.** $x(x + 2)(x + 5) < 0$

79. $\dfrac{x(x - 4)}{x + 2} > 0$ **80.** $\dfrac{(x - 2)(x - 5)}{x + 3} < 0$

81. $\dfrac{x - 3}{(x + 2)(x - 5)} \geq 0$ **82.** $\dfrac{x(x - 5)}{x + 3} \leq 0$

Solve each inequality and graph the solution on the number line.

83. $\dfrac{3}{x + 4} \geq -1$

84. $\dfrac{2x}{x - 2} \leq 1$

85. $\dfrac{2x + 3}{x - 5} < 2$

86. $\dfrac{4x}{x + 5} \leq 3$

[9.4] *(a) Determine whether the parabola opens upward or downward, (b) find the y intercept, (c) find the vertex, (d) find the x intercepts if they exist, and (e) sketch the graph.*

87. $y = x^2 + 6x$

88. $y = x^2 + 2x - 8$

89. $y = 2x^2 + 4x - 16$

90. $y = -x^2 - 9$

91. $y = -2x^2 - x + 15$

Write an equation in two variables that has the given x intercepts.

93. $3, -2$

94. $\dfrac{2}{3}, -3$

95. $-3, -3$

96. $\dfrac{1}{2}, \dfrac{2}{3}$

[9.1–9.4]

97. The product of two consecutive positive integers is 90. Find the integers.

98. The larger of two positive numbers is 4 greater than the smaller. Find the two numbers if their product is 45.

99. The length of a rectangle is 1 inch less than twice its width. Find the sides of the rectangle if its area is 66 square inches.

100. The value, V, of a wheat crop per acre, in dollars, d days after planting is given by the formula $V = 12d - 0.05d^2$, $20 < d < 80$. Find the value of an acre of wheat after it has been planted 50 days.

101. The distance an object is from the ground, in feet, t seconds after being dropped from an airplane is given by the formula $d = -16t^2 + 1800$. Find the distance the object is from the ground 3 seconds after it has been dropped.

102. If an object is dropped from the top of a 100-foot-tall building, its height above the ground, h, at any time, t, can be found by the formula $h = -16t^2 + 100$, $t < 2.25$ seconds. Find the height of the object at 2 seconds.

103. A tractor has an oil leak. The amount of oil in milliliters that leaks out per hour is a function of the tractor's operating temperature, t, in degrees Celsius. The formula that can be used to determine the amount of oil that leaks out, L, is $L = 0.0004t^2 + 0.16t + 20$, $100°C \leq t \leq 160°C$. (a) How many milliliters of oil will leak out in one hour if the operating temperature of the tractor is $100°C$? (b) If oil is leaking out at a rate of 53 milliliters per hour, what is the operating temperature of the tractor?

104. Two molding machines can complete an order in 12 hours. The larger machine could complete the order by itself in 1 hour less time than the smaller machine could by itself. How long will it take each machine to complete the order working by itself.

105. When a cannon is fired at a certain angle, the height of its shell above the ground, in feet, at any time, t, can be calculated by the formula $h = -5t^2 + 26t + 8$.

(a) Find the time at which the shell reaches its maximum height.

(b) Find the maximum height.

106. Ruben throws a rock upward from the top of a building 64 feet tall. The rock's distance from the ground, d, in feet, can be found by the formula $d = -16t^2 + 32t + 64$.

(a) Find the time, t, at which the rock attains its maximum height.

(b) Find the maximum height obtained by the rock.

[9.5] *Given* $f(x) = x^2 - 3x + 4$ *and* $g(x) = 2x - 5$, *find the following:*

107. $(f + g)(x)$ **108.** $(f + g)(3)$ **109.** $(g - f)(x)$ **110.** $(g - f)(-1)$

111. $(fg)(x)$ **112.** $(fg)(5)$ **113.** $\left(\dfrac{f}{g}\right)(x)$ **114.** $\left(\dfrac{f}{g}\right)(2)$

115. $(f \circ g)(x)$ **116.** $(f \circ g)(2)$ **117.** $(g \circ f)(x)$ **118.** $(g \circ f)(-3)$

Given $f(x) = 3x + 2$ *and* $g(x) = \sqrt{x - 4}$, $x \geq 4$, *find the following:*

119. $(f + g)(x)$ **120.** $(f - g)(x)$ **121.** $(fg)x$ **122.** $\left(\dfrac{g}{f}\right)(x)$

123. $(f \circ g)(x)$ **124.** $(g \circ f)(x)$

[9.6] *Determine which functions are one-to-one.*

125.   **126.**   **127.**   **128.**

129. $\{(2, 3), (4, 0), (-5, 7), (3, 8)\}$ **130.** $\left\{(0, -2), (5, 6), (3, -2), \left(\dfrac{1}{2}, 4\right)\right\}$

Give the domain and range of $f(x)$ *and* $f^{-1}(x)$.

131. Function: $\{(5, 3), (6, 2), (-4, -3), (0, 7)\}$ **132.** Function: $\left\{\left(\dfrac{1}{2}, 2\right), (-3, 8), (-1, 3), (\sqrt{5}, \sqrt{7})\right\}$

Find $f^{-1}(x)$ *and graph* $f(x)$ *and* $f^{-1}(x)$ *on the same set of axes.*

133. $y = f(x) = 4x - 2$ **134.** $y = f(x) = -3x - 5$

135. $y = f(x) = \dfrac{2x + 5}{3}$ **136.** $y = f(x) = \dfrac{3}{5} - \dfrac{2}{3}x$

137. Show that for $f(x) = \dfrac{5x - 4}{2}$ and $f^{-1}(x) = \dfrac{2x + 4}{5}$, $(f \circ f^{-1})(x) = x$ and $(f^{-1} \circ f)(x) = x$.

Practice Test

Solve by completing the square.

1. $x^2 = -x + 12$ **2.** $4x^2 + 8x = -12$

Solve by the quadratic formula.

3. $x^2 - 5x - 6 = 0$ **4.** $x^2 + 5 = -8x$

Solve by the method of your choice.

5. $3x^2 - 5x = 0$ **6.** $-2x^2 = 9x - 5$

7. Solve the formula $P = 3a - b^2$ for b. Use only the positive root.

8. Determine whether the following equation has two distinct real solutions, a single unique solution, or no real solution: $5x^2 = 4x + 2$.

Graph the solution to the inequality on the number line.

9. $x^2 - x \geq 42$

10. $\dfrac{(x + 3)(x - 4)}{x + 1} \geq 0$

Solve the inequality. Write the answer in (a) interval notation and (b) set builder notation.

11. $\dfrac{x + 3}{x + 2} \leq -1$

In problems 12 and 13, (a) determine whether the parabola opens upward or downward, (b) find the y intercept, (c) find the vertex, (d) find the x intercepts if they exist, and (e) sketch the graph.

12. $y = x^2 - 2x - 8$

13. $y = -2x^2 - 3x + 9$

14. Write an equation in two variables that has x intercepts of $-6, \frac{1}{2}$.

15. The length of a rectangle is 4 feet greater than twice its width. Find the length and width of the rectangle if the area of the rectangle is 48 square feet.

16. Kerry throws a ball upward from the top of a building. The distance, d, of the ball from the ground at any time t is $d = -16t^2 + 64t + 80$.

    (a) Find the time the object reaches its maximum height.

    (b) Find the maximum height.

Given $f(x) = x^2 - x + 8$ and $g(x) = 2x - 4$, find the following:

17. $(g - f)(x)$

18. $(f \circ g)(x)$

19. Find the inverse function of $f(x)$.

$$y = f(x) = 2x + 4$$

10

Conic Sections

See Section 10.5, Exercise 38.

10.1

The Circle

► **1**   Identify and describe the conic sections.

► **2**   Graph circles with the center at the origin.

► **3**   Graph circles with the center at (h, k).

Conic Sections

► **1**   In Chapter 9 we discussed parabolas. A parabola is one type of conic section. Parabolas will be discussed further in Section 10.3. Other conic sections are circles, ellipses, and hyperbolas. Each of these shapes is called a conic section because each can be made by slicing a cone and observing the shape of the resulting slice. The methods used to slice the cone to obtain each individual conic section are illustrated in Fig. 10.1.

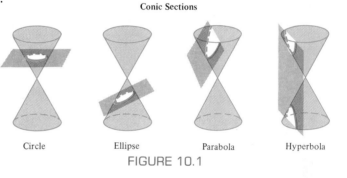

Conic Sections

Circle   Ellipse   Parabola   Hyperbola

FIGURE 10.1

The Circle

► **2**   A **circle** may be defined as the set of points in a plane that are the same distance from a fixed point called its **center.**

The formula for the **standard form** of a circle may be derived using the distance formula discussed in section 8.7. Let (x, y) be a point on a circle of radius r with the center at $(0, 0)$ (see Fig. 10.2). Then, using the distance formula,

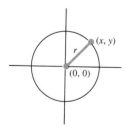

FIGURE 10.2

$$d = \sqrt{(x_2 - x_1)^2 + (y_2 - y_1)^2}$$
$$\text{or} \quad r = \sqrt{(x - 0)^2 + (y - 0)^2}$$
$$r = \sqrt{x^2 + y^2}$$
$$r^2 = x^2 + y^2 \quad \text{or} \quad x^2 + y^2 = r^2$$

Circle with Center at the Origin and Radius r

$$x^2 + y^2 = r^2$$

For example, $x^2 + y^2 = 16$ is a circle with center at the origin and a radius of 4, and $x^2 + y^2 = 7$ is a circle with center at the origin and a radius of $\sqrt{7}$. Note that $4^2 = 16$ and $(\sqrt{7})^2 = 7$.

EXAMPLE 1   Sketch the graph of $x^2 + y^2 = 64$.

Solution: If we rewrite the equation as

$$x^2 + y^2 = 8^2$$

We see that the radius is 8. The graph is illustrated in Fig. 10.3.

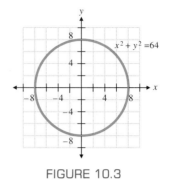

FIGURE 10.3 ■

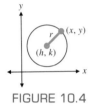

FIGURE 10.4

▶ **3**   The standard form of a circle with center at (h, k) and radius r can also be derived using the distance formula. Let (h, k) be the center of the circle and let (x, y) be any point on the circle (see Fig. 10.4). If the radius r represents the distance between points (x, y) and (h, k), then by the distance formula

$$r = \sqrt{(x - h)^2 + (y - k)^2}$$

We now square both sides of the equation to obtain the standard form of a circle with center at (h, k) and radius r.

$$r^2 = (x - h)^2 + (y - k)^2$$

Circle with Center at (h, k) and Radius r

$$(x - h)^2 + (y - k)^2 = r^2$$

EXAMPLE 2   (a) Determine the equation of the circle with center at $(3, -1)$ with radius 4.
(b) Sketch the circle.

Solution: (a) The center is $(3, -1)$. Thus h has a value of 3 and k is -1. The radius, r, is 4.

$$(x - h)^2 + (y - k)^2 = r^2$$
$$(x - 3)^2 + [y - (-1)]^2 = 4^2$$
$$(x - 3)^2 + (y + 1)^2 = 16$$

(b)

Note that each point on the circle is four units from the center.   ■

EXAMPLE 3   Determine the equation of the circle shown in Fig. 10.5.

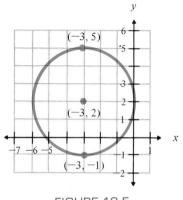

FIGURE 10.5

Solution: The center is $(-3, 2)$ and the radius is 3. The equation is therefore

$$[x - (-3)]^2 + (y - 2)^2 = 3^2$$
$$(x + 3)^2 + (y - 2)^2 = 9$$ ■

EXAMPLE 4   (a) Show that the equation $x^2 + y^2 + 6x - 2y + 6 = 0$ is an equation of a circle by using the procedure for completing the square discussed in Section 9.1 to rewrite this equation in standard form.
(b) Determine the center and radius of the circle and then sketch the circle.

Solution: (a) First rewrite the equation, placing all the terms containing like variables together.

$$x^2 + 6x + y^2 - 2y + 6 = 0$$

Then move the constant to the right side of the equation.

$$x^2 + 6x + y^2 - 2y = -6$$

Now we complete the square twice, once for each variable. We will first work with the variable x.

$$x^2 + 6x \boxed{+ 9} + y^2 - 2y = -6 \boxed{+ 9}$$

Now work with the variable y.

$$x^2 + 6x + 9 + y^2 - 2y \boxed{+ 1} = -6 + 9 \boxed{+ 1}$$

or

$$\underbrace{x^2 + 6x + 9}_{(x + 3)^2} + \underbrace{y^2 - 2y + 1}_{(y - 1)^2} = 4$$
$$(x + 3)^2 \ + \ (y - 1)^2 \ = 4$$
$$(x + 3)^2 \ + \ (y - 1)^2 \ = 2^2$$

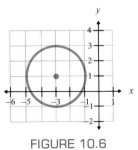

FIGURE 10.6

(b) The center of the circle is at $(-3, 1)$ and the radius is 2. The circle is sketched in Fig. 10.6.   ■

Exercise Set 10.1

Write the equation of the circle with the given center and radius; then sketch the graph of the equation.

1. Center (0, 0), radius 3

2. Center (0, 0), radius 5

3. Center (3, 0), radius 1

4. Center (0, −2), radius 7

5. Center (−6, 5), radius 5

6. Center (−4, −1), radius 4

7. Center (4, 7), radius $\sqrt{8}$

8. Center (0, −2), radius $\sqrt{12}$

Sketch the graph of each equation.

9. $x^2 + y^2 = 16$

10. $x^2 + y^2 = 9$

11. $x^2 + y^2 = 3$

12. $x^2 + y^2 = 10$

13. $x^2 + (y - 3)^2 = 4$

14. $(x + 4)^2 + y^2 = 25$

15. $(x - 2)^2 + (y + 3)^2 = 16$

16. $(x + 8)^2 + (y + 2)^2 = 9$

17. $(x + 1)^2 + (y - 4)^2 = 36$

Determine the equation of the circle.

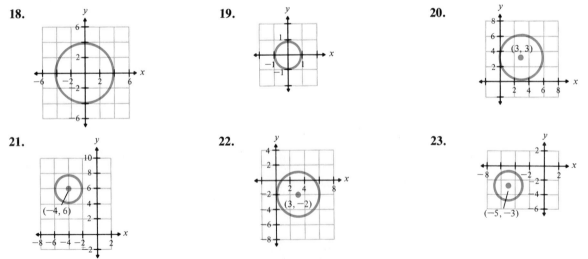

18.

19.

20.

21.

22.

23.

Use the completing the square procedure to write the equation in standard form; then sketch the graph. See Example 4.

24. $x^2 + 4x + y^2 - 12 = 0$

25. $x^2 + y^2 + 10y - 75 = 0$

26. $x^2 + y^2 - 4y = 0$

27. $x^2 + 8x - 9 + y^2 = 0$

28. $x^2 + y^2 + 6x - 4y + 9 = 0$

29. $x^2 + y^2 + 2x - 4y - 4 = 0$

30. $x^2 + y^2 + 4x - 6y - 3 = 0$

31. $x^2 + y^2 + 6x - 2y + 6 = 0$

32. $x^2 + y^2 + 8x - 4y + 4 = 0$

33. $x^2 + y^2 - 8x + 2y + 13 = 0$

34. What is the definition of a circle?

Cumulative Review Exercises

[6.3] **35.** Factor $8x^3 - 64$.

Solve the formula for the indicated variable.

[7.5] **36.** $E = 1 - \dfrac{T_1}{T_2}$ for T_2.

[9.1] **37.** $E = \dfrac{V + P^2}{2}$ for P.

[9.2] **38.** Solve using the quadratic formula
$$3x^2 - 5x + 5 = 0$$

10.2

The Ellipse

▸**1** Graph ellipses.

▸**1** An **ellipse** may be defined as a set of points in a plane, the sum of whose distances from two fixed points is a constant. The two fixed points are called the **foci** (each is a focus) of the ellipse (see Fig. 10.7).

FIGURE 10.7

We can construct an ellipse using a loop of string and two thumbtacks. Place the two thumbtacks fairly close together (Fig. 10.8). Then place the loop of string around the two thumbtacks. With a pencil or pen pull the string taut, and while keeping the string taut, draw the ellipse by moving the pencil around the thumbtacks.

FIGURE 10.8

The standard form of an ellipse with its center at the origin (Fig. 10.9) follows.

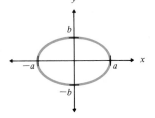

FIGURE 10.9

Ellipse with Center at the Origin

$$\frac{x^2}{a^2} + \frac{y^2}{b^2} = 1$$

where a and $-a$ are the x intercepts and b and $-b$ are the y intercepts.

In Fig. 10.9 the line segment from $-a$ to a on the x axis is the *longer* or **major axis** and the line segment from $-b$ to b is the *shorter* or **minor axis** of the ellipse.

EXAMPLE 1   Sketch the graph of $\dfrac{x^2}{9} + \dfrac{y^2}{4} = 1$.

Solution: We can rewrite the equation as

$$\frac{x^2}{3^2} + \frac{y^2}{2^2} = 1$$

Thus $a = 3$ and the x intercepts are ± 3. Since $b = 2$, the y intercepts are ± 2. The ellipse is illustrated in Fig. 10.10.

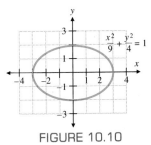

FIGURE 10.10

An equation may be camouflaged so that it may not be obvious that its graph is an ellipse. This is illustrated in Example 2.

EXAMPLE 2   Sketch the graph of $20x^2 + 9y^2 = 180$.

Solution: If we divide both sides of the equation by 180 to make the right side of the equation equal to 1, we obtain an equation that we can recognize as an ellipse.

$$\frac{20x^2 + 9y^2}{180} = \frac{180}{180}$$

$$\frac{20x^2}{180} + \frac{9y^2}{180} = 1$$

$$\frac{x^2}{9} + \frac{y^2}{20} = 1$$

The equation can now be recognized as an ellipse in standard form.

$$\frac{x^2}{a^2} + \frac{y^2}{b^2} = 1$$

Since $a^2 = 9$, $a = 3$. We know that $b^2 = 20$; thus $b = \sqrt{20}$ (or approximately 4.47).

$$\frac{x^2}{3^2} + \frac{y^2}{(\sqrt{20})^2} = 1$$

The x intercepts are at ± 3. The y intercepts are at $\pm\sqrt{20}$. The graph is illustrated in Fig. 10.11. Note that the major axis lies along the y axis instead of along the x axis.

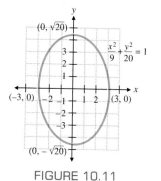

FIGURE 10.11

EXAMPLE 3 Write the equation of the ellipse illustrated in Fig. 10.12.

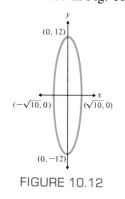

FIGURE 10.12

Solution: The x intercepts are $\pm\sqrt{10}$; thus $a = \sqrt{10}$ and $a^2 = 10$. The y intercepts are ± 12; thus $b = 12$ and $b^2 = 144$.

$$\frac{x^2}{a^2} + \frac{y^2}{b^2} = 1$$

$$\frac{x^2}{10} + \frac{y^2}{144} = 1$$

Exercise Set 10.2

Sketch the graph of each equation.

1. $\dfrac{x^2}{4} + \dfrac{y^2}{1} = 1$

2. $\dfrac{x^2}{9} + \dfrac{y^2}{4} = 1$

3. $\dfrac{x^2}{4} + \dfrac{y^2}{9} = 1$

4. $\dfrac{x^2}{25} + \dfrac{y^2}{9} = 1$

5. $\dfrac{x^2}{16} + \dfrac{y^2}{25} = 1$

6. $\dfrac{x^2}{9} + \dfrac{y^2}{121} = 1$

7. $9x^2 + 12y^2 = 108$

8. $9x^2 + 4y^2 = 36$

9. $25x^2 + 16y^2 = 400$

10. $x^2 + 36y^2 = 36$

11. $9x^2 + 25y^2 = 225$

12. $x^2 + 2y^2 = 8$

13. What is the definition of an ellipse?

14. In your own words, discuss the graphs of $\dfrac{x^2}{a^2} + \dfrac{y^2}{b^2} = 1$ when $a > b$, $a < b$, and $a = b$.

15. Consider the graph of the equation $\dfrac{x^2}{a^2} + \dfrac{y^2}{b^2} = 1$, where $a > b$. Explain what will happen to the shape of the graph as the value of b gets closer to the value of a. What is the shape of the graph when $a = b$?

Cumulative Review Exercises

Solve the following inequalities. Indicate the solution on the number line.

[2.5] **16.** $-3 \le 4 - \frac{1}{2}x < 6$

[2.6] **17.** $|2x - 4| = 8$

18. $|2x - 4| \le 8$

19. $|2x - 4| > 8$

JUST FOR FUN

The standard form of an ellipse with center at (h, k) is

$$\frac{(x - h)^2}{a^2} + \frac{(y - k)^2}{b^2} = 1$$

where a and b are distances from the center to the vertices, as shown.

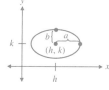

Sketch a graph of the following equations.

1. $\dfrac{x^2}{16} + \dfrac{(y - 2)^2}{9} = 1$

2. $\dfrac{(x - 4)^2}{9} + \dfrac{(y + 3)^2}{25} = 1$

3. Write the following equation in standard form. Determine the center of the ellipse and then sketch the ellipse.

$$x^2 + 4y^2 - 4x - 8y - 92 = 0$$

10.3

The Parabola

▶ **1** Derive the equation $y = a(x - h)^2 + k$.

▶ **2** Graph parabolas of the forms $y = a(x - h)^2 + k$ and $x = a(y - k)^2 + h$.

▶ **3** Convert equations from $y = ax^2 + bx + c$ form to $y = a(x - h)^2 + k$ form.

▶**1**   We have discussed the parabola in Chapters 5 and 9. In this section we will discuss the parabola further. In Section 9.4 we began with a quadratic equation of the form $y = ax^2 + bx + c$. By completing the square, we obtained

$$y = a\left(x + \frac{b}{2a}\right)^2 + \frac{4ac - b^2}{4a}$$

We then showed that the coordinates of the vertex of the parabola are

$$\left(\frac{-b}{2a}, \frac{4ac - b^2}{4a}\right)$$

Since a, b, and c are all constants, the expressions $-b/2a$ and $(4ac - b^2)/4a$ will also be constants. Let's do a little simplification by letting $h = -b/2a$ and $k = (4ac - b^2)/4a$. Then, by substitution, we get

$$y = a\left[x - \left(\frac{-b}{2a}\right)\right]^2 + \frac{4ac - b^2}{4a}$$
$$y = a(x - h)^2 + k$$

We can therefore reason that an equation of the form $y = a(x - h)^2 + k$ will be a parabola with its vertex at the point (h, k). If a in the equation $y = a(x - h)^2 + k$ is a positive number, the parabola will open upward, and if a is a negative number, the parabola will open downward.

   Parabolas can also open to the right or left. The graph of an equation of the form $x = a(y - k)^2 + h$ will be a parabola whose vertex is at the point (h, k). If a is a positive number, the parabola will open to the right, and if a is a negative number, the parabola will open to the left. The four different forms of a parabola are shown in Fig. 10.13.

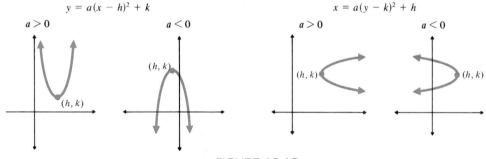

FIGURE 10.13

Parabola with Vertex at (h, k)
1. $y = a(x - h)^2 + k, a > 0$ (opens upward)
2. $y = a(x - h)^2 + k, a < 0$ (opens downward)
3. $x = a(y - k)^2 + h, a > 0$ (opens to the right)
4. $x = a(y - k)^2 + h, a < 0$ (opens to the left)

   Note that equations of the form $y = a(x - h)^2 + k$ are functions since their graphs pass the vertical line test. However, equations of the form $x = a(y - k)^2 + h$ are not functions since their graphs do not pass the vertical line test.

▸**2**   Now we will sketch some parabolas.

EXAMPLE 1   Sketch the graph of $y = (x - 2)^2 + 3$.

Solution: The graph opens upward since the equation is of the form $y = a(x - h)^2 + k$ and $a = 1$, which is greater than 0. The vertex is at $(2, 3)$ (see Fig. 10.14). Note that when $x = 0$, the y intercept is $(0 - 2)^2 + 3 = 4 + 3$ or 7.

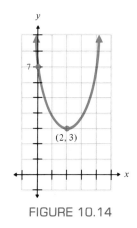

FIGURE 10.14   ■

EXAMPLE 2   Sketch the graph of $x = -2(y + 4)^2 - 1$.

Solution: The graph opens to the left since the equation is of the form $x = a(y - k)^2 + h$ and $a = -2$, which is less than 0 (Fig. 10.15). The equation can be expressed as $x = -2[y - (-4)]^2 - 1$. Thus $h = -1$ and $k = -4$. The vertex of the graph is $(-1, -4)$. When $y = 0$, we see that the x intercept is at $-2(0 + 4)^2 - 1 = -2(16) - 1$ or -33.

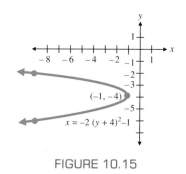

$$x = -2\,(y + 4)^2 - 1$$

FIGURE 10.15   ■

▸**3**   In Examples 3 and 4, we will convert an equation from $y = ax^2 + bx + c$ form to $y = a(x - h)^2 + k$ form before graphing.

EXAMPLE 3   (a) Write the equation $y = x^2 - 6x + 8$ in $y = a(x - h)^2 + k$ form.
(b) Sketch the graph of $y = x^2 - 6x + 8$.

Solution: (a) We convert $y = x^2 - 6x + 8$ to $y = a(x - h)^2 + k$ form by completing the square.

$$y = x^2 - 6x + 8$$

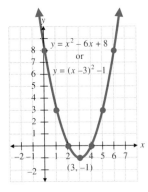

$y = x^2 - 6x + 8$
or
$y = (x - 3)^2 - 1$

$(3, -1)$

FIGURE 10.16

Take one-half the coefficient of the x term; then square it.

$$\frac{-6}{2} = -3, \qquad (-3)^2 = 9$$

Now add $+9$ and -9 to the equation to obtain

$$y = x^2 - 6x + 9 - 9 + 8$$

By doing this, we have created a perfect square trinomial plus some constant.

$$y = \underbrace{x^2 - 6x + 9}\ \underbrace{- 9 + 8}$$
$$y = \quad (x - 3)^2 \qquad -1$$

(b) The vertex of the parabola is at $(3, -1)$ and the parabola opens upward since $a = 1$. The y intercept is at $(0 - 3)^2 - 1 = 9 - 1$ or 8. The graph is sketched in Fig. 10.16.   ■

EXAMPLE 4   (a) Write the equation $y = 2x^2 + 4x - 6$ in $y = a(x - h)^2 + k$ form.
(b) Sketch the graph of $y = 2x^2 + 4x - 6$.

Solution: (a) First, factor 2 from the two terms containing the variable to make the coefficient of the squared term equal to 1. Do not factor 2 from the constant, -6.

$$y = 2x^2 + 4x - 6$$
$$y = 2(x^2 + 2x) - 6$$

Now complete the square by taking one-half the coefficient of the x term and squaring it.

$$\frac{2}{2} = 1, \qquad 1^2 = 1$$

Now add $+1$ inside the parentheses. Since the terms inside the parentheses are multiplied by 2, we are really adding $2(1)$ or 2. Therefore, we must also add a -2 to the right of the parentheses. In doing this, we are not changing the equation, since we are adding 2 and -2, which sums to zero.

$$y = 2(x^2 + 2x + 1) - 2 - 6$$
$$y = 2(x + 1)^2 - 8$$

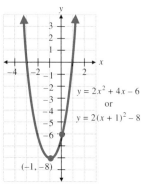

$y = 2x^2 + 4x - 6$
or
$y = 2(x + 1)^2 - 8$

$(-1, -8)$

FIGURE 10.17

(b) This parabola opens upward since $a = 2$, which is greater than 0. Its vertex is at $(-1, -8)$. The y intercept is at $2(0 + 1)^2 - 8 = 2(1) - 8$ or -6 (see Fig. 10.17).   ■

EXAMPLE 5   (a) Write the equation $x = -2y^2 + 4y + 5$ in $x = a(y - k)^2 + h$ form.
(b) Sketch the graph of $x = -2y^2 + 4y + 5$.

Solution: (a)
$$x = -2y^2 + 4y + 5$$
$$x = -2(y^2 - 2y) + 5$$
$$x = -2(y^2 - 2y + 1) + 2 + 5$$
$$x = -2(y - 1)^2 + 7$$

(b) Since $a < 0$, this parabola opens to the left. Note that when $y = 0$, $x = 5$. Therefore, the x intercept is 5. The vertex of the parabola is $(7, 1)$. The graph is shown in Fig. 10.18 on page 478.

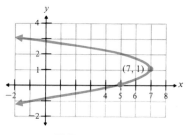

FIGURE 10.18

Exercise Set 10.3

Sketch the graph of each equation.

1. $y = (x - 2)^2 + 3$

2. $y = (x + 1)^2 - 2$

3. $y = -(x - 3)^2 - 6$

4. $x = (y + 6)^2 + 1$

5. $x = (y - 4)^2 - 3$

6. $x = -(y - 5)^2 + 4$

7. $y = 2(x + 6)^2 - 4$

8. $y = -3(x - 5)^2 + 3$

9. $x = -5(y + 3)^2 - 6$

10. $x = -(y - 7)^2 + 8$

11. $y = -2\left(x + \dfrac{1}{2}\right)^2 + 6$

12. $y = -\left(x - \dfrac{5}{2}\right)^2 + \dfrac{1}{2}$

Write each equation in the form $y = a(x - h)^2 + k$ or $x = a(y - k)^2 + h$, and then sketch the graph of the equation.

13. $y = x^2 + 2x$

14. $y = x^2 - 4x$

15. $x = y^2 + 6y$

16. $y = x^2 - 6x + 8$

17. $y = x^2 + 2x - 15$

18. $x = y^2 + 8y + 7$

19. $x = -y^2 + 6y - 9$

20. $y = -x^2 + 4x - 4$

21. $y = x^2 + 7x + 10$

22. $x = -y^2 + 3y - 4$

23. $y = 2x^2 - 4x - 4$

24. $x = 3y^2 - 12y - 36$

25. Explain how to determine the direction the parabola will open by observing the equation.

26. When functions of the form $y = a(x - h)^2 + k$ are graphed, their graphs will be parabolas with vertex at the point (h, k). By studying the equation, explain why the vertex occurs at (h, k).

27. Will all parabolas of the form $y = a(x - h)^2 + k$,

$a > 0$, be functions? Explain your answer. What will be the domain and range of $y = a(x - h)^2 + k$, $a > 0$?

28. Will all parabolas of the form $x = a(y - k)^2 + h$, $a > 0$, be functions? Explain your answer. What will be the domain and range of $x = a(y - k)^2 + h$, $a > 0$?

29. How will the graphs of $y = 2(x - 3)^2 + 4$ and $y = -2(x - 3)^2 + 4$ compare with each other? Explain your answer.

Cumulative Review Exercises

[3.4] **30.** Write the equation, in slope–intercept form, of the graph that passes through the points $(-6, 4)$ and $(-2, 2)$.

[4.4] **31.** Evaluate the determinant: $\begin{vmatrix} 4 & 0 & 3 \\ 5 & 2 & -1 \\ 3 & 6 & 4 \end{vmatrix}$.

[5.7] **32.** $f(x) = -2x^2 - x + 2$; find (a) $f(\frac{1}{3})$; (b) $f(a + b)$.

[7.7] **33.** T varies jointly as m_1 and m_2 and inversely as the square of R. If $T = \frac{3}{2}$ when $m_1 = 6$, $m_2 = 4$, and $R = 4$, find T when $m_1 = 6$, $m_2 = 10$, and $R = 2$.

10.4

The Hyperbola

▸ **1** Identify hyperbolas.

▸ **2** Graph hyperbolas in standard form using asymptotes.

▸ **3** Graph hyperbolas in nonstandard form.

▸ **4** Review conic sections.

▸ **5** Identify nonstandard forms of conic sections.

▸ **1** A **hyperbola** is the set of all points in a plane the difference of whose distances from two fixed points (called foci) is a constant. A hyperbola looks like a pair of parabolas (Fig. 10.19). However, the shapes are actually quite different. A hyperbola has two **vertices.** The point halfway between the two vertices is the **center** of the hyperbola. The dashed lines in the figure are called **asymptotes.** The asymptotes are not a part of the hyperbola, but are used as an aid in graphing the hyperbola. We will discuss asymptotes further shortly.

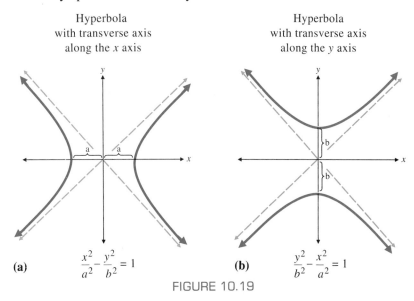

FIGURE 10.19

Also given in Fig. 10.19 is the standard form of the equation for each hyperbola. The axis through the center that intersects the vertices is called the **transverse axis.** The axis perpendicular to the transverse axis is the **conjugate axis.** In Fig. 10.19a, the transverse axis lies along the x axis and the conjugate axis lies along the y axis. Both vertices are a units from the origin. In Fig. 10.19b, the transverse axis lies along the y axis and the conjugate axis lies along the x axis. Both vertices are b units from the origin. Note that in the standard form of the equation the denominator of the x^2 term is always a^2 and the denominator of the y^2 term is always b^2.

A hyperbola whose transverse and conjugate axes are parallel to the coordinate axes has either x intercepts (Fig. 10.19a) or y intercepts (Fig. 10.19b), but not both. When written in standard form, the intercepts will be on the axis indicated by the variable with the positive coefficient. For example, $\frac{x^2}{25} - \frac{y^2}{9} = 1$ will intersect the x axis, and $\frac{y^2}{9} - \frac{x^2}{25} = 1$ will intersect the y axis. In either case the intercepts will be the positive and the negative square root of the denominator of the positive term (which will be the first term when written in standard form). Thus $\frac{x^2}{25} - \frac{y^2}{9} = 1$ has x intercepts of ±5 and $\frac{y^2}{9} - \frac{x^2}{25} = 1$ has y intercepts of ±3. Note that the intercepts are the vertices of the hyperbola.

▶ **2 Asymptotes** can be used as an aid in graphing hyperbolas. The asymptotes are two straight lines that go through the center of the hyperbola. As the values of x and y get larger, the graph of the hyperbola will approach the asymptotes. The asymptotes are an aid in graphing the hyperbola. The equations of the asymptotes are determined using a and b. The equations of the asymptotes of a hyperbola whose center is the origin are

$$y = \frac{b}{a}x \quad \text{and} \quad y = -\frac{b}{a}x$$

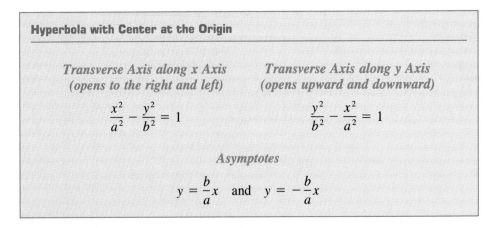

Hyperbola with Center at the Origin

Transverse Axis along x Axis
(opens to the right and left)

$$\frac{x^2}{a^2} - \frac{y^2}{b^2} = 1$$

Transverse Axis along y Axis
(opens upward and downward)

$$\frac{y^2}{b^2} - \frac{x^2}{a^2} = 1$$

Asymptotes

$$y = \frac{b}{a}x \quad \text{and} \quad y = -\frac{b}{a}x$$

EXAMPLE 1 (a) Determine the equations of the asymptotes of the hyperbola with equation

$$\frac{x^2}{9} - \frac{y^2}{16} = 1$$

(b) Sketch the hyperbola using the asymptotes as an aid.

Solution: (a) The value of a^2 is 9; the positive square root of 9 is 3. The value of b^2 is 16; the positive square root of 16 is 4. The asymptotes are

$$y = \frac{b}{a}x \quad \text{and} \quad y = \frac{-b}{a}x$$

or

$$y = \frac{4}{3}x \quad \text{and} \quad y = \frac{-4}{3}x$$

(b) To graph the hyperbola, we first graph the asymptotes as illustrated in Fig. 10.20. Since the x term is positive, the graph intersects the x axis. Since the denominator of the positive term is 9, the vertices will be at 3 and -3. Now draw the hyperbola by letting the hyperbola approach its asymptotes (Fig. 10.21). Note that the asymptotes are drawn using dashed lines since they are not part of the hyperbola. They are used merely to help sketch the graph.

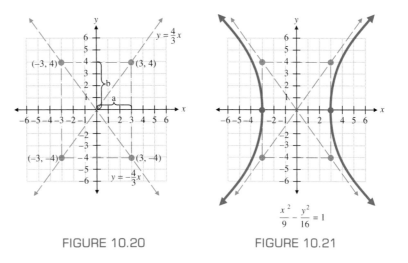

FIGURE 10.20   FIGURE 10.21   ■

EXAMPLE 2   (a) Show that the equation $-25x^2 + 4y^2 = 100$ is a hyperbola by expressing the equation in standard form.
(b) Determine the equation of the asymptotes of the graph.
(c) Sketch the graph.

Solution: (a) Divide both sides of the equation by 100 to obtain a 1 on the right side of the equation.

$$\frac{-25x^2 + 4y^2}{100} = \frac{100}{100}$$

$$\frac{-25x^2}{100} + \frac{4y^2}{100} = 1$$

$$\frac{-x^2}{4} + \frac{y^2}{25} = 1$$

Rewriting the equation in standard form (positive term first), we get

$$\frac{y^2}{25} - \frac{x^2}{4} = 1$$

(b) The equations of the asymptotes are

$$y = \frac{5}{2}x \quad \text{and} \quad y = \frac{-5}{2}x$$

(c) The graph intersects the y axis at 5 and -5. Figure 10.22a on p. 482 illustrates the asymptotes, and Fig. 10.22b illustrates the graph of the hyperbola.

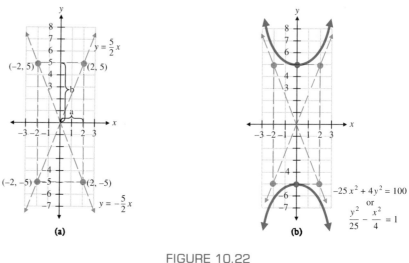

FIGURE 10.22

Nonstandard Form of the Hyperbola

▶ 3   Another form of the hyperbola is $xy = c$, where c is a nonzero constant. The following equations are examples of hyperbolas in nonstandard form:

$$xy = 4, \qquad xy = -6, \qquad x = \frac{1}{y}, \qquad y = -\frac{3}{x}$$

Note that $x = \dfrac{1}{y}$ is equivalent to $xy = 1$ and $y = \dfrac{-3}{x}$ is equivalent to $xy = -3$.

Equations of the form $xy = c$ will be hyperbolas with the x and y axes as asymptotes.

EXAMPLE 3   (a) Sketch the graph of $xy = 6$.
(b) Sketch the graph of $xy = -6$.

Solution   (a) $xy = 6$ or $y = \dfrac{6}{x}$. We will graph the equation $y = \dfrac{6}{x}$ in two parts. First we will make a table of values for x less than 0. Then we will make a table of values for x greater than 0. This will allow us to see what happens to the graph as the values of x approach 0 from the left and from the right, respectively. It is important that you realize that the equation is not defined when x is 0 since division by 0 is not permitted.

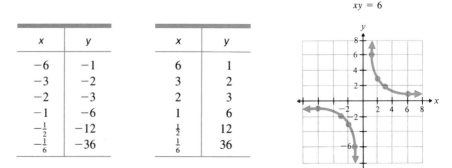

x	y
-6	-1
-3	-2
-2	-3
-1	-6
$-\frac{1}{2}$	-12
$-\frac{1}{6}$	-36

x	y
6	1
3	2
2	3
1	6
$\frac{1}{2}$	12
$\frac{1}{6}$	36

By looking at the tables and the graphs, we can see that as x gets closer and closer to 0 coming from the left (values of x less than 0), the values of y decrease.

As x gets closer and closer to zero coming from the right (values of x greater than 0), the values of y increase. The graph does not cross the y axis, which is what we expect since the equation $y = \dfrac{6}{x}$ is not defined when $x = 0$.

(b) $xy = -6$ or $y = \dfrac{-6}{x}$. We will follow the same procedure as in part (a) to graph $y = \dfrac{-b}{x}$.

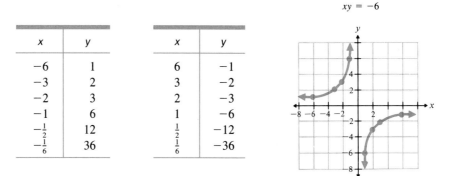

x	y
-6	1
-3	2
-2	3
-1	6
$-\frac{1}{2}$	12
$-\frac{1}{6}$	36

x	y
6	-1
3	-2
2	-3
1	-6
$\frac{1}{2}$	-12
$\frac{1}{6}$	-36

Summary of Conic Sections

▶ **4** The following chart summarizes conic sections.

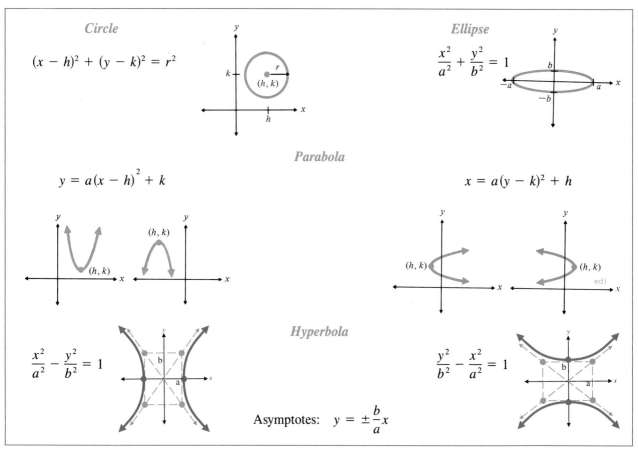

Circle

$$(x - h)^2 + (y - k)^2 = r^2$$

Ellipse

$$\frac{x^2}{a^2} + \frac{y^2}{b^2} = 1$$

Parabola

$$y = a(x - h)^2 + k$$

$$x = a(y - k)^2 + h$$

Hyperbola

$$\frac{x^2}{a^2} - \frac{y^2}{b^2} = 1$$

$$\frac{y^2}{b^2} - \frac{x^2}{a^2} = 1$$

Asymptotes: $y = \pm\dfrac{b}{a}x$

Nonstandard Forms of Conic Sections

▶ **5** Often, a conic section will be given in a nonstandard form. Conic sections given in nonstandard form can be recognized with a little practice. The circle, ellipse, and hyperbola can be discussed using the equation

$$ax^2 + by^2 = c^2$$

If a and b are both positive and $a = b$, the equation is a circle. For example, $4x^2 + 4y^2 = 16$ is a circle.

$$4x^2 + 4y^2 = 16$$

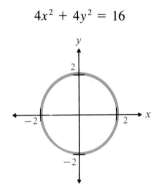

If a and b are both positive and $a \neq b$, the equation is an ellipse. If $a < b$, the major axis is along the x axis. If $a > b$, the major axis is along the y axis. For example,

$$4x^2 + 9y^2 = 36 \qquad\qquad 9x^2 + 4y^2 = 36$$

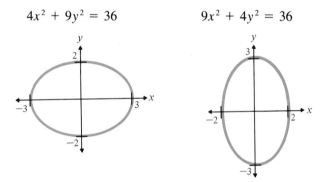

If a and b have opposite signs, the equation is a hyperbola. If $a > 0$, the hyperbola will have its transverse axis along the x axis and will intersect the x axis. If $b > 0$, the hyperbola will have its transverse axis along the y axis and will intersect the y axis. For example,

$$4x^2 - 9y^2 = 36 \qquad\qquad -4x^2 + 9y^2 = 36$$

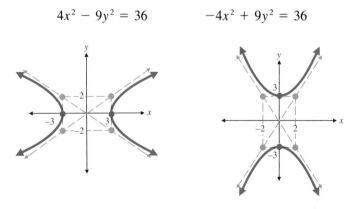

Nonstandard forms of parabolas are $y = ax^2 + bx + c$ (parabola opens up when $a > 0$, down when $a < 0$), and $x = ay^2 + by + c$ (parabola opens right when $a > 0$, left when $a < 0$).

Exercise Set 10.4

Determine the equations of the asymptotes, and then sketch the graph of the equation.

1. $\dfrac{x^2}{4} - \dfrac{y^2}{1} = 1$

2. $\dfrac{x^2}{9} - \dfrac{y^2}{4} = 1$

3. $\dfrac{y^2}{9} - \dfrac{x^2}{16} = 1$

4. $\dfrac{y^2}{25} - \dfrac{x^2}{4} = 1$

5. $\dfrac{y^2}{25} - \dfrac{x^2}{36} = 1$

6. $\dfrac{x^2}{9} - \dfrac{y^2}{25} = 1$

7. $\dfrac{x^2}{4} - \dfrac{y^2}{4} = 1$

8. $\dfrac{y^2}{49} - \dfrac{x^2}{100} = 1$

9. $\dfrac{y^2}{16} - \dfrac{x^2}{81} = 1$

10. $\dfrac{x^2}{25} - \dfrac{y^2}{16} = 1$

11. $\dfrac{y^2}{25} - \dfrac{x^2}{16} = 1$

12. $\dfrac{y^2}{4} - \dfrac{x^2}{36} = 1$

Write each equation in standard form, determine the equation of the asymptotes, and then sketch the graph.

13. $16x^2 - 4y^2 = 64$

14. $25x^2 - 16y^2 = 400$

15. $9y^2 - x^2 = 9$

16. $4y^2 - 25x^2 = 100$

17. $4y^2 - 36x^2 = 144$

18. $x^2 - 25y^2 = 25$

19. $25x^2 - 4y^2 = 100$

20. $25x^2 - 9y^2 = 225$

21. $81x^2 - 9y^2 = 729$

22. $64y^2 - 25x^2 = 1600$

Graph each hyperbola.

23. $xy = 10$

24. $xy = 1$

25. $xy = -8$

26. $xy = -4$

27. $y = \dfrac{12}{x}$

28. $y = -\dfrac{3}{x}$

29. What are asymptotes? How do you find the equations of the asymptotes of a hyperbola?

30. If you are only given the equations of the two asymptotes of a hyperbola, can you determine if the hyperbola will have its transverse axis along the x axis or y axis? Explain your answer.

31. Discuss the graph of $\dfrac{x^2}{a^2} - \dfrac{y^2}{b^2} = 1$ for non-zero real numbers a and b. Mention transverse and conjugate axes, vertices, and asymptotes.

32 Discuss the graph of $\dfrac{y^2}{b^2} - \dfrac{x^2}{a^2} = 1$ for non-zero real numbers a and b. Mention transverse and conjugate axes, vertices, and asymptotes.

33. Given a hyperbola with equation $\dfrac{x^2}{a^2} - \dfrac{y^2}{b^2} = 1$, where $a > b$; if this equation is graphed, and then the values

of a and b are switched and the new equation is graphed, how will the two graphs compare? Explain your answer.

34. Given a hyperbola with equation $\dfrac{x^2}{a^2} - \dfrac{y^2}{b^2} = 1$, where $a > b$; if this equation is graphed, and then the signs of each term on the left side of the equation are changed and the new equation is graphed, how will the two graphs compare? Explain your answer.

Indicate whether the equation when graphed will be a parabola, circle, ellipse, or hyperbola.

35. $y = 6x^2 + 4x + 3$

36. $4x^2 + 4y^2 = 16$

37. $5x^2 - 5y^2 = 25$

38. $9x^2 - 16y^2 = 36$

39. $x = y^2 + 6y - 7$

40. $x^2 - 4y^2 = 36$

41. $-2x^2 + 4y^2 = 16$

42. $3x^2 + 3y^2 = 12$

43. $5x^2 + 10y^2 = 12$

44. $9x^2 + 16y^2 = 144$

45. $x = 3y^2 - y + 7$

46. $6x^2 - 9y^2 = 36$

47. $6x^2 + 6y^2 = 36$

48. $-y^2 + 4x^2 = 16$

49. $-3x^2 - 3y^2 = -27$

50. $-6x^2 + 2y^2 = -6$

51. $-6y^2 + x^2 = -9$

52. $4x^2 - 9y^2 = 36$

53. Consider the equation $ax^2 + by^2 = c^2$, where a, b, and c represent real numbers. Write a report explaining under what conditions the graph of this equation will be (a) a circle, (b) an ellipse with major axis along the x axis, (c) an ellipse with major axis along the y axis,

(d) a hyperbola with transverse axis along the x axis, and (e) a hyperbola with transverse axis along the y axis. In your report, discuss the characteristics of each graph using a, b, and c to represent real numbers.

Cumulative Review Exercises

[2.1] **54.** Solve the equation $\dfrac{x}{2} + \dfrac{2}{3}(x - 6) = x + 4$.

[3.5] **55.** What are the range and the domain of a function?

[4.1] **56.** Determine the solution to the system of inequalities graphically.

$$6x - 2y < 12$$
$$y \geq -2x + 3$$

[9.2] **57.** Solve the quadratic equation $-2x^2 + 6x - 5 = 0$.

JUST FOR FUN

The standard forms of a hyperbola with center at (h, k) are

$$\dfrac{(x - h)^2}{a^2} - \dfrac{(y - k)^2}{b^2} = 1 \qquad \dfrac{(y - k)^2}{b^2} - \dfrac{(x - h)^2}{a^2} = 1$$

The asymptotes of both graphs are

$$y - k = \pm\dfrac{b}{a}(x - h)$$

First graph the asymptotes, and then sketch the graph.

1. $\dfrac{(x - 3)^2}{9} - \dfrac{(y + 2)^2}{4} = 1$ **2.** $\dfrac{(y + 5)^2}{16} - \dfrac{(x - 1)^2}{9} = 1$

3. Write the following equation in standard form, and then sketch the graph.

$$y^2 - 4x^2 + 2y + 8x - 7 = 0$$

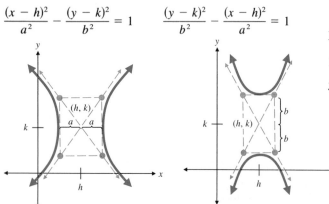

Nonlinear Systems of Equations and Their Applications

▶ **1** Identify a nonlinear system of equations.

▶ **2** Solve a nonlinear system using substitution.

▶ **3** Solve a nonlinear system using addition.

▶ **4** Use nonlinear systems to solve applied problems.

▶ **1** In Chapter 4 we discussed systems of linear equations. Here, we discuss nonlinear systems of equations. A **nonlinear system of equations** is a system of equations containing at least one that is not a linear equation (that is, one whose graph is not a straight line).

The solution to a system of equations is the point or points that satisfy all equations in the system. Consider the system of equations

$$x^2 + y^2 = 25$$
$$3x + 4y = 0$$

Both equations are graphed on the same set of axes in Fig. 10.23. Note that the graphs appear to intersect at the points $(-4, 3)$ and $(4, -3)$. The check shows these points satisfy all equations in the system and are therefore solutions to the system.

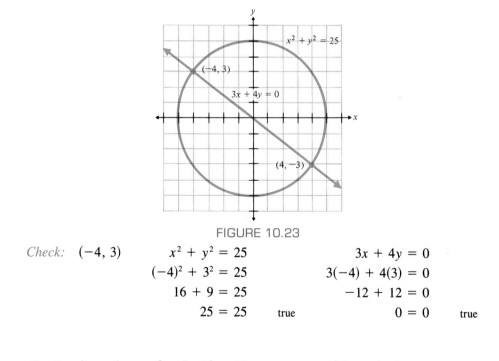

FIGURE 10.23

Check: $(-4, 3)$

$x^2 + y^2 = 25$	$3x + 4y = 0$
$(-4)^2 + 3^2 = 25$	$3(-4) + 4(3) = 0$
$16 + 9 = 25$	$-12 + 12 = 0$
$25 = 25$ true	$0 = 0$ true

Check: $(4, -3)$

$4^2 + (-3)^2 = 25$	$3(4) + 4(-3) = 0$
$16 + 9 = 25$	$12 - 12 = 0$
$25 = 25$ true	$0 = 0$ true

▶ **2** The graphical procedure for solving a system of equations may result in an inaccurate result since you will have to estimate the point or points of intersection. An exact answer may be obtained algebraically.

To solve a system of equations algebraically, we often solve one or more of the equations for one of the variables and then use the substitution principle. This procedure is illustrated in Examples 1 through 3.

EXAMPLE 1 Solve the previous system of equations algebraically using substitution.

$$x^2 + y^2 = 25$$
$$3x + 4y = 0$$

Solution: Solve the linear equation $3x + 4y = 0$ for either x or y. We will select to solve for y.

$$3x + 4y = 0$$
$$4y = -3x$$
$$y = -\frac{3x}{4}$$

Now substitute $-\dfrac{3x}{4}$ for y in the equation $x^2 + y^2 = 25$ and solve for the remaining variable, x.

$$x^2 + y^2 = 25$$
$$x^2 + \left(\frac{-3x}{4}\right)^2 = 25$$
$$x^2 + \frac{9x^2}{16} = 25$$
$$16\left(x^2 + \frac{9x^2}{16}\right) = 16(25)$$
$$16x^2 + 9x^2 = 400$$
$$25x^2 = 400$$
$$x^2 = \frac{400}{25} = 16$$
$$x = \pm\sqrt{16} = \pm 4$$

Next, find the corresponding value of y for each value of x by substituting each value of x (one at a time) in the equation solved for y.

$x = 4$	$x = -4$
$y = -\dfrac{3x}{4}$	$y = \dfrac{-3x}{4}$
$y = \dfrac{-3(4)}{4}$	$y = \dfrac{-3(-4)}{4}$
$y = -3$	$y = 3$

Solutions: $(4, -3)$ and $(-4, 3)$

This answer checks with the answer obtained graphically.   ■

Note that our objective in the substitution method is to make a substitution that will result in obtaining a single equation containing only one variable.

EXAMPLE 2   Solve the following system of equations using substitution.

$$y = x^2 - 3$$
$$x^2 + y^2 = 9$$

Solution: Since both equations contain an x^2 term, we will solve one of the equations for x^2. We will chose to solve $y = x^2 - 3$ for x^2.

$$y = x^2 - 3$$
$$y + 3 = x^2$$

Now substitute $y + 3$ for x^2 in the equation $x^2 + y^2 = 9$.

$$x^2 + y^2 = 9$$
$$(y + 3) + y^2 = 9$$
$$y^2 + y + 3 = 9$$
$$y^2 + y - 6 = 0$$
$$(y + 3)(y - 2) = 0$$
$$y + 3 = 0 \quad \text{or} \quad y - 2 = 0$$
$$y = -3 \qquad\qquad y = 2$$

Now find the corresponding values of x.

$$\begin{array}{ll} y = -3 & y = 2 \\ y = x^2 - 3 & y = x^2 - 3 \\ -3 = x^2 - 3 & 2 = x^2 - 3 \\ 0 = x^2 & 5 = x^2 \\ 0 = x & \pm\sqrt{5} = x \end{array}$$

Solutions: $(0, -3)$ $(\sqrt{5}, 2), (-\sqrt{5}, 2)$

This system has three solutions: $(0, -3)$, $(\sqrt{5}, 2)$, and $(-\sqrt{5}, 2)$.   ■

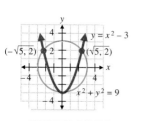

FIGURE 10.24

Note that in Example 2 the graph of the equation $y = x^2 - 3$ is a parabola and the graph of the equation $x^2 + y^2 = 9$ is a circle. The graphs of both equations are illustrated in Fig. 10.24.

HELPFUL HINT   Students will sometimes solve for one variable and assume that they have found the answer. Remember that the solution, if one exists, consists of one or more *ordered pairs*.

▶ **3**   We can often solve systems of equations more easily using the addition method that was discussed in Section 4.1. As with the substitution method, our objective will be to obtain a single equation containing only one variable.

EXAMPLE 3   Solve the system of equations by the addition method.

$$x^2 + y^2 = 9$$
$$2x^2 - y^2 = -6$$

Solution: If we add the two equations, we will obtain one equation containing only one variable.

$$x^2 + y^2 = 9$$
$$\underline{2x^2 - y^2 = -6}$$
$$3x^2 = 3$$
$$x^2 = 1$$
$$x = \pm 1$$

Now solve for the variable y by substituting $x = \pm 1$ in either the original equations.

$x = 1$	$x = -1$
$x^2 + y^2 = 9$	$x^2 + y^2 = 9$
$1^2 + y^2 = 9$	$(-1)^2 + y^2 = 9$
$1 + y^2 = 9$	$1 + y^2 = 9$
$y^2 = 8$	$y^2 = 8$
$y = \pm\sqrt{8} = \pm 2\sqrt{2}$	$y = \pm\sqrt{8} = \pm 2\sqrt{2}$

Solution: $(1, 2\sqrt{2}), (1, -2\sqrt{2})$ $(-1, 2\sqrt{2}), (-1, -2\sqrt{2})$

There are four solutions to this system of equations: $(1, 2\sqrt{2}), (1, -2\sqrt{2}), (-1, 2\sqrt{2})$, and $(-1, -2\sqrt{2})$ ∎

The graphs of the equations in the system solved in Example 3 are given in Fig. 10.25. Notice there are 4 points of intersection of the two graphs.

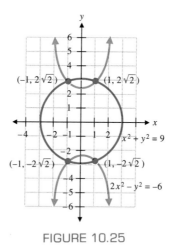

FIGURE 10.25

It is possible that a system of equations has no real solution (therefore, the graphs do not intersect). Example 4 illustrates such a case.

EXAMPLE 4   Solve the system of equations using the addition method.

$$x^2 + 4y^2 = 16$$
$$x^2 + y^2 = 1$$

Solution: Recall from Section 4.1 that in this text we place an equation within brackets, [ ], to indicate that the entire equation is to be multiplied by the value to the left of the brackets.

$$x^2 + 4y^2 = 16 \qquad\qquad x^2 + 4y^2 = 16$$
$$-1[x^2 + y^2 = 1] \quad \text{gives} \quad \underline{-x^2 - y^2 = -1}$$
$$3y^2 = 15$$
$$y^2 = 5$$
$$y = \pm\sqrt{5}$$

Now solve for x.

$$y = \sqrt{5} \qquad\qquad\qquad y = -\sqrt{5}$$
$$x^2 + y^2 = 1 \qquad\qquad\quad x^2 + y^2 = 1$$
$$x^2 + (\sqrt{5})^2 = 1 \qquad\qquad x^2 + (-\sqrt{5})^2 = 1$$
$$x^2 + 5 = 1 \qquad\qquad\quad x^2 + 5 = 1$$
$$x^2 = -4 \qquad\qquad\qquad x^2 = -4$$
$$x = \pm\sqrt{-4} \qquad\qquad\quad x = \pm\sqrt{-4}$$
$$x = \pm 2i \qquad\qquad\qquad x = \pm 2i$$

Since x is an imaginary number for both values of y, this system of equations has no real solution.   ■

The graphs of the equations in the system solved in Example 4 are given in Fig. 10.26. Notice that the two graphs do not intersect; therefore, there is no real solution. This is in agreement with the answer we obtained in Example 4.

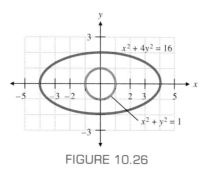

FIGURE 10.26

EXAMPLE 5   Solve the system of equations by the addition method.

$$6x^2 + y^2 = 10$$
$$2x^2 + 4y^2 = 40$$

Solution: We can choose to eliminate either the variable x or the variable y. Here we will eliminate the variable x.

$$6x^2 + y^2 = 10 \quad \text{gives} \quad 6x^2 + y^2 = 10$$
$$-3[2x^2 + 4y^2 = 40] \qquad\qquad \underline{-6x^2 - 12y^2 = -120}$$
$$-11y^2 = -110$$
$$y^2 = \frac{-110}{-11} = 10$$
$$y = \pm\sqrt{10}$$

Now find the value of x by substituting $y = \pm\sqrt{10}$ in either of the original equations.

$$y = \sqrt{10} \qquad\qquad\qquad y = -\sqrt{10}$$
$$6x^2 + y^2 = 10 \qquad\qquad 6x^2 + y = 10$$
$$6x^2 + (\sqrt{10})^2 = 10 \qquad 6x^2 + (-\sqrt{10})^2 = 10$$
$$6x^2 + 10 = 10 \qquad\qquad 6x^2 + 10 = 10$$
$$6x^2 = 0 \qquad\qquad\qquad 6x^2 = 0$$
$$x^2 = 0 \qquad\qquad\qquad x^2 = 0$$
$$x = 0 \qquad\qquad\qquad x = 0$$

Solutions: $(0, \sqrt{10})$ $(0, -\sqrt{10})$ ■

Applications of Nonlinear Systems

▶ **4**   Now we will study some applications of nonlinear systems.

EXAMPLE 6   The area, A, of a rectangle is 80 square feet and its perimeter, P, is 36 feet. Find the length and width of the rectangle.

Solution: Begin by drawing a rectangle as illustrated in Fig. 10.27.

Let x = length
y = width

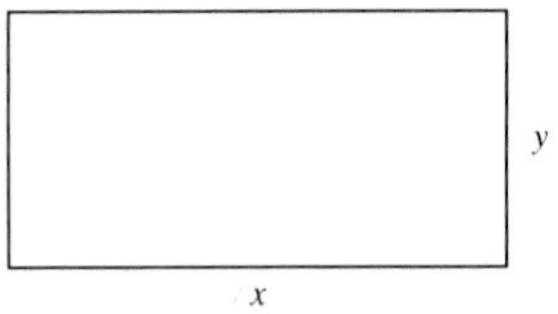

FIGURE 10.27

Since $A = xy$ and $P = 2x + 2y$, the system of equations is

$$xy = 80$$
$$2x + 2y = 36$$

We will solve this system using substitution. Since $2x + 2y = 36$ is a linear equation, we solve this equation for y (we could have also solved for x).

$$2x + 2y = 36$$
$$2y = 36 - 2x$$
$$y = \frac{36 - 2x}{2} = \frac{36}{2} - \frac{2x}{2} = 18 - x$$

Now substitute $18 - x$ for y in $xy = 80$.

$$xy = 80$$
$$x(18 - x) = 80$$
$$18x - x^2 = 80$$

or $$x^2 - 18x + 80 = 0$$
$$(x - 10)(x - 8) = 0$$
$$x - 10 = 0 \quad \text{or} \quad x - 8 = 0$$
$$x = 10 \qquad\qquad x = 8$$

If x is 10, then $y = 18 - x = 18 - 10 = 8$. And if $x = 8$, then $y = 18 - 8 = 10$. Thus the dimensions of the rectangle are 8 feet by 10 feet.   ■

Note in Example 6 that the graph of $xy = 80$ is a hyperbola, and the graph of $2x + 2y = 36$ is a straight line. If you graph these two equations, the graphs will intersect at (8, 10) and (10, 8). Try solving the system given in Example 6 graphically at this time.

EXAMPLE 7   A company breaks even when its cost equals its revenue. When its cost is greater than its revenue, the company has a loss. When its revenue exceeds its cost, the company makes a profit.

Hike 'n' Bike Company produces and sells bicycles. Its weekly cost equation is $C = 50x + 400$, $0 \le x \le 160$, and its weekly revenue equation is $R = 100x - 0.3x^2$, $0 \le x \le 160$, where x is the number of bicycles produced and sold each week. Find the number of bicycles that must be produced and sold for Hike 'n' Bike to break even.

Solution: The system of equations is

$$C = 50x + 400$$
$$R = 100x - 0.3x^2$$

For Hike 'n' Bike to break even, its cost must equal its revenue. Thus we write

$$C = R$$
$$50x + 400 = 100x - 0.3x^2$$

Writing this quadratic equation in standard form, we obtain

$$0.3x^2 - 50x + 400 = 0, \qquad 0 \le x \le 160$$

We will solve this equation using the quadratic formula.

$$a = 0.3, \qquad b = -50, \qquad c = 400$$

$$x = \frac{-b \pm \sqrt{b^2 - 4ac}}{2a}$$

$$= \frac{-(-50) \pm \sqrt{(-50)^2 - 4(0.3)(400)}}{2(0.3)}$$

$$= \frac{50 \pm \sqrt{2020}}{0.6}$$

$$x = \frac{50 \pm \sqrt{2020}}{0.6} \approx 158.2 \quad \text{or} \quad x = \frac{50 - \sqrt{2020}}{0.6} \approx 8.4$$

The cost will equal the revenue and the company will break even when approximately 8 bicycles are sold. The cost will also equal the revenue when approximately 158 bicycles are sold. The company will make a profit when between 9 and 158 bicycles are sold. When fewer than 9 or more than 158 bicycles are sold, the company will take a loss (see Fig. 10.28).

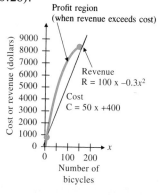

FIGURE 10.28

Exercise Set 10.5

Solve the system of equations using the substitution method.

1. $x^2 + y^2 = 4$
  $x - 2y = 4$

2. $x^2 + y^2 = 9$
  $x + 2y - 3 = 0$

3. $y = x^2 - 5$
  $3x + 2y = 10$

4. $x + y = 4$
  $x^2 - y^2 = 4$

5. $2x^2 - y^2 = -8$
  $x - y = 6$

6. $y^2 = -x + 4$
  $x^2 + y^2 = 6$

7. $x^2 - 4y^2 = 16$
  $x^2 + y^2 = 1$

8. $2x^2 + y^2 = 16$
  $x^2 - y^2 = -4$

9. $xy = 4$
  $6x - y = 5$

10. $xy = 4$
  $y = 5 - x$

11. $y = x^2 - 3$
  $x^2 + y^2 = 9$

12. $x^2 + y^2 = 25$
  $x - 3y = -5$

Solve the system of equations using the addition method.

13. $x^2 - y^2 = 4$
  $x^2 + y^2 = 4$

14. $x^2 + y^2 = 25$
  $x^2 - 2y^2 = 7$

15. $x^2 + y^2 = 13$
  $2x^2 + 3y^2 = 30$

16. $3x^2 - y^2 = 4$
  $x^2 + 4y^2 = 10$

17. $4x^2 + 9y^2 = 36$
  $2x^2 - 9y^2 = 18$

18. $5x^2 - 2y^2 = -13$
  $3x^2 + 4y^2 = 39$

19. $2x^2 + 3y^2 = 21$
  $x^2 + 2y^2 = 12$

20. $2x^2 + y^2 = 11$
  $x^2 + 3y^2 = 28$

21. $-x^2 - 2y^2 = 6$
  $5x^2 + 15y^2 = 20$

22. $x^2 - 2y^2 = 7$
  $x^2 + y^2 = 34$

23. $x^2 + y^2 = 9$
  $16x^2 - 4y^2 = 64$

24. $3x^2 + 4y^2 = 35$
  $2x^2 + 5y^2 = 42$

Solve exercises 25 through 41.

25. The sum of two numbers is 12 and the sum of their squares is 74. Find the two numbers.

26. The sum of two numbers is 14 and the sum of their squares is 106. Find the numbers.

27. The sum of the squares of two numbers is 34. The difference of their squares is 16. Find the two numbers.

28. The difference of the squares of two numbers is 80. The sum of their squares is 208. Find the two numbers.

29. The square of one number is 11 less than the square of another. The sum of their squares is 61. Find the numbers.

30. The product of two numbers is 135. The sum of the two numbers is 24. Find the numbers.

31. The area of a rectangle is 84 square feet and its perimeter is 38 ft. Find the length and width of the rectangle.

32. The area of a rectangle is 80 square feet and its perimeter is 36 feet. Find the length and width of the rectangle.

33. A garden is shaped like a right triangle with a perimeter of 30 ft and a hypotenuse of 13 feet. Find the length of the legs of the triangle.

34. A sail on a sailboat is shaped like a right triangle with a perimeter of 36 meters and a hypotenuse 15 meters. Find the length of the legs of the triangle.

35. The area of a rectangle is 300 square meters and its length is 5 feet longer than its width. Find the length and width of the rectangle.

36. The area of a rectangle is 72 square yards and its length is twice its width. Find the length and width of the rectangle.

37. A rectangular area is to be fenced in along a river bank as illustrated. If 20 feet of fencing encloses an area of 48 square feet, find the length and width of the enclosed area.

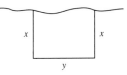

38. An football is thrown up from the ground. Its distance from the ground at any time t is given by the formula $d = -16t^2 + 64t$. At the same time that the football is thrown, a baseball is thrown up from the top of an 80-foot building. Its distance from the ground at any time t is given by the formula $d = -16t^2 + 16t + 80$. Find the time at which the two balls will be the same distance from the ground.

39. A quarter is thrown down from an airplane flying at a height of 2000 feet. The distance of the quarter above the ground at any time t is found by the formula $d = -16t^2 - 10t + 2000$. At the instant the quarter is thrown from the plane, a snowball is thrown upward from the top of a 100-foot-tall building. The distance from the ground of the snowball at any time t is found by the formula $d = -16t^2 + 800t + 100$. At what time will the quarter and snowball be the same distance from the ground?

40. Simple interest is calculated using the simple interest formula, interest = principal · rate · time or $i = prt$. When a certain principal is invested at a certain interest rate for 1 year, the interest obtained is $7.50. If the principal is increased by $25 and the interest rate is decreased by 1%, the interest remains the same. Find the principal and the interest rate.

41. When a certain principal is invested at a certain interest rate for 1 year, the interest obtained is $72. If the principal is increased by $120 and the interest rate is decreased by 2%, the interest remains the same. Find the principal and the interest rate. Use $i = prt$.

Cost and revenue equations are given below. Find the break-even point(s). See Example 7.

42. $C = 10x + 300, R = 30x - 0.1x^2$

43. $C = 80x + 900, R = 120x - 0.2x^2$

44. $C = 12.6x + 150, R = 42.8x - 0.3x^2$

45. $C = 0.6x^2 + 9, R = 12x - 0.2x^2$

46. What is a system of nonlinear equations?

47. Make up your own nonlinear system of equations whose solution is the empty set. Explain how you know the system has no solution.

48. If a system of equations consists of an ellipse and a hyperbola, what is the maximum number of points of intersection? Make a sketch to illustrate this.

49. If a system of equations consists of two hyperbolas, what is the maximum number of points of intersection? Make a sketch to illustrate this.

Cumulative Review Exercises

[1.6] **50.** List the order of operations we follow when evaluating an expression.

[2.2] **51.** Use the compound interest formula $A = P\left(1 + \dfrac{r}{n}\right)^{nt}$ to find the amount, A, when the principal is $P = 5000$, the rate r is 8%, the number of compounding periods, n, is 2, and the number of years, t, is 2.

52. Solve the equation $\dfrac{3}{5}(2x - y) = \dfrac{3}{4}(2x - 3y) + 6$ for y.

[2.5] **53.** Solve the inequality $\dfrac{3 - 4y}{3} \geq \dfrac{2y - 6}{4} - \dfrac{7}{6}$.

JUST FOR FUN

1. Find the length of the legs of a right triangle if the hypotenuse is 13 meters and the area is 30 square meters.

2. Find the length of the legs of a right triangle if the hypotenuse is 7 ft and the area is 11.76 square feet.

10.6

Nonlinear Systems of Inequalities

▶ **1** Graph second-degree inequalities.

▶ **2** Solve a system of inequalities graphically.

▶ **1** In Sections 10.1 through 10.5 we graphed second-degree equations such as $x^2 + y^2 = 25$. Now we will graph second-degree inequalities, such as $x^2 + y^2 \leq 25$. To graph second-degree inequalities, we use a procedure similar to the one we used to graph linear inequalities. That is, graph the equation using a solid line if the inequality is $\leq$ or $\geq$, and a dashed line if the inequality is $<$ or $>$, and then shade in the region that satisfies the inequality. Examples 1 and 2 illustrate this procedure.

EXAMPLE 1   Graph the inequality.

$$\frac{x^2}{4} + \frac{y^2}{9} \geq 1$$

Solution: First, graph the equation $\frac{x^2}{4} + \frac{y^2}{9} = 1$. Use a solid line when drawing the ellipse since the inequality contains $\geq$ (Fig. 10.29a). Next select a test point not on the graph and test it in the original inequality. We will select the test point $(0, 0)$.

$$\frac{x^2}{4} + \frac{y^2}{9} \geq 1$$

$$\frac{0^2}{4} + \frac{0^2}{9} \geq 1$$

$$0 + 0 \geq 1$$

$$0 \geq 1 \qquad \text{false}$$

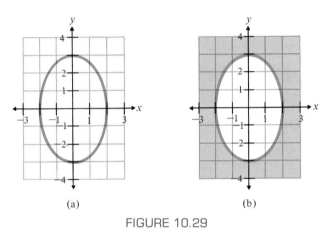

(a)   (b)

FIGURE 10.29

The point $(0, 0)$, which is within the ellipse, is not a solution. Thus all the points outside the ellipse will satisfy the inequality. Shade in this outer area (Fig. 10.29b). The shaded area, and the ellipse itself, form the solution to the inequality.   ∎

EXAMPLE 2   Graph the inequality.

$$\frac{y^2}{25} - \frac{x^2}{9} < 1$$

Solution: Graph the equation $\frac{y^2}{25} - \frac{x^2}{9} = 1$. Use a dashed line when drawing the hyperbola since the inequality contains $<$ (Fig. 10.30a). Next select the test point. We will use $(0, 0)$.

$$\frac{y^2}{25} - \frac{x^2}{9} < 1$$

$$\frac{0^2}{25} - \frac{0^2}{9} < 1$$

$$0 - 0 < 1$$

$$0 < 1 \qquad \text{true}$$

Since $(0, 0)$ is a solution, we shade in the region containing the point $(0, 0)$. The solution is indicated in Fig. 10.30b.

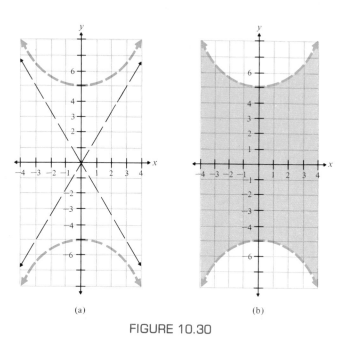

(a)   (b)

FIGURE 10.30

▸**2**   Now we will find the graphical solution to a system of inequalities in which at least one inequality is not linear. The solution to a system of inequalities is the set of ordered pairs that satisfy all inequalities in the system.

To Solve a System of Inequalities Graphically

1. Graph one inequality.
2. On the same set of axes, draw the second inequality. Use a different type of shading than was used in the first inequality.
3. The solution is the area containing the shaded area from both inequalities.

EXAMPLE 3   Solve the system of inequalities graphically.

$$x^2 + y^2 < 25$$
$$2x + y \geq 4$$

Solution: First, graph the inequality, $x^2 + y^2 < 25$. The inner region of the circle satisfies the inequality (Fig. 10.31). On the same set of axes, graph the second inequality, $2x + y \geq 4$. The solution to the system is the area containing both types of shading and that part of the straight line within the boundaries of the circle (see Fig. 10.32).

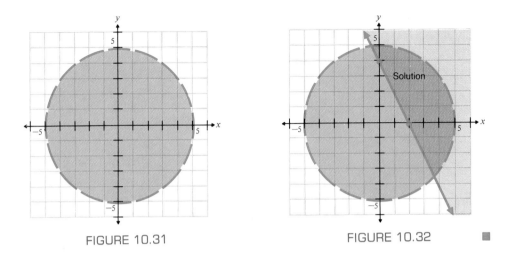

FIGURE 10.31   FIGURE 10.32

EXAMPLE 4   Solve the system of inequalities graphically.

$$\frac{x^2}{4} - \frac{y^2}{9} \le 1$$
$$y > (x + 2)^2 - 4$$

Solution: Graph each inequality on the same set of axes (Fig. 10.33). The solution is the area containing both types of shading and the solid line within the parabola.

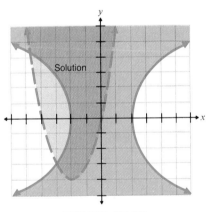

FIGURE 10.33

Exercise Set 10.6

Graph each system of inequalities.

1. $x^2 + y^2 \ge 16$
   $x + y < 5$

2. $\frac{x^2}{25} - \frac{y^2}{9} < 1$
   $2x + 3y > 6$

3. $4x^2 + y^2 > 16$
   $y \ge 2x + 2$

4. $x^2 - y^2 > 1$
$\dfrac{x^2}{9} + \dfrac{y^2}{4} \le 1$

5. $x^2 + y^2 \le 36$
$y < (x + 1)^2 - 5$

6. $y \ge x^2 - 2x + 1$
$x^2 + y^2 > 4$

7. $\dfrac{y^2}{16} - \dfrac{x^2}{4} \ge 1$
$\dfrac{x^2}{4} - \dfrac{y^2}{1} < 1$

8. $4x^2 + 9y^2 \le 36$
$x > (y - 2)^2 + 1$

9. $xy \le 6$
$2x - y \le 8$

10. $25x^2 - 4y^2 > 100$
$5x + 3y \le 15$

11. $(x - 3)^2 + (y + 2)^2 \ge 16$
$y \le 4x - 2$

12. $x^2 + (y - 3)^2 \le 25$
$y > (x - 2)^2 + 3$

Cumulative Review Exercises

[5.2] **13.** Simplify $\left(\dfrac{4x^{-2}y^3}{2xy^{-4}}\right)^2 \left(\dfrac{3xy^{-1}}{6x^4y^{-3}}\right)^{-2}$

Solve the equation $x^2 - 2x - 4 = 0$:

[9.1] **14.** By completing the square.

[9.2] **15.** Using the quadratic formula.

[9.3] **16.** Graph the solution to the inequality $(x + 4)(x - 2)(x - 4) \le 0$ on the number line.

JUST FOR FUN

Graph the following systems of inequalities.

1. $y > 4x - 6$
$x^2 + y^2 \ge 36$
$2x + y \le 8$

2. $x^2 + y^2 > 25$
$\dfrac{x^2}{25} + \dfrac{y^2}{9} > 1$
$2x - 3y < 12$

SUMMARY

GLOSSARY

Asymptotes of a hyperbola *(480):* Two lines through the center of the hyperbola that are used as an aid in graphing the hyperbola.

Circle *(467):* The set of points in a plane that are the same distance from a fixed point called the center.

Conic sections *(467):* Circles, ellipses, hyperbolas, and parabolas.

Conjugate axis of a hyperbola *(479):* The axis perpendicular to the transverse axis.

Ellipse *(471):* The set of points in a plane the sum of whose distance from two fixed points, called foci, is a constant.

Hyperbola *(479):* The set of points in a plane the difference of whose distance from two fixed points, called **foci,** is a constant.

Major axis of ellipse *(472):* The longer axis through the center of the ellipse.

Minor axis of ellipse *(472):* The shorter axis through the center of the ellipse.

Nonlinear system of equations *(487):* A system of equations containing at least one equation that is not a linear equation.

Parabola *(475):* The shape of the graph of a quadratic equation of the form $y = a(x - h)^2 + k$ or $x = a(y - k)^2 + h$.

Transverse axis of a hyperbola *(479):* The axis through the center of the hyperbola that intersects the vertices.

IMPORTANT FACTS

Circle	*Ellipse*	*Parabola*	*Hyperbola*

Circle

$(x - h)^2 + (y - k)^2 = r^2$
center at (h, k)
radius r

Ellipse

$\dfrac{x^2}{a^2} + \dfrac{y^2}{b^2} = 1$
center at $(0, 0)$

Parabola

$y = a(x - h)^2 + k$
vertex at (h, k)
opens up when $a > 0$
opens down when $a < 0$

$x = a(y - k)^2 + h$
vertex at (h, k)
opens right when $a > 0$
opens left when $a < 0$

Hyperbola

$\dfrac{x^2}{a^2} - \dfrac{y^2}{b^2} = 1$
x axis transverse axis

$\dfrac{y^2}{b^2} - \dfrac{x^2}{a^2} = 1$
y axis transverse axis

asymptotes:
$y = \pm\dfrac{b}{a}x$

Review Exercises

[10.1] *Write the equation of the circle in standard form; then sketch the graph of the circle.*

1. Center $(0, 0)$, radius 5   **2.** Center $(-3, 4)$, radius 3   **3.** Center $(4, 2)$, radius $\sqrt{8}$

Sketch the graph of each equation.

4. $x^2 + y^2 = 16$ **5.** $(x - 2)^2 + (y + 3)^2 = 25$ **6.** $(x + 1)^2 + (y + 3)^2 = 6$

Determine the equation of the circle.

7.
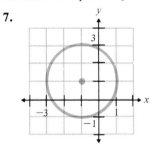

8.
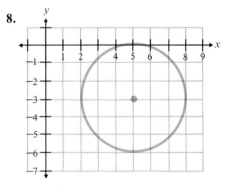

Write the equation in standard form; then sketch the graph.

9. $x^2 + y^2 - 4y = 0$ **10.** $x^2 + y^2 - 2x + 6y + 1 = 0$

11. $x^2 - 8x + y^2 - 10y + 40 = 0$ **12.** $x^2 + y^2 - 4x + 10y + 17 = 0$

[10.2] *Sketch the graph of each equation.*

13. $\dfrac{x^2}{9} + \dfrac{y^2}{4} = 1$ **14.** $\dfrac{x^2}{16} + \dfrac{y^2}{1} = 1$ **15.** $\dfrac{x^2}{9} + \dfrac{y^2}{64} = 1$

16. $4x^2 + 9y^2 = 36$ **17.** $9x^2 + 16y^2 = 144$ **18.** $16x^2 + y^2 = 16$

[10.3] *Graph each equation.*

19. $y = (x - 3)^2 + 4$ **20.** $y = (x + 4)^2 - 5$ **21.** $x = (y - 1)^2 + 4$

22. $x = -2(y + 4)^2 - 3$ **23.** $y = -3(x + 5)^2$

Write each equation in the form $y = a(x - h)^2 + k$ or $x = a(y - k)^2 + h$; then graph the equation.

24. $y = x^2 - 6x$ **25.** $y = x^2 - 2x - 3$ **26.** $x = -y^2 - 2y + 8$

27. $x = y^2 + 5y + 4$ **28.** $y = 2x^2 - 8x - 24$

[10.4] *Determine the equations of the asymptotes: then sketch the graph.*

29. $\dfrac{x^2}{4} - \dfrac{y^2}{9} = 1$ **30.** $\dfrac{y^2}{16} - \dfrac{x^2}{4} = 1$ **31.** $\dfrac{y^2}{9} - \dfrac{x^2}{25} = 1$ **32.** $\dfrac{x^2}{4} - \dfrac{y^2}{36} = 1$

Write each equation in standard form and determine the equations of the asymptotes; then sketch the graph.

33. $9y^2 - 4x^2 = 36$ **34.** $x^2 - 16y^2 = 16$ **35.** $25x^2 - 16y^2 = 400$ **36.** $49y^2 - 9x^2 = 441$

Graph each equation.

37. $xy = 6$ **38.** $xy = -8$ **39.** $y = \dfrac{3}{x}$ **40.** $y = -\dfrac{10}{x}$

[10.1–10.4] *Identify the graph as a circle, ellipse, parabola, or hyperbola; then sketch the graph.*

41. $\dfrac{x^2}{4} - \dfrac{y^2}{16} = 1$ **42.** $16x^2 + 9y^2 = 144$

43. $(x - 4)^2 + (y + 2)^2 = 16$ **44.** $x^2 - 25y^2 = 25$

45. $\dfrac{x^2}{64} + \dfrac{y^2}{9} = 1$ **46.** $y = (x - 2)^2 - 5$

47. $4x^2 + 9y^2 = 36$ **48.** $y = x^2 + 2x - 3$

49. $x = -2(y + 3)^2 - 1$ **50.** $25y^2 - 9x^2 = 225$

51. $x^2 - 4x + y^2 + 6y = -4$ **52.** $x = -y^2 - 6y - 7$

[10.5] *Solve the systems of equations using the substitution method.*

53. $x^2 + y^2 = 9$
$v = 3x + 9$

54. $xy = 5$
$3x + y = 16$

55. $x^2 + y^2 = 4$
$x^2 - y^2 = 4$

56. $x^2 + 4y^2 = 4$
$x^2 - 6y^2 = 12$

Solve the systems of equations using the addition method.

57. $x^2 + y^2 = 16$
$\quad\ x^2 - y^2 = 16$

58. $x^2 + \ y^2 = \ 25$
$\quad\ x^2 - 2y^2 = -2$

59. $-4x^2 + \ y^2 = -12$
$\quad\ 8x^2 + 2y^2 = \ -8$

60. $-2x^2 - 3y^2 = -6$
$\quad\ 5x^2 + 4y^2 = \ 15$

61. The square of one number is 9 less than the square of another. The sum of their squares is 41. Find the numbers.

62. A right triangle with a hypotenuse of 13 meters has a perimeter of 30 meters. Find the length of the legs of the triangle.

[10.6] *Graph each system of nonlinear inequalities.*

63. $2x + y \geq 6$
$\quad\ x^2 + y^2 < 9$

64. $xy > 5$
$\quad\ y < 3x + 4$

65. $4x^2 + 9y^2 \leq 36$
$\quad\ x^2 + y^2 > 25$

66. $\dfrac{x^2}{4} - \dfrac{y^2}{9} > 1$
$\quad\ y \geq (x - 3)^2 - 4$

Practice Test

1. Write the equation of the circle with center at $(-3, -1)$ and radius 9. Then sketch the circle.

2. Write the equation in standard form; then sketch the graph.
$$x^2 + y^2 - 2x - 6y + 1 = 0$$

3. Sketch the graph of $9x^2 + 16y^2 = 144$.

4. Sketch the graph of $y = -2(x - 3)^2 - 9$.

5. Sketch the graph of $\dfrac{y^2}{25} - \dfrac{x^2}{1} = 1$.

6. Sketch the graph of $y = \dfrac{8}{x}$.

7. Solve the system of equations by the method of your choice.
$$x^2 + y^2 = 16$$
$$2x^2 - y^2 = 2$$

8. Sketch the system of inequalities.
$$\frac{x^2}{9} - \frac{y^2}{25} < 1$$
$$x^2 + y^2 \leq 4$$

9. The product of two numbers is 54. The sum of the two numbers is 29. Find the numbers.

Cumulative Review Test

1. Solve the system of equations graphically:
$$2x + 3y = 12$$
$$x + 3y = 9$$

2. Solve the system of equations algebraically:
$$2x - y = 6$$
$$3x + 2y = 4$$

3. Simplify $\left(\dfrac{2x^{1/2}y^{2/3}}{x^2}\right)^2\left(\dfrac{3x^{1/3}y^{1/2}}{y^{-3/2}}\right)$.

4. If $f(x) = x^2 + 2x + 5$, find $f(a + 3)$.

5. Multiply $\dfrac{6x^2 + 5x - 4}{2x^2 - 3x + 1} \cdot \dfrac{4x^2 - 1}{8x^3 + 1}$.

6. Subtract $\dfrac{x}{x + 3} - \dfrac{x + 1}{2x^2 - 2x - 24}$.

7. Solve the equation $\dfrac{y + 1}{y + 3} + \dfrac{y - 3}{y - 2} = \dfrac{2y^2 - 15}{y^2 + y - 6}$.

8. Simplify $\sqrt{\dfrac{12x^5y^3}{8z}}$.

9. Simplify $\dfrac{6}{\sqrt{3} - \sqrt{5}}$.

10. Solve the equation $3\sqrt[3]{2x + 2} = \sqrt[3]{80x - 24}$.

11. Evaluate $(5 - 4i)(5 + 4i)$.

12. Solve the equation $(x - 3)^2 = 28$.

13. Solve the equation $3x^2 - 4x - 8 = 0$ by the quadratic formula.

14. Solve the inequality $\dfrac{3x - 2}{x + 4} \geq 0$ and graph the solution on the number line.

15. Consider the equation $y = x^2 - 4x + 4$.

    (a) Determine whether the parabola opens upward or downward.

    (b) Find the y intercept.

    (c) Find the vertex.

    (d) Find the x intercepts if they exist.

    (e) Sketch the graph.

16. If $f(x) = x^2 + 6x$ and $g(x) = 2x - 3$, find (a) $(f \circ g)(x)$ and (b) $(g \circ f)(x)$.

17. Graph $9x^2 + 4y^2 = 36$.

18. Graph $\dfrac{y^2}{25} - \dfrac{x^2}{16} = 1$.

19. A suit is reduced in price by 20% and then reduced an additional \$25. If the sale price of the suit is \$155, find the original cost of the suit.

20. The Nut Shop sells cashews for \$7 per pound and peanuts for \$5 per pound. If a customer wants a 4-pound mixture of these two nuts and the mixture is to cost a total of \$25 before tax, how many pounds of each nut should be mixed?

CHAPTER

11

Exponential and Logarithmic Functions

See Section 11.1, Exercise 68.

11.1

Exponential and Logarithmic Functions

▶ **1** Graph exponential functions.

▶ **2** Evaluate exponential equations.

▶ **3** Change an exponential equation to logarithmic form.

▶ **4** Graph logarithmic functions.

▶ **5** Graph inverse functions of the form $y = a^x$ and $y = \log_a x$ on the same set of axes.

▶ **6** Solve logarithmic equations.

In this section we introduce the exponential function and the logarithmic function. We present both functions in the same section because the two functions are closely related. We will show that the exponential function and the logarithmic function are inverse functions as defined in Section 9.6.

Exponential Functions

▶ **1** An **exponential function** is a function of the form $y = f(x) = a^x$, where a is a positive real number not equal to 1. Examples of exponential functions are

$$y = 2^x, \qquad y = 5^x, \qquad y = \left(\frac{1}{2}\right)^x$$

Exponential Function

For any real number $a > 0$ and $a \neq 1$,

$$f(x) = a^x$$

is an exponential function.

Exponential functions can be graphed by selecting values for x, finding the corresponding values of y [or $f(x)$], and plotting the points.

EXAMPLE 1 Graph the exponential function $y = 2^x$. State the domain and range of the function.

Solution: First, construct a table of values.

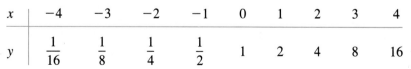

x	-4	-3	-2	-1	0	1	2	3	4
y	$\dfrac{1}{16}$	$\dfrac{1}{8}$	$\dfrac{1}{4}$	$\dfrac{1}{2}$	1	2	4	8	16

Now plot these points and connect them with a smooth curve (Fig. 11.1).

Domain: $\mathbb{R}$

Range: $\{y \mid y > 0\}$

The domain of this function is all real numbers, $\mathbb{R}$. The range is the set of values greater than 0. If you study the equation $y = 2^x$, you should realize that y must always be greater than 0 for any real number x. ∎

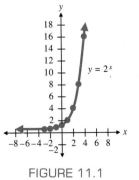

FIGURE 11.1

EXAMPLE 2   Graph $y = \left(\dfrac{1}{2}\right)^x$. State the domain and range of the function.

Solution: We will construct a table of values and plot the curve (Fig. 11.2).

x	-4	-3	-2	-1	0	1	2	3	4
y	16	8	4	2	1	$\dfrac{1}{2}$	$\dfrac{1}{4}$	$\dfrac{1}{8}$	$\dfrac{1}{16}$

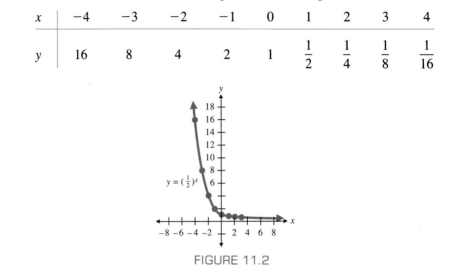

FIGURE 11.2

The domain is the set of real numbers. The range is $\{\, y \mid y > 0 \,\}$.

Note that the graphs in Figures 11.1 and 11.2 are both one-to-one functions. Both graphs appear to be horizontal next to the x axis, but in fact they are not.

In Figure 11.1, when $x = -2$, $y = 1/4$; when $x = -3$, $y = 1/8$; when $x = -10$, $y = 1/1024$ (about 0.001); and so on. As the values of x decrease, the closer the values of y get to 0 and the x axis.

In Figure 11.2, when $x = 2$, $y = 1/4$; when $x = 3$, $y = 1/8$; when $x = 10$, $y = 1/1024$; and so on. As the values of x increase, the closer the values of y get to 0 and the x axis.

Exponential functions of the form $y = a^x$ will have a shape similar to that in Fig. 11.1 when $a > 1$ and a shape similar to Fig. 11.2 when $0 < a < 1$. Note that $y = 1^x$ is not a one-to-one function, so we exclude it from our discussion of exponential functions.

▶ **2**   An **exponential equation** is one that has a variable as an exponent. Exponential equations are often used in applications describing the growth and decay of some quantities. Example 3 illustrates the use of an exponential equation in the study of heredity.

EXAMPLE 3   The number of gametes, g, in a certain species of plant is determined by the equation $g = 2^n$, where n is the number of cells in the species. Determine the number of gametes in a species that has 12 cells.

Solution: By evaluating 2^{12} on a calculator, we can determine that the species with 12 cells has 4096 gametes.

EXAMPLE 4   The formula $A = P(1 + r)^n$ is called the **compound interest formula.** When interest is compounded periodically (yearly, monthly, daily), this formula can be used to find the amount, A, in the account after n periods. In the formula, P is the principal, or the original amount invested, r is the interest rate per compounding period, and n is the number of compounding periods. Suppose that $10,000 is invested at 8% interest compounded quarterly for 6 years. Then $P = 10,000$; the interest rate per compounding period, r, is $\dfrac{8\%}{4}$ or 2% and the number of compounding periods, n, is $6 \cdot 4$ or 24. Find the amount in the account after 6 years given these values.

Solution:
$$A = P(1 + r)^n$$
$$= 10,000(1 + .02)^{24}$$
$$= 10,000(1.02)^{24}$$
$$= 10,000(1.61) \qquad \text{from a calculator (to the nearest hundredth).}$$
$$= 16,100$$

Therefore, the original $10,000 investment has grown to $16,100.   ▨

EXAMPLE 5   Carbon 14 dating is used by scientists to find the age of fossils and other items. The formula used in carbon dating is

$$A = A_0 \cdot 2^{-t/5600}$$

where A_0 represents the original amount of carbon 14 present and A represents the amount of carbon 14 present after t years. If 500 grams of carbon 14 are present originally, how many grams will remain after 2000 years?

Solution:
$$A = A_0 \cdot 2^{-t/5600}$$
$$= 500(2)^{-2000/5600}$$
$$= 500(2)^{-0.36}$$
$$= 500(0.78)$$
$$= 390 \text{ grams}$$
   ▨

Logarithms

▶**3**   Consider the exponential function $y = 2^x$. Recall from Section 9.6 that to find the inverse function, we interchange the x and y and solve the equation for y. Interchanging x and y gives the equation $x = 2^y$, which is the inverse of $y = 2^x$. But at this time we have no way of solving the equation $x = 2^y$ for y. To solve this equation for y, we introduce a new definition.

Definition of Logarithm

For all positive numbers a, where $a \neq 1$,

$$y = \log_a x \quad \text{means} \quad x = a^y$$

In the expression $y = \log_a x$, the word *log* is an abbreviation for the word *logarithm*. $y = \log_a x$ is read "y is the logarithm of x to the base a." The letter y represents the logarithm, the letter a represents the base, and the letter x represents the number.

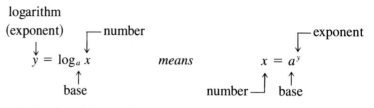

In words, the logarithm of the number x to the base a is the exponent to which the base a must be raised to obtain the number x. For example,

$$2 = \log_{10} 100 \quad \text{means} \quad 100 = 10^2$$

The logarithm is 2, the base is 10, and the number is 100. The logarithm, 2, is the exponent to which the base, 10, must be raised to give the number, 100. Note $10^2 = 100$.

The following are some examples of how an exponential expression can be converted to a logarithmic expression.

Exponential Form	*Logarithmic Form*		
$10^0 = 1$	$0 = \log_{10} 1$	or	$\log_{10} 1 = 0$
$4^2 = 16$	$2 = \log_4 16$	or	$\log_4 16 = 2$
$\left(\dfrac{1}{2}\right)^5 = \dfrac{1}{32}$	$5 = \log_{1/2} \dfrac{1}{32}$	or	$\log_{1/2} \dfrac{1}{32} = 5$
$5^{-2} = \dfrac{1}{25}$	$-2 = \log_5 \dfrac{1}{25}$	or	$\log_5 \dfrac{1}{25} = -2$

Now let us do a few examples involving conversion from exponential form to logarithmic form, and vice versa.

EXAMPLE 6 Write each of the following in logarithmic form.

(a) $3^4 = 81$ (b) $\left(\dfrac{1}{5}\right)^3 = \dfrac{1}{125}$ (c) $2^{-4} = \dfrac{1}{16}$

Solution: (a) $\log_3 81 = 4$ (b) $\log_{1/5} \dfrac{1}{125} = 3$ (c) $\log_2 \dfrac{1}{16} = -4$ ■

EXAMPLE 7 Write each of the following logarithms in exponential form.

(a) $\log_6 36 = 2$ (b) $\log_3 9 = 2$ (c) $\log_{1/3} \dfrac{1}{81} = 4$

Solution: (a) $6^2 = 36$ (b) $3^2 = 9$ (c) $\left(\dfrac{1}{3}\right)^4 = \dfrac{1}{81}$ ■

EXAMPLE 8 Write each logarithm in exponential form; then find the missing value.

(a) $y = \log_5 25$ (b) $2 = \log_a 16$ (c) $3 = \log_{1/2} x$

Solution: (a) $5^y = 25$
Since $5^2 = 25$, $y = 2$.

(b) $a^2 = 16$
Since $4^2 = 16$, $a = 4$.

(c) $\left(\dfrac{1}{2}\right)^3 = x$

Since $\left(\dfrac{1}{2}\right)^3 = \dfrac{1}{8}$, $x = \dfrac{1}{8}$. ■

▶ **4**   Now that we know how to convert from exponential form to logarithmic form, and vice versa, we will graph logarithmic functions. Equations of the form $y = \log_a x$, $a > 0$, $a \neq 1$, and $x > 0$, are called **logarithmic functions** since their graphs pass the vertical line test. To graph a logarithmic function, change it to exponential form. This procedure is illustrated in Examples 9 and 10.

EXAMPLE 9   Graph $y = \log_2 x$. State the domain and range of the function.

Solution: $y = \log_2 x$ means $x = 2^y$. Using $x = 2^y$, construct a table of values. The table will be easier to develop by selecting values for y and finding the corresponding values for x.

x	$\dfrac{1}{16}$	$\dfrac{1}{8}$	$\dfrac{1}{4}$	$\dfrac{1}{2}$	1	2	4	8	16
y	-4	-3	-2	-1	0	1	2	3	4

Now draw the graph (Fig. 11.3). The domain, the x values, is $\{x \mid x > 0\}$. The range, the y values, is all real numbers, $\mathbb{R}$.

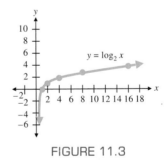

FIGURE 11.3   ■

EXAMPLE 10   Graph $y = \log_{1/2} x$. State the domain and the range of the function.

Solution: $y = \log_{1/2} x$ means $x = \left(\frac{1}{2}\right)^y$. Construct a table of values by selecting values for y and finding the corresponding values of x.

x	16	8	4	2	1	$\dfrac{1}{2}$	$\dfrac{1}{4}$	$\dfrac{1}{8}$	$\dfrac{1}{16}$
y	-4	-3	-2	-1	0	1	2	3	4

The graph is illustrated in Fig. 11.4. The domain is $\{x \mid x > 0\}$. The range is all real numbers, $\mathbb{R}$.

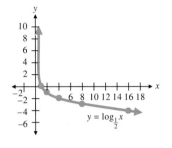

FIGURE 11.4

If we study the domains in Examples 9 and 10, we see that both $y = \log_2 x$ and $y = \log_{1/2} x$ have domains $\{x \mid x > 0\}$. In fact, **for any expression of the form $y = \log_a x$, the domain will be $\{x \mid x > 0\}$.** Also note that the graphs in Examples 9 and 10 are both one-to-one functions.

▶ **5**   Earlier we stated that functions of the form $y = 2^x$ and $x = 2^y$ are inverse functions. Since $x = 2^y$ means $y = \log_2 x$, the functions $y = 2^x$ and $y = \log_2 x$ are inverse functions. In fact, for any $a > 0$, $a \neq 1$, the functions $y = a^x$ and $y = \log_a x$ will be inverse functions.

The graphs of $y = 2^x$ and $y = \log_2 x$ are illustrated in Fig. 11.5. Note that they are symmetric with respect to the line $y = x$. Since these graphs are inverses of each other, the symmetry is expected.

The graphs of $y = (\frac{1}{2})^x$ and $y = \log_{1/2} x$ are illustrated in Fig. 11.6. They are also inverses of each other and are symmetric with respect to the line $y = x$.

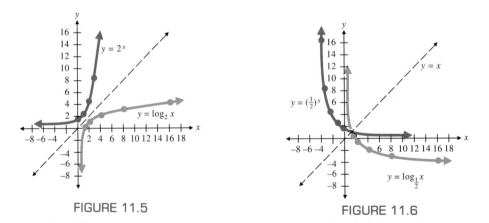

FIGURE 11.5   FIGURE 11.6

▶ **6** When finding the logarithm of an expression, the expression is called the **argument** of the logarithm. For example, in $\log_{10} 3$ the 3 is the argument, and in $\log_{10}(2x + 4)$ the $(2x + 4)$ is the argument. When the argument contains a variable, we assume that the argument represents a number that is greater than 0. *Remember, only the logarithms of positive numbers exist.*

A **logarithmic equation** is one in which a variable appears in the argument of some logarithm. Exponential and logarithmic equations are discussed in more detail in Sections 11.4 and 11.5. One application of logarithms is to measure the intensity of earthquakes. Example 11 illustrates this use of logarithms.

EXAMPLE 11   The magnitude of an earthquake on the Richter scale, R, is given by the formula

$$R = \log_{10} I$$

The I in the formula represents the number of times greater (or more intense) the earthquake is than the smallest measurable activity that can be measured on the seismograph.

(a) If an earthquake measures 4 on the Richter scale, how many times more intense is it than the smallest measurable activity?

(b) How many times more intense is an earthquake that measures 5 than an earthquake that measures 4?

Solution: (a) $R = \log_{10} I$

$4 = \log_{10} I$

or

$10^4 = I$

$10{,}000 = I$

Therefore, an earthquake that measures 4 is 10,000 times greater than the smallest measurable activity.

(b) $5 = \log_{10} I$

$10^5 = I$

$100{,}000 = I$

Since $(10{,}000)(10) = 100{,}000$, an earthquake that measures 5 is 10 times more intense than an earthquake that measures 4.   ∎

Exercise Set 11.1

Graph the exponential function.

1. $y = 2^x$

2. $y = 3^x$

3. $y = \left(\dfrac{1}{2}\right)^x$

4. $y = \left(\dfrac{1}{3}\right)^x$

5. $y = 4^x$

6. $y = 5^x$

Graph the logarithmic function.

7. $y = \log_2 x$

8. $y = \log_3 x$

9. $y = \log_{1/2} x$

10. $y = \log_{1/3} x$

11. $y = \log_5 x$

12. $y = \log_{1/4} x$

Graph each pair of functions on the same set of axes.

13. $y = 2^x,\ y = \log_{1/2} x$

14. $y = \left(\dfrac{1}{2}\right)^x,\ y = \log_2 x$

15. $y = 2^x,\ y = \log_2 x$

16. $y = \left(\dfrac{1}{2}\right)^x,\ y = \log_{1/2} x$

Write each expression in logarithmic form.

17. $2^3 = 8$

18. $5^2 = 25$

19. $3^5 = 243$

20. $9^{1/2} = 3$

21. $8^{1/3} = 2$

22. $\left(\dfrac{1}{2}\right)^5 = \dfrac{1}{32}$

23. $\left(\dfrac{1}{4}\right)^2 = \dfrac{1}{16}$

24. $2^{-3} = \dfrac{1}{8}$

25. $5^{-2} = \dfrac{1}{25}$

26. $4^3 = 64$

27. $4^{-3} = \dfrac{1}{64}$

28. $(64)^{1/3} = 4$

29. $16^{-1/2} = \dfrac{1}{4}$

30. $36^{1/2} = 6$

31. $8^{-1/3} = \dfrac{1}{2}$

32. $81^{-1/4} = \dfrac{1}{3}$

Write each expression in exponential form.

33. $\log_2 8 = 3$

34. $\log_3 9 = 2$

35. $\log_4 64 = 3$

36. $\log_{1/3} \dfrac{1}{9} = 2$

37. $\log_{1/2} \dfrac{1}{16} = 4$

38. $\log_4 \dfrac{1}{16} = -2$

39. $\log_5 \dfrac{1}{125} = -3$

40. $\log_9 3 = \dfrac{1}{2}$

41. $\log_{125} 5 = \dfrac{1}{3}$

42. $\log_8 \dfrac{1}{64} = -2$

43. $\log_{27} \dfrac{1}{3} = -\dfrac{1}{3}$

44. $\log_{10} 100 = 2$

45. $\log_{10} 1000 = 3$

46. $\log_6 216 = 3$

Write each logarithm in exponential form; then find the missing value.

47. $y = \log_6 36$

48. $y = \log_3 27$

49. $3 = \log_2 x$

50. $5 = \log_3 x$

51. $y = \log_{64} 8$

52. $\dfrac{1}{2} = \log_{100} x$

53. $3 = \log_a 64$

54. $2 = \log_a 16$

55. $4 = \log_{1/2} x$

56. $3 = \log_{1/4} x$

57. $5 = \log_a 32$

58. $y = \log_2 \dfrac{1}{4}$

59. $\dfrac{1}{3} = \log_8 x$

60. $-\dfrac{1}{4} = \log_{16} x$

61. Use the formula $g = 2^n$ to determine the number of gametes the specified plant has if it has 8 cells. See Example 3.

62. If José invests \$5000 at 10% interest compounded quarterly, find the amount after 4 years. Use $A = p(1 + r)^n$. See Example 4.

63. If Marsha invests \$8000 at 12% interest compounded quarterly, find the amount after 5 years.

64. If 12 grams of carbon 14 are originally present in a certain animal bone, how much will remain at the end of 1000 years? Use $A = A_0 \cdot 2^{-t/5600}$. See Example 5.

65. If 60 grams of carbon 14 are originally present in the fossil Jonas found at the archeological site, how much will remain after 10,000 years?

66. The expected future population of Ackworth, which presently has 2000 residents, can be approximated by the formula $y = 2000 (1.2)^{0.1x}$, where x is the number of years in the future. Find the expected population of the town in

(a) 10 years;   **(b)** 100 years.

67. The amount of a radioactive substance present, in grams, at time t is given by the formula $y = 80(2)^{-0.4t}$, where t is measured in years. Find the number of grams present in

(a) 10 years;   **(b)** 100 years.

68. The number of a certain type of bacteria present in a culture is determined by the equation $y = 5000(3)^x$, where x is the number of days the culture has been growing. Find the number of bacteria in

(a) 5 days;   **(b)** 7 days.

69. If an earthquake has a magnitude of 7 on the Richter scale, how many times greater is the earthquake than the smallest measurable activity? Use $R = \log_{10} I$. See Example 11.

70. How many times greater is an earthquake that measures 3 on the Richter scale than an earthquake that measures 1?

71. For the logarithmic function $y = \log_a x$:

(a) What are the restrictions on a?

(b) What is the domain of the function?

(c) What is the range of the function?

72. For the logarithmic function $y = \log_a(x - 3)$, what must be true about x? Explain your answer.

73. We stated earlier that, for exponential functions $f(x) = a^x$, the value of a cannot equal 1. (a) What does the graph of $f(x) = a^x$ look like when $a = 1$? (b) Is $f(x) = a^x$ a function when $a = 1$? (c) Does $f(x) = a^x$ have an inverse function when $a = 1$? Explain your answer.

74. If some points on the exponential function $f(x) = a^x$ are $(-3, \frac{1}{27})$, $(-2, \frac{1}{9})$, $(-1, \frac{1}{3})$, $(0, 1)$, $(1, 3)$, $(2, 9)$, and $(3, 27)$, list some points on the logarithmic function $f(x) = \log_a x$. Explain how you determined your answer.

75. Discuss the relation between the graphs $y = a^x$ and $y = \log_a x$ for any real number a, $a \neq 1$.

76. Discuss the graphs of $y = a^x$ (a) for $0 < a < 1$ and (b) for $a > 1$.

77. How will the graphs of $y = a^x$ and $y = a^x + k$ compare? Explain your answer.

78. Explain why $y = \log_a x$, $a > 0$, $a \neq 1$, is not defined for $x \leq 0$. *Hint:* Change the equation to exponential form.

Cumulative Review Exercises

[6.1–6.4] *Factor each of the following.*

79. $24x^2 - 6xy + 16xy - 4y^2$

80. $2(a - 3)^2 + 7(a - 3) - 15$

81. $4x^4 - 36x^2$

82. $8x^3 + \dfrac{1}{27}$

11.2

Properties of Logarithms

▶ **1** Use the product rule for logarithms.

▶ **2** Use the quotient rule for logarithms.

▶ **3** Use the power rule for logarithms.

▶ **1** To be able to do calculations using logarithms, you must have an understanding of their properties. We use these properties in Section 11.3 when we solve exponential and logarithmic equations. The first property we discuss is the product rule for logarithms.

> **Product Rule for Logarithms**
>
> For positive real numbers x, y, and a, $a \neq 1$,
>
> $$\log_a xy = \log_a x + \log_a y \qquad \text{Property 1}$$

To prove this property, let $\log_a x = m$ and $\log_a y = n$. Now write each logarithm in exponential form.

$$\log_a x = m \quad \text{means} \quad a^m = x$$
$$\log_a y = n \quad \text{means} \quad a^n = y$$

By substitution and using the rules of exponents, we see that

$$xy = a^m \cdot a^n = a^{m+n}$$

We can now convert $xy = a^{m+n}$ back to logarithmic form.

$$xy = a^{m+n} \quad \text{means} \quad \log_a xy = m + n$$

Finally, substituting $\log_a x$ for m and $\log_a y$ for n, we obtain

$$\log_a xy = \log_a x + \log_a y$$

which is property 1.

Examples of Property 1

$$\log_3 5 \cdot 7 = \log_3 5 + \log_3 7$$
$$\log_4 3x = \log_4 3 + \log_4 x$$
$$\log_8 x^2 = \log_8 (x \cdot x) = \log_8 x + \log_8 x$$

▶ **2** Now we give the quotient rule for logarithms, which we refer to as property 2.

> **Quotient Rule for Logarithms**
>
> For positive real numbers x, y, and a, $a \neq 1$,
>
> $$\log_a \frac{x}{y} = \log_a x - \log_a y \qquad \text{Property 2}$$

Examples of Property 2

$$\log_3 \frac{12}{4} = \log_3 12 - \log_3 4$$

$$\log_6 \frac{x}{3} = \log_6 x - \log_6 3$$

$$\log_5 \frac{x}{x+2} = \log_5 x - \log_5 (x + 2)$$

▶ **3** The last property we discuss in this section is the power rule for logarithms.

> **Power Rule for Logarithms**
>
> If x and a are positive real numbers, $a \neq 1$, and n is any real number, then
>
> $$\log_a x^n = n \log_a x \qquad \text{Property 3}$$

Examples of Property 3

$$\log_2 4^3 = 3 \log_2 4$$

$$\log_{10} x^2 = 2 \log_{10} x$$

$$\log_5 \sqrt{12} = \log_5 (12)^{1/2} = \frac{1}{2} \log_5 12$$

$$\log_8 \sqrt[5]{x+3} = \log_8 (x+3)^{1/5} = \frac{1}{5} \log_8 (x+3)$$

Properties 2 and 3 can be proved in a manner similar to that given for property 1.

EXAMPLE 1 Use properties 1 through 3 to expand each of the following.

(a) $\log_8 \dfrac{27}{43}$ (b) $\log_4 (64)(180)$ (c) $\log_{10} (32)^{1/5}$

Solution: (a) $\log_8 \dfrac{27}{43} = \log_8 27 - \log_8 43$

(b) $\log_4 (64)(180) = \log_4 64 + \log_4 180.$

(c) $\log_{10} (32)^{1/5} = \dfrac{1}{5} \log_{10} 32$ ∎

Often we will have to use two or more of these properties in the same problem.

EXAMPLE 2 Expand each of the following.

(a) $\log_{10} 4(x+2)^3$ (b) $\log_5 \dfrac{(4-x)^2}{3}$

(c) $\log_5 \left(\dfrac{4-x}{3}\right)^2$ (d) $\log_5 \dfrac{[x(x+4)]^3}{2}$

Solution: (a) $\log_{10} 4(x+2)^3 = \log_{10} 4 + \log_{10} (x+2)^3$

$$= \log_{10} 4 + 3 \log_{10} (x+2)$$

(b) $\log_5 \dfrac{(4-x)^2}{3} = \log_5 (4-x)^2 - \log_5 3$

$$= 2 \log_5 (4-x) - \log_5 3$$

(c) $\log_5 \left(\dfrac{4-x}{3}\right)^2 = 2 \log_5 \left(\dfrac{4-x}{3}\right)$

$$= 2[\log_5 (4-x) - \log_5 3]$$
$$= 2 \log_5 (4-x) - 2 \log_5 3$$

(d) $\log_5 \dfrac{[x(x+4)]^3}{2} = \log_5 [x(x+4)]^3 - \log_5 2$

$$= 3 \log_5 x(x+4) - \log_5 2$$
$$= 3[\log_5 x + \log_5 (x+4)] - \log_5 2$$
$$= 3 \log_5 x + 3 \log_5 (x+4) - \log_5 2$$ ∎

Note that property 1 can be expanded to evaluate the product of 3 or more quantities. Thus, for example, $\log_5 xyz = \log_5 x + \log_5 y + \log_5 z$.

HELPFUL HINT

In Example 2(b), when we expanded $\log_5 \dfrac{(4 - x)^2}{3}$, we first used the quotient rule.

In Example 2(c), when we expanded $\log_5 \left(\dfrac{4 - x}{3}\right)^2$, we first used the power rule.

Do you see the difference in the two problems? In $\log_5 \dfrac{(4 - x)^2}{3}$ just the numerator of the argument is squared; therefore, we use the quotient rule first. In $\log_5 \left(\dfrac{4 - x}{3}\right)^2$ the entire argument is squared, and so we use the power rule first. If we wished, we could rewrite $\log_5 \left(\dfrac{4 - x}{3}\right)^2$ as $\log_5 \dfrac{(4 - x)^2}{3^2}$. Then we could use the quotient rule to expand the expression. Show that $\log_5 \dfrac{(4 - x)^2}{3^2}$ gives the same answer as $\log_5 \left(\dfrac{4 - x}{3}\right)^2$.

EXAMPLE 3   Write each of the following as the logarithm of a single expression.

    (a) $3 \log_8 (x + 2) - \log_8 x$ (b) $\log_7 (x + 1) + 2 \log_7 (x + 4) - 3 \log_7 (x - 5)$

Solution: (a) $3 \log_8 (x + 2) - \log_8 x = \log_8 (x + 2)^3 - \log_8 x$

$$= \log_8 \frac{(x + 2)^3}{x}$$

(b) $\log_7 (x + 1) + 2 \log_7 (x + 4) - 3 \log_7 (x - 5) = \log_7 (x + 1) + \log_7 (x + 4)^2 - \log_7 (x - 5)^3$

$$= \log_7 (x + 1)(x + 4)^2 - \log_7 (x - 5)^3$$

$$= \log_7 \frac{(x + 1)(x + 4)^2}{(x - 5)^3}$$

COMMON STUDENT ERROR

The correct rules are:

$$\log (A \cdot B) = \log A + \log B$$

$$\log \frac{A}{B} = \log A - \log B$$

A common mistake made by students is to use the following *incorrect procedures*.

Wrong	*Wrong*
$\log (A + B) = \log A + \log B$	$\dfrac{\log A}{\log B} = \log A - \log B$
$\log (A - B) = \log A - \log B$	$\dfrac{\log A}{\log B} = \log \dfrac{A}{B}$
$\log (A \cdot B) = (\log A)(\log B)$	

For example,

$$\log (x + 2) \neq \log x + \log 2 \qquad \frac{\log x}{\log 2} \neq \log x - \log 2$$

$$\log (3x) \neq (\log 3)(\log x) \qquad \frac{\log 10}{\log x} \neq \log \frac{10}{x}$$

Exercise Set 11.2

Use properties 1 through 3 to expand each of the following.

1. $\log_3 7 \cdot 12$

2. $\log_5 8 \cdot 29$

3. $\log_8 7(x + 3)$

4. $\log_{10} x(x - 2)$

5. $\log_4 \dfrac{15}{7}$

6. $\log_9 \dfrac{\sqrt{x}}{12}$

7. $\log_{10} \dfrac{\sqrt{x}}{x - 3}$

8. $\log_5 3^{12}$

9. $\log_8 x^4$

10. $\log_5 (x + 4)^3$

11. $\log_{10} 3(8^2)$

12. $\log_8 x^2(x - 2)$

13. $\log_4 \sqrt{\dfrac{x^5}{x + 4}}$

14. $\log_{10} (x - 3)^2 x^3$

15. $\log_{10} \dfrac{x^4}{(x + 2)^3}$

16. $\log_7 x^2(x - 2)$

17. $\log_8 \dfrac{x(x - 6)}{x^3}$

18. $\log_{10} \left(\dfrac{x}{6}\right)^2$

19. $\log_{10} \dfrac{2x}{3}$

20. $\log_5 \dfrac{(3x)^2}{x + 5}$

Write each of the following as a logarithm of a single expression.

21. $2 \log_{10} x - \log_{10} (x - 2)$

22. $3 \log_8 x + 2 \log_8 (x + 1)$

23. $2(\log_5 x - \log_5 4)$

24. $2[\log_6 (x - 1) - \log_6 3]$

25. $\log_{10} x + \log_{10}(x - 4) - \log_{10} (x + 1)$

26. $2 \log_5 x + \log_5 (x - 4) + \log_5 (x - 2)$

27. $\dfrac{1}{2}[\log_7 (x - 2) - \log_7 x]$

28. $5 \log_7 (x + 3) + 2 \log_7 (x - 1) - \dfrac{1}{2} \log_7 x$

29. $2 \log_9 5 - 4 \log_9 6 + \log_9 3$

30. $5 \log_6 (x + 3) - [2 \log_6 (x - 4) + 3 \log_6 x]$

31. $4 \log_6 3 - [2 \log_6 (x + 3) + 4 \log_6 x]$

32. $2 \log_7 (x - 6) + 3 \log_7 (x + 3) - [5 \log_7 2 + 3 \log_7 (x - 2)]$

Find the value of each of the following by writing each argument using the numbers 2 and/or 5 and using the values $\log_{10} 2 = 0.3010$ and $\log_{10} 5 = 0.6990$.

33. $\log_{10} 10$

34. $\log_{10} 0.4$

35. $\log_{10} 2.5$

36. $\log_{10} 4$

37. $\log_{10} 25$

38. $\log_{10} 8$

Determine whether each of the following statements is true or false.

39. $\log_{10} 3 + \log_{10} 4 = \log_{10}(3 + 4)$

40. $(\log_4 5)(\log_4 8) = \log_4(5 \cdot 8)$

41. $\log_3 8 - \log_3 4 = \log_3 \frac{8}{4}$

42. $\dfrac{\log_6 9}{\log_6 2} = \log_6 9 - \log_6 2$

43. $\dfrac{\log_5 8}{\log_5 2} = \log_5 \dfrac{8}{2}$

44. $(\log_4 5)(\log_4 8) = \log_4 5 + \log_4 8$

45. $2 \log_{10} 10 = (\log_{10} 10)^2$

46. $3 \log_{10} 10 = \log_{10} 10^3$

✎ **47.** In your own words explain the product rule for logarithms.

✎ **48.** In your own words explain the quotient rule for logarithms.

✎ **49.** In your own words explain the power rule for logarithms.

Cumulative Review Exercises

Perform the indicated operations.

[7.2] **50.** $\dfrac{2x + 5}{x^2 - 7x + 12} \div \dfrac{x - 4}{2x^2 - x - 15}$.

[7.3] **51.** $\dfrac{2x + 5}{x^2 - 7x + 12} - \dfrac{x - 4}{2x^2 - x - 15}$.

[7.6] **52.** Mike can paint a house by himself in 4 days and Jill can paint the same house by herself in 5 days. How long would it take them to paint the house together?

[8.3] **53.** Multiply and then simplify $\sqrt[3]{4x^4y^7} \cdot \sqrt[3]{12x^7y^{10}}$

11.3

Common Logarithms

▶ **1** Identify common logarithms.

▶ **2** Find common logarithms of powers of 10.

▶ **3** Find common logarithms on a calculator.

▶ **4** Find antilogarithms on a calculator.

▶ **1** The properties discussed in Section 11.2 can be used with any valid base (greater than 0 and not equal to 1). However, since we are used to working in base 10, we will often use the base 10 when computing with logarithms. Base 10 logarithms are called **common logarithms.** When we are working with common logarithms, it is not necessary to list the base. Thus $\log x$ is the same as $\log_{10} x$.

The properties of logarithms written as common logarithms follow. For positive real numbers x and y, and any real number, n;

1. $\log xy = \log x + \log y$

2. $\log \dfrac{x}{y} = \log x - \log y$

3. $\log x^n = n \log x$.

▶ **2** In Chapter 5 we learned that

$$10^{-2} = \frac{1}{10^2} = \frac{1}{100} = 0.01$$

$$10^{-1} = \frac{1}{10^1} = \frac{1}{10} = 0.1$$

$$10^0 = 1$$

$$10^1 = 10$$

$$10^2 = 100$$

We could enlarge this chart by using exponents less than -2 or greater than 2.

Note that the number 1 can be expressed as 10^0 and the number 10 can be expressed as 10^1. Since, for example, the number 5 is between 1 and 10, it must be a number between 10^0 and 10^1.

$$1 < 5 < 10$$
$$10^0 < 5 < 10^1$$

The number 5 can be expressed as the base 10 raised to an exponent between 0 and 1. The number 5 is equal to $10^{0.69897}$. The common logarithm of 5 is 0.69897.

$$\log 5 = 0.69897$$

Common Logarithms

The **common logarithm** of a number is the **exponent** to which the base 10 is raised to obtain the number.

$$\text{If } \log N = L, \quad \text{then } 10^L = N.$$

For example, if $\log 5 = 0.69897$, then $10^{0.69897} = 5$.
    Now consider the number 50.

$$10 < 50 < 100$$
$$10^1 < 50 < 10^2$$

The number 50 can be expressed as the base 10 raised to an exponent between 1 and 2. The number $50 = 10^{1.69897}$; thus $\log 50 = 1.69897$.

▸**3**   To find common logarithms of numbers, we can use a calculator that has a logarithm key, ⎡log⎤. If you have access to such a calculator and your instructor permits its use, you should use it to find logarithms. If you do not have such a calculator and plan to take further mathematics courses, you may wish to consider purchasing one. If a calculator is not available, you can find common logarithms of numbers with the use of the common logarithm table given in Appendix D. Illustrations of how the table is used are also given in the appendix.
    The logarithms of most numbers are irrational numbers. Even the answers given by calculators are usually only approximations of the actual values. Even though we are working with approximations when evaluating most logarithms, we generally write the logarithm with an equal sign. Thus, rather than writing $\log 6 \approx 0.7781513$, we will write $\log 6 = 0.7781513$.

Calculator Corner

Finding Common Logarithms Using a Calculator
Common logarithms can be found using a calculator that contains a ⎡log⎤ key.
    To find the common logarithm of a real number greater than or equal to zero, press the clear key, ⎡c⎤; then enter the number; then press the logarithm key. The answer will then be displayed.

Example	*Keys to Use*	*Answer Displayed*
Find log 400	⎡c⎤ 400 ⎡log⎤	2.60206
Find log 0.0538	⎡c⎤ 0.0538 ⎡log⎤	−1.2692177

EXAMPLE 1   Indicate the keys you would use to find the following logarithms using a calculator and then give the answer obtained.

(a) log 962   (b) log 1   (c) log 0.00046   (d) log −5.2

Solution: (a) ⎡c⎤ 962 ⎡log⎤ 2.9831751
    (b) ⎡c⎤ 1 ⎡log⎤ 0

(c) $\boxed{\text{c}}$ 0.00046 $\boxed{\text{log}}$ −3.3372422

(d) $\boxed{\text{c}}$ 5.2 $\boxed{^{+}/_{-}}$ $\boxed{\text{log}}$ Error

Recall in part (d) that the logarithms of negative numbers do not exist.   ■

EXAMPLE 2   Find the exponent to which the base 10 must be raised to obtain the number 43,600.

Solution: We are asked to find the exponent, which is a logarithm. We need to determine log 43,600.

$$\boxed{\text{c}}\ 43{,}600\ \boxed{\text{log}}\ 4.6394865$$

Thus the exponent is 4.6394865. Note that $10^{4.6394865} = 43{,}600$.   ■

The question that should now be asked is, "if we know the common logarithm of a number, how do we find the number?" For example, if log $N = 3.406$, what is N? When we find the value of the number from the logarithm, we say we are finding the **antilogarithm** or **inverse logarithm.***

If the logarithm of N is L, then N is the antilogarithm or inverse logarithm of L.

Antilogarithm

$$\text{If } \log N = L, \quad \text{then } N = \text{antilog } L.$$

▶ **4**   When we are given the common logarithm, which is the exponent on the base 10, the *antilog is the number* obtained when the base 10 is raised to that exponent. For example, in Example 1 we found log $962 = 2.9831751$. If we are given the logarithm, or exponent, 2.9831751, the antilog will be the value of $10^{2.9831751}$, which is 962. Thus, if log $962 = 2.9831751$, then antilog $2.9831751 = 962$. In Example 1 we showed that log $0.00046 = -3.3372422$. Thus, antilog $-3.3372422 = 0.00046$. When finding an antilog, we are converting from a logarithm, or exponent, to a number.

Calculator Corner

Finding Antilogarithms on a Calculator

To find antilogarithms on a calculator, we use a combination of the inverse key, $\boxed{\text{inv}}$, and the $\boxed{\text{log}}$ key. First, press the clear key, $\boxed{\text{c}}$, and then enter the logarithm whose antilog you wish to find. Press the inverse key, $\boxed{\text{inv}}$, followed by the logarithm key, $\boxed{\text{log}}$. The antilog will then be displayed.

Example	*Keys to Use*	*Answer Display*
Find antilog 2.9831751	$\boxed{\text{c}}$ 2.9831751 $\boxed{\text{inv}}$ $\boxed{\text{log}}$	962.00006*
Find antilog −3.3372422	$\boxed{\text{c}}$ 3.3372422 $\boxed{^{+}/_{-}}$ $\boxed{\text{inv}}$ $\boxed{\text{log}}$	0.00046**

When you are finding the antilog of a negative value, enter the value and then press the $\boxed{^{+}/_{-}}$ key before pressing the inverse and logarithm keys.

* Some calculators will give slightly different answers, depending upon their electronics.
** Some calculators may give the answers in scientific notation form.

---

* The word antilogarithm is slowly being replaced by the term inverse logarithm because we use the inverse key on the calculator to find these values.

EXAMPLE 3   Indicate the keys you would use to find the following antilogarithms on a calculator and give the answer obtained on a calculator.

(a) antilog 6.827   (b) antilog 0
(c) antilog -2.35 (d) antilog -5.822

Solution: (a) $\boxed{c}$ 6.827 $\boxed{\text{inv}}$ $\boxed{\text{log}}$ 6714288.4
(b) $\boxed{c}$ 0 $\boxed{\text{inv}}$ $\boxed{\text{log}}$ 1
(c) $\boxed{c}$ 2.35 $\boxed{+/-}$ $\boxed{\text{inv}}$ $\boxed{\text{log}}$ 0.0044668
(d) $\boxed{c}$ 5.822 $\boxed{+/-}$ $\boxed{\text{inv}}$ $\boxed{\text{log}}$ 0.0000015   ■

EXAMPLE 4   Find the value obtained when the base 10 is raised to the -1.052 power.

Solution: We are asked to find the value of $10^{-1.052}$. Since we are given the exponent, or logarithm, we can find the answer by taking the antilog of -1.052.

$$\text{Antilog } -1.052 = 0.0887156$$

Thus $10^{-1.052} = 0.0887156$.   ■

EXAMPLE 5   Find N if log $N = 3.742$.

Solution: We are given the logarithm and asked to find the antilog, or the number N.

$$\text{Antilog } 3.742 = 5520.7743$$

Thus $N = 5520.7743$.   ■

Since we generally do not need the eight-place accuracy given by most calculators, in the exercise set that follows we will round logarithms to four decimal places and antilogarithms to three *significant digits*. In a number written in decimal form, the leading zeros preceding the first nonzero digit are not considered significant digits. The first nonzero digit in a number, moving left to right, will be the first significant digit.

Example

0.0063402	first significant digit is shaded
3.0424080	first three significant digits are shaded
0.0000138483	first three significant digits are shaded
206,435.05	first four significant digits are shaded

EXAMPLE 6   Find the following antilogs rounded to three significant digits: (a) antilog 6.827; (b) antilog -2.35.

Solution: In Example 3a, we found antilog $6.827 = 6,714,288.4$. Rounded to three significant digits, we get antilog $6.827 = 6,710,000$. In Example 3c, we found antilog $-2.35 = 0.0044668$. Rounded to three significant digits, we get antilog $-2.35 = 0.00447$.   ■

Exercise Set 11.3

Find the common logarithm of the number. Round the answer to four decimal places.

1. 870 **2.** 36 **3.** 8 **4.** 19,200
5. 1000 **6.** 0.00152 **7.** 0.0000857 **8.** 27,700
9. 100 **10.** 0.000835 **11.** 1.74 **12.** 3.75
13. 0.375 **14.** 0.0000375 **15.** 0.00872 **16.** 960

Find the antilog of the logarithm. Round the answer to three significant digits.

17. 0.5416 **18.** 2.6464 **19.** 2.3201 **20.** 5.8149
21. −1.0585 **22.** −2.3382 **23.** 0.0000 **24.** 5.5922
25. 2.5011 **26.** −4.4306 **27.** −0.1543 **28.** −1.2549

Find the number N. Round N to three significant digits.

29. $\log N = 2.0000$ **30.** $\log N = 1.6730$ **31.** $\log N = -3.104$ **32.** $\log N = 1.9330$
33. $\log N = 3.8202$ **34.** $\log N = 2.7404$ **35.** $\log N = -1.06$ **36.** $\log N = -1.1469$
37. $\log N = -0.3686$ **38.** $\log N = 1.5159$ **39.** $\log N = -0.3936$ **40.** $\log N = -1.3206$

To what exponent must the base 10 be raised to obtain each of the following values? Round your answer to four decimal places.

41. 2370 **42.** 817,000 **43.** 0.0410 **44.** 0.00612
45. 102 **46.** 8.92 **47.** 0.00128 **48.** 73,700,000

Find the value obtained to three significant digits, when 10 is raised to the following exponents.

49. 2.5866 **50.** 3.7118 **51.** −0.158 **52.** −2.2351
53. −1.6091 **54.** 4.8537 **55.** 1.1903 **56.** −2.1918

By changing the logarithm to exponential form, determine the common logarithm without the use of a calculator or a table.

57. $\log 1$ **58.** $\log 100$ **59.** $\log 0.1$ **60.** $\log 1000$
61. $\log 0.01$ **62.** $\log 10$ **63.** $\log 0.001$ **64.** $\log 10,000$

Find the solution. Use a calculator if available and round the answer to the nearest hundredth.

65. The magnitude of an earthquake on the Richter scale is given by the formula $R = \log I$, where I is the number of times more intense the quake is than the minimum level for comparison. (a) Find the Richter scale number for an earthquake that is 12,000 times more intense than the minimum level for comparison. (b) If an earthquake has a Richter scale number of 4.29, how many times more intense is the earthquake than the minimum level for comparison?

66. In astronomy, a formula used to find the diameter, in kilometers, of minor planets (also called asteroids) is $\log d = 3.7 - 0.2g$, where g is a quantity called the absolute magnitude of the minor planet. Find the diameter of a minor planet that has an absolute magnitude of (a) 11; (b) 20. (c) Find the absolute magnitude of the minor planet that has a diameter of 5.8 kilometers.

67. A formula that is sometimes used to estimate the seismic energy released by an earthquake is $\log E = 11.8 + 1.5m_s$, where E is the seismic energy and m_s is the surface wave magnitude. (a) Find the energy released in an earthquake with a surface wave magnitude of 6. (b) If the energy released during an earthquake is 1.2×10^{15}, what was the magnitude?

68. The sound pressure level, s_p, is given by the formula

$s_p = 20 \log \dfrac{p_r}{0.0002}$, where p_r is the sound pressure in dynes/cm². (a) Find the sound pressure level if the sound pressure is 0.0036 dynes/cm². (b) If the sound pressure level is 10.0, find the sound pressure.

69. The pH is a measure of the acidity or alkalinity of a solution. The pH of water, for example, is 7. In general, acids have pH numbers less than 7 while alkaline solutions have pH numbers greater than 7. The pH of a solution is defined to be

$$pH = -\log[H_3O^+]$$

where H_3O^+ represents the hydronium ion concentration of the solution. Find the pH of a solution with a hydronium ion concentration of 2.8×10^{-3}.

70. On your calculator, you find log 462 and obtain the answer 1.6646. Can this answer be correct? Explain your answer.

71. On your calculator, you find log 6250 and obtain the answer 2.7589. Can this answer be correct? Explain your answer.

72. On your calculator, you find log 0.163 and obtain the answer -2.7878. Can this answer be correct? Explain your answer.

73. On your calculator, you find log 0.0024 and obtain the answer -1.6198. Can this answer be correct? Explain your answer.

74. On your calculator, you find log -1.23 and obtain the answer 0.08991. Can this answer be correct? Explain your answer.

Cumulative Review Exercises

[9.2] **75.** Solve the quadratic equation using the quadratic formula.

$$-3x^2 - 4x - 8 = 0$$

76. In 4 hours the Simpsons first travel 15 miles down river in their motorboat, and then turn around and return home. If the river current is 5 miles per hour, find the speed of their boat in still water.

[9.3] **77.** Graph the solution to $\dfrac{2x - 3}{5x + 10} < 0$ on the number line.

[10.3] **78.** Sketch the graph of $x = 3(y - 2)^2 + 1$.

11.4

Exponential and Logarithmic Equations

▶**1** Solve exponential and logarithmic equations.

▶**2** Solve some practical problem using exponential and logarithmic equations.

▶**1** In Section 11.1 we introduced exponential and logarithmic equations. In this section we give more examples of their use and discuss further procedures for solving such equations.

To solve exponential and logarithmic equations, we often use properties 5a through 5d.

To Solve Exponential and Logarithmic Equations

We may use these properties: Properties 5a–5d

a. If $x = y$, then $a^x = a^y$.

b. If $a^x = a^y$, then $x = y$.

c. If $x = y$, then $\log x = \log y$ $(x > 0, y > 0)$.

d. If $\log x = \log y$, then $x = y$ $(x > 0, y > 0)$.

EXAMPLE 1   Solve the equation $8^x = \dfrac{1}{2}$.

Solution:

$$8^x = \dfrac{1}{2}$$

$$(2^3)^x = \dfrac{1}{2}$$

$$(2^3)^x = 2^{-1}$$

$$2^{3x} = 2^{-1}$$

Using property 5b, we can write

$$3x = -1$$

$$x = -\dfrac{1}{3}$$

When both sides of the exponential equation cannot be written as a power of the same base, we often begin by taking the logarithm of both sides of the equation, as illustrated in Example 2. In the following examples, we will use logarithms rounded to the nearest ten-thousandth.

EXAMPLE 2   Solve the equation $5^n = 20$.

Solution: Take the logarithm of both sides of the equation and solve for the variable n.

$$\log 5^n = \log 20$$

$$n \log 5 = \log 20$$

$$n = \dfrac{\log 20}{\log 5} = \dfrac{1.3010}{0.6990} \approx 1.861$$

Some logarithmic equations can be solved by expressing the equation in exponential form. **It is necessary to check logarithmic equations for extraneous solutions.** When checking an answer, if you obtain the logarithm of a nonpositive number, that answer is not a solution to the equation.

EXAMPLE 3   Solve the equation $\log_2 (x + 1)^3 = 4$.

Solution: Writing the equation in exponential form gives

$$(x + 1)^3 = 2^4$$

$$(x + 1)^3 = 16$$

$$x + 1 = \sqrt[3]{16}$$

$$x = -1 + \sqrt[3]{16}$$

Check:

$$\log_2 (x + 1)^3 = 4$$
$$\log_2 [(-1 + \sqrt[3]{16}) + 1]^3 = 4$$
$$\log_2 (\sqrt[3]{16})^3 = 4$$
$$\log_2 16 = 4$$
$$2^4 = 16$$
$$16 = 16 \qquad \text{true}$$

■

Other logarithmic equations can be solved using the properties of logarithms given in earlier sections.

EXAMPLE 4 Solve the equation $\log (3x + 2) + \log 9 = \log (x + 5)$.

Solution:
$$\log (3x + 2) + \log 9 = \log (x + 5)$$
$$\log (3x + 2)(9) = \log (x + 5)$$
$$(3x + 2)(9) = (x + 5) \qquad \text{Property 5d}$$
$$27x + 18 = x + 5$$
$$26x + 18 = 5$$
$$26x = -13$$
$$x = -\frac{1}{2}$$

A check will show that the solution is $-\frac{1}{2}$.

■

EXAMPLE 5 Solve the equation $\log x + \log(x + 1) = \log 12$.

Solution:
$$\log x + \log (x + 1) = \log 12$$
$$\log x (x + 1) = \log 12$$
$$x (x + 1) = 12$$
$$x^2 + x = 12$$
$$x^2 + x - 12 = 0$$
$$(x + 4)(x - 3) = 0$$
$$x + 4 = 0 \quad \text{or} \quad x - 3 = 0$$
$$x = -4 \quad \text{or} \quad x = 3$$

Check:

$x = 3$	$x = -4$
$\log x + \log (x + 1) = \log 12$	$\log x + \log (x + 1) = \log 12$
$\log 3 + \log 4 = \log 12$	$\log (-4) + \log (-3) = \log 12$
$\log (3)(4) = \log 12$	↑ ↑
$\log 12 = \log 12 \qquad \text{true}$	Logarithms of negative numbers are not real numbers

Thus -4 is an extraneous solution. The only solution is 3.

■

EXAMPLE 6   Solve the equation $\log(3x - 5) - \log 5x = 1.23$.

Solution:

$$\log \frac{3x - 5}{5x} = 1.23 \qquad \text{Property 2}$$

$$\frac{3x - 5}{5x} = \text{antilog } 1.23$$

$$\frac{3x - 5}{5x} = 17.0$$

$$3x - 5 = 5x(17.0)$$

$$3x - 5 = 85x$$

$$-5 = 82x$$

$$x = -\frac{5}{82} \approx -0.061$$

Check:
$$\log(3x - 5) - \log 5x = 1.23$$
$$\log[3(-0.061) - 5] - \log[(5)(-0.061)] = 1.23$$
$$\log(-5.183) - \log(-0.305) = 1.23$$

Since we are evaluating the logs of negative numbers, -0.061 is an extraneous solution. Thus the answer to this equation is no solution, or the empty set, $\{\ \}$. ∎

▶ **2**   In Section 11.1 we introduced the compound interest formula, $A = P(1 + r)^n$. We found A at that time by using a calculator. The value of A can also be found by using logarithms, as illustrated in Example 7.

EXAMPLE 7   The amount of money, A, accumulated in a savings account for a given principal P, interest rate r, and number of compounding periods n can be found by the formula

$$A = P(1 + r)^n$$

For example, if $1000 is invested in a savings account giving 8% interest compounded annually for 5 years, the amount accumulated at the end of 5 years is

$$A = 1000(1 + 0.08)^5$$

Use logarithms to find the amount accumulated.

Solution: Begin by taking the logarithms of both sides of the equation,

$$\log A = \log 1000(1 + 0.08)^5$$
$$= \log 1000 + \log(1 + 0.08)^5 \qquad \text{Property 1}$$
$$= \log 1000 + \log(1.08)^5$$
$$= \log 1000 + 5 \log 1.08 \qquad \text{Property 3}$$
$$= 3.00 + 5(0.0334)$$
$$= 3.00 + 0.167$$
$$= 3.167$$
$$A = \text{antilog } 3.167$$
$$A \approx 1470$$

In 5 years the $1000 would grow to about $1470. This includes the $1000 principal and $470 interest. ∎

EXAMPLE 8   If there are initially 1000 bacteria in a culture, and the number of bacteria double each hour, the number of bacteria after t hours can be found by the formula

$$N = 1000(2^t)$$

How long will it take for the culture to grow to 30,000 bacteria?

Solution:

$$N = 1000(2^t)$$
$$30,000 = 1000(2^t)$$

Now take the log of both sides of the equation.

$$\log 30,000 = \log 1000(2^t)$$
$$\log 30,000 = \log 1000 + \log 2^t$$
$$\log 30,000 = \log 1000 + t(\log 2)$$
$$4.4771 = 3.000 + t(0.3010)$$
$$4.4771 = 3.000 + 0.3010t$$
$$1.4771 = 0.3010t$$
$$\frac{1.4771}{0.3010} = t$$
$$4.91 \approx t$$

In approximately 4.91 hours there will be 30,000 bacteria in the culture. ∎

Exercise Set 11.4

Solve the exponential equation. Use a calculator if available and round the answer to nearest hundredth. Note that answers may be slightly different if tables are used, due to round-off error.

1. $3^x = 243$

2. $16^x = \dfrac{1}{4}$

3. $5^x = 125$

4. $8^x = 60$

5. $1.05^x = 15$

6. $4^{x-1} = 20$

7. $1.63^{x+1} = 25$

8. $2^{2x-1} = 50.6$

Find the solutions to the logarithmic equation. Round the answer to the nearest hundredth. If the equation has no real solution, so state.

9. $\log_4 (x + 1)^3 = 3$

10. $\log_3 (x - 2)^2 = 2$

11. $\log_2 (x + 4)^2 = 4$

12. $\log_2 x + \log_2 3 = 1$

13. $\log (2x - 3)^3 = 3$

14. $\log_3 3 + \log_3 x = 3$

15. $\log (x + 2) = \log (3x - 1)$

16. $\log x = \log (1 - x)$

17. $\log (3x - 1) + \log 4 = \log (9x + 2)$

18. $\log (x + 3) + \log x = \log 4$

19. $\log x + \log (3x - 5) = \log 2$

20. $\log (x + 4) - \log x = \log (x + 1)$

21. $\log x + \log 4 = 0.56$

22. $\log (x + 4) - \log x = 1.22$

23. $2 \log x - \log 4 = 2$

24. $\log 6000 - \log (x + 2) = 3.15$

25. $\log x + \log (x - 3) = 1$

26. $2 \log_2 x = 2$

27. $\log x = \frac{1}{3} \log 27$

28. $\log_7 x = \frac{3}{2} \log_7 64$

29. $\log_8 x = 3 \log_8 2 - \log_8 4$

30. $\log_4 x + \log_4 (6x - 7) = \log_4 5$

Find the solution. Use a calculator if available and round the answer to the nearest hundredth.

31. Find the amount accumulated if Marlina puts $1200 in a savings account offering 10% interest compounded annually for 5 years. Use $A = P(1 + r)^n$. See Example 7.

32. If the initial amount of bacteria in the culture in Example 8 is 4500, how long will it take for the number of bacteria in the culture to reach a count of 50,000? Use $N = 4500(2^t)$.

33. If after 4 hours the culture in Example 8 contains 2224 bacteria, how many bacteria were present initially?

34. The amount, A, of 200 grams of a certain radioactive material remaining after t years can be found by the equation $A = 200(0.800)^t$. After how long will there be 40 grams of the material remaining?

35. A machine purchased for business can be depreciated to reduce income tax. The value of the machine at the end of its useful life is called its *scrap value*. When the machine depreciates by a constant percentage annually, its scrap value, S, is

$$S = c(1 - r)^n$$

where c is the original cost, r is the annual rate of depreciation as a decimal, and n is the useful life in years. Find the scrap value of a machine that costs $50,000, has a useful life of 12 years, and has an annual depreciation rate of 15%.

36. If the machine in Exercise 35 costs $100,000, has a useful life of 15 years, and has an annual depreciation rate of 8%, find its scrap value.

37. The power gain, P, of an amplifier is defined to be

$$P = \log 10 \frac{P_{out}}{P_{in}}$$

where P_{out} is the power output in watts and P_{in} is the power input in watts. If an amplifier has an output power of 12.6 watts and an input power of 0.146 watts, find the power gain.

38. Measured by the Richter scale, the magnitude, R, of an earthquake of intensity I is defined by

$$R = \log I$$

where I is the number of times greater (or more intense) the earthquake is than the minimum level for comparison.

(a) How many times more intense was the 1906 San Francisco earthquake, which measured 8.25 on the Richter scale, than the minimum level for comparison?

(b) How many times more intense is an earthquake that measures 6.4 on the Richter scale than one that measures 4.7?

39. The decibel scale is used to measure the magnitude of sound. The magnitude of a sound, d, in decibels, is defined to be $d = 10 \log I$

where I is the number of times greater (or more intense) the given sound is than the minimum intensity of audible sound. An airplane engine (nearby) measures 120 decibels.

(a) How many times the minimum level of audible sound is the airplane engine?

(b) A busy city street has an intensity of 70 decibels. How many times greater is the intensity of the sound of the airplane engine than the sound of the city street?

40. After solving a logarithmic equation, what is it necessary to do?

Cumulative Review Exercises

[2.5] **41.** Solve the inequality $\dfrac{x - 4}{2} - \dfrac{2x - 5}{5} > 3$ and indicate the solution in (a) set builder notation and (b) interval notation.

[3.5, 11.1] **42.** Indicate which of the following graphs are functions. If the graph is a function, indicate if it is a one-to-one function.

(a)   (b)   (c)

[5.4] **43.** Multiply $(x^2 - 4x + 3)(2x - 3)$

[5.5] **44.** Divide $2x^2 + 11x + 15$ by $x + 4$

11.5

The Natural Exponential Function and Natural Logarithms

▶ **1** Identify the natural exponential function.

▶ **2** Identify the natural logarithmic function.

▶ **3** Show that the natural exponential function and the natural logarithmic functions are inverse functions.

▶ **4** Use a calculator to find natural logarithms and natural exponential values.

▶ **5** Find natural logarithms using the change of base formula.

▶ **6** Solve natural logarithmic and natural exponential equations.

▶ **7** Study applications of natural exponential functions and natural logarithms.

The natural exponential function and *its inverse* natural logarithmic function that we present in this section are in fact exponential functions and logarithmic functions of the type presented in the previous sections. They share all the properties of exponential functions and logarithmic functions discussed earlier. The importance of these special functions lies in the many varied applications in real life involving a unique irrational number designated by the letter e.

▶ **1** In Section 11.1 we indicated that exponential functions were of the form $f(x) = a^x$, $a > 0$ and $a \neq 1$. Now we introduce a very special exponential function. It is called the **natural exponential function** and it makes use of the letter e. Like the irrational number π, the number e is an irrational number whose value can only be approximated by a decimal number. The number e plays a very important role in higher-level mathematics courses. The number e has a value of approximately 2.7183. Now we define the natural exponential function.

The Natural Exponential Function

$$f(x) = e^x$$

where $e \approx 2.7183$.

▶ **2** We discussed common logarithms in Section 11.3. Now we will discuss natural logarithms. **Natural logarithms** are logarithms to the base e. Natural logarithms are indicated by the letters ln. The notation *ln x* is read the "natural log of x."

Natural Logarithms

$$\log_e x = \ln x$$

You must remember that the base of the natural logarithm is e. Thus, when you change a natural logarithm to exponential form, the base of the exponential expression will be e.

> For $x > 0$ if $y = \ln x$, then $e^y = x$.

EXAMPLE 1   Find the value of each of the following by changing the natural logarithm to exponential form.

(a) $\ln 1$ (b) $\ln e$

Solution: (a) Let $y = \ln 1$; then $e^y = 1$. Since any nonzero value to the 0th power equals 1, y must equal 0. Thus $\ln 1 = 0$.

(b) Let $y = \ln e$; then $e^y = e$. For e^y to equal e, y must have a value of 1. Thus $\ln e = 1$.   ∎

▶ **3**   The functions $y = a^x$ and $y = \log_a x$ are inverse functions. Similarly, the functions $y = e^x$ and $y = \ln x$ are inverse functions. (Remember, $y = \ln x$ means $y = \log_e x$.) The graphs of $y = e^x$ and $y = \ln x$ are illustrated in Fig. 11.7. Notice that the graphs are symmetric about the line $y = x$, which is what we expect of inverse functions.

Note that the graph of $y = e^x$ is similar to graphs of the form $y = a^x$, $a > 1$, and that the graph of $y = \ln x$ is similar to graphs of the form $y = \log_a x$, $a > 1$.

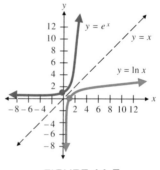

FIGURE 11.7

▶ **4**   Now we will learn how to find natural logarithms on a calculator.

🖩 Calculator Corner

Find Natural Logarithms on a Calculator

Natural logarithms can be found using a calculator that has a $\boxed{\ln}$ key. Natural logarithms are found in the same manner that we found common logarithms on a calculator, except we use the natural log key, $\boxed{\ln}$, instead of the common log key, $\boxed{\log}$.

Example	*Keys to Push*	*Display*
Find $\ln 242$	$\boxed{c}$ 242 $\boxed{\ln}$	5.4889377
Find $\ln 0.85$	$\boxed{c}$ 0.85 $\boxed{\ln}$	-0.1625189

When finding the natural logarithm of a number, we are finding an exponent. The natural logarithm of a number is the exponent to which the base e must be raised to obtain that number. For example,

$$\text{if } \ln 242 = 5.4889377, \text{ then } e^{5.4889377} = 242$$
$$\text{if } \ln 0.85 = -0.1625189, \text{ then } e^{-0.1625189} = 0.85$$

EXAMPLE 2 Indicate the keys to press to evaluate each of the natural logarithms; then give the answer obtained on a calculator.

(a) ln 1462   (b) ln 0.000381

Solution: (a) $\boxed{c}$ 1462 $\boxed{\ln}$ 7.2875606
(b) $\boxed{c}$ 0.000381 $\boxed{\ln}$ -7.8727112

Since $y = \ln x$ and $y = e^x$ are inverse functions of one another, we can use the inverse key $\boxed{\text{inv}}$ in combination with the natural log key $\boxed{\ln}$ to obtain values for e^x.

Calculator Corner

Find Values of e^x on a Calculator

To find the value of e^x on a calculator, press the clear key $\boxed{c}$, enter the value of exponent x, press the inverse key $\boxed{\text{inv}}$, and then press the natural log key $\boxed{\ln}$. After the $\boxed{\ln}$ key is pressed, the value of e^x will be displayed.

Example	*Keys to Push*	*Display*
Find $e^{5.24}$	$\boxed{c}$ 5.24 $\boxed{\text{inv}}$ $\boxed{\ln}$	188.6701
Find $e^{-1.639}$	$\boxed{c}$ 1.639 $\boxed{+/-}$ $\boxed{\text{inv}}$ $\boxed{\ln}$	0.1941741

Since e is about 2.7183, if we evaluated $(2.7183)^{5.24}$ we would obtain an answer close to 188.6701. Also, if we found ln 188.6701 on a calculator, we would obtain the answer 5.24. What do you think we would get as an answer if we evaluated ln 0.1941741 on a calculator? If you answered -1.639, you answered correctly.

EXAMPLE 3 Indicate the keys you would push to evaluate each of the following on a calculator, and then indicate the answer obtained.

(a) $e^{7.214}$ (b) $e^{-1.245}$

Solution: (a) $\boxed{c}$ 7.214 $\boxed{\text{inv}}$ $\boxed{\ln}$ 1358.3147
(b) $\boxed{c}$ 1.245 $\boxed{+/-}$ $\boxed{\text{inv}}$ $\boxed{\ln}$ 0.2879409

EXAMPLE 4 If $\ln N = 4.92$, find N.

Solution: If we write $\ln N = 4.92$ in exponential form, we get $e^{4.92} = N$. Thus we simply need to evaluate $e^{4.92}$ to determine N. We do this by pressing the following keys on the calculator.

$$\boxed{c} \ 4.92 \ \boxed{\text{inv}} \ \boxed{\ln} \ 137.00261$$

Thus $N = 137.00261$. Note that $e^{4.92} = 137.00261$.

EXAMPLE 5 If $\ln N = -0.0253$, find N.

Solution:

$$\boxed{c} \ 0.0253 \ \boxed{+/-} \ \boxed{\text{inv}} \ \boxed{\ln} \ 0.9750174$$

Thus $N = 0.9750174$.

▶ **5** If you do not have a calculator with $\boxed{\log}$ and $\boxed{\ln}$ keys, you can find natural logs by using the common logarithm table (Appendix D). To do so, you use the change of base formula.

Change of Base Formula

If $a > 0$, $a \neq 1$, $b > 0$, $b \neq 1$, and $x > 0$, then

$$\log_a x = \frac{\log_b x}{\log_b a}$$

In the change of base formula, 10 is often used in place of base b because we know how to find common logarithms. Replacing base b with 10 we get

$$\log_a x = \frac{\log_{10} x}{\log_{10} a} \qquad \text{or} \qquad \log_a x = \frac{\log x}{\log a}$$

EXAMPLE 6 Use the change of base formula to find $\log_3 24$.

Solution: If we substitute 3 for a and 24 for x in $\log_a x = \dfrac{\log x}{\log x}$ we obtain

$$\log_3 24 = \frac{\log 24}{\log 3} \approx \frac{1.3802}{0.4771} \approx 2.8928$$

Note that $3^{2.8929} \approx 24$. ■

We can use the same basic procedure illustrated in Example 6 to find natural logarithms using the change of base formula. For example to evaluate $\ln 20$ (or $\log_e 20$), we can substitute e for a and 20 for x in the formula $\log_a x = \dfrac{\log x}{\log a}$.

$$\log_e 20 = \frac{\log 20}{\log e} \approx \frac{1.3010}{0.4343} \approx 2.9956$$

Thus, $\ln 20 \approx 2.9956$. If you find $\ln 20$ on a calculator, you will obtain a very close answer.

Since $\log e \approx 0.4343$, to evaluate natural logarithms using common logarithms, we use the formula

$$\ln x \approx \frac{\log x}{0.4343}$$

EXAMPLE 7 Use the change of base formula to find $\ln 95$.

Solution: $\ln 95 = \dfrac{\log 95}{\log e} \approx \dfrac{1.9778}{0.4343} \approx 4.5540$ ■

▶ **6** The properties of logarithms discussed in Section 11.2 still hold true for natural logarithms. Following is a summary of the properties.

Properties for Natural Logarithms

$$\ln xy = \ln x + \ln y \qquad (x > 0 \text{ and } y > 0)$$

$$\ln \frac{x}{y} = \ln x - \ln y \qquad (x > 0 \text{ and } y > 0)$$

$$\ln x^n = n \ln x \qquad (x > 0)$$

Consider the expression $\ln e^x$. To what is this equal? Let us call this expression y and write the expression in exponential form.

$$y = \ln e^x$$
$$e^y = e^x$$

For this statement to be true, y must equal x. Substituting x for y in $y = \ln e^x$, we obtain $x = \ln e^x$. We can also show that, for any positive real value x, $e^{\ln x} = x$.

Additional Properties for Natural Logarithms and Natural Exponential Expressions

(a) $\ln e^x = x$ **Property 6**

(b) $e^{\ln x} = x, \qquad x > 0$

Using the property $\ln e^x = x$, we can state, for example, that $\ln e^{kt} = kt$, and $\ln e^{-2.06t} = -2.06t$. Using the property $e^{\ln x} = x$, we can state, for example, that $e^{\ln(t+2)} = t + 2$ and $e^{\ln kt} = kt$.

EXAMPLE 8   Solve the equation $\ln y - \ln(x + 6) = t$ for y.

Solution:

$$\ln y - \ln(x + 6) = t.$$

$$\ln \frac{y}{x + 6} = t$$

To eliminate the natural logarithm on the left side of the equation, we rewrite both sides of the equation with the base e, as follows:

$$e^{\ln[y/(x+6)]} = e^t$$

Since $\ln[y/(x + 6)] = t$ we are permitted to write $e^{\ln[y/(x+6)]} = e^t$ by property 5a given on page 523 (if $x = y$, then $a^x = a^y$). By Property 6b, we know that $e^{\ln[y/(x+6)]} = y/(x + 6)$. Replacing $e^{\ln[y/(x+6)]}$ with $y/(x + 6)$ we obtain

$$\frac{y}{x + 6} = e^t$$

Now proceed to solve the equation for y.

$$y = (x + 6)e^t$$

EXAMPLE 9   Solve the equation $450 = 225e^{-0.4t}$ for t.

Solution: Begin by dividing both sides of the equation by 225 to isolate the $e^{-0.4t}$.

$$\frac{450}{225} = \frac{\cancel{225}e^{-0.4t}}{\cancel{225}}$$

$$2 = e^{-0.4t}$$

Now take the natural log of both sides of the equation. We do this to eliminate the exponential expression on the right side of the expression.

$$\ln 2 = \ln e^{-0.4t}$$

$$\ln 2 = -0.4t \qquad (\ln e^{-0.4t} = -0.4t \text{ by Property } 6a)$$

$$0.6931472 = -0.4t$$

$$\frac{0.6931472}{-0.4} = t$$

$$-1.732868 = t$$ ■

EXAMPLE 10   Solve the equation $P = P_o e^{kt}$ for t.

Solution: We can follow the same general procedure as given in Example 9.

$$P = P_o e^{kt}$$

$$\frac{P}{P_o} = \frac{\cancel{P_o} e^{kt}}{\cancel{P_o}}$$

$$\frac{P}{P_o} = e^{kt}$$

$$\ln \frac{P}{P_o} = \ln e^{kt}$$

$$\ln P - \ln P_o = \ln e^{kt}$$

$$\ln P - \ln P_o = kt$$

$$\frac{\ln P - \ln P_o}{k} = t$$ ■

▶ **7**   Now let us look at some applications that involve the natural exponential function and natural logarithms.

When something increases or decreases at an *exponential rate,* a formula that is often used to find the value P after time t is

$$P = P_o e^{kt},$$

where P_o is the initial or starting value, and k is the constant rate of growth or decay.

EXAMPLE 11   Banks often credit compound interest on a continuous basis. When interest is compounded continuously, the principal in the account, P, at any time t can be calculated by the exponential growth formula $P = P_o e^{kt}$, where P_o is the initial principal invested, and k is the interest rate.

(a) Suppose the interest rate is 10% compounded continuously and $1000 is initially invested. Determine the balance in the account after 1 year.

(b) How long will it take the account to double in value?

Solution: (a) $P = P_o e^{kt}$

$$P = 1000e^{(0.10)(1)} = 1000e^{0.10} = 1000(1.1051709) = 1105.1709$$

Thus, after 1 year, the balance in the account is $1105.17. Since the initial amount invested was $1000, the interest that has accumulated in 1 year is $105.17.

(b) For the value of the account to double, the balance in the account would have to reach $2000. Therefore, we substitute 2000 for P and solve for t.

$$P = P_o e^{kt}$$
$$2000 = 1000e^{0.10t}$$
$$2 = e^{0.10t}$$
$$\ln 2 = \ln e^{0.10t}$$
$$\ln 2 = 0.10t$$
$$\frac{\ln 2}{0.10} = t$$
$$\frac{0.6931472}{0.10} = t$$
$$6.931472 = t$$

Thus with an interest rate of 10% compounded continuously, the account would double in slightly over 6.9 years. ■

EXAMPLE 12 Assume that the value of the island of Manhattan has grown at an exponential rate of 8% per year since 1626 when Peter Minuit of the Dutch West India Company purchased the island for $24. What is the value of the island of Manhattan in 1992, 366 years later?

Solution: Since the value is increasing exponentially, we can use the exponential growth formula. We will use V_o to represent initial value, and V to represent the value in 1992.

$$V = V_o e^{kt}$$
$$V = 24e^{(0.08)(366)}$$
$$= 24e^{29.28}$$
$$= 24(5.2017 \times 10^{12}) \quad \text{from a calculator}$$
$$= 1.2484 \times 10^{14}$$
$$= \$124,840,000,000,000$$

Thus, if the value grows exponentially at a rate of 8%, the value of Manhattan after 366 years is $124,840,000,000,000. ■

EXAMPLE 13 Strontium 90 is a radioactive material that decays exponentially at a decay rate of 2.8% per year. The amount of strontium 90 left after t years can be found by the formula $P = P_o e^{-0.028t}$. If there are initially 1000 grams of strontium 90:

(a) Find the number of grams of strontium 90 left after 50 years.
(b) Find the half-life of strontium 90.

Solution: (a) $P = P_o e^{-0.028t}$

$$P = 1000e^{-0.028(50)} = 1000e^{-1.4} = 1000(0.246597) = 246.597$$

Thus, after 50 years, there are still 246.597 grams of strontium 90 remaining.

(b) To find the half-life, we need to determine when there are 500 grams of strontium 90 left.

$$P = P_o e^{-0.028t}$$
$$500 = 1000 e^{-0.028t}$$
$$0.5 = e^{-0.028t}$$
$$\ln 0.5 = \ln e^{-0.028t}$$
$$-0.6931472 = -0.028t$$
$$\frac{-0.6931472}{-0.028} = t$$
$$24.755257 = t$$

Thus the half-life of strontium 90 is about 24.8 years.   ■

EXAMPLE 14   The formula for estimating the number of units sold, N, of a particular toy is $N = 400 + 250 \ln a$, where a is the amount of money spent on advertising.

(a) If $2000 is spent on advertising, what are the estimated sales of the toy?
(b) How much money would need to be spent on advertising for sales to reach 1500 units?

Solution: (a) $N = 400 + 250 \ln a$
$$= 400 + 250 \ln 2000$$
$$= 400 + 250(7.6009025) = 400 + 1900.2256 = 2300.2256$$

Thus approximately 2300 units of the toy will be sold.

(b)
$$N = 400 + 250 \ln a$$
$$1500 = 400 + 250 \ln a$$
$$1100 = 250 \ln a$$
$$\frac{1100}{250} = \ln a$$
$$4.4 = \ln a$$
$$81.45 = a$$

Thus about $81 needs to be spent on advertising to sell 1500 units of the toy.   ■

Exercise Set 11.5

Use a calculator if available to find the following values. Round your answer off to four decimal places.

1. $\ln 50$
2. $\ln 0.432$
3. $\ln 302$
4. $\ln 0.0038$

Use a calculator if available to find the value of N. Round your answer to three significant digits.

5. $\ln N = 1.6$
6. $\ln N = 4.96$
7. $\ln N = -2.63$
8. $\ln N = 0.632$

Use the change of base formula to find the value of the following logarithms.

9. $\ln 40$
10. $\ln 562$
11. $\ln 0.046$
12. $\ln 198$
13. $\log_3 25$
14. $\log_7 96$
15. $\log_2 20$
16. $\log_5 0.463$
17. $\ln 2700$
18. $\log_6 4000$
19. $\log_3 0.0049$
20. $\ln 8462$

Solve the following logarithmic equations.

21. $\ln x + \ln(x - 1) = \ln 12$

22. $\ln(x + 3) + \ln(x + 2) = \ln 6$

23. $\ln x = 5 \ln 2 - \ln 8$

24. $\ln x + \ln(x - 1) = \ln 2$

25. $\ln(x^2 - 4) - \ln(x + 2) = \ln 1$

26. $\ln x = \dfrac{3}{2} \ln 4$

Each of the following equations is in the form $P = P_o e^{kt}$. Solve each equation for the remaining variable. Remember, e is a constant.

27. $P = 500e^{1.6(1.2)}$

28. $1000 = V_o e^{0.6(2)}$

29. $50 = P_o e^{-0.05(3)}$

30. $120 = 60e^{2t}$

31. $90 = 30e^{1.4t}$

32. $20 = 40e^{-0.5t}$

33. $100 = 50e^{k(3)}$

34. $10 = 50e^{k(4)}$

35. $20 = 40e^{k(2.4)}$

36. $100 = A_o e^{-0.02(3)}$

37. $A = 6000e^{-0.08(3)}$

38. $75 = 100e^{-0.04t}$

Solve each of the equations for the indicated variable.

39. $V = V_o e^{kt}$ for V_o

40. $P = P_o e^{kt}$ for P_o

41. $P = 150e^{4t}$ for t

42. $200 = P_o e^{kt}$ for t

43. $A = A_o e^{kt}$ for k

44. $140 = R_o e^{kt}$ for k

45. $\ln y - \ln x = 2.3$ for y

46. $\ln y + 5 \ln x = \ln 2$ for y

47. $\ln y - \ln(x + 3) = 6$ for y

48. $\ln(x + 2) - \ln(y - 1) = \ln 5$ for y

49. The intensity of light as it passes through a certain medium is found by the formula $x = k(\ln I_o - \ln I)$. Solve this equation for I_o.

50. The distance traveled by a train moving at velocity v_o after the engine is shut off can be calculated by the formula $x = \dfrac{1}{k} \ln(kv_o t + 1)$. Solve this equation for v_o.

51. A formula used in studying the action of a protein molecule is $\ln M = \ln Q - \ln(1 - Q)$. Solve this equation for Q.

52. An equation relating the current and time in an electric circuit is $\ln i - \ln I = \dfrac{-t}{RC}$. Solve this equation for i.

Use a calculator if available to solve the following exercises.

53. If $5000 is invested at a rate of 8% compounded continuously, (a) determine the balance in the account after 2 years, and (b) how long would it take the value of the account to double (see Example 11)?

54. Refer to Example 13. Determine the volume of strontium 90 remaining after 20 years if there are originally 70 grams.

55. For a certain soft drink the percentage of a target market, $f(t)$, that buys the soft drink after t days of advertising is a function of the time the soft drink is advertised. The function that describes this relationship is $f(t) = 1 - e^{-0.04t}$. (a) What percentage of the target market buys the soft drink after 50 days of advertising? (b) How many days of advertising are needed if 75% of the target market is to buy the soft drink?

56. For a certain type of tie, the number of ties, $N(T)$, sold is a function of the dollar amount spent on advertising, a (in thousands of dollars). The function that describes this relationship is $N(T) = 800 + 300 \ln a$. (a) How many ties were sold after $1500 (or 1.5 thousand) was spent on advertising? (b) How much money must be spent on advertising to sell 1000 ties?

57. It was found in a psychological study that the average walking speeds, $f(P)$, of a person living in a certain city is a function of the population of the city. For a city of population P, the average walking speed in feet per second is given by $f(P) = 0.37 \ln P + 0.05$. The population of Nashville, Tennessee, is 972,000. (a) What is the average walking speed of a person living in Nashville? (b) What is the average walking speed of a person living in New York City, population 8,567,000? (c) If the average walking speed of the people in a certain city is 5.0 feet per second, what is the population of the city?

58. The percentage of doctors who accept and prescribe a new medicine is given by the formula $P(t) = 1 - e^{-0.22t}$, where t is the time in months since the medicine has been placed on the market. What percentage of doctors accept a new medicine 2 months after it is placed on the market?

59. The world population in 1988 is estimated to be 5.026 billion people. It is estimated that the world's population continues to grow exponentially at the rate of 1.8% per year. The world's expected population, in billions, in t years is given by the formula $P(t) = 5.026e^{0.018t}$.

(a) Find the expected world's population in the year 2000. (b) In how many years will the world's population double?

60. Plutonium, which is commonly used in nuclear reactors, decays at a rate of 0.003% per year. The formula $A = A_o e^{kt}$ can be used to find the amount of plutonium remaining from an initial amount, A_o, after t years. In the formula the k is replaced with -0.00003. (a) If 1000 grams of plutonium is present in 1990, how many grams of plutonium will remain in the year 2090, 100 years later? (b) Find the half-life of plutonium.

61. Carbon dating is used to determine the age of many ancient plants and objects. The radioactive element, carbon 14, is most often used for this purpose. Carbon 14 decays at a rate of 0.01205% per year. The amount of carbon 14 remaining in an object after t years can be found by the function $f(t) = v_o e^{-0.0001205t}$, where v_o is the initial amount present. (a) If an ancient animal bone originally had 20 grams of carbon 14, and when found it had 9 grams of carbon 14, how old is the bone? (b) How

old is an item that has 50% of its original carbon 14 remaining?

62. The power supply of a satellite is a radioisotope. The power output P, in watts, remaining in the power supply is a function of the time the satellite is in space. If there are 50 grams of the isotope originally, the power remaining after t days is $P = 50e^{-0.002t}$ (a) Find the power remaining after 100 days. (b) When will the power supply drop to 10 watts?

63. What is the approximate value of e?

64. If $e^x = 12.183$, find the value of x. Explain how you obtained your answer.

65. To what exponent must the base e be raised to obtain the value 184.93? Explain how you obtained your answer.

66. Find the value of $e^{4.32}$. Explain how you obtained your answer.

67. Find the value of e raised to the exponent -1.73. Explain how you obtained your answer.

Cumulative Review Exercises

[5.2] **68.** Simplify $\dfrac{(x^2y^{-2})^{-1}(4xy^3)^2}{(x^{-2}y^3)^{-2}}$.

[7.4] **69.** Simplify $\dfrac{\frac{3}{x^2} - \frac{2}{x}}{\frac{x}{4}}$.

[7.5] **70.** Solve the formula $\dfrac{1}{f} = \dfrac{1}{p} + \dfrac{1}{q}$ for q.

[8.3] **71.** Simplify $\sqrt[3]{128x^7y^9z^{13}}$.

SUMMARY

GLOSSARY

Antilogarithm *(520):* If log $N = L$, then $N = $ antilog L.

Argument of a logarithm *(511):* When finding the logarithm of an expression, the expression is called the argument. In $y = \log(x + 3)$, the $x + 3$ is the argument.

Common logarithm *(518):* Logarithms to the base 10 are common logarithms.

Exponential equation *(506):* An equation that has a variable as an exponent.

Exponential function *(505):* A function of the form $f(x) = a^x$, $a > 0$, $a \neq 1$.

Logarithm *(507):* $y = \log_a x$ means $x = a^y$, $a > 0$, $a \neq 1$.

Logarithmic equation *(511):* An equation in which a variable appears in the argument of the logarithm.

Logarithmic functions *(509):* Equations of the form $y = \log_a x$.

Natural exponential function *(529):* A function of the form $y = e^x$.

Natural logarithm *(529):* A logarithm to the base e, $\log_e x = \ln x$.

IMPORTANT FACTS

$y = a^x$ and $y = \log_a x$ are inverse functions.

$y = e^x$ and $y = \ln x$ are inverse functions.

The domain of a logarithm of the form $y = \log_a x$ is $x > 0$.

Properties of Logarithms

1. $\log_a xy = \log_a x + \log_a y$

2. $\log_a \dfrac{x}{y} = \log_a x - \log_a y$

3. $\log_a x^n = n \log_a x$

4. $\log 10^n = n$

To solve exponential and logarithmic equations, we may use these properties:

1. If $x = y$, then $a^x = a^y$.

2. If $a^x = a^y$, then $x = y$.

3. If $x = y$, then $\log x = \log y$ $(x > 0, y > 0)$.

4. If $\log x = \log y$, then $x = y$ $(x > 0, y > 0)$.

Natural logarithms

1. $\ln e^x = x$

2. $e^{\ln x} = x, \quad x > 0$

Review Exercises

[11.1] *Graph the following functions.*

1. $y = 2^x$

2. $y = \left(\dfrac{1}{2}\right)^x$

3. $y = \log_2 x$

4. $y = \log_{1/2} x$

5. On the same set of axes, graph $y = 3^x, \quad y = \log_3 x$.

Write in logarithmic form.

6. $4^2 = 16$

7. $8^{1/3} = 2$

8. $6^{-2} = \dfrac{1}{36}$

9. $25^{1/2} = 5$

Write in exponential form.

10. $\log_5 25 = 2$

11. $\log_{1/3} \dfrac{1}{9} = 2$

12. $\log_3 \dfrac{1}{9} = -2$

13. $\log_2 32 = 5$

Write in exponential form and find the missing value.

14. $3 = \log_4 x$

15. $2 = \log_4 x$

16. $3 = \log_a 8$

17. $-3 = \log_{1/4} x$

[11.2] *Use the properties of logarithms to expand each expression.*

18. $\log_8 \sqrt{12}$

19. $\log (x - 8)^5$

20. $\log \dfrac{2(x - 3)}{x}$

21. $\log \dfrac{x^4}{39(2x + 8)}$

Write as the logarithm of a single expression.

22. $\log_5 (x - 2) - 2 \log_5 x$

23. $2 \log x - 3 \log (x + 1)$

24. $3[\log x + \log 2] - \log x$

25. $2 \log_8 (x + 3) + 4 \log_8(x - 1) - \frac{1}{2} \log_8 x$

26. $\frac{1}{2}[\ln x - \ln (x + 2)] - \ln 2$

27. $3 \ln x + \frac{1}{2} \ln (x + 1) - 3 \ln (x + 4)$

[11.3] *Use a calculator to find the common logarithm. Round your answer to four decimal places.*

28. 8200

29. 0.000716

30. 0.00189

31. 17,600

Use a calculator to find the antilog. Give the answer to three significant digits.

32. 1.7528

33. 2.9186

34. -1.0991

35. -1.3747

Use a calculator to find N. Round the answer to three significant digits.

36. $\log N = 2.3304$

37. $\log N = -1.2262$

[11.4] *Solve the exponential equation.*

38. $4^x = 37$

39. $(3.2)^x = 187$

40. $(10.9)^{x+1} = 492$

41. $49^x = \frac{1}{7}$

Solve the logarithmic equation.

42. $\log_5(x + 2) = 3$

43. $\log(3x + 2) = \log 300$

44. $\log x - \log(3x - 5) = \log 2$

45. $\log_3 x + \log_3(2x + 1) = 1$

46. $\ln x = \frac{2}{3} \ln 8$

47. $\ln(x + 1) - \ln(x - 2) = \ln 4$

[11.5] *Solve each exponential equation for the remaining variable.*

48. $40 = 20e^{0.6t}$

49. $P_o = 80e^{-0.02(10)}$

50. $100 = A_o e^{-0.42(3)}$

Solve each equation for the indicated variable.

51. $w = w_o e^{kt}$ for w_o

52. $A = A_o e^{kt}$ for t

53. $150 = 600e^{kt}$ for k

54. $\ln y - \ln x = 2$ for y

55. $\ln(y + 3) - \ln(x + 1) = \ln 5$ for y

56. $\ln(x - 3) - \ln y = 1.4$ for y

Use the change of base formula to evaluate the following.

57. $\ln 450$

58. $\log_3 50$

59. $\log_5 0.0862$

60. Find the amount of money accumulated if Mrs. Elwood puts $12,000 in a savings account yielding 10% interest per year for 8 years. Use $A = p(1 + r)^n$.

61. Plutonium is a radioactive element that decays, or disintegrates, over time. If there are originally 1000 mg of plutonium, the amount remaining, R, after t years is

$$R = 1000(0.5)^{0.000041t}$$

Calculate the amount of plutonium present after 20,000 years.

62. The atmospheric pressure, P, in pounds per square inch at an elevation of x feet above sea level can be found by the formula $p = 14.7e^{-0.00004x}$. Find the atmospheric pressure at an elevation of 2000 feet.

63. If $10,000 is placed in a savings account paying 7% interest compounded continuously, find the time needed for the account to double in value.

Practice Test

1. Graph $y = 2^x$.

2. Graph $y = \log_2 x$.

3. Write $4^{-3} = \frac{1}{64}$ in logarithmic form.

4. Write $\log_3 243 = 5$ in exponential form.

Write in exponential form and find the missing value.

5. $4 = \log_2 x$

6. $y = \log_{27} 3$

7. Expand $\log_3 \dfrac{x(x - 4)}{x^2}$.

8. Write as the logarithm of a single expression.

$$3 \log_8(x - 4) + 2 \log_8(x + 1) - \frac{1}{2} \log_8 x$$

9. Find $\log 4620$.

10. Find $\log 0.000638$.

11. $\text{Log } N = -2.3002$; find N.

12. Solve for x: $3^x = 123$.

13. Solve for x: $\log 4x = \log(x + 3) + \log 2$.

14. What amount of money accumulates if Say-Chun puts $1500 in a savings account yielding 12% interest per year for 10 years? Use $A = p(1 + r)^n$.

15. The amount of carbon 14 remaining after t years is found by the formula $v = v_o e^{-0.0001205t}$, where v_o is the original amount of carbon 14 present. If a fossil originally contained 60 grams of carbon 14, and now contains 40 grams of carbon 14, how old is the fossil?

12

Sequences, Series, and the Binomial Theorem

See Section 12.3, Exercise 41.

Sequences and Series

▶**1** Find the terms of a sequence given the general term.

▶**2** Write a series.

▶**3** Find partial sums of a series.

▶**4** Use summation notation, Σ.

▶**1** Consider the following two lists of numbers

$$5, 10, 15, 20, 25, 30, \ldots$$

$$2, 4, 8, 16, 32, \ldots$$

The three dots at the ends of the lists indicate that the lists continue in the same manner. Can you determine the next two numbers in the first list? In the second list? The next numbers in the first list would seem to be 35 and 40. In the second list of numbers they would seem to be 64 and 128. The two lists of numbers above are examples of sequences. A **sequence (or progression)** of numbers is a list of numbers arranged in a specified order. Consider the sequence 5, 10, 15, 20, 25, 30, ... The first term is 5. We indicate this by writing $a_1 = 5$. Since the second term is 10, $a_2 = 10$, and so on.

An **infinite sequence** is a function whose domain is the set of natural numbers.

Consider the infinite sequence 5, 10, 15, 20, 25, 30, 35, . . .

Domain: $\{1, \ 2, \ 3, \ 4, \ 5, \ 6, \ 7, \ \ldots, \ n, \ \ldots\}$

$\downarrow \ \downarrow \ \downarrow \ \downarrow \ \downarrow \ \downarrow \ \downarrow \qquad \downarrow$

Range: $\{5, 10, 15, 20, 25, 30, 35, \ldots, 5n, \ldots\}$

Note that the terms of the sequence 5, 10, 15, 20, . . . , are found by multiplying each natural number by 5. For any natural number n, the corresponding term in the sequence is $5 \cdot n$ or $5n$. The **general term of the sequence,** a_n, which defines the sequence, is $a_n = 5n$.

$$a_n = f(n) = 5n$$

To find the twelfth term of the sequence, substitute 12 for n in the general term of the sequence, $a_{12} = 5 \cdot 12 = 60$. Thus, the twelfth term of the sequence is 60. Note that the terms in the sequence are the function values, or the range of the function. When writing the sequence, we do not use set braces. The general form of a sequence is

$$a_1, a_2, a_3, a_4, \ldots, a_n, \ldots$$

For the infinite sequence 2, 4, 8, 16, 32, . . . , 2^n, . . . we can write

$$a_n = f(n) = 2^n$$

Notice that when $n = 1$, $a_1 = 2^1 = 2$; when $n = 2$, $a_2 = 2^2 = 4$; when $n = 3$, $a_3 = 2^3 = 8$; when $n = 4$, $a_4 = 2^4 = 16$; and so on. What is the seventh term of this sequence? The answer is $a_7 = 2^7 = 128$.

A sequence may also be finite.

A **finite sequence** is a function whose domain includes only the first n natural numbers.

A finite sequence has only a finite number of terms.

Examples of Finite Sequences

5, 10, 15, 20   domain is {1, 2, 3, 4}

2, 4, 8, 16, 32   domain is {1, 2, 3, 4, 5}

EXAMPLE 1 Write the finite sequence defined by $a_n = 2n + 1$, for $n = 1, 2, 3, 4$.

Solution:
$$a_1 = 2(1) + 1 = 3$$
$$a_2 = 2(2) + 1 = 5$$
$$a_3 = 2(3) + 1 = 7$$
$$a_4 = 2(4) + 1 = 9$$

Thus the sequence is 3, 5, 7, 9. ∎

Since each term of the sequence in Example 1 is larger than the preceding term, it is called an **increasing sequence**.

EXAMPLE 2 Given $a_n = (2n + 3)/n^2$, find the following:
(a) The first term in the sequence.
(b) The third term in the sequence.
(c) The fifth term in the sequence.

Solution: (a) When $n = 1$, $a_1 = \dfrac{2(1) + 3}{1^2} = \dfrac{5}{1} = 5$.

(b) When $n = 3$, $a_3 = \dfrac{2(3) + 3}{3^2} = \dfrac{9}{9} = 1$.

(c) When $n = 5$, $a_5 = \dfrac{2(5) + 3}{5^2} = \dfrac{13}{25}$. ∎

Note in Example 2 that since there is no restriction on n, a_n is the general term of an infinite sequence.

In Example 2, since each term of the sequence generated by $a_n = (2n + 3)/n^2$ will be smaller than the preceding term, the sequence is called a **decreasing sequence**.

EXAMPLE 3 Find the first four terms of the sequence whose general term is given by $a_n = (-1)^n(n)$.

Solution:
$$a_n = (-1)^n(n)$$
$$a_1 = (-1)^1(1) = -1$$
$$a_2 = (-1)^2(2) = 2$$
$$a_3 = (-1)^3(3) = -3$$
$$a_4 = (-1)^4(4) = 4$$
∎

If we write the sequence in Example 3, we get $-1, 2, -3, 4, \ldots, (-1)^n(n)$. Notice that each term alternates in sign. We call this an **alternating sequence**.

▶ **2**   A **series** is the sum of a sequence. A series may be finite or infinite depending on whether the sequence it is based on is finite or infinite. Examples of a finite sequence and a finite series are

$$a_1, a_2, a_3, a_4, a_5 \qquad \textbf{finite sequence}$$

$$a_1 + a_2 + a_3 + a_4 + a_5 \qquad \textbf{finite series}$$

Examples of an infinite sequence and an infinite series are

$$a_1, a_2, a_3, a_4, a_5, \ldots, a_n, \ldots \qquad \textbf{infinite sequence}$$

$$a_1 + a_2 + a_3 + a_4 + a_5 + \cdots + a_n + \cdots \qquad \textbf{infinite series}$$

EXAMPLE 4   Write the first eight terms of the sequence; then write the series that represents the sum of that sequence if:

(a) $a_n = \left(\dfrac{1}{2}\right)^n$ (b) $a_n = \left(\dfrac{1}{2}\right)^{n-1}$

Solution: (a) We begin with $n = 1$; thus the first eight terms of the sequence whose general term is $a_n = \left(\frac{1}{2}\right)^n$ are

$$\left(\frac{1}{2}\right)^1, \left(\frac{1}{2}\right)^2, \left(\frac{1}{2}\right)^3, \left(\frac{1}{2}\right)^4, \left(\frac{1}{2}\right)^5, \left(\frac{1}{2}\right)^6, \left(\frac{1}{2}\right)^7, \left(\frac{1}{2}\right)^8$$

or

$$\frac{1}{2}, \frac{1}{4}, \frac{1}{8}, \frac{1}{16}, \frac{1}{32}, \frac{1}{64}, \frac{1}{128}, \frac{1}{256}$$

The series that represents the sum of the sequence is

$$\frac{1}{2} + \frac{1}{4} + \frac{1}{8} + \frac{1}{16} + \frac{1}{32} + \frac{1}{64} + \frac{1}{128} + \frac{1}{256}$$

(b) We again begin with $n = 1$; thus the first eight terms of the sequence whose general term is $a_n = \left(\frac{1}{2}\right)^{n-1}$ are

$$\left(\frac{1}{2}\right)^{1-1}, \left(\frac{1}{2}\right)^{2-1}, \left(\frac{1}{2}\right)^{3-1}, \left(\frac{1}{2}\right)^{4-1}, \left(\frac{1}{2}\right)^{5-1}, \left(\frac{1}{2}\right)^{6-1}, \left(\frac{1}{2}\right)^{7-1}, \left(\frac{1}{2}\right)^{8-1}$$

or

$$\left(\frac{1}{2}\right)^0, \left(\frac{1}{2}\right)^1, \left(\frac{1}{2}\right)^2, \left(\frac{1}{2}\right)^3, \left(\frac{1}{2}\right)^4, \left(\frac{1}{2}\right)^5, \left(\frac{1}{2}\right)^6, \left(\frac{1}{2}\right)^7$$

or

$$1, \frac{1}{2}, \frac{1}{4}, \frac{1}{8}, \frac{1}{16}, \frac{1}{32}, \frac{1}{64}, \frac{1}{128}$$

The series that represents the sum of this sequence is

$$1 + \frac{1}{2} + \frac{1}{4} + \frac{1}{8} + \frac{1}{16} + \frac{1}{32} + \frac{1}{64} + \frac{1}{128}$$ ■

▶ **3** A **partial sum of a series** is the sum of a finite number of consecutive terms of the series, beginning with the first term.

$$s_1 = a_1 \qquad\qquad \text{first partial sum}$$
$$s_2 = a_1 + a_2 \qquad\qquad \text{second partial sum}$$
$$s_3 = a_1 + a_2 + a_3 \qquad\qquad \text{third partial sum}$$
$$\vdots$$
$$s_n = a_1 + a_2 + a_3 + \cdots + a_n \qquad n\text{th partial sum}$$

EXAMPLE 5 Given the infinite sequence defined by $a_n = (1 + n^2)/n$, find the following:

(a) The first partial sum.
(b) The third partial sum.

Solution: (a) $s_1 = a_1 = \dfrac{1 + 1^2}{1} = \dfrac{1 + 1}{1} = 2$

(b) $s_3 = a_1 + a_2 + a_3$

$$= \frac{1 + 1^2}{1} + \frac{1 + 2^2}{2} + \frac{1 + 3^2}{3}$$

$$= 2 + \frac{5}{2} + \frac{10}{3}$$

$$= \frac{12}{6} + \frac{15}{6} + \frac{20}{6} = \frac{47}{6} \quad \text{or} \quad 7\frac{5}{6}$$

▶ **4** In Section 2.2 we introduced the **summation symbol, Σ**, which is the Greek letter *sigma*. This symbol plays a very important role in mathematics and statistics. The summation notation gives a compact way to write a series from the general term of the corresponding sequence. For example, the first five terms of the sequence with general term $a_n = 2n + 3$, where $n = 1, 2, 3, 4, 5$, are

$$5, 7, 9, 11, 13$$

The series representing the sum of this sequence is

$$5 + 7 + 9 + 11 + 13$$

The series representing the sum of this five-term sequence can be indicated using the summation notation:

$$\sum_{n=1}^{5}(2n + 3)$$

To list and evaluate the series represented by $\sum_{n=1}^{5}(2n + 3)$, we first substitute 1 for n in $2n + 3$ and list the value obtained. Then we substitute 2 for n in $2n + 3$ and list the value. We follow this procedure for the values 1 through 5. We then sum these values to obtain the series value.

$$\sum_{n=1}^{5}(2n + 3) = (2 \cdot 1 + 3) + (2 \cdot 2 + 3) + (2 \cdot 3 + 3) + (2 \cdot 4 + 3) + (2 \cdot 5 + 3)$$

$$= 5 + 7 + 9 + 11 + 13 = 45$$

Notice that the series generated by $\sum_{n=1}^{5}(2n + 3)$ is $5 + 7 + 9 + 11 + 13$, and the sum of the series is 45.

The letter n used in the summation notation is called the **index of summation** or simply the *index*. Any letter can be used for the index. Thus, for example, $\sum_{n=1}^{5} (2n + 3) = \sum_{i=1}^{5} (2i + 3)$. The 1 below the summation symbol is called the **lower limit,** and the 5 above the summation symbol is called the **upper limit** of the summation. The expression $\sum_{n=1}^{5} (2n + 3)$ is read "the sum of the terms $2n + 3$ as n varies from 1 to 5."

EXAMPLE 6 Write out the series $\sum_{k=1}^{6} (k^2 + 1)$ and find the sum of the series.

Solution: $\sum_{k=1}^{6}(k^2 + 1) = (1^2 + 1) + (2^2 + 1) + (3^2 + 1) + (4^2 + 1) + (5^2 + 1) + (6^2 + 1)$

$$= 2 + 5 + 10 + 17 + 26 + 37 = 97 \qquad \blacksquare$$

EXAMPLE 7 Consider the general term of a sequence $a_n = 2n^2 - 4$. Represent the third partial sum in summation notation.

Solution: The third partial sum will be the sum of the first three terms, $a_1 + a_2 + a_3$. We can represent the third partial sum as $\sum_{n=1}^{3} (2n^2 - 4)$. $\blacksquare$

EXAMPLE 8 For the following set of values $x_1 = 3$, $x_2 = 4$, $x_3 = 5$, $x_4 = 6$, and $x_5 = 7$, does $\sum_{i=1}^{5} x_i^2 = (\sum_{i=1}^{5} x_i)^2$?

Solution:

$$\sum_{i=1}^{5} x_i^2 = x_1^2 + x_2^2 + x_3^2 + x_4^2 + x_5^2$$

$$= 3^2 + 4^2 + 5^2 + 6^2 + 7^2$$

$$= 9 + 16 + 25 + 36 + 49$$

$$= 135$$

$$\left(\sum_{i=1}^{5} x_i\right)^2 = (x_1 + x_2 + x_3 + x_4 + x_5)^2$$

$$= (3 + 4 + 5 + 6 + 7)^2$$

$$= (25)^2$$

$$= 625$$

Since $135 \neq 625$, $\sum_{i=1}^{5} x_i^2 \neq (\sum_{i=1}^{5} x_1)^2$. $\blacksquare$

When a summation symbol is written without any upper and lower limits, it means that all the given data are to be summed.

EXAMPLE 9 A formula used to find the arithmetic mean, $\bar{x}$ (read x bar), of a set of data is $\bar{x} = \dfrac{\sum x}{n}$, where n is the number of pieces of data.

Sally's five test grades are 70, 90, 83, 74, and 92. Find the mean of her grades.

Solution: $\bar{x} = \dfrac{\sum x}{n} = \dfrac{70 + 90 + 83 + 74 + 92}{5} = \dfrac{409}{5} = 81.8$ $\blacksquare$

Exercise Set 12.1

Write the first five terms of the sequence whose nth term is shown.

1. $a_n = 2n$ **2.** $a_n = 2n + 3$ **3.** $a_n = \dfrac{n + 5}{n}$ **4.** $a_n = \dfrac{n}{n^2}$

5. $a_n = \dfrac{1}{n}$ **6.** $a_n = n^2 - n$ **7.** $a_n = \dfrac{n + 2}{n + 1}$ **8.** $a_n = \dfrac{n + 2}{n + 3}$

9. $a_n = (-1)^n$ **10.** $a_n = (-1)^{2n}$ **11.** $a_n = (-2)^{n+1}$ **12.** $a_n = n(n + 2)$

Find the indicated term of the sequence whose nth term is shown.

13. $a_n = 2n + 3$, twelfth term   **14.** $a_n = 2^n$, seventh term

15. $a_n = 2n - 4$, fifth term   **16.** $a_n = (-1)^n$, eighth term

17. $a_n = (-2)^n$, fourth term   **18.** $a_n = n(n + 5)$, eighth term

19. $a_n = \dfrac{n^2}{(2n + 1)}$, ninth term   **20.** $a_n = \dfrac{n(n + 1)}{n^2}$, tenth term

Find the first and third partial sums, s_1 and s_3, for the given sequences.

21. $a_n = 2n + 5$ **22.** $a_n = \dfrac{3n}{n + 2}$ **23.** $a_n = 2^n + 1$ **24.** $a_n = \dfrac{n - 1}{n + 1}$

25. $a_n = (-1)^{2n}$ **26.** $a_n = \dfrac{2n^2}{n + 1}$ **27.** $a_n = \dfrac{n^2}{2}$ **28.** $a_n = \dfrac{n + 3}{2n}$

Write the next three terms of each sequence.

29. $2, 4, 8, 16, 32, \ldots$ **30.** $\dfrac{1}{2}, \dfrac{1}{3}, \dfrac{1}{4}, \dfrac{1}{5}, \ldots$

31. $3, 5, 7, 9, 11, 13, \ldots$ **32.** $5, 10, 15, 20, 25, \ldots$

33. $1, \dfrac{1}{2}, \dfrac{1}{3}, \dfrac{1}{4}, \dfrac{1}{5}, \ldots$ **34.** $\dfrac{2}{3}, \dfrac{3}{4}, \dfrac{4}{5}, \dfrac{5}{6}, \dfrac{6}{7}, \ldots$

35. $-1, 1, -1, 1, -1, \ldots$ **36.** $-10, -20, -30, -40, \ldots$

37. $1, \dfrac{1}{3}, \dfrac{1}{9}, \dfrac{1}{27}, \ldots$ **38.** $\dfrac{1}{3}, \dfrac{2}{3}, \dfrac{3}{3}, \dfrac{4}{3}, \ldots$

39. $1, -\dfrac{1}{2}, \dfrac{1}{4}, -\dfrac{1}{8}, \ldots$ **40.** $\dfrac{2}{3}, \dfrac{1}{3}, \dfrac{1}{6}, \dfrac{1}{12}, \ldots$

41. $7, -1, -9, -17, \ldots$ **42.** $37, 32, 27, 22, \ldots$

Write out the series, and then find the sum of the series.

43. $\displaystyle\sum_{n=1}^{5} 3n - 1$ **44.** $\displaystyle\sum_{k=1}^{4} k^2 - 1$ **45.** $\displaystyle\sum_{k=1}^{6} 2k^2 - 3$ **46.** $\displaystyle\sum_{i=1}^{4} \dfrac{i^2}{2}$

47. $\displaystyle\sum_{n=2}^{4} \dfrac{n^2 + n}{n + 1}$ **48.** $\displaystyle\sum_{n=2}^{5} \dfrac{n^3}{n + 1}$

For the general term a_n, write an expression using Σ to represent the indicated partial sum.

49. $a_n = n + 3$, fifth partial sum   **50.** $a_n = n^2 + 1$, fourth partial sum

51. $a_n = \dfrac{n^2}{4}$, third partial sum   **52.** $a_n = \dfrac{n^2 + 1}{n + 1}$, third partial sum

For the set of values $x_1 = 2$, $x_2 = 3$, $x_3 = 5$, $x_4 = -1$, and $x_5 = 4$, find each of the following

53. $\displaystyle\sum_{i=1}^{5} x_i$

54. $\displaystyle\sum_{i=1}^{5} x_i^2$

55. $\displaystyle\left(\sum_{i=1}^{5} x_i\right)^2$

56. $\displaystyle\sum_{i=1}^{4} (x_i^2 + 3)$

57. $\displaystyle\sum_{i=3}^{5} \frac{2x_i}{3}$

58. $\displaystyle\sum_{i=2}^{4} x_i^2 + x_i$

Find the mean, $\bar{x}$, of the following sets of data. See Example 9.

59. 15, 20, 25, 30, 35

60. 16, 20, 96, 18, 25

61. 72, 83, 4, 60, 18, 20

62. 5, 12, 9, 12, 17, 36, 70

63. Write $\displaystyle\sum_{i=1}^{n} x_i$ as a sum of terms.

64. Solve the formula $\bar{x} = \dfrac{\sum x}{n}$ for (a) $\sum x$; (b) n.

65. What is a sequence?

66. What is a series?

67. What is the nth partial sum of a series?

68. How is the following notation read:

Cumulative Review Exercises

[6.5] **69.** Solve the equation $2x^2 + 15 = 13x$ using factoring.

[9.2] **70.** How many real solutions does the equation $6x^2 - 3x - 4 = 2$ have? Explain how you obtained your answer.

[10.2] **71.** Sketch the graph of $\dfrac{x^2}{4} + \dfrac{y^2}{1} = 1$.

[10.5] **72.** Solve the system of equations $x^2 + y^2 = 5$
$\quad x = 2y$.

JUST FOR FUN

1. When no upper and lower limits are placed on a summation symbol, it indicates that the sum of all values is to be found. Consider the following values:

$$x_1 = 1, \quad x_2 = 3, \quad x_3 = 5, \quad x_4 = 7, \quad x_5 = 9$$
$$f_1 = 3, \quad f_2 = 4, \quad f_3 = 5, \quad f_4 = 0, \quad f_5 = 2$$

Then

$$\sum x = x_1 + x_2 + x_3 + x_4 + x_5 = 1 + 3 + 5 + 7 + 9 = 25$$
$$\sum f = f_1 + f_2 + f_3 + f_4 + f_5 = 3 + 4 + 5 + 0 + 2 = 14$$
$$\sum xf = x_1f_1 + x_2f_2 + x_3f_3 + x_4f_4 + x_5f_5$$
$$= 1(3) + 3(4) + 5(5) + 7(0) + 9(2) = 58$$

A formula used to calculate **standard deviation** in statistics is

$$s = \sqrt{\frac{n(\sum x^2 f) - (\sum xf)^2}{n(n-1)}}$$

where $n = \sum f$. For the values of x and f given above, find the value of s.

12.2

Arithmetic Sequences and Series

▸**1** Find the common difference in an arithmetic sequence.

▸**2** Find the nth term of an arithmetic sequence.

▸**3** Find the nth partial sum of an arithmetic series.

▸**1** A sequence in which each term after the first differs from the preceding term by a constant amount is called an **arithmetic sequence** or **arithmetic progression.** The constant amount by which each pair of successive terms differs is called the **common difference,** d. The common difference can be found by subtracting any term from the term that directly follows it.

Arithmetic Sequence	*Common Difference*
$1, 4, 7, 10, 13, 16, \ldots$	$d = 4 - 1 = 3$
$-7, -2, 3, 8, 13, \ldots$	$d = -2 - (-7) = -2 + 7 = 5$
$\dfrac{7}{2}, \dfrac{2}{2}, -\dfrac{3}{2}, -\dfrac{8}{2}, -\dfrac{13}{2}, -\dfrac{18}{2}, \ldots$	$d = \dfrac{2}{2} - \dfrac{7}{2} = -\dfrac{5}{2}$

EXAMPLE 1   Write the first five terms of the arithmetic sequence with the first term 6 and the common difference 3.

Solution: 6, 9, 12, 15, 18   ■

EXAMPLE 2   Write the first six terms of the arithmetic sequence with the first term 3 and the common difference -2.

Solution: $3, 1, -1, -3, -5, -7$ ■

▶ **2**   In general, an arithmetic sequence with the first term a_1 and the common difference d will have the following terms:

$$a_1 = a_1, \quad a_2 = a_1 + d, \quad a_3 = a_1 + 2d, \quad a_4 = a_1 + 3d, \quad \text{and so on}$$

If we continue this process, we can see that the nth term, a_n, can be found by

---

nth Term of an Arithmetic Sequence

$$a_n = a_1 + (n - 1)d$$

EXAMPLE 3   (a) Write an expression for the general (or nth) term, a_n, of the arithmetic sequence with a first term of -3 and a common difference of 4.
(b) Find the twelfth term of the sequence.

Solution: (a) The nth term of the sequence is $a_n = a_1 + (n - 1)d$. Substituting $a_1 = -3$ and $d = 4$, we obtain

$$\begin{aligned} a_n &= a_1 + (n - 1)d \\ &= -3 + (n - 1)4 \\ &= -3 + 4(n - 1) \end{aligned}$$

Thus $a_n = -3 + 4(n - 1)$.
(b) $a_n = -3 + 4(n - 1)$
$a_{12} = -3 + 4(12 - 1) = -3 + 4(11) = -3 + 44 = 41$

The twelfth term in the sequence will be 41.   ■

EXAMPLE 4 Find the number of terms in the arithmetic sequence 5, 9, 13, 17, . . . , 41.

Solution: The first term, a_1, is 5; the nth term is 41, and the common difference, d, is 4.

$$a_n = a_1 + (n - 1)d$$
$$41 = 5 + (n - 1)4$$
$$41 = 5 + 4n - 4$$
$$41 = 4n + 1$$
$$40 = 4n$$
$$10 = n$$

Thus the sequence has a total of 10 terms. ■

▶ **3** An **arithmetic series** is the sum of the terms of an arithmetic sequence. A finite arithmetic series can be written

$$s_n = a_1 + (a_1 + d) + (a_1 + 2d) + (a_1 + 3d) + \cdots + (a_n - 2d) + (a_n - d) + a_n$$

If we consider the last term as a_n, the term before the last term will be $a_n - d$, the second before the last term will be $a_n - 2d$, and so on.

A formula for the nth partial sum, s_n, can be obtained by adding the reverse of s_n to itself.

$$s_n = a_1 + (a_1 + d) + (a_1 + 2d) + \cdots + (a_n - 2d) + (a_n - d) + a_n$$
$$s_n = a_n + (a_n - d) + (a_n - 2d) + \cdots + (a_1 + 2d) + (a_1 + d) + a_1$$
$$\overline{2s_n = (a_1 + a_n) + (a_1 + a_n) + (a_1 + a_n) + \cdots + (a_1 + a_n) + (a_1 + a_n) + (a_1 + a_n)}$$

Since the right side of the equation contains n terms of $(a_1 + a_n)$, we can write

$$2s_n = n(a_1 + a_n)$$

Therefore,

---

nth Partial Sum of an Arithmetic Series

$$s_n = \frac{n(a_1 + a_n)}{2}$$

EXAMPLE 5 Find the sum of the first 25 natural numbers.

Solution: The arithmetic sequence is 1, 2, 3, 4, 5, 6, . . . , 25. The first term, a_1, is 1; the last term, a_n, is 25. There are 25 terms; thus $n = 25$.

$$s_n = \frac{n(a_1 + a_n)}{2} = \frac{25(1 + 25)}{2} = \frac{25(26)}{2} = 25(13) = 325$$

The sum of the first 25 natural numbers is 325. Thus $s_{25} = 325$. ■

EXAMPLE 6 The first term of an arithmetic sequence is 4, and the last term is 31. If $s_n = 175$, find the number of terms in the sequence and the common difference.

Solution: $a_1 = 4$, $a_n = 31$, and $s_n = 175$.

$$s_n = \frac{n(a_1 + a_n)}{2}$$

$$175 = \frac{n(4 + 31)}{2}$$

$$175 = \frac{35n}{2}$$

$$350 = 35n$$

$$10 = n$$

Thus there are 10 terms in the sequence. We can now find the common difference.

$$a_n = a_1 + (n - 1)d$$
$$31 = 4 + (10 - 1)d$$
$$31 = 4 + 9d$$
$$27 = 9d$$
$$3 = d$$

The common difference is 3. The sequence is 4, 7, 10, 13, 16, 19, 22, 25, 28, 31. ∎

Examples 7 and 8 illustrate some applications of arithmetic sequences and series.

EXAMPLE 7 Donna Stansell is given a starting salary of $25,000 and is promised a $1200 raise after each of the next 8 years. Find her salary during her eighth year of work.

Solution: Her salaries after the first few years would be

$$25,000, \ 26,200, \ 27,400, \ 28,600, \ \ldots$$

Since we are adding a constant amount each year, this is an arithmetic sequence. The general term of an arithmetic sequence is $a_n = a_1 + (n - 1)d$. In this example, $a_1 = 25,000$ and $d = 1200$. Thus, for $n = 8$, Donna's salary would be

$$a_8 = 25,000 + (8 - 1)1200$$
$$= 25,000 + 7(1200)$$
$$= 25,000 + 8400$$
$$= 33,400$$

If we listed all the salaries for the 8-year period, they would be

$$25,000, \ 26,200, \ 27,400, \ 28,600, \ 29,800, \ 31,000, \ 32,200, \ 33,400$$ ∎

EXAMPLE 8 Each swing of a pendulum is 3 inches shorter than its preceding swing. The first swing is 8 feet.

(a) Find the length of the twelfth swing.
(b) Determine the total distance traveled by the pendulum during the first 12 swings.

Solution: (a) Since each swing is decreasing by a constant amount, this problem can be represented as an arithmetic series. Since the first swing is given in feet and the decrease in swing in inches, we will change 3 inches to 0.25 feet $(3 \div 12 = 0.25)$. The twelfth swing can be considered a_{12}. The difference, d, is negative since the distance is decreasing with each swing.

$$a_n = a_1 + (n - 1)d$$
$$a_{12} = 8 + (12 - 1)(-0.25)$$
$$= 8 + 11(-0.25)$$
$$= 8 - 2.75$$
$$= 5.25 \text{ feet}$$

The twelfth swing will travel 5.25 feet.

(b) The total distance traveled during the first 12 swings can be found using the formula

$$s_n = \frac{n(a_1 + a_n)}{2}$$

$$s_{12} = \frac{12(a_1 + a_{12})}{2}$$

$$= \frac{12(8 + 5.25)}{2} = \frac{12(13.25)}{2} = 6(13.25) = 79.5 \text{ feet}$$

The pendulum travels a total of 79.5 feet during its first 12 swings.   ■

Exercise Set 12.2

Write the first five terms of the arithmetic sequence with the given first term and common difference. Write the expression for the general (or nth) term, a_n, of the arithmetic sequence.

1. $a_1 = 3, d = 4$

2. $a_1 = 8, d = 2$

3. $a_1 = -5, d = 2$

4. $a_1 = -8, d = -3$

5. $a_1 = \frac{1}{2}, d = \frac{3}{2}$

6. $a_1 = -\frac{5}{3}, d = -\frac{1}{3}$

7. $a_1 = 100, d = -5$

8. $a_1 = \frac{5}{4}, d = -\frac{3}{4}$

Find the desired quantity of the arithmetic sequence.

9. $a_1 = 4, d = 3$; find a_7

10. $a_1 = 8, d = -2$; find a_6

11. $a_1 = -6, d = -1$; find a_{18}

12. $a_1 = -15, d = 3$; find a_{20}

13. $a_1 = -2, d = \frac{5}{3}$; find a_{10}

14. $a_1 = 5, a_8 = -21$; find d

15. $a_1 = 3, a_9 = 19$; find d

16. $a_1 = \frac{1}{2}, a_7 = \frac{19}{2}$; find d

17. $a_1 = 4, a_n = 28, d = 3$; find n

18. $a_1 = -2, a_n = -20, d = -3$; find n

19. $a_1 = -\frac{7}{3}, a_n = -\frac{17}{3}, d = -\frac{2}{3}$; find n

20. $a_1 = 100, a_n = 60, d = -8$; find n

Find the sum, s_n, and common difference, d.

21. $a_1 = 1, a_n = 19, n = 10$

22. $a_1 = -5, a_n = 13, n = 7$

23. $a_1 = \frac{3}{5}, a_n = 2, n = 8$

24. $a_1 = 12, a_n = -23, n = 8$

25. $a_1 = \dfrac{12}{5}$, $a_n = \dfrac{28}{5}$, $n = 5$

26. $a_1 = -3$, $a_n = 15.5$, $n = 6$

27. $a_1 = 7$, $a_n = 67$, $n = 11$

Write the first four terms of each sequence; then find a_{10} and s_{10}.

28. $a_1 = 5$, $d = 3$

29. $a_1 = -4$, $d = -2$

30. $a_1 = \dfrac{7}{2}$, $d = \dfrac{5}{2}$

31. $a_1 = -8$, $d = -5$

32. $a_1 = -15$, $d = 4$

33. $a_1 = 100$, $d = -7$

34. $a_1 = 35$, $d = 3$

35. $a_1 = \dfrac{9}{5}$, $d = \dfrac{3}{5}$

Find the number of terms in each sequence and s_n.

36. $1, 4, 7, 10, \ldots , 43$

37. $-8, -6, -4, -2, \ldots , 42$

38. $-9, -5, -1, 3, \ldots , 27$

39. $\dfrac{1}{2}, \dfrac{2}{2}, \dfrac{3}{2}, \dfrac{4}{2}, \dfrac{5}{2}, \ldots , \dfrac{17}{2}$

40. $-\dfrac{5}{6}, -\dfrac{7}{6}, -\dfrac{9}{6}, -\dfrac{11}{6}, \ldots , -\dfrac{21}{6}$

41. $7, 14, 21, 28, \ldots , 63$

42. $-12, -16, -20, \ldots , -52$

43. $9, 12, 15, 18, \ldots , 93$

44. Mr. Baudean is given a starting salary of $20,000 and is told he will receive a $1000 raise at the end of each year for the next 5 years.
 (a) Write a sequence showing his salary for the next 5 years
 (b) What is the general term for this sequence?
 (c) If this procedure is extended, find his salary after 12 years.

45. Find the sum of the first 1000 positive integers.

46. Find the sum of the numbers between 50 and 200 inclusive.

47. Determine how many numbers between 7 and 1610 are divisible by 6.

48. An object falls 16.0 feet during the first second, 48.0 feet during the second second, 80.0 feet during the third second, and so on.
 (a) How far will it fall in its tenth second?

 (b) How far will the object fall, in total, during the first 10 seconds?

49. Each time a ball bounces the height attained is 6 inches less than the previous height reached. If its first bounce reaches a height of 6 feet, find the height attained on the eleventh bounce.

50. If you are given $1 on January 1, $2 on January 2, $3 on the 3rd, and so on, how much money will you have received in total by January 31?

51. Jack piles logs so that there are 20 logs in the bottom layer, and each layer contains one log less than the layer below it. How many logs are on the pile?

52. What is an arithmetic sequence?

53. What is an arithmetic series?

54. How can the common difference in an arithmetic sequence be found?

Cumulative Review Exercises

[3.4] **55.** Consider the system of equations
$$2x + 3y = -4$$
$$-x - y = -1$$
Without solving the system, determine how many solutions the system has. Explain how you determined your answer.

[4.1] **56.** Determine the solution to the system of equations in problem 55.

[4.5] **57.** Graph $|x - 2| < 4$ in the Cartesian coordinate system.

[9.3] **58.** Graph the inequality $y \geq 2x^2 - 4x + 6$.

Geometric Sequences and Series

▶ **1** Find the common ratio in a geometric sequence.

▶ **2** Find the nth term of a geometric sequence.

▶ **3** Find the nth partial sum of a geometric series.

▶ **1** A **geometric sequence** (or **geometric progression**) is a sequence in which each term after the first is the same multiple of the preceding term. The common multiple is called the **common ratio.** The common ratio, r, in any geometric sequence can be found by taking any term, except the first, and dividing that term by the preceding term. Consider the geometric sequence

$$1, 3, 9, 27, 81, \ldots , 3^{n-1}, \ldots$$

The common ratio is 3 since $3 \div 1 = 3$ (or $9 \div 3 = 3$, and so on).

The common ratio of the geometric sequence

$$4, 8, 16, 32, 64, \ldots , 4(2^{n-1}), \ldots \text{ is } 2.$$

The common ratio of the geometric sequence

$$3, 12, 48, 192, 576, \ldots , 3(4^{n-1}), \ldots \text{ is } 4.$$

The common ratio of the geometric sequence

$$7, \frac{7}{2}, \frac{7}{4}, \frac{7}{8}, \frac{7}{16}, \ldots , 7\left(\frac{1}{2}\right)^{n-1}, \ldots \text{ is } \frac{1}{2}.$$

EXAMPLE 1 Determine the first five terms of the geometric sequence if $a_1 = 4$ and $r = \frac{1}{2}$.

Solution: $a_1 = 4, \quad a_2 = 4 \cdot \frac{1}{2} = 2, \quad a_3 = 2 \cdot \frac{1}{2} = 1, \quad a_4 = 1 \cdot \frac{1}{2} = \frac{1}{2}, \quad a_5 = \frac{1}{2} \cdot \frac{1}{2} = \frac{1}{4}$

Thus the first five terms of the geometric sequence are

$$4, 2, 1, \frac{1}{2}, \frac{1}{4}$$

▶ **2** In general, a geometric sequence with first term a_1 and common ratio r has the following terms:

$$a_1, \quad a_1r, \quad a_1r^2, \quad a_1r^3, \quad a_1r^4, \ldots , a_1r^{n-1}, \ldots$$

$$\uparrow \qquad \uparrow \qquad \uparrow \qquad \uparrow \qquad \uparrow \qquad\qquad \uparrow$$

1st   2nd   3rd   4th   5th nth
term   term   term   term   term   term

By observing the preceding geometric sequence, we can see that the nth term of a geometric sequence is

---

***n*th Term of a Geometric Sequence**

$$a_n = a_1 r^{n-1}$$

EXAMPLE 2   (a) Write an expression for the general (or *n*th) term, a_n, of the geometric sequence with $a_1 = 3$ and $r = -2$
            (b) Find the twelfth term of this sequence.

Solution: (a) The *n*th term of the sequence is $a_n = a_1 n^{r-1}$. Substituting $a_1 = 3$ and $r = -2$, we obtain

$$a_n = a_1 r^{n-1} = 3(-2)^{n-1}$$

Thus $a_n = 3(-2)^{n-1}$.

(b)

$$a_n = 3(-2)^{n-1}$$
$$a_{12} = 3(-2)^{12-1} = 3(-2)^{11} = 3(-2048) = -6144$$

The twelfth term of the sequence is -6144. The first twelve terms of the sequence are $3, -6, 12, -24, 48, -96, 192, -384, 768, -1536, 3072, -6144$. ∎

EXAMPLE 3   Find r and a_1 for the geometric sequence with $a_2 = 24$ and $a_5 = 648$.

Solution: The sequence looks like

$$\underline{}, \; 24, \; \underline{}, \; \underline{}, \; 648$$
$$\qquad\quad \uparrow \qquad\qquad\qquad \uparrow$$
$$\qquad\quad a_2 \qquad\qquad\qquad a_5$$

If we assume that a_2 is the first term of a sequence with the same common ratio, we obtain

$$24, \; \underline{}, \; \underline{}, \; 648$$
$$\;\; \uparrow \qquad\qquad\quad \uparrow$$
$$\;\text{1st} \qquad\qquad\;\; \text{4th}$$
$$\text{term} \qquad\qquad \text{term}$$

Now use the formula

$$a_n = a_1 r^{n-1}$$
$$648 = 24 r^{4-1}$$
$$648 = 24 r^3$$
$$\frac{648}{24} = r^3$$
$$27 = r^3$$
$$3 = r$$

Thus the common ratio is 3.

The first term of the original sequence must be $24 \div 3$ or 8. Thus $a_1 = 8$. The first term could also be found using

$$a_n = a_1 r^{n-1}$$
$$648 = a_1 3^4$$
$$648 = a_1(81)$$
$$\frac{648}{81} = a_1$$
$$8 = a_1$$
∎

▶ **3**   A **geometric series** is the sum of the terms of a geometric sequence. The sum of the first n terms, s_n, of a geometric sequence can be expressed as

$$s_n = a_1 + a_1 r + a_1 r^2 + a_1 r^3 + \cdots + a_1 r^{n-1}$$

If we multiply both sides of the equation by r, we obtain

$$rs_n = a_1r + a_1r^2 + a_1r^3 + \cdots + a_1r^n$$

Now subtract the second equation from the first.

$$
\begin{array}{l}
s_n = a_1 + a_1r + a_1r^2 + a_1r^3 + \cdots + a_1r^{n-1} \\
\underline{rs_n = a_1r + a_1r^2 + a_1r^3 + \cdots + a_1r^{n-1} + a_1r^n} \\
s_n - rs_n = a_1 \phantom{+ a_1r + a_1r^2 + a_1r^3 + \cdots + a_1r^{n-1} +} - a_1r^n
\end{array}
$$

or

$$s_n - rs_n = a_1 - a_1r^n$$
$$s_n(1 - r) = a_1 - a_1r^n$$
$$s_n = \frac{a_1 - a_1r^n}{1 - r}$$

Thus we have the following formula for the nth partial sum of a geometric series.

---

nth Partial Sum of a Geometric Series

$$s_n = \frac{a_1(1 - r^n)}{1 - r}, \qquad r \neq 1$$

EXAMPLE 4   Find the seventh partial sum of a geometric series whose first term is 8 and whose common ratio is $\frac{1}{2}$.

Solution:

$$s_n = \frac{a_1(1 - r^n)}{1 - r}$$

$$s_7 = \frac{8\left[1 - \left(\frac{1}{2}\right)^7\right]}{1 - \frac{1}{2}} = \frac{8\left(1 - \frac{1}{128}\right)}{\frac{1}{2}} = \frac{8\left(\frac{127}{128}\right)}{\frac{1}{2}} = \frac{127}{16} \cdot \frac{2}{1} = \frac{127}{8}$$

Thus $s_7 = \frac{127}{8}$. ∎

EXAMPLE 5   Given $s_n = 93$, $a_1 = 3$, and $r = 2$, find n.

Solution:

$$S_n = \frac{a_1(1 - r^n)}{1 - r}$$

$$93 = \frac{3(1 - 2^n)}{1 - 2}$$

$$93 = \frac{3(1 - 2^n)}{-1}$$

$$-93 = 3(1 - 2^n)$$

$$-31 = 1 - 2^n$$

$$-32 = -2^n$$

$$32 = 2^n$$

$$2^5 = 2^n$$

Therefore, $n = 5$. ∎

Examples 6 and 7 illustrate applications of geometric sequences and series.

EXAMPLE 6 A certain substance decomposes and loses 20% of its weight each hour. If there is originally 300 grams of the substance, how much remains after 7 hours?

Solution: This problem can be considered as a geometric sequence since the substance is decreasing by a certain rate (or percent) each hour. Often, when working with a sequence, it is helpful to write out the first few terms of the sequence. In this problem, since we are concerned with the amount of the substance *remaining,* the rate, r, is 100% - 20% = 80% or 0.80. To obtain the terms in the sequence, the preceding term is multiplied by 0.80, giving the amount of the substance left in each succeeding hour.

Amount remaining at beginning of:

1st hour	2nd hour	3rd hour	4th hour
$a_1 = 300$	$a_2 = 300(0.80)$	$a_3 = 300(0.80)^2$	$a_4 = 300(0.80)^3$

In this geometric sequence, $a_1 = 300$ and $r = 0.8$. In general, the amount of substance remaining *at the beginning* of the nth hour is $a_n = 300(0.80)^{n-1}$.

We are asked to find the amount remaining after 7 hours. Thus we must find a_8, since the amount remaining after 7 hours is the same as the amount remaining at the beginning of the eighth hour.

$$a_8 = 300(0.8)^{8-1}$$
$$= 300(0.8)^7$$
$$= 300(0.2097)$$
$$= 62.91 \text{ grams}$$

Note that $a_n = 300(0.80)^n$ could also be used to find the amount *remaining after n hours*. Thus, after 7 hours, $a_7 = 300(0.80)^7$. In explaining the solution to this example, we used the form $a_n = a_1 r^{n-1}$ rather than $a_n = a_1 r^n$, since the first form was presented in the section. ■

EXAMPLE 7 Mary Foster invests $1000 at 8% interest compounded annually in a savings account. Determine the amount in her account at the end of 6 years.

Solution: At the beginning of the second year, the amount is $1000 + 0.08(1000) = 1000(1 + 0.8) = 1000(1.08)$. At the beginning of the third year, this new amount increases by 8% to give $1000(1.08)^2$, and so on.

Amount in account at beginning of:

1st year	2nd year	3rd year	4th year
$a_1 = 1000$	$a_2 = 1000(1.08)$	$a_3 = 1000(1.08)^2$	$a_4 = 1000(1.08)^3$

This is a geometric sequence with $a_1 = 1000$ and $r = 1.08$. Since we are seeking the amount at the end of 6 years, which is the same as the beginning of the seventh year, we will find a_7.

$$a_n = 1000(1.08)^{n-1}$$
$$a_7 = 1000(1.08)^{7-1}$$
$$= 1000(1.08)^6$$
$$= 1000(1.58687)$$
$$= 1586.87$$

After 6 years the amount in the account is $1586.87. The amount of interest gained is $1586.87 − $1000 = $586.87. ■

Exercise Set 12.3

Determine the first five terms of the geometric sequence.

1. $a_1 = 5, r = 3$

2. $a_1 = 6, r = \dfrac{1}{2}$

3. $a_1 = 90, r = \dfrac{1}{3}$

4. $a_1 = -12, r = -1$

5. $a_1 = -15, r = -2$

6. $a_1 = 1, r = -\dfrac{1}{2}$

7. $a_1 = 3, r = \dfrac{3}{2}$

8. $a_1 = 60, r = \dfrac{1}{3}$

Find the indicated term of the geometric sequence.

9. $a_1 = 5, r = 2$; find a_6

10. $a_1 = -12, r = \dfrac{1}{2}$; find a_{10}

11. $a_1 = 18, r = 3$; find a_7

12. $a_1 = -20, r = -2$; find a_{10}

13. $a_1 = 2, r = \dfrac{1}{2}$; find a_8

14. $a_1 = 5, r = \dfrac{2}{3}$; find a_9

15. $a_1 = -3, r = -2$; find a_{12}

16. $a_1 = 80, r = \dfrac{1}{3}$; find a_{12}

Find the sum indicated.

17. $a_1 = 3, r = 4$; find s_5

18. $a_1 = 9, r = \dfrac{1}{2}$; find s_6

19. $a_1 = 80, r = 2$; find s_7

20. $a_1 = 1, r = -2$; find s_{12}

21. $a_1 = -30, r = -\dfrac{1}{2}$; find s_9

22. $a_1 = \dfrac{3}{5}, r = 3$; find s_7

23. $a_1 = -9, r = \dfrac{2}{5}$; find s_5

24. $a_1 = 35, r = \dfrac{1}{5}$; find s_{12}

Find the common ratio, r; then write an expression for the general (or nth) term, a_n, for the geometric sequence.

25. $5, \dfrac{5}{2}, \dfrac{5}{4}, \dfrac{5}{8}, \ldots$

26. $3, 9, 27, 81, \ldots$

27. $2, -6, 18, -54, \ldots$

28. $\dfrac{3}{4}, \dfrac{6}{12}, \dfrac{12}{36}, \dfrac{24}{108}$

29. $-1, -3, -9, -18, \ldots$

30. $\dfrac{4}{3}, \dfrac{8}{3}, \dfrac{16}{3}, \dfrac{32}{3}, \ldots$

31. In the geometric series, $a_3 = 28$ and $a_5 = 112$; find r and a_1.

32. In the geometric series, $a_2 = 27$ and $a_5 = 1$; find r and a_1.

33. In the geometric series, $a_2 = 15$ and $a_5 = 405$; find r and a_1.

34. In a geometric series, $a_2 = 12$ and $a_5 = -324$; find r and a_1.

35. Your salary increases at a rate of 15% per year. If your initial salary is $20,000, what will your salary be at the start of your twenty-fifth year?

36. A ball, when dropped, rebounds to four-fifths of its original height. How high will the ball rebound after the fourth bounce if it is dropped from a height of 30 feet?

37. A substance loses half its mass each day. If there is initially 300 grams of the substance, find:
   (a) The number of days after which only 37.5 grams of the substance remain.
   (b) The amount of the substance remaining after 8 days.

38. A certain type of bacteria doubles every hour. If there are initially 1000 bacteria, how long will it take for the number of bacteria to reach 64,000?

39. The population of the United States is 217.3 million. If the population grows at a rate of 6% per year, find:
   (a) The population in 12 years.
   (b) The number of years for the population to double.

40. A piece of farm equipment that costs $75,000 new decreases in value by 15% per year. Find the value of the equipment after 4 years.

41. The amount of light filtering through a lake diminishes by one-half for each meter of depth.
   (a) Write a sequence indicating the amount of light remaining at depths of 1, 2, 3, 4, and 5 meters.
   (b) What is the general term for this sequence?
   (c) What is the remaining light at a depth of 7 meters.

42. You invest $10,000 in a savings account paying 6% interest annually. Find the amount in your account at the end of 8 years.

43. A tracer dye is injected into Mark for medical reasons. After each hour, two-thirds of the previous hour's dye remains. How much dye remains in Mark's system after 10 hours?

44. If you start with $1 and double your money each day, how many days will it take to surpass $1,000,000?

45. One method of depreciating an item on an income tax return is the declining balance method. With this method, a given percentage of the original value of the item is depreciated each year. Suppose an item has a 5-year life and is depreciated using the declining balance method. Then at the end of its first year it loses $\frac{1}{5}$ of its

value and $\frac{4}{5}$ of its value remains. At the end of the second year it loses $\frac{1}{5}$ of the remaining $\frac{4}{5}$ of its value, and so on. A car has a 5-year life expectancy and costs $9800.

(a) Write a sequence showing the value of the car remaining for each of the first 3 years.

(b) What is the general term of this sequence?

(c) Find its value at the end of 5 years.

46. What is a geometric sequence?

47. What is a geometric series?

48. Explain how to find the common ratio in a geometric sequence.

Cumulative Review Exercises

[7.6] **49.** It takes Mrs. Donovan twice as long to load a truck as Mr. Donovan. If together they can load the truck in 8 hours, how long would it take Mr. Donovan to load the truck by himself?

[8.2] **50.** Simplify $\left(\dfrac{x^{1/2}y^{1/3}}{4x^{-3/2}y^{3/5}}\right)^{1/2}$.

[8.3] **51.** Simplify $\sqrt[3]{9x^2y}\,(\sqrt[3]{3x^4y^6} - \sqrt[3]{8xy^4})$.

[8.5] **52.** Simplify $x\sqrt{y} - 2\sqrt{x^2y} + \sqrt{4x^2y}$.

JUST FOR FUN

1. In Exercise Set 11.4, number 35, a formula for scrap value was given. The scrap value, S, is found by

$$S = c(1 - r)^n$$

where c is the original cost, r is the annual depreciation rate, and n is the number of years the object is depreciated.

(a) If you have not already done so, do Exercise 45 of this section to find the value of the car remaining at the end of 5 years.

(b) Use the formula above to find the scrap value of the car at the end of 5 years and compare this answer with the answer found in part (a).

2. Find the sum of the sequence 1, 2, 4, 8, . . . , 1,048,576 and the number of terms in the sequence.

12.4

Infinite Geometric Series

▶ **1** Identify infinite geometric series.

▶ **2** Find the sum of an infinite geometric series.

▶ **1** All the geometric sequences that we have examined thus far have been finite since they have had a last term. The following sequence is an example of an infinite geometric sequence.

$$1, \frac{1}{2}, \frac{1}{4}, \frac{1}{8}, \frac{1}{16}, \ldots, \left(\frac{1}{2}\right)^{n-1}, \ldots$$

Note that the three dots at the end of the sequence indicate that the sequence continues indefinitely in the same manner. The sum of the terms in an infinite geometric sequence form an **infinite geometric series.** For example,

$$1 + \frac{1}{2} + \frac{1}{4} + \frac{1}{8} + \frac{1}{16} + \cdots + \left(\frac{1}{2}\right)^{n-1} + \cdots$$

is an infinite geometric series. Let's find some partial sums.

Sum of first two terms, s_2: $1 + \dfrac{1}{2} = \dfrac{3}{2} = 1.5$

Sum of first three terms, s_3: $1 + \dfrac{1}{2} + \dfrac{1}{4} = \dfrac{7}{4} = 1.75$

Sum of first four terms, s_4: $1 + \dfrac{1}{2} + \dfrac{1}{4} + \dfrac{1}{8} = \dfrac{15}{8} = 1.875$

Sum of first five terms, s_5: $1 + \dfrac{1}{2} + \dfrac{1}{4} + \dfrac{1}{8} + \dfrac{1}{16} = \dfrac{31}{16} = 1.9375$

Sum of first six terms, s_6: $1 + \dfrac{1}{2} + \dfrac{1}{4} + \dfrac{1}{8} + \dfrac{1}{16} + \dfrac{1}{32} = \dfrac{63}{32} = 1.96875$

Since each term of the geometric sequence is smaller than the preceding term, each additional term adds less and less to the sum. Also, the sum seems to be getting closer and closer to 2.00.

▶ **2**   Consider the formula for the sum of an n-term geometric series:

$$s_n = \frac{a_1(1 - r^n)}{1 - r}, \qquad r \neq 1$$

What happens to r^n if $|r| < 1$ and n gets larger and larger? Suppose that $r = \frac{1}{2}$; then

$$\left(\frac{1}{2}\right)^1 = \frac{1}{2} = 0.5$$

$$\left(\frac{1}{2}\right)^2 = \frac{1}{4} = 0.25$$

$$\left(\frac{1}{2}\right)^3 = \frac{1}{8} = 0.125$$

$$\left(\frac{1}{2}\right)^{20} \approx 0.000001$$

We can see that when $|r| < 1$ the value of r^n gets exceedingly close to 0 as n gets larger and larger. Thus, when considering the sum of an infinite series, symbolized s_∞, the expression r^n approaches 0 when $|r| < 1$. Therefore, replacing r^n with a 0 in the formula $s_n = \dfrac{a_1(1 - r^n)}{1 - r}$ gives

---

Sum of an Infinite Series

$$s_\infty = \frac{a_1}{1 - r}, \qquad |r| < 1$$

EXAMPLE 1   Find the sum of the terms of the infinite sequence $1, \frac{1}{2}, \frac{1}{4}, \frac{1}{16}, \ldots$.

Solution: $a_1 = 1$ and $r = \frac{1}{2}$.

$$s_\infty = \frac{a_1}{1 - r} = \frac{1}{1 - \dfrac{1}{2}} = \frac{1}{\dfrac{1}{2}} = 2$$

Thus $1 + \frac{1}{2} + \frac{1}{4} + \frac{1}{8} + \frac{1}{16} + \cdots + \left(\frac{1}{2}\right)^{n-1} + \cdots = 2$. ∎

EXAMPLE 2   Find the sum of the infinite geomertric series

$$5 - 2 + \frac{4}{5} - \frac{8}{25} + \frac{16}{125} - \frac{32}{625} + \cdots$$

Solution: The terms of the corresponding sequence are

$$5, -2, \frac{4}{5}, -\frac{8}{25}, \ldots$$

$$r = -2 \div 5 = -\frac{2}{5} \quad \text{and} \quad a_1 = 5$$

$$s_\infty = \frac{a_1}{1 - r}$$

$$= \frac{5}{1 - \left(-\dfrac{2}{5}\right)} = \frac{5}{1 + \dfrac{2}{5}} = \frac{5}{\dfrac{7}{5}} = \frac{25}{7}$$

EXAMPLE 3   Write a fraction equivalent to $0.343434 \ldots$.

Solution: We can write this decimal as

$$0.34 + 0.0034 + 0.000034 + \cdots + (0.34)(0.01)^{n-1} + \cdots$$

This is an infinite geometric series with $r = 0.01$. Since $|r| < 1$,

$$s_\infty = \frac{a_1}{1 - r} = \frac{0.34}{1 - 0.01} = \frac{0.34}{0.99} = \frac{34}{99}$$

If you divide 34 by 99 on a calculator, you will see .34343434 displayed.

EXAMPLE 4   Each swing of a certain pendulum travels 90% of its previous swing. For example, if the swing to the right is 10 feet, then the swing back to the left is $0.9 \cdot 10 = 9$ feet (see Fig. 12.1). If the first swing is 10 feet long, determine the total distance traveled by the pendulum by the time it comes to rest.

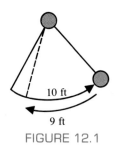

10 ft

9 ft

FIGURE 12.1

Solution: This problem may be considered an infinite geometric series with $a_1 = 10$ and $r = 0.9$. We can therefore use the formula $s_\infty = \dfrac{a_1}{1 - r}$ to find the total distance traveled by the pendulum.

$$s_\infty = \frac{a_1}{1 - r} = \frac{10}{1 - 0.9} = \frac{10}{0.1} = 100 \text{ feet}$$

What is the sum of a geometric series where $|r| > 1$? Consider the geometric sequence where $a_1 = 1$ and $r = 2$.

$$1, 2, 4, 8, 16, 32, \ldots, 2^{n-1}, \ldots$$

The sum of its terms is

$$1 + 2 + 4 + 8 + 16 + 32 + \cdots + 2^{n-1}$$

What is the sum of this series? As n gets larger and larger, the sum gets larger and larger. We therefore say that the sum "does not exist." For $|r| > 1$, the sum of an infinite geometric series does not exist.

Exercise Set 12.4

Find the sum of the terms in each sequence.

1. $6, 3, \dfrac{3}{2}, \dfrac{3}{4}, \dfrac{3}{8}, \cdots$

2. $\dfrac{1}{3}, \dfrac{1}{9}, \dfrac{1}{27}, \dfrac{1}{81}, \cdots$

3. $5, 2, \dfrac{4}{5}, \dfrac{8}{25}, \cdots$

4. $-\dfrac{4}{3}, -\dfrac{4}{9}, -\dfrac{4}{27}, -\dfrac{4}{81}, \cdots$

5. $\dfrac{1}{3}, \dfrac{4}{15}, \dfrac{16}{75}, \dfrac{64}{375}, \cdots$

6. $6, -2, \dfrac{2}{3}, -\dfrac{2}{9}, \dfrac{2}{27}, \cdots$

7. $9, -1, \dfrac{1}{9}, -\dfrac{1}{81}, \cdots$

8. $\dfrac{5}{3}, 1, \dfrac{3}{5}, \dfrac{9}{25}, \cdots$

Find the sum of each infinite series.

9. $1 + \dfrac{1}{2} + \dfrac{1}{4} + \dfrac{1}{8} + \cdots$

10. $4 + 2 + 1 + \dfrac{1}{2} + \cdots$

11. $8 + \dfrac{16}{3} + \dfrac{32}{9} + \dfrac{64}{27} + \cdots$

12. $10 - 5 + \dfrac{5}{2} - \dfrac{5}{4} + \cdots$

13. $-60 + 20 - \dfrac{20}{3} + \dfrac{20}{9} - \cdots$

14. $4 + \dfrac{8}{3} + \dfrac{16}{9} + \dfrac{32}{27} + \cdots$

15. $-12 - \dfrac{12}{5} - \dfrac{12}{25} - \dfrac{12}{125} - \cdots$

16. $5 - 1 + \dfrac{1}{5} - \dfrac{1}{25} + \cdots$

Write a fraction equivalent to the repeating decimal.

17. $0.2727\ldots$

18. $0.454545\ldots$

19. $0.5555\ldots$

20. $0.375375\ldots$

21. $0.515151\ldots$

22. $0.742742\ldots$

23. Each swing of a pendulum travels 80% of its previous swing. If the first swing is 8 feet long, determine the total distance traveled by the pendulum by the time it comes to rest.

24. What is an infinite geometric series?

25. What is the sum of an infinite geometric series when $|r| > 1$?

Cumulative Review Exercises

[5.4] **26.** Multiply $(4x^2 - 3x - 6)(2x - 3)$.

[5.5] **27.** Divide $(16x^2 + 10x - 18) \div (2x + 5)$.

[8.2] **28.** Evaluate $\left(\dfrac{9}{100}\right)^{-1/2}$

[8.6] **29.** Solve the equation $\sqrt{a^2 + 9a + 3} = -a$.

JUST FOR FUN

1. A ball is dropped from a height of 10 feet. The ball bounces to a height of 9 feet. On each successive bounce the ball reaches a height of 90% of its previous height. Find the total vertical distance traveled by the ball when it comes to rest (see the figure).

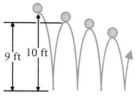

2. A particle follows the path indicated by the wave shown. Find the total vertical distance traveled by the particle.

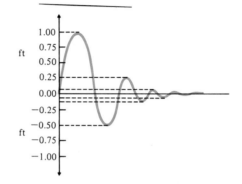

12.5

The Binomial Theorem (Optional)

▸**1**   Evaluate factorials.

▸**2**   Use Pascal's triangle to find the numerical coefficients of an expanded binomial.

▸**3**   Use the binomial theorem to expand binomials.

▸**1**   To understand the binomial theorem, you must have an understanding of what **factorials** are. The notation $n!$ is read "n factorial" and is defined by

---

n Factorial

$$n! = n(n - 1)(n - 2)(n - 3) \cdots (1)$$

for any positive integer n.

For example,

$$6! = 6 \cdot 5 \cdot 4 \cdot 3 \cdot 2 \cdot 1 = 720$$
$$8! = 8 \cdot 7 \cdot 6 \cdot 5 \cdot 4 \cdot 3 \cdot 2 \cdot 1 = 40{,}320$$

Note that **0! is defined to be 1.**

▶ **2**   Using direct multiplication, we can obtain the following expansions of the binomial $a + b$:

$$(a + b)^0 = 1$$
$$(a + b)^1 = a + b$$
$$(a + b)^2 = a^2 + 2ab + b^2$$
$$(a + b)^3 = a^3 + 3a^2b + 3ab^2 + b^2$$
$$(a + b)^4 = a^4 + 4a^3b + 6a^2b^2 + 4ab^3 + b^4$$
$$(a + b)^5 = a^5 + 5a^4b + 10a^3b^2 + 10a^2b^3 + 5ab^4 + b^5$$
$$(a + b)^6 = a^6 + 6a^5b + 15a^4b^2 + 20a^3b^3 + 15a^2b^4 + 6ab^5 + b^6$$

Note that when expanding a binomial of the form $(a + b)^n$:

1. There are $n + 1$ terms in the expansion.
2. The first term is a^n and the last term is b^n.
3. Progressing from the term on the left to the term on the right, the exponents on a decrease by 1 from term to term, while the exponents on b increase by 1 from term to term.
4. The sum of the exponents on the variables in any one term is n.
5. The coefficients of the terms equidistant from the ends are the same.

If we examine just the variables in $(a + b)^5$, we have a^5, a^4b, a^3b^2, a^2b^3, ab^4, and b^5.

The numerical coefficients of each term in the expansion of $(a + b)^n$ can be found by using **Pascal's triangle.** Consider the case where $n = 5$, then we can determine the numerical coefficients of $(a + b)^5$ as follows.

Exponent on Binomial	*Pascal's Triangle*

$n = 0$						1					
$n = 1$					1		1				
$n = 2$				1		2		1			
$n = 3$			1		3		3		1		
$n = 4$		1		4		6		4		1	
$n = 5$	1		5		10		10		5		1
$n = 6$	1	6		15		20		15		6	1

Examine row 5 ($n = 4$) and row 6 ($n = 5$).

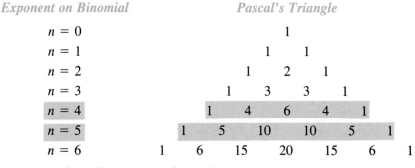

Notice that the first and last numbers in each row are 1, and the inner numbers are obtained by adding the two numbers in the row above (to the right and left). The numerical coefficients of $(a + b)^5$ are 1, 5, 10, 10, 5, and 1. Combining the numerical coefficients and the variables, we obtain

$$(a + b)^5 = a^5 + 5a^4b + 10a^3b^2 + 10a^2b^3 + 5ab^4 + b^5$$

This method of expanding a binomial is not practical when n is large.

▶ **3**   We will shortly introduce a more practical method, called the binomial theorem, to expand expressions of the form $(a + b)^n$. However, before we introduce this formula, we need to introduce the procedure for finding binomial coefficients of the form $\binom{n}{r}$.

Binomial Coefficients

For n and r nonnegative integers, $n > r$,

$$\binom{n}{r} = \frac{n!}{r! \cdot (n - r)!}$$

The binomial coefficient $\binom{n}{r}$ is read "the number of *combinations* of n items taken r at a time." Combinations are used in the many areas of mathematics, including the study of probability. Lack of space prohibits us from discussing them further here.

EXAMPLE 1 Evaluate $\binom{5}{2}$.

Solution: By the definition, if we substitute 5 for n and 2 for r, we obtain

$$\binom{5}{2} = \frac{5!}{2! \cdot (5 - 2)!} = \frac{5!}{2! \cdot 3!} = \frac{5 \cdot 4 \cdot 3 \cdot 2 \cdot 1}{(2 \cdot 1) \cdot (3 \cdot 2 \cdot 1)} = 10$$

Thus $\binom{5}{2}$ has a value of 10. ■

EXAMPLE 2 Evaluate: (a) $\binom{7}{4}$ (b) $\binom{4}{4}$ (c) $\binom{5}{0}$

Solution: (a) $\binom{7}{4} = \dfrac{7!}{4! \cdot (7 - 4)!} = \dfrac{7!}{4! \cdot 3!} = \dfrac{7 \cdot 6 \cdot 5 \cdot 4 \cdot 3 \cdot 2 \cdot 1}{(4 \cdot 3 \cdot 2 \cdot 1)(3 \cdot 2 \cdot 1)} = 35$

(b) $\binom{4}{4} = \dfrac{4!}{4! \cdot (4 - 4)!} = \dfrac{4!}{4! \cdot 0!} = \dfrac{1}{1} = 1$ remember that $0! = 1$

(c) $\binom{5}{0} = \dfrac{5!}{0! \cdot (5 - 0)!} = \dfrac{5!}{0! \cdot 5!} = \dfrac{1}{1} = 1$ ■

By observing Examples 2b and 2c, you can reason that, for any positive integer n,

$$\binom{n}{n} = 1 \quad \text{and} \quad \binom{n}{0} = 1$$

Now we introduce the binomial theorem.

Binomial Theorem

For any positive integer n,

$$(a + b)^n = \binom{n}{0}a^n b^0 + \binom{n}{1}a^{n-1}b^1 + \binom{n}{2}a^{n-2}b^2$$

$$+ \binom{n}{3}a^{n-3}b^3 + \cdots + \binom{n}{n}a^0 b^n$$

Notice in the binomial theorem that the sum of the exponents on the variables in any one term is n, and the bottom number in the combination is always the same as the exponent on the second variable in the term. For example, if we consider the term $\binom{n}{3}a^{n-3}b^3$, the sum of the exponents on the variables is $(n - 3) + 3 = n$. Also, the exponent on the variable b is 3, and the bottom number in the combination is also 3.

Now we will expand $(a + b)^5$ using the binomial theorem and see if we get the same results we did when we used direct multiplication and Pascal's triangle to obtain the expansion.

$$(a + b)^5 = \binom{5}{0}a^5b^0 + \binom{5}{1}a^{5-1}b^1 + \binom{5}{2}a^{5-2}b^2 + \binom{5}{3}a^{5-3}b^3 + \binom{5}{4}a^{5-4}b^4 + \binom{5}{5}a^{5-5}b^5$$

$$= \binom{5}{0}a^5b^0 + \binom{5}{1}a^4b^1 + \binom{5}{2}a^3b^2 + \binom{5}{3}a^2b^3 + \binom{5}{4}a^1b^4 + \binom{5}{5}a^0b^5$$

$$= \frac{5!}{0! \cdot 5!}a^5 + \frac{5!}{1! \cdot 4!}a^4b + \frac{5!}{2! \cdot 3!}a^3b^2 + \frac{5!}{3! \cdot 2!}a^2b^3 + \frac{5!}{4! \cdot 1!}ab^4 + \frac{5!}{5! \cdot 0!}b^5$$

$$= a^5 + 5a^4b + 10a^3b^2 + 10a^2b^3 + 5ab^4 + b^5$$

This answer is the same as the answer obtained earlier.

In the binomial theorem, the first and last terms of the expansion contain a factor raised to the zero power. Since any non zero number raised to the 0th power equals one, we could have omitted those factors. These factors were initially included so that you could see the pattern better.

EXAMPLE 3   Use the binomial theorem to expand $(2x + 3)^6$.

Solution: If we use $2x$ for a and 3 for b, we obtain

$$(2x + 3)^6 = \binom{6}{0}(2x)^6(3)^0 + \binom{6}{1}(2x)^5(3)^1 + \binom{6}{2}(2x)^4(3)^2 + \binom{6}{3}(2x)^3(3)^3 + \binom{6}{4}(2x)^2(3)^4 + \binom{6}{5}(2x)^1(3)^5 + \binom{6}{6}(2x)^0(3)^6$$

$$= 1(2x)^6 + 6(2x)^5(3) + 15(2x)^4(9) + 20(2x)^3(27) + 15(2x)^2(81) + 6(2x)(243) + 1(3)^6$$

$$= 64x^6 + 576x^5 + 2160x^4 + 4320x^3 + 4860x^2 + 2916x + 729 \qquad \blacksquare$$

EXAMPLE 4   Use the binomial theorem to expand $(5x - 2y)^4$.

Solution: Write $(5x - 2y)^4$ as $[5x + (-2y)]^4$. Use $5x$ in place of a and $-2y$ in place of b in the binomial theorem.

$$[5x + (-2y)]^4 = \binom{4}{0}(5x)^4(-2y)^0 + \binom{4}{1}(5x)^3(-2y)^1 + \binom{4}{2}(5x)^2(-2y)^2 + \binom{4}{3}(5x)^1(-2y)^3 + \binom{4}{4}(5x)^0(-2y)^4$$

$$= 1(5x)^4 + 4(5x)^3(-2y) + 6(5x)^2(-2y)^2 + 4(5x)(-2y)^3 + 1(-2y)^4$$

$$= 625x^4 - 1000x^3y + 600x^2y^2 - 160xy^3 + 16y^4 \qquad \blacksquare$$

Exercise Set 12.5

Evaluate each of the following combinations.

1. $\binom{4}{2}$

2. $\binom{6}{3}$

3. $\binom{5}{3}$

4. $\binom{9}{3}$

5. $\binom{7}{0}$

6. $\binom{10}{8}$

7. $\binom{8}{4}$

8. $\binom{12}{8}$

9. $\binom{8}{2}$

10. $\binom{7}{5}$

Use the binomial theorem to expand each expression.

11. $(x + 4)^3$

12. $(2x + 3)^3$

13. $(a - b)^4$

14. $(2r + s^2)^4$

15. $(3a - b)^5$

16. $(x + 2y)^5$

17. $\left(2x + \dfrac{1}{2}\right)^4$

18. $\left(\dfrac{2}{3}x + \dfrac{3}{2}\right)^4$

19. $\left(\dfrac{x}{2} - 3\right)^4$

20. $(3x^2 + y)^5$

Write the first four terms of each expansion.

21. $(x + y)^{10}$

22. $(2x + 3)^8$

23. $(3x - y)^7$

24. $(3p + 2q)^{11}$

25. $(x^2 - 3y)^8$

26. $\left(2x + \dfrac{y}{5}\right)^9$

27. Explain how to construct Pascal's Triangle. Construct the first five rows of Pascal's triangle.

28. Explain in your own words how to find n factorial for any positive integer n.

29. Is $n!$ equal to $n \cdot (n - 1)!$? Explain your answer and give an example to support your answer.

30. Is $(n + 1)!$ equal to $(n + 1) \cdot n!$? Explain your answer and give an example to support your answer.

31. Is $(n - 3)!$ equal to $(n - 3)(n - 4)(n - 5)!$, for $n \geq 5$? Explain your answer and give an example to support your answer.

32. Under what conditions will $\binom{n}{m}$ where n and m are non-negative integers, have a value of 1?

33. What are the first, second, next to last and last terms of the expansion $(x + 3)^8$?

34. What are the first, second, next to last, and last terms of the expansion $(2x + 5)^6$?

Cumulative Review Exercises

Factor

[6.2] **35.** $16x^2 - 8x - 3$.

[6.4] **36.** $3a^2b^2 - 24ab^2 + 48b^2$.

[6.6] *Solve* $S_n - S_n r = a_1 - a_1 r^n$:

37. For S_n.

38. For a_1.

JUST FOR FUN

1. Express the binomial theorem using summation notation.
2. **(a)** Explain how you can find the rth term in a binomial expansion of the form $(x + y)^n$, where r is any positive integer less than or equal to n without expanding the binomial. **(b)** Find the third term in the expansion of $(3x + 2)^4$ without expanding the expression.

SUMMARY

GLOSSARY

Alternating sequence *(544):* A sequence in which the signs of the terms alternate.

Arithmetic sequence (or **arithmetic progression**) *(548):* A sequence in which each term after the first differs from the preceding term by a constant amount.

Arithmetic series *(550):* The sum of the terms of an arithmetic sequence.

Common difference *(549):* The amount by which each pair of terms differs in an arithmetic sequence.

Common ratio *(554):* The common multiple in a geometric series.

Decreasing sequence *(543):* A sequence in which each term is less than the term that precedes it.

Finite sequence *(543):* A function whose domain includes only the first n natural numbers.

General term of a sequence *(542):* An expression that defines the nth term of the sequence.

Geometric sequence (or **geometric progression**) *(555):* A sequence in which each term after the first is the same multiple of the preceding term.

Geometric series *(555):* The sum of the terms of a geometric sequence.

Increasing sequence *(543):* A sequence in which each term is greater than the term that precedes it.

Infinite geometric series *(559):* The sum of the terms in an infinite geometric sequence.

Infinite sequence *(542):* A function whose domain is the set of natural numbers.

Partial sum of a series *(545):* The sum of a finite number of consecutive terms of a series, beginning with the first term.

Sequence (or **progression**) *(542):* A list of numbers arranged in a specific order.

Series *(544):* The sum of a sequence.

IMPORTANT FACTS

$$\sum_{i=1}^{n} x_i = x_1 + x_2 + x_3 + \cdots + x_n$$

n Factorial

$$n! = n(n-1)(n-2)\cdots(2)(1)$$

nth Term of an arithmetic sequence

$$a_n = a_1 + (n-1)d$$

nth Partial sum of an arithmetic series

$$S_n = \frac{n(a_1 + a_n)}{2}$$

nth Term of a geometric sequence

$$a_n = a_1 r^{n-1}$$

nth Partial sum of a geometric series

$$S_n = \frac{a_1(1 - r^n)}{1 - r}, \; r \neq 1$$

Sum of an infinite series

$$S_\infty = \frac{a_1}{1 - r}, \; |r| < 1$$

Binomial coefficients

$$\binom{n}{r} = \frac{n!}{r! \cdot (n - r)!}$$

Binomial theorem

$$(a + b)^n = \binom{n}{0}a^n b^0 + \binom{n}{1}a^{n-1}b^1 + \binom{n}{2}a^{n-2}b^2$$
$$+ \binom{n}{3}a^{n-3}b^3 + \cdots + \binom{n}{n}a^0 b^n$$

Review Exercises

[12.1] *Write the first five terms of the sequence.*

1. $a_n = n + 2$ **2.** $a_n = \dfrac{1}{n}$ **3.** $a_n = n(n + 1)$ **4.** $a_n = \dfrac{n^2}{n + 4}$

Find the indicated term of the sequence.

5. $a_n = 3n + 4$, seventh term   **6.** $a_n = (-1)^n + 3$, seventh term

7. $a_n = \dfrac{n + 7}{n^2}$, ninth term   **8.** $a_n = (n)(n - 3)$, eleventh term

Find the first and third partial sums, s_1 and s_3.

9. $a_n = 3n + 2$ **10.** $a_n = 2n^2$ **11.** $a_n = \dfrac{n + 3}{n + 2}$ **12.** $a_n = (-1)^n(n + 2)$

Write the next three terms of each sequence; then write an expression for the general term, a_n.

13. 1, 2, 4, 8, . . .   **14.** $-8, 4, -2, 1, \ldots$ **15.** $\dfrac{2}{3}, \dfrac{4}{3}, \dfrac{8}{3}, \dfrac{16}{3}, \ldots$ **16.** 9, 6, 3, 0, . . .

17. $-1, 1, -1, 1, \ldots$ **18.** 6, 12, 18, 24, . . .

Write out the series, and then find the sum of the series.

19. $\displaystyle\sum_{n=1}^{3}(n^2 + 2)$ **20.** $\displaystyle\sum_{k=1}^{4} k(k + 2)$ **21.** $\displaystyle\sum_{k=1}^{5} \dfrac{k^2}{3}$ **22.** $\displaystyle\sum_{n=1}^{4} \dfrac{n}{n + 1}$

For the set of values $x_1 = 3$, $x_2 = 9$, $x_3 = 5$, $x_4 = 10$, find:

23. $\displaystyle\sum_{i=1}^{4} x_i$ **24.** $\displaystyle\sum_{i=1}^{4} x_i^2$ **25.** $\displaystyle\sum_{i=2}^{3}(x_i^2 + 1)$ **26.** $\left(\displaystyle\sum_{i=1}^{4} x_i\right)^2$

[12.2] *Write the first five terms of the arithmetic sequence with first term and common difference as given.*

27. $a_1 = 5$, $d = 2$ **28.** $a_1 = \dfrac{1}{2}$, $d = -2$

29. $a_1 = -12$, $d = -\dfrac{1}{2}$ **30.** $a_1 = -100$, $d = \dfrac{1}{5}$

Find the desired quantity of the arithmetic sequence.

31. $a_1 = 2$, $d = 3$; find a_9

32. $a_1 = -12$, $d = -\dfrac{1}{2}$; find a_7

33. $a_1 = 50$, $a_5 = 34$; find d

34. $a_1 = -3$, $a_7 = 0$; find d

35. $a_1 = 12$, $a_n = -13$, $d = -5$; find n

36. $a_1 = 80$, $a_n = 152$, $d = 12$; find n

Find s_n and d for each arithmetic sequence.

37. $a_1 = 7$, $a_n = 21$, $n = 8$

38. $a_1 = -12$, $a_n = -48$, $n = 7$

39. $a_1 = \dfrac{3}{5}$, $a_n = 3$, $n = 7$

40. $a_1 = -\dfrac{10}{3}$, $a_n = -6$, $n = 9$

Write the first four terms of each arithmetic sequence; then find a_{10} and s_{10}.

41. $a_1 = 2$, $d = 4$

42. $a_1 = -8$, $d = -3$

43. $a_1 = \dfrac{5}{6}$, $d = \dfrac{2}{3}$

44. $a_1 = -80$, $d = 4$

Find the number of terms in each arithmetic sequence and s_n.

45. $3, 8, 13, \ldots , 53$

46. $-16, -11, -6, -1, \ldots , 24$

47. $\dfrac{6}{10}, \dfrac{9}{10}, \dfrac{12}{10}, \dfrac{15}{10}, \ldots , \dfrac{36}{10}$

48. $-5, 0, 5, 10, \ldots , 85$

[12.3] *Determine the first five terms of each geometric sequence.*

49. $a_1 = 5$, $r = 2$

50. $a_1 = -12$, $r = \dfrac{1}{2}$

51. $a_1 = 20$, $r = -\dfrac{2}{3}$

52. $a_1 = -100$, $r = \dfrac{1}{5}$

Find the indicated term of the geometric sequence.

53. $a_1 = 12$, $r = \dfrac{1}{3}$; find a_7

54. $a_1 = 25$, $r = 2$; find a_9

55. $a_1 = -8$, $r = -2$; find a_9

56. $a_1 = \dfrac{5}{12}$, $r = \dfrac{2}{3}$; find a_8

Find the indicated sum.

57. $a_1 = 12$, $r = 2$; find s_8

58. $a_1 = \dfrac{3}{5}$, $r = \dfrac{5}{3}$; find s_7

59. $a_1 = -84$, $r = -\dfrac{1}{4}$; find s_5

60. $a_1 = 9$, $r = \dfrac{3}{2}$; find s_9

Find the common ratio, r; then write an expression for the general term, a_n, for the geometric sequences.

61. $6, 12, 24, \ldots$

62. $8, \dfrac{8}{3}, \dfrac{8}{9}, \ldots$

63. $-4, -20, -100, \ldots$

64. $\dfrac{9}{5}, \dfrac{18}{15}, \dfrac{36}{45}, \ldots$

[12.4] *Find the sum of the terms in each infinite sequence.*

65. $7, \dfrac{7}{2}, \dfrac{7}{4}, \dfrac{7}{8}, \ldots$

66. $-8, \dfrac{8}{3}, -\dfrac{8}{9}, \dfrac{8}{27}, \ldots$

67. $-5, -\dfrac{10}{3}, -\dfrac{20}{9}, -\dfrac{40}{27}, \ldots$

68. $\dfrac{7}{2}, 1, \dfrac{2}{7}, \dfrac{4}{49}, \ldots$

Find the sum of each infinite series.

69. $2 + 1 + \dfrac{1}{2} + \dfrac{1}{4} + \cdots$

70. $7 + \dfrac{7}{3} + \dfrac{7}{9} + \dfrac{7}{27} + \cdots$

71. $-12 - \dfrac{24}{3} - \dfrac{48}{9} - \dfrac{96}{27} - \cdots$

72. $5 - 1 + \dfrac{1}{5} - \dfrac{1}{25} + \cdots$

Write a fraction equivalent to the repeating decimal.

73. $0.5252\ldots$

74. $0.375375\ldots$

[12.5] *Use the binomial theorem to expand the following.*

75. $(3x + y)^4$

76. $(2x - 3y^2)^3$

Write the first four terms of the expansion.

77. $(x - 2y)^9$

78. $(2a^2 + 3b)^8$

[12.2–12.4]

79. Find the sum of the integers between 100 and 200 inclusive.

80. Professor Gayvert is offered a job with a starting salary of \$30,000 with the agreement that his salary will increase \$1000 at the end of each of the next seven years.
   (a) Write a sequence showing his salary for the first five years.
   (b) What is the general term of this sequence?
   (c) If this process were continued, what would his salary be after nine years?

81. You begin with \$100, double that to get \$200, double that again to get \$400, and so on. How much will you have after you perform this process 10 times?

82. If the inflation rate remains constant at 8% per year (each year the cost of living is 8% greater than the previous year), how much would an object that presently costs \$200 cost after 12 years?

83. Each successive swing of a pendulum travels 92% of the length of its previous swing. If the first swing is 8 feet in length, find the distance traveled by the pendulum by the time it comes to rest.

Practice Test

1. Write the first five terms of the sequence with $a_n = \dfrac{n + 2}{n^2}$.

2. Find the first and third partial sums of $a_n = \dfrac{2n + 3}{n}$.

3. Write out the series and find the sum of the series $\displaystyle\sum_{n=1}^{5} (2n^2 + 3)$

4. Write the general term for the arithmetic sequence
$$\dfrac{1}{3}, \dfrac{2}{3}, \dfrac{3}{3}, \dfrac{4}{3}, \ldots$$

5. Write the general term for the geometric sequence 5, 10, 20, 40, . . .

Write the first four terms of the sequence.

6. $a_1 = 12, d = -3$

7. $a_1 = \dfrac{5}{8}, r = \dfrac{2}{3}$

8. Find a_8 when $a_1 = 100$ and $d = -12$.

9. Find s_8 when $a_1 = 3$, $a_8 = -11$.

10. Find the number of terms in the arithmetic sequence $-4, -16, -28, \ldots, -148$.

11. Find a_7 when $a_1 = 8$ and $r = \dfrac{2}{3}$.

12. Find s_7 when $a_1 = \dfrac{3}{5}$ and $r = -5$.

13. Find the common ratio and write an expression for the general term of the sequence $12, 6, 3, \dfrac{3}{2}, \ldots$.

14. Find the sum of the infinite geometric series $3 + \dfrac{6}{3} + \dfrac{12}{9} + \dfrac{24}{27} + \cdots$.

15. Use the binomial theorem to expand $(x + 2y)^4$.

Cumulative Review Test

1. Solve the equation $\frac{1}{2}x + \frac{1}{3}(x - 2) = \frac{3}{4}(x - 5)$.

2. Solve the inequality $|2x - 3| - 4 > 10$.

3. Graph $3x - 5y > 10$.

4. Solve the system of equations:

$$5x - 2y = 8$$
$$x - y = 4$$

5. Solve the formula $A = \frac{pt}{p + t}$ for p.

6. Add $\frac{x}{x^2 + x - 12} + \frac{x + 2}{3x^2 + 16x + 16}$.

7. Simplify $\frac{\sqrt[3]{24x^6y^3}}{\sqrt[3]{2x^2y^5}}$.

8. Simplify $\sqrt{28} - 3\sqrt{7} + \sqrt{63}$.

9. Solve the equation $\sqrt{5x + 1} - \sqrt{2x - 2} = 2$.

10. Simplify $\frac{5 - 2i}{3 + 4i}$.

11. Solve the equation $-4x^2 - 2x + 8 = 0$.

12. Solve the inequality $\frac{(2x - 3)(x - 4)}{x + 1} < 0$.

13. Graph $4x^2 + 4y^2 = 36$.

14. Write the following equation in the form $y = a(x - h)^2 + k$ and then sketch the graph of the equation.

$$y = x^2 + 2x - 3$$

15. Graph $y = 2^x$ and $y = \log_2 x$ on the same set of axes.

16. Solve the equation $\log(4x - 1) + \log 3 = \log(8x + 13)$.

17. In an arithmetic series, if $a_1 = 8$ and $a_6 = 28$, find d.

18. In a geometric series, if $a_1 = 5$ and $r = 3$, find s_3.

19. A train and car leave from South Point heading for the same destination. The car averaged 40 mph and the train averaged 60 mph. If the train arrives 2 hours ahead of the car, find the distance between the two points.

20. The Donovan's garden is in the shape of a rectangle. If the area of their garden is 300 square feet and the length of the garden is 20 feet greater than the width, find the dimensions of their garden.

Appendices

A A Review of Fractions
B Geometric Formulas
C Squares and Square Roots
D Common Logarithms

A Review of Fractions

To be successful in algebra, you must have a thorough understanding of fractions. This appendix gives a brief review of addition, subtraction, multiplication, and division of fractions. For a more complete explanation of fractions, see any arithmetic text.

The top number of a fraction is called the **numerator.** The bottom number is called the **denominator.**

$$\frac{3}{5} \begin{array}{l} \leftarrow numerator \\ \leftarrow denominator \end{array}$$

Multiplication of Fractions

The product of two or more fractions is obtained by multiplying their numerators together and then multiplying their denominators together, as follows:

$$\frac{a}{b} \cdot \frac{c}{d} = \frac{a \cdot c}{b \cdot d}, \ b \neq 0, d \neq 0$$

EXAMPLE 1   (a) $\dfrac{3}{5} \cdot \dfrac{4}{7} = \dfrac{3 \cdot 4}{5 \cdot 7} = \dfrac{12}{35}$

(b) $\dfrac{5}{12} \cdot \dfrac{3}{10} = \dfrac{5 \cdot 3}{10 \cdot 12} = \dfrac{15}{120} = \dfrac{1}{8}$ ∎

In Example 1(b) the fraction $\frac{15}{120}$ was reduced to $\frac{1}{8}$. To help avoid having to reduce fractions after multiplication, we often divide out common factors. When any nu-

573

merator and any denominator in a **multiplication problem** have a common factor, divide both the numerator and the denominator by the common factor prior to multiplying. This process is illustrated in Example 2.

EXAMPLE 2 $\dfrac{5}{12} \cdot \dfrac{3}{10} = \dfrac{\overset{1}{\cancel{5}}}{12} \cdot \dfrac{3}{\underset{2}{\cancel{10}}}$ Divide both 5 and 10 by 5.

$= \dfrac{\overset{1}{\cancel{5}}}{\underset{4}{\cancel{12}}} \cdot \dfrac{\overset{1}{\cancel{3}}}{\underset{2}{\cancel{10}}}$ Divide both 3 and 12 by 3.

$= \dfrac{1 \cdot 1}{4 \cdot 2} = \dfrac{1}{8}$ ∎

EXAMPLE 3 $\dfrac{25}{36} \cdot \dfrac{8}{15} = \dfrac{25}{36} \cdot \dfrac{8}{\underset{3}{\overset{5}{\cancel{15}}}}$ Divide both 25 and 15 by 5.

$= \dfrac{\overset{5}{\cancel{25}}}{\underset{9}{\cancel{36}}} \cdot \dfrac{\overset{2}{\cancel{8}}}{\underset{3}{\cancel{15}}}$ Divide both 8 and 36 by 4.

$= \dfrac{5 \cdot 2}{9 \cdot 3} = \dfrac{10}{27}$ ∎

Division of Fractions

To divide fractions, invert the divisor and proceed as in multiplication.

$$\frac{a}{b} \div \frac{c}{d} = \frac{a}{b} \cdot \frac{d}{c} = \frac{a \cdot d}{b \cdot c}, \ b \neq 0, \ c \neq 0, \ d \neq 0$$

EXAMPLE 4 $\dfrac{3}{8} \div \dfrac{5}{9} = \dfrac{3}{8} \cdot \dfrac{9}{5} = \dfrac{27}{40}$ ∎

EXAMPLE 5 $\dfrac{7}{15} \div \dfrac{3}{5} = \dfrac{7}{\underset{3}{\cancel{15}}} \cdot \dfrac{\overset{1}{\cancel{5}}}{3} = \dfrac{7}{9}$ ∎

EXAMPLE 6 $4 \div \dfrac{3}{5} = \dfrac{4}{1} \cdot \dfrac{5}{3} = \dfrac{20}{3}$ ∎

EXAMPLE 7 $\dfrac{9}{16} \div 6 = \dfrac{9}{16} \div \dfrac{6}{1} = \dfrac{\overset{3}{\cancel{9}}}{16} \cdot \dfrac{1}{\underset{2}{\cancel{6}}} = \dfrac{3}{32}$ ∎

Addition and Subtraction of Fractions

Only fractions with the same denominators can be added or subtracted. To add (or subtract) fractions with the same denominators, add (or subtract) the numerators while maintaining the common denominator.

$$\frac{a}{c} + \frac{b}{c} = \frac{a+b}{c}, \qquad \frac{a}{c} - \frac{b}{c} = \frac{a-b}{c}, c \neq 0$$

EXAMPLE 8 (a) $\dfrac{3}{7} + \dfrac{2}{7} = \dfrac{3+2}{7} = \dfrac{5}{7}$

(b) $\dfrac{8}{12} - \dfrac{5}{12} = \dfrac{8-5}{12} = \dfrac{3}{12} = \dfrac{1}{4}$

To add or subtract fractions with different denominators, rewrite the fractions so that they have the same, or a common, denominator; then proceed as above.

EXAMPLE 9 (a) $\dfrac{3}{4} + \dfrac{5}{6} = \left(\dfrac{3}{4} \cdot \dfrac{3}{3}\right) + \left(\dfrac{5}{6} \cdot \dfrac{2}{2}\right) = \dfrac{9}{12} + \dfrac{10}{12} = \dfrac{19}{12}$

(b) $\dfrac{5}{12} - \dfrac{7}{18} = \left(\dfrac{5}{12} \cdot \dfrac{3}{3}\right) - \left(\dfrac{7}{18} \cdot \dfrac{2}{2}\right) = \dfrac{15}{36} - \dfrac{14}{36} = \dfrac{1}{36}$

EXAMPLE 10 $4 + \dfrac{5}{7} = \dfrac{4}{1} + \dfrac{5}{7} = \dfrac{4}{1} \cdot \dfrac{7}{7} + \dfrac{5}{7} = \dfrac{28}{7} + \dfrac{5}{7} = \dfrac{33}{7}$

COMMON STUDENT ERROR

Dividing out common factors can be be performed only when multiplying fractions; it cannot be performed when adding or subtracting fractions.

Correct *Wrong*

$\dfrac{\overset{1}{\cancel{3}}}{5} \cdot \dfrac{4}{\underset{3}{\cancel{9}}} = \dfrac{1 \cdot 4}{5 \cdot 3} = \dfrac{4}{15}$ $\dfrac{\overset{1}{\cancel{3}}}{5} + \dfrac{\overset{1}{\cancel{1}}}{\underset{1}{\cancel{3}}}$ cannot divide out common factors when adding or subtracting

$\dfrac{\overset{1}{\cancel{4}} \cdot 7}{\underset{2}{\cancel{8}}} = \dfrac{1 \cdot 7}{2} = \dfrac{7}{2}$ $\dfrac{\overset{1}{\cancel{4}} + 7}{\underset{2}{\cancel{8}}}$

Appendix B

Geometric Formulas

Areas and Perimeters

Figure	Sketch	Area	Perimeter
Square		$A = s^2$	$P = 4s$
Rectangle		$A = lw$	$P = 2l + 2w$
Parallelogram		$A = lh$	$P = 2l + 2w$
Trapezoid		$A = \frac{1}{2}h(b_1 + b_2)$	$P = s_1 + s_2 + b_1 + b_2$
Triangle		$A = \frac{1}{2}bh$	$P = s_1 + s_2 + b$

Area and Circumference of Circle

Circle	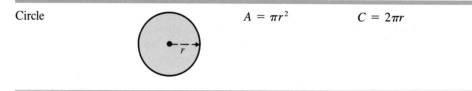	$A = \pi r^2$	$C = 2\pi r$

Volume and Surface Area of Three-dimensional Figures

Figure	Sketch	Volume	Surface Area
Rectangular solid		$V = lwh$	$s = 2lh + 2wh + 2wl$
Right circular cylinder		$V = \pi r^2 h$	$s = 2\pi rh + 2\pi r^2$
Sphere		$V = \dfrac{4}{3}\pi r^3$	$s = 4\pi r^2$
Right circular cone		$V = \dfrac{1}{3}\pi r^2 h$	$s = \pi r \sqrt{r^2 + h^2}$
Square or rectangular pyramid		$V = \dfrac{1}{3}lwh$	

Appendix C

Squares and Square Roots

Number	Square	Square Root	Number	Square	Square Root
1	1	1.000	51	2,601	7.141
2	4	1.414	52	2,704	7.211
3	9	1.732	53	2,809	7.280
4	16	2.000	54	2,916	7.348
5	25	2.236	55	3,025	7.416
6	36	2.449	56	3,136	7.483
7	49	2.646	57	3,249	7.550
8	64	2.828	58	3,364	7.616
9	81	3.000	59	3,481	7.681
10	100	3.162	60	3,600	7.746
11	121	3.317	61	3,721	7.810
12	144	3.464	62	3,844	7.874
13	169	3.606	63	3,969	7.937
14	196	3.742	64	4,096	8.000
15	225	3.873	65	4,225	8.062
16	256	4.000	66	4,356	8.124
17	289	4.123	67	4,489	8.185
18	324	4.243	68	4,624	8.246
19	361	4.359	69	4,761	8.307
20	400	4.472	70	4,900	8.367
21	441	4.583	71	5,041	8.426
22	484	4.690	72	5,184	8.485
23	529	4.796	73	5,329	8.544
24	576	4.899	74	5,476	8.602
25	625	5.000	75	5,625	8.660
26	676	5.099	76	5,776	8.718
27	729	5.196	77	5,929	8.775
28	784	5.292	78	6,084	8.832
29	841	5.385	79	6,241	8.888
30	900	5.477	80	6,400	8.944
31	961	5.568	81	6,561	9.000
32	1,024	5.657	82	6,724	9.055
33	1,089	5.745	83	6,889	9.110
34	1,156	5.831	84	7,056	9.165
35	1,225	5.916	85	7,225	9.220
36	1,296	6.000	86	7,396	9.274
37	1,369	6.083	87	7,569	9.327
38	1,444	6.164	88	7,744	9.381
39	1,521	6.245	89	7,921	9.434
40	1,600	6.325	90	8,100	9.487
41	1,681	6.403	91	8,281	9.539
42	1,764	6.481	92	8,464	9.592
43	1,849	6.557	93	8,649	9.644
44	1,936	6.633	94	8,836	9.695
45	2,025	6.708	95	9,025	9.747
46	2,116	6.782	96	9,216	9.798
47	2,209	6.856	97	9,409	9.849
48	2,304	6.928	98	9,604	9.899
49	2,401	7.000	99	9,801	9.950
50	2,500	7.071	100	10,000	10.000

Appendix D

Common Logarithms

n	0	1	2	3	4	5	6	7	8	9
1.0	.0000	.0043	.0086	.0128	.0170	.0212	.0253	.0294	.0334	.0374
1.1	.0414	.0453	.0492	.0531	.0569	.0607	.0645	.0682	.0719	.0755
1.2	.0792	.0828	.0864	.0899	.0934	.0969	.1004	.1038	.1072	.1106
1.3	.1139	.1173	.1206	.1239	.1271	.1303	.1335	.1367	.1399	.1430
1.4	.1461	.1492	.1523	.1553	.1584	.1614	.1644	.1673	.1703	.1732
1.5	.1761	.1790	.1818	.1847	.1875	.1903	.1931	.1959	.1987	.2014
1.6	.2041	.2068	.2095	.2122	.2148	.2175	.2201	.2227	.2253	.2279
1.7	.2304	.2330	.2355	.2380	.2405	.2430	.2455	.2480	.2504	.2529
1.8	.2553	.2577	.2601	.2625	.2648	.2672	.2695	.2718	.2742	.2765
1.9	.2788	.2810	.2833	.2856	.2878	.2900	.2923	.2945	.2967	.2989
2.0	.3010	.3032	.3054	.3075	.3096	.3118	.3139	.3160	.3181	.3201
2.1	.3222	.3243	.3263	.3284	.3304	.3324	.3345	.3365	.3385	.3404
2.2	.3424	.3444	.3464	.3483	.3502	.3522	.3541	.3560	.3579	.3598
2.3	.3617	.3636	.3655	.3674	.3692	.3711	.3729	.3747	.3766	.3784
2.4	.3802	.3820	.3838	.3856	.3874	.3892	.3909	.3927	.3945	.3962
2.5	.3979	.3997	.4014	.4031	.4048	.4065	.4082	.4099	.4116	.4133
2.6	.4150	.4166	.4183	.4200	.4216	.4232	.4249	.4265	.4281	.4298
2.7	.4314	.4330	.4346	.4362	.4378	.4393	.4409	.4425	.4440	.4456
2.8	.4472	.4487	.4502	.4518	.4533	.4548	.4564	.4579	.4594	.4609
2.9	.4624	.4639	.4654	.4669	.4683	.4698	.4713	.4728	.4742	.4757
3.0	.4771	.4786	.4800	.4814	.4829	.4843	.4857	.4871	.4886	.4900
3.1	.4914	.4928	.4942	.4955	.4969	.4983	.4997	.5011	.5024	.5038
3.2	.5051	.5065	.5079	.5092	.5105	.5119	.5132	.5145	.5159	.5172
3.3	.5185	.5198	.5211	.5224	.5237	.5250	.5263	.5276	.5289	.5302
3.4	.5315	.5328	.5340	.5353	.5366	.5378	.5391	.5403	.5416	.5428
3.5	.5441	.5453	.5465	.5478	.5490	.5502	.5514	.5527	.5539	.5551
3.6	.5563	.5575	.5587	.5599	.5611	.5623	.5635	.5647	.5658	.5670
3.7	.5682	.5694	.5705	.5717	.5729	.5740	.5752	.5763	.5775	.5786
3.8	.5798	.5809	.5821	.5832	.5843	.5855	.5866	.5877	.5888	.5899
3.9	.5911	.5922	.5933	.5944	.5955	.5966	.5977	.5988	.5999	.6010
4.0	.6021	.6031	.6042	.6053	.6064	.6075	.6085	.6096	.6107	.6117
4.1	.6128	.6138	.6149	.6160	.6170	.6180	.6191	.6201	.6212	.6222
4.2	.6232	.6243	.6253	.6263	.6274	.6284	.6294	.6304	.6314	.6325
4.3	.6335	.6345	.6355	.6365	.6375	.6385	.6395	.6405	.6415	.6425
4.4	.6435	.6444	.6454	.6464	.6474	.6484	.6493	.6503	.6513	.6522
4.5	.6532	.6542	.6551	.6561	.6571	.6580	.6590	.6599	.6609	.6618
4.6	.6628	.6637	.6646	.6656	.6665	.6675	.6684	.6693	.6702	.6712
4.7	.6721	.6730	.6739	.6749	.6758	.6767	.6776	.6785	.6794	.6803
4.8	.6812	.6821	.6830	.6839	.6848	.6857	.6866	.6875	.6884	.6893
4.9	.6902	.6911	.6920	.6928	.6937	.6946	.6955	.6964	.6972	.6981
5.0	.6990	.6998	.7007	.7016	.7024	.7033	.7042	.7050	.7059	.7067
5.1	.7076	.7084	.7093	.7101	.7110	.7118	.7126	.7135	.7143	.7152
5.2	.7160	.7168	.7177	.7185	.7193	.7202	.7210	.7218	.7226	.7235
5.3	.7243	.7251	.7259	.7267	.7275	.7284	.7292	.7300	.7308	.7316
5.4	.7324	.7332	.7340	.7348	.7356	.7364	.7372	.7380	.7388	.7396
5.5	.7404	.7412	.7419	.7427	.7435	.7443	.7451	.7459	.7466	.7474
5.6	.7482	.7490	.7497	.7505	.7513	.7520	.7528	.7536	.7543	.7551
5.7	.7559	.7566	.7574	.7582	.7589	.7597	.7604	.7612	.7619	.7627
5.8	.7634	.7642	.7649	.7657	.7664	.7672	.7679	.7686	.7694	.7701
5.9	.7709	.7716	.7723	.7731	.7738	.7745	.7752	.7760	.7767	.7774

Table continues on page 580

COMMON LOGARITHMS (continued)

n	0	1	2	3	4	5	6	7	8	9
6.0	.7782	.7789	.7796	.7803	.7810	.7818	.7825	.7832	.7839	.7846
6.1	.7853	.7860	.7868	.7875	.7882	.7889	.7896	.7903	.7910	.7917
6.2	.7924	.7931	.7938	.7945	.7952	.7959	.7966	.7973	.7980	.7987
6.3	.7993	.8000	.8007	.8014	.8021	.8028	.8035	.8041	.8048	.8055
6.4	.8062	.8069	.8075	.8082	.8089	.8096	.8102	.8109	.8116	.8122
6.5	.8129	.8136	.8142	.8149	.8156	.8162	.8169	.8176	.8182	.8189
6.6	.8195	.8202	.8209	.8215	.8222	.8228	.8235	.8241	.8248	.8254
6.7	.8261	.8267	.8274	.8280	.8287	.8293	.8299	.8306	.8312	.8319
6.8	.8325	.8331	.8338	.8344	.8351	.8357	.8363	.8370	.8376	.8382
6.9	.8388	.8395	.8401	.8407	.8414	.8420	.8426	.8432	.8439	.8445
7.0	.8451	.8457	.8463	.8470	.8476	.8482	.8488	.8494	.8500	.8506
7.1	.8513	.8519	.8525	.8531	.8537	.8543	.8549	.8555	.8561	.8567
7.2	.8573	.8579	.8585	.8591	.8597	.8603	.8609	.8615	.8621	.8627
7.3	.8633	.8639	.8645	.8651	.8657	.8663	.8669	.8675	.8681	.8686
7.4	.8692	.8698	.8704	.8710	.8716	.8722	.8727	.8733	.8739	.8745
7.5	.8751	.8756	.8762	.8768	.8774	.8779	.8785	.8791	.8797	.8802
7.6	.8808	.8814	.8820	.8825	.8831	.8837	.8842	.8848	.8854	.8859
7.7	.8865	.8871	.8876	.8882	.8887	.8893	.8899	.8904	.8910	.8915
7.8	.8921	.8927	.8932	.8938	.8943	.8949	.8954	.8960	.8965	.8971
7.9	.8976	.8982	.8987	.8993	.8998	.9004	.9009	.9015	.9020	.9025
8.0	.9031	.9036	.9042	.9047	.9053	.9058	.9063	.9069	.9074	.9079
8.1	.9085	.9090	.9096	.9101	.9106	.9112	.9117	.9122	.9128	.9133
8.2	.9138	.9143	.9149	.9154	.9159	.9165	.9170	.9175	.9180	.9186
8.3	.9191	.9196	.9201	.9206	.9212	.9217	.9222	.9227	.9232	.9238
8.4	.9243	.9248	.9253	.9258	.9263	.9269	.9274	.9279	.9284	.9289
8.5	.9294	.9299	.9304	.9309	.9315	.9320	.9325	.9330	.9335	.9340
8.6	.9345	.9350	.9355	.9360	.9365	.9370	.9375	.9380	.9385	.9390
8.7	.9395	.9400	.9405	.9410	.9415	.9420	.9425	.9430	.9435	.9440
8.8	.9445	.9450	.9455	.9460	.9465	.9469	.9474	.9479	.9484	.9489
8.9	.9494	.9499	.9504	.9509	.9513	.9518	.9523	.9528	.9533	.9538
9.0	.9542	.9547	.9552	.9557	.9562	.9566	.9571	.9576	.9581	.9586
9.1	.9590	.9595	.9600	.9605	.9609	.9614	.9619	.9624	.9628	.9633
9.2	.9638	.9643	.9647	.9652	.9657	.9661	.9666	.9671	.9675	.9680
9.3	.9685	.9689	.9694	.9699	.9703	.9708	.9713	.9717	.9722	.9727
9.4	.9731	.9736	.9741	.9745	.9750	.9754	.9759	.9763	.9768	.9773
9.5	.9777	.9782	.9786	.9791	.9795	.9800	.9805	.9809	.9814	.9818
9.6	.9823	.9827	.9832	.9836	.9841	.9845	.9850	.9854	.9859	.9863
9.7	.9868	.9872	.9877	.9881	.9886	.9890	.9894	.9899	.9903	.9908
9.8	.9912	.9917	.9921	.9926	.9930	.9934	.9939	.9943	.9948	.9952
9.9	.9956	.9961	.9965	.9969	.9974	.9978	.9983	.9987	.9991	.9996

Examples 1 through 4 illustrate how to use the common logarithm table.

EXAMPLE 1

$$\log 4610 = \log (4.61 \times 10^3)$$
$$= \log 4.61 + \log 10^3$$
$$= 0.6637 + 3$$
$$= 3.6637$$

0.6637 is found in the body of the table by first finding 4.61 along the margins of the table.

EXAMPLE 2
$$\begin{aligned}
\log 0.000547 &= \log(5.47 \times 10^{-4}) \\
&= \log 5.47 + \log 10^{-4} \\
&= 0.7380 + (-4) \\
&= 0.7380 - 4 \\
&= -3.262
\end{aligned}$$

0.7380 is found in the body of the table by first finding 5.47 along the margins of the table. ∎

EXAMPLE 3
$$\begin{aligned}
\log x &= 2.8476 \\
x &= 7.04 \times 10^{2} \\
&= 704
\end{aligned}$$

7.04 is found along the margins of the table by first finding .8476 in the body of the table. ∎

EXAMPLE 4
$$\begin{aligned}
\log x &= -3.0696 \\
&= 4 - 3.0696 - 4 \\
&= 0.9304 - 4 \\
x &= 8.52 \times 10^{-4} \\
&= 0.000852
\end{aligned}$$

8.52 is found along the margins of the table by first finding 9304 in the body of the table. When the logarithm is negative, add and subtract the same constant to obtain a positive number that can be looked up in the body of the text. ∎

Answers

CHAPTER 1

Exercise Set 1.2

1. A = {4, 5, 6, 7} **3.** C = {6, 8, 10} **5.** E = {0, 1, 2, 3, 4, 5, 6} **7.** G = { } **9.** I = {−4, −3, −2, −1, . . .}
11. K = { } **13.** ∉ **15.** ∈ **17.** ∉ **19.** ∉ **21.** ⊆ **23.** ⊄ **25.** ⊄ **27.** ⊄ **29.** ⊄ **31.** ⊆
33. ⊆ **35.** ⊄ **37.** ⊆ **39.** ⊄ **41.** ⊄ **43.** ⊆ **45.** True **47.** False **49.** True **51.** False
53. True **55.** True **57.** True **59.** False **61.** True **63.** 4 **65.** −6, 4, 0 **67.** $\sqrt{7}, \sqrt{5}$ **69.** 2, 4
71. 2, 4, −5.33, $\frac{9}{2}$, −100, −7, 4.7 **73.** 2, 4, −5.33, $\frac{9}{2}$, $\sqrt{7}$, $\sqrt{2}$, −100, −7, 4.7
75. A ∪ B = {1, 2, 3, 4}, A ∩ B = {2, 3} **77.** A ∪ B = {−1, −2, −4, −5, −6}, A ∩ B = {−2, −4}
79. A ∪ B = {0, 1, 2, 3}, A ∩ B = { } **81.** A ∪ B = {2, 4, 6, 8, . . .}, A ∩ B = {2, 4, 6}
83. A ∪ B = {0, 1, 2, 3, 4, 5, 6, 7, 8}, A ∩ B = { } **85.** A ∪ B = {1, 2, 3, 4, . . .}, A ∩ B = {2, 4, 6, 8, . . .}
87. The set of odd natural numbers greater than or equal to 5. **89.** The set of letters in the English alphabet.
91. The set of states in the United States.
93. (a) Set B is the set of all x such that x is one of the last five capital letters in the English alphabet.
(b) B = {V, W, X, Y, Z}

Exercise Set 1.3

1. Commutative property of addition **3.** Distributive property **5.** Associative property of addition
7. Commutative property of addition **9.** Commutative property of multiplication **11.** Identity property of addition
13. Commutative property of addition **15.** Commutative property of addition **17.** Commutative property of addition
19. Distributive property **21.** Double negative property **23.** Identity property of multiplication
25. Inverse property of addition **27.** Inverse property of addition **29.** Inverse property of addition
31. Identity property of multiplication **33.** Inverse property of multiplication **35.** Double negative property
37. Multiplication property of 0. **39.** Double negative property **41.** 3 + x **43.** x + (2 + 3) **45.** x **47.** x
49. x **51.** 0 **53.** 1x + 1y or x + y **55.** 3 **57.** 0 **59.** −4, $\frac{1}{4}$ **61.** 3, −$\frac{1}{3}$ **63.** −$\frac{2}{3}$, $\frac{3}{2}$ **65.** 6, −$\frac{1}{6}$
67. $\frac{3}{7}$, −$\frac{7}{3}$ **69. (a)** $a + b = b + a$ **71. (a)** $a(b + c) = ab + ac$

Cumulative Review Exercises

73. True **74.** ⊆ **75. (a)** 3, 4, −2, 0 **(b)** 3, 4, −2, $\frac{5}{6}$, 0 **(c)** $\sqrt{3}$ **(d)** 3, 4, −2, $\frac{5}{6}$, $\sqrt{3}$, 0
76. A ∪ B = {1, 4, 7, 9, 12, 15}, A ∩ B = {4, 7}

Exercise Set 1.4

1. > **3.** > **5.** > **7.** < **9.** < **11.** > **13.** < **15.** > **17.** > **19.** < **21.** > **23.** >
25. 6 **27.** 4 **29.** 2 **31.** $\frac{1}{2}$ **33.** 0 **35.** 45 **37.** 13.84 **39.** -7 **41.** -3 **43.** $-\frac{5}{9}$ **45.** =
47. > **49.** > **51.** > **53.** > **55.** > **57.** < **59.** < **61.** $-1, 2, |3|, |-5|, 6$
63. $-|3|, -2, 4, |-6|, 8$ **65.** $-3, |0|, |-5|, |7|, |-12|$ **67.** $|-9|, 12, 24, |36|, |-45|$
69. $-|-6|, -4, -2, 6, |-8|$ **71.** $-|2.9|, -2.4, -2.1, -2, |-2.8|$ **73.** $-2, \frac{1}{3}, |-\frac{1}{2}|, |\frac{3}{5}|, |-\frac{3}{4}|$ **75.** 6
77. $4, -4$ **79.** $5, -17$ **81.** All real numbers, $\mathbb{R}$ **83.** $a \le 0$ **85.** $5, -5$
87. The absolute value of 7 is greater than 4. **89.** Negative four is equal to the negative of the absolute value of 4.
91. For any real number a, $|a| = \begin{cases} a, a \ge 0 \\ -a, a < 0 \end{cases}$ **93.** No, for example $|2| - |3| = -1$

Cumulative Review Exercises

94. A rational number can be written as a quotient of two integers, denominator not 0; an irrational number cannot.
95. Commutative property of addition **96.** Distributive Property **97.** Associative Property of addition.

Exercise Set 1.5

1. 1 **3.** 10 **5.** 5 **7.** 8 **9.** -11 **11.** 39 **13.** -5.78 **15.** $\frac{19}{56}$ **17.** 7 **19.** 11.4 **21.** $-\frac{21}{2}$
23. -5 **25.** 1 **27.** 2 **29.** -10 **31.** -2 **33.** -2 **35.** -48 **37.** $\frac{5}{4}$ **39.** 6 **41.** -3 **43.** 1
45. -144 **47.** 1 **49.** -24 **51.** 1 **53.** $-\frac{3}{16}$ **55.** $-\frac{1}{9}$ **57.** -2 **59.** -16 **61.** $-\frac{52}{45}$ **63.** 10
65. -96 **67.** -27 **69.** -2 **71.** $\frac{12}{5}$ **73.** 5 **75.** -18 **77.** -14 **79.** 121
81. 170 ft below sea level (or -170 ft) **83.** \$4313 **85.** a) Owes \$83,000 (or $-$\$83,000) b) receive \$1,050,000
87. True **89.** False **91.** True **93.** True **95.** False **97.** False **99.** True

Cumulative Review Exercises

105. Identity property of multiplication **106.** Identity property of addition **107.** >
108. $-|-7|, -5, -1, |-4|, |-5|, |6|$

Just for Fun **1.** -50 **2.** 84 **3.** -1 **4.** 1

Exercise Set 1.6

1. 9 **3.** 25 **5.** 16 **7.** 81 **9.** -32 **11.** $\frac{16}{81}$ **13.** 0.09 **15.** 0.008 **17.** 1 **19.** 4 **21.** -3
23. -1 **25.** 4 **27.** 8 **29.** $\frac{5}{6}$ **31.** $\frac{15}{9} = \frac{5}{3}$ **33.** 0.2 **35.** 0.5 **37.** 4 **39.** -2 **41.** -4 **43.** 1
45. 5 **47.** -6 **49.** $\frac{1}{4}$ **51.** 0.1 **53.** 32 **55.** 3 **57.** $\frac{2}{3}$ **59.** -3 **61.** -5 **63.** $\frac{1}{256}$ **65.** $9, -9$
67. $1, -1$ **69.** $1, -1$ **71.** $\frac{1}{9}, -\frac{1}{9}$ **73.** $27, -27$ **75.** $-27, 27$ **77.** $-8, 8$ **79.** $\frac{8}{27}, -\frac{8}{27}$ **81.** 21
83. 29 **85.** 0 **87.** -19 **89.** 16 **91.** -12 **93.** -12.76 **95.** $-\frac{49}{72}$
97. No real number when squared gives a negative number.
99. A negative number raised to an odd power is a negative number.

Cumulative Review Exercises

100. All real numbers, $\mathbb{R}$ **101.** $a \ge 0$ **102.** $4, -4$ **103.** 6

Exercise Set 1.7

1. 26 **3.** 50 **5.** $\frac{27}{4}$ **7.** 24 **9.** -64 **11.** $\frac{81}{40}$ **13.** 178 **15.** 2 **17.** 294 **19.** 1156 **21.** 81
23. $-\frac{1}{6}$ **25.** $-\frac{47}{96}$ **27.** $\frac{27}{5}$ **29.** $\frac{5}{6}$ **31.** -2 **33.** $-\frac{10}{3}$ **35.** 3 **37.** 17 **39.** 1 **41.** $-\frac{3}{4}$
43. -7 **45.** 44 **47.** $\frac{147}{16}$ **49.** -40 **51.** 0 **53.** 21 **55.** 33 **57.** 42 **59.** $\frac{25}{2}$ **61.** 8 **63.** $\frac{4}{9}$
65. (a) 169 (b) $\sqrt{169} = 13$ (c) 8 (d) $\frac{4}{3}$ (e) -3 **67.** $(3x + 6)^2, 225$ **69.** $6(3x + 6) - 9, 81$
71. $\left(\dfrac{x + 3}{2y}\right)^2 - 3, 1$ **73.** (b) $-\frac{3}{2}$ **75.** (b) 24

Cumulative Review Exercises

76. $A \cap B = \{b, c, f\}, A \cup B = \{a, b, c, d, f, g, h\}$ **77.** Associative property of addition
78. $-|6|, -4, -|-2|, 0, |-5|$ **79.** 4 **80.** -12

Just for Fun   **1.** $\frac{883}{48}$ **2.** $\frac{131,072}{5}$ **3.** $-\frac{15}{19}$

Review Exercises

1. {3, 4, 5, 6}   **2.** {0, 3, 6, 9, . . .}   **3.** $\in$ **4.** $\notin$ **5.** $\notin$ **6.** $\in$ **7.** $\subseteq$ **8.** $\not\subseteq$ **9.** $\not\subseteq$
10. $\subseteq$ **11.** $\subseteq$ **12.** $\subseteq$ **13.** $\subseteq$ **14.** $\subseteq$ **15.** $\subseteq$ **16.** $\not\subseteq$ **17.** 4, 6   **18.** 4, 6, 0   **19.** $-3, 4, 6, 0$
20. $-3, 4, 6, \frac{1}{2}, 0, \frac{15}{27}, -\frac{1}{5}, 1.47$ **21.** $\sqrt{5}, \sqrt{3}$ **22.** $-3, 4, 6, \frac{1}{2}, \sqrt{5}, \sqrt{3}, 0, \frac{15}{27}, -\frac{1}{5}, 1.47$ **23.** True
24. True   **25.** False   **26.** True   **27.** True   **28.** A $\cup$ B = {1, 2, 3, 4, 5}, A $\cap$ B = {2, 3, 4, 5}
29. A $\cup$ B = {2, 3, 4, 5, 6, 7, 8, 9}, A $\cap$ B = $\varnothing$ **30.** A $\cup$ B = {1, 2, 3, 4, . . .}, A $\cap$ B = {2, 4, 6, . . .}
31. A $\cup$ B = {3, 4, 5, 6, 9, 10, 11, 12}, A $\cap$ B = {9, 10}   **32.** Commutative property of addition
33. Commutative property of addition   **34.** Distributive property   **35.** Commutative property of multiplication
36. Associative property of addition   **37.** Identity property of addition   **38.** Associative property of multiplication
39. Identity property of multiplication   **40.** Double negative property   **41.** Multiplication property of zero.
42. Identity property of addition   **43.** Identity property of multiplication   **44.** Inverse property of addition
45. Inverse property of multiplication   **46.** Inverse property of multiplication   **47.** Identity property of multiplication
48. $3 + x$ **49.** $3x + 15$ **50.** $x + [6 + (-4)]$ **51.** $x \cdot 3$ **52.** $9 \cdot (x \cdot y)$ **53.** $4x - 4y + 20$ **54.** a
55. a **56.** 0   **57.** 1   **58.** a **59.** $>$ **60.** $<$ **61.** $<$ **62.** $>$ **63.** $<$ **64.** $<$ **65.** $<$
66. $>$ **67.** $=$ **68.** $=$ **69.** $<$ **70.** $<$ **71.** $>$ **72.** $>$ **73.** $>$ **74.** $>$ **75.** $-5, -2, 4, |7|$
76. $0, \frac{3}{5}, 2.3, |-3|$ **77.** $-2, 3, |-5|, |-7|$ **78.** $-4, -|3|, -2.1, -2$ **79.** $-4, -|-3|, 5, 6$
80. $-3, 0, |1.6|, |-2.3|$ **81.** $\frac{22}{5}$ **82.** -63 **83.** -12 **84.** 8   **85.** -1 **86.** -2 **87.** -1 **88.** 16
89. 21   **90.** 21   **91.** -47 **92.** 15   **93.** 31   **94.** 6   **95.** 512   **96.** $-\frac{8}{5}$ **97.** $\frac{8}{3}$ **98.** 5   **99.** $-\frac{16}{19}$
100. 15   **101.** 10   **102.** $\frac{26}{3}$ **103.** 39   **104.** 7   **105.** -50 **106.** -249 **107.** 204

Practice Test

1. {6, 7, 8, 9, . . .}   **2.** $\not\subseteq$ **3.** $\subseteq$ **4.** True   **5.** False   **6.** True   **7.** $-\frac{3}{5}, 2, -4, 0, \frac{19}{12}, 2.57, -1.92$
8. $-\frac{3}{5}, 2, -4, 0, \frac{19}{12}, 2.57, \sqrt{8}, \sqrt{2}, -1.92$ **9.** A $\cup$ B = {5, 7, 8, 9, 10, 11, 14}, A $\cap$ B = {8, 10}
10. A $\cup$ B = {1, 3, 5, 7, . . .}, A $\cap$ B = {3, 5, 7, 9, 11}   **11.** $<$ **12.** $>$ **13.** $-|4|, -2, |3|, 6$
14. Distributive property   **15.** Associative property of addition   **16.** Commutative property of addition
17. Inverse property of multiplication   **18.** Identity property of addition   **19.** -5 **20.** 23   **21.** $-\frac{1}{4}$ **22.** $-\frac{37}{22}$
23. $\frac{64}{5}$ **24.** 17   **25.** 39

CHAPTER 2

Exercise Set 2.1

1. Reflexive property   **3.** Symmetric property   **5.** Transitive property   **7.** Reflexive property
9. Addition property   **11.** Multiplication property   **13.** Multiplication property   **15.** Addition property
17. Multiplication property   **19.** Symmetric property   **21.** First   **23.** Second   **25.** Fifth   **27.** Zero
29. Thirteenth   **31.** Twelfth   **33.** $15x - 5$ **35.** $5x^2 - x - 5$ **37.** $-4x^2 - 8x + 7$
39. Cannot be simplified   **41.** $4xy + y^2 - 2$ **43.** $-10x + 47$ **45.** $\frac{8}{3}x + \frac{13}{2}$ **47.** $-17x - 4$ **49.** $6x - 3y$
51. 1   **53.** $-\frac{15}{4}$ **55.** -32 **57.** $-\frac{8}{3}$ **59.** 3.67   **61.** 3   **63.** $\frac{11}{7}$ **65.** -3 **67.** -8 **69.** 20
71. 1   **73.** -5 **75.** -4 **77.** 5   **79.** 1.05   **81.** -4 **83.** -5 **85.** No solution   **87.** $-\frac{119}{25}$
89. All real numbers   **91.** -2 **93.** $\frac{7}{9}$ **95.** 4   **97.** $\frac{96}{5}$ **99.** -6 **101. (b)** $\frac{14}{5}$
103. An equation that is true for all real numbers.   **105.** An equation that has no solution.
107. $2x = 8, x + 3 = 7, x - 2 = 2$

Cumulative Review Exercises

109. $|a| = \begin{cases} a, a \geq 0 \\ -a, a < 0 \end{cases}$ **110.** -9 **111.** $-\frac{27}{64}$ **112.** -4

Just for Fun   **1.** $\frac{306}{157}$ **2.** $-\frac{115}{31}$ **3.** $\frac{1524}{131}$

Exercise Set 2.2

1. 42   **3.** 250   **5.** 1100   **7.** 532   **9.** 310.88   **11.** $\frac{7}{4}$ **13.** 3.33   **15.** $\sqrt{145} \approx 12.04$ **17.** 66.67
19. 4   **21.** 1252   **23.** 65.94   **25.** 20   **27.** 0.29   **29.** 84   **31.** 119.10   **33.** $y = -2x + 3$

35. $y = \dfrac{-2x + 6}{3}$ **37.** $y = x - 8$ **39.** $y = \dfrac{x - 3}{2}$ **41.** $y = \dfrac{8x - 3}{2}$ **43.** $y = \dfrac{-9x + 15}{5}$

45. $y = \dfrac{2x + 5}{10}$ **47.** $y = \dfrac{4x - 90}{23}$ **49.** $r = \dfrac{d}{t}$ **51.** $b = \dfrac{2A}{h}$ **53.** $t = \dfrac{i}{pr}$ **55.** $l = \dfrac{P - 2w}{2}$

57. $h = \dfrac{3V}{B}$ **59.** $h = \dfrac{V}{\pi r^2}$ **61.** $\sigma = \dfrac{x - \mu}{z}$ **63.** $R = \dfrac{P}{I^2}$ **65.** $h = \dfrac{2A}{b_1 + b_2}$ **67.** $m = \dfrac{y - b}{x}$

69. $m = \dfrac{y - y_1}{x - x_1}$ **71.** $n = \dfrac{2S}{f + l}$ **73.** $F = \frac{9}{5}C + 32$ **75.** $z = \dfrac{kx}{y}$ **77.** $m_1 = \dfrac{Fd^2}{km_2}$ **79. (b)** $A = \dfrac{0.5cV^2}{\mu d}$

Cumulative Review Exercises

81. -22 **82.** $\frac{3}{2}$ **83.** 15   **84.** 78

Exercise Set 2.3

1. $x + 6x = 56$; 8, 48   **3.** $x + (x + 1) = 51$; 25, 26   **5.** $2x - 8 = 38$; 23   **7.** $x + 3x = 48$; 12, 36
9. $x + (x + 2) + (x + 4) = 66$; 20, 22, 24   **11.** $10 - \frac{3}{5}x = 4$; 10   **13.** $5x = \frac{1}{2}(4x + 2) + 2$; 1, 6
15. $x + x + 2x = 180$; 45°, 45°, 90°   **17.** $x + x + (x + 15) = 45$; 10 in, 10 in, 25 in
19. $2.89x = 37.99$; 13.15 months   **21.** $1.25x = 30$; over 24   **23.** $5200 + 300x = 8800$; 12 yrs
25. $35 + 0.20x = 80$; 225 mi   **27.** $240 + 0.12x = 540$; $2500   **29.** $x + 0.04x = 10,000$; 9615.38
31. $x + 0.07x = 499$; $466.36   **33.** $x + 0.07x + 0.15x = 9.25$; $7.58   **35.** $150 + 6x = 510$; 60 days

37. $2x + 2(2x + 2) = 40$; $w = 6$ ft, $l = 14$ ft   **39.** $2\left(\dfrac{x}{2} + 1\right) + 2x = 20$; $l = 6$ m, $w = 4$ m

41. $4x + 2(x + 3) = 30$; $w = 4$ ft, $h = 7$ ft   **43.** $x - 0.10x - 5 = 49$; $60   **45.** $x - \frac{1}{4}x - 10 = 290$; $400

47. $x + 2x + (3x - 4) = 512$; 86 acres, 172 acres, 254 acres   **49. (a)** $\dfrac{87 + 93 + 97 + 96 + x}{5} = 90$ **(c)** 77

Cumulative Review Exercises

53. $\frac{2}{19}$ **54.** $\frac{12}{5}$ **55.** $\frac{40}{63}$ **56.** $y = \dfrac{4x - 9}{6}$ or $y = \dfrac{2}{3}x - \dfrac{3}{2}$

Just for Fun   **1.** $x + 0.05x - 0.05(x + 0.05x) = 59.85$; $60
2. $3(28) + 0.15x + 0.04[(3)(28) + 0.15x] = 121.68$; 220 mi
3. Let n be the number.

Multiply by 2	$2n$
Add 33	$2n + 33$
Subtract 13	$2n + 33 - 13 = 2n + 20$
Divide by 2	$\dfrac{2n + 20}{2} = n + 10$
Subtract number started with	$n + 10 - n = 10$

Exercise Set 2.4

1. 4 mph   **3. (a)** 1.91 mph **(b)** 1.78 mph   **5.** 16.39 min   **7.** 0.5 yr or 6 months
9. (a) 26.07 days   **(b)** 3,345,991.5 mi   **11. (a)** 2.29 min   **(b)** 1.83 min   **(c)** 1.02 min   **(d)** 0.81 min
13. $550t + 650t = 3000$; 2.5 hrs   **15.** $330 = 60(3) + 3r$; 50 mph   **17.** $520 = 60t + 70t$; 4 hrs
19. $3(x + 20) = 4.2x$; Freight train, 50 mph; Passenger train, 70 mph
21. $18 = 2(4x) - 2(x)$;   **(a)** 12 mph   **(b)** 24 mi   **23.** $2.6t = 1.2(16 - t)$ **(a)** 5.05 hrs   **(b)** 26.26 mi
25. $600(x + 2) + 400x = 15,000$; 13.8 hr   **27.** $0.09x + 0.10(11,000 - x) = 1050$; $5000 at 9%, $6000 at 10%
29. (a) $45x + 14(5x) = 9000$; 78 shares GM, 390 shares Reebok   **(b)** $30
31. $0.10x + 0.25(33 - x) = 4.50$; 25 dimes, 8 quarters   **33.** $6.20x + 5.80(18) = 6.10(x + 18)$; 54 lbs
35. $0.10(40) + 1.00x = 0.25(x + 40)$; 8 oz

37. $0.20x + 0.50(12 - x) = 0.30(12)$; 8 l of 20% solution, 4 l of 50% solution   **39.** $12x + 0.76(16) = 0.82(28)$; 90%
41. $28,000 - x = 32,450 - (6400 - x)$; $1075 for Mr. Clar, $5325 for Mrs. Clar
43. $8.2 = 2x + 2(x + 0.4)$; 1.85 mph, 2.25 mph
45. **(a)** $34(3x) + 23(x) = 6000$; 48 shares US Steel, 144 shares Sears   **(b)** no money left over
47. $10x + 20x = 15,000$; 500 min or $8\frac{1}{3}$ hrs   **49.** $0.09(2500) + x(1500) = 315$; 6%
51. $300t = 220(11.2 - t)$; **(a)** 4.74 hr   **(b)** 1422 mi   **53.** $6x + 6.50(18 - x) = 114$; 6 hr at $6, 12 hr at $6.50
55. $20t = 12(8 - t)$; **(a)** 3 hr   **(b)** 120 nautical miles   **57.** $0.03(16) + 0.07(64) = x(80)$; 6.2%
59. $0.80x + 0.00(128 - x) = 0.06(128)$; 9.6 oz 80% sol, 118.4 oz water
61. $0.05(400) + 0.015x = 0.02(x + 400)$; 2400 qt   **63.** $70x = 50x + 200$; 10 min

Cumulative Review Exercises

69. -5.7 **70.** $\frac{126}{35} = \frac{18}{5}$ **71.** $y = \dfrac{x - 42}{30}$ **72.** 140 mi

Just for Fun   **1. (a)** 121.9 mph   **(b)** 0.0509 hr or 183.1 sec   **2.** 300,000,000 ft
3. 82,944 stitches require 10,368 minutes or 172.8 hours   **4.** 6 qt

Exercise Set 2.5

1. (a)   **(b)** $(-\infty, -3)$ **(c)** $\{x \mid x < -3\}$ **3. (a)**   **(b)** $[5, \infty)$ **(c)** $\{x \mid x \geq 5\}$

5. (a)   **(b)** $[2, \frac{12}{5})$ **(c)** $\{x \mid 2 \leq x < \frac{12}{5}\}$ **7. (a)**   **(b)** $(-6, -4]$ **(c)** $\{x \mid -6 < x \leq -4\}$

9. (a)   **(b)** $[-4, 5]$ **(c)** $\{x \mid -4 \leq x \leq 5\}$

11.   **13.**   **15.**   **17.**

19.   **21.**   **23.**   **25.** No solution

27. $[-19, \infty)$ **29.** $(-\infty, \frac{4}{3})$ **31.** $\varnothing$ **33.** $(-\infty, \infty)$ **35.** $(1, 6)$ **37.** $(-\frac{3}{5}, \frac{8}{5}]$ **39.** $[\frac{7}{2}, 5)$ **41.** $(-\frac{7}{6}, \frac{2}{3})$
43. $[-\frac{35}{2}, \frac{21}{2})$ **45.** $\{x \mid -\frac{56}{3} \leq x < \frac{14}{3}\}$ **47.** $\{x \mid 0 < x \leq 1\}$ **49.** $\{x \mid -23 < x \leq 2\}$ **51.** $\{x \mid -1 < x < \frac{1}{3}\}$
53. $\{x \mid 2 < x < 4\}$ **55.** $\varnothing$ **57.** $\{x \mid -3 < x < 1\}$ **59.** $\{x \mid x \leq 2 \text{ or } x > 8\}$ **61.** $(-\infty, 4)$ **63.** $[0, 2]$
65. $(\frac{13}{5}, \infty)$ **67.** $(-\infty, 0) \cup (6, \infty)$ **69.** $80x \leq 900$; 11 boxes   **71.** $4.25 + 0.48x \leq 9.50$; 13 minutes
73. $10,025 + 1.09x < 6.42x$; 1881 books   **75.** $60 + 0.084x < 0.167x$; 723 pieces
77. $\dfrac{90 + 87 + 96 + 95 + x}{5} \geq 90$; 82   **79.** $\dfrac{2.7 + 3.42 + x}{3} < 3.2$; any value less than 3.48

81. $7.2 < \dfrac{7.48 + 7.85 + x}{3} < 7.8$; $6.27 < x < 8.07$

83. (a) Solve individually $x + 5 < 3x - 8$ and $3x - 8 \leq 2(x + 7)$. Then find the intersection of their solution sets.
(b) $(\frac{13}{2}, 22]$

Cumulative Review Exercises

84. (a) $\{1, 3, 4, 5, 6, 7, 9\}$ **(b)** $\{1, 4, 6\}$ **85. (a)** 4   **(b)** 0, 4   **(c)** $-3, 4, \frac{5}{2}, 0, -\frac{29}{80}$ **(d)** $-3, 4, \frac{5}{2}, \sqrt{7}, 0, -\frac{29}{80}$
86. Associative Property of addition   **87.** Commutative Property of addition   **88.** $V = \dfrac{R - L + Dr}{r}$

Just for Fun   **1.** $84 \leq x \leq 100$

Exercise Set 2.6

1. $\{-5, 5\}$ **3.** $\{-12, 12\}$ **5.** $\varnothing$ **7.** $\{-12, 2\}$ **9.** $\{-5, 1\}$ **11.** $\{\frac{3}{2}, \frac{11}{6}\}$ **13.** $\{-\frac{5}{2}, \frac{13}{2}\}$ **15.** $\{-\frac{59}{3}, \frac{49}{3}\}$
17. $\{-1, \frac{11}{5}\}$ **19.** $\{y \mid -5 \leq y \leq 5\}$ **21.** $\{x \mid -2 \leq x \leq 16\}$ **23.** $\{z \mid 0 \leq z \leq \frac{10}{3}\}$ **25.** $\{x \mid -9 \leq x \leq 6\}$
27. $\{x \mid \frac{9}{2} \leq x \leq \frac{11}{2}\}$ **29.** $\varnothing$ **31.** $\{x \mid -4 < x < \frac{52}{3}\}$ **33.** $\{x \mid x < -3 \text{ or } x > 3\}$ **35.** $\{x \mid x < -9 \text{ or } x > 1\}$
37. $\{x \mid x < -\frac{5}{3} \text{ or } x > 1\}$ **39.** $\{z \mid z < -6 \text{ or } z > 0\}$ **41.** $\{x \mid x < -\frac{1}{2} \text{ or } x > 2\}$ **43.** $\mathbb{R}$
45. $\{x \mid x \leq -18 \text{ or } x \geq 2\}$ **47.** $\{7, -7\}$ **49.** $\{x \mid -2 < x < 8\}$ **51.** $\{x \mid x < -14 \text{ or } x > 4\}$

53. $\{y \mid -\frac{5}{2} < y < -\frac{3}{2}\}$ **55.** $\{-\frac{11}{4}, \frac{7}{4}\}$ **57.** $\{x \mid x \le -4 \text{ or } x \ge -1\}$ **59.** $\{x \mid -\frac{13}{3} \le x \le \frac{5}{3}\}$ **61** $\varnothing$

63. $\{-\frac{22}{3}, \frac{26}{3}\}$ **65.** $\{w \mid -16 < w < 8\}$ **67.** $\mathbb{R}$ **69.** $\{x \mid x \le 0 \text{ or } x \ge \frac{4}{3}\}$ **71.** $\{x \mid -\frac{3}{2} < x < \frac{15}{2}\}$

73. (a) Write $ax + b = c$ or $ax + b = -c$, then solve each equation for x. **(b)** $\left\{x \mid x = \dfrac{c - b}{a} \text{ or } x = \dfrac{-b - c}{a}\right\}$

75. (a) Write $ax + b < -c$ or $ax + b > c$, then solve both inequalities for x. **(b)** $\left\{x \mid x < \dfrac{-c - b}{a} \text{ or } x > \dfrac{c - b}{a}\right\}$

Cumulative Review Exercises

77. $\frac{29}{72}$ **78.** 25 **79.** 1.33 mi **80.** $\{x \mid x < 4\}$

Just for Fun **1.** All real numbers **2.** All x and all y **3.** $\{-1, 9\}$ **4.** $\{-1\}$

Review Exercises

1. Tenth **2.** First **3.** Seventh **4.** Cannot be simplified **5.** $7x^2 + 2xy - 4$ **6.** 8 **7.** $4x - 3y + 6$
8. $\frac{49}{6}$ **9.** 20 **10.** $-\frac{13}{3}$ **11.** -3 **12.** $-\frac{9}{2}$ **13.** No solution **14.** 3 **15.** 200 **16.** $\frac{1}{4}$ **17.** 176
18. -20 **19.** $l = \dfrac{A}{w}$ **20.** $h = \dfrac{A}{\pi r^2}$ **21.** $w = \dfrac{P - 2l}{2}$ **22.** $r = \dfrac{d}{t}$ **23.** $m = \dfrac{y - b}{x}$ **24.** $y = \dfrac{2x - 5}{3}$
25. $V_2 = \dfrac{P_1 V_1}{P_2}$ **26.** $a = \dfrac{2S - b}{3}$ **27.** $l = \dfrac{K - 2d}{2}$ **28.** $t = \dfrac{I - P}{Pr}$ **29.** $b_1 = \dfrac{2A - hb_2}{h}$ **30.** $t = \dfrac{w + 2l}{V_0}$
31. 16 **32.** sister 16, Paul 20 **33.** 5 **34.** 20 **35.** 50 **36.** 13, 15 **37.** 200 **38.** 10, 19
39. 30.6 rolls per hour **40.** \$6000 at 8%, \$4000 at 5% **41.** $2\frac{2}{3}$ hrs **42. (a)** 3000 mph **(b)** 16,500 mi
43. 15 lbs at \$6, 25 lbs at \$6.80 **44.** \$25 **45. (a)** 1 hr **(b)** 14.4 mi **46.** 40°, 65°, 75°
47. 300 gal/hr, 450 gal/hr **48.** 24, 25 **49.** 7.5 oz **50.** \$4500 at 10%, \$7500 at 6% **51.** more than 5
52. 90 mph **53.** ——₇ **54.** ——₋₃ **55.** ——_{5/2} **56.** ——_{21/4}

57. ——_{-9/2} **58.** ——₋₁₀ **59.** ——_{2/5} **60.** ——_{20/9} **61.** 13 boxes **62.** 7 min

63. 17 weeks **64.** $(5, 11)$ **65.** $[-3, 3)$ **66.** $(\frac{7}{2}, 6)$ **67.** $(\frac{8}{3}, 6)$ **68.** $[-\frac{1}{2}, \frac{23}{2})$ **69.** $(2, 14)$
70. $\{x \mid 81 \le x \le 100\}$ **71.** $\{x \mid -3 < x < 3\}$ **72.** $\mathbb{R}$ **73.** $\{x \mid x > -1\}$ **74.** $\{x \mid 2 \le x \le \frac{5}{2}\}$
75. $\{x \mid x \le -4\}$ **76.** $\{x \mid x \le -4 \text{ or } x > \frac{17}{5}\}$ **77.** $\{-4, 4\}$ **78.** $\{x \mid -3 < x < 3\}$ **79.** $\{x \mid x \le -4 \text{ or } x \ge 4\}$
80. $\{-5, 13\}$ **81.** $\{x \mid x \le -3 \text{ or } x \ge 7\}$ **82.** $\{-\frac{1}{2}, \frac{9}{2}\}$ **83.** $\{x \mid -2 < x < 5\}$ **84.** $\{-1, 4\}$
85. $\{x \mid -14 < x < 22\}$ **86.** $\{x \mid x \le \frac{5}{2} \text{ or } x \ge \frac{11}{2}\}$ **87.** $\{x \mid x < \frac{3}{4} \text{ or } x > \frac{13}{4}\}$ **88.** $\mathbb{R}$ **89.** $(3, \infty)$
90. $[3, \infty)$ **91.** $\varnothing$ **92.** $[-17, 23]$ **93.** $(-\frac{17}{2}, \frac{27}{2}]$ **94.** $[-2, 4]$ **95.** $[-\frac{7}{2}, -\frac{1}{2}]$ **96.** $(-\infty, -11] \cup (14, \infty)$
97. $(-12, 6)$ **98.** $(-\infty, -\frac{1}{3}] \cup [3, \infty)$ **99.** $(-\infty, -12) \cup (20, \infty)$ **100.** $(\frac{29}{10}, \frac{43}{8}]$

Practice Test

1. Sixth **2.** $\frac{27}{7}$ **3.** $-\frac{34}{5}$ **4.** 68 **5.** $\frac{13}{5}$ **6.** $b = \dfrac{a - 2c}{3}$ **7.** $b_2 = \dfrac{2A - hb_1}{h}$ **8.** 23, 24 **9.** 200 mi
10. 11.56 mi **11.** 6.25 l **12.** \$7000 at 8%, \$5000 at 7% **13.** ←→₃₃ **14.** $(-12, 12)$ **15.** $\{-1, 9\}$
16. $\{x \mid x < -1 \text{ or } x > 4\}$ **17.** $\{x \mid \frac{1}{2} < x < \frac{5}{2}\}$

Cumulative Review Test

1. (a) $\{1, 2, 3, 4, 5, 6, 7, 9, 10, 12\}$ **(b)** $\{4, 6, 9, 12\}$
2. (a) Commutative Property of addition **(b)** Associative Property of multiplication **(c)** Distributive property
3. $<$ **4.** -80 **5.** -14 **6.** -29 **7.** 7 **8.** $\frac{1}{5}$ **9.** 1.15 **10.** $-\frac{56}{33}$ **11.** 10
12. Conditional linear equation is true for only one value of the variable; Identity is true for all values of the variable; Inconsistent equation is never true.
13. 3 **14.** $t = \dfrac{I - p}{pr}$ **15. (a)** ○——○_{-2 8/5} **(b)** $\{x \mid -2 < x < \frac{8}{5}\}$ **(c)** $(-2, \frac{8}{5})$ **16.** $\{1, -5\}$
17. $\{x \mid x \le -10 \text{ or } x \ge 14\}$ **18.** \$2250 **19.** 40 mph, 50 mph **20.** $\frac{2}{3} l$ of 50%, $1\frac{1}{3} l$ of 20%

CHAPTER 3

Exercise Set 3.1

1. $A(3, 1), B(-6, 0), C(2, -4), D(-2, -4), E(0, 3), F(-8, 1), G(\frac{3}{2}, -1)$ **3.**
5. 3 **7.** 9 **9.** 5

11. $\sqrt{90} \approx 9.49$ **13.** $\sqrt{74} \approx 8.60$ **15.** $\sqrt{34.33} \approx 5.86$ **17.** $\sqrt{\frac{281}{16}} \approx 4.19$ **19.** (2, 3) **21.** (0, 0)
23. $(-4, -5)$ **25.** $(-\frac{7}{2}, -5)$ **27.** $(-3.05, 9.575)$ **29.** $(\frac{9}{4}, \frac{15}{4})$ **31.** 24 **33.** $\sqrt{53} + \sqrt{116} + \sqrt{13} \approx 21.66$
35. The square of any nonzero number is positive and the square root of a positive number is positive.

Cumulative Review Exercises

36. $\frac{3}{2}$ **37.** 140 mi **38.** $\{x \mid -2 < x \le 4\}$ **39.** $\{x \mid x < -\frac{7}{3} \text{ or } x > 1\}$

Exercise Set 3.2

1. **3.** **5.** **7.**

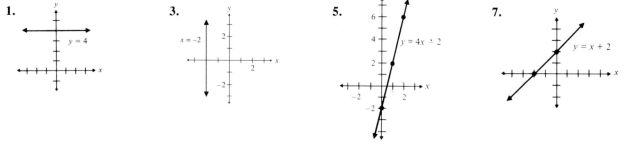

9. **11.** **13.** **15.**

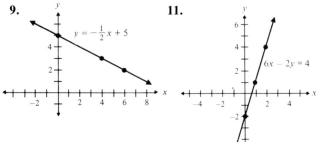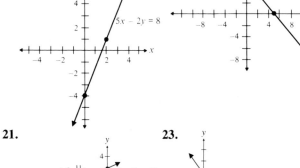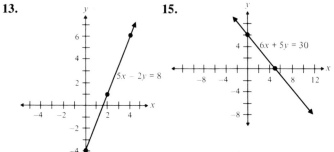

17. **19.** **21.** **23.**

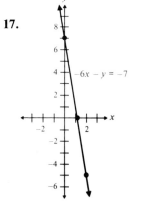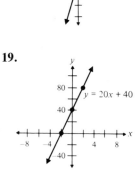

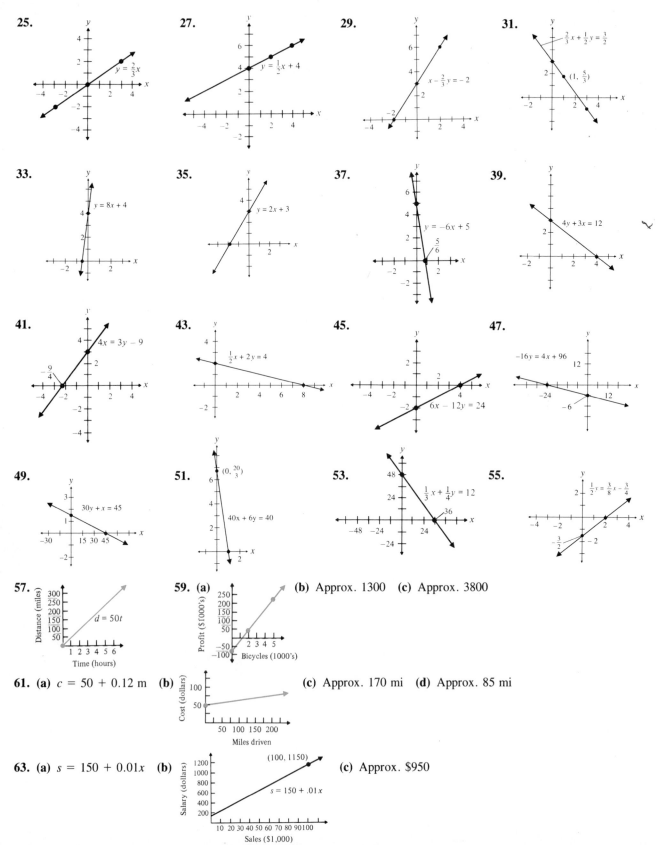

25. $y = \frac{2}{3}x$

27. $y = \frac{1}{2}x + 4$

29. $x - \frac{2}{3}y = -2$

31. $\frac{2}{3}x + \frac{1}{2}y = \frac{3}{2}$ $(1, \frac{5}{3})$

33. $y = 8x + 4$

35. $y = 2x + 3$

37. $y = -6x + 5$ $\frac{5}{6}$

39. $4y + 3x = 12$

41. $4x = 3y - 9$ $-\frac{9}{4}$

43. $\frac{1}{2}x + 2y = 4$

45. $6x - 12y = 24$

47. $-16y = 4x + 96$

49. $30y + x = 45$

51. $40x + 6y = 40$ $(0, \frac{20}{3})$

53. $\frac{1}{3}x + \frac{1}{4}y = 12$

55. $\frac{1}{2}x = \frac{3}{8}x - \frac{3}{4}$ $-\frac{3}{2}$

57. Distance (miles) vs Time (hours), $d = 50t$

59. **(a)** Profit ($1000's) vs Bicycles (1000's)   **(b)** Approx. 1300   **(c)** Approx. 3800

61. (a) $c = 50 + 0.12m$ **(b)** Cost (dollars) vs Miles driven   **(c)** Approx. 170 mi   **(d)** Approx. 85 mi

63. (a) $s = 150 + 0.01x$ **(b)** Salary (dollars) vs Sales ($1,000), $(100, 1150)$, $s = 150 + .01x$ **(c)** Approx. $950

Cumulative Review Exercises

67. $\frac{1}{2}$ **68.** $p_2 = \dfrac{E - a_1 p_1 - a_3 p_3}{a_2}$ **69. (a)**    **(b)** $(3, \infty)$ **(c)** $\{x \mid x > 3\}$ **70.** $-2, 10$

Just for Fun

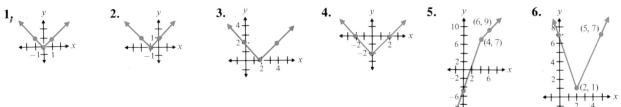

1.   **2.**   **3.**   **4.**   **5.**   **6.**

Exercise Set 3.3

1. -8 **3.** $-\frac{1}{2}$ **5.** -1 **7.** Undefined   **9.** -5 **11.** 0 **13.** $-\frac{1}{7}$ **15.** $\frac{1}{3}$ **17.** $\frac{1}{5}$ **19.** 0
21. undefined   **23.** 1 **25.** Parallel   **27.** Perpendicular   **29.** Neither   **31.** Perpendicular
33. Parallel   **35.** $a = 7$ **37.** $b = 2$ **39.** $c = -4$ **41.** $x = 6$ **43.** $x = 0$
45. Select two points on the line; then find the ratio of the change in y to the change in x.
47. The line is falling going from left to right.   **49.** Because the change in x is zero and you cannot divide by zero.

Cumulative Review Exercises

51. $7, 9, 11$ **52.** $x = a + b$ or $x = a - b$ **53.** $a - b < x < a + b$
54. $x < a - b$ or $x > a + b$

Just for Fun   **1. (a)** 1292.2 in   **(b)** 582.4 in   **(c)** 2.21875
2. Positive slope in intervals (b, c) and (d, e); slope 0 at points b, c, and d; Negative slope in intervals (a, b) and (c, d).

Exercise Set 3.4

1. $y = x + 2$ **3.** $y = -\frac{3}{2}x + 15$ **5.** $y = -4$ **7.**   **9.**   **11.**

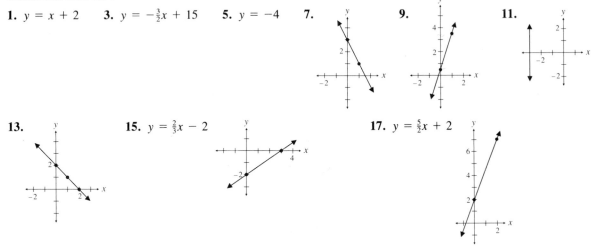

13.   **15.** $y = \frac{2}{3}x - 2$ **17.** $y = \frac{5}{2}x + 2$

19. Parallel   **21.** Parallel   **23.** Neither   **25.** Perpendicular   **27.** Perpendicular   **29.** Parallel
31. Perpendicular   **33.** $y = 4x - 5$ **35.** $y = -x + 6$ **37.** $y = -\frac{2}{3}x - \frac{8}{3}$ **39.** $y = \frac{5}{6}x + \frac{8}{3}$
41. $y = x - 3$ **43.** $y = -\frac{4}{3}x + \frac{4}{3}$ **45.** $y = 2x + 2$ **47.** $2x + 3y = 10$ **49.** $y = \frac{2}{3}x + \frac{41}{12}$
51. $x + 2y = 6$ **53.** $y = -2x - \frac{16}{3}$ **55.** $2x - 4y = 1$ **57.** $y = \frac{2}{3}x - \frac{14}{3}$ **59.** $y = -\frac{2}{3}x + 6$
61. $12x + 5y = 62$ **63. (a)** $-x + y = 2$ **(b)** $y = x + 2$ **(c)** $y - 2 = 1(x - 0)$
65. (a) If the slopes are the same the lines are parallel.   **(b)** If the product of the slopes is -1 the lines are perpendicular.

Cumulative Review Exercises

67. change the sense, or direction, of the inequality symbol.   **68.** distance $= \sqrt{162} \approx 12.73$, midpoint $\left(-\frac{1}{2}, -\frac{3}{2}\right)$

69. (a) To find the y intercept set $x = 0$ and solve for y. To find the x intercept set $y = 0$ and solve for x.
(b)   **70. (a)** $c = 15 + 0.10$ m   **(b)**

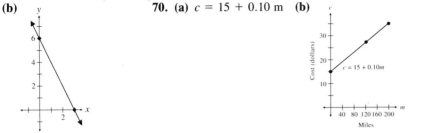

Exercise Set 3.5

1. Function, domain $\{1, 2, 3, 4, 5\}$, range $\{1, 2, 3, 4, 5\}$ **3.** Function, domain $\{1, 2, 3, 4, 5, 7\}$, range $\{-1, 0, 2, 4, 5\}$
5. Relation, domain $\{1, 2, 3, 5\}$, range $\{-4, -1, 0, 1, 2\}$
7. Function, domain $\{-2, \frac{1}{2}, 0, 2, 3, 5\}$, range $\{-3, -1, 0, \frac{2}{3}, 2, 5\}$ **9.** Relation, domain $\{1, 2, 6\}$, range $\{-3, 0, 2, 5\}$
11. Relation, domain $\{0, 1, 2\}$, range $\{-7, -1, 2, 3\}$ **13.** Relation, domain $\{x \mid -2 \le x \le 2\}$, range $\{y \mid -2 \le y \le 2\}$
15. Function, domain $\mathbb{R}$, range $\{y \mid y \ge 0\}$ **17.** Function, domain $\{-1, 0, 1, 2, 3\}$, range $\{-1, 0, 1, 2, 3\}$
19. Function, domain $\mathbb{R}$, range $\mathbb{R}$ **21.** Relation, domain $\{-2\}$, range $\mathbb{R}$
23. Function, domain $\mathbb{R}$, range $\{y \mid -5 \le y \le 5\}$ **25. (a)** 13   **(b)** 3   **27. (a)** 4   **(b)** 9   **29. (a)** -1 **(b)** 2

31. (a) -1 **(b)** -11 **33.**   **35.**   **37.**   **39.**
41.

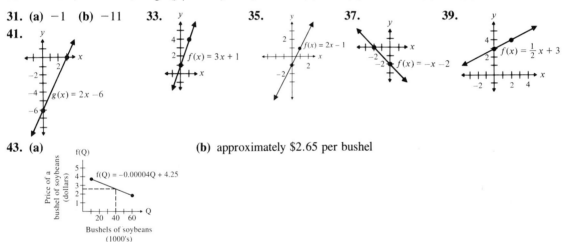

43. (a)   **(b)** approximately \$2.65 per bushel

45. Any set of ordered pairs.   **47.** No, a function has no two distinct ordered pairs with the same first coordinate.
49. If a vertical line can be drawn at any value of x that intersects the graph more than once, then each x does not have a unique y value, and the graph is not a function.
51. The set of second coordinates of the ordered pairs.   **53.** Domain: $\mathbb{R}$, Range: $\mathbb{R}$

Cumulative Review Exercise

54. 19   **55.** $-\frac{92}{5}$ **56.** 2.4   **57.** 60 mph, 75 mph

Exercise Set 3.6

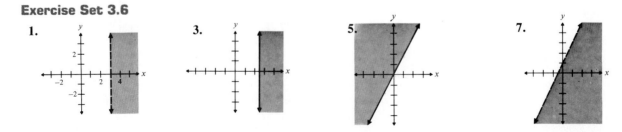

1.   **3.**   **5.**   **7.**

9.
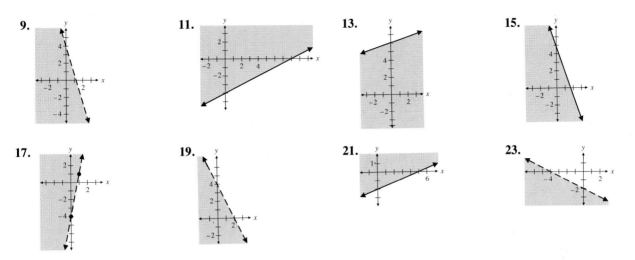

11.

13.

15.

17.

19.

21.

23.

25. Because ≥ means greater than or *equal to* and ≤ means less than or *equal to*.

Cumulative Review Exercises

27. -56 **28.** 81.176 **29.** \$15.72 **30.** $x + 2y = 2$ (other answers are possible)

Just for Fun

1.

2.

Review Exercises

1.

2. $5, (\frac{3}{2}, -2)$ **3.** $5, (4, \frac{1}{2})$ **4.** $13, (\frac{1}{2}, 3)$ **5.** $\sqrt{8} \approx 2.83, (-3, 4)$ **6.** $2, (4, 4)$

7. $13, (-3, -\frac{3}{2})$ **8.**

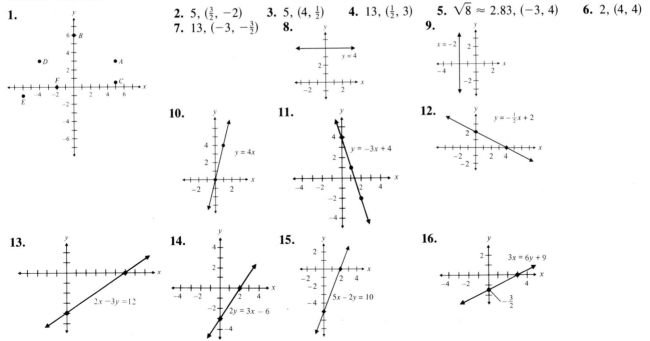

9.

10. $y = 4x$

11. $y = -3x + 4$

12. $y = -\frac{1}{2}x + 2$

13. $2x - 3y = 12$

14. $2y = 3x - 6$

15. $5x - 2y = 10$

16. $3x = 6y + 9$

17.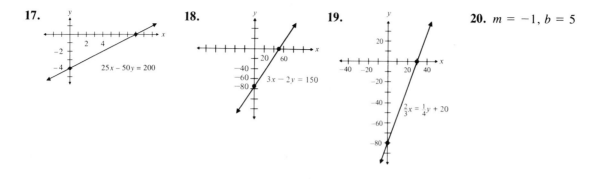

25 *x* − 50 *y* = 200

18.

3 *x* − 2 *y* = 150

19.

$\frac{2}{3}x = \frac{1}{4}y + 20$

20. $m = -1, b = 5$

21. $m = -4, b = \frac{1}{2}$ **22.** $m = -\frac{1}{2}, b = \frac{3}{2}$ **23.** $m = -\frac{3}{5}, b = \frac{12}{5}$ **24.** $m = -\frac{9}{7}, b = \frac{15}{7}$ **25.** $m = \frac{1}{2}, b = -2$
26. *m* is undefined, no *y* intercept   **27.** $m = 0, b = 6$ **28.** $m = -4, b = 0$ **29.** -7 **30.** $-\frac{1}{3}$ **31.** $\frac{1}{3}$
32. 7   **33.** Neither   **34.** Parallel   **35.** Perpendicular   **36.** Neither   **37.** $a = 3$ **38.** $y = 6$
39. $y = -37$ **40.** $x = 7$ **41.** $y = 3$ **42.** $x = 2$ **43.** $y = -\frac{1}{2}x + 2$ **44.** Parallel   **45.** Perpendicular
46. Parallel   **47.** Parallel   **48.** Perpendicular   **49.** Neither   **50.** $y = 2x - 2$ **51.** $y = -\frac{2}{3}x + 4$
52. $y = x - 1$ **53.** $y = -\frac{7}{2}x - 4$ **54.** $y = 3x + 20$ **55.** $y = \frac{2}{5}x - \frac{18}{5}$ **56.** $y = -\frac{5}{3}x - 4$
57. $y = -\frac{1}{2}x + 4$

58. (a)

Profit ($1000)

$p = .1x - 5000$

Sales (1000's)

(b) Approx. 50,000 bagels   **(c)** Approx. 250,000 bagels   **59.**

Interest (dollars)

$I = 12{,}000r$

Rate

60. Domain: $\{-2, 0, 3, 6\}$, range: $\{-1, 4, 5, 9\}$ **61.** Domain $\{\frac{1}{2}, 2, 4, 5\}$, range: $\{-6, -1, 2, 3\}$
62. Domain $\{x \mid -1 \le x \le 1\}$, range $\{y \mid -1 \le y \le 1\}$
63. Domain $\{x \mid -2 \le x \le 2\}$, range $\{y \mid -1 \le y \le 1\}$ **64.** Domain $\mathbb{R}$, range $\{y \mid y \le 0\}$ **65.** Domain $\mathbb{R}$, range $\mathbb{R}$
66. Function   **67.** Function   **68.** Function   **69.** Not a function   **70.** Function   **71.** Function   **72.** Function
73. Function   **74.** Not a function   **75.** Not a function   **76. (a)** 10   **(b)** -5 **77. (a)** 7   **(b)** 4
78. (a) 2   **(b)** $\frac{5}{2}$ **79. (a)** 6   **(b)** 16   **80.**

$f(x) = 2x + 4$

81.

$f(x) = 3x - 5$

82.

$f(x) = -4x + 2$

83.

$f(x) = \frac{1}{2}x + 2$

84. (a)

f(n)

Profit or loss ($ 1000's)

$f(n) = 24n - 200{,}000$

Copies (1000's)

(b) 8,300   **(c)** 12,500

85.   **86.**   **87.**   **88.**   **89.**

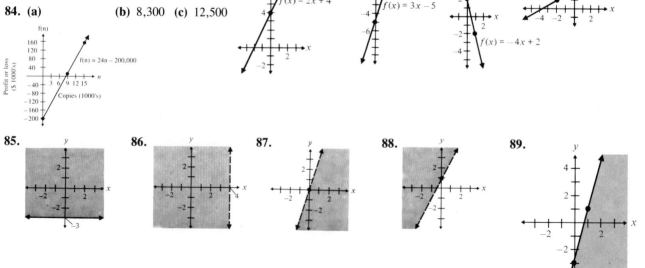

90. 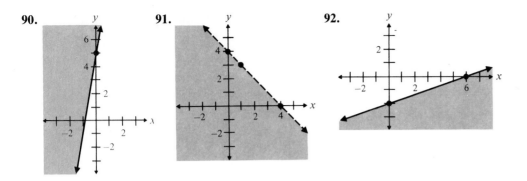 **91.** **92.**

Practice Test

1. 5, $\left(-\frac{1}{2}, 1\right)$ **2.** $m = \frac{4}{9}, b = -\frac{5}{3}$ **3.** $y = 3x - 3$ **4.** $y = 4x + 7$ **5.** $y = -\frac{3}{7}x + \frac{2}{7}$ **6.** $y = \frac{3}{2}x + \frac{11}{2}$

7. **8.** **9. (a)** **(b)** ≈ 1900 **(c)** ≈ 320

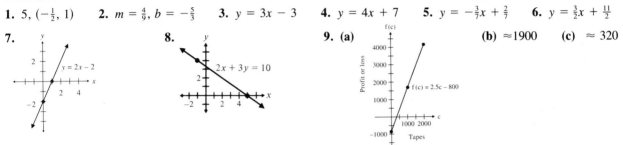

10. Domain $\left\{\frac{1}{2}, 2, 4, 6\right\}$, range $\{-3, 0, 2, 9\}$ **11.** a, c **12. (a)** 1 **(b)** -17

13. **14.** **15.**

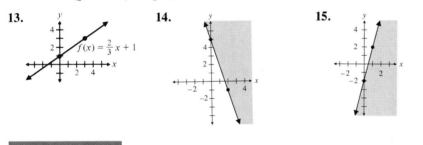

CHAPTER 4

Exercise Set 4.1

1. c **3.** None **5.** b, c **7.** c **9.** None **11.** $y = -\frac{1}{2}x + \frac{5}{2}$; consistent—one
$y = \frac{1}{2}x - \frac{1}{2}$
13. $y = \frac{2}{3}x + 1$; inconsistent—none **15.** $y = \frac{2}{3}x - \frac{4}{3}$; consistent—one
$y = \frac{2}{3}x - 2$ $\qquad\qquad y = \frac{3}{2}x + 1$
17. $y = \frac{2}{3}x - \frac{4}{3}$; dependent—infinite number **19.** $y = \frac{3}{2}x + \frac{1}{2}$; inconsistent—no solution
$y = \frac{2}{3}x - \frac{4}{3}$ $\qquad\qquad\qquad y = \frac{3}{2}x + \frac{1}{4}$

21. **23.** **25.** **27.** **29.**

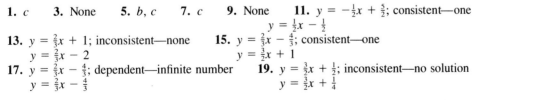
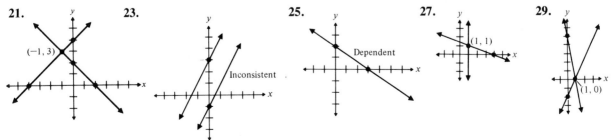

31. (5, 2)   **33.** (3, 3)   **35.** No solution   **37.** $(\frac{1}{2}, -\frac{39}{2})$ **39.** Infinite number of solutions   **41.** (−1, 2)
43. (5, −3)   **45.** (−3, −4)   **47.** $(-\frac{13}{2}, -\frac{23}{6})$ **49.** (8, 6)   **51.** (1, −3)   **53.** (−3, 2)   **55.** $(2, \frac{9}{2})$
57. No solution   **59.** (4, −2)   **61.** (4, −1)   **63.** Infinite number of solutions   **65.** $(\frac{59}{7}, \frac{60}{7})$ **67.** (2, −1)
69. $(-\frac{15}{43}, -\frac{27}{43})$ **71.** $(\frac{29}{22}, -\frac{5}{11})$ **73.** (3, 2)   **75.** (4, 0)   **77.** (14, 66)   **79.** $(\frac{192}{25}, \frac{144}{25})$ **81.** (10, 4)
83. (1, 2)   **85.** Write both equations in slope intercept form and compare their slopes and y intercepts.
87. At some point during the solution process, both sides of the equation will be the same.

Cumulative Review Exercises

88. Rational numbers can be expressed as the quotient of two integers; denominator not 0, irrational numbers cannot.
89. (a) yes **(b)** yes   **90.** $\mathbb{R}$ **91.** 520   **92.** No, no two ordered pairs can have the same first element and a different second element (each x must have a unique y).

Just for Fun   **1.** (8, −1)   **2.** $(-\frac{105}{41}, \frac{447}{82})$ **3.** (−1, 2)   **4.** (−3, 1)   **5.** $(\frac{1}{a}, 5)$ **6.** $(\frac{1}{a}, \frac{1}{b})$

Exercise Set 4.2

1. (1, 2, 5)   **3.** $(-7, -\frac{35}{4}, -3)$ **5.** (0, 3, 6)   **7.** (−1, 1, 3)   **9.** (−1, 3, 2)   **11.** (2, −1, 3)
13. $(\frac{2}{3}, -\frac{1}{3}, 1)$ **15.** (5, −3, −2)   **17.** $(-\frac{11}{17}, \frac{7}{34}, -\frac{49}{17})$ **19.** (4, 6, 8)   **21.** $(\frac{2}{3}, \frac{23}{15}, \frac{37}{15})$
23. No point is common to all 3 planes. Therefore the system is inconsistent.
25. A straight line is common to all 3 planes. Therefore there are an infinite number of points common to all 3 planes and the system is dependent.   **27.** dependent   **29.** inconsistent

Cumulative Review Exercises

31. (a) $\frac{5}{12}$ hr (or 25 minutes) after Cameron starts **(b)** 1.25 miles
32. $\{x \mid x < -\frac{3}{2}$ or $x > \frac{27}{2}\}$ **33.** $\{x \mid -\frac{8}{3} < x < \frac{16}{3}\}$ **34.**

Just for Fun   **1.** (−1, 2, 1, 5)

Exercise Set 4.3

1. $x + y = 73$ **3.** $x - y = 25$ **5.** $x + y = 180$ **7.** $x + y = 50$
   $y = 3x - 15$ $x = 3y - 1$ $y = 3x - 28$ $y = 3x + 2$
   22, 51   13, 38   52°, 128°   12 ft, 38 ft
9. $2x + 100y = 60$ **11.** $x + y = 1000$ **13.** $x + y = 100$
   $3x + 400y = 115$ $0.05x + 0.25y = 0.10(1000)$ $0.05x + 0.00y = 0.035(100)$
   \$25 day and 10¢ mi   750 ml of 5%, 250 ml of 25%   70 gal of 5%, 30 gal skim
15. $x + y = 16$ **17.** $4x + 4y = 420$ **19.** $x = 3y$
   $0.36x + 0.105y = 0.20(16)$ $y = x + 5$ $35x + 20y = 6250$
   5.96 oz cream, 10.04 oz half and half   50 mph, 55 mph   150 shares Apple; 50 shares Loews
21. $x + y = 30$ **23.** $x + y = 25$ **25.** $x + y = 12,400$
   $6x + 5.40y = 30(5.80)$ $0.10x + 0.25y = 3.55$ $26,200 - x = 22,450 - y$
   20 lbs almonds, 10 lbs walnuts   18 dimes, 7 quarters   \$8075 for Mr., \$4325 for Mrs.
27. $7 = -a + b$ **29.** $2x + \frac{1}{2}y = 35$ **31.** $0.10A + 0.20B = 20$
   $4 = \frac{1}{2}a + b$ $\frac{1}{2}x + \frac{1}{3}y = 15$ $0.06A + 0.02B = 6$
   $a = -2, b = 5$ 10, 30   80 gm A, 60 gm B

33. $\left. \begin{array}{l} x + y = 300 \\ 0.70x + 0.40y = 0.6(300) \end{array} \right\}$ or $\left\{ \begin{array}{l} 0.70x + 0.40y = 0.60(300) \\ 0.30x + 0.60y = 0.40(300) \end{array} \right.$ **35.** $x = 2y$
        $z = y - 6$
        $z = x - 21$
   200 gm 1st alloy, 100 gm 2nd alloy   Balcony \$9, floor seats further back \$15, up front seats \$30

37. $x + y + z = 180$ **39.** $6 = 4a + 2b + c$ **41.** $x + y + z = 8$
   $x = \frac{2}{3}y$ $17 = 9a + 3b + c$ $z = x - 2$
   $z = 3y - 30$ $-3 = a - b + c$ $0.10x + 0.12y + 0.20z = 0.13(8)$
   30°, 45°, 105° $a = 2, b = 1, c = -4$ $4l$ of 10%, $2l$ of 12%, $2l$ of 20%

43. $5x + 4y + 7z = 154$
$3x + 2y + 5z = 94$
$2x + 2y + 4z = 76$
10 children's, 12 standard, 8 executive

45. $I_A = \frac{27}{38}$, $I_B = -\frac{15}{38}$, $I_c = -\frac{6}{19}$

Cumulative Review Exercises

47. $-\frac{35}{8}$ **48.** $\sqrt{32} \approx 5.66$, $(4, -6)$ **49.** $y = x - 10$
50. Use the vertical line test. If a vertical line cannot be drawn to intersect the graph in more than one point, the graph is a function.

Just for Fun   1. (a) $\frac{2400}{17} \approx 141.2$ cm   **(b)** Pull away

Exercise Set 4.4

1. $(3, 1)$ **3.** $(3, 2)$ **5.** $\left(\frac{60}{17}, -\frac{11}{17}\right)$ **7.** $\left(\frac{1}{2}, -4\right)$ **9.** $(2, -3)$ **11.** $(2, 5)$ **13.** $(-1, 1, 3)$
15. $\left(-\frac{1}{2}, \frac{1}{2}, -\frac{3}{2}\right)$ **17.** $\left(\frac{165}{14}, -\frac{153}{14}, -\frac{6}{7}\right)$ **19.** $(-1, 0, 2)$ **21.** Dependent, infinite number of solutions
23. $(1, -1, 2)$
25. A square array of numbers enclosed between two vertical bars. A determinant that has two rows and two columns of elements. A determinant that has three rows and three columns of elements.   **27.** It will have the opposite sign.

Cumulative Review Exercises

28. $\left(-\infty, \frac{14}{11}\right)$ **29.**   **30.**   **31.**

Exercise Set 4.5

1.   **3.**   **5.**   **7.**   **9.**

11.   **13.**   **15.**   **17.**   **19.**

21.   **23.**   **25.**   **27.**   **29.**

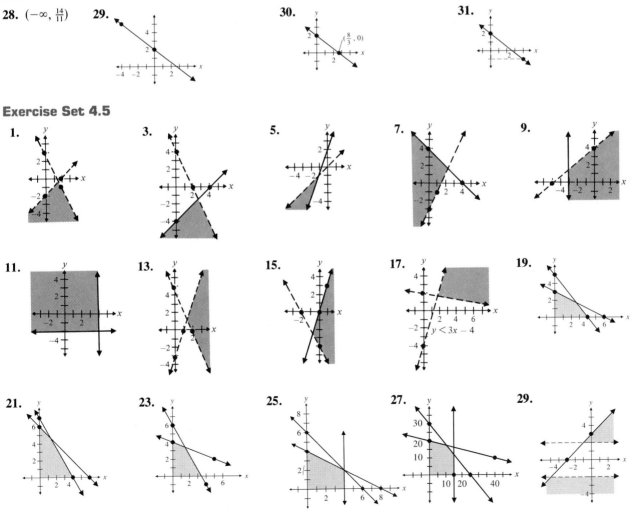

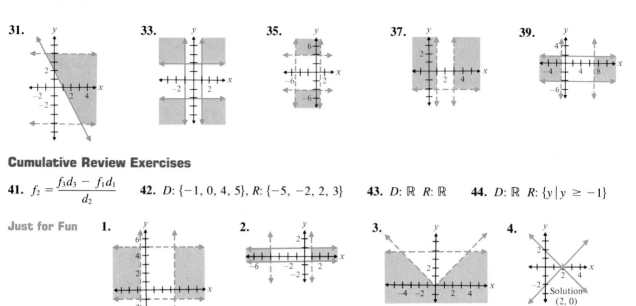

31. **33.** **35.** **37.** **39.**

Cumulative Review Exercises

41. $f_2 = \dfrac{f_3 d_3 - f_1 d_1}{d_2}$ **42.** $D: \{-1, 0, 4, 5\}$, $R: \{-5, -2, 2, 3\}$ **43.** $D: \mathbb{R}$ $R: \mathbb{R}$ **44.** $D: \mathbb{R}$ $R: \{y \mid y \geq -1\}$

Just for Fun   **1.**   **2.**   **3.**   **4.**

Solution (2, 0)

Review Exercises

1. $y = -\frac{1}{2}x + 4$
$y = -\frac{1}{2}x + 2$
inconsistent, none

2. $y = -3x - 6$
$y = -\frac{2}{3}x + \frac{8}{3}$
consistent, one

3. $y = \frac{1}{2}x + 4$
$y = -\frac{1}{2}x + 4$
consistent, one

4. $y = \frac{3}{2}x + 2$
$y = \frac{2}{3}x - \frac{4}{3}$
consistent, one

5.

$y = 2x + 5$
$y = x + 3$
$(-2, 1)$

6.

$y = 3$
$(-2, 3)$
$x = -2$

7.

$(0, 4)$
$2x - y = -4$
$2x + 2y = 8$

8.

$\frac{1}{2}x - \frac{1}{2}y = \frac{3}{2}$
Dependent
$2y = 2x - 6$

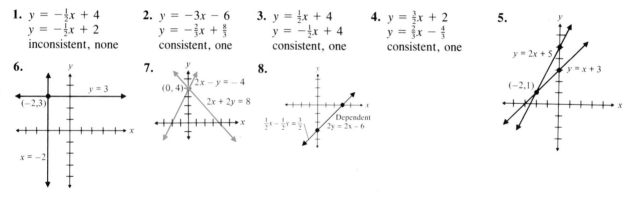

9. $(3, 7)$ **10.** $(2, 3)$ **11.** $(5, 2)$ **12.** $(-3, 2)$ **13.** $(2, 1)$ **14.** $(4, 2)$ **15.** $(5, 2)$ **16.** $(-18, 6)$
17. $(8, -2)$ **18.** $(1, -2)$ **19.** $(26, -16)$ **20.** $(5, -2)$ **21.** $\left(\frac{32}{13}, \frac{8}{13}\right)$ **22.** $\left(-1, \frac{13}{3}\right)$ **23.** $(1, 2)$
24. $\left(\frac{7}{5}, \frac{13}{5}\right)$ **25.** $(6, -2)$ **26.** $\left(-\frac{78}{7}, -\frac{48}{7}\right)$ **27.** $\left(2, 5, \frac{34}{5}\right)$ **28.** $\left(5, -\frac{15}{4}, -2\right)$ **29.** $(1, 2, -1)$ **30.** $(3, 1, 2)$
31. $\left(\frac{8}{3}, \frac{2}{3}, 3\right)$ **32.** $(0, 2, -3)$

33. $x + y = 48$
$y = 2x - 3$
17, 31

34. $x - y = 18$
$x = 4y$
6, 24

35. $x + y = 600$
$x - y = 530$
565 mph plane, 35 mph wind

36. $x + y = 6$
$0.3x + 0.5y = 0.4(6)$
3 l of each

37. $x + y = 650$
$7.50x + 5.50y = 4395$
410 adults, 240 children

38. $x + y + z = 17$
$x = y + z + 1$
$y = 3z$
$(9, 6, 2)$

39. $x + y + z = 40{,}000$
$y = x - 5{,}000$
$0.10x + 0.08y + 0.06z = 3500$
$20,000 at 10%, $15,000 at 8%
$5,000 at 6%

40. $(4, -1)$

41. $(1, -1)$ **42.** $(-1, 2)$ **43.** $(4, 1, 3)$ **44.** $(-1, 5, -2)$ **45.** No solution
46.   **47.**   **48.**   **49.**

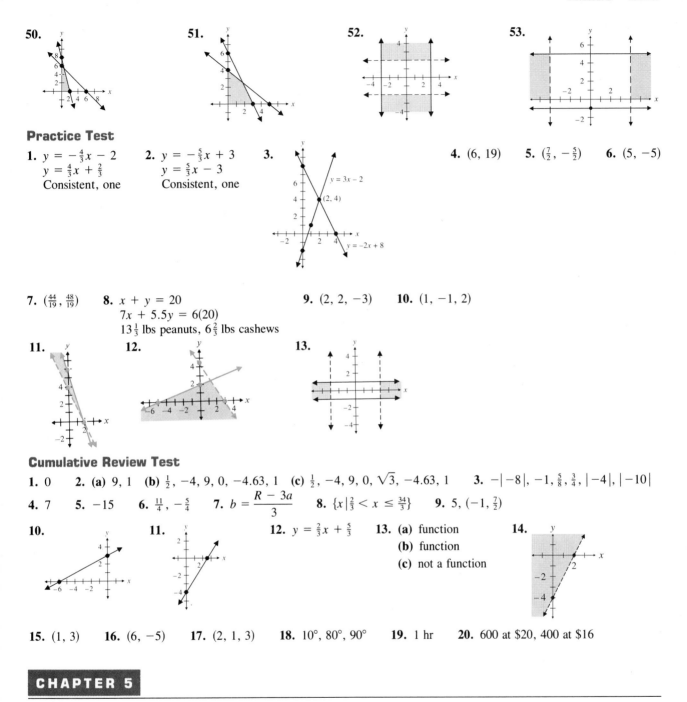

50.

51.

52.

53.

Practice Test

1. $y = -\frac{4}{3}x - 2$
$y = \frac{4}{3}x + \frac{2}{3}$
Consistent, one

2. $y = -\frac{5}{3}x + 3$
$y = \frac{5}{3}x - 3$
Consistent, one

3.

$y = 3x - 2$
$(2, 4)$
$y = -2x + 8$

4. $(6, 19)$ **5.** $(\frac{7}{2}, -\frac{5}{2})$ **6.** $(5, -5)$

7. $(\frac{44}{19}, \frac{48}{19})$ **8.** $x + y = 20$
$7x + 5.5y = 6(20)$
$13\frac{1}{3}$ lbs peanuts, $6\frac{2}{3}$ lbs cashews

9. $(2, 2, -3)$ **10.** $(1, -1, 2)$

11.

12.

13.

Cumulative Review Test

1. 0 **2. (a)** $9, 1$ **(b)** $\frac{1}{2}, -4, 9, 0, -4.63, 1$ **(c)** $\frac{1}{2}, -4, 9, 0, \sqrt{3}, -4.63, 1$ **3.** $-|-8|, -1, \frac{5}{8}, \frac{3}{4}, |-4|, |-10|$

4. 7 **5.** -15 **6.** $\frac{11}{4}, -\frac{5}{4}$ **7.** $b = \dfrac{R - 3a}{3}$ **8.** $\{x \mid \frac{2}{3} < x \le \frac{34}{3}\}$ **9.** $5, (-1, \frac{7}{2})$

10.

11.

12. $y = \frac{2}{3}x + \frac{5}{3}$ **13. (a)** function
(b) function
(c) not a function

14.

15. $(1, 3)$ **16.** $(6, -5)$ **17.** $(2, 1, 3)$ **18.** $10°, 80°, 90°$ **19.** 1 hr **20.** 600 at \$20, 400 at \$16

CHAPTER 5

Exercise Set 5.1

1. $\frac{1}{9}$ **3.** 1 **5.** $\dfrac{5}{y^3}$ **7.** x^4 **9.** $2xy^3$ **11.** $\dfrac{5z}{2x^2y^3}$ **13.** $\dfrac{5z^4}{x^2y^3}$ **15.** $\dfrac{1}{4xy}$ **17.** 1 **19.** 4 **21.** -1

23. -3 **25.** -1 **27.** 7 **29.** $\frac{1}{6}$ **31.** x^3 **33.** x^4 **35.** 9 **37.** 625 **39.** $\dfrac{1}{x^{11}}$ **41.** x^3 **43.** $3y^5$

45. $\dfrac{4}{x^2}$ **47.** $\dfrac{12}{x^7}$ **49.** $3y$ **51.** $\dfrac{12y^3}{x^7}$ **53.** $-10x^7z^5$ **55.** $\dfrac{8x^7y^2}{z^3}$ **57.** $\dfrac{3x^2}{y^6}$ **59.** $\dfrac{3x^3}{y^5}$ **61.** $3x^5$

63. $\dfrac{-x^2}{y^9}$ **65.** $\dfrac{1}{2x^6y^3}$ **67.** $6x^6yz^3$ **69.** $\dfrac{2y^5}{3x}$ **71.** $x^6y^6z^3$ **73.** x^{7a+4} **75.** w^{7b+1} **77.** x^{w+7} **79.** x^{p+3}

81. 3.7×10^3 **83.** 9×10^2 **85.** 4.7×10^{-2} **87.** 1.9×10^4 **89.** 1.86×10^{-6} **91.** 9.14×10^{-6}
93. 5200 **95.** 40,000,000 **97.** 0.0000213 **99.** 0.312 **101.** 9,000,000 **103.** 535 **105.** 150,000,000
107. 0.000064 **109.** 0.021 **111.** 20 **113.** 4,200,000,000,000 **115.** 4.5×10^{-7} **117.** 2.0×10^3
119. 2.13×10^{-7} **121.** 30,000 hr **123.** $\frac{1}{5}$

Cumulative Review Exercises

125. -5 **126.** -0.75 **127.** 20 **128.** $y = \frac{1}{2}x - 2$

Just for Fun **1. (a)** 0 gives $10^0 = 1$, 1 gives $10^1 = 10$, 2 gives $10^2 = 100$, 3 gives $10^3 = 1000$, 4 gives $10^4 = 10,000$,
5 gives $10^5 = 100,000$, 6 gives $10^6 = 1,000,000$, 7 gives $10^7 = 10,000,000$, 8 gives $10^8 = 100,000,000$, 9 gives $10^9 =$
1,000,000,000, 10 gives $10^{10} = 10,000,000,000$ **(b)** 10,000 **(c)** $10^{1.4} \approx 25.1$
2. (a) about 5.87×10^{12} miles **(b)** 500 sec or $8\frac{1}{3}$ min. **(c)** 6.72×10^{11} sec or 21,309 years

Exercise Set 5.2

1. x^4 **3.** 81 **5.** $\frac{1}{64}$ **7.** 1 **9.** x^2 **11.** $-\frac{1}{x^3}$ **13.** $\frac{3}{x^8}$ **15.** $27x^6$ **17.** 8 **19.** $81x^8y^4$ **21.** $\frac{16x^4}{y^4}$

23. $\frac{1}{8x^9y^3}$ **25.** $-\frac{x^{12}}{64y^{15}}$ **27.** $\frac{36x^2}{y^4}$ **29.** $8x^6y^{15}$ **31.** $\frac{125y^9}{x^3}$ **33.** 4 **35.** $64x^9y^3$ **37.** $\frac{z^3}{8x^3y^3}$

39. $72x^{11}y^7$ **41.** $96x^8y^{13}$ **43.** $\frac{81x^2}{y^{14}}$ **45.** $\frac{8z^{13}}{27x^2y^{11}}$ **47.** $\frac{2}{27x^{11}y^{36}}$ **49.** $\frac{x^9y^{32}}{z^7}$ **51.** $\frac{-2}{9x^{10}y^3z^8}$ **53.** $\frac{x^7y^6}{576}$

55. $\frac{3y^3}{10}$ **57.** x^{5m+4} **59.** b^{5y^2+3y} **61.** m^{3-7y} **63.** x^{-3}, y^1 **65.** x^{-4}, y^{33}, z^{-22}

Cumulative Review Exercises

67. $(1, 3)$ **68.** $(2, 4)$ **69.** **70.** 15 l of 40%, 5 l of 60%

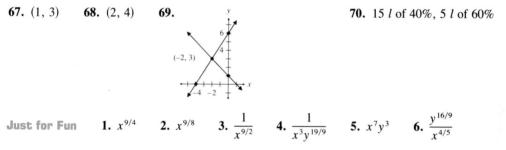

Just for Fun **1.** $x^{9/4}$ **2.** $x^{9/8}$ **3.** $\frac{1}{x^{9/2}}$ **4.** $\frac{1}{x^3y^{19/9}}$ **5.** x^7y^3 **6.** $\frac{y^{16/9}}{x^{4/5}}$

Exercise Set 5.3

1. Monomial **3.** Monomial **5.** Binomial **7.** Trinomial **9.** Not polynomial **11.** Binomial
13. $-x^2 - 4x - 8$, second **15.** $10x^2 + 3xy + 6y^2$, second **17.** $2x^3 + x^2 - 3x + 5$, third
19. In descending order, fourth **21.** $-2x^3 + 3x^2y + 5xy^2 - 6$, third **23.** $7x - 2$ **25.** $x - 6$ **27.** $-7x + 4$
29. $x^2 - 8x - 2$ **31.** $3x - z + 3$ **33.** $2x^2 + 8x + 5$ **35.** $8y^2 - 5y - 3$ **37.** $-7x^2 + x - 12$
39. $5x^2 - 2x + 7$ **41.** $-x^3 + 3x^2y + 4xy^2$ **43.** $5xy^2 - 8$ **45.** $3x^2 - 2xy$ **47.** $4x^3 - 8x^2 - x - 4$
49. $7x^2y - 3x + 2$ **51.** $9x^3 - x^2 - 9$ **53.** $x^2 + x + 16$ **55.** $-x + 11$ **57.** $-3x^2 + 2x - 12$
59. $-3x^2 - 4x + 14$ **61.** $7x^2 + 7x - 13$ **63.** $15y^2 - 6y + 4$ **65.** $4x^2 + 7xy + 3y$
67. $x^3 + 4x^2 - 6x - 6$ **69.** $4x^2 + 7x - 4y^2 - 3y$ **71.** $-7x^2y + 6xy^2$
73. The signs of all the terms in the polynomial being subtracted will change.

Cumulative Review Exercises

74. $-\frac{15}{72}$ **75.** $\{x \mid -\frac{4}{3} \le x < \frac{14}{3}\}$
76. A function is a relation in which no two ordered pairs have the same first coordinate and a different second coordinate.
77. $(2, -1, 6)$

Exercise Set 5.4

1. $24x^2y^5$ **3.** $\frac{1}{9}x^7y^8z^2$ **5.** $-2x^3 + 4x^2 - 10x$ **7.** $-20x^4 + 30x^3 - 20x^2$ **9.** $6x^6y^3 - 9x^3y^4 - 12x^2y$
11. $2xyz + \frac{8}{3}y^2z - 6y^3z$ **13.** $x^2 + x - 20$ **15.** $3x^2 - 8x - 3$ **17.** $x^2 - y^2$ **19.** $4x^3 - 6x^2 - 6x + 9$

21. $-2x^3 + 8x^2 - 3x + 12$ **23.** $x^2 + \frac{23}{6}xy - \frac{2}{3}y^2$ **25.** $-12x^3 + 8x^2y^2 + 9xy - 6y^3$ **27.** $x^3 - 7x^2 + 14x - 8$
29. $-2y^3 + 9y^2 - 17y + 12$ **31.** $-14x^3 - 22x^2 + 19x - 3$ **33.** $x^3y - x^2y^2 - 6xy^3$
35. $2a^3 - 7a^2b + 5ab^2 - 6b^3$ **37.** $2x^5 - 7x^4 + 20x^3 - 35x^2 + 38x - 24$ **39.** $27x^3 - 27x^2 + 9x - 1$
41. $x^2 - 16$ **43.** $4x^2 - 1$ **45.** $4x^2 - 12xy + 9y^2$ **47.** $4x^2 + 20xy + 25y^2$ **49.** $4y^4 - 20y^2w + 25w^2$
51. $25m^4 - 4n^2$ **53.** $y^2 + 8y - 4xy + 16 - 16x + 4x^2$ **55.** $16 - 8x + 24y + x^2 - 6xy + 9y^2$
57. $x^2 + 2xy + y^2 + 8x + 8y + 16$ **59.** $x^2 - 6xy + 9y^2 - 10x + 30y + 25$ **61.** $-24r^6s^{13}$
63. $3x^3 + 9x^2 - 3x$ **65.** $\frac{1}{6}x^2 - \frac{8}{15}xy + \frac{2}{5}y^2$ **67.** $6y^2 - y - 12$ **69.** $\frac{2}{5}x^3y^5 - \frac{1}{15}x^3y^2 - \frac{9}{5}x^2y$ **71.** $4x^2 - \frac{9}{16}$
73. $16x^2 - 40xy + 25y^2$ **75.** $2x^3 + 10x^2 + 9x - 9$ **77.** $5x^3 - x^2 + 16x + 16$ **79.** $6x^3 - x^2y - 16xy^2 + 6y^3$
81. $\frac{1}{4}r^2 + \frac{1}{4}rs + \frac{1}{16}s^2$ **83.** $\frac{2}{5}x^3y^7 - \frac{1}{6}x^6y^5 + \frac{4}{3}x^3y^7z^5$ **85.** $w^2 - 9x^2 - 24x - 16$ **87.** $a^3 + b^3$
89. $a^3 + 8b^3$ **91.** $x^3 + 9x^2 + 27x + 27$ **93.** $9m^2 + 12m + 4 - n^2$ **95.** $25x^2 + 10x + 1 - 36y^2$

Cumulative Review Exercises

97. $-\frac{31}{5}$ **98.** $0.05x + 0.06(10,000 - x) = 560$; $6000 at 6%, $4000 at 5%
99. $x + y = 10,000$; $6000 at 6%, $4000 at 5% **100.** $3x - 4y = 8$
$\quad 0.05x + 0.06y = 560$

Just for Fun **1.** $y^2 - 2y - 2xy + 2x + x^2 + 1$ **2.** $x^4 - 12x^3y + 54x^2y^2 - 108xy^3 + 81y^4$

Exercise Set 5.5

1. $3x + 4$ **3.** $2x + 1$ **5.** $3x^2 - x - 2$ **7.** $x^3 - \frac{3}{2}x^2 + 3x$ **9.** $2x^2 - 4xy + \frac{3}{2}y^2$ **11.** $\frac{3x}{y} - 6x^2 + \frac{9y}{2x}$

13. $x + 3$ **15.** $2x + 3$ **17.** $3x + 2$ **19.** $2x + 3$ **21.** $x^2 + 2x + 3 + \frac{1}{x + 1}$

23. $3x^2 - 3x + 1 + \frac{2}{3x + 2}$ **25.** $2x^2 + x - 2 - \frac{2}{2x - 1}$ **27.** $2x^3 - 6x + 4$ **29.** $3x + \frac{3}{2} + \frac{6}{x}$ **31.** $2x + 5$

33. $2x + 1 + \frac{1}{2x} + \frac{3}{2x^2}$ **35.** $\frac{-x^2y^2}{2} + y - \frac{3}{5x}$ **37.** $3x^2 + 2x + 1 + \frac{5}{3x - 2}$ **39.** $\frac{z}{2} + z^2 - \frac{3}{2}x^2y^4z^7$

41. $2x^2 - 6x + 3$ **43.** $x^3 + x^2 - 6$

Cumulative Review Exercises

45. (a) $ax + by = c$ (b) $y = mx + b$ (c) $y - y_1 = m(x - x_1)$
46. Every function is a relation, but not every relation is a function. A relation is any set of ordered pairs. A function is a set of ordered pairs no two of which have the same first coordinate.
47. $\frac{2}{21}$ **48.** 400 bulk, 150 first class

Just for Fun **1.** $2x^2 + 3xy - y^2$ **2.** $x^2 - xy + y^2$ **3.** $x^2 + \frac{2}{3}x + \frac{4}{9} - \frac{37}{9(3x - 2)}$

Exercise Set 5.6

1. $x + 3$ **3.** $x - 1$ **5.** $x + 8 + \frac{12}{x - 3}$ **7.** $3x + 5 + \frac{10}{x - 4}$ **9.** $4x^2 + x + 3 + \frac{3}{x - 1}$

11. $3x^2 - 2x + 2 + \frac{6}{x + 3}$ **13.** $5x^2 - 11x + 14 - \frac{20}{x + 1}$ **15.** $x^3 - 4x^2 + 16x - 64 + \frac{272}{x + 4}$

17. $y^4 - \frac{10}{y + 1}$ **19.** $3x^2 + 3x - 3$ **21.** $2x^3 + 2x - 2$

Cumulative Review Exercises

23. **24.** $\sqrt{34} + \sqrt{50} + \sqrt{104}$ (or ≈ 23.1) **25.** **26.** $\frac{x^9}{18y^{12}}$

Just for Fun   **1.** $0.2x^2 - 3.92x - 1.248 - \dfrac{1.1392}{x - 0.4}$

2. (a) $3x^2 - 2x + 5 - \dfrac{13}{3x + 5}$ **(b)** Because we are expressing the remainder in terms of $3x + 5$ rather than $x + \frac{5}{3}$, the denominator of the remainder is altered rather than the numerator.

Exercise Set 5.7

1. (a) 2   **(b)** -10 **(c)** $3a - 1$ **3. (a)** 11   **(b)** -1 **(c)** $2a + 2b + 3$ **5. (a)** 4   **(b)** $\frac{11}{2}$ **(c)** $\frac{11}{2}$ **7. (a)** 2
(b) 17   **(c)** $\frac{79}{27}$ **9. (a)** -35 **(b)** -1.89 **(c)** $-c^2 + 4c - 3$ **11. (a)** 23   **(b)** $\frac{107}{16}$ **(c)** $3a^2 + 16a + 28$
13. (a) 5   **(b)** $-c^2 - 2c + 5$ **(c)** $-c^2 - 6c - 3$ **15. (a)** 40   **(b)** $2x^2 + 3x + 5$ **(c)** $8x^2 - 6x + 5$
17. (a) $h^2 + 3h - 4$ **(b)** $h^2 + 11h + 24$ **(c)** $a^2 + 2ah + h^2 + 3a + 3h - 4$ **19. (a)** $2h^2 - 3h + 1$
(b) $2x^2 + 9x + 10$ **(c)** $2x^2 + 4xh + 2h^2 - 3x - 3h + 1$ **21. (a)** 110   **(b)** 240
23. (a) $v = 19.65$ m/s, h $= 60.3$ m   **(b)** $v = 8.45$ m/s, h $= 11.125$ m   **25. (a)** 34 m   **(b)** 10.75 m   **27. (a)** 5.6 m
(b) 8.988 cm   **29. (a)** 91   **(b)** 204   **31. (a)** Speed 19 ft/sec, height 46.8 ft
(b) Speed 40.6 ft/sec, height 0 feet (at 5 sec the last car is at the bottom of the drop).
33. No, polynomial functions must have exponents that are whole numbers.

Cumulative Review Exercises

35. less than or equal to 65.6 thousand   **36.** An illustration of the set of points that satisfy an equation.
37. The ratio of the vertical change to the horizontal change between any two points on a line.   **38.** $(-2, 2, 3)$

Just for Fun   **1.** $x^3 + 7x^2 + 21x + 30$ **2. (a)** $2x^2 + 4xh + 2h^2 + 3x + 3h - 4$ **(b)** $4xh + 2h^2 + 3h$
(c) $4x + 2h + 3$ **3. (a)** $A = d^2 - (\pi d^2/4)$ **(b)** 3.44 sq ft   **(c)** 7.74 sq ft

Exercise Set 5.8

1. $x = -1$, $(-1, -8)$, up   **3.** $x = 2$, $(2, -2)$, down   **5.** $x = 1$, $(1, 11)$, down   **7.** $x = -1$, $(-1, -8)$, down
9. $x = \frac{1}{2}$, $(\frac{1}{2}, \frac{7}{4})$, up   **11.** $x = -\frac{3}{2}$, $(-\frac{3}{2}, -14)$, up
13. Domain: $\mathbb{R}$ **15.** Domain: $\mathbb{R}$ **17.** Domain: $\mathbb{R}$ **19.** Domain: $\mathbb{R}$
    Range: $\{y \mid y \geq -1\}$ Range: $\{y \mid y \leq 3\}$ Range: $\{y \mid y \geq -16\}$ Range: $\{y \mid y \leq -1\}$

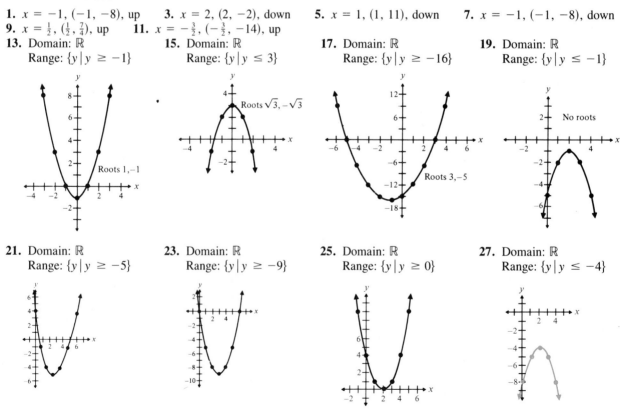

21. Domain: $\mathbb{R}$ **23.** Domain: $\mathbb{R}$ **25.** Domain: $\mathbb{R}$ **27.** Domain: $\mathbb{R}$
    Range: $\{y \mid y \geq -5\}$ Range: $\{y \mid y \geq -9\}$ Range: $\{y \mid y \geq 0\}$ Range: $\{y \mid y \leq -4\}$

29. Domain: $\mathbb{R}$
Range: $\{y \mid y \geq -16\}$

31. Domain: $\mathbb{R}$
Range: $\{y \mid y \geq -2\}$

33. Domain: $\mathbb{R}$
Range: $\{y \mid y \geq -2\}$

35.

37.

39.

41.

43.

45.

47.

49.

51.

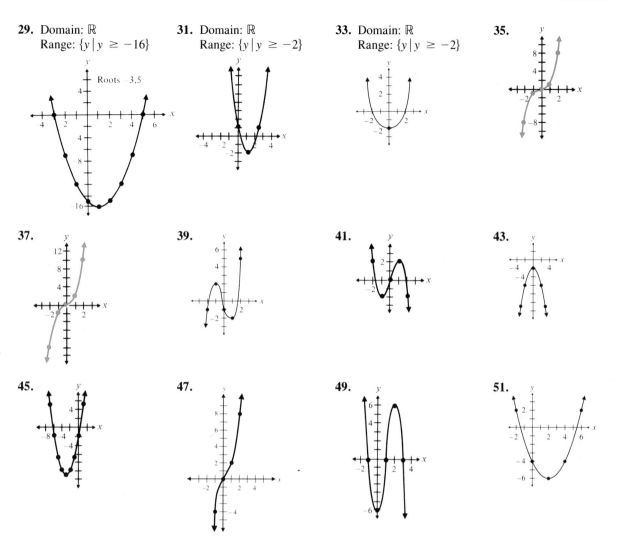

53. When the squared term is positive, the graph opens up; if negative, it opens down.
55. As x increases, y increases; as x decreases, y decreases
57. y decreases as x goes from -3 to 0; then y increases as x goes from 0 to 3. (y is a minimum when $x = 0$.)

Cumulative Review Exercises
61. $(-8, 2)$ **62.** $(-8, 2)$ **63.** $(1, 5, 6)$ **64.** $15x^3 + 21x^2 - 38x + 12$

Just for Fun   **1.**   **2.**

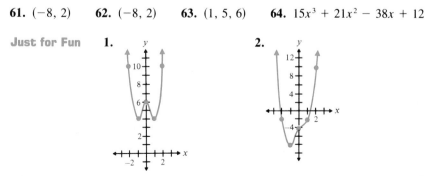

Review Exercises

1. 64 **2.** x^8 **3.** y^7 **4.** 27 **5.** x^4 **6.** y^9 **7.** $\dfrac{1}{y}$ **8.** x^7 **9.** $\dfrac{1}{x^3}$ **10.** $\frac{1}{27}$ **11.** $\frac{1}{32}$ **12.** 3

13. $9x^4$ **14.** $\frac{3}{2}$ **15.** $\frac{16}{9}$ **16.** $\dfrac{y^2}{x}$ **17.** $-12x^2y^6$ **18.** $-21x^3y^9$ **19.** $\dfrac{8}{x^2y}$ **20.** $\dfrac{30x^5}{y^6}$ **21.** $\dfrac{3y^7}{x^5}$

22. $\dfrac{3}{xy^9}$ **23.** $\dfrac{y^8}{2}$ **24.** $\dfrac{x^7}{y^8}$ **25.** $125x^3y^3$ **26.** $\dfrac{x^{10}}{9y^2}$ **27.** $\dfrac{4x^2}{y^2}$ **28.** x^6y^8 **29.** $\dfrac{-125y^3}{x^6z^9}$ **30.** $\dfrac{z^4}{36x^2y^6}$

31. $\dfrac{x^9}{27}$ **32.** $\dfrac{x^{12}}{16y^8}$ **33.** $\dfrac{125}{x^9y^3}$ **34.** $\dfrac{8x^{10}y^{16}}{z^6}$ **35.** $\dfrac{64y^3}{x^3z^{15}}$ **36.** $\dfrac{4x^5}{z^6}$ **37.** $\dfrac{27x^{21}}{32y^{19}}$ **38.** $\dfrac{x^9y^{19}}{8z}$ **39.** $\dfrac{x^{26}}{64y^6}$

40. $\dfrac{x^{11}}{125y^{23}}$ **41.** 7.42×10^{-5} **42.** 2.6×10^5 **43.** 1.83×10^5 **44.** 1×10^{-6} **45.** 30,000 **46.** 0.02

47. 200,000,000 **48.** 2000 **49.** Binomial, $-x + 5$, first **50.** Trinomial, $x^2 + 5x - 3$, second
51. Trinomial, $x^2 + xy - y^2$, second **52.** Trinomial, $x^5y^3 + x^4y - 6xy^3$, eighth
53. Polynomial, $2x^4 - 9x^2y + 6xy^3 - 3$, fourth **54.** Not polynomial **55.** $10x - 5$ **56.** $7x^2 - 4x - 15$
57. $4x^3 + 8x^2 + 12x$ **58.** $x^2 + 10x + 25$ **59.** $5y^2 + 2$ **60.** $2x^2 + 4x + 4$ **61.** $4x^2 - 9$
62. $6x^2y + 9xy^2 + 2x + 3y$ **63.** $2x - 3$ **64.** $x^2 - y^2$ **65.** $2x^3 - 8x^2 - 9$ **66.** $-2x^4y^2 - 2x^3y^7 + 12xy^3$
67. $9x^2 - 12xy + 4y^2$ **68.** $3x^2y + 3xy - 9y^2$ **69.** $25x^2y^2 - 36$ **70.** $3x - 2y + 1$ **71.** $4x^2 - 25y^4$

72. $x + 4 - \dfrac{5}{x - 3}$ **73.** $\dfrac{x^2}{2y} + x + \dfrac{3y}{2}$ **74.** $x^2 + 6xy + 9y^2 + 4x + 12y + 4$ **75.** $x^2 + 6xy + 9y^2 - 4$

76. $-9xy - 9x + 5y^2$ **77.** $6x^3 - x^2 - 24x + 18$ **78.** $2x^3 + x^2 - 3x - 4$

79. $4x^4 + 12x^3 + 6x^2 + 13x - 15$ **80.** $2x^2 + 3x - 4 + \dfrac{2}{2x + 3}$ **81.** $x^3y + 6x^2y + 7xy^2 + x^2y^2 + y^3$

82. $3x^2 + 7x + 21 + \dfrac{73}{x - 3}$ **83.** $2y^4 - 2y^3 - 8y^2 + 8y - 7 + \dfrac{6}{y + 1}$ **84.** $x^4 + 2x^3 + 4x^2 + 8x + 16 + \dfrac{12}{x - 2}$

85. $2x^2 + 2x + 6$ **86.** (a) 3 (b) 94 **87.** (a) $\frac{17}{8}$ (b) 7 **88.** (a) 5 (b) 30
89. (a) $a^2 + 2a - 1$ (b) $a^2 + 6a + 7$ **90.** (a) $a^2 - a + 3$ (b) $a^2 + 2ab + b^2 - a - b + 3$
91. Domain: $\mathbb{R}$ **92.** Domain: $\mathbb{R}$ **93.** Domain: $\mathbb{R}$ **94.** **95.** **96.**
Range: $\{y \mid y \geq 0\}$ Range: $\{y \mid y \geq -1\}$ Range: $\{y \mid y \geq 1\}$

97. (a) 720 (b) 1500 **98.** (a) 84 ft (b) 36 ft

Practice Test

1. $\dfrac{y^{10}}{9x^6}$ **2.** $\dfrac{9}{x^9y^{11}}$ **3.** Trinominal, $-6x^4 - 4x^2y^3 + 2x$, fifth **4.** $4x^3 - 2x^2 + 2x + 8$ **5.** $4x^5 - 2y^2 + \dfrac{5}{x}$

6. $-6x^2 + xy + y^2$ **7.** $4x^3 + 8x^2y - 9xy^2 - 6y^3$ **8.** $x - 5 + \dfrac{25}{2x + 3}$ **9.** $2x^2y + 3x + 7y^2$

10. $2x^2 + 5x + 20 + \dfrac{53}{x - 3}$ **11.** $-6x^7y^6 + 18x^4y^7 - 9x^3y^4$ **12.** $4x^2 + 12xy + 9y^2$

13. $3x^3 + 3x^2 + 15x + 15 + \dfrac{79}{x - 5}$ **14.** -18 **15.** Domain: $\mathbb{R}$ **16.** 6848
Range: $\{y \mid y \geq -2\}$

CHAPTER 6

Exercise Set 6.1

1. $8(n + 1)$ **3.** Cannot be factored   **5.** $2(8x^2 - 6x - 3)$ **7.** $x^3(7x^2 - 9x + 3)$ **9.** $3y(8y^{14} - 3y^2 + 1)$
11. $x(1 + 3y^2)$ **13.** Cannot be factored   **15.** $8xy^2(5x + 2y^2 + 8y)$ **17.** $9xyz(4yz^2 + 4x^2y + x)$
19. $4x^3(6x^3 + 2x - y)$ **21.** $2(26x^2y^2 + 8xy^3 + 13z)$ **23.** $(5x + 3)(2x - 5)$ **25.** $(4x - 5)^2(12x^2 - 15x + 1)$
27. $3(2x + 5)(-3x - 10)$ **29.** $3(3p - q)(p - q)$ **31.** $(x - 2)(-2x + 9)$ **33.** $(2x + 5)(6x^3 - 2x^2 - 1)$
35. $(2r - 3)^6(8pr - 12p - 3)$ **37.** $(x - 5)(x + 3)$ **39.** $(3x + 1)(x + 3)$ **41.** $(2x - 1)^2$
43. $2(2x - 5)(2x - 1)$ **45.** $(b + c)(2 + a)$ **47.** $(5x + 3y)(7x - 8y)$ **49.** $(x^2 + 4)(x - 3)$
51. $(2x - 5y)(5x - 6y)$ **53.** $(x^2 - 2)(x + 3)$ **55.** $(a^3b - c^2)(2a - 3c)$ **57.** $(3p + 2q)(p^2 + q^2)$
59. $2p(10p^2 - 9p + 6)$ **61.** $4(4xy^2z + x^3y - 2)$ **63.** $(x - 2)(5x + 3)$ **65.** $y^2z^2(14yz^3 - 28yz^4 - 9x)$
67. $7y^7(x^4y^2 - 3x^3z^5 - 5yz^9)$ **69.** $(3a - 4b)(5a - 6b)$ **71.** $(3x - 2)(7x + 1)$ **73.** $(3x + y)(2x - 3y)$
75. $(x + 3)(5x^2 + 15x - 3)$ **77.** $(3c - d)(2d)$ **79.** $x(3x^2 + 2)(x^2 - 5)$
81. Determine if all the terms have a gcf, and if so factor it out.

Cumulative Review Exercises

83. $25   **84.** 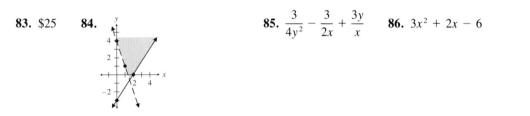   **85.** $\dfrac{3}{4y^2} - \dfrac{3}{2x} + \dfrac{3y}{x}$ **86.** $3x^2 + 2x - 6$

Just for Fun   **1.** $2(x - 3)(2x^4 - 12x^3 + 15x^2 + 9x + 2)$ **2. (a)** $(x + 1)[(x + 1) + 1] = (x + 1)(x + 2)$
(b) $(x + 1)^2[(x + 1) + 1] = (x + 1)^2(x + 2)$ **(c)** $(x + 1)^{n-1}[(x + 1) + 1] = (x + 1)^{n-1}(x + 2)$
(d) $(x + 1)^n[(x + 1) + 1] = (x + 1)^n(x + 2)$ **3.** $2x(x + 5)^{-2}[2 + (x + 5)^1] = 2x(x + 5)^{-2}(x + 7)$
4. $(2x - 3)^{-1/2}[x + 6(2x - 3)^1] = (2x - 3)^{-1/2}(13x - 18)$

Exercise Set 6.2

1. $(x + 6)(x + 1)$ **3.** $(p - 5)(p + 2)$ **5.** Cannot be factored   **7.** $(x - 32)(x - 2)$ **9.** $(a - 15)(a - 3)$
11. Cannot be factored   **13.** $(x - y)(x - 3y)$ **15.** $(z - 2y)(z - 5y)$ **17.** $5(x + 1)(x + 3)$
19. $x(x - 6)(x + 3)$ **21.** $x(x - 8)(x + 3)$ **23.** $(4w + 1)(w + 3)$ **25.** Cannot be factored
27. $(w - 2)(3w + 4)$ **29.** $(3y - 5)(y + 1)$ **31.** $(2x + 3y)(2x - y)$ **33.** $2(4x^2 + x - 10)$
35. $2(2x - 3y)(2x + y)$ **37.** $xy(x - 6)(x + 3)$ **39.** $ab(a + 7)(a - 5)$ **41.** $6pq^2(p - 5q)(p + q)$
43. $(7x - 3)(5x + 4)$ **45.** $2(4x - 5)(x - 3)$ **47.** $(x^2 + 3)(x^2 - 2)$ **49.** $(x^2 + 2)(x^2 + 3)$
51. $(2a^2 + 5)(3a^2 - 5)$ **53.** $(2x + 5)(2x + 3)$ **55.** $(3a + 1)(2a + 5)$ **57.** $(ab - 3)(ab - 5)$
59. $(3xy - 5)(xy + 1)$ **61.** $(2a - 5)(a - 1)(5 - a)$ **63.** $(2x + 3)(x + 2)(x - 3)$ **65.** $(x + 8)^2$
67. $3(y - 9)(y - 2)$ **69.** $(y^2 - 10)(y^2 + 3)$ **71.** $(3z^2 + 1)(z^2 - 5)$ **73.** $(6x - 1)(2x + 3)$
75. $(xy + 3)(2xy - 3)$ **77.** $3(x - 3)(3x + 4)$ **79.** $(x + 2)(x + 1)(x + 3)$ **81.** $5(x + 4y)(x + y)$
83. $(5a - 3)(3a + 5)$ **85.** $(5y - 3)(4y + 5)$ **87.** Factor out the GCF if there is one.
89. $24x^2 - 66x + 45$ **91.** Divide $(x^2 - xy - 6y^2)$ by $(x - 3y)$, $x + 2y$ **93. (b)** $2(4x - 1)(x - 3)$

Cumulative Review Exercises

94. 0, change in y is 0.   **95.** Slope is undefined, change in x is 0 and you cannot divide by 0.
96. 9.0×10^{10} **97.** $-x^2y - 8xy^2 + 6$ **98. (a)** $x = 2$ **(b)**

Just for Fun   **1.** Cannot divide both sides of the equation by $a - b$ since it equals 0.   **2.** $(2a^n + 3)(2a^n - 5)$
3. $2(3x^ny^n - 1)(2x^ny^n + 1)$

Exercise Set 6.3

1. $(x + 9)(x - 9)$ **3.** Cannot be factored   **5.** $(1 + 2x)(1 - 2x)$ **7.** $(x + 6y)(x - 6y)$
9. $(x^3 - 12y^2)(x^3 + 12y^2)$ **11.** $(x^3 + 2)(x^3 - 2)$ **13.** $(ab - 7c)(ab + 7c)$ **15.** $x^2(3y + 2)(3y - 2)$
17. $x(12 - x)$ **19.** $(a + 3b + 2)(a - 3b - 2)$ **21.** $(x + 5)^2$ **23.** $(2 + x)^2$ **25.** $(2x - 5y)^2$
27. $(3a + 2)^2$ **29.** $(w^2 + 8)^2$ **31.** $(x + y + 1)^2$ **33.** $(a + b)^2(a - b)^2$ **35.** $(x + 3 + y)(x + 3 - y)$
37. $(x + 7)(-x + 3)$ **39.** $(3a - 2b + 3)(3a - 2b - 3)$ **41.** $(x - 3)(x^2 + 3x + 9)$
43. $(x + y)(x^2 - xy + y^2)$ **45.** $(x - 2a)(x^2 + 2ax + 4a^2)$ **47.** $(y + 1)(y^2 - y + 1)$
49. $(3y - 2x)(9y^2 + 6xy + 4x^2)$ **51.** $3(2x - 3y)(4x^2 + 6xy + 9y^2)$ **53.** $5(x - 5y)(x^2 + 5xy + 25y^2)$
55. $(x + 2)(x^2 + x + 1)$ **57.** $(x - y - 3)(x^2 - 2xy + y^2 + 3x - 3y + 9)$ **59.** $(y^2 - 7x)(y^2 + 7x)$
61. $(4y - 9x)(4y + 9x)$ **63.** $(5x^2 + 9y^3)(5x^2 - 9y^3)$ **65.** $(a - 2)(a^2 + 2a + 4)$ **67.** $(x - 4)(x^2 + 4x + 16)$
69. $(a^2 + 6)^2$ **71.** $(a^2 + b^2)^2$ **73.** $(x - 1 + y)(x - 1 - y)$ **75.** $(x + y + 1)(x^2 + 2xy + y^2 - x - y + 1)$
79. $16, -16$ **81.** 9

Cumulative Review Exercises

83. (a) $3, 6$ **(b)** $-2, \frac{5}{9}, -1.67, 0, 3, 6$ **(c)** $\sqrt{3}, -\sqrt{6}$ **(d)** $-2, \frac{5}{9}, -1.67, 0, \sqrt{3}, -\sqrt{6}, 3, 6$ **84.** $\in$ **85.** $\subseteq$
86. ≈ 0.44 **87.** $w = 3$ ft, $l = 8$ ft

Just for Fun   **1.** $(x + \sqrt{7})(x - \sqrt{7})$ **2.** $(x\sqrt{2} + \sqrt{15})(x\sqrt{2} - \sqrt{15})$ **3.** $-3(2x - 13)$ **4.** $(a^n - 8)^2$

Exercise Set 6.4

1. $3(x + 4)(x - 3)$ **3.** $(5s - 3)(2s + 5)$ **5.** $(4r - 3)(2r - 5)$ **7.** $2(x + 6)(x - 6)$ **9.** $5x(x^2 + 3)(x^2 - 3)$
11. $3x(x - 4)(x - 1)$ **13.** $5x^2y^2(x - 3)(x + 4)$ **15.** $x^2(x + y)(x - y)$ **17.** $x^4y^2(x - 1)(x^2 + x + 1)$
19. $x(x^2 + 4)(x + 2)(x - 2)$ **21.** $4(x^2 + 2y)(x^4 - 2x^2y + 4y^2)$ **23.** $2(a + b + 3)(a + b - 3)$ **25.** $(x + 3y)^2$
27. $x(x + 4)$ **29.** $(2a + b)(a - 2b)$ **31.** $(y + 5)^2$ **33.** $5a^2(3a - 1)^2$ **35.** $(x + \frac{1}{3})(x^2 - \frac{1}{3}x + \frac{1}{9})$
37. $(x + 3)(x - 3)(3x + 2)$ **39.** $ab(a + 4b)(a - 4b)$ **41.** $(3 + x + y)(3 - x - y)$ **43.** $2(3x - 2)(4x - 3)$
45. $(7x - 6)(x - 1)$ **47.** $(x^2 + 9)(x + 3)(x - 3)$ **49.** $(5c - 6y)(b - 2x)$ **51.** $(3x^2 - 4)(x^2 + 1)$
53. $(y + x - 4)(y - x + 4)$ **55.** $3(2x + 3y)(4a + 3)$ **57.** $(x^3 - 6)(x^3 - 5)$ **59.** $y(1 + y)(1 - y)$
61. $(2xy + 3)^2$ **63.** $(3rs - 1)(2rs + 1)$

Cumulative Review Exercises

66. -8 **67.** No solution   **68.** $\{x \mid x > 6\}$ **69.** $\mathbb{R}$

Exercise Set 6.5

1. $0, -5$ **3.** $0, -9$ **5.** $-\frac{5}{2}, 3, -2$ **7.** 3   **9.** $0, 12$ **11.** $0, -2$ **13.** $-4, 3$ **15.** $2, 10$ **17.** $3, -6$
19. $-4, 6$ **21.** $0, 2, -21$ **23.** $-5, -6$ **25.** $-\frac{1}{6}, \frac{3}{2}$ **27.** $\frac{1}{4}, \frac{2}{7}$ **29.** $0, -\frac{1}{3}, 3$ **31.** $\frac{1}{3}, 7$ **33.** $\frac{2}{3}, -1$
35. $0, -4, 3$ **37.** $0, \frac{8}{3}$ **39.** $0, \frac{4}{5}, -\frac{4}{5}$ **41.** $0, -8$ **43.** $-3, 5$ **45.** $\frac{5}{2}, \frac{4}{3}$ **47.** $-\frac{1}{3}, 2$ **49.** $0, 1$
51. $0, -3, -5$ **53.** $8, 9$ **55.** $9, 11$ **57.** $5, 7$ **59.** $w = 3$ ft, $l = 12$ ft   **61.** $h = 10$ cm, $b = 16$ cm
63. 7 m   **65. (a)** 80 ft   **(b)** 60 ft   **(c)** $\sqrt{6} \approx 2.45$ sec   **67.** 5 sec
69. (a) Zero factor property only holds when one side of the equation is 0.   **(b)** $-2, -5$

Cumulative Review Exercises

70. (a) Carmen 4.8 mph, Bob 6 mph   **(b)** 24 miles   **71.** $(3, 0)$ **72.** $\varnothing$ **73.** $3x + 4$

Just for Fun   **1.** $x^2 + 3x - 10 = 0$ (other answers are possible for 1 through 3).   **2.** $2x^2 - 7x + 3 = 0$
3. $x^3 - 2x^2 - 3x = 0$ **4.** $(x^3 + 5)(x^3 - 12) = 0, x = \sqrt[3]{-5}, x = \sqrt[3]{12}$
5. $1, -9$ **6.** $\pm 1, \pm 2$

Exercise Set 6.6

1. $y = \dfrac{3x - 4}{x - 2}$ **3.** $y = -\dfrac{x}{x - 2}$ **5.** $x = \dfrac{y}{y - 1}$ **7.** $x = \dfrac{2y + 4}{y + 1}$ **9.** $z = \dfrac{5x - 2}{3y - 1}$ **11.** $z = \dfrac{-6x}{2x + 3}$

13. $y = \dfrac{9}{x - 24}$ **15.** $r = \dfrac{5s}{6s - 2}$ **17.** $x = \dfrac{33}{3a - 4}$ **19.** $y = \dfrac{9x - 20}{90x + 30}$ **21.** $P = \dfrac{I}{1 + rt}$

23. $m = \dfrac{y_2 - y_1}{x_2 - x_1}$ **25.** $d = \dfrac{a_n - a_1}{n - 1}$ **27.** $r = \dfrac{O + R}{V + D}$ **29.** $S_n = \dfrac{a_1 - a_1 r^n}{1 - r}$ **31.** $a_1 = \dfrac{S_n - S_n r}{1 - r^n}$

33. $q_H = \dfrac{q_c}{e - 1}$ **35.** $b = \dfrac{3a + 4}{2a + 5}$, the two equations are the same except that the variables differ.

Cumulative Review Exercises

36.

37. $(3x + 4)(x + 2)$ **38.** $(3x - 7)(2x + 3)$ **39.** $3(6x - 1)(x - 3)$

Just for Fun **1.** $y' = \dfrac{1}{x + y}$ **2.** $y' = \dfrac{xy - 2}{x + 3}$ **3.** $y' = \dfrac{x + xy}{2xy + 3}$

Review Exercises

1. $4x^2$ **2.** $6xy$ **3.** $3xy^2$ **4.** $2x - 5$ **5.** $x + 5$ **6.** 1 **7.** $4(3x - 2)(x + 1)$ **8.** $6x^4(10 + x^5 - 3xy^2)$
9. Cannot be factored **10.** $(x - 2)(5x + 3)$ **11.** $(2x + 1)(4x - 3)$ **12.** $(x - 1)(3x^2 - 3x - 2)$
13. $(2x - 1)(10x - 3)$ **14.** $3xy^2z^2(4y^2z + 2xy - 5x^2z)$ **15.** $(x + 2)(x + 3)$ **16.** $(x + 3)(x - 5)$
17. $(x + 7)(x - 7)$ **18.** $(x + 3)(3x + 1)$ **19.** $(5x - 1)(x + 4)$ **20.** $(x + 4y)(5x - y)$
21. $(4x + 5y)(3x - 2y)$ **22.** $(3x - y)(-4x + 9y)$ **23.** $3(a^2 + 3b)(a^2 - 4b)$ **24.** $(x + 5)(x + 3)$
25. $(x - 5)(x - 3)$ **26.** $(x - 9)(x + 3)$ **27.** $(x - 15)(x + 3)$ **28.** $(x - 10y)(x + 5y)$
29. $(x - 18y)(x + 3y)$ **30.** $2(x + 4)^2$ **31.** $3(x - 3)^2$ **32.** $x(x - 6)(x + 3)$ **33.** $x(4x - 5)(2x + 5)$
34. $(3x + 1)(x + 4)$ **35.** $(4x - y)(x + 3y)$ **36.** $x(4x - 5)(x - 1)$ **37.** $x(12x + 1)(x + 5)$
38. $(x^2 - 5)(x^2 + 2)$ **39.** $(x^2 + 4)(x^2 - 5)$ **40.** $(x + 9)(x + 11)$ **41.** $(3x + 8)(x - 4)$ **42.** $(x + 6)(x - 6)$
43. $4(x + 2y^2)(x - 2y^2)$ **44.** $(x^2 + 9)(x - 3)(x + 3)$ **45.** $(x - 1)(x + 5)$ **46.** $(x - 1)(x - 5)$
47. $(2x - 3)^2$ **48.** $(3y + 4)^2$ **49.** $(w^2 - 8)^2$ **50.** $(a + 3b + 2c)(a + 3b - 2c)$ **51.** $(x - 2)(x^2 + 2x + 4)$
52. $(2x + 3)(4x^2 - 6x + 9)$ **53.** $(3x - 2y)(9x^2 + 6xy + 4y^2)$ **54.** $(3x - 2y^2)(9x^2 + 6xy^2 + 4y^4)$
55. $(2y^2 - 5x)(4y^4 + 10xy^2 + 25x^2)$ **56.** $(x - 1)(x^2 + 4x + 7)$ **57.** $y^2(x + 3)(x - 5)$ **58.** $3x(x - 4)(x - 2)$
59. $3xy^4(x - 2)(x + 6)$ **60.** $3y(y^2 + 3)(y^2 - 3)$ **61.** $2y(x + 2)(x^2 - 2x + 4)$ **62.** $5x^2y(x + 2)^2$
63. $3x(2x + 1)(x - 4)$ **64.** $(x + 5 + y)(x + 5 - y)$ **65.** $3(x + 2y)(x^2 - 2xy + 4y^2)$ **66.** $(x + 4)^2(x - 1)$
67. $(4x + 1)(4x + 5)$ **68.** $(2x^2 - 1)(2x^2 + 3)$ **69.** $(x + 1)(x - 2)(x - 1)$ **70.** $3(5x - 2)(2x + 1)$
71. $(3x + 4y)(3a - b)$ **72.** $(2pq - 3)(3pq + 2)$ **73.** $(3x^2 - 2)^2$ **74.** $(2y + x + 2)(2y - x - 2)$
75. $2(3a + 5)(4a + 3)$ **76.** $3x^2y^4(x + 3)(2x - 3)$ **77.** $(x - \tfrac{2}{3}y^2)(x^2 + \tfrac{2}{3}xy^2 + \tfrac{4}{9}y^4)$ **78.** $5, -\tfrac{2}{3}$ **79.** $0, \tfrac{3}{2}$
80. $0, -\tfrac{4}{3}$ **81.** $6, -4$ **82.** $-3, -5$ **83.** $-4, 2$ **84.** $-2, -5$ **85.** $0, 2, 4$ **86.** $0, \tfrac{4}{3}, -\tfrac{1}{4}$ **87.** $\tfrac{1}{4}, -\tfrac{3}{2}$
88. $2, -2$ **89.** $2, -3$ **90.** $5, 6$ **91.** $w = 7$ ft, $l = 9$ ft **92.** $h = 4$ m, $b = 11$ m **93.** 5 in, 9 in

94. 9 sec **95.** $y = \dfrac{3x - 6}{x - 4}$ **96.** $z = \dfrac{6x}{2x - 3}$ **97.** $z = \dfrac{3x}{6xy + 2}$ **98.** $y = \dfrac{12x + 10}{5x + 12}$ **99.** $y = \dfrac{3x + 2}{x - 1}$

100. $l = \dfrac{Rw}{2 - R}$ **101.** $\beta = \dfrac{-w}{3\alpha + 6}$ **102.** $w_1 = \dfrac{15p}{2x_2 - 2x_1}$

Practice Test

1. $4x(xy - 1)$ **2.** $(3x + 2)(x + 4)$ **3.** $3xy^2(3x^2 + 4xy^3 - 9y^2)$ **4.** $5(x - 2)(x + 1)$ **5.** $(2x + 3y)(x + 2y)$
6. $(x - 4y)(x - 3y)$ **7.** $3x(x - 3)(x + 1)$ **8.** $(3x - 2)(2x - 1)$ **9.** $(5x + 2)(x + 3)$
10. $(9x + 4y^2)(9x - 4y^2)$ **11.** $y^6(3x - 2)(9x^2 + 6x + 4)$ **12.** $(x + 2)(x + 6)$ **13.** $(2x^2 + 9)(x^2 - 2)$
14. $5, -\tfrac{2}{3}$ **15.** $-\tfrac{3}{4}, 6$ **16.** $0, -5, 1$ **17.** $y = \dfrac{8x}{x - 6}$ **18.** $p = \dfrac{8w + 45}{15w - 8}$ **19.** $h = 4$ m, $b = 14$ m
20. 7 sec

Cumulative Practice Test

1. $-\frac{9}{8}$ **2.** $-\frac{15}{14}$ **3.** $L = \dfrac{12P + W}{2}$ **4.** 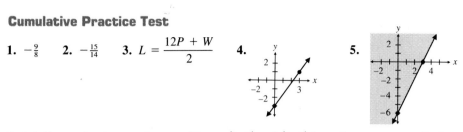   **5.**

6. **(a)** Yes, each x has a unique y **(b)** no, $(1, 2)$ and $(1, 0)$ have the same x coordinate   **7.** $\left(\frac{20}{11}, -\frac{14}{11}\right)$ **8.** $4xy^9$

9. $\dfrac{27p^{11}q^9}{8}$ **10.** $7x^2 - 9x$ **11.** $2x^3 - 11x^2 + 3x + 30$ **12.** $3xy^4 - \dfrac{8y^3}{3} - \dfrac{4}{x}$ **13.** -5 **14.**

15. $x(x^2 + 2)(x - 3)$ **16.** $3y(x - 2)(4x - 1)$ **17.** $(y + 2)(y - 2)(y^2 + 6)$
18. $(2x - 3y^2)(4x^2 + 6xy^2 + 9y^4)$ **19.** 620 pages   **20.** $34 \le x < 84$

CHAPTER 7

Exercise Set 7.1

1. $\{x \mid x \ne 0\}$ **3.** $\{x \mid x \ne 3\}$ **5.** $\mathbb{R}$ **7.** $\{x \mid x \ne 2\}$ **9.** $\{r \mid r \ne 4, r \ne -4\}$ **11.** $\{x \mid x \ne -1, x \ne -6\}$
13. $\{y \mid y \ne 1\}$ **15.** $\{z \mid z \ne \frac{15}{8}\}$ **17.** $\{x \mid x \ne 4, x \ne -4\}$ **19.** $\{x \mid x \ne 0\}$ **21.** $1 - y$ **23.** $\dfrac{x(x - 5)}{2}$
25. $x - 2$ **27.** $\dfrac{x - 2y}{5}$ **29.** $\dfrac{2x + 6 + 9x^2y^2}{4y}$ **31.** -1 **33.** $\dfrac{1}{p + 5}$ **35.** $-(x + 4)$
37. $\dfrac{2 - 8x^2 + 3x^3y}{4xy}$ **39.** $\dfrac{y - 6}{y - 1}$ **41.** x^3 **43.** $\dfrac{2x - 5}{2}$ **45.** $\dfrac{x - 8}{x - 2}$ **47.** $\dfrac{x - w}{x + w}$ **49.** $\dfrac{a^2 + ab + b^2}{a + b}$
51. $x - 2$ **53.** $\dfrac{5x}{x^2}$ **55.** $\dfrac{(w + 3)^2}{w^2 - 9}$ or $\dfrac{w^2 + 6w + 9}{w^2 - 9}$ **57.** $\dfrac{2xy^2z^2}{2x^3y^5z^2}$ **59.** $\dfrac{4(x + 1)(x + 2)}{8(x - 4)(x + 2)}$
61. $\dfrac{y(3y + 1)}{9y^2 + 15y + 4}$ **63.** **(a)** $x + 5$ **65.** **(a)** $y^2 - 4y - 5$ **67.** $\sqrt{x}$ is not a polynomial
69. **(a)** 1   **(b)** $\frac{1}{10}$ **(c)** $\frac{1}{100}$ **(d)** It decreases   **71.** **(a)** Factor out -1 from either the numerator or denominator.   **(b)** -1

Cumulative Review Exercises

72. $h = \dfrac{3V}{4\pi r^2}$ **73.** slope $= -\dfrac{1}{3}$, y intercept $= \dfrac{14}{3}$ **74.** $(4, 0)$ **75.** $\dfrac{2x}{3}$

Just for Fun   **1.** **(a)** Domain: $\{x \mid x \ne 2\}$ **(b)**   **2.** **(a)** Domain: $\{x \mid x \ne 1\}$ **(b)**

Exercise Set 7.2

1. $\dfrac{xy^2}{4}$ **3.** $12x^3y^2$ **5.** $\dfrac{27x}{8}$ **7.** 1   **9.** $\dfrac{36x^9y^2}{25z^7}$ **11.** $\dfrac{z^2}{2y}$ **13.** x **15.** 1   **17.** 1   **19.** 1
21. $\dfrac{1}{x - 3}$ **23.** $\dfrac{x + 3}{2(x + 2)}$ **25.** $\dfrac{2(x + 2)}{x - 2}$ **27.** $\dfrac{1}{2}$ **29.** $\dfrac{1}{x^2 + 2x + 4}$ **31.** 1   **33.** 1
35. $(2a - 1)(a - 1)$ **37.** $\dfrac{x^2(2x - 1)}{(x - 1)(x - 3)}$ **39.** $\dfrac{r + 5s}{2r + 5s}$ **41.** $\dfrac{p - q}{p + q}$ **43.** $\dfrac{2p - q^2}{p^2 + q^2}$ **45.** $\dfrac{2x(x + 2)}{x - 5}$
47. $x(x^2 - 4x + 16)$ **49.** $\dfrac{(a + b)^2}{ab}$ **51.** $\dfrac{x - 1}{x + 3}$ **53.** **(a)** $x^2 + x - 2$ **(b)** Factors must be $(x - 1)(x + 2)$
55. **(a)** $2x^2 + x - 6$ **(b)** Factors must be $(x + 2)(2x - 3)$

Cumulative Review Exercises

57. 2   **58.** (0, 4)   **59.** $\left(-\frac{2}{7}, \frac{15}{7}\right)$ **60.** $3x^2y - 7xy - 4y^2 - 2x$

Exercise Set 7.3

1. $\dfrac{2x - 11}{3}$ **3.** $\dfrac{-8}{x}$ **5.** $\dfrac{1}{t - 4}$ **7.** -1 **9.** $\dfrac{y}{x - 2y}$ **11.** $x - 5$ **13.** $\dfrac{x + 5}{x + 3}$ **15.** $(x + 1)(x + 2)$

17. $48x^3y$ **19.** $z - 3$ **21.** $(a - 8)(a + 3)(a + 8)$ **23.** $(x + 3)(x - 3)(x - 1)$

25. $(x + y)(3x + 2y)(2x + 3y)$ **27.** $(x - 1)(x + 4)(4x + 9)$ **29.** $\dfrac{10}{3x}$ **31.** $\dfrac{3x + 12}{2x^2}$ **33.** $\dfrac{10y + 9}{12y^2}$

35. $\dfrac{25y^2 - 12x^2}{60x^4y^3}$ **37.** $\dfrac{4 + 6y}{3y}$ **39.** $\dfrac{5 - 3x}{b - 2}$ **41.** $\dfrac{a^2}{b(a - b)}$ **43.** $\dfrac{20z}{(z + 5)(z - 5)}$ **45.** $\dfrac{-4x^2 + 25x - 36}{(x + 3)(x - 3)}$

47. $\dfrac{2m^2 + 5m - 2}{(m - 5)(m + 2)}$ **49.** $\dfrac{1}{x - 5}$ **51.** $\dfrac{2x^2 + 5x + 8}{(x - 1)(x + 4)(x - 2)}$ **53.** $\dfrac{5x^2 + 14x - 49}{(x + 5)(x - 2)}$ **55.** $\dfrac{3a - 1}{4a + 1}$

57. $\dfrac{x - 2}{(x - 1)(2x - 3)}$ **59.** $\dfrac{2x^2 - 4xy + 4y^2}{(x - 2y)^2(x + 2y)}$ **61.** $\dfrac{12}{x - 3}$ **63.** 0   **65.** 1   **67.** $\dfrac{x - 6}{x - 4}$

69. $\dfrac{x^2 - 18x - 30}{(5x + 6)(x - 2)}$ **71.** $\dfrac{3x + 10}{x - 2}$ **73.** $\dfrac{1}{(a - 5)(a + 3)}$ **75.** $\dfrac{x^2 + 5x + 8}{(x + 1)(x + 2)}$ **77. (a)** $-5x^2 + x + 6$

79. (a) $r^2 - r - 3$

Cumulative Review Exercises

80. \$180   **81. (a)** 6 min   **(b)** 960 bottles   **82.** $3x^3y^2 - \dfrac{x^2}{y^2} + \dfrac{5y}{3}$ **83.** $2x - 1$

Just for Fun   **1.** $\dfrac{(x + 6)(x - 1)}{(x + 2)(x - 3)(x - 2)(x + 3)}$ **2.** $\dfrac{4x^2 - 9x + 27}{(x + 3)(x - 3)(x^2 - 3x + 9)}$

Exercise Set 7.4

1. $\frac{8}{11}$ **3.** $\frac{57}{32}$ **5.** $\frac{25}{1224}$ **7.** $\dfrac{x^3y}{8}$ **9.** $6xz^2$ **11.** $\dfrac{xy + 1}{x}$ **13.** $\dfrac{3}{x}$ **15.** $\dfrac{3y - 1}{2y - 1}$ **17.** $\dfrac{x - y}{y}$ **19.** $\dfrac{-a}{b}$

21. -1 **23.** $\dfrac{2(x + 2)}{x^3}$ **25.** $\dfrac{x - 2}{8}$ **27.** $\dfrac{x + 1}{1 - x}$ **29.** $\dfrac{-4a}{a^2 + 4}$ **31.** $\dfrac{2 + a^2b}{a^2}$ **33.** $\dfrac{ab}{b + a}$ **35.** $a + b$

37. $\dfrac{a^2 + b}{b(b + 1)}$ **39.** $\dfrac{y - x}{x + y}$ **41.** $\dfrac{(a + b)^2}{ab}$ **43.** $\dfrac{6y - x}{3xy}$ **45. (a)** $\dfrac{2}{7}$ **(b)** $\dfrac{4}{13}$ **47.** $R_T = \dfrac{R_1 R_2 R_3}{R_2 R_3 + R_1 R_3 + R_1 R_2}$

49. A complex fraction is one that has a fractional expression in its numerator or denominator or both its numerator and denominator.

Cumulative Review Exercises

51. $\frac{13}{48}$ **52.** $\left\{x \,\middle|\, x \le -\frac{5}{2} \text{ or } x \ge \frac{13}{2}\right\}$ **53.**

54. $x^2 - 5x - 23 - \dfrac{37}{x - 2}$

Just for Fun   **1.** $\dfrac{4a^2 + 1}{4a(2a^2 + 1)}$ **2.** $\dfrac{x^2 + x + 1}{x^3 + x^2 + 2x + 1}$

Exercise Set 7.5

1. 4   **3.** -30 **5.** -1 **7.** 4   **9.** 3   **11.** $-\frac{1}{5}$ **13.** $\frac{1}{4}$ **15.** $\frac{14}{3}$ **17.** No solution   **19.** 2   **21.** $-\frac{12}{7}$

23. 8   **25.** $-\frac{4}{3}$, 1   **27.** -14 **29.** $-2, -3$ **31.** 4   **33.** $-\frac{5}{2}$ **35.** 5   **37.** No solution   **39.** No solution

41. $\frac{17}{4}$ **43.** 12, 2   **45.** 12, 4   **47.** $w = \dfrac{fl - df}{d}$ **49.** $p = \dfrac{qf}{q - f}$ **51.** $R_1 = \dfrac{R_T R_2}{R_2 - R_T}$ **53.** $\bar{x} = \mu + \dfrac{z\sigma}{\sqrt{n}}$

55. $q = q' + E\epsilon_0 A$ **57.** $C_T = \dfrac{C_1 C_2 C_3}{C_2 C_3 + C_1 C_3 + C_1 C_2}$ **59.** 12 in   **61.** 17.14 cm

63. object: 16 cm, image: 48 cm   **65.** 155.6 ohms   **67.** 176.47 ohms
69. A number obtained when solving an equation which is not a true solution to the original equation.
71. (a) Multiply both sides of the equation by the LCD, 12. This removes fractions from the equation.   **(b)** -24
(c) Write each term with the common denominator, 12. This step will allow the fractions to be added and subtracted.
(d) $\dfrac{-x + 24}{12}$

Cumulative Review Exercises

73. $\frac{3}{2}$ **74.**   **75.**   **76.** $(2x - 4y^2)(4x^2 + 8xy^2 + 16y^4)$

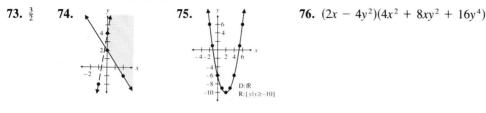

D: $\mathbb{R}$
R: $\{y \mid y \ge -10\}$

Just for Fun   1. (a) 9.71%   **(b)** Tax free money market since 9.71% > 7.68%   **2.** 115.96 days

Exercise Set 7.6

1. $\frac{20}{9}$ or 2.22 hr   **3.** $\frac{40}{13}$ or 3.08 hr   **5.** 100 hr   **7.** $\frac{160}{6}$ or 26.67 hr   **9.** 7.30 hr   **11.** $\frac{60}{7}$ or 8.57 hr
13. 18 hr   **15.** $\frac{160}{3}$ or 53.33 hr   **17.** -4 **19.** 5, 6   **21.** 3   **23.** all real numbers except 3   **25.** 1 or 2
27. 4 mph   **29.** 12 mi   **31. (a)** $\frac{300}{8}$ or 37.5 mi   **(b)** 9:30 AM   **33.** ≈ 6.33 mph   **35.** train, 60 mph; plane, 300 mph
37. 108,000 mi

Cumulative Review Exercises

38. $-\frac{5}{2}x^2 + \frac{7}{5}xy - 6y^2$ **39.** $12x^3 - 34x^2 + 21x + 4$ **40.** $4x + 5 + \dfrac{22}{3x - 2}$

41. $(4x + 3)(2x + 5)$

Exercise Set 7.7

1. Direct   **3.** Direct   **5.** Direct   **7.** Inverse   **9.** Direct   **11.** Direct   **13.** Inverse   **15.** Direct
17. $C = kZ^2$; 60.75   **19.** $x = k/y$; 0.2   **21.** $L = k/P^2$; 6.25   **23.** $A = kR_1R_2/L^2$; 57.6   **25.** $x = ky$; 18
27. $y = kR^2$; 20   **29.** $C = k/J$; 0.41   **31.** $F = kM_1M_2/d$; 40   **33.** $S = kIT^2$; 0.2   **35.** $S = kF$; 0.7 in
37. $I = k/d^2$; 31.25 footcandles   **39.** $W = k/d^2$; 133.25 lb   **41.** $R = kL/A$; 25 ohms

Cumulative Review Exercises

43. $d = -5$ **44.** weekly salary $400, commission rate 3%   **45.**   **46.**

Just for Fun   1. $\frac{1}{9}$

Review Exercises

1. $\{x \mid x \ne 4\}$ **2.** $\{x \mid x \ne -1\}$ **3.** $\mathbb{R}$ **4.** $\{x \mid x \ne -3\}$ **5.** $\{x \mid x \ne 0\}$ **6.** $\{x \mid x \ne 5, x \ne -2\}$ **7.** x
8. $x - 3$ **9.** -1 **10.** $\dfrac{x - 1}{x - 2}$ **11.** $\dfrac{x - 3}{x + 1}$ **12.** $\dfrac{a^2 + 2a + 4}{a + 2}$ **13.** $x(x + 1)$ **14.** $(x + y)(x - y)$
15. $(x + 7)(x - 5)(x + 2)$ **16.** $(x + 2)^2(x - 2)(x + 1)$ **17.** $6xz^2$ **18.** $-\dfrac{1}{2}$ **19.** $\dfrac{32z}{x^3}$ **20.** $\dfrac{3}{x}$
21. $\dfrac{(x + y)y^2}{2x^3}$ **22.** $3x + 2$ **23.** 1   **24.** $\dfrac{3}{x - y}$ **25.** $\dfrac{6x^2 - 3x - 16}{2x - 3}$ **26.** $\dfrac{5x^2 - 12y}{3x^2y}$ **27.** $\dfrac{7x + 12}{x + 2}$
28. $\dfrac{5x + 12}{x + 3}$ **29.** $\dfrac{x + y}{x}$ **30.** $\dfrac{1}{3(a + 3)}$ **31.** $\dfrac{-4x - 2}{2x - 3}$ **32.** $\dfrac{a^2 + c^2}{ac}$ **33.** $\dfrac{x + 1}{2x - 1}$ **34.** 1
35. $2x(x - 5y)$ **36.** $\dfrac{-3x^2 + 2x - 4}{3x(x - 2)}$ **37.** $\dfrac{x^2 - 2x - 5}{(x + 5)(x - 5)}$ **38.** $\dfrac{2(2x + 13)}{(x + 5)^2}$ **39.** $\dfrac{3(x - 1)}{(x + 3)(x - 3)}$

40. $\dfrac{1}{a+2}$ **41.** $\dfrac{16(x-2y)}{3(x+2y)}$ **42.** $\dfrac{4}{(x+2)(x-3)(x-2)}$ **43.** $\dfrac{2}{x(x+2)}$ **44.** $\dfrac{2(x-4)}{(x-3)(x-5)}$

45. $\dfrac{22x+5}{(x-5)(x-10)(x+5)}$ **46.** $\dfrac{x^3+y^2}{x^3-y^2}$ **47.** $\dfrac{-1}{x-3}$ **48.** $\dfrac{(x+3)(x-1)}{4(x+1)}$ **49.** $\dfrac{x+3}{x+5}$ **50.** $\dfrac{x-4}{x-6}$

51. $\dfrac{x^2+6x-24}{(x-1)(x+9)}$ **52.** $\dfrac{55}{26}$ **53.** $\dfrac{5yz}{6}$ **54.** $\dfrac{16x^3z^2}{y^3}$ **55.** $\dfrac{xy+1}{y^3}$ **56.** $\dfrac{xy-x}{x+1}$ **57.** $\dfrac{4x+2}{x(6x-1)}$

58. $\dfrac{2a+1}{2}$ **59.** $\dfrac{x+1}{-x+1}$ **60.** 9   **61.** 1   **62.** 6   **63.** 6   **64.** 52   **65.** 20   **66.** 18   **67.** $\frac{1}{2}$

68. -6 **69.** -18 **70.** $a_1=\dfrac{S_n(r-1)}{r^n-1}$ **71.** $P_2=\dfrac{V_1P_1T_2}{T_1V_2}$ **72.** $f=\dfrac{pq}{p+q}$ **73.** $R_2=\dfrac{R_1R_T}{R_1-R_T}$

74. 120 ohms   **75.** 900 ohms, 1800 ohms   **76.** 3 cm   **77.** 15 cm   **78.** $\frac{12}{7}$ or 1.71 hrs   **79.** $\frac{3000}{35}$ or 85.71 hrs

80. 2   **81.** $\frac{5}{6}$ **82.** 5 mph   **83.** 50 mph, 150 mph   **84.** 75   **85.** 20   **86.** 1   **87.** 20   **88.** 426.7

89. 5 in   **90.** \$119.88   **91.** 400 ft   **92.** 200.96   **93.** 2.38 min

Practice Test

1. $\{x\mid x\neq7,\;x\neq-4\}$ **2.** $-(x+4)$ **3.** $\dfrac{4x^3z}{3}$ **4.** $a+3$ **5.** $\dfrac{x-3y}{3}$ **6.** x^2+y^2 **7.** $\dfrac{10x+3}{2x^2}$

8. $\dfrac{-1}{(x+4)(x-4)}$ **9.** $\dfrac{x(x+10)}{(2x-1)^2(x+3)}$ **10.** $\dfrac{y+x}{y-x}$ **11.** $\dfrac{x^2(y+1)}{y+x}$ **12.** 60   **13.** 12   **14.** 10

15. 1.125   **16.** $\dfrac{40}{13}$ or 3.08 hrs

CHAPTER 8

1. 5   **3.** -3 **5.** 5   **7.** not a real number   **9.** -2 **11.** 12   **13.** 1   **15.** 7   **17.** not a real number

19. not a real number   **21.** -6 **23.** $\frac{1}{3}$ **25.** not a real number   **27.** -1 **29.** 6   **31.** 1   **33.** 43

35. 147   **37.** 83   **39.** 179   **41.** $|y-8|$ **43.** $|x-3|$ **45.** $|3x+5|$ **47.** $|6-3x|$

49. $|y^2-4y+3|$ **51.** $|8a-b|$

53. (a) Two, positive or principal square root and negative square root. (b) $6,\,-6$ (c) the positive square root (d) 6

55. No real number when squared gives -49.   **57.** No, example $\sqrt{-2}$ is not a real number.   **59.** $x=a^n$

63. $x\geq-4$ **65.** $x\geq1$ **67.** when n is an even integer and $x<0$

69. when n is an even integer, m is an odd integer, and $x<0$.

Cumulative Review Exercises

70. $3(y-3+z)(y-3-z)$ **71.** $(x+\frac{1}{3})(x^2-\frac{1}{3}x+\frac{1}{9})$ **72.** $(x-2)(x+5)$ **73.** $x(x^2-3)(2x-3)$

Exercise Set 8.2

1. $x^{3/2}$ **3.** $4^{5/2}$ **5.** $x^{4/5}$ **7.** $x^{3/2}$ **9.** $5^{3/4}$ **11.** $y^{2/7}$ **13.** $\sqrt{x}$ **15.** $\sqrt{z^3}$ **17.** $\sqrt[4]{2}$ **19.** $\sqrt[4]{z^9}$

21. $\sqrt[5]{x^4}$ **23.** $\sqrt[3]{7}$ **25.** y^3 **27.** z^4 **29.** x^3 **31.** $\sqrt{z}$ **33.** 5   **35.** y **37.** $\sqrt[4]{x}$ **39.** x^5 **41.** $\sqrt{y}$

43. $\sqrt[3]{y}$ **45.** 2   **47.** 9   **49.** not a real number   **51.** $\frac{3}{5}$ **53.** -4 **55.** -3 **57.** $\frac{1}{3}$ **59.** $\frac{1}{8}$ **61.** $\frac{7}{2}$

63. $\frac{9}{4}$ **65.** 11   **67.** $\frac{5}{6}$ **69.** $x^{11/2}$ **71.** $x^{1/6}$ **73.** $x^{2/15}$ **75.** $\dfrac{1}{x}$ **77.** 1   **79.** $y^{5/3}$ **81.** $\dfrac{1}{x^{11/6}}$ **83.** $\dfrac{4}{x^{1/3}}$

85. $\dfrac{1}{x^{1/7}}$ **87.** $\dfrac{1}{y^{66/5}}$ **89.** $\dfrac{x^{1/4}}{y^4}$ **91.** $a^{5/3}b^{2/15}$ **93.** $x^{1/2}(x+1)$ **95.** $y^{1/3}(1-y)$ **97.** $\dfrac{1+y}{y^{3/5}}$

99. $\dfrac{1-y^2}{y}$ **101.** $\dfrac{1+x^2}{x^7}$ **103.** $\dfrac{x^2+1}{x^{5/2}}$ **105.** $\dfrac{2(x-3)}{x^5}$ **107.** $\dfrac{5(x^2-2)}{x}$ **109.** $(x^{1/3}+3)(x^{1/3}-1)$

111. $(x^{1/2}+3)(x^{1/2}+3)$ **113.** $(2x^{1/7}+3)(x^{1/7}-1)$ **115.** $(2x^{2/5}+3)(2x^{2/5}+1)$ **117.** $(5x^{1/6}-3)(3x^{1/6}-1)$

119. n is odd, or n is even and $a\geq0$ **121.** $(4^{1/2}+9^{1/2})^2\neq4+9;\;25\neq13$ **123.** They are equal; both equal $x^{1/6}$ when converted to exponential form.

Cumulative Review Exercises

124. (c) is a function; (a), (b) and (c) are relations.   **125.** $\dfrac{b^2 + a^3b}{a^3 - b}$ **126.** 0, 3   **127.** 441.67 mph

Just for Fun   **1.** $\dfrac{2(3x - 2)}{(6x - 5)^3}$ **2.** $\dfrac{2x + 4}{(2x + 3)^{1/3}}$

Exercise Set 8.3

1. $5\sqrt{2}$ **3.** $4\sqrt{2}$ **5.** $2\sqrt[3]{2}$ **7.** $3\sqrt[3]{2}$ **9.** $x\sqrt{x}$ **11.** $x^5\sqrt{x}$ **13.** $b^{13}\sqrt{b}$ **15.** $y^2\sqrt[4]{y}$ **17.** $2x\sqrt{6x}$
19. $2y^2\sqrt[3]{3y}$ **21.** $xy^3\sqrt{xy}$ **23.** $3x^2y^2\sqrt[3]{3y^2}$ **25.** $3x^4y^4\sqrt[3]{2y}$ **27.** $2x^2y\sqrt[5]{2x^2y^2}$ **29.** $2cw^3\sqrt[3]{4cz}$
31. $3x^2y^7z^{16}\sqrt[3]{3xz^2}$ **33.** 5   **35.** 2   **37.** $3\sqrt[3]{6}$ **39.** $3xy^3\sqrt{10y}$ **41.** $4x^3y^2$ **43.** $5xy^4\sqrt[3]{x^2y^2}$
45. $x^2y^2\sqrt[4]{4y^2}$ **47.** $xy^3z\sqrt[4]{24yz^3}$ **49.** $x^7y^7z^3\sqrt[5]{x^2y^3z}$ **51.** $2\sqrt{3} + 2$ **53.** $6 - 3\sqrt{2}$ **55.** 10
57. $9y\sqrt{y} - y\sqrt{3}$ **59.** $2xy^2 + 2x^4\sqrt[3]{3y^2}$ **61.** $4x^5y^3\sqrt[3]{x} + 4xy^4\sqrt[3]{2x^2y^2}$ **63.** $6x^2y^6\sqrt{10x} - 12xy^6\sqrt{3y}$
65. $2\sqrt{6}$ **67.** $2\sqrt[3]{4}$ **69.** $x\sqrt[3]{x^2}$ **71.** $6x^2\sqrt{x}$ **73.** $x^2y^6\sqrt{x}$ **75.** $2b^4c^2\sqrt[4]{abc}$ **77.** $15\sqrt{2}$
79. $3x^3\sqrt{10x}$ **81.** $2x^3y^5\sqrt{30y}$ **83.** $x + 3\sqrt{x}$ **85.** $2y^2\sqrt[3]{2x^2}$ **87.** $2\sqrt[3]{y^2} - y^3$ **89.** $xy\sqrt[3]{12x^2y^2} - 2x^2y^2\sqrt[3]{3}$
91. If n is even and a or b is negative the numbers are not real numbers, and the rule does not apply.

Cumulative Review Exercises

92. A number that can be expressed as a quotient of two integers, denominator not 0.
93. A number that can be represented on the real number line.
94. A real number that cannot be expressed as a quotient of two integers, denominator not 0.
95. $|a| = \begin{cases} a, & a \geq 0 \\ -a, & a < 0 \end{cases}$ **96.** $m = \dfrac{2E}{v^2}$ **97.** ○━━━━● $-\frac{1}{2}$ 4   **(b)** $(-\frac{1}{2}, 4]$ **(c)** $\{x \mid -\frac{1}{2} < x \leq 4\}$

Exercise Set 8.4

1. 3   **3.** $\frac{1}{3}$ **5.** $\frac{1}{2}$ **7.** $2\sqrt{2}$ **9.** $\dfrac{x^2}{5}$ **11.** $2x^2$ **13.** $\dfrac{1}{2x^2}$ **15.** $\dfrac{3}{y}$ **17.** $\dfrac{\sqrt{3}}{3}$ **19.** $\dfrac{\sqrt{2}}{2}$ **21.** $\dfrac{x\sqrt{5}}{5}$
23. $\dfrac{x\sqrt{y}}{y}$ **25.** $\dfrac{\sqrt{2x}}{2}$ **27.** $\dfrac{\sqrt[3]{10}}{4}$ **29.** $\dfrac{2\sqrt{15}}{5}$ **31.** $\dfrac{\sqrt{6}}{4}$ **33.** $\dfrac{\sqrt[3]{4}}{2}$ **35.** $\dfrac{\sqrt[3]{9}}{3}$ **37.** $\dfrac{\sqrt[3]{5xy^2}}{y}$ **39.** $\dfrac{\sqrt[3]{10xy}}{2y}$
41. $\dfrac{5x\sqrt[4]{8}}{2}$ **43.** $\dfrac{\sqrt[4]{8xy^2}}{2y}$ **45.** $\dfrac{2x^2\sqrt{xyz}}{z}$ **47.** $\dfrac{y^2\sqrt{10xz}}{2z}$ **49.** $\dfrac{y^3\sqrt{30xz}}{6z}$ **51.** $\dfrac{3x^2y\sqrt{yz}}{z}$ **53.** $\dfrac{x^2y^2\sqrt[3]{15yz}}{z}$
55. $\dfrac{2y^3\sqrt[4]{x^2}}{x}$ **57.** 6   **59.** 31   **61.** $x - 25$ **63.** $x - y^2$ **65.** $x^2 - y$ **67.** $25 - y$ **69.** $3\sqrt{2} - 3$
71. $\dfrac{-3\sqrt{6} - 15}{19}$ **73.** $\dfrac{-4\sqrt{2} - 28}{47}$ **75.** $-5 - \sqrt{30}$ **77.** $\dfrac{\sqrt{17} + 2\sqrt{2}}{9}$ **79.** $\dfrac{5\sqrt{x} + 15}{x - 9}$ **81.** $\dfrac{4\sqrt{x} + 4y}{x - y^2}$
83. $3 - 2\sqrt{2}$ **85.** $\dfrac{x + \sqrt{xy} - \sqrt{2xy} - y\sqrt{2}}{x - y}$ **87.** $\dfrac{x\sqrt{y} - y\sqrt{x}}{x - y}$ **89.** $\dfrac{\sqrt{x}}{3}$ **91.** $\dfrac{\sqrt{10}}{5}$ **93.** -1
95. $\dfrac{2xy^3\sqrt{30xz}}{5z}$ **97.** $\dfrac{\sqrt{6}}{x}$ **99.** $x - 9$ **101.** $\dfrac{\sqrt{2x}}{2}$ **103.** $\dfrac{\sqrt[4]{24x^3}}{2x}$ **105.** $\dfrac{2y^4z^3\sqrt[3]{2x^2z}}{x}$ **107.** $\dfrac{\sqrt{2pq}}{q}$
109. $y - 9$ **111.** $\dfrac{y^3z\sqrt[4]{54x^2}}{3x}$ **113.** (a), (b), and (c) $\dfrac{\sqrt{2}}{6}$ **115.** $\dfrac{3}{\sqrt{3}}$ **117.** Yes, they are equal
119. 1) No perfect powers in any radicand.   2) No radicand contains fractions.   3) No radicals in any denominator.

Cumulative Review Exercises

121. 40 mph, 50 mph

122. **(a)** Solve equation for y. Select values for x and find corresponding values for y. Plot points.

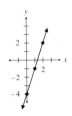

(b) Set $x = 0$ and solve for y. Set $y = 0$ and solve for x.

(c) Mark y intercept, then use the slope to determine a second point.

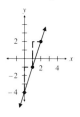

123. $8x^3 - 18x^2 + 5x + 6$ **124.** $4, \frac{3}{2}$

Exercise Set 8.5

1. $2\sqrt{3}$ **3.** $9\sqrt{10} + 2$ **5.** $9\sqrt[3]{15}$ **7.** $-3\sqrt{y}$ **9.** $7\sqrt{5} + 2\sqrt[3]{x}$ **11.** $5 - 4\sqrt[3]{x}$ **13.** $2(\sqrt{2} - \sqrt{3})$
15. $-30\sqrt{3} + 20\sqrt{5}$ **17.** $-4\sqrt{10}$ **19.** $29\sqrt{2}$ **21.** $-16\sqrt{5x}$ **23.** $-27x\sqrt{2}$ **25.** $-6\sqrt[3]{5}$ **27.** $7\sqrt[3]{2}$
29. $(9x + 1)\sqrt{5x}$ **31.** $10a^2b\sqrt{5a}$ **33.** $(6xy - x^2y)\sqrt[4]{3x}$ **35.** $4x^2\sqrt[3]{x^2y^2}$ **37.** 0 **39.** $\sqrt{2}$ **41.** $\dfrac{2\sqrt{3}}{3}$
43. $\dfrac{13\sqrt{6}}{6}$ **45.** $\dfrac{13\sqrt{2}}{2}$ **47.** $2\sqrt{x}\left(2 + \dfrac{1}{x}\right)$ **49.** $-15\sqrt{2}$ **51.** $23 + 9\sqrt{3}$ **53.** $11 + 7\sqrt{5}$ **55.** $18 - \sqrt{2}$
57. $8 + 2\sqrt{15}$ **59.** $1 - \sqrt{6}$ **61.** $7 - 4\sqrt{3}$ **63.** $x - \sqrt{3x} - 6$ **65.** $18x - \sqrt{3xy} - y$
67. $8 - 2\sqrt[3]{18} - \sqrt[3]{12}$ **69.** $3\sqrt{5}$ **71.** $7\sqrt{5}$ **73.** $\dfrac{2\sqrt{6}}{3}$ **75.** $3\sqrt[4]{x}$ **77.** $7 - 3\sqrt{y}$ **79.** $-14 + 11\sqrt{2}$
81. $(4x - 2)\sqrt{3x}$ **83.** $\sqrt{y}$ **85.** $16ay\sqrt[3]{3y}$ **87.** $(5xy^2 + 2x)\sqrt[3]{xy}$ **89.** $\dfrac{-301\sqrt{2}}{20}$ **91.** ≈ 5.97
93. $2 + \sqrt{3}$

Cumulative Review Exercises

95. $\dfrac{4y^7}{x^3}$ **96.** $\dfrac{3}{5}, -\dfrac{3}{4}$ **97.** $\dfrac{x^{1/2}}{y^{2/3}}$ **98.** $3x^2y\sqrt[3]{y} - x^4y^2\sqrt[3]{3y^2}$

Just for Fun   **1.** $\dfrac{\sqrt[5]{81x^3y}}{x^2y^2}$ **2.** $\dfrac{\sqrt[4]{27x^3y^2z^3}}{3x^2y^2z^4}$

Exercise Set 8.6

1. 25 **3.** 8 **5.** 81 **7.** 16 **9.** 8 **11.** 9 **13.** No solution   **15.** No solution   **17.** 1 **19.** $-\frac{1}{3}$
21. 5 **23.** -1 **25.** $\frac{1}{4}$ **27.** 1 **29.** -7 **31.** 2 **33.** 2 **35.** $\frac{9}{16}$ **37.** 9 **39.** 4 **41.** 4 **43.** 6
45. $v = \dfrac{p^2}{2}$ **47.** $g = \dfrac{v^2}{2h}$ **49.** $F = \dfrac{v^2m}{R}$ **51.** $m = \dfrac{x^2k}{v_0^2}$ **53.** $\sqrt{x + 3}$ cannot equal a negative number.
55. 0 is the only solution. For any nonzero value, the left side of the equation is positive and the right side is negative.

Cumulative Review Exercises

57. $\dfrac{3a}{2b(2a + 3b)}$ **58.** $t(t - 4)$ **59.** $\dfrac{3}{x + 3}$ **60.** 2

Just for Fun   **1.** 6 **2.** 6 **3.** no solution   **4.** 16 **5.** $n = \dfrac{z^2\sigma^2}{(\bar{x} - \mu)^2}$ **6.** $n = \dfrac{z^2pq}{(p' - p)^2}$

Exercise Set 8.7

1. 5 in   **3.** $\sqrt{45} \approx 6.71$ cm   **5.** $\sqrt{18.25} \approx 4.27$ m   **7.** 13 in   **9.** $\sqrt{16{,}200} \approx 127.28$ ft   **11.** $\sqrt{60} \approx 7.75$ in
13. $\sqrt{5120} \approx 71.55$ ft/sec   **15.** 3.14 sec   **17.** $\sqrt{576} = 24$ in.^{2} **19.** $0.2(\sqrt{149.4})^3 \approx 365.2$ days
21. $\sqrt{1{,}000{,}000} = 1000$ lb   **23.** $\sqrt{320} \approx 17.89$ ft/sec   **25. (a)** $2, -2$ **(b)** $3, -4$ **(c)** $-4, \frac{3}{2}$ **(d)** $5, -1$

Cumulative Review Exercises

26. $-2x - 3$ **27.** All real numbers except 4 and $-\frac{2}{3}$.   **28.** $P_2 = \dfrac{P_1 P_3}{P_1 - P_3}$ **29.** -5

Just for Fun   **1.** $\sqrt{1066.67} \approx 32.66$ ft/sec   **2.** 1.59 cps

Exercise Set 8.8

1. $3 + 0i$ **3.** $3 + 2i$ **5.** $(6 + \sqrt{3}) + 0i$ **7.** $0 + 5i$ **9.** $4 + 2i\sqrt{3}$ **11.** $0 + 3i$ **13.** $9 - 3i$
15. $0 + (2 - 4\sqrt{5})i$ **17.** $15 - 4i$ **19.** $8 + \frac{47}{36}i$ **21.** $18 - 5i$ **23.** $(4\sqrt{2} + \sqrt{3}) - 2i\sqrt{2}$
25. $19 + 12i\sqrt{3}$ **27.** $(2\sqrt{3} - 7) + (7 + 2\sqrt{3})i$ **29.** $-6 - 4i$ **31.** $-1 + 6i$ **33.** $-6.3 - 22.4i$
35. $-4 + 2i\sqrt{3}$ **37.** $-6 + 3i\sqrt{2}$ **39.** $1 + 5i$ **41.** $6 - 22i$ **43.** $\left(\dfrac{1}{2} + \dfrac{\sqrt{3}}{4}\right) + (2\sqrt{3} + 3)i$
45. $\dfrac{21}{4} + \dfrac{9}{2}i\sqrt{2}$ **47.** $-\dfrac{5i}{3}$ **49.** $\dfrac{3 - 2i}{2}$ **51.** $\dfrac{5 - 2i}{5}$ **53.** $\dfrac{49 + 14i}{53}$ **55.** $\dfrac{9 - 12i}{10}$ **57.** $\dfrac{3 + i}{5}$
59. $\dfrac{\sqrt{2} + i\sqrt{6}}{4}$ **61.** $\dfrac{(5\sqrt{10} - 2\sqrt{15}) + (10\sqrt{2} + 5\sqrt{3})i}{45}$ **63.** $7 - 7i$ **65.** $6 + i\sqrt{6}$ **67.** $20.8 - 16.64i$
69. $2\sqrt{15} + 3i\sqrt{2}$ **71.** $7\sqrt{2}$ **73.** $\dfrac{-4 - 5i}{2}$ **75.** $\dfrac{4\sqrt{3} + 8i}{7}$ **77.** $3 + \dfrac{2}{45}i$ **79.** $\dfrac{1}{4} - \dfrac{31}{50}i$ **81.** $\dfrac{3 + 5i}{5}$
83. $-3.33 - 9.01i$ **85.** $0.83 - 3i$ **87.** $1.5 - 0.33i$ **89.** $i^{23} = (i^4)^5 \cdot i^3 = i^3 = -i$ **91.** $\dfrac{1}{i} = \dfrac{1}{i} \cdot \dfrac{-i}{-i} = \dfrac{-i}{1} = -i$
93. false   **95.** false   **97.** true

Cumulative Review Exercises

100. 15 lbs at \$5, 25 lbs at \$5.80   **101.**   **102.** $(2x^2 + 3)^2$ **103.** $(3rs + 2)(5rs - 3)$

Exercise Set 8.9

1. $\{x \mid x \geq -2\}$ **3.** $\{x \mid x \geq 0\}$ **5.** $\{x \mid x \leq 4\}$ **7.** $\{x \mid x \geq -4\}$ **9.** $\{x \mid x \geq -\frac{5}{3}\}$ **11.** $\{x \mid x \geq \frac{1}{14}\}$
13. $\{x \mid x \leq \frac{5}{12}\}$
15. Domain: $\{x \mid x \geq -4\}$
Range: $\{y \mid y \geq 0\}$

17. Domain: $\{x \mid x \geq 2\}$
Range: $\{y \mid y \geq 0\}$

19. Domain: $\{x \mid x \geq -3\}$
Range: $\{y \mid y \leq 0\}$

21. Domain: $\{x \mid x \leq 3\}$
Range: $\{y \mid y \geq 0\}$

23. Domain: $\{x \mid x \leq 2\}$
Range: $\{y \mid y \leq 0\}$

25. Domain: $\{x \mid x \geq 0\}$
Range: $\{y \mid y \geq 0\}$

27. Domain: $\{x \mid x \geq -\frac{1}{2}\}$
Range: $\{y \mid y \geq 0\}$

29. Domain: $\{x \mid x \geq -2\}$
Range: $\{y \mid y \geq 0\}$

31. (a) Domain: $\mathbb{R}$
(c) Range: $\{y \mid y \geq 0\}$
(b)

33. No, since $\sqrt{-4}$ is not a real number. **35.** $\mathbb{R}$ **37.** No, it is a cube root function.
Domain: $\mathbb{R}$
Range: $\mathbb{R}$

Cumulative Review Exercise

38. $1.67\ l$ **39.** $-\dfrac{5}{a^2}$ **40.** 9 min **41.** -2

Just for Fun

1. (a) Domain: $\{x \mid -2 \le x \le 2\}$ **(b)** **2. (a)** Domain: $\{x \mid x \le -2 \text{ or } x \ge 2\}$
(c) Range: $\{y \mid 0 \le y \le 2\}$ **(c)** Range: $\{y \mid y \ge 0\}$

3.

(b)

Review Exercises

1. 3 **2.** 5 **3.** -2 **4.** 4 **5.** 3 **6.** -3 **7.** 6 **8.** 19 **9.** 7 **10.** 93 **11.** $|x|$ **12.** $|x-2|$
13. $|x-y|$ **14.** $|2x-3|$ **15.** $x^{5/2}$ **16.** $x^{5/3}$ **17.** $y^{3/4}$ **18.** $5^{2/7}$ **19.** $\sqrt{x}$ **20.** $\sqrt[5]{y^3}$ **21.** $\sqrt[4]{2}$
22. $\sqrt[4]{8^3}$ **23.** 125 **24.** 9 **25.** $\sqrt[3]{y}$ **26.** x^5 **27.** 16 **28.** 81 **29.** $\sqrt[5]{x}$ **30.** x **31.** 2 **32.** -5
33. not a real number **34.** $\frac{3}{2}$ **35.** $x^{9/10}$ **36.** $\dfrac{4}{y^2}$ **37.** $\dfrac{1}{y^{8/15}}$ **38.** $\dfrac{x^9}{y^{7/2}z^3}$ **39.** $x^{1/5}(1+x)$ **40.** $\dfrac{x-1}{x^4}$
41. $\dfrac{x^{1/6}+1}{x^{2/3}}$ **42.** $\dfrac{2(2+3x^2)}{x^6}$ **43.** $(x^{1/4}-1)(x^{1/4}-4)$ **44.** $(x^{1/5}+3)(x^{1/5}-5)$ **45.** $(2x^{1/2}-3)(3x^{1/2}+2)$
46. $(4x^{1/3}-1)(2x^{1/3}+3)$ **47.** $2\sqrt{6}$ **48.** $4\sqrt{5}$ **49.** $2\sqrt[3]{2}$ **50.** $3\sqrt[3]{2}$ **51.** $5xy^3\sqrt{2xy}$ **52.** $x^2y\sqrt[3]{9y^2}$
53. $2x^2y^3\sqrt[4]{x}$ **54.** $5x^2y^3\sqrt[3]{xy}$ **55.** $2x^2yz\sqrt[3]{x^2y^2z^2}$ **56.** 10 **57.** $2x^3\sqrt{10}$ **58.** $2x^2y^3\sqrt[3]{4x^2}$
59. $xy^2\sqrt[3]{25x}$ **60.** $2x^2y^4\sqrt[4]{x}$ **61.** $6x-2\sqrt{15x}$ **62.** $4y-y^3\sqrt[3]{y^2}$ **63.** $2x^2y^3\sqrt[3]{y}+xy^5\sqrt[3]{18}$
64. $x^2y^2\sqrt[4]{6y^3}+3x^3y\sqrt[4]{y}$ **65.** $\frac{1}{2}$ **66.** $\frac{6}{5}$ **67.** $\frac{1}{3}$ **68.** $\dfrac{x}{2}$ **69.** $\dfrac{x}{2}$ **70.** $\dfrac{4y^2}{x}$ **71.** $3xy$ **72.** $\dfrac{5}{xy}$
73. $\dfrac{\sqrt{2}}{2}$ **74.** $\dfrac{\sqrt{3}}{3}$ **75.** $\dfrac{x\sqrt{7}}{7}$ **76.** $\dfrac{\sqrt{10}}{5}$ **77.** $\dfrac{\sqrt{xy}}{x}$ **78.** $\dfrac{\sqrt{15xy}}{5y}$ **79.** $\dfrac{2\sqrt[3]{x^2}}{x}$ **80.** $\dfrac{\sqrt[3]{75xy^2}}{5y}$
81. $\dfrac{x\sqrt{3y}}{y}$ **82.** $\dfrac{x^2y^2\sqrt{6yz}}{z}$ **83.** $\dfrac{5xy^2\sqrt{3yz}}{3z}$ **84.** $\dfrac{xy^2\sqrt{10xy}}{2}$ **85.** $\dfrac{3x^2z^2\sqrt{15yz}}{5y}$ **86.** $\dfrac{y^3z^4\sqrt{30xz}}{3x^2}$
87. $\dfrac{y\sqrt[3]{4x^2}}{x}$ **88.** $\dfrac{y^2\sqrt[3]{4x}}{2x}$ **89.** 7 **90.** -2 **91.** $x-y^2$ **92.** x^2-y **93.** $28+10\sqrt{3}$ **94.** 5
95. $x+\sqrt{5xy}-\sqrt{3xy}-y\sqrt{15}$ **96.** $2x+\sqrt{xy}-3y$ **97.** $\sqrt[3]{6x^2}-\sqrt[3]{4xy}-\sqrt[3]{9xy}+\sqrt[3]{6y^2}$
98. $-10+5\sqrt{5}$ **99.** $\dfrac{12+3\sqrt{2}}{14}$ **100.** $\dfrac{3x-x\sqrt{x}}{9-x}$ **101.** $\dfrac{x-\sqrt{xy}}{x-y}$ **102.** $\dfrac{\sqrt{30}+\sqrt{15}+2\sqrt{3}+\sqrt{6}}{3}$
103. $\dfrac{x-6\sqrt{x}+9}{x-9}$ **104.** $\dfrac{x-\sqrt{xy}-2y}{x-y}$ **105.** $\dfrac{4\sqrt{y+2}+12}{y-7}$ **106.** $6\sqrt{3}$ **107.** $2\sqrt[3]{x}$ **108.** $-\sqrt[3]{2}$
109. $-4\sqrt{3}$ **110.** $8-13\sqrt[3]{2}$ **111.** $\dfrac{69\sqrt{2}}{8}$ **112.** $(3x^2y^3-4x^3y^4)\sqrt{x}$ **113.** $(2x^2y^2-x+3x^3)\sqrt[3]{xy^2}$
114. 64 **115.** 7 **116.** 8 **117.** 2 **118.** -3 **119.** 2 **120.** 0, 9 **121.** 5 **122.** $L=\dfrac{V^2w}{2}$
123. $A=\pi r^2$ **124.** $\sqrt{29}\approx 5.39$ m **125.** $\sqrt{1280}\approx 35.78$ **126.** $2\pi\sqrt{2}\approx 2.83\pi\approx 8.89$ **127.** $5+0i$
128. $-6+0i$ **129.** $2-2i$ **130.** $3+4i$ **131.** $7+i$ **132.** $1-2i$ **133.** $2+5i$
134. $3\sqrt{3}+(\sqrt{5}-\sqrt{7})i$ **135.** $12+8i$ **136.** $-2\sqrt{3}+2i$ **137.** $6\sqrt{2}+4i$ **138.** $-6+6i$
139. $17-6i$ **140.** $(24+3\sqrt{5})+(4\sqrt{3}-6\sqrt{15})i$ **141.** $-2i/3$ **142.** $-3i/5$ **143.** $(-2-\sqrt{3})i/2$
144. $5(3-2i)/13$ **145.** $(5\sqrt{3}+3i\sqrt{2})/31$ **146.** $[(\sqrt{10}-6\sqrt{2})+(3\sqrt{2}+2\sqrt{10})i]/10$

147. $\{x \mid x \geq 6\}$ **148.** $\{x \mid x \geq -\frac{5}{3}\}$ **149.** $\{x \mid x \leq \frac{5}{2}\}$ **150.** $\{x \mid x \geq 6\}$

151. Domain: $\{x \mid x \geq 0\}$ **152.** Domain: $\{x \mid x \geq -2\}$ **153.** Domain: $\{x \mid \leq 5\}$ **154.** Domain: $\{x \mid x \geq 3\}$
    Range: $\{y \mid y \geq 0\}$ Range: $\{y \mid y \geq 0\}$ Range: $\{y \mid y \geq 0\}$ Range: $\{y \mid y \leq 0\}$

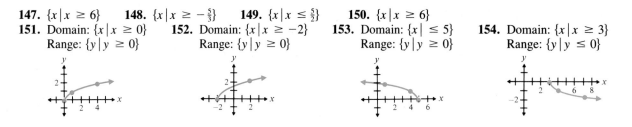

Practice Test

1. 26   **2.** $|3x - 4|$ **3.** $\dfrac{1}{y^{7/6}}$ **4.** $(2x^{1/3} + 5)(x^{1/3} - 2)$ **5.** $5x^2y^4\sqrt{2x}$ **6.** $2x^3y^3\sqrt[3]{5x^2y}$ **7.** $\dfrac{x^2y^2\sqrt{yz}}{2z}$

8. $\dfrac{\sqrt[3]{x^2}}{x}$ **9.** $\dfrac{2 - \sqrt{2}}{2}$ **10.** $-20\sqrt{3}$ **11.** $(2xy + 2x^2y^2)\sqrt[3]{y^2}$ **12.** $2\sqrt{5} - 2\sqrt{10} - 6 + 6\sqrt{2}$ **13.** 13

14. -3 **15.** 16   **16.** $h = \dfrac{8w^2}{g}$ **17.** 13 ft   **18.** $18 + 2\sqrt{2} + (6\sqrt{2} - 6)i$ **19.** $\dfrac{\sqrt{5} + i\sqrt{10}}{6}$

20. Domain: $\{x \mid x \geq -2\}$
     Range: $\{y \mid y \geq 0\}$

Cumulative Review Test

1. $\frac{57}{9}$ **2.**

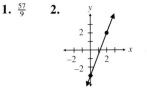

3. (a) A set of ordered pairs   **(b)** A set of ordered pairs no two of which have the same first coordinate.   **4.** $\left(\frac{2}{11}, -\frac{16}{11}\right)$

5. $\dfrac{3}{2xy^3}$ **6.** $6x^3 - 23x^2 + 8x + 30$ **7.** $3x + 4 + \dfrac{2}{x + 2}$ **8.**   **9.** $2(x - 3 + y)(x - 3 - y)$

10. $x \neq \frac{3}{5}$ **11.** $\dfrac{2x + 1}{3}$ **12.** $\dfrac{x + 1}{2x + 3}$ **13.** $\dfrac{x^2 + 8x + 5}{(x + 3)(x - 1)(2x + 5)}$ **14.** -5 **15.** $x^{7/2}y$ **16.** $2x^4y^9\sqrt[3]{2xy^2}$

17. $3, -3$ **18.** $\dfrac{6 - 2i\sqrt{6}}{15}$ **19.** $1\frac{1}{5}$ hr   **20.** $\sqrt{1300} \approx 36.1$ ft

CHAPTER 9

Exercise Set 9.1

1. $5, -5$ **3.** $5\sqrt{3}, -5\sqrt{3}$ **5.** $2\sqrt{7}, -2\sqrt{7}$ **7.** $8, 0$ **9.** $\frac{1}{3}, -1$ **11.** $-0.9, -2.7$

13. $\dfrac{5 + 2\sqrt{3}}{2}, \dfrac{5 - 2\sqrt{3}}{2}$ **15.** $-\frac{1}{20}, -\frac{9}{20}$ **17.** $s = \sqrt{A}$ **19.** $r = \sqrt{\dfrac{A}{\pi}}$ **21.** $F = \sqrt{F_x^2 + F_y^2}$

23. $V_x = \sqrt{V^2 - V_y^2}$ **25.** $b = \sqrt{a^2 - L}$ **27.** $t = \sqrt{\dfrac{v - m}{n}}$ **29.** $b = \sqrt{d + 4ac}$ **31.** $d = \sqrt{\dfrac{w - 3l}{2}}$

33. 4   **35.** $\sqrt{175} \approx 13.23$ **37.** $\sqrt{41} \approx 6.40$ **39.** $\sqrt{128} \approx 11.31$ **41.** 10   **43.** 1, -3 **45.** 5, -1

47. $-2, -1$ **49.** 5, 3   **51.** $-1 + i\sqrt{14}, -1 - i\sqrt{14}$ **53.** $-2, -3$ **55.** $-3, -6$ **57.** 7, 8

59. 6, -2 **61.** $-1 + \sqrt{7}, -1 - \sqrt{7}$ **63.** $-3 + \sqrt{3}, -3 - \sqrt{3}$ **65.** $(5 + \sqrt{57})/2, (5 - \sqrt{57})/2$

67. 0, -2 **69.** 0, $\frac{1}{3}$ **71.** 1, -3 **73.** $(-9 + \sqrt{73})/2, (-9 - \sqrt{73})/2$ **75.** $-8, -3$

77. $(-2 + i\sqrt{2})/2, (-2 - i\sqrt{2})/2$ **79.** $1 + i, 1 - i$ **81.** $(-3 + \sqrt{59})/10, (-3 - \sqrt{59})/10$

83. $(-1 + i\sqrt{191})/12, (-1 - i\sqrt{191})/12$ **85.** $-a \pm \sqrt{k}$ **87.** 7, 9   **89.** 5 ft, 12 ft

91. $12 + 12\sqrt{2} \approx 28.97$ ft   **93.** Make the coefficient of the squared term equal to 1.

Cumulative Review Exercises

96. $z = \dfrac{3xy}{3y + 1}$ **97.** $|x^2 - 4x|$ **98.** $\frac{1}{5}$ **99.** $\dfrac{x^{1/2}}{y^{3/2}}$

Just for Fun   **1. (a)** $S = 32 + 80\sqrt{\pi} \approx 173.80$ sq. in.   **(b)** $r = \dfrac{4\sqrt{\pi}}{\pi} \approx 2.26$ in   **3.** $(x + 2)^2 + (y - 3)^2 = 16$

4. $4(x - 6)^2 + 9(y + 4)^2 = 144$ **5.** $(x + 2)^2 - 4(y + 2)^2 = 16$

Exercise Set 9.2

1. Two real solutions   **3.** No real solution   **5.** Two real solutions   **7.** One real solution   **9.** Two real solutions

11. No real solution   **13.** Two real solutions   **15.** Two real solutions   **17.** 1, 2   **19.** 4, 5   **21.** 3, -8

23. 4,9   **25.** 5, -5 **27.** 0, 3   **29.** $(7 + i\sqrt{159})/8, (7 - i\sqrt{159})/8$ **31.** $(7 + \sqrt{17})/4, (7 - \sqrt{17})/4$

33. $\frac{1}{3}, -\frac{1}{2}$ **35.** $(2 + \sqrt{6})/2, (2 - \sqrt{6})/2$ **37.** $(2 + i\sqrt{2})/2, (2 - i\sqrt{2})/2$ **39.** $\frac{5}{4}, -1$ **41.** $\frac{9}{2}, -1$

43. 3, $\frac{5}{2}$ **45.** $(-1 + i\sqrt{23})/4, (-1 - i\sqrt{23})/4$ **47.** 0, -3 **49.** $\sqrt{7}/2, -\sqrt{7}/2$

51. $(-6 + 2\sqrt{6})/3, (-6 - 2\sqrt{6})/3$ **53.** $(11 + \sqrt{241})/6, (11 - \sqrt{241})/6$

55. $(-0.6 + \sqrt{0.84})/0.2, (-0.6 - \sqrt{0.84})/0.2$ or $-3 + \sqrt{21}, -3 - \sqrt{21}$ if decimals eliminated first

57. $(-0.94 + \sqrt{32.3116})/3.24, (-0.94 - \sqrt{32.3116})/3.24$ or $(-47 + \sqrt{80,779})/162, (-47 - \sqrt{80,779})/162$, if decimals eliminated first

59. 2   **61.** $w = 3$ ft, $l = 7$ ft   **63.** 2 in.   **65.** ≈ 2.18 sec   **67.** ≈ 16.37 sec

69. 50 mph going, 60 mph returning   **71.** larger heater 81.1 min, smaller heater 87.1 min   **73. (a)** 0.5 thousand

(b) 11.3 thousand   **(c)** $1.55   **75. (a)** 1.2   **(b)** 2.16   **(c)** 7.3 hr

77. Yes, if both sides of either equation are multiplied by -1 you obtain two identical equations.

Cumulative Review Exercises

79. $2xy^2z^4\sqrt[5]{2x^4y^2}$ **80.** $2x^2y^3\sqrt[3]{x} + 2xy^3\sqrt[3]{3}$ **81.** $\dfrac{x^2 + 2x\sqrt{y} + y}{x^2 - y}$ **82.** 6

Just for Fun   **1.** $2\sqrt{5}, -\sqrt{5}$ **2.** $-2\sqrt{6}, -3\sqrt{6}$ **3.** 1, -1, 2, -2 **4.** 800 mph

5. $(-0.12 + \sqrt{14.3952})/1.2 \approx 3.0617$ **6.** $x = \dfrac{-1 \pm \sqrt{1 + 4y}}{2}$ **7.** $x = \dfrac{-3y \pm \sqrt{9y^2 - 16}}{2}$

8. $x = \dfrac{-3 \pm \sqrt{9 + 20y}}{2y}$

Exercise Set 9.3

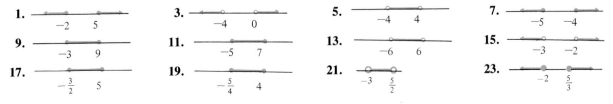

25. $(-4, -1) \cup (1, \infty)$ **27.** $(-\infty, -3] \cup [1, 4]$ **29.** $[-3, 0] \cup [3, \infty)$ **31.** $(-6, -3) \cup (2, \infty)$ **33.** $[3, \infty)$
35. $(-\infty, -3) \cup (-3, 4)$ **37.** $\{x \mid x < -3 \text{ or } x > 1\}$ **39.** $\{x \mid 1 < x \le 4\}$ **41.** $\{x \mid 1 < x < 2\}$
43. $\{x \mid x \le -2 \text{ or } x > \frac{1}{2}\}$ **45.** $\{x \mid -4 \le x < 4\}$ **47.** $\{x \mid x < \frac{1}{2} \text{ or } x > 2\}$
49. $(-\infty, -6) \cup (-2, 4)$, $\{x \mid x < -6 \text{ or } -2 < x < 4\}$ **51.** $[1, 3) \cup [6, \infty)$, $\{x \mid 1 \le x < 3 \text{ or } x \ge 6\}$
53. $(-\infty, -4) \cup (1, 6]$, $\{x \mid x < -4 \text{ or } 1 < x \le 6\}$ **55.** $(-\frac{5}{2}, 3) \cup (6, \infty)$, $\{x \mid -\frac{5}{2} < x < 3 \text{ or } x > 6\}$
57. $[-2, 0) \cup [\frac{3}{2}, \infty)$, $\{x \mid -2 \le x < 0 \text{ or } x \ge \frac{3}{2}\}$ **59.**   **61.**   **63.**
65.   **67.**   **69.**   **71.** All real numbers
73. No real solution   **75.** All real numbers except 0 and 3

Cumulative Review Exercises

76. $x \ne \pm 2$ **77.** $x \le 4$ **78.** $\dfrac{a^3 - ab^3}{b^2 + a^3 b}$ **79.** $-8 - 6i$

Just for Fun   **1.**   **2.**   **3.**

Exercise Set 9.4

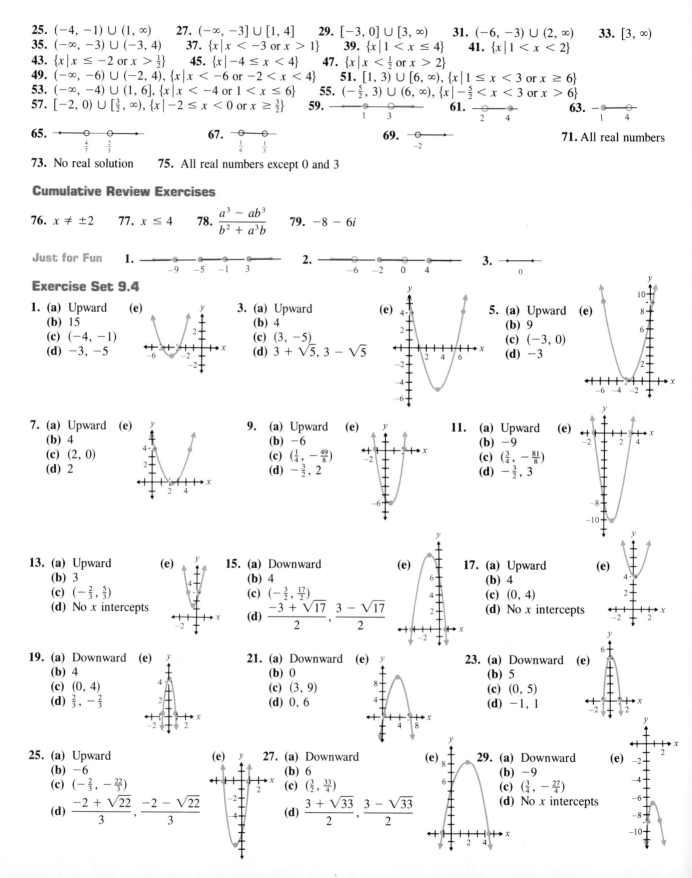

1. (a) Upward **(b)** 15 **(c)** $(-4, -1)$ **(d)** $-3, -5$ **(e)**

3. (a) Upward **(b)** 4 **(c)** $(3, -5)$ **(d)** $3 + \sqrt{5}, 3 - \sqrt{5}$ **(e)**

5. (a) Upward **(b)** 9 **(c)** $(-3, 0)$ **(d)** -3 **(e)**

7. (a) Upward **(b)** 4 **(c)** $(2, 0)$ **(d)** 2 **(e)**

9. (a) Upward **(b)** -6 **(c)** $(\frac{1}{4}, -\frac{49}{8})$ **(d)** $-\frac{3}{2}, 2$ **(e)**

11. (a) Upward **(b)** -9 **(c)** $(\frac{3}{4}, -\frac{81}{8})$ **(d)** $-\frac{3}{2}, 3$ **(e)**

13. (a) Upward **(b)** 3 **(c)** $(-\frac{2}{3}, \frac{5}{3})$ **(d)** No x intercepts **(e)**

15. (a) Downward **(b)** 4 **(c)** $(-\frac{3}{2}, \frac{17}{2})$ **(d)** $\dfrac{-3 + \sqrt{17}}{2}, \dfrac{3 - \sqrt{17}}{2}$ **(e)**

17. (a) Upward **(b)** 4 **(c)** $(0, 4)$ **(d)** No x intercepts **(e)**

19. (a) Downward **(b)** 4 **(c)** $(0, 4)$ **(d)** $\frac{2}{3}, -\frac{2}{3}$ **(e)**

21. (a) Downward **(b)** 0 **(c)** $(3, 9)$ **(d)** 0, 6 **(e)**

23. (a) Downward **(b)** 5 **(c)** $(0, 5)$ **(d)** $-1, 1$ **(e)**

25. (a) Upward **(b)** -6 **(c)** $(-\frac{2}{3}, -\frac{22}{3})$ **(d)** $\dfrac{-2 + \sqrt{22}}{3}, \dfrac{-2 - \sqrt{22}}{3}$ **(e)**

27. (a) Downward **(b)** 6 **(c)** $(\frac{3}{2}, \frac{33}{4})$ **(d)** $\dfrac{3 + \sqrt{33}}{2}, \dfrac{3 - \sqrt{33}}{2}$ **(e)**

29. (a) Downward **(b)** -9 **(c)** $(\frac{3}{4}, -\frac{27}{4})$ **(d)** No x intercepts **(e)**

31. $y = x^2 - 10x + 24$ **33.** $y = x^2 + x - 12$ **35.** $y = x^2 + x - 6$ **37.** $y = x^2 - 4x + 4$
39. $y = 3x^2 + 4x - 4$ **41.** $y = 6x^2 - x - 2$ **43.** 28   **45.** 38.8 ft   **47.** 11.2 in.
49. (a) 432 ft   **(b)**
(c) 576 ft   **(d)** 6 sec   **51. (a)**   **(b)** $2 **(c)** $22 **(d)** $12
(e) 12 sec   **(e)** $10,000

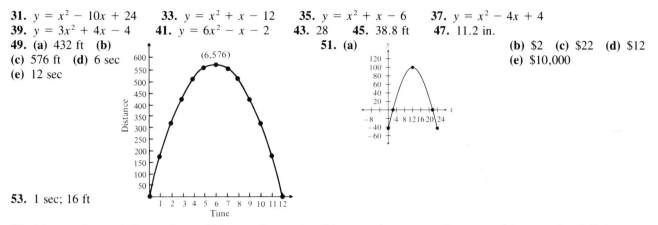

53. 1 sec; 16 ft

55. (a) no real roots indicates there will be no x intercepts   **(b)** one real root, one x intercept   **(c)** two real and distinct roots, 2 distinct x intercepts.
57. If the vertex is in the first or second quadrant and $a > 0$, there are no x intercepts. If the vertex is in the third or fourth quadrant and $a < 0$, there are no x intercepts. If the vertex is on the x axis, there is a single x intercept. In all other cases there are two distinct x intercepts.
59. (b) Yes   **(c)** No   **(d)**

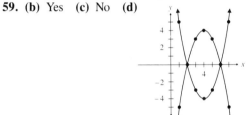

Cumulative Review Exercises

60. $\left(\frac{56}{11}, -\frac{8}{11}\right)$ **61.** $1153.85 at 12%, $13,846.15 at 5.5%   **62.** fifth   **63.** $(3x - 5)(2x + 9)$

Just for Fun   **1.** $w = 16$ ft, $l = 16$ ft   **2.** $w = 25$ ft, $l = 50$ ft

Exercise Set 9.5

1. (a) $x^2 + 3x - 10$ **(b)** 0   **3. (a)** $x^3 - 12x + 16$ **(b)** 27   **5. (a)** $x^2 - 2x - 8$ **(b)** -5
7. (a) $x^2 + x - 2$ **(b)** 0   **9. (a)** $x^3 + 2x^2 - 4x - 8$ **(b)** -24 **11. (a)** $x^2 - 2$ **(b)** 14
13. (a) $x - 4 + \sqrt{x + 6}$ **(b)** 2   **15. (a)** $(x - 4)\sqrt{x + 6}$ **(b)** 24   **17. (a)** $\sqrt{x + 6} - 4$ **(b)** $\sqrt{13} - 4$
19. yes   **21.** yes   **23.** No, composition of functions is not commutative
25. (a) $(f \circ g)(x) = x$ and $(g \circ f)(x) = x$ **(b)** $(f \circ g)(x) = \sqrt{x^2}$, without specifying that $x \geq 0$, $\sqrt{x^2} = |x|$, not x.

Cumulative Review Exercises

26. All functions are relations, but not all relations are functions.
27. Yes, both linear and quadratic functions are polynomial functions.   **28.** $f(x) = x^3$
29. All polynomial functions are rational functions, but not all rational functions are polynomial functions.
30. No, if a function is a polynomial function, it cannot be a square root function, and vice versa.

Exercise Set 9.6

1. No   **3.** Yes   **5.** No   **7.** Yes   **9.** No   **11.** $f(x)$: Domain $\{-2, -1, 2, 4, 9\}$ Range $\{0, 3, 4, 6, 7\}$
$f^{-1}(x)$: Domain $\{0, 3, 4, 6, 7\}$ Range $\{-2, -1, 2, 4, 9\}$
13. $f(x)$: Domain $\{-2.9, 0, 1.7, 5.7\}$ Range $\{-3.4, 3, 4, 9.76\}$
$f^{-1}(x)$: Domain $\{-3.4, 3, 4, 9.76\}$ Range $\{-2.9, 0, 1.7, 5.7\}$

15. $f^{-1}(x) = (x - 8)/2$ **17.** $f^{-1}(x) = -(x + 10)/3$ **19.** $f^{-1}(x) = (5x + 3)/10$ **21.** $f^{-1}(x) = (-x + 6)/3$

23. $f^{-1}(x) = (3x - 6)/4$ **25.** $f^{-1}(x) = (24x + 16)/15$

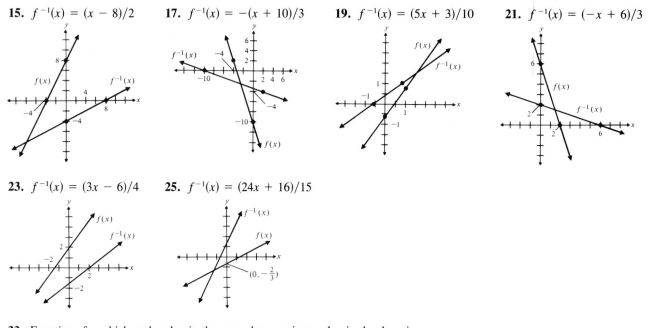

33. Functions for which each value in the range has a unique value in the domain.
35. The domain of the function is the range of the inverse function, and the range of the funciton is the domain of the inverse function.

Cumulative Review Exercises

37. $(4, 2, -1)$ **38.** $x^2 + 4x - 2 - \dfrac{4}{x + 2}$ **39.** $\dfrac{2x\sqrt{2y}}{y}$ **40.** $x = -1 \pm \sqrt{7}$

Just for Fun

1. $f^{-1}(x) = \sqrt{x^2 + 9}$

$f(x) \begin{cases} \text{Domain } \{x \mid x \geq 3\} \\ \text{Range } \{y \mid y \geq 0\} \end{cases}$ $f^{-1}(x) \begin{cases} \text{Domain } \{x \mid x \geq 0\} \\ \text{Range } \{y \mid y \geq 3\} \end{cases}$

2. $f^{-1}(x) = -\sqrt{x^2 + 9}$

$f(x) \begin{cases} \text{Domain } \{x \mid x \leq -3\} \\ \text{Range } \{y \mid y \geq 0\} \end{cases}$ $f^{-1}(x) \begin{cases} \text{Domain } \{x \mid x \geq 0\} \\ \text{Range } \{y \mid y \leq -3\} \end{cases}$

Review Exercises

1. $4 + 2\sqrt{5}, 4 - 2\sqrt{5}$ **2.** $\dfrac{4 + 2\sqrt{15}}{3}, \dfrac{4 - 2\sqrt{15}}{3}$ **3.** $1, \frac{1}{3}$ **4.** $\frac{5}{4}, -\frac{3}{4}$ **5.** $F_b = \sqrt{F_T^2 - F_a^2}$
6. $s = \sqrt{\dfrac{L - 3r}{2}}$ **7.** $\sqrt{128} \approx 11.31$ **8.** $\sqrt{128} \approx 11.31$ **9.** $2, 8$ **10.** $3, 5$ **11.** $1, 13$ **12.** $2, -3$

13. $9, -6$ **14.** $1, -6$ **15.** $-1 + \sqrt{6}, -1 - \sqrt{6}$ **16.** $(3 + i\sqrt{23})/2, (3 - i\sqrt{23})/2$
17. $2 + 2i\sqrt{7}, 2 - 2i\sqrt{7}$ **18.** $5, -3$ **19.** $(1 + i\sqrt{47})/4, (1 - i\sqrt{47})/4$ **20.** $-3 + \sqrt{19}, -3 - \sqrt{19}$
21. Two real solutions **22.** No real solution **23.** No real solution **24.** No real solution **25.** One real solution
26. No real solution **27.** Two real solutions **28.** Two real solutions **29.** $2, 7$ **30.** $3, -10$ **31.** $2, 5$
32. $2, -\frac{3}{5}$ **33.** $-2, 9$ **34.** $(1 + i\sqrt{119})/2, (1 - i\sqrt{119})/2$ **35.** $\frac{3}{2}, -\frac{5}{3}$
36. $(-2 + \sqrt{10})/2, (-2 - \sqrt{10})/2$ **37.** $(3 + \sqrt{57})/4, (3 - \sqrt{57})/4$ **38.** $3 + \sqrt{2}, 3 - \sqrt{2}$

39. $(2 + i\sqrt{14})/3$, $(2 - i\sqrt{14})/3$ **40.** $(3 + \sqrt{33})/3$, $(3 - \sqrt{33})/3$ **41.** 0, $-\frac{3}{2}$ **42.** 0, $\frac{5}{2}$ **43.** 3, 8
44. 7, 9 **45.** 5, -8 **46.** 3, -9 **47.** 10, -6 **48.** $(1 + i\sqrt{167})/2$, $(1 - i\sqrt{167})/2$
49. $(-11 + \sqrt{73})/2$, $(-11 - \sqrt{73})/2$ **50.** 5, -5 **51.** 0, -6 **52.** $\frac{1}{2}$, -3
53. $(9 + i\sqrt{39})/6$, $(9 - i\sqrt{39})/6$ **54.** $\frac{2}{3}$, $-\frac{3}{2}$ **55.** $(-3 + \sqrt{33})/2$, $(-3 - \sqrt{33})/2$ **56.** 2, $\frac{5}{3}$ **57.** 1, $-\frac{8}{3}$
58. $(3 + 3\sqrt{3})/2$, $(3 - 3\sqrt{3})/2$ **59.** 0, $\frac{5}{2}$ **60.** 0, $-\frac{5}{3}$ **61.** $\frac{1}{4}$, $-\frac{3}{2}$ **62.** $\frac{5}{2}$, $-\frac{5}{3}$
63. ─────── -5 -1 **64.** ─────── -5 3 **65.** ─────── 5 6 **66.** ─────── -3 0

67. ─────── -4 $\frac{4}{3}$ **68.** ─────── -2 2 **69.** ─────── $-\sqrt{5}$ $\sqrt{5}$ **70.** ─────── $-\frac{5}{3}$ $\frac{5}{3}$

71. $(-\infty, -2) \cup (3, \infty)$, $\{x \mid x < -2 \text{ or } x > 3\}$ **72.** $(-2, 5]$, $\{x \mid -2 < x \le 5\}$
73. $(-\infty, -1) \cup [2, \infty)$, $\{x \mid x < -1 \text{ or } x \ge 2\}$ **74.** $(-\frac{5}{3}, 6)$, $\{x \mid -\frac{5}{3} < x < 6\}$
75. $(-3, -1) \cup (2, \infty)$, $\{x \mid -3 < x < -1 \text{ or } x > 2\}$ **76.** $(-\infty, 0] \cup [3, 5]$, $\{x \mid x \le 0 \text{ or } 3 \le x \le 5\}$
77. $[-4, 1] \cup [3, \infty)$, $\{x \mid -4 \le x \le 1 \text{ or } x \ge 3\}$ **78.** $(-\infty, -5) \cup (-2, 0)$, $\{x \mid x < -5 \text{ or } -2 < x < 0\}$
79. $(-2, 0) \cup (4, \infty)$, $\{x \mid -2 < x < 0 \text{ or } x > 4\}$ **80.** $(-\infty, -3) \cup (2, 5)$, $\{x \mid x < -3 \text{ or } 2 < x < 5\}$
81. $(-2, 3] \cup (5, \infty)$, $\{x \mid -2 < x \le 3 \text{ or } x > 5\}$ **82.** $(-\infty, -3) \cup [0, 5]$, $\{x \mid x < -3 \text{ or } 0 \le x \le 5\}$
83. ─────── -7 -4 **84.** ─────── -2 2 **85.** ─────── 5 **86.** ─────── -5 15

87. (a) Upward (e) **88.** (a) Upward (e) **89.** (a) Upward (e)
(b) 0 (b) -8 (b) -16
(c) $(-3, -9)$ (c) $(-1, -9)$ (c) $(-1, -18)$
(d) 0, -6 (d) -4, 2 (d) -4, 2

90. (a) Downward (e) **91.** (a) Downward (e) **92.** (a) Upward (e)
(b) -9 (b) 15 (b) 8
(c) $(0, -9)$ (c) $(-\frac{1}{4}, \frac{121}{8})$ (c) $(-\frac{3}{2}, \frac{23}{4})$
(d) No x intercepts (d) $\frac{5}{2}$, -3 (d) No x intercepts

93. $y = x^2 - x - 6$ **94.** $y = 3x^2 + 7x - 6$ **95.** $y = x^2 + 6x + 9$ **96.** $y = 6x^2 - 7x + 2$ **97.** 9, 10
98. 5, 9 **99.** $w = 6$ in., $l = 11$ in. **100.** \$475 **101.** 1656 ft **102.** 36 ft **103.** (a) 40 ml (b) 150°C
104. larger machine 23.51 hr, smaller machine 24.51 hrs. **105.** (a) $t = 2.6$ sec (b) $h = 41.8$ ft
106. (a) 1 sec (b) 80 ft **107.** $x^2 - x - 1$ **108.** 5 **109.** $-x^2 + 5x - 9$ **110.** -15

111. $2x^3 - 11x^2 + 23x - 20$ **112.** 70 **113.** $\dfrac{x^2 - 3x + 4}{2x - 5}$, $x \ne \frac{5}{2}$ **114.** -2 **115.** $4x^2 - 26x + 44$

116. 8 **117.** $2x^2 - 6x + 3$ **118.** 39 **119.** $3x + 2 + \sqrt{x - 4}$ **120.** $3x + 2 - \sqrt{x - 4}$

121. $(3x + 2)\sqrt{x - 4}$ **122.** $\dfrac{\sqrt{x - 4}}{3x + 2}$, $x \ne -\frac{2}{3}$ **123.** $3\sqrt{x - 4} + 2$ **124.** $\sqrt{3x - 2}$, $x \ge \frac{2}{3}$

125. One to one **126.** One to one **127.** Not one to one **128.** Not one to one **129.** One to one
130. Not one to one
131. $f(x)$, domain $\{-4, 0, 5, 6\}$, range $\{-3, 2, 3, 7\}$ **132.** $f(x)$, domain $\{-3, -1, \frac{1}{2}, \sqrt{5}\}$, range $\{2, \sqrt{7}, 3, 8\}$
$f^{-1}(x)$, domain $\{-3, 2, 3, 7\}$, range $\{-4, 0, 5, 6\}$ $f^{-1}(x)$, domain $\{2, \sqrt{7}, 3, 8\}$, range $\{-3, -1, \frac{1}{2}, \sqrt{5}\}$
133. $f^{-1}(x) = (x + 2)/4$ **134.** $f^{-1}(x) = -\frac{1}{3}(x + 5)$ **135.** $f^{-1}(x) = (3x - 5)/2$ **136.** $f^{-1}(x) = (-15x + 9)/10$

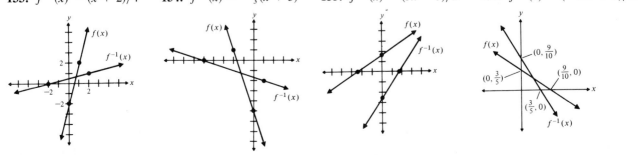

Practice Test

1. 3, −4 **2.** −1 + $i\sqrt{2}$, −1 − $i\sqrt{2}$ **3.** −1, 6 **4.** −4 + $\sqrt{11}$, −4 − $\sqrt{11}$ **5.** 0, $\frac{5}{3}$ **6.** $\frac{1}{2}$, −5
7. $b = \sqrt{3a - P}$ **8.** Two real solutions **9.** —— **10.** —— **11.** **(a)** $[-\frac{5}{2}, -2)$ **(b)** $\{x \,|\, -\frac{5}{2} \le x < -2\}$
12. **(a)** Upward **(e)** **13.** **(a)** Downward **(e)** **14.** $y = 2x^2 + 11x - 6$
(b) −8
(c) $(1, -9)$
(d) 4, −2

(b) 9
(c) $(-\frac{3}{4}, \frac{81}{8})$
(d) $\frac{3}{2}$, −3

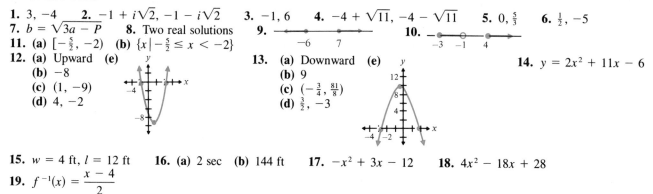

15. $w = 4$ ft, $l = 12$ ft **16.** **(a)** 2 sec **(b)** 144 ft **17.** $-x^2 + 3x - 12$ **18.** $4x^2 - 18x + 28$
19. $f^{-1}(x) = \dfrac{x - 4}{2}$

CHAPTER 10

Exercise Set 10.1

1. $x^2 + y^2 = 9$ **3.** $(x - 3)^2 + y^2 = 1$ **5.** $(x + 6)^2 + (y - 5)^2 = 25$

7. $(x - 4)^2 + (y - 7)^2 = 8$ **9.**

11. **13.** **15.** **17.**

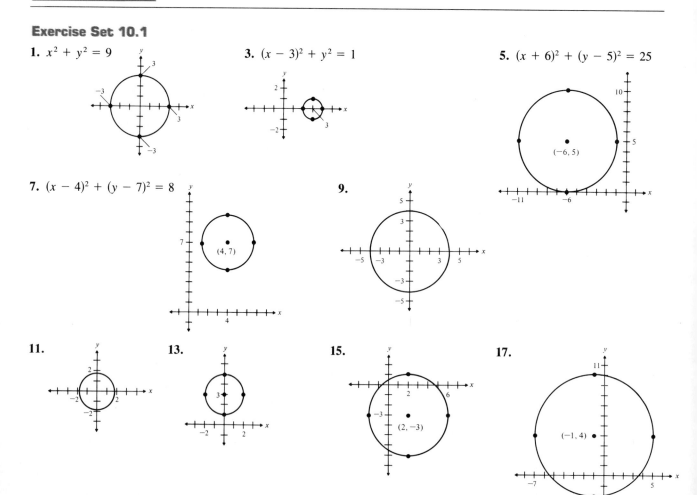

19. $x^2 + y^2 = 1$ **21.** $(x + 4)^2 + (y - 6)^2 = 4$ **23.** $(x + 5)^2 + (y + 3)^2 = 4$

25. $x^2 + (y + 5)^2 = 10^2$ **27.** $(x + 4)^2 + y^2 = 5^2$ **29.** $(x + 1)^2 + (y - 2)^2 = 3^2$ **31.** $(x + 3)^2 + (y - 1)^2 = 2^2$

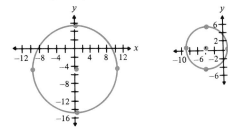 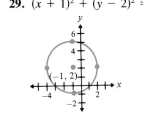 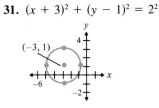

33. $(x - 4)^2 + (y + 1)^2 = 2^2$

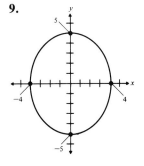

Cumulative Review Exercises

35. $8(x - 2)(x^2 + 2x + 4)$ **36.** $T_2 = \dfrac{T_1}{1 - E}$ **37.** $P = \sqrt{2E - V}$ **38.** $\dfrac{5 + i\sqrt{35}}{6}, \dfrac{5 - i\sqrt{35}}{6}$

Exercise Set 10.2

1.

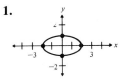

3.

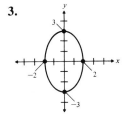

5.

7.

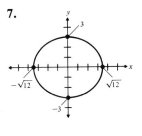

9.

11.

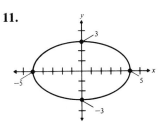

13. The set of points in a plane, the sum of whose distance from two fixed points is a constant.

15. The graph becomes more circular as the value of b approaches the value of a. When $a = b$ the graph is a circle.

Cumulative Review Exercises

16. ○———● -4 14

17. ●———● -2 6

18. ●———● -2 6

19. ←———○———○———→ -2 6

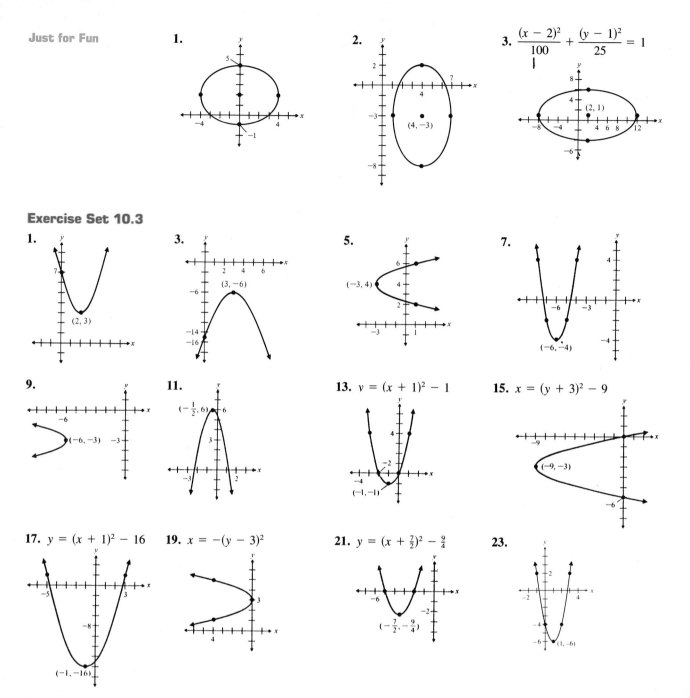

Just for Fun

1.

2.

3. $\dfrac{(x-2)^2}{100} + \dfrac{(y-1)^2}{25} = 1$

(2, 1)

Exercise Set 10.3

1.

(2, 3)

3.

(3, −6)

5.

(−3, 4)

7.

(−6, −4)

9.

(−6, −3)

11.

$(-\frac{1}{2}, 6)$

13. $v = (x + 1)^2 - 1$

(−1, −1)

15. $x = (y + 3)^2 - 9$

(−9, −3)

17. $y = (x + 1)^2 - 16$

(−1, −16)

19. $x = -(y - 3)^2$

21. $y = (x + \frac{7}{2})^2 - \frac{9}{4}$

$(-\frac{7}{2}, -\frac{9}{4})$

23.

(1, −6)

25. $y = a(x - h)^2 + k$ opens up if $a > 0$ and opens down if $a < 0$, $x = a(y - k)^2 + h$ opens to the right if $a > 0$ and opens to the left if $a < 0$.

27. Yes, they will all pass the vertical line test. D: $\mathbb{R}$, R: $\{y \mid y \geq k\}$

29. Both graphs will have their vertex at (3, 4). The graph of $y = 2(x - 3)^2 + 4$ will open upward and have no x intercepts. The graph of $y = -2(x - 3)^2 + 4$ will open downward and have two x intercepts.

Cumulative Review Exercises

30. $y = -\frac{1}{2}x + 1$ **31.** 128   **32.** (a) $\frac{13}{9}$ (b) $-2a^2 - 4ab - 2b^2 - a - b + 2$ **33.** 15

Exercise Set 10.4

1. $y = \pm\frac{1}{2}x$

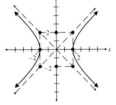

3. $y = \pm\frac{3}{4}x$

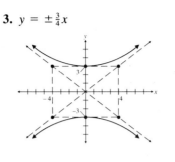

5. $y = \pm\frac{5}{6}x$

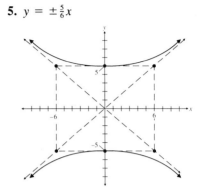

7. $y = \pm x$

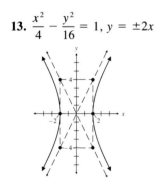

9. $y = \pm\frac{4}{9}x$

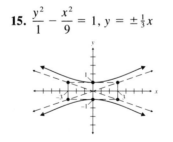

11. $y = \pm\frac{5}{4}x$

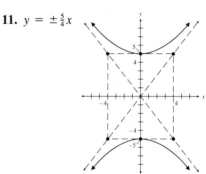

13. $\dfrac{x^2}{4} - \dfrac{y^2}{16} = 1$, $y = \pm 2x$

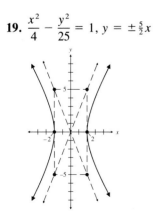

15. $\dfrac{y^2}{1} - \dfrac{x^2}{9} = 1$, $y = \pm\frac{1}{3}x$

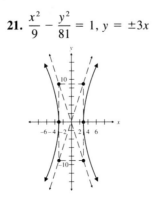

17. $\dfrac{y^2}{36} - \dfrac{x^2}{4} = 1$, $y = \pm 3x$

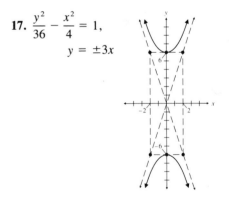

19. $\dfrac{x^2}{4} - \dfrac{y^2}{25} = 1$, $y = \pm\frac{5}{2}x$

21. $\dfrac{x^2}{9} - \dfrac{y^2}{81} = 1$, $y = \pm 3x$

23.

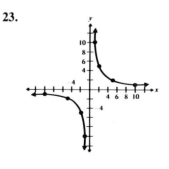

25. 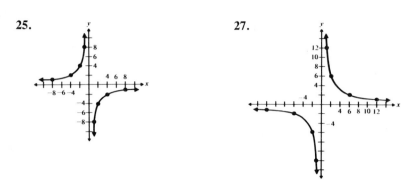   **27.**

29. Asymptotes are lines that the hyperbola approaches, $y = \pm \frac{b}{a}x$

31. Hyperbola with vertices at $(a, 0)$ and $(-a, 0)$. Transverse axis along x axis. Conjugate axis along y axis. Asymptotes, $y = \pm \frac{b}{a}x$.

33. The transverse axis of both graphs would be along the x axis. The vertices of the second graph will be closer to the origin, and the hyperbola of the second graph will open wider than that of the first graph.

35. Parabola   **37.** Hyperbola   **39.** Parabola   **41.** Hyperbola   **43.** Ellipse   **45.** Parabola   **47.** Circle
49. Circle   **51.** Hyperbola

Cumulative Review Exercises

54. 48   **55.** Domain: the set of values that can be used for the independent variable.   **56.**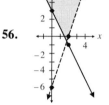
Range: the set of values that are obtained for the dependent variable.

57. $\dfrac{3 + i}{2}, \dfrac{3 - i}{2}$

Just for Fun   **1.** $y + 2 = \pm \frac{2}{3}(x - 3)$ **2.** $y + 5 = \pm \frac{4}{3}(x - 1)$ **3.**

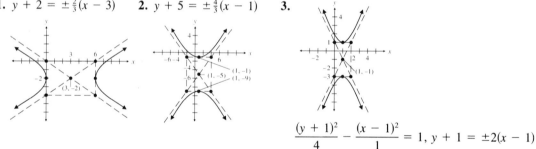

$$\frac{(y + 1)^2}{4} - \frac{(x - 1)^2}{1} = 1, \quad y + 1 = \pm 2(x - 1)$$

Exercise Set 10.5

1. $(0, -2), (\frac{8}{5}, -\frac{6}{5})$ **3.** $(-4, 11), (\frac{5}{2}, \frac{5}{4})$ **5.** $(2, -4), (-14, -20)$ **7.** No real solution   **9.** $(\frac{4}{3}, 3), (-\frac{1}{2}, -8)$
11. $(0, -3), (\sqrt{5}, 2), (-\sqrt{5}, 2)$ **13.** $(2, 0), (-2, 0)$ **15.** $(3, 2), (3, -2), (-3, 2), (-3, -2)$ **17.** $(3, 0), (-3, 0)$
19. $(\sqrt{6}, \sqrt{3}), (\sqrt{6}, -\sqrt{3}), (-\sqrt{6}, \sqrt{3}), (-\sqrt{6}, -\sqrt{3})$ **21.** No real solution
23. $(\sqrt{5}, 2), (\sqrt{5}, -2), (-\sqrt{5}, 2), (-\sqrt{5}, -2)$ **25.** 5, 7   **27.** 3, 5 or 3, -5 or -3, 5 or -3, -5
29. 5, 6 or 5, -6, or -5, 6 or -5, -6 **31.** 7 ft, 12 ft   **33.** 5 ft, 12 ft   **35.** 15 ft, 20 ft
37. 6 ft by 8 ft or 4 ft by 12 ft   **39.** 2.35 sec   **41.** $r = 12\%, P = \$600$ **43.** ≈ 26 and ≈ 174
45. ≈ 1 and ≈ 14 **49.** 8 points

Cumulative Review Exercises

50. Parentheses, exponents, multiplication or division from left to right, addition or subtraction from left to right.

51. $\$5849.29$ **52.** $y = \dfrac{6x + 120}{33}$ **53.** $y \le 2$

Just for Fun   **1.** 5 m, 12 m   **2.** 4.2 ft, 5.6 ft

Exercise Set 10.6

1. 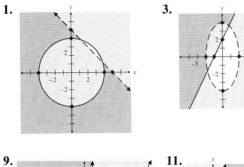   **3.**   **5.** 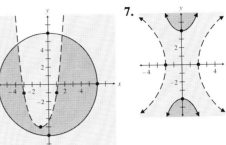   **7.**

9. 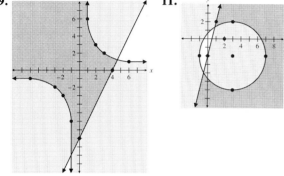   **11.**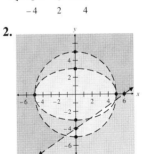

Cumulative Review Exercises

13. $16y^{10}$ **14.** $1 + \sqrt{5}, 1 - \sqrt{5}$ **15.** $1 + \sqrt{5}, 1 - \sqrt{5}$ **16.**

Just for Fun   **1.**   **2.**

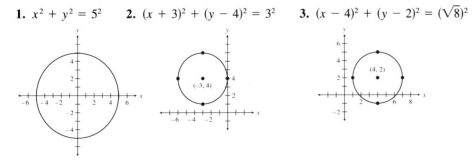

Review Exercises

1. $x^2 + y^2 = 5^2$ **2.** $(x + 3)^2 + (y - 4)^2 = 3^2$ **3.** $(x - 4)^2 + (y - 2)^2 = (\sqrt{8})^2$ **4.**

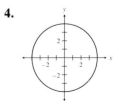

5.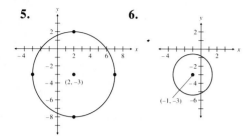

6.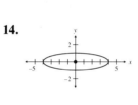

7. $(x + 1)^2 + (y - 1)^2 = 4$ **9.** $x^2 + (y - 2)^2 = 2^2$

8. $(x - 5)^2 + (y + 3)^2 = 9$

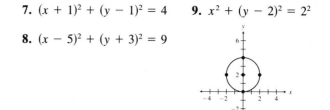

10. $(x - 1)^2 + (y + 3)^2 = 3^2$ **11.** $(x - 4)^2 + (y - 5)^2 = 1^2$ **12.** $(x - 2)^2 + (y + 5)^2 = (\sqrt{12})^2$

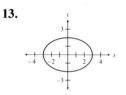

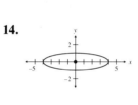

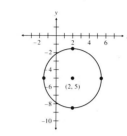

13.   **14.**   **15**   **16.**

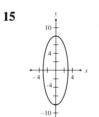

17.   **18.**   **19.**   **20.**

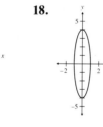

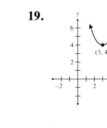

21.   **22.**   **23.**   **24.** $y = (x - 3)^2 - 9$

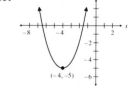

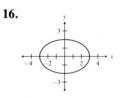

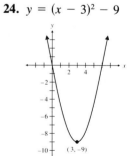

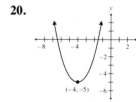

25. $y = (x - 1)^2 - 4$ **26.** $x = -(y + 1)^2 + 9$ **27.** $x = \left(y + \frac{5}{2}\right)^2 - \frac{9}{4}$

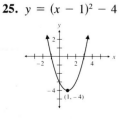

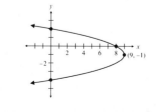

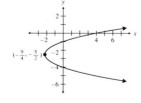

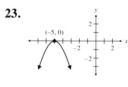

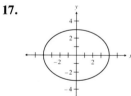

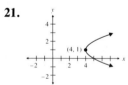

28. $y = 2(x - 2)^2 - 32$

29. $y = \pm\frac{3}{2}x$

30. $y = \pm2x$

31. $y = \pm\frac{3}{5}x$

32. $y = \pm3x$

33. $\dfrac{y^2}{4} - \dfrac{x^2}{9} = 1$, $y = \pm\frac{2}{3}x$

34. $\dfrac{x^2}{16} - \dfrac{y^2}{1} = 1$, $y = \pm\frac{1}{4}x$

35. $\dfrac{x^2}{16} - \dfrac{y^2}{25} = 1$, $y = \pm\frac{5}{4}x$

36. $\dfrac{y^2}{9} - \dfrac{x^2}{49} = 1$, $y = \pm\frac{3}{7}x$

37.

38.

39.

40.

41. Hyperbola

42. Ellipse

43. Circle

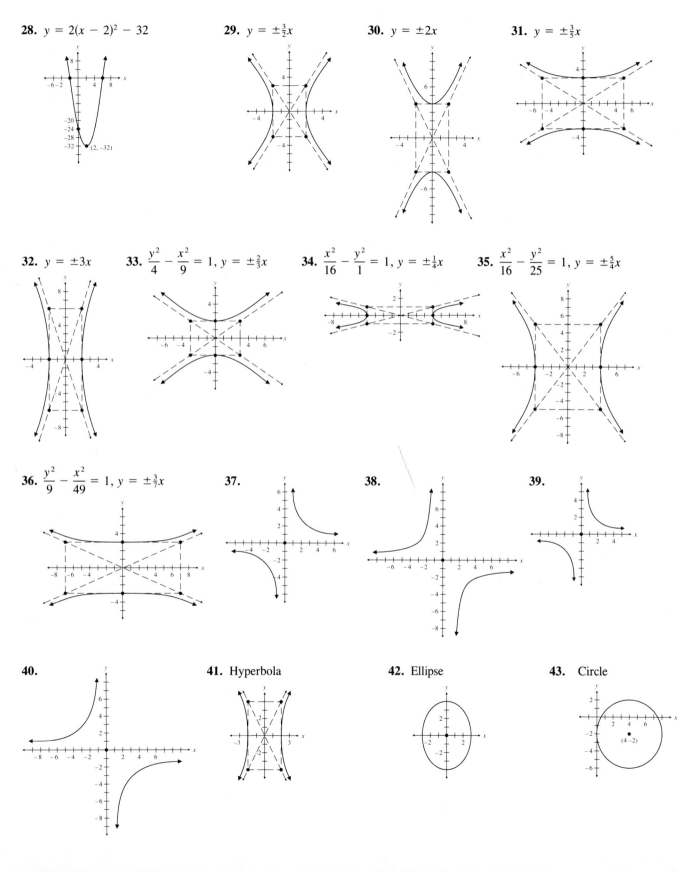

44. Hyperbola   **45.** Ellipse   **46.** Parabola   **47.** Ellipse   **48.** Parabola

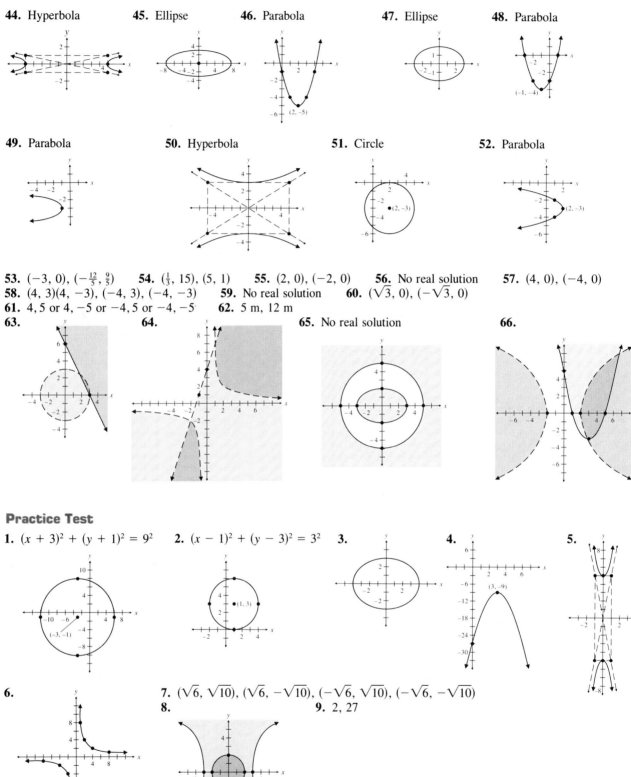

53. $(-3, 0), \left(-\frac{12}{5}, \frac{9}{5}\right)$ **54.** $\left(\frac{1}{3}, 15\right), (5, 1)$ **55.** $(2, 0), (-2, 0)$ **56.** No real solution   **57.** $(4, 0), (-4, 0)$
58. $(4, 3)(4, -3), (-4, 3), (-4, -3)$ **59.** No real solution   **60.** $(\sqrt{3}, 0), (-\sqrt{3}, 0)$
61. $4, 5$ or $4, -5$ or $-4, 5$ or $-4, -5$ **62.** 5 m, 12 m
63.   **64.**   **65.** No real solution   **66.**

Practice Test

1. $(x + 3)^2 + (y + 1)^2 = 9^2$ **2.** $(x - 1)^2 + (y - 3)^2 = 3^2$ **3.**   **4.**   **5.**

6.   **7.** $(\sqrt{6}, \sqrt{10}), (\sqrt{6}, -\sqrt{10}), (-\sqrt{6}, \sqrt{10}), (-\sqrt{6}, -\sqrt{10})$
8.   **9.** $2, 27$

Cumulative Review Test

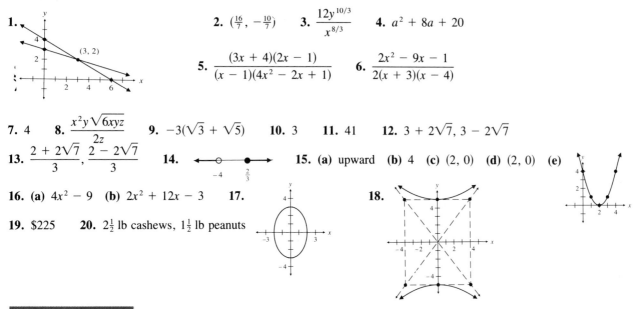

1.

2. $(\frac{16}{7}, -\frac{10}{7})$ **3.** $\dfrac{12y^{10/3}}{x^{8/3}}$ **4.** $a^2 + 8a + 20$

5. $\dfrac{(3x + 4)(2x - 1)}{(x - 1)(4x^2 - 2x + 1)}$ **6.** $\dfrac{2x^2 - 9x - 1}{2(x + 3)(x - 4)}$

7. 4   **8.** $\dfrac{x^2y\sqrt{6xyz}}{2z}$ **9.** $-3(\sqrt{3} + \sqrt{5})$ **10.** 3   **11.** 41   **12.** $3 + 2\sqrt{7}, 3 - 2\sqrt{7}$

13. $\dfrac{2 + 2\sqrt{7}}{3}, \dfrac{2 - 2\sqrt{7}}{3}$ **14.**   **15. (a)** upward   **(b)** 4   **(c)** $(2, 0)$ **(d)** $(2, 0)$ **(e)**

16. (a) $4x^2 - 9$ **(b)** $2x^2 + 12x - 3$ **17.**

18.

19. \$225   **20.** $2\frac{1}{2}$ lb cashews, $1\frac{1}{2}$ lb peanuts

CHAPTER 11

Exercise Set 11.1

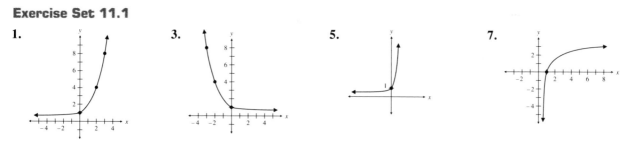

1.   **3.**   **5.**   **7.**

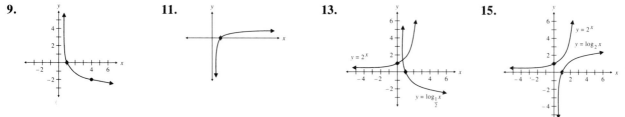

9.   **11.**   **13.**   **15.**

17. $\log_2 8 = 3$ **19.** $\log_3 243 = 5$ **21.** $\log_8 2 = \frac{1}{3}$ **23.** $\log_{1/4} \frac{1}{16} = 2$ **25.** $\log_5 \frac{1}{25} = -2$ **27.** $\log_4 \frac{1}{64} = -3$
29. $\log_{16} \frac{1}{4} = -\frac{1}{2}$ **31.** $\log_8 \frac{1}{2} = -\frac{1}{3}$ **33.** $2^3 = 8$ **35.** $4^3 = 64$ **37.** $(\frac{1}{2})^4 = \frac{1}{16}$ **39.** $5^{-3} = \frac{1}{125}$
41. $125^{1/3} = 5$ **43.** $27^{-1/3} = \frac{1}{3}$ **45.** $10^3 = 1000$ **47.** $6^y = 36, y = 2$ **49.** $2^3 = x, x = 8$
51. $64^y = 8, y = \frac{1}{2}$ **53.** $a^3 = 64, a = 4$ **55.** $(\frac{1}{2})^4 = x, x = \frac{1}{16}$ **57.** $a^5 = 32, a = 2$ **59.** $8^{1/3} = x, x = 2$
61. 256   **63.** \$14,448.89   **65.** 17.4 grams   **67. (a)** 5 grams   **(b)** 7.28×10^{-11} grams   **69.** 10,000,000
71. (a) $a > 0, a \neq 1$ **(b)** $\{x \mid x > 0\}$ **(c)** all real numbers
73. (a) A horizontal line through $y = 1$.   **(b)** Yes   **(c)** No, $f(x)$ is not a one-to-one function
75. The graphs are symmetric about the line $y = x$. For each ordered pair (x, y) on the graph of $y = a^x$, the ordered pair (y, x) is on the graph of $y = \log_a x$
77. $y = a^x + k$ will have the same shape as the graph of $y = a^x$. However $y = a^x + k$ will be k units higher or lower than $y = a^x$. If $k > 0$ the graph will be raised k units and if $k < 0$ the graph will be lowered k units. The y intercept of $y = a^x + k$ will be $(0, 1 + k)$.

Cumulative Review Exercises

79. $2(3x + 2y)(4x - y)$　　**80.** $(2a - 9)(a + 2)$　　**81.** $4x^2(x + 3)(x - 3)$　　**82.** $(2x + \frac{1}{3})(4x^2 - \frac{2}{3}x + \frac{1}{9})$

Exercise Set 11.2

1. $\log_3 7 + \log_3 12$　　**3.** $\log_8 7 + \log_8 (x + 3)$　　**5.** $\log_4 15 - \log_4 7$　　**7.** $\frac{1}{2}\log_{10} x - \log_{10} (x - 3)$　　**9.** $4 \log_8 x$
11. $\log_{10} 3 + 2 \log_{10} 8$　　**13.** $\frac{1}{2}[5 \log_4 x - \log_4 (x + 4)]$　　**15.** $4 \log_{10} x - 3 \log_{10} (x + 2)$

17. $\log_8 x + \log_8 (x - 6) - 3 \log_8 x = -2 \log_8 x + \log_8 (x - 6)$　　**19.** $\log_{10} 2 + \log_{10} x - \log_{10} 3$　　**21.** $\log_{10} \dfrac{x^2}{x - 2}$

23. $\log_5 \left(\dfrac{x}{4}\right)^2$　　**25.** $\log_{10} \dfrac{x(x - 4)}{x + 1}$　　**27.** $\log_7 \left[\dfrac{(x - 2)}{x}\right]^{1/2}$　　**29.** $\log_9 \left(\dfrac{5^2}{6^4}\right)(3)$　　**31.** $\log_6 \dfrac{3^4}{(x + 3)^2 x^4}$　　**33.** 1
35. 0.3980　　**37.** 1.3980　　**39.** false　　**41.** true　　**43.** false　　**45.** false

Cumulative Review Exercises

50. $\dfrac{(2x + 5)^2}{(x - 4)^2}$　　**51.** $\dfrac{3x^2 + 28x + 9}{(x - 4)(x - 3)(2x + 5)}$　　**52.** $2\frac{2}{9}$ days　　**53.** $2x^3 y^5 \sqrt[3]{6x^2 y^2}$

Exercise Set 11.3

1. 2.9395　　**3.** 0.9031　　**5.** 3.0000　　**7.** -4.0671　　**9.** 2.0000　　**11.** 0.2405　　**13.** -0.4260　　**15.** -2.0595
17. 3.48　　**19.** 209　　**21.** 0.0874　　**23.** 1.00　　**25.** 317　　**27.** 0.701　　**29.** 100　　**31.** 0.000787　　**33.** 6610
35. 0.0871　　**37.** 0.428　　**39.** 0.404　　**41.** 3.3747　　**43.** -1.3872　　**45.** 2.0086　　**47.** -2.8928　　**49.** 386.0
51. 0.695　　**53.** 0.0246　　**55.** 15.5　　**57.** 0　　**59.** -1　　**61.** -2　　**63.** -3　　**65. (a)** 4.08　　**(b)** $19{,}500$
67. (a) 6.31×10^{20}　　**(b)** 2.19　　**69.** 2.55　　**71.** No, Since 10^3 is 1000, log 6250 must be greater than 3.
73. No, since $10^{-2} = 0.01$ and 10^{-3} is 0.001, log 0.0024 must be between -2 and -3.

Cumulative Review Exercises

75. $\dfrac{-2 + 2i\sqrt{5}}{3}, \dfrac{-2 - 2i\sqrt{5}}{3}$　　**76.** 10 mph　　**77.** ⟜─○　　**78.**

Exercise Set 11.4

1. 5　　**3.** 3　　**5.** 55.50　　**7.** 5.59　　**9.** 3　　**11.** $0, -8$　　**13.** $\frac{13}{2}$　　**15.** $\frac{3}{2}$　　**17.** 2　　**19.** 2　　**21.** 0.91
23. 20　　**25.** 5　　**27.** 3　　**29.** 2　　**31.** $\$1932.61$　　**33.** 139　　**35.** $\$7112.10$　　**37.** 2.94
39. (a) $1{,}000{,}000{,}000{,}000$　　**(b)** $100{,}000$

Cumulative Review Exercises

41. (a) $\{x \mid x > 40\}$　　**(b)** $(40, \infty)$　　**42. (b)** and **(c)** are functions, only **(b)** is a one-to-one function
43. $2x^3 - 11x^2 + 18x - 9$　　**44.** $2x + 3 + \dfrac{3}{x + 4}$

Exercise Set 11.5

1. 3.9120　　**3.** 5.7104　　**5.** 4.95　　**7.** 0.0721　　**9.** 3.6889　　**11.** -3.0791　　**13.** 2.9300　　**15.** 4.3219
17. 7.9010　　**19.** -4.8411　　**21.** 4　　**23.** 4　　**25.** 3　　**27.** $P = 3410.48$　　**29.** $P_0 = 58.09$　　**31.** $t = 0.7847$

33. $k = 0.2310$　　**35.** $k = -0.2888$　　**37.** $A = 4719.77$　　**39.** $V_0 = \dfrac{V}{e^{kt}}$　　**41.** $t = \dfrac{\ln P - \ln 150}{4}$

43. $k = \dfrac{\ln A - \ln A_0}{t}$　　**45.** $y = xe^{2.3}$　　**47.** $y = (x + 3)e^6$　　**49.** $I_0 = Ie^{x/k}$　　**51.** $Q = \dfrac{M}{1 + M}$
53. (a) $\$5867.55$　　**(b)** 8.66 yr　　**55. (a)** 86.47%　　**(b)** 34.66 days　　**57. (a)** 5.15 ft/sec　　**(b)** 5.96 ft/sec　　**(c)** 646,000
59. (a) 6.24 billion　　**(b)** 38.51 yr　　**61. (a)** 6626.62 yr　　**(b)** 5752.26 yr　　**63.** ≈ 2.718
65. 5.2200, find ln 184.93 to obtain the answer.　　**67.** 0.1773, press $\boxed{c}$　　$\boxed{+/-}$ $\boxed{\text{inv}}$ $\boxed{\ln}$

Cumulative Review Exercises

68. $\dfrac{16y^{14}}{x^4}$　　**69.** $\dfrac{12 - 8x}{x^3}$　　**70.** $q = \dfrac{fp}{p - f}$　　**71.** $4x^2 y^3 z^4 \sqrt[3]{2xz}$

Review Exercises

1. 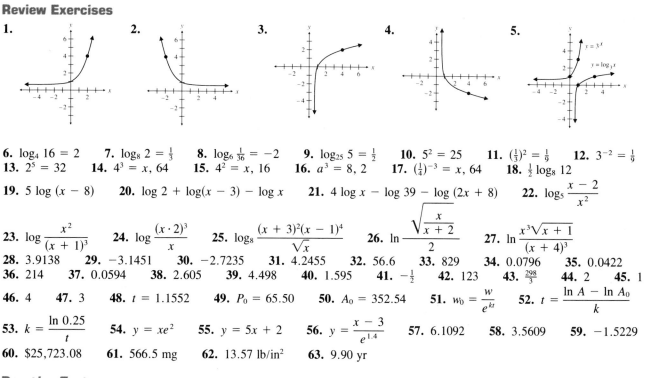 **2.** **3.** **4.** **5.**

6. $\log_4 16 = 2$ **7.** $\log_8 2 = \frac{1}{3}$ **8.** $\log_6 \frac{1}{36} = -2$ **9.** $\log_{25} 5 = \frac{1}{2}$ **10.** $5^2 = 25$ **11.** $\left(\frac{1}{3}\right)^2 = \frac{1}{9}$ **12.** $3^{-2} = \frac{1}{9}$
13. $2^5 = 32$ **14.** $4^3 = x, 64$ **15.** $4^2 = x, 16$ **16.** $a^3 = 8, 2$ **17.** $\left(\frac{1}{4}\right)^{-3} = x, 64$ **18.** $\frac{1}{2}\log_8 12$
19. $5\log(x - 8)$ **20.** $\log 2 + \log(x - 3) - \log x$ **21.** $4\log x - \log 39 - \log(2x + 8)$ **22.** $\log_5 \dfrac{x - 2}{x^2}$

23. $\log \dfrac{x^2}{(x + 1)^3}$ **24.** $\log \dfrac{(x \cdot 2)^3}{x}$ **25.** $\log_8 \dfrac{(x + 3)^2(x - 1)^4}{\sqrt{x}}$ **26.** $\ln \dfrac{\sqrt{\dfrac{x}{x + 2}}}{2}$ **27.** $\ln \dfrac{x^3\sqrt{x + 1}}{(x + 4)^3}$
28. 3.9138 **29.** -3.1451 **30.** -2.7235 **31.** 4.2455 **32.** 56.6 **33.** 829 **34.** 0.0796 **35.** 0.0422
36. 214 **37.** 0.0594 **38.** 2.605 **39.** 4.498 **40.** 1.595 **41.** $-\frac{1}{2}$ **42.** 123 **43.** $\frac{298}{3}$ **44.** 2 **45.** 1
46. 4 **47.** 3 **48.** $t = 1.1552$ **49.** $P_0 = 65.50$ **50.** $A_0 = 352.54$ **51.** $w_0 = \dfrac{w}{e^{kt}}$ **52.** $t = \dfrac{\ln A - \ln A_0}{k}$
53. $k = \dfrac{\ln 0.25}{t}$ **54.** $y = xe^2$ **55.** $y = 5x + 2$ **56.** $y = \dfrac{x - 3}{e^{1.4}}$ **57.** 6.1092 **58.** 3.5609 **59.** -1.5229
60. \$25,723.08 **61.** 566.5 mg **62.** 13.57 lb/in^2 **63.** 9.90 yr

Practice Test

1. 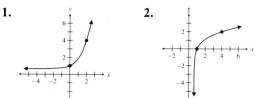 **2.** **3.** $\log_4 \frac{1}{64} = -3$ **4.** $3^5 = 243$ **5.** $2^4 = x, 16$ **6.** $27^y = 3, \frac{1}{3}$
7. $\log_3 x + \log_3 (x - 4) - 2\log_3 x$ **8.** $\log_8 \dfrac{(x - 4)^3(x + 1)^2}{\sqrt{x}}$
9. 3.6646 **10.** -3.1952 **11.** 0.00501 **12.** 4.38 **13.** 3
14. \$4658.77 **15.** 3364.86 yr

CHAPTER 12

Exercise Set 12.1

1. 2, 4, 6, 8, 10 **3.** 6, $\frac{7}{2}, \frac{8}{3}, \frac{9}{4}, 2$ **5.** 1, $\frac{1}{2}, \frac{1}{3}, \frac{1}{4}, \frac{1}{5}$ **7.** $\frac{3}{2}, \frac{4}{3}, \frac{5}{4}, \frac{6}{5}, \frac{7}{6}$ **9.** $-1, 1, -1, 1, -1$
11. 4, -8, 16, -32, 64 **13.** 27 **15.** 6 **17.** 16 **19.** $\frac{81}{19}$ **21.** $s_1 = 7, s_3 = 27$ **23.** $s_1 = 3, s_3 = 17$
25. $s_1 = 1, s_3 = 3$ **27.** $s_1 = \frac{1}{2}, s_3 = 7$ **29.** 64, 128, 256 **31.** 15, 17, 19 **33.** $\frac{1}{6}, \frac{1}{7}, \frac{1}{8}$ **35.** 1, -1, 1
37. $\frac{1}{81}, \frac{1}{243}, \frac{1}{729}$ **39.** $\frac{1}{16}, -\frac{1}{32}, \frac{1}{64}$ **41.** $-25, -33, -41$ **43.** 2, 5, 8, 11, 14; 40 **45.** -1, 5, 15, 29, 47, 69; 164
47. 2, 3, 4; 9 **49.** $\displaystyle\sum_{n=1}^{5} (n + 3)$ **51.** $\displaystyle\sum_{n=1}^{3} \dfrac{n^2}{4}$ **53.** 13 **55.** 169 **57.** $\dfrac{16}{3}$ **59.** 25 **61.** 42.83
63. $x_1 + x_2 + x_3 + \cdots + x_n$ **65.** A list of numbers arranged in a specific order
67. The sum of the first consecutive n terms of a series

Cumulative Review Exercises

69. 5, $\frac{3}{2}$ **70.** Two, $b^2 - 4ac > 0$ **71.** **72.** $(2, 1), (-2, -1)$

Just for Fun **1.** $S \approx \sqrt{6.59} \approx 2.57$

Exercise Set 12.2

1. 3, 7, 11, 15, 19; $a_n = 3 + 4(n - 1)$ **3.** $-5, -3, -1, 1, 3; a_n = -5 + 2(n - 1)$
5. $\frac{1}{2}, 2, \frac{7}{2}, 5, \frac{13}{2}; a_n = \frac{1}{2} + \frac{3}{2}(n - 1)$ **7.** 100, 95, 90, 85, 80; $a_n = 100 - 5(n - 1)$ **9.** 22   **11.** -23 **13.** 13
15. 2   **17.** 9   **19.** 6   **21.** $s_{10} = 100, d = 2$ **23.** $s_8 = \frac{52}{5}, d = \frac{1}{5}$ **25.** $s_5 = 20, d = \frac{4}{5}$
27. $s_{11} = 407, d = 6$ **29.** $-4, -6, -8, -10; a_{10} = -22, s_{10} = -130$
31. $-8, -13, -18, -23; a_{10} = -53, s_{10} = -305$ **33.** 100, 93, 86, 79; $a_{10} = 37, s_{10} = 685$
35. $\frac{9}{5}, \frac{12}{5}, \frac{15}{5}, \frac{18}{5}; a_{10} = \frac{36}{5}, s_{10} = 45$ **37.** $n = 26, s_n = 442$ **39.** $n = 17, s_n = \frac{153}{2}$ **41.** $n = 9, s_n = 315$
43. $n = 29, s_n = 1479$ **45.** 500,500   **47.** 267   **49.** 1 ft   **51.** 210 logs
53. The sum of the terms of an arithmetic sequence.

Cumulative Review Exercises

55. One, the slopes of the lines are different. Therefore the lines must intersect once.   **56.** $(7, -6)$
57.   **58.**

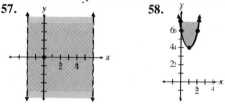

Exercise Set 12.3

1. 5, 15, 45, 135, 405   **3.** 90, 30, 10, $\frac{10}{3}, \frac{10}{9}$ **5.** $-15, 30, -60, 120, -240$ **7.** 3, $\frac{9}{2}, \frac{27}{4}, \frac{81}{8}, \frac{243}{16}$ **9.** 160
11. 13,122   **13.** $\frac{1}{64}$ **15.** 6144   **17.** 1023   **19.** 10,160   **21.** $\frac{2565}{128}$ **23.** $-\frac{9279}{625}$ **25.** $r = \frac{1}{2}, a_n = 5(\frac{1}{2})^{n-1}$
27. $r = -3, a_n = 2(-3)^{n-1}$ **29.** $r = 3, a_n = -1(3)^{n-1}$ **31.** $r = 2, a_1 = 7$ or $r = -2, a_1 = 7$ **33.** $r = 3, a_1 = 5$
35. \$572,503.52   **37.** (a) 3 days (b) 1.172 grams   **39.** (a) 437.25 million (b) 11.9 years
41. (a) $\frac{1}{2}, \frac{1}{4}, \frac{1}{8}, \frac{1}{16}, \frac{1}{32}$ (b) $a_n = a_1 r^{n-1} = \frac{1}{2}(\frac{1}{2})^{n-1} = (\frac{1}{2})^n$ (c) amount remaining $= \frac{1}{128}$ or 0.78%   **43.** $(\frac{2}{3})^{10}$ or 1.7%
45. (a) \$7840, \$6272, \$5017.60 (b) $a_n = 7840(\frac{4}{5})^{n-1}$ (c) \$3211.26   **47.** The sum of the terms of a geometric sequence.

Cumulative Review Exercises

49. 12 hrs   **50.** $\dfrac{x}{y^{2/15}}$ **51.** $3x^2y^2\sqrt[3]{y} - 2xy\sqrt[3]{9y^2}$ **52.** $x\sqrt{y}$

Just for Fun   **1.** (a) \$3211.26 (b) \$3211.26   **2.** $n = 21, s = 2{,}097{,}151$

Exercise Set 12.4

1. 12   **3** $\frac{25}{3}$ **5.** $\frac{5}{3}$ **7.** $\frac{81}{10}$ **9.** 2   **11.** 24   **13.** -45 **15.** -15 **17.** $\frac{3}{11}$ **19.** $\frac{5}{9}$ **21.** $\frac{17}{33}$ **23.** 40 ft
25. The sum does not exist.

Cumulative Review Exercises

26. $8x^3 - 18x^2 - 3x + 18$ **27.** $8x - 15 + \dfrac{15}{2x + 5}$ **28.** $\dfrac{10}{3}$ **29.** $-\dfrac{1}{3}$

Just for Fun   **1.** 190 ft   **2.** 4 ft

Exercise Set 12.5

1. 6   **3.** 1   **5.** 1   **7.** 70   **9.** 28   **11.** $x^3 + 12x^2 + 48x + 64$ **13.** $a^4 - 4a^3b + 6a^2b^2 - 4ab^3 + b^4$
15. $243a^5 - 405a^4b + 270a^3b^2 - 90a^2b^3 + 15ab^4 - b^5$ **17.** $16x^4 + 16x^3 + 6x^2 + x + \frac{1}{16}$
19. $(x^4/16) - (3x^3/2) + (27x^2/2) - 54x + 81$ **21.** $x^{10} + 10x^9y + 45x^8y^2 + 120x^7y^3$
23. $2187x^7 - 5103x^6y + 5103x^5y^2 - 2835x^4y^3$ **25.** $x^{16} - 24x^{14}y + 252x^{12}y^2 - 1512x^{10}y^3$ **29.** yes, $4! = 4 \cdot 3!$
31. yes, $(7 - 3)! = (7 - 3)(7 - 4)(7 - 5)!$ **33.** $x^8, 24x^7, 17{,}496x, 6561$
$4! = 4 \cdot 3 \cdot 2!$

Cumulative Review Exercise

35. $(4x - 3)(4y + 1)$ **36.** $3b^2(a - 4)^2$ **37.** $S_n = \dfrac{a_1 - a_1 r^n}{1 - r}$ or $S_n = \dfrac{a_1(1 - r^n)}{1 - r}$ **38.** $a_1 = \dfrac{S_n - S_n r}{1 - r^n}$

Just for Fun   **1.** $(a + b)^n = \sum\limits_{i=0}^{n} \binom{n}{i} a^{n-i} b^i$ **2. (a)** r^{th} term $= \dfrac{n!}{[n - (r - 1)]!(r - 1)!} x^{n-(r-1)} y^{r-1}$ **(b)** $216x^2$

Review Exercises

1. 3, 4, 5, 6, 7   **2.** $1, \frac{1}{2}, \frac{1}{3}, \frac{1}{4}, \frac{1}{5}$ **3.** 2, 6, 12, 20, 30   **4.** $\frac{1}{5}, \frac{2}{3}, \frac{9}{7}, 2, \frac{25}{9}$ **5.** 25   **6.** 2   **7.** $\frac{16}{81}$ **8.** 88
9. $s_1 = 5, s_3 = 24$ **10.** $s_1 = 2, s_3 = 28$ **11.** $s_1 = \frac{4}{3}, s_3 = \frac{227}{60}$ **12.** $s_1 = -3, s_3 = -4$
13. 16, 32, 64; $a_n = 2^{n-1}$ **14.** $-\frac{1}{2}, \frac{1}{4}, -\frac{1}{8}; a_n = -8(-\frac{1}{2})^{n-1}$ **15.** $\frac{32}{3}, \frac{64}{3}, \frac{128}{3}; a_n = \frac{2}{3}(2)^{n-1}$
16. $-3, -6, -9; a_n = 9 - 3(n - 1) = 12 - 3n$ **17.** $-1, 1, -1; a_n = (-1)^n$
18. 30, 36, 42; $a_n = 6 + 6(n - 1) = 6n$ **19.** 3, 6, 11; 20   **20.** 3, 8, 15, 24; 50   **21.** $\frac{1}{3}, \frac{4}{3}, \frac{9}{3}, \frac{16}{3}, \frac{25}{3}; \frac{55}{3}$
22. $\frac{1}{2}, \frac{2}{3}, \frac{3}{4}, \frac{4}{5}; \frac{163}{60}$ **23.** 27   **24.** 215   **25.** 108   **26.** 729   **27.** 5, 7, 9, 11, 13   **28.** $\frac{1}{2}, -\frac{3}{2}, -\frac{7}{2}, -\frac{11}{2}, -\frac{15}{2}$
29. $-12, -\frac{25}{2}, -13, -\frac{27}{2}, -14$ **30.** $-100, -\frac{499}{5}, -\frac{498}{5}, -\frac{497}{5}, -\frac{496}{5}$ **31.** 26   **32.** -15 **33.** -4 **34.** $\frac{1}{2}$
35. 6   **36.** 7   **37.** $s = 112, d = 2$ **38.** $s = -210, d = -6$ **39.** $s = \frac{63}{5}, d = \frac{2}{5}$ **40.** $s = -42, d = -\frac{1}{3}$
41. 2, 6, 10, 14; $a_{10} = 38, s_{10} = 200$ **42.** $-8, -11, -14, -17; a_{10} = -35, s_{10} = -215$
43. $\frac{5}{6}, \frac{9}{6}, \frac{13}{6}, \frac{17}{6}, a_{10} = \frac{41}{6}, s_{10} = \frac{115}{3}$ **44.** $-80, -76, -72, -68; a_{10} = -44, s_{10} = -620$ **45.** $n = 11, s_n = 308$
46. $n = 9, s_n = 36$ **47.** $n = 11, s_n = \frac{231}{10}$ **48.** $n = 19, s_n = 760$ **49.** 5, 10, 20, 40, 80
50. $-12, -6, -3, -\frac{3}{2}, -\frac{3}{4}$ **51.** $20, -\frac{40}{3}, \frac{80}{9}, -\frac{160}{27}, \frac{320}{81}$ **52.** $-100, -20, -4, -\frac{4}{5}, -\frac{4}{25}$ **53.** $\frac{4}{243}$ **54.** 6400
55. -2048 **56.** $\frac{160}{6561}$ **57.** 3060   **58.** $\frac{37,969}{1215}$ **59.** $-\frac{4305}{64}$ **60.** $\frac{172,539}{256}$ **61.** $r = 2, a_n = 6(2)^{n-1}$
62. $r = \frac{1}{3}, a_n = 8(\frac{1}{3})^{n-1}$ **63.** $r = 5, a_n = -4(5)^{n-1}$ **64.** $r = \frac{2}{3}, a_n = \frac{9}{5}(\frac{2}{3})^{n-1}$ **65.** 14   **66.** -6 **67.** -15
68. $\frac{49}{10}$ **69.** 4   **70.** $\frac{21}{2}$ **71.** -36 **72.** $\frac{25}{6}$ **73.** $\frac{52}{99}$ **74.** $\frac{125}{333}$ **75.** $81x^4 + 108x^3 y + 54x^2 y^2 + 12xy^3 + y^4$
76. $8x^3 - 36x^2 y^2 + 54xy^4 - 27y^6$ **77.** $x^9 - 18x^8 y + 144x^7 y^2 - 672x^6 y^3$
78. $256a^{16} + 3072a^{14} b + 16{,}128a^{12} b^2 + 48{,}384a^{10} b^3$ **79.** 15,150
80. (a) 30,000, 31,000, 32,000, 33,000, 34,000   **(b)** $a_n = 30{,}000 + (n - 1)1000$ **(c)** 39,000   **81.** $102,400
82. $503.63   **83.** 100 ft

Practice Test

1. $3, 1, \frac{5}{9}, \frac{3}{8}, \frac{7}{25}$ **2.** $s_1 = 5, s_3 = \frac{23}{2}$ **3.** 5, 11, 21, 35, 33, 53; 125   **4.** $a_n = \frac{1}{3} + \frac{1}{3}(n - 1) = \frac{1}{3}n$
5. $a_n = 5(2)^{n-1}$ **6.** 12, 9, 6, 3   **7.** $\frac{5}{8}, \frac{10}{24}, \frac{20}{72}, \frac{40}{216}$ **8.** 16   **9.** -32 **10.** 13   **11.** $\frac{512}{729}$ **12.** $\frac{39,063}{5}$
13. $r = \frac{1}{2}, a_n = 12(\frac{1}{2})^{n-1}$ **14.** 9   **15.** $x^4 + 8x^3 y + 24x^2 y^2 + 32xy^3 + 16y^4$

Cumulative Review Test

1. -37 **2.** $x < -\frac{11}{2}$ or $x > \frac{17}{2}$ **3.**   **4.** $(0, -4)$ **5.** $p = \dfrac{At}{t - A}$ **6.** $\dfrac{4x^2 + 3x - 6}{(x + 4)(x - 3)(3x + 4)}$

7. $\dfrac{x\sqrt[3]{12xy}}{y}$ **8.** $2\sqrt{7}$ **9.** 3   **10.** $\dfrac{7 - 26i}{25}$ **11.** $\dfrac{-1 + \sqrt{33}}{4}, \dfrac{-1 - \sqrt{33}}{4}$ **12.** $x < -1$ or $\frac{3}{2} < x < 4$
13.   **14.** $y = (x + 1)^2 - 4$ **15.**

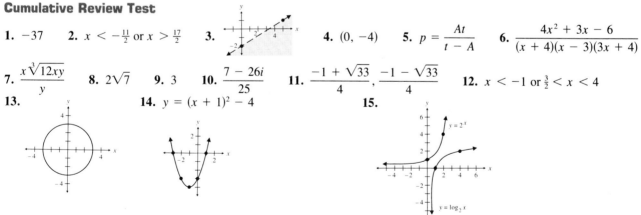

16. 4   **17.** 4   **18.** 65   **19.** 240 mi   **20.** $w = 10$ ft, $l = 30$ ft

Index

Chapter 7   Rational Expressions and Equations

To Multiply Rational Expressions:

1. Factor all numerators and denominators.
2. Divide out the common factors.
3. Multiply numerators together and multiply denominators together.
4. Reduce the answer when possible.

To Divide Fractional Expressions:

1. Invert the divisor.
2. Multiply the numerator by the inverted divisor.

To Add or Subtract Rational Expressions:

1. Write each fraction with a common denominator.
2. Add or subract the numerators while keeping the common denominator.
3. When possible factor the remaining numerator and reduce the fraction.

Similar Figures: Corresponding angles have the same measure and corresponding sides are in proportion.

Variation: direct, $x = ky$; inverse, $x = \frac{k}{y}$ or $xy = k$;

joint, $x = kyz$

Chapter 8   Roots, Radicals, and Complex Numbers

$\sqrt{a^2} = |a|$

$\sqrt[n]{x^n} = x, \; x > 0$

$\sqrt[n]{x} = x^{\frac{1}{n}}, \; x^n \geq 0$

$\sqrt[n]{x^m} = (\sqrt[n]{x})^m = x^{\frac{m}{n}}, \; x \geq 0$

$\sqrt[n]{ab} = \sqrt[n]{a}\sqrt[n]{b}, \; a \geq 0, \; b \geq 0$

$\sqrt[m]{a/b} = \sqrt[m]{a} / \sqrt[m]{b}, \; a \geq 0 \text{ and } b > 0$

A radical is simplified when: (a) there are no perfect powers that are factors of any radicand, (b) no radicand contains fractions, and (c) there are no radicals in any denominator.

Complex numbers: $i = \sqrt{-1}$ and $i^2 = -1$

Square root function: $f(x) = \sqrt{x}, \; x \geq 0$

Chapter 9   Quadratic Functions and the Algebra of Functions

Standard form of a quadratic equation: $ax^2 + bx + c = 0$

A quadratic equation may be solved by factoring, completing the square, or the quadratic formula

Quadratic Formula: $x = \dfrac{-b \pm \sqrt{b^2 - 4ac}}{2a}$

Pythagorean Theorem: $a^2 + b^2 = c^2$

Discriminant: $b^2 - 4ac$

If $b^2 - 4ac > 0$ then equation has two distinct real solutions.
If $b^2 - 4ac = 0$ then equation has a single real solution (double root).
If $b^2 - 4ac < 0$ then equation has no real solution.

Vertex of a parabola: $\left(\dfrac{-b}{2a}, \dfrac{4ac - b^2}{4a} \right)$

Functions
Sum: $(f + g)(x) = f(x) + g(x)$
difference: $(f - g)(x) = f(x) - g(x)$
Product: $(fg)(x) = f(x) \cdot g(x)$
quotient: $\left(\dfrac{f}{g} \right)(x) = \dfrac{f(x)}{g(x)}, \; g(x) \neq 0$

Composite function of function f with function g: $(f \circ g)(x) = f[g(x)]$

To find the **inverse function**, $f^{-1}(x)$, interchange all x's and y's and solve the resulting equation for y.